T0211642

Lecture Notes in Computer Science   12708

More information about this subseries at http://www.springer.com/series/7412

Joakim Lindblad · Filip Malmberg ·
Nataša Sladoje (Eds.)

Discrete Geometry
and Mathematical Morphology

First International Joint Conference, DGMM 2021
Uppsala, Sweden, May 24–27, 2021
Proceedings

 Springer

Editors
Joakim Lindblad
Uppsala University
Uppsala, Sweden

Filip Malmberg
Uppsala University
Uppsala, Sweden

Nataša Sladoje
Uppsala University
Uppsala, Sweden

ISSN 0302-9743   ISSN 1611-3349   (electronic)
Lecture Notes in Computer Science
ISBN 978-3-030-76656-6   ISBN 978-3-030-76657-3   (eBook)
https://doi.org/10.1007/978-3-030-76657-3

LNCS Sublibrary: SL6 – Image Processing, Computer Vision, Pattern Recognition, and Graphics

This Springer imprint is published by the registered company Springer Nature Switzerland AG
The registered company address is: Gewerbestrasse 11, 6330 Cham, Switzerland

Preface

This volume contains the papers presented at DGMM 2021: IAPR International Conference on Discrete Geometry and Mathematical Morphology held during May 24–27, 2021, in Uppsala, Sweden.

DGMM is sponsored by the International Association of Pattern Recognition (IAPR), and is associated with the IAPR Technical Committee on Discrete Geometry and Mathematical Morphology (TC18). This is the first joint event between the two main conference series of IAPR TC18, the International Conference on Discrete Geometry for Computer Imagery (DGCI), with 21 successful previous editions, and the International Symposium on Mathematical Morphology (ISMM), with 14 successful previous editions.

This first DGMM edition attracted 59 submissions by authors from 15 countries: Austria, Brazil, Canada, China, Denmark, England, France, Germany, Greece, Hungary, Italy, Serbia, Sweden, Turkey, and the USA. Out of these 59 submissions, 36 were selected for presentation at the conference after a review and rebuttal process where each submission received, on average, 3.3 reports, and a meta-review. The DGMM 2021 papers highlight the current trends and advances in discrete geometry and mathematical morphology, be they purely theoretical contributions, algorithmic developments, or novel applications in image processing, computer vision, and pattern recognition.

In addition, three internationally well-known researchers were invited for keynote lectures:

- María-José Jiménez, on "On topological analysis of cells organization in biological images"
- Jesús Angulo, on "Some open questions on morphological operators and representations in the deep learning era"
- Cecilia Holmgren, on "Split trees — A unifying model for many important random trees of logarithmic height"

Each keynote speaker contributed a written article that can be found in this volume.

Following the tradition of both DGCI and ISMM, the DGMM 2021 proceedings appear in Springer's LNCS series and a special issue of the Journal of Mathematical Imaging and Vision, with extended versions of selected outstanding contributions, is planned.

We are thankful to the IAPR for its sponsorship. We would like to thank all contributors, the keynote speakers, the Program and Steering Committees of DGMM, the Organizing Committee of DGMM 2021, and all those who made this conference happen. Last but not least, we thank all participants and we hope that everyone found great interest in DGMM 2021.

April 2021

Filip Malmberg
Joakim Lindblad
Nataša Sladoje

Organization

DGMM 2021 was organized by the Centre for Image Analysis, Department of Information Technology, Uppsala University, Sweden.

Organizing Committee

Filip Malmberg (General Chair)
Nataša Sladoje (Program Chair)
Joakim Lindblad (Program Chair)
Johan Öfverstedt (Local Chair)
Eva Breznik (Webmaster)
Ingela Nyström (Advisor)
Robin Strand (Advisor)
Gunilla Borgefors (Honorary Chair)
Christer Kiselman (Honorary Chair)

Steering Committee

Discrete Geometry

Eric Andres	XLIM-SIC, University of Poitiers, France
Gunilla Borgefors	Uppsala University, Sweden
Srećko Brlek	Université du Québec à Montréal, Canada
David Coeurjolly	CNRS, Université de Lyon, France
Isabelle Debled-Rennesson	Université de Lorraine, Nancy, France
Andrea Frosini	Università di Firenze, Italy
María-José Jiménez	University of Seville, Spain
Yukiko Kenmochi	CNRS, Université Paris-Est, France
Bertrand Kerautret	Université de Lyon, France
Walter G. Kropatsch	Vienna University of Technology, Austria
Jacques-Olivier Lachaud	LAMA, University Savoie Mont Blanc, France
Nicolas Normand	Université de Nantes, France

Mathematical Morphology

Jesús Angulo	Ecole des Mines de Paris, France
Junior Barrera	University of São Paulo, Brazil
Jón Atli Benediktsson	University of Iceland
Isabelle Bloch	LTCI, Tlcom Paris, France
Gunilla Borgefors	Uppsala University, Sweden
Bernhard Burgeth	Universitt des Saarlandes, Germany
David Coeurjolly	CNRS, Université de Lyon, France
Renato Keshet	GE Global Res. Center, Tirat Carmel, Israel

Ron Kimmel	Technion, Israel
Cris Luengo Hendricks	Flagship Biosciences, USA
Petros Maragos	National Technical University of Athens, Greece
Laurent Najman	ESIEE Paris, Université Paris-Est, France
Nicolas Passat	Université de Reims Champagne-Ardenne, France
Dan Schonfeld	University of Illinois, Chicago, USA
Pierre Soille	Joint Research Centre, European Commission, Ispra, Italy
Hugues Talbot	Université Paris-Est, France
Michael H. F. Wilkinson	University of Groningen, Netherlands

Program Committee

Eric Andres	XLIM-SIC, University of Poitiers, France
Jesús Angulo	Ecole des Mines de Paris, France
Péter Balázs	University of Szeged, Hungary
Partha Bhowmick	Indian Institute of Technology, Kharagpur, India
Isabelle Bloch	LTCI, Tlcom Paris, France
Sara Brunetti	University of Siena, Italy
David Coeurjolly	CNRS, Université de Lyon, France
Guillaume Damiand	CNRS, Université de Lyon, France
Yan Gerard	Université Clermont Auvergne, France
Rocio Gonzalez-Diaz	University of Seville, Spain
Atsushi Imiya	IMIT, Chiba University, Japan
María José Jiménez	University of Seville, Spain
Yukiko Kenmochi	CNRS, Université Paris-Est, France
Bertrand Kerautret	Université de Lyon, France
Walter G. Kropatsch	Vienna University of Technology, Austria
Jacques-Olivier Lachaud	LAMA, University Savoie Mont Blanc, France
Sébastien Lefèvre	Université de Bretagne Sud, France
Joakim Lindblad	Uppsala University, Sweden
Cris Luengo	Flagship Biosciences, USA
Filip Malmberg	Uppsala University, Sweden
Loïc Mazo	ICube, CNRS, Université de Strasbourg, France
Laurent Najman	ESIEE Paris, Université Paris-Est, France
Nicolas Passat	Université de Reims Champagne-Ardenne, France
Gabriella Sanniti di Baja	ICAR-CNR, Italy
Isabelle Sivignon	GIPSA-lab, CNRS, France
Nataša Sladoje	Uppsala University, Sweden
Pierre Soille	Joint Research Centre, European Commission, Ispra, Italy
Robin Strand	Uppsala University, Sweden

Hugues Talbot   Université Paris Est, France

Antoine Vacavant   Université Clermont Auvergne, France

Additional Reviewers

Almeida Carneiro, Sarah
Alpers, Andreas
Anosova, Olga
Araujo, Arnaldo
Ascolese, Michela
Asplund, Teo
Banaeyan, Majid
Baudrier, Etienne
Belém, Felipe
Bertrand, Gilles
Bilodeau, Michel
Blusseau, Samy
Borgefors, Gunilla
Bosilj, Petra
Boutry, Nicolas
Bruckstein, Alfred
Brunetti, Sara
Buzer, Lilian
Carlinet, Edwin
Chanda, Bhabatosh
Chaussard, John
Chazalon, Joseph
Comic, Lidija
Cousty, Jean
Crombez, Loic
Debled-Rennesson, Isabelle
Decenciere, Etienne
Detterfelt, Åsa
Dey, Tamal
Didas, Stephan
Dokladal, Petr
Domenjoud, Eric
Ehrle, Steffen
Ericsson, Daniel
Falcão, Alexandre
Fehri, Amin
Feschet, Fabien
Frosini, Andrea

Genctav, Asli
Geraud, Thierry
Godehardt, Michael
Gonzalez-Lorenzo, Aldo
Guimaraes, Silvio
Kiselman, Christer Oscar
Kocay, William
Kurlin, Vitaliy
Kurtz, Camille
Lascabettes, Paul
Leborgne, Aurélie
Lindeberg, Tony
Lohou, Christophe
Lomenie, Nicolas
Magillo, Paola
Malmberg, Filip
Maragos, Petros
Marcotegui, Beatriz
Mari, Jean-Luc
Martins Antunes, Daniel
Mazo, Loïc
Merciol, François
Meyron, Jocelyn
Miranda, Paulo
Monasse, Pascal
Moreaud, Maxime
Mukherjee, Sohan
Nagy, Benedek
Ngo, Phuc
Normand, Nicolas
Noyel, Guillaume
Obara, Boguslaw
Öfverstedt, Johan
Pagani, Silvia
Pal, Shyamosree
Palagyi, Kalman
Palma, Giulia
Paluzo-Hidalgo, Eduardo

Patrocínio Jr., Zenilton K. G.
Perret, Benjamin
Pham, Minh-Tan
Phelippeau, Harold
Pianta, Silvia
Pino Perez, Ramon
Pratihar, Sanjoy
Puybareau, Elodie
Ragnemalm, Ingemar
Redenbach, Claudia
Riva, Mateus
Roussillon, Tristan
Saadat, Mohammadreza
Sangalli, Mateus
Santana Maia, Deise

Schladitz, Katja
Strand, Robin
Tankyevych, Olena
Tarsissi, Lama
Telea, Alexandru
Tochon, Guillaume
Tsilivis, Nikolaos
Vacavant, Antoine
Valle, Marcos Eduardo
Welk, Martin
Xu, Yongchao
Yger, Florian
Zheng, Xiqiang
Zrour, Rita

Contents

Discrete Tomography and Inverse Problems

Hierarchical and Graph-Based Models, Analysis and Segmentation

Learning-Based Approaches to Mathematical Morphology

**Multivariate and PDE-Based Mathematical Morphology,
Morphological Filtering**

Invited Papers

Some Open Questions on Morphological Operators and Representations in the Deep Learning Era
A Personal Vision

Jesús Angulo[✉]

MINES ParisTech, PSL-Research University,
CMM-Centre de Morphologie Mathématique, Fontainebleau, France
`jesus.angulo@mines-paristech.fr`

Abstract. *"Work on deep learning or perish"*: folklore wisdom in 2021.

During recent years, the renaissance of neural networks as the major machine learning paradigm and more specifically, the confirmation that deep learning techniques provide state-of-the-art results for most of computer vision tasks has been shaking up traditional research in image processing. The same can be said for research in communities working on applied harmonic analysis, information geometry, variational methods, etc. For many researchers, this is viewed as an existential threat. On the one hand, research funding agencies privilege mainstream approaches especially when these are unquestionably suitable for solving real problems and for making progress on artificial intelligence. On the other hand, successful publishing of research in our communities is becoming almost exclusively based on a quantitative improvement of the accuracy of any benchmark task.

As most of my colleagues sharing this research field, I am confronted with the dilemma of continuing to invest my time and intellectual effort on mathematical morphology as my driving force for research, or simply focussing on how to use deep learning and contributing to it. The solution is not obvious to any of us since our research is not fundamental, it is just oriented to solve challenging problems, which can be more or less theoretical. Certainly, it would be foolish for anyone to claim that deep learning is insignificant or to think that one's favourite image processing domain is productive enough to ignore the state-of-the-art. I fully understand that the labs and leading people in image processing communities have been shifting their research to almost exclusively focus on deep learning techniques. My own position is different: I do think there is room for progress on mathematically grounded image processing branches, under the condition that these are rethought in a broader sense from the deep learning paradigm. Indeed, I firmly believe that the convergence between mathematical morphology and the computation methods which gravitate around deep learning (fully connected networks, convolutional neural networks, residual neural networks, recurrent neural networks, etc.) is worthwhile.

The goal of this talk is to discuss my personal vision regarding these potential interactions. Without any pretension of being exhaustive, I

© Springer Nature Switzerland AG 2021
J. Lindblad et al. (Eds.): DGMM 2021, LNCS 12708, pp. 3–19, 2021.
https://doi.org/10.1007/978-3-030-76657-3_1

want to address it with a series of open questions, covering a wide range of specificities of morphological operators and representations, which could be tackled and revisited under the paradigm of deep learning. An expected benefit of such convergence between morphology and deep learning is a cross-fertilization of concepts and techniques between both fields. In addition, I think the future answer to some of these questions can provide some insight on understanding, interpreting and simplifying deep learning networks.

1 Mathematical Morphology Is Powerful and Still Attractive Despite Its Age

Mathematical morphology is *not only* a mathematical theory of shape. Its corpus nowadays provides a vast theoretical and practical machinery to address fundamental problems arising from the fields of computer vision and structured-data analysis.

Unfortunately, the significant scope of morphology is overshadowed by its widely use as mainly a post-processing tool to regularize binary images as well as the progressive shift of interest of practitioners towards deep learning techniques, which require little knowledge on image processing theory and provide impressive results. A "win-win" game. In the case of theorists, a progressive fading of interest has been caused by at least three possible reasons: i) the apparent exotic mathematical formulation of morphology, ii) a theoretical apparatus which is aged, or worse, already depleted it of new discoveries, iii) the perception of morphology as a useless theory for the future of signal/image and data processing since, once again, deep learning can solve everything and the mathematics underlying deep learning cannot interact with the algebraic and geometrical formulation of morphology. I challenge these three arguments. I believe, on the contrary, that the arrival of real progress on artificial intelligence based on deep learning provides a mind frame to push the boundaries of morphology and to prove that it is one of the appropriate nonlinear machineries to address some open issues on understanding, interpreting and simplifying deep learning networks. But also to introduce new layers and architectures inspired from morphological operators and representations. Let me enumerate the major themes on mathematical morphology which are relevant in the context of this talk:

- An abstract algebraic formulation of the theory on complete lattices, which requires very little assumptions to be instantiated into a specific lattice structure of the space of interest.
- A common representation theory for the Boolean and the semicontinuous function cases, in which, for instance, any translation-invariant increasing, upper semicontinuous operator can be represented exactly as a minimal superposition of morphological erosions or dilations.
- An intimate relationship with the random set theory via the notion of Choquet capacity from stochastic geometry.

- Strong connections with idempotent mathematics (max-plus and max-min algebra and calculus) and tropical geometry.
- Continuous models which correspond to Hamilton–Jacobi PDEs, relevant also in optics and optimal control.
- A powerful extension to the case of morphology on groups, which bring a proper dealing with space symmetries and provide equivariant operators to the groups of transforms relevant in computer vision.
- Multiscale operators and semigroups formulated in Riemannian, metric and ultrametric spaces.
- Multiple morphological representations that provide a rich family of shape-based and geometrical descriptions and decompositions: skeletons, pattern spectra and size distributions, topological description of functions using maxima-minima extinction values, etc.
- A privileged mathematical tool for Lipschitz characterization and regularization.
- A counterpart of the perceptron which yields to the scope of morphological neural networks, morphological associative memories.

The previous list is not exhaustive and of course, it is based on my personal research interests. Nevertheless, I do believe it illustrates which I mean by a central position on the field of nonlinear mathematics for visual computing.

2 A Selection of Themes Where Morphology and Deep Learning Can Meet

I discuss now in a rather informal way a few fields of potential interaction between deep learning and mathematical morphology and the questions which arise from. The bibliography is not exhaustive since I am covering a large scope of topics. Some of the subjects that I mention below are already the object of current research. Due to prospective nature of these reflections, this should be considered more like a personal roadmap than a systematic review of the state-of-the-art.

2.1 Lattice Theory and Algebraic Models for Deep Learning

Current mathematical models for deep learning networks are based on approximation theory and harmonic analysis [26,76]. Other approaches explore the relevance of tropical geometry [50] to describe networks with Rectified Linear Units (ReLUs) [8,84]. The Matheron-Maragos-Banon-Barrera (MMBB) [10,48,49] representation theorems provide an astonishing general formulation for any nonlinear operator between complete lattices, based on combinations of infimum of dilations and supremum of erosions. The theory is relevant when the basis (minimal kernel) of the operators can be learnt. In the case of non-increasing or non-translation-invariant operators the constructive decomposition of operators become more complex but still it would based on basic morphological dilation, erosion, anti-dilation and anti-erosion which can be the minimal bricks to construct architectures of networks which mimic the MMBB representations.

How effective MMBB networks would be to learn the minimal basis of structuring functions approximating any nonlinear image transform? How the idea of hierarchical architectures from deep learning can be used in the case of MMBB networks?

Can MMBB networks be combined with standard layers in deep learning pipelines, providing relevant learnable models?

Any network architecture combining convolution, down/up-sampling, ReLUs, etc. could be seen at first sight as incompatible with lattice theory formulation. In fact, as it was shown by Keshet [40,41], low-pass filters, decimation/interpolation, Gaussian/Laplacian pyramids and other typical image processing operators, admit an interpretation as erosions and adjunctions in the framework of (semi)-lattices. In addition, max-pooling and ReLUs are just dilation operators. The notion of deepness or recurrence in a network can be seen as the iteration of basic operators, which yields to questions on the convergence to idempotency or, at least, to study order stability in the corresponding lattice [34].

What kind of unified algebraic models, integrating standard layers and morphological layers can be used to mathematically study deep learning architectures? Is there any information on order continuity, on invariance and fixed points, on decomposition and simplification, etc., which can be inferred from these unified algebraic models?

What is the expressiveness of deep MMBB networks and the hybrid deep networks?

This last question is related to study of the capacity of neural networks to be universal approximators for smooth functions. For instance, both maxout networks [27] and max-plus networks [85] can approximate arbitrarily well any continuous function on a compact domain. The proofs are based on the fact that [75] continuous pricewise linear (PWL) functions can be expressed as a difference of two convex PWL functions, and each convex PWL can be seen as maximum of affine terms. Alternative theory by Ovchinnikov [56,57] shows that a PWL (or a smooth) function can be represented as a max-min polynomial of its linear components, and the theory is also valid in a Boolean representation. The representation formulas by Ovchinnikov are equivalent to MMBB theorems which justify the potential interest of the latter. Tropical formulation of ReLU networks has shown that a deeper network is exponentially more expressive than a shallow network [84]. To explore the expressiveness of complex morphological networks with respect to the deepness is therefore a fundamental relevant topic.

2.2 Lipschitz Regularity in Neural Networks

Lipschitz regularity has been proven to be a fundamental property to deal with robustness of the predictions made by deep neural networks when their input is subject to an adversarial perturbation [29]. This mathematical topic of Lipschitz regularity is quite important in deep learning since adversarial attacks against machine learning models are a proof of their limited resilience to small perturbations. Training neural networks under a strict Lipschitz constraint is useful also for generalization bounds and interpretable gradients [72]. By the composition property of Lipschitz functions, it suffices to ensure that each individual affine transformation or nonlinear activation is 1-Lipschitz. That can be achieved by constraining the spectral norm of the weights in the layers: for instance, maintaining during the training weight matrices of linear and convolutional layers to be approximately Parseval tight frames (extensions of orthogonal matrices to non-square matrices) [20]. This approach satisfies the Lipschitz constraint, but comes at a cost in expressive power [32]. Other techniques replace the ReLU layers by more elaborated functions based of ordering the inputs and computing max-min operations (GroupSort activation function) [7]. Morphological operators using multiscale convex structuring functions are a powerful tool to deal with Lipschitz extension of functions [3], which is connected to Lasry-Lions regularization [43].

What kind of Lipschitz morphological layers can be introduced into deep networks to control their Lipschitz constant and therefore their regularity? Can these morphological regularizing layers replace the standard nonlinearities like pooling+ReLU without degrading their expressiveness?

The later question is also the object of related work [23], where it has been proven that Lipschitz constraint models using FullSort activation functions are universal Lipschitz function approximators.

2.3 Group Equivariance and Integration of Data Symmetries and Topology

Motivated by the Gestalt pattern theory, some studies investigated by synthetic experiments the ability of deep learning to infer simple (at least for human) visual concepts, such as shape or symmetry, from examples [80]. Humans can often infer a semantic geometric/morphological concept quickly after looking at only a very small number of examples; on the contrary, deep convolutional neural networks approximate some concepts statistically, but only after seeing many (thousands or millions) more examples. It seems reasonable that the use of morphological layers which deal more naturally with the notion of shape could improve some visual taks.

Are networks integrating morphological layers more "intelligent" (i.e., requires less training samples) than standard deep learning architectures to learn tasks inspired from Gestalt pattern theory?

I think in particular about the potential role to be played by the theory of group morphology [61], which extends the construction of morphological operators to be invariant under, for instance, the motion group, the roto-translation group, the affine group, the projective group, etc. Indeed, the notion of group equivariance of networks [22] is central nowadays in the field of deep learning. It provides a sound approach to deal with the explicit introduction of the desired symmetries into the network, without the need to approach them by means of costly techniques such as data augmentation.

A combinatorial shape problem called generalized Tailor problem [60], which is connected to the one of finding the decomposition of any shape according to a set of templates, can be a relevant to assess the interest of group morphology in deep learning based part-based decomposition.

> How efficiently can be solved the generalized tailor problem using networks inspired from the morphological iterative algorithm?

Another interesting problem is to design models for machine learning tasks defined on sets [83]. In contrast to traditional approaches, which operate on fixed dimensional vectors, the idea is to consider objective functions on sets that are invariant to permutations. The issue is also relevant on graphs. This equivariance to the permutation of the elements of the input requires specific pooling strategies across set-members. In mathematical morphology and discrete geometry, there are compact representations of sets by a minimal number of points, typically based on the notion of skeleton, maxima/minima of the signed distance function, etc. Other shape morphological representations such as shape-size distribution from granulometries, provided set descriptors which are invariant to many transforms of the sets. It was proved in the past the interest of skeletons and shape-size representations when they are combined with neural networks [81].

> Are morphological compact representations of sets more efficient to deal with permutation invariance in machine learning? Can we introduce loss functions based on morphological descriptions which are inherently permutation-invariant and scale-invariant?

In a similar way, considering that an image is a function whose relevant information is associated to its topology, namely the location of the maxima/minima and the features associated to them, provides a representation which is invariant to many isometries. Dealing with maxima and minima can be addressed using morphological representations based on residues of morphological reconstruction (and iterative algorithm) and the appropriate markers [74]. The idea of topology-preserving has been considered for the problem of deep image segmentation [31], basically to learn continuous-valued loss functions that enforces a segmentation to have the same topology as the ground truth. That is done using the notion of persistence diagrams from computational topology [24]. However, the integration of computational topology and deep learning is not natural.

> Can we learn persistence (extinction values) features to be used in topological image classification and segmentation using morphological operators based on reconstruction? Which architecture to be used for the iterative-based reconstruction (residual, recurrent, other)?

2.4   Interpretability and "Small" Parametric Models

Due to the black-box nature of deep learning, it is inherently difficult to understand which aspects of the input data are contributing to the decisions on a complex network. It is also difficult to identify which combinations of data features are appropriate in the context of the deployment of networks as a decision support system in critical domains. Understating better by humans why a deep neural network is taking a particular decision is the object of the so-called explainable deep learning [79].

A way to move towards explainable networks is to have layers which are easy to interpret. For instance, if a part of the network is learning patterns, as in a template matching problem, and those patterns are easy to visualize, this part of the networks can be explainable. One of the most studied and rather simple morphological operator (i.e., the intersection of an erosion and an anti-dilation), the hit-or-miss transform, can be seen as a powerful template matching approach.

The use of the hit-or-miss transform as part of a neural network for object recognition was pioneered in [77] and some recent attempts of extending its use in the context of deep neural networks are promising [36]. However, to have robust to noise template detection [13] or the multiple ways to extend the hit-or-miss transform to grey-scale images [42,54], as well as the fact that the patterns to be matched can appear at different scales or at different rotations, yield interesting topics to be explored.

What is the best formulation for the hit-or-miss transform to provide easy learnable and robust template extraction layers? How efficient is the integration of hit-or-miss layers into a complex architecture of deep learning? Only as the first layers?

How to deal with equivariance in pattern detection by means of group morphology-based hit-or-miss transforms?

What other morphological template extraction operators [62] are relevant as interpretable layers in deep learning?

An alternative in the quest for a better interpretability of deep learning is to replace regular convolutional neural networks filters by parametric families of canonical or well-known filters and scales spaces. That reduces significantly the number of parameters and make them more interpretable. These hybrid approaches are constructed by coupling parameterized scale-space operations in cascade [44], or circular harmonics banks of filters [78] or Gabor filters [46], etc. with other neural networks layers.

In the case of morphological operators, we can also consider the use of parametric models. For instance, by defining an architecture mimicking the notion of granulometry: a series of multiscale openings followed by a global integral pooling operator, such that the parameters to be learn would be the structuring function, which would shared by all the openings, and the scale parameter for each opening. We can also consider the interest of learning parametric structuring functions (typically quadratic ones) or to consider pipelines of multiscale dilations/erosions used to predict quantitative parameters like the fractal dimension or the Hölder exponent [6].

Are there architectures based on parametric families of morphological multiscale operators which can be efficiently learned and provide a better interpretation on tasks of shape or texture recognition?

What parametric families of structuring functions are the most fruitful in deep learning: quadratic ones defined by a shape covariance matrix? Convex ones defined by Minkowski addition of oriented segments?

2.5   Image Generation and Simulation of Microstructures

Morphological operators are the fundamental computational tool for the characterization and simulation of random sets (for instance, the binary images associated to a random microstructure) in the theory developed by Matheron [48]. The notion of Choquet capacity of a random set relies on computing how the integral of the set changes when it is dilated or eroded by a particular structuring element. By considering specific families of structuring elements (i.e., pairs or triplets of points, segments, disks, etc.), the random set is characterized and, by working on well-studied stochastic models of random sets, the corresponding parameters of the model are fit thanks to the morphological measurements. Then, it is possible to simulate new images following the model and test if the new images have the prescribed morphological measurements. This theory has been of significant success in the characterisation and simulation of microstructure images in material sciences [38].

In the field of deep learning, Generative Adversarial Networks (GANs) are an approach to generate images from an illustrative dataset [28,65]. GANs involve automatically discovering and learning the regularities or patterns in the input data, then the model is used to generate new examples that plausibly could have been drawn from the initial dataset. GANs consider the problem as a supervised learning framework with two sub-models: the generator model that one trains to generate new examples, and the discriminator model that tries to classify examples as either real (from the domain) or fake (generated). The two models are trained together until the discriminator model is fooled about half the time, meaning the generator model is producing plausible fake examples. The generator is typically a deconvolutional neural network, and the discriminator is a convolutional neural network.

GANs are nowadays used in many domains, where image simulation or image synthesis is required, with impressive visual results. GANs are also being explored in the field of virtual material simulation in physics [21,82]. The additional dimension in material science is the fact that the generated image should satisfy some physical or mathematical constraints. For instance, the model can explicitly enforces known physical invariances by replacing the traditional discriminator in a GAN with an invariance checker [68].

Can the notion of Choquet capacity play a role in the GAN discriminators to explicitly impose morphological constraints learned from the empirical data? Can one incorporate a "deconvolution" image simulation closer to the morphological random set models into the GAN generators, thanks to the use of morphological layers?

2.6 Ultrametric Convolutional Neural Networks

Many scientific fields work on data with an underlying mathematical structure modelled as a non-Euclidean space. Some examples include social networks, sensor networks, biological networks (functional networks in brain imaging or regulatory networks in genetics), and meshed surfaces in computer vision. Geometric deep learning is a generic term for techniques attempting to generalize structured deep neural models to non-Euclidean domains such as graphs and manifolds [14]. The area of deep learning on graphs is particularly active [9,86], with many alternative approaches seeking to generalize to graphs fundamental deep learinng notions as convolution, pooling, coding/decoding, loss functions, etc. Research topics for instance deal with the role played by the nodes/edges, the use of graph spectral techniques or kernel methods, etc.

Another discrete powerful setting for structured data (or unstructured data which can be first embedded into a graph) are the hierarchical representations associated to dendrograms (rooted trees). In that case, the mathematical structure corresponds to an ultrametric space. Ultrametric spaces are ubiquitous in mathematical morphology, ranging from watershed segmentation (minimum spanning tree [52,55]) to connected-components preserving filtering (notion of max/min-tree [64]). When an image, or any other element in a dataset embedded into an edge-weighted graph, is represented by a dendrogram, the use of deep learning techniques and in particular of ultrametric convolutional neural networks requires to be have a specific definition of the typical layers for dendrograms: their structure is very different from metric graphs. Classical image/data processing opeartors and transforms, including Gaussian and Laplacian operators, convolution, morphological semigroups, etc., have been formulated on ultrametric spaces [4,5], and the basic ingredients are the ultrametric distance and the distribution of diameters of the ultrametric balls.

How to efficiently learn the convolution operation on ultrametric convolutional neural networks? How to define the max-pooling and unpooling-layers on ultrametric convolutional neural networks? Are there specific layers useful in this kind of neural networks?

Note that this is different from the problem of learning an ultrametric distance from a dissimilarity graph, an optimization problem that can be now efficiently solved by gradient descent methods [19], useful in learning watershed image segmentation or in other hierarchical clustering methods. Both approaches could be potentially integrated into end-to-end learnable frameworks, where graphs can be embedded into dendrograms and then, ultrametric deep learning techniques could be used for classification or prediction.

2.7   The Difficulty of Training Morphological Neural Networks

The training of morphological neural networks has been, and still is, the object of active research. The dendrite morphological neurons have been trained using geometrical-based algorithms (enclosing patterns in hyper- boxes) [59,71]. The non-differentiability of the lattice-based operations makes training morphological neural networks more difficult for gradient-based algorithms, but the backpropagation can be achieved without difficulty [85]: in fact, maxpooling and ReLU are just examples of morphological layers widely used in deep learning.

More recently, inspired from the tropical mathematics framework, training algorithms not rooted in stochastic optimization but rather on Difference-of-Convex Programming have been used for training dilation and erosion layers [18]. These techniques have been also adapted to train other generalized dilation-erosion perceptrons combined with linear transformations [73]. Additional progress have been made the use of tropical geometry tools for neural network pruning [70]. The property of sparsity induced by max-plus layers [70,85] seems of great potential, however to integrate morphological layers into complex deep learning architectures is not always straightforward from the viewpoint of the optimization.

Is there any efficient network learning technique which can combine the optimization techniques inspired from tropical geometry and stochastic gradient descent?

What are the best gradient-descent optimizers in the case of hybrid networks including morphological layers? Is there any strategy of alternate optimization between the convolution layers and the morphological layers?

An alternative is to use smooth approximations of the max-min based morphological operators, either using counter-harmonic means [47] or Log-Sum-Exp terms [17] (related to the Maslov dequantization [2,45]). In these differentiable frameworks, the morphological ones are the limit cases and it allows, if the nonlinearity parameter is learned too, to provide layers which after training can behave as standard convolutions or as morphological ones.

Which smooth approximation to morphological operators is more relevant for training deep learning hybrid networks?

Discrete geometry focusses on the mathematically sound definitions and efficient algorithms to study binary discrete objects, like lines, circles, convex shapes, etc., as well as to work on discrete functions. Discrete convolution and its equations have been also studied [39]. Deep Learning with limited numerical precision [30] and in particular with integer-arithmetic-only [37] is relevant in the field of resource-efficient AI, especially for its deployment in embedded systems. The issue of neural network training with constrained integer weights was considered also in the past [58]. The limit case of discrete representation and computation corresponds to the binary or ternary neural networks [1,25,87]. Efficiently training these systems uses rounding off, and other numerical tricks, which do not always guarantees some properties related to discrete or binary convolution, down/up-sampling, etc.

Is it possible to incorporate genuine discrete layers and to train them using gradient-descent approaches? Are there other approaches from discrete optimization better adapted to these networks?

Connections between convolution, non-linear operators and PDEs have been the object of major research in image processing during many years. Image data is interpreted as the discretization of multivariate functions and the output of image processing algorithms as solutions to some PDEs. A few works have considered network layers as a set of PDE solvers, where the geometrically significant coefficients of the equation become the trainable weights of the layer [63,67]. The well-established theory of PDEs allows the introduction of neural network layers with good approximation properties (relevant for the problem of equivariance, for instance). The potential interest of the interpretation of some non-linear layers from the viewpoint of morphological PDEs has been considered in [69], even if the formulated PDEs are solved using the viscosity solutions, which correspond to the morphological convolutions. It would be interesting to explore the interest of the numerical solvers for Hamilton–Jacobi PDEs which can be plugged into a deep learning pipelines to learn non-linearities and morphological layers.

What numerical schemes for morphological PDEs are relevant in order to learn morphological operators or another non-linear layer? How to deal with the iterative nature of the approximations?

I think about the case of deep learning for positive definite matrices (SPD) [35]. Riemannian geometry-based tools are used to formulate neural network layers which allow computing and learning in that setting. The non-linearity layers, like ReLU and max-pooling, are not well formulated in the SPD case. There is a well-established theory of numerical solution schemes of PDE-based morphology for matrix fields [15,16].

Can the numerical schemes for morphological PDEs of matrix fields be used to learn non-linearities for SPD, or other matrices, in Riemannian deep learning?

2.8 Morphological AI

More than thirty year ago, Schmitt [66] showed the possibility of creating an automatic programming system of artificial intelligence morphology for image processing, in a rule-based paradigm of geometric reasoning. The starting point is the obvious fact that complex morphological transforms are just based on a few bricks. Those primitives can be seen as the words of morphological language and the possible combinations making sense as the grammar of the language.

Typical primitives in [66] are dilations, erosions, hit-or-mis transforms, thinnings and thickennigs, with their corresponding structuring elements. The appropiate constructive combinations provide the grammar. The solution for a given problem is solved using combinational optimization. Other paradigms producing successful automatic design of morphological operators were based on genetic algorithms [33] or on PAC (Probably Approximately Correct) learning [12]. All these approaches are based on learning the transformations by collections of observed-ideal pairs of images and the result of the desirable operators from these data, which fit with training from datasets as it is done nowadays in deep learning. In the field of deep learning, the first tentative of learning pipelines of morphological operators was [47]. Using the counter-harmonic mean as an asymptotic approximation to both dilation and erosion, it was proved that stochastic gradient descent-based convolutional neural networks can learn both the structuring element and the composition of operators, including compositions of openings and closings which approximate TV-regularization. However, this kind of approach does not exploit the vast complexity of morphological language.

An alternative would be to use natural language processing (NLP) deep learning techniques. In that context, the training dataset will be composed of examples of morphological programs, considered as a morphological text, written by experts to solve specific tasks, illustrated with input-output images too. Supervised NLP tasks are based on building pretrained representations of the distribution of words (called word embeddings), such word2vec or context2vect [51,53].

What is the most appropiate coding of the morphological language to be use with NLP techniques? Is it efficient to use a word for the operator and a declination for the structuring element? Or to use different words for operator and structuring element?

What is the optimal granularity on the decomposition of the operators to obtain a language with a high enough semantic interpretation and which can be still learnt?

In general NLP, training the word embeddings required a (relatively) large amount of data, which reduced the amount of labeled data necessary for training on the supervised tasks.

Do we have a large enough corpus of morphological programs available to learn the operator2vect embeddings? How to extract and parse morphological programs available in multiple repositories to train the algorithms?

Another source of inspiration for developing morphological AI is the field of deep coding [11]. This time, the perspective from the computational morphology viewpoint is to start from a toolbox of programmed morphological functions and the goal would be to learn to write programs using that basic functions. Solving automatic programming problems from input-output examples using deep learning is quite challenge and works only on domain specific languages, which is the case of a morphological language toolbox. It requires to find consistent programs by searching over a suitable set of possible ones, and a ranking of them if there are multiple programs consistent with the input-output examples. The current paradigm provides only limited and short programs. These techniques can be more efficiently use eventually to rewrite programs and to decompose them into subprograms which can be learn and optimize separately.

How can one introduce syntactical transformations to simplify and rewrite a morphological program?

How to integrate combinatorial and deep learning techniques to explore large enough program spaces which can provide an efficient morphological AI?

3 My Conclusion: Invest Yourself in Studying the Theory

I anticipate that the main reaction of the reader may be that of frustration. I have not provided many more reasons for the choice of the previous topics than my own intuition. I concede that the answer to my open questions will not always lead to any methodological or algorithmic breakthroughs. However, working on the program I discussed above, or on alternative problems, will definitely advance our discipline and keep it flourishing.

I am open to collaborate and discuss with anyone interested in some of these topics. Joining efforts to work on challenging problems is fundamental in a research field quickly moving forward and in multiple directions.

A last take-home message. Successful interaction between morphology and deep learning is not only related to computational based aspects. It requires, about all, an intimate understating of the theoretical aspects of both fields. My main advice for young researchers starting a PhD thesis on exploring these interactions is the following: invest part of your time in studying the theoretical papers, read your *classics*, enlarge the scope of your theoretical interests. It will be worthwhile.

References

1. Alemdar, H., Leroy, V., Prost-Boucle, A., Pétrot, F.: Ternary Neural Networks for Resource-Efficient AI Applications. arXiv:1609.00222 (2017)
2. Angulo, J., Velasco-Forero, S.: Stochastic morphological filtering and Bellman-Maslov chains. In: Hendriks, C.L.L., Borgefors, G., Strand, R. (eds.) ISMM 2013. LNCS, vol. 7883, pp. 171–182. Springer, Heidelberg (2013). https://doi.org/10.1007/978-3-642-38294-9_15
3. Angulo, J.: Lipschitz Regularization of Images supported on Surfaces using Riemannian Morphological Operators. HAL hal-01108130v2 (2014)
4. Angulo, J., Velasco-Forero, S.: Morphological semigroups and scale-spaces on ultrametric spaces. In: Angulo, J., Velasco-Forero, S., Meyer, F. (eds.) ISMM 2017. LNCS, vol. 10225, pp. 28–39. Springer, Cham (2017). https://doi.org/10.1007/978-3-319-57240-6_3

5. Angulo, J.: Hierarchical laplacian and its spectrum in ultrametric image processing. In: Burgeth, B., Kleefeld, A., Naegel, B., Passat, N., Perret, B. (eds.) ISMM 2019. LNCS, vol. 11564, pp. 29–40. Springer, Cham (2019). https://doi.org/10.1007/978-3-030-20867-7_3

6. Angulo, J.: Hölder Exponents and Fractal Analysis on Metric Spaces using Morphological Operators. HAL hal-03108997 (2021)

7. Anil, C., Lucas, J., Grosse, R.: Sorting out Lipschitz function approximation. arXiv:1811.05381 (2019)

8. Arora, R., Basu, A., Mianjy, P., Mukherjee, A.: Understanding Deep Neural Networks with Rectified Linear Units. arXiv. 1611.01491 (2018)

9. Bacciu, D., Errica, F., Micheli, A., Podda, M.: A gentle introduction to deep learning for graphs. Neural Netw. **129**, 203–221 (2020)

10. Banon, G.J.F., Barrera, J.: Minimal representations for translation-invariant set mappings by mathematical morphology. SIAM J. Appl. Math. **51**(6), 1782–1798 (1991)

11. Balog, M., Gaunt, A.L., Brockschmidt, M., Nowozin, S., Tarlow, D.: DeepCoder: Learning to Write Programs. arXiv:1611.01989 (2017)

12. Barrera, J., Terada, R., Hirata Jr., R., Hirata, N.S.T.: Automatic programming of morphological machines by PAC learning. Fund. Inform. **41**(1–2), 229–258 (2000)

13. Bloomberg, D.S., Vincent, L.: Pattern matching using the blur hit-or-miss transform. J. Electron. Imaging **9**, 140–150 (2000)

14. Bronstein, M., Bruna, J., LeCun, Y., Szlam, A., Vandergheynst, P.: Geometric deep learning: going beyond Euclidean data. IEEE Signal Process. Mag. **34**(4), 18–42 (2017)

15. Burgeth, B., Breuß, M., Didas, S., Weickert, J.: PDE-based morphology for matrix fields: numerical solution schemes. In: Aja-Fernández, S., de Luis García, R., Tao, D., Li, X. (eds.) Tensors in Image Processing and Computer Vision, pp. 125–150. Springer, London (2009). https://doi.org/10.1007/978-1-84882-299-3_6

16. Burgeth, B., Kleefeld, A.: A unified approach to PDE-driven morphology for fields of orthogonal and generalized doubly-stochastic matrices. In: Angulo, J., Velasco-Forero, S., Meyer, F. (eds.) ISMM 2017. LNCS, vol. 10225, pp. 284–295. Springer, Cham (2017). https://doi.org/10.1007/978-3-319-57240-6_23

17. Calafiore, G.C., Gaubert, S., Possieri, C.: Log-sum-exp neural networks and posynomial models for convex and log-log-convex data. arXiv:1806.07850 (2018)

18. Charisopoulos, V., Maragos, P.: Morphological perceptrons: geometry and training algorithms. In: Angulo, J., Velasco-Forero, S., Meyer, F. (eds.) ISMM 2017. LNCS, vol. 10225, pp. 3–15. Springer, Cham (2017). https://doi.org/10.1007/978-3-319-57240-6_1

19. Chierchia, G., Perret, B.: Ultrametric fitting by gradient descent. J. Stat. Mech: Theory Exp. **12**, 124004 (2020)

20. Cisse, M., Bojanowski, P., Grave, E., Dauphin, Y., Usunier, N.: Parseval Networks: Improving Robustness to Adversarial Examples. arXiv:1704.08847 (2017)

21. Chun, S., Roy, S., Nguyen, Y.T., et al.: Deep learning for synthetic microstructure generation in a materials-by-design framework for heterogeneous energetic materials. Sci. Rep. **10**, 13307 (2020)

22. Cohen, T.S., Welling, M.: Group equivariant convolutional networks. arXiv:1602.07576 (2016)

23. Cohen, J.E.J., Huster, T., Cohen, R.: Universal Lipschitz Approximation in Bounded Depth Neural Networks. arXiv:1904.04861 (2019)

24. Cohen-Steiner, D., Edelsbrunner, H., Harer, J.: Stability of persistence diagrams. Discrete Comput. Geom. **37**(1), 103–120 (2007)

25. Courbariaux, M., Hubara, I., Soudry, D., El-Yaniv, R., Bengio, Y.: Binarized neural networks: training deep neural networks with weights and activations constrained to +1 or −1. arXiv:1602.02830 (2016)
26. Daubechies, I., DeVore, R., Foucart, S., Hanin, B., Petrova, G.: Nonlinear Approximation and (Deep) ReLU Networks. arXiv:1905.02199 (2019)
27. Goodfellow, I., Warde-Farley, D., Mirza, M., Courville, A., Bengio, Y.: Maxout networks. In: Proceedings of ICML 2013, III, pp. 1319–1327 (2013)
28. Goodfellow, I., et al.: Generative adversarial networks. In: Proceedings of NIPS 2014 (2014)
29. Goodfellow, I., Shlens, J., Szegedy, C.: Explaining and harnessing adversarial examples. In: Proceedings of ICLR 2015 (2015)
30. Gupta, S., Agrawal, A., Gopalakrishnan, K., Narayanan, P.: Deep Learning with Limited Numerical Precision. arXiv:1502.02551 (2015)
31. Hu, X., Fuxin, L., Samaras, D., Chen, C.: Topology-Preserving Deep Image Segmentation. arXiv:1906.05404 (2019)
32. Huster, T., Chiang, C.-Y.J., Chadha, R.: Limitations of the Lipschitz constant as a defense against adversarial examples. arXiv:1807.09705 (2018)
33. Harvey, N.R., Marshall, S.: The use of genetic algorithms in morphological filter design. Signal Process. Image Commun. 8(1), 55–71 (1996)
34. Hejmans, H.J.A.M., Serra, J.: Convergence, continuity, and iteration in mathematical morphology. J. Vis. Commun. Image Represent. 3(1), 84–102 (1992)
35. Huang, Z., Gool, L.V.: A Riemannian Network for SPD Matrix Learning. arXiv:1608.04233 (2016)
36. Islam, M.A., et al.: Extending the Morphological Hit-or-Miss Transform to Deep Neural Networks. arXiv:1912.02259 (2020)
37. Jacob, B., et al.: Quantization and Training of Neural Networks for Efficient Integer-Arithmetic-Only Inference. arXiv:1712.05877 (2017)
38. Jeulin, D.: Morphological models. In: Altenbach, H., Öchsner, A. (eds.) Encyclopedia of Continuum Mechanics. Springer, Heidelberg (2018). https://doi.org/10.1007/978-3-662-53605-6
39. Kiselman, C.O.: Estimates for solutions to discrete convolution equations. Mathematika 61, 295–308 (2015)
40. Keshet, R.: A morphological view on traditional signal processing. In: Goutsias, J., Vincent, L., Bloomberg, D.S. (eds.) Mathematical Morphology and its Applications to Image and Signal Processing. Computational Imaging and Vision, vol. 18, pp. 3–12. Springer, Boston (2002). https://doi.org/10.1007/0-306-47025-X_2
41. Keshet, R., Heijmans, H.J.A.M.: Adjunctions in pyramids, curve evolution and scale-spaces. Int. J. Comput. Vision 52, 139–151 (2003)
42. Khosravi, M., Schafer, R.W.: Template matching based on a grayscale hit-or-miss transform. IEEE Trans. Image Process. 5(5), 1060–1066 (1996)
43. Lasry, J.M., Lions, P.-L.: A remark on regularization in Hilbert spaces. Israel J. Math. 55, 257–266 (1986)
44. Lindeberg, T.: Scale-covariant and scale-invariant Gaussian derivative networks. arXiv:2011.14759 (2021)
45. Litvinov, G.L.: Maslov dequantization, idempotent and tropical mathematics: a brief introduction. J. Math. Sci. 140(3), 426–444 (2007)
46. Luan, S., Chen, C., Zhang, B., Han, J., Liu, J.: Gabor convolutional networks. IEEE Trans. Image Process. 27(9), 4357–4366 (2018)

47. Masci, J., Angulo, J., Schmidhuber, J.: A learning framework for morphological operators using counter–harmonic mean. In: Hendriks, C.L.L., Borgefors, G., Strand, R. (eds.) ISMM 2013. LNCS, vol. 7883, pp. 329–340. Springer, Heidelberg (2013). https://doi.org/10.1007/978-3-642-38294-9_28
48. Matheron, G.: Random Sets and Integrad Geometry. Wiley, NewYork (1975)
49. Maragos, P.: A representation theory for morphological image and signal processing. IEEE Trans. Pattern Anal. Mach. Intell. **11**(6), 586–599 (1989)
50. Maragos, P., Theodosis, E.: Tropical Geometry and Piecewise-Linear Approximation of Curves and Surfaces on Weighted Lattices. arXiv:1912.03891 (2019)
51. Melamud, O., Goldberger, J., Dagan, I.: context2vec: learning generic context embedding with bidirectional LSTM. In: Proceedings of the 20th SIGNLL Conference on Computational Natural Language Learning, pp. 51–61 (2016)
52. Meyer, F.: Watersheds on weighted graphs. Pattern Recogn. Lett. **47**, 72–79 (2014)
53. Mikolov, T., Sutskever, I., Chen, K., Corrado, G., Dean, J.: Distributed representations of words and phrases and their compositionality. In: Proceedings of NIPS 2013, pp. 3111–3119 (2013)
54. Naegel, B., Passat, N., Ronse, C.: Grey-level hit-or-miss transforms-part i: unified theory. Pattern Recogn. **40**(2), 635–647 (2007)
55. Najman, L., Cousty, J., Perret, B.: Playing with Kruskal: algorithms for morphological trees in edge-weighted graphs. In: Hendriks, C.L.L., Borgefors, G., Strand, R. (eds.) ISMM 2013. LNCS, vol. 7883, pp. 135–146. Springer, Heidelberg (2013). https://doi.org/10.1007/978-3-642-38294-9_12
56. Ovchinnikov, S.: Boolean representation of manifolds functions. J. Math. Anal. Appl. **263**, 294–300 (2001)
57. Ovchinnikov, S.: Max-min representations of piecewise linear functions. Beiträge Algebra Geom. **43**, 297–302 (2002)
58. Plagianakos, V.P., Vrahatis, M.N.: Neural network training with constrained integer weights. In: Proceedings of the IEEE 1999 Congress on Evolutionary Computation-CEC 1999, vol. 3, pp. 2007–2013 (1999)
59. Ritter, G.X., Urcid, G.: Lattice algebra approach to single-neuron computation. IEEE Trans. Neural Networks **14**(2), 282–295 (2003)
60. Roerdink, J.B.T.M.: The generalized tailor problem. In: Maragos, P., Schafer, R.W., Butt, M.A. (eds.) Mathematical Morphology and its Applications to Image and Signal Processing. Computational Imaging and Vision, vol. 5. Springer, Boston (1996). https://doi.org/10.1007/978-1-4613-0469-2_8
61. Roerdink, J.B.T.M.: Group morphology. Pattern Recogn. **33**(6), 877–895 (2000)
62. Ronse, C.: A lattice-theoretical morphological view on template extraction in images. J. Vis. Commun. Image Represent. **7**(3), 273–295 (1996)
63. Ruthotto, L., Haber, E.: Deep neural networks motivated by partial differential equations. J. Math. Imaging Vision 1–13 (2018)
64. Salembier, P., Garrido, L.: Binary partition tree as an efficient representation for image processing, segmentation, and information retrieval. IEEE Trans. Image Process. **9**(4), 561–576 (2000)
65. Salimans, T., Goodfellow, I., Zaremba, W., Cheung, V., Radford, A., Chen, X.: Improved Techniques for Training GANs. arXiv:1606.03498 (2016)
66. Schmitt, M.: Mathematical morphology and artificial intelligence: an automatic programming system. Signal Process. **16**(4), 389–401 (1989)
67. Shen, Z., He, L., Lin, Z., Ma, J.: PDO-eConvs: Partial Differential Operator Based Equivariant Convolutions. arXiv:2007.10408 (2020)

68. Singh, R., Shah, V., Pokuri, B., Sarkar, S., Ganapathysubramanian, B., Hegde, Ch.: Physics-aware Deep Generative Models for Creating Synthetic Microstructures. arXiv:1811.09669 (2018)
69. Smets, B., Portegies, J., Bekkers, E., Duits, R.: PDE-based Group Equivariant Convolutional Neural Networks. arXiv:2001.09046 (2020)
70. Smyrnis, G., Maragos, P.: Tropical Polynomial Division and Neural Networks. arXiv:1911.12922 (2019)
71. Sossa, H., Guevara, E.: Efficient training for dendrite morphological neural networks. Neurocomputing **131**, 132–142 (2014)
72. Tsipras, D., Santurkar, S., Engstrom, L., Turner, A., Madry, A.: There is no free lunch in adversarial robustness (but there are unexpected benefits). arXiv:1805.12152 (2018)
73. Valle, M.E.: Reduced dilation-erosion perceptron for binary classification. Mathematics **8**(4), 512 (2020)
74. Vincent, L.: Morphological grayscale reconstruction in image analysis: applications and efficient algorithms. IEEE Trans. Image Process. **2**(2), 176–201 (1993)
75. Wang, S.: General constructive representations for continuous piecewise-linear functions. IEEE Trans. Circ. Syst. I **51**(9), 1889–1896 (2004)
76. Wiatowski, T., Bölcskei, H.: A mathematical theory of deep convolutional neural networks for feature extraction. IEEE Trans. Inf. Theory **64**(3), 1845–1866 (2018)
77. Won, Y., Gader, P.D., Coffield, P.C.: Morphological shared-weight networks with applications to automatic target recognition. IEEE Trans. Neural Networks **8**(5), 1195–1203 (1997)
78. Worrall, D.E., Garbin, S.J., Turmukhambetov, D., Brostow, G.J.: Harmonic Networks: Deep Translation and Rotation Equivariance. arXiv:1612.04642 (2017)
79. Xie, N., Ras, G., van Gerven, M., Doran, D.: Explainable Deep Learning: A Field Guide for the Uninitiated. arXiv:2004.14545 (2020)
80. Yan, Zh., Zhou, X.S.: How intelligent are convolutional neural networks? arXiv. 1709.06126 (2017)
81. Yang, P.-F., Maragos, P.: Morphological systems for character image processing and recognition. In: IEEE International Conference on Acoustics, Speech, and Signal Processing, vol. 5, pp. 97–100 (1993)
82. Yang, Z., Li, X., Brinson, C.L., Choudhary, A.N., Chen, W., Agrawal, A.: Microstructural materials design via deep adversarial learning methodology. J. Mech. Des. **140**(11) (2018)
83. Zaheer, M., Kottur, S., Ravanbakhsh, S., Poczos, B., Salakhutdinov, R., Smola, A.: Deep Sets. arXiv:1703.06114 (2018)
84. Zhang, L., Naitzat, G., Lim, L.-H.: Tropical Geometry of Deep Neural Networks. arXiv. 1805.07091 (2018)
85. Zhang, Y., Blusseau, S., Velasco-Forero, S., Bloch, I., Angulo, J.: Max-plus operators applied to filter selection and model pruning in neural networks. In: Burgeth, B., Kleefeld, A., Naegel, B., Passat, N., Perret, B. (eds.) ISMM 2019. LNCS, vol. 11564, pp. 310–322. Springer, Cham (2019). https://doi.org/10.1007/978-3-030-20867-7_24
86. Zhang, Z., Cui, P., Zhu, W.: Deep Learning on Graphs: A Survey. arXiv:1812.04202 (2020)
87. Zhu, C., Han, S., Mao, H., Dally, W.J.: Trained Ternary Quantization. arXiv:1612.01064 (2017)

Split Trees – A Unifying Model for Many Important Random Trees of Logarithmic Height: A Brief Survey

Cecilia Holmgren[✉][iD]

Department of Mathematics, Uppsala University, Uppsala, Sweden
cecilia.holmgren@math.uu.se
https://katalog.uu.se/empinfo/?id=N5-824

Abstract. Split trees were introduced by Devroye [28] as a novel approach for unifying many important random trees of logarithmic height. They are interesting not least because of their usefulness as models of sorting algorithms in computer science; for instance the well-known Quicksort algorithm (introduced by Hoare [35,36]) can be depicted as a binary search tree (which is one example of a split tree). A split tree of cardinality n is constructed by distributing n balls (which often represent data items) to a subset of nodes of an infinite tree. In [39], renewal theory was introduced by the author as a novel approach for studying split trees. This approach has proved to be highly useful for investigating such trees and has lead (often in combination with other methods) to several general results valid for all split trees. In this brief survey, we will present split trees, give an introduction to renewal theory in relation to split trees and describe some of the characteristics of split trees including results on the depths for the balls and nodes in the tree, the height (maximal depth) of the tree and the size of the tree in terms of the number of nodes; see [11,17,18,28,39]. Furthermore, we will briefly describe some of our later results for this large class of random trees, e.g. on the total path length [19], number of cuttings [12,21,38] and number of inversions (and more general permutations) [2,20] as well as on the size of the giant and other components after bond percolation [11,13].

Keywords: Random trees · Split trees · Sorting algorithms · Quicksort · Binary search tree · Renewal theory · Contraction method · Limit laws · Size · Depths · Height · Total path length · Inversions · Cuttings · Bond percolation

MSC 2020 Subject Classifications: Primary 60C05 · Secondary 05C05 · 05C80 · 60K05 · 60J80 · 60J85 · 68P05 · 68P10 · 60F05 · 60K35 · 05A05

This work is supported by grants from the Ragnar Söderberg Foundation, the Swedish Research Council and the Knut and Alice Wallenberg Foundation.

J. Lindblad et al. (Eds.): DGMM 2021, LNCS 12708, pp. 20–57, 2021.
https://doi.org/10.1007/978-3-030-76657-3_2

1 Introduction

Random trees are often used to model data structures and algorithms in computer science and also have a wide variety of other applications in e.g., biology, sociology, physics, chemistry, literature and economics. During the last decades mathematical research related to random trees has been a hot topic and probabilistic techniques have been developed for describing characteristics and properties of trees in different settings (often in an asymptotic sense as the size of the tree grows to infinity) [29,45].

1.1 Deterministic and Random Trees

A *rooted tree* T is a connected graph with one node r defined as the *root*. We also require that there is a unique path of edges from each node v to the root r and there are no cycles in the graph. It is common to draw a tree with the root in the top and the other nodes below as in a family tree. Thus, it is natural to define the *children*, the *descendants* and the *ancestors* to a node v in T (in the same way as for a family tree). It is also natural to define the *depth* $d(v)$ of a node v in T as the distance of v to the root r, i.e., the number of edges in the unique path from v to r. The *height* is the maximal depth of the tree, i.e., $\max_{v \in T} d(v)$. Nodes without any children are called *leaves*. A *subtree* T_v in a tree T is a smaller tree inside T that has its root as node v and contains all descendants of v; sometimes one instead use the notation fringe subtree for such a T_v. A random tree is a tree that is generated from some probability procedure. Rooted trees can be divided into two main different classes, trees of *logarithmic height* and trees of *non-logarithmic height*. An important parameter is the (total) number of nodes of the tree T, which we for the moment denote by n (and which is often referred to as the size of T). Then we can also define the *subtree size* n_v as the (total) number of nodes in a subtree T_v rooted at v. Trees of logarithmic height, or equivalently $\mathcal{O}(\log n)$ trees, are classes of trees where the height for large n is bounded by $C \log n$ for some constant C (and are in particular important in computer science applications). Trees of non-logarithmic height, has typically a height of order $\mathcal{O}(n^c)$ for some constant $c < 1$, an important large class of random trees of this type are the critical conditioned Galton-Watson tree (with $c = 1/2$). In this brief survey, we will focus on random trees of logarithmic height, and more precisely on the large class of *split trees* that unifies many important random trees of this type. Many types of split trees also represent models of efficient data structures or *sorting algorithms*. Examples of split trees include e.g., *binary search trees, m-ary search trees, median-of-$(2k+1)$ trees, quad trees, simplex trees* and *tries* [28].

1.2 The Quicksort Sorting Algorithm and the Binary Search Tree

Quicksort is a well-known *sorting algorithm* (introduced by Hoare [35,36]) that is represented by the so-called *binary search tree*, which is one of the most studied examples of a random tree.

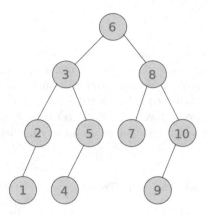

Fig. 1. An example of a binary search tree constructed from the set of keys $\{1, 2, \ldots, 10\}$ by drawing them in e.g., the order 6, 3, 5, 8, 10, 4, 7, 2, 1 and 9.

The Binary Search Tree: Draw a number σ_1, which we call a *key*, uniformly and independently from a set $\{1, 2 \ldots, n\}$, and associate the key to the root node r. Draw another key independently from the remaining numbers and place it at the left child of the root r if it is smaller than σ_1 and place it at the right child of r if it is larger than σ_1. Then one proceeds to grow the tree incrementally: When the k'th key is to be placed, one starts the comparison at the root r, and recursively finds the subtree to which the key is going to belong to by comparing the key to the current root's key, then go to the right child if it is larger and to the left child otherwise. Eventually one locates an empty node (subtree) and then one places the key at this position. When all keys are placed one gets a binary tree of n nodes that each contain one key, this is called the binary search tree (see Fig. 1).

1.3   Defining the Binary Search Tree Through the Subtree Method

There is another equivalent description of the binary search tree which is often more useful in applications, see e.g., [27]. Since the rank σ_1 of the root's key is equally likely to be $\{1, 2, \ldots, n\}$, the size of the left subtree (rooted at the left child of the root) is uniformly distributed on the set $0, 1, \ldots, n-1$ (i.e., there are n possibilities for the size which are equally likely). Thus, it follows that the size of the left subtree of the root is distributed as $\lfloor nU \rfloor$ (i.e., the integer part of nU), where U is a *uniform* $U(0, 1)$ random variable. Similarly the right subtree (rooted at the right child of the root) is distributed as $\lfloor n(1 - U) \rfloor$. This can be considered as if all n keys are placed in the root r, and by a uniform split of an interval of length n, it is determined how many of the $n - 1$ keys that are sent to the right and left subtree respectively (one key is kept in r). All subtree sizes (i.e., the number of keys in each subtree) can be explained in this manner by associating each node with an independent uniform random variable U_v. If a subtree rooted at v has size V, the size of its left subtree is given by $\lfloor V U_v \rfloor$.

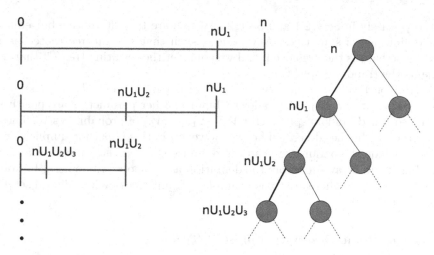

Fig. 2. Since the rank σ_1 of the root's key is equally likely to be $\{1, 2, \ldots, n\}$, the size of its left subtree is distributed as $\lfloor nU_1 \rfloor \approx nU_1$, where U_1 is a uniform $U(0,1)$ random variable. This can be regarded as performing a uniform split of an interval of length n. Other subtree sizes in the tree can be described in the same manner, for instance the left subtree of the root is distributed as $\lfloor \lfloor nU_1 \rfloor U_2 \rfloor \approx nU_1U_2$, where U_2 is a uniform $U(0,1)$ random variable independent from U_1.

Thus, given all U_v's, the subtree size n_v for a node v at depth k is given by

$$\lfloor \ldots \lfloor \lfloor nU_1 \rfloor U_2 \rfloor \ldots U_k \rfloor, \tag{1}$$

where U_i, $i \in \{1, \ldots, k\}$, are independent $U(0,1)$ random variables. Note that at least for "larger" n_vs (i.e., when k is not too large) the product in (1) is well approximated by the simple product

$$nU_1U_2 \ldots U_k;$$

see the illustration of this fact in Fig. 2. Note that since (in the final tree) each node holds exactly one key the number of keys in each subtree is the same as the number of nodes in the subtree.

2   Defining Split Trees

In this section we introduce the *split tree model* of [28]. A (random) *split tree* T^n can (vaguely described) be constructed as follows. Consider a rooted infinite b-ary tree where each node is a bucket of finite capacity s. We place n balls at the root, and the balls individually trickle down the tree in a random fashion until no bucket is above capacity. Each node draws a *split vector* $\mathcal{V} = (V_1, \ldots, V_b)$ from a common distribution, where V_i describes the probability that a ball passing through the node continues to the ith child. The trickle-down procedure is defined precisely in Sect. 2.1 and Sect. 2.2, which are two alternative but equivalent definitions of split trees. Any node u such that the subtree rooted as u contains no balls is then removed, and we consider the resulting tree T^n that we define as the (random) split tree.

A (random) *split tree* is defined by several parameters n, b, s_0, s that are non-negative integers ($s_0 = 0$ is allowed but the other parameters are positive) and the so-called random split vector $V = (V_1, \ldots, V_b)$ of probabilities V_i. These parameters are further described below. However, in this paper for simplicity we will not include the parameter s_1 in the definition. Often the parameter $s_1 = 0$ (and thus it can be avoided from the definition as we choose to do here) but one should observe that there are some examples of split trees such as digital search trees when $s_1 > 0$.

2.1   Algorithmic Description of Split Trees

We will first define split trees through an algorithm that describes incremental insertion of data items into an initially empty data structure, see e.g., [20, 28, 39]. The items are represented as *balls* and labelled using $\{1, 2, \ldots, n\}$ in the order of insertion.

Consider an infinite rooted b-ary tree $\mathcal{U}$ (every node has b children, this is called the *branch factor*). A split tree T^n of *cardinality* n is constructed by distributing n items (pieces of data, for instance the *keys* in the binary search tree) to the nodes $v \in \mathcal{U}$.

Let C_v denote the number of balls in node v. A *leaf* is defined as a node v that holds at least one ball (i.e., $C_v > 0$) but whose descendants are devoid of any balls.

Below there is a description of the algorithm, which determines how the n balls are distributed over the nodes. Initially there are no balls, i.e. $C_v = 0$ for each node v. Let s be the maximal number of balls a node can hold, we call s the *node capacity*; hence, for each node v, $C_v \leq s$.

Let $V = (V_1, \ldots, V_b)$ be a random vector satisfying $V_i \geq 0$ and $\sum_i V_i = 1$. Each node $u \in \mathcal{U}$ receives an independent copy $\mathcal{V}_u$ of the random vector V (chosen randomly according to some distribution); this is the so-called *split vector* and is the most important parameter in the definition of split trees. (Thus, different nodes can get different vectors assigned to them, but the vectors have the same distribution.) In the following, we always assume that $\mathbf{P}(\exists i : V_i = 1) < 1$ for technical reasons[1].

Add balls, one by one, to the root by the following recursive procedure for adding a ball to the subtree rooted at v.

After having added the first ball to the root:

1. If v is not a leaf, choose child i with probability V_i, and recursively add the ball to the subtree rooted at child i, by the rules given in steps 1, 2 and 3.
2. If v is a leaf and $Cv < s$, (where s is the node capacity) then add the ball to v and stop. Thus, C_v increases by 1.

---

[1] This is crucial, to avoid that the tree becomes a path!

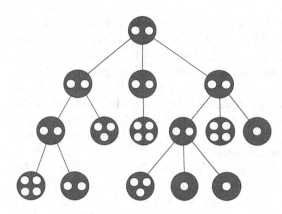

- Branch factor b

- Cardinality n

- Node capacity s>0

- Internal node capacity $s_0 \leq s$

- An independent copy of the random split vector $\mathcal{V} = (V_1, V_2, ..., V_b)$ is attached to each node.

Fig. 3. A split tree with $n = 35$, $b = 3$, $s = 4$ and $s_0 = 2$. The split vector $\mathcal{V}$ is the most important parameter and its components are probabilities

3. If v is a leaf and $C_v = s$, the ball cannot be placed in v since it is occupied by the maximal number of balls it can hold. In this case we move the now $s + 1$ balls from node v, by first selecting $s_0 \leq s$ of the balls uniformly at random to stay at v. (Hence, C_v decreases from $C_v = s$ which it was before the new ball was added to $C_v = s_0$.) The remaining of the $s + 1 - s_0$ balls are then moved to the children of v. This is done by choosing a child for each ball independently according to the probability vector $\mathcal{V}_v = (V_1, V_2 \ldots, V_b)$, i.e., each ball is added to the ith child with probability V_i. If a child receives more than s balls the procedure is repeated. Note however, that if $s_0 > 0$, this can not happen since no child could reach the node capacity s, whereas in the case $s_0 = 0$ this procedure may have to be repeated several times. (Note that v is no longer a leaf after this step.)

Note from the algorithm just described, that every nonleaf (internal) node has $C_v = s_0$ balls (this is why we denote s_0 as the *internal node capacity* and every leaf has $0 < C_v \leq s$ balls. In Fig. 3 we illustrate an example of a split tree.

2.2 Defining Split Trees Through Subtree Sizes

In this section we will instead use a definition (similarly explained as in [19]) that is similar to the description of the binary search tree in terms of subtree sizes in Sect. 1.3.

Let n_v denote the total number of balls that the nodes in the subtree rooted at node v hold together. Note that n_v is for split trees defined as the number of balls in a subtree, rather than the number of nodes which is the more common definition of a subtree size. (The reason for this is that the number of balls n in a split tree is a more natural parameter than the number of nodes N, since N is in fact a random variable, while n is deterministic.)

To describe the split tree T^n, it suffices to define the number of balls n_u in the subtree rooted at each node $u \in \mathcal{U}$. (Recall that we also did this observation in Sect. 1.3 concerning the case of the binary search tree, which is an example of a split tree as we will see below.) The tree T^n is then defined as the smallest relevant tree, i.e., the subset of nodes u such that $n_u > 0$ (which is indeed a tree). We keep s_0 balls in the root and then distribute $n - s_0$ balls to its children (according to the probabilities V_i of the random split vector $\mathcal{V} = (V_1, \ldots, V_b)$ as we will describe below). Distributing the balls is equivalent to determine the subtree sizes of the children of the root, i.e., we want to determine $(n_1, n_2, \ldots, n_b)$, where $i \in \{1, \ldots, b\}$ are the b children of the root. The construction of the remaining subtree sizes is then done inductively.

Choose an independent copy $\mathcal{V}_u$ of $\mathcal{V} = (V_1, \ldots, V_b)$ for every node $u \in \mathcal{U}$. We can now describe $(n_u, u \in \mathcal{U})$. Note that in particular $n_r = n$, where r is the root node. Given the subtree size n_v and the split vector $\mathcal{V}_v = (V_1, V_2, \ldots, V_b)$ of a node v, the subtree sizes $(n_{v_1}, n_{v_2}, \ldots, n_{v_b})$ for the b subtrees rooted at $v_1, v_2, \ldots, v_b$ (the children of v) are distributed as

$$\text{Mult}(n_v - s_0, V_1, V_2, \ldots, V_b); \tag{2}$$

recall that a *multinomial distribution* $\text{Mult}(m, p_1, \ldots, p_k)$ describes the distribution of sending m items into k boxes, where an item is sent to box i with probability p_i. Hence, the components V_i in the split vector describes the probability of sending a ball to the ith child. Here the children of v are the boxes, the balls are the items and V_i is the probability of sending a ball to the ith child.

2.3   Some Final Remarks on Defining Split Trees

Remark 1. One can without loss of generality (for most applications) assume that the *components* V_i of the split vector $\mathcal{V}$ are identically distributed. If this were not the case they can anyway be made identically distributed by using a random permutation, see [28]. Let V be a random variable with this distribution. This gives (because $\sum_i V_i = 1$) that $\mathbf{E}(V) = \frac{1}{b}$.

Remark 2. As a final remark for the definition of split trees, we mention that it is possible to include trees with $b = \infty$ as described by Janson [43] who showed that *preferential attachment trees*, which is an important class of random trees of logarithmic height with infinite branching (including e.g., *random recursive trees* and *plane oriented recursive tree*) can be regarded as split trees with a split vector that obtains the *Poisson-Dirichlet distribution*. The preferential attachment model (for which the tree is a special case of more general graphs) was made popular by Barabasi and Albert [4] to describe the *World Wide Web* (WWW), whose nodes are the HTML documents connected by links (edges) pointing from one page to another. Preferential attachment means that the more connected a node is, there is a higher chance to receive new links. Nodes with a higher degree are more likely to grab new links added to the network. However, although it is possible to extend split trees to include trees where $b = \infty$, such as preferential attachment trees, we will assume that b is finite since extensions of proofs of results for split trees would often be needed if one would allow $b = \infty$.

2.4   Examples of Split Trees

Depending on the choice of parameters b, s_0, s and the distribution of $\mathcal{V} = (V_1, \ldots, V_b)$ many important data structures may be modelled, such as *binary search trees, m-ary search trees, median-of-$(2k + 1)$ trees, quad trees, simplex trees* and *tries* [28]. To make sure that the model is clear and give a hint of the wide applicability of the model, we illustrate with three canonical examples.

EXAMPLE 1: BINARY SEARCH TREE. The *binary search tree* as described above is a simple example of a split tree. Here we assume that the data set is $\{1, \ldots, n\}$; these keys correspond to the balls that we are distributing to the nodes of the split tree T^n. A first (uniformly) random key is drawn σ_1, and stored at the root of a binary tree. The remaining keys are then divided into two subgroups, depending on whether they are smaller or larger than σ_1. The left and right subtrees are then binary search trees built from the two subgroups $\{i : i < \sigma_1\}$ and $\{i : i > \sigma_1\}$, respectively. The sizes of the two subtrees of the root are $\sigma_1 - 1$ and $n - \sigma_1$. One easily verifies (similarly as in the description of the binary search tree with the so-called subtree method above) that, since σ_1 is uniform in $\{1, 2, \ldots, n\}$, one has

$$(\sigma_1 - 1, n - \sigma_1) \overset{d}{=} \mathrm{Mult}(n - 1; U, 1 - U),$$

where $\overset{d}{=}$ denotes equality in distribution and U is a uniform $U(0, 1)$ random variable. Hence, $\sigma_1 - 1 \overset{d}{=} Bin(n - 1, U)$. Here $Bin(m, p)$ denotes a binomial distribution; recall that this is the special case of a multinomial distribution when there are only two boxes, such that each of the m items go to the first box with probability p and to the second box with probability $1 - p$. Thus, a binary search tree can be described as a split tree with parameters $b = 2$, $s_0 = 1$, $s = 1$, and $\mathcal{V}$ is distributed as $(U, 1 - U)$ for a random variable U that is uniform on the interval $[0, 1]$, i.e., U is a $U(0, 1)$ random variable.

EXAMPLE 2: M-ARY SEARCH TREES. The *m-ary search trees* for integers $m \geq 2$ generalize the binary search trees where $m = 2$. Here we again assume that the data set is $\{1, \ldots, n\}$; these keys correspond to the balls that we are distributing to the nodes of the split tree T^n. Instead of one key σ_1 draw $m - 1$ keys $\sigma_1, \sigma_2, \ldots, \sigma_{m-1}$ from the set $\{1, 2, \ldots, n\}$, and associate them to the root in increasing order $\sigma_1' < \sigma_2' < \ldots < \sigma_{m-1}'$. Note that this results in m different intervals. Now sort the remaining keys into m sets in accordance to which of the intervals that resulted from $\sigma_1' < \sigma_2' < \ldots < \sigma_{m-1}'$ that they belong to. Proceed recursively in each node as long as a node holds more than $m - 1$ keys, by first drawing $m - 1$ keys to be placed in the node in increasing order and then split the remaining keys into m sets (where some could be empty). When the procedure is ended all internal nodes hold exactly $m - 1$ keys and the leaves hold between 1 and $m - 1$ keys; see Fig. 4 for two examples when $m = 3$ and $m = 4$, respectively.

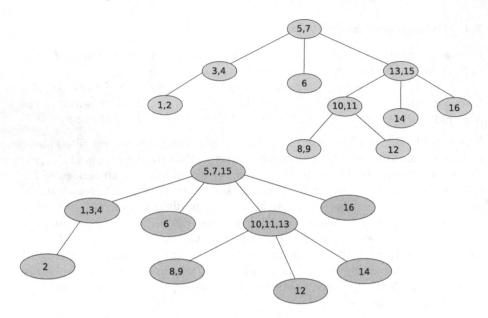

Fig. 4. The m-ary search trees are generalisations of the binary search tree where $m = 2$. The figure shows a 3-ary and a 4-ary search tree constructed from e.g., the sequence of keys 7, 5, 15, 3, 4, 6, 1, 13, 11, 10, 2, 16, 8, 9, 14 and 12.

Let $(n_1, n_2, \ldots, n_m)$ be the vector of the subtree sizes (of the number of keys) for the m children of the root. Then

$$(n_1, n_2, \ldots, n_m) \stackrel{d}{=} \text{Mult}(n - m + 1, V_1, \ldots, V_m),$$

where the V_i, $i \in \{1, \ldots, m\}$ are the proportion of the keys in these m subtrees and are given by the lengths (or spacings) induced by $m - 1$ independent uniform $U(0, 1)$ random variables on the interval $[0, 1]$, see [26]. Hence, the $V_i \stackrel{d}{=} V$ are distributed as the minimum of $(U_1, U_2, \ldots, U_{m-1})$, where U_i are independent uniform $U(0, 1)$ random variables; this is in fact equal to a beta$(1, m - 1)$ distribution. The m-ary search trees are split trees with parameters $b = m$, $s_0 = m - 1$, $s = m - 1$, and $V \stackrel{d}{=} \min(U_1, U_2, \ldots, U_{b-1}) \stackrel{d}{=} \text{beta}(1, m - 1)$. Note that for $m > 2$ in contrast to the binary search tree with $m = 2$ (where the number of balls is the same as the number of keys), the number of nodes in the whole tree is a random number N although the number of keys (balls) is a deterministic number n.

EXAMPLE 3: TRIES. We are given n (infinite) strings $X_1, \ldots, X_n$ on the alphabet $\{1, \ldots, b\}$; here the strings correspond to the balls in the split tree T^n. The strings are drawn independently, and the symbols in each string are also independent with distribution on $\{1, \ldots, b\}$ given by $p_1, \ldots, p_b$. Each string naturally corresponds to an infinite path in the infinite complete b-ary tree, where the sequence of symbols indicate the sequence of directions to take as one walks away from the root r. The *trie* is then defined as the smallest tree so that all the paths corresponding to the infinite strings are eventually distinguished; see Friedkin [32] who invented this model. Thus, for a node u let n_u denote the number of strings that pass through node u. Then eliminate all nodes with $n_u = 0$ and also those with $n_u = 1$ whose parents also have $n_u = 1$. The internal nodes store no data, each leaf stores a unique string, and thus there are n leaves. Note that each leaf u has $n_u = 1$ and every internal node v has $n_v > 1$; see Fig. 5 which shows a binary trie (i.e., $b = 2$) built from eight strings.

For the b children of the root it clearly holds that

$$(n_1, \ldots, n_b) \stackrel{d}{=} \mathrm{Mult}(n; p_1, \ldots, p_b).$$

The trie is thus a random split tree with parameters $s = 1, s_0 = 0$ and $\mathcal{V} = (p_1, p_2, \ldots, p_b)$.

3   An Introduction to Renewal Theory for Split Trees

Renewal theory is a widely used branch of probability theory that generalizes *Poisson processes* (which have *holding times* that have *exponential distributions*) to processes with arbitrary holding times. Renewal theory is a powerful tool to study sums of independent and identically distributed random variables. A classic in this field is Feller [31] on recurrent events.

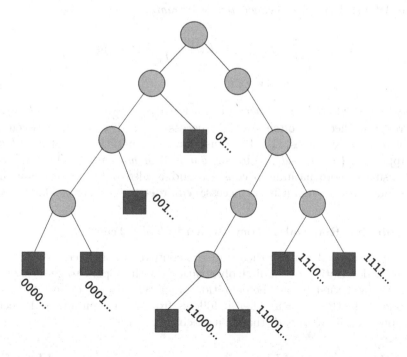

Fig. 5. A binary trie (with eight leaves) built from the eight strings $0000\ldots$, $0001\ldots$, $001\ldots$, $01\ldots$, $11000\ldots$, $11001\ldots$, $1110\ldots$ and $1111\ldots$.

3.1   Renewal Theory 101

Let X_k, $k \geq 1$, be independent and identically distributed (*i.i.d.*), non-negative random variables distributed as X. Let S_n, $n \geq 1$, be the *partial sums* i.e., $S_n = \sum_{k=1}^{n} X_k$. Let $F(x)$ denote the *distribution function* of X i.e., $F(x) = P(X \leq k)$, and let $F_n(x)$ be the distribution function of S_n, $n \geq 1$ i.e., $F_n(x) = P(S_n \leq x)$. The *renewal counting process* $\{\mathcal{N}(t),\ t \geq 0\}$ is defined by

$$\mathcal{N}(t) := \max\{n : S_n \leq t\},$$

which can be considered as the number of renewals before time t of an object with a lifetime distributed as the random variable X. A natural application of the renewal counting process is to consider the number of lightbulbs that needs to be changed in a lamp during say 5 years time, where each lightbulb has a lifetime (in years) distributed as the random variable X; hence, this number is given by $\mathcal{N}(5)$. In the specific case when X is an *exponential random variable*, $\{\mathcal{N}(t),\ t \geq 0\}$ is a *Poisson process*. An important well studied function is the so called *standard renewal function* defined as

$$V(t) := \sum_{k=0}^{\infty} F_k(t) = \sum_{k=1}^{\infty} \mathbf{P}(S_k \leq t),$$

which obviously is equal to the expected value of $\mathcal{N}(t)$ i.e., $\mathbf{E}\left[\mathcal{N}(t)\right]$. The renewal function $V(t)$ satisfies the so called *renewal equation*

$$V(t) = \sum_{k=1}^{\infty} \mathbf{P}(S_k \leq t) = F(t) + \int_0^t V(t-s)dF(s)$$
$$= F(t) + (V * dF)(t) \tag{3}$$

and has by the *law of large numbers* a solution $V(t) = \frac{t}{\mathbf{E}[X]} + o(t)$, as $t \to \infty$.

In renewal theory it often becomes necessary to distinguish between two kinds of renewal processes, i.e., if the distribution function $F(x)$ of X is *lattice* (has support on $\{0, \pm d, \pm 2d, \ldots\}$ for some $d > 0$) or *non-lattice*. The non-lattice case is usually more natural to consider and is also much more common in applications, even though it is not necessary harder to consider the lattice case.

3.2   Applying Renewal Theory to Study Split Trees

In [39] we introduced renewal theory in the context of split trees. We will now give a brief description of the first observations with respect to this that were obtained in [39]. First, recall the definition of subtree size n_v in Sect. 2.2. Let v be a node at depth $d(v) = d$. Then it follows from (2) (given all the $\mathcal{V}$ vectors for all nodes u with $d(u) \leq d$) that in a stochastic sense,

$$\mathrm{Bin}(n, \prod_{i=1}^{d} V_i) - \mathrm{Bin}(s_0, \prod_{i=2}^{d} V_i) - \ldots - \mathrm{Bin}(s_0, V_d) \leq n_v \leq \mathrm{Bin}(n, \prod_{i=1}^{d} V_i), \quad (4)$$

where $V_i, i \in \{1, \ldots d\}$ are i.i.d. random variables (distributed as a random variable V, see Sect. 2.3) given by the split vectors associated with the nodes in the unique path from v to the root. Recall that $Bin(n, p)$ denotes a random variable with a *binomial distribution* and that for a $Bin(n, p)$ random variable, the expected value is np and the variance is $np(1 - p)$. Thus, *Chebyshev's inequality* applied to the dominating term $Bin(n, \prod_{i=1}^{d} V_i)$ in (4) gives that n_v for v at depth d (at least for n_v:s that are large enough, which is true if d is not too large) is close to

$$nV_1 V_2 \ldots V_d; \tag{5}$$

see an illustration of this fact in Fig. 6 also compare with Fig. 2 for the special case of the binary search tree. More precisely, *Chebyshev's inequality* gives for v with $d(v) = d$, that

$$\mathbf{P}\Big(| n_v - n \prod_{i=1}^{d} V_i | > n^{0.6} \Big) \leq 2 \frac{\mathbf{E}\Big(\mathbf{Var}\Big(Bin(n, \prod_{i=1}^{d} V_i) \Big) \Big)}{n^{1.2}} \leq \frac{1}{n^{0.1}}. \tag{6}$$

Now let $Y_k := -\sum_{i=1}^{k} \ln V_i$. Note that $nV_1 V_2 \ldots V_k = ne^{-Y_k}$. Recall that in a binary search tree, the split vector $\mathcal{V} = (V_1, V_2)$ is distributed as $(U, 1 - U)$ where U is a uniform $U(0, 1)$ random variable. For this specific case of a split tree, the sum Y_k (where $V_i, i \in \{1, \ldots, k\}$ in this case are i.i.d. uniform $U(0, 1)$ random variables) is distributed as a $\Gamma(k, 1)$ random variable (where $\Gamma(\alpha, \beta)$ denotes a Gamma random variable with parameters $\alpha, \beta > 0$). This fact is used by, for example, Devroye in [24] to determine the height (maximal depth) of a binary search tree. For general split trees, there is no simple common distribution function of $\sum_{i=1}^{k} \ln V_i$, instead renewal theory can be used as first applied in [39] to obtain novel characteristics of split trees.

Let $\nu_k(t) := b^k \mathbf{P}(Y_k \leq t)$. We define the *renewal function*

$$U(t) := \sum_{k=1}^{\infty} \nu_k(t) = \sum_{k=1}^{\infty} b^k \mathbf{P}(Y_k \leq t). \tag{7}$$

We also denote by $\nu(t) := \nu_1(t) = b\mathbf{P}(-\ln V_1 \leq t)$. For $U(t)$ we obtain the following *renewal equation*

$$U(t) = \nu(t) + \sum_{k=1}^{\infty} (\nu_k * \nu)(t) = \nu(t) + (U * \nu)(t), \tag{8}$$

where $(U * d\nu)(t) = \int_0^t U(t - z) d\nu(z)$ (compare with Eq. (3)). The measure $d\nu(t)$ is not a *probability measure* (i.e., a measure that can be regarded as a probability with a value between 0 and 1). To work with more convenient renewal equations, involving probability measures, we introduce the *tilted measure* $d\omega(t) = e^{-t} d\nu(t)$. It is easily seen that $d\omega(t)$ is probability measure and defines a random variable X by $\mathbf{P}(X \in dt) = d\omega(t)$. In fact ω is the distribution

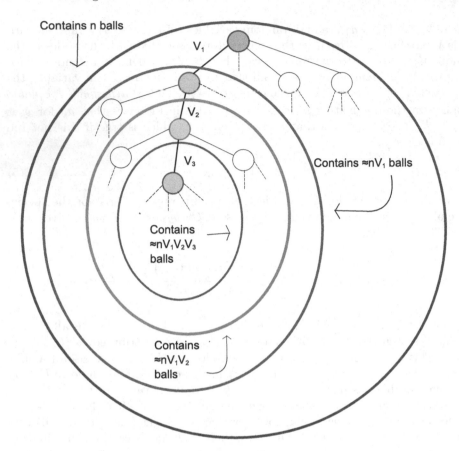

Fig. 6. Given all split vectors in the tree, n_v for v at depth d is close to $nL_v = n \prod_{j=1}^{d} V_j$ (at least for n_v:s that are large enough), where the V_j's are *i.i.d.* random variables distributed as the components in the split vector $\mathcal{V}$.

function of $-\ln \Delta$, where Δ is the *size-biased random variable* which is defined to be the component of $(V_1, \ldots, V_b)$ picked with probability proportional to its size, i.e., given $(V_1, \ldots, V_b)$, let $\Delta = V_j$ with probability V_j. Recall from Sect. 2.3 that V is a random variable distributed as the components V_i of the split vector $\mathcal{V}$. We write

$$\mu := \mathbf{E}[-\ln \Delta] = b\mathbf{E}[-V \ln V], \qquad \text{and}$$
$$\sigma^2 := \mathbf{Var}(\ln \Delta) = b\mathbf{E}[V \ln^2 V] - \mu^2. \tag{9}$$

Then from (9), X obviously satisfies

$$\mathbf{E}[X] = \mathbf{E}[-\ln \Delta] = \mu \qquad \text{and} \qquad \mathbf{E}[X^2] = \sigma^2 + \mu^2.$$

The renewal Eq. (8) can then be rewritten as

$$\widehat{U}(t) = \widehat{\nu}(t) + (\widehat{U} * d\omega)(t), \tag{10}$$

where $\widehat{U}(t) := e^{-t}U(t)$ and $\widehat{\nu}(t) := e^{-t}\nu(t)$. From the renewal equation in (10) we can obtain asymptotic results for $U(t)$ as we will present below. Recall that V is distributed as in the components of the split vector; see Sect. 2.3. For simplicity, we only state the results for $U(t)$ below under the assumption that $-\ln V$ is non-lattice, which holds true for most types of split trees (however, for the tries in Example 3 of Sect. 2.4 $-\ln V$ is lattice). The first order asymptotics for $U(t)$ as $t \to \infty$ follows from the *key renewal theorem* [3, Theorem IV.4.3] applied to $\widehat{U}(t)$. The result we get is presented in the following crucial lemma (see [39, Lemma 2.1] for a formal proof), which we have applied frequently to study split trees.

Lemma 1. *The renewal function $U(t)$ in (7) satisfies*

$$U(t) = (\mu^{-1} + o(1))e^t \ \ as \ t \to \infty, \tag{11}$$

where μ is the constant in (9).

The following result from [39, Corollary 2.2] gives some useful information about the second order behaviour of $U(t)$.

Lemma 2. *As $x \to \infty$, we have*

$$\int_0^x e^{-t}(U(t) - \mu^{-1}e^t)dt = \left\{ \frac{\sigma^2 - \mu^2}{2\mu^2} - \mu^{-1} + o(1).\right.$$

where μ and σ^2 are the constants in (9).

One can show similar results for $U(t)$ also if $-\ln V$ is lattice; see [19, Lemma 4.2].

4   Characteristics of Split Trees

In this section we summarize some important properties of split trees including results on depths for the balls and nodes in the tree, the height (maximal depth) of the tree and the size of the tree in terms of the number of nodes; see [11,17,18,28,39]. These characteristics give corresponding properties to the search algorithms. Understanding the characteristics of split trees can enable us to better understand the corresponding sorting algorithms, where the characteristics of the trees corresponds to properties of the algorithm. In Sect. 5 we will explain this correspondence in the case of the total path length which corresponds to the total running time of the sorting algorithm.

4.1   Results for the Depths of Balls and Nodes in a Split Tree

Devroye [28, Theorem 2] showed a *central limit theorem* for the depth D_n of the last ball, where we consider the labelling of the balls as in the order they were added to the tree in the algorithmic definition of split trees in Sect. 2.1. From now on we always assume that μ and σ^2 are the constants defined in (9).

Theorem 1. *For the depth D_n of the last ball it holds that if $\sigma > 0$ (i.e., assuming that V is not monoatomic with $\mathcal{V} = (\frac{1}{b}, \ldots \frac{1}{b})$) then*

$$\frac{D_n - \mu^{-1} \ln n}{\sqrt{\sigma^2 \mu^{-3} \ln n}} \xrightarrow{d} N(0,1).$$

For example for the case of the binary search tree $\mu^{-1} = 2$ and $\sigma^2 = \frac{1}{4}$. As earlier also shown by Devroye in [25] (in a previous study where he only considered the binary search tree), this implies that for the specific case of the binary search tree it holds that

$$\frac{D_n - 2\ln n}{\sqrt{2\ln n}} \xrightarrow{d} N(0,1).$$

In [39, see Proposition 1.1 and Proof of Theorem 1.3] (by using *coupling arguments*) we extended the central limit law in Theorem 1 to hold for other balls (ordered as in the split tree generating algorithm described in Sect. 2.1).

Proposition 1. *For the depth of the k:th ball D_k such that $\frac{n}{\ln n} \le k \le n$ it holds that*

$$\frac{D_k - \mu^{-1} \ln n}{\sqrt{\sigma^2 \mu^{-3} \ln n}} \xrightarrow{d} N(0,1).$$

Hence, it follows that the depth of a random ball (or node) in T^n obtains a central limit law. Thus, most nodes in T^n are close to $\mu^{-1} \ln n + \mathcal{O}\left(\sqrt{\ln n}\right)$. The next theorem [39, Theorem 1.2] sharpens this result.

Definition 1. *For a given $\epsilon > 0$ we say that a node v in T^n is "good" if for its depth $d(v)$ it holds*

$$\mu^{-1} \ln n - \ln^{0.5+\epsilon} n \le d(v) \le \mu^{-1} \ln n + \ln^{0.5+\epsilon} n,$$

and "bad" otherwise.

Theorem 2. *For any choice of $\epsilon > 0$, the expected number of bad nodes in T^n is bounded by $\mathcal{O}\left(\frac{n}{\ln^k n}\right)$ for any constant k.*

The basic idea of the proof was to use so-called *large deviations*, and using the fact that the subtree size n_v is "often" close to $n \prod_{i=1}^{d(v)} V_i$ (i.e., when n_v is large enough). In [39, Remark 3.4] we also note that the proof of Theorem 2 implies that the expected number of nodes with large depths is very small as stated in the corollary below.

Corollary 41. *For any constant r there is a constant $C > 0$ so that the expected number of nodes with depth $d(v) \ge C \ln n$ is bounded by $\mathcal{O}\left(\frac{1}{n^r}\right)$.*

Further, in [39, Theorem 1.3] we also showed the first asymptotic for the variances of the depths of the balls in the tree:

Theorem 3. *For all k such that $\frac{n}{\ln n} \leq k \leq n$, the variance of D_k (the depth of the k:th ball) satisfies*

$$\frac{\mathbf{Var}(D_k)}{\ln n} \overset{n\to\infty}{\to} \mathbf{E}\left(N(0, \sigma^2 \mu^{-3})^2\right) = \sigma^2 \mu^{-3}.$$

From the central limit laws in Theorem 1 and Proposition 1 this is a natural guess, but one needs to be prove *uniform integrability* of $(D_k - \mu^{-1} \ln n)^2 / \ln n$, which we did in the proof of [39, Theorem 1.3].

We complete this section by stating the result in [11, Corollary 1] that shows in particular that most pairs of balls (and pairs of nodes) in split trees are far away from each other, i.e., the latest common ancestor of two (randomly chosen) balls (or nodes) is with high probability close to the root. Let u_1 and u_2 be two independent uniformly chosen nodes in T^n. We write $d_n(u_1, u_2)$ for the number of edges of T^n which are needed to connect the root, u_1 and u_2. Note that $d_n(u_1, u_2) \leq d_n(u_1) + d_n(u_2)$, where $d_n(u_1)$ and $d_n(u_2)$ are the depths of u_1 and u_2, respectively. Similarly, let b_1 and b_2 be two independent uniformly chosen balls in T^n. We write $D_n(b_1, b_2)$ for the number of edges of T^n which are needed to connect the root, and nodes where the balls b_1 and b_2 are stored. Again note that $D_n(b_1, b_2) \leq D_n(b_1) + D_n(b_2)$, where $D_n(b_1)$ and $D_n(b_2)$ are the depths of b_1 and b_2 respectively.

Proposition 1. *We have that*

$$\lim_{n\to\infty} \frac{D_n(b_1, b_2)}{\mu^{-1} \ln n} = 2, \text{ in probability.}$$

If we further assume that Condition 1 is also satisfied (see Sect. 6). We have that

$$\lim_{n\to\infty} \frac{d_n(u_1, u_2)}{\mu^{-1} \ln n} = 2, \text{ in probability.}$$

Remark 3. Recall that almost all balls and nodes in a split tree are close to the depth $\mu^{-1} \ln n$. Note that Proposition 1 implies that for the two independently chosen balls b_1 and b_2 (respectively the two independently chosen nodes u_1 and u_2) most ancestors are different; hence their latest common ancestor is close to the root.

4.2 Results for the Height of a Split Tree

Devroye [24, Theorem 5.1] studied the height (maximal depth) of the binary search tree and showed that

Theorem 4. *Let T^n be a random binary search tree. Let H_n be its height. Then $H_n \sim c \ln n$, in probability as n tends to infinity, where $c = 1/\rho_0 = 4.311\ldots$ and $\rho_0 = \inf\{\rho : \rho - 1 - \log \rho \leq \log 2\}$.*

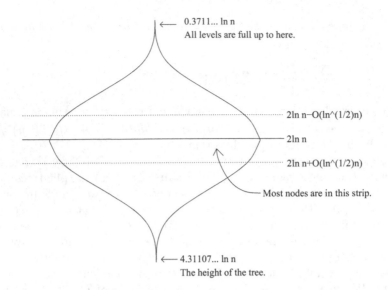

Fig. 7. This figure illustrates the shape of the binary search tree. The root is at the top. The horizontal width represents the number of nodes at each level. Most nodes are in a strip of width $\mathcal{O}(\sqrt{\ln n})$ around $2 \ln n$. The height is almost surely $H_n \sim 4.311 \ldots \ln n$. General split trees turn out to have a similar structure.

Devroye [28] extended this result to split trees and proved that the behavior of the height H_n of a split tree is related to the *moment function* $m(t) = E(V^t)$ $t \geq 0$, where $V \stackrel{d}{=} V_1$ is distributed as the components in the split vector $\mathcal{V}$ (see Sect. 2.3). Devroye [28, Theorem 1] showed that under some conditions of $m(t)$ it holds that.

Theorem 5. *Let T^n be a split tree. Let H_n be its height. Then $H_n \sim \gamma \ln n$, in probability as n tends to infinity, where $\gamma > 0$ is a constant only depending on b and the distribution of V.*

Furthermore, in Devroye [28, Theorem 1] there is also a definition of $\gamma > 0$ that uses the moment function $m(t)$, but since it is rather involved we choose to not write the full expression here. The proof of Theorem 5 also used key properties of $m(t)$ obtained from Devroye [28, Lemma 1] and again using the crucial fact that the subtree sizes n_v:s are "often" close to $n \prod_{i=1}^{d(v)} V_i$. In [17, Theorem 1] and [18, Theorem 2], by using large deviations the result of Devroye [28, Theorem 1] were extended to hold for an even more general class of trees in which split trees were included and also they expressed $\gamma > 0$ in a somewhat different way and also provided with geometric interpretations.

Figure 7 shows an illustration of the shape of the binary search tree, in the sense how the nodes (or keys) are distributed at different levels (depths) in the tree; and the results discussed in Sect. 4.1 and Sect. 4.2 show that split trees in general have a similar structure.

4.3   A Result for the Number of Nodes in a Split Tree

Except for studying depths of T^n an important remaining natural property to study is the (total) *number of nodes* N in T^n, which is (as mentioned in Sect. 2.2) a different parameter for the size of T^n than the cardinality n which refers to the (total) number of balls in T^n. Note that for split trees the number of nodes N is usually random (as remarked in the example of m-ary search trees in Sect. 2), whereas the number of balls n is a deterministic number. In [39, Theorem 1.1] we provided a relationship between the (random) number of nodes N and the (deterministic) number of balls n. Since we proved this result by using renewal theory (where it is necessary to distinguish between lattice and non-lattice distributions as explained above) for simplicity we assumed that $-\ln V$ is non-lattice, where we recall from Sect. 2.3 that V is distributed as the components V_i in the split vector. (However, we could have applied renewal theory for lattice distributions to obtain a similar result also if $-\ln V$ is lattice.)

Theorem 6. *Let N be the (random) number of nodes in a split tree with n balls. Then it holds that there is a constant C depending on the type of the split tree such that*

$$\mathbf{E}\,[N] = \alpha n + o(n) \ and \ \mathrm{Var}(N) = o(n^2).$$

For some applications of split trees we need to assume that Theorem 6 can be sharpened a little so that the $o(n)$ term is bounded by $O\left(\frac{n}{\ln^{1+\varepsilon} n}\right)$. This is known to hold for most type of split trees; see e.g., [38, Assumption 3. (A3) and its remark].

Assumption 1. *Suppose that $-\ln V$ is non-lattice. Furthermore, for some $\alpha > 0$ and $\varepsilon > 0$, assume that*

$$\mathbb{E}[N] = \alpha n + O\left(\frac{n}{\ln^{1+\varepsilon} n}\right);$$

hence the error term of $|E(N) - \alpha n|$ is assumed to be a little bit smaller than the upper bound $o(n)$, which we obtained from Theorem 6.

5   The Total Path Length of Split Trees

Sorting algorithms sort a collection of data items (often called *keys*) by *comparisons* of the input data. The number of comparisons for a certain key is given by its depth in the tree. The *total number of comparisons* (or equivalently the sum of all depths of keys) is the *total path length*, which therefore represents a natural cost measure or *running time* of these algorithms.

5.1   Introduction to the Total Path Length

The total path length can be defined for any type of rooted tree:

Definition 2. *Let T be a rooted tree and $d(v)$ denote the depth of node v, i.e., the distance from v to the root r. Define $\Psi(T) \overset{d}{=} \sum_{v \in T} d(v)$ as the total path length of T.*

Effective sorting algorithms are represented by trees of logarithmic height with a total path length (or equivalently, a running time of the corresponding algorithm), which is of order $\mathcal{O}(n \log n)$. The total path length of tree data structures have been studied by many authors, but in most cases the analyses and proofs are very much tied to a specific case. The case that has been mostly studied is the analysis of the total path length for the binary search tree (or equivalently the running time of the sorting algorithm Quicksort), see e.g., [48, 49] who proved that the total path length for binary search trees does not obtain a central limit law as for the depths of the nodes as discussed in Sect. 4. Instead they proved that the limit law is characterized by a *fixed point equation*. Figure 8 illustrates the total number of comparisons in Quicksort.

5.2   A Fixed-Point Equation for the Normalized Total Path Length of a Split Tree

The main result of [19] was to show that for general split trees, the total path length (after normalization) converges in distribution to a random variable characterized by some fixed point equation. In that sense our result extended the earlier studies of [48, 49] and [47] who used the so-called *contraction method* to

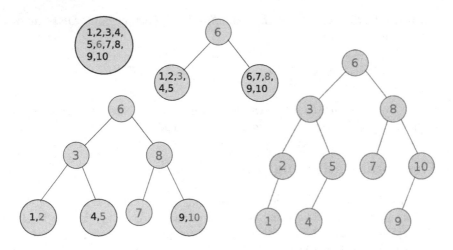

Fig. 8. This figure illustrates the total number of comparisons in Quicksort. The number of comparisons for a certain key is given by the total number of black digits in the figure, which is equivalent to its depth in the final tree (to the right). Thus, the total number of comparisons of Quicksort is equivalent to the total number of black digits in the figure, or equivalently the sum of all depths in the final tree (to the right) i.e., the total path length of the binary search tree.

show convergence in distribution of the total path length for the specific examples of the binary search trees, the median-of-$(2k+1)$ trees and quad trees. By applying renewal theory to the contraction method we proved the following result [19, Theorem 2.1]:

Theorem 7. *Let $\Psi(T^n)$ be the total path length in a general split tree with split vector $\mathcal{V} = (V_1, \ldots, V_b)$. Suppose that $\mathbf{P}\,(\exists i : V_i = 1) < 1$. Let*

$$X_n := \frac{\Psi(T^n) - \mathbf{E}[\Psi(T^n)]}{n} \qquad and \qquad C(\mathcal{V}) = 1 + \frac{1}{\mu}\sum_{i=1}^{b} V_i \ln V_i,$$

where μ is the constant in (9). If $C(\mathcal{V}) \neq 0$ with positive probability, then $X_n \to X$ in distribution, where X is the unique solution of the fixed point equation

$$X \stackrel{d}{=} \sum_{i=1}^{b} V_i X^{(i)} + C(\mathcal{V}), \tag{12}$$

satisfying $\mathbf{E}[X] = 0$ and $\mathbf{Var}\,(X) < \infty$, and $X^{(i)}$ are i.i.d. copies of X.

Remark 4. From the proof of Theorem 7 (which is based on the contraction method) it follows that the variance also needs to converge. In particular

$$\mathbf{Var}(\Psi(T^n)) \sim Kn^2, \tag{13}$$

where the constant $K = \mathbf{Var}\,(X) = \mathbf{E}\,[X^2] - (\mathbf{E}\,[X])^2$. Computing $\mathbf{E}[X^2]$ using the fixed point equation, one easily obtains the following expression for the constant K in (13),

$$K = \mathbf{Var}\,(X) = \frac{\mu^{-2}\mathbf{E}[(\sum_{i=1}^{b} V_i \log V_i)^2] - 1}{1 - \sum_{i=1}^{b}\mathbf{E}[V_i^2]}. \tag{14}$$

Remark 5. When the split vector $\mathcal{V}$ is deterministic, i.e. $\mathcal{V}$ is a permutation of some fixed vector $(p_1, \ldots, p_b)$, the cost function $C(\mathcal{V}) = 0$. In some sense, part of Theorem 7 still holds true, but the limit X is trivial since $X = 0$ almost surely. Instead one should use a different normalization since the variance is no longer of order n^2 but smaller, for instance in the case of binary tries where the variance is at most of order $\mathcal{O}(n \log n)$ it was shown in [40] that the total path length obtains a normal central limit law.

For the result in Theorem 7 to hold one needs in particular that the mean of $\Psi(T^n)$ satisfies a precise asymptotic form, which we showed by using renewal theory in the following theorem. (The result below also holds in the case when $C(\mathcal{V}) = 0$ as was discussed in Remark 5.) Recall that V is a random variable distributed as the components V_i in the split vector $\mathcal{V}$. Thus, we proved in [19, Theorem 3.1]:

Theorem 8. *The expected value of the total path length $\Psi(T^n)$ exhibits the following asymptotics, as $n \to \infty$,*

$$\mathbf{E}[\Psi(T^n)] = \mu^{-1} n \ln n + n\varpi(\ln n) + o(n), \tag{15}$$

where μ is the constant in (9) and ϖ is a continuous periodic function of period $d = \sup\{a \geq 0 : \mathbf{P}(-\ln V \in a\mathbb{Z}) = 1\}$. In particular, if $-\ln V$ is non-lattice, then $d = 0$ and $\varpi = \zeta$ is a constant.

Remark 6. Note that the first order term in (15) is obvious from Theorem 2, which implies that most nodes are *good*, see Definition 1. However, the second order term in (15) is certainly non-trivial. Note also that for most cases of split trees $-\ln V$ is non-lattice so that $\varpi(\ln n) = \zeta$ is just a constant (and the period is then $d = 0$).

5.3   Brief Description of Proofs of Theorems 7 and 8

A key part of the contraction method (see e.g., [47–49]) involves a *recursive identity* that can easily be adapted so that it can be used also in the case of split trees, since it holds true for any type of random tree e.g., for a binary search tree or another type of split tree (see the argument of the equality in (16) below). The hard part of the proof is actually to show that the precise asymptotic expansion in Theorem 8 holds since otherwise the convergence part of the contraction method can not be used. A brief sketch of how the contraction method can be used to prove Theorem 7 (under the condition that Theorem 8 i.e., [19, Theorem 3.1] holds true) is explained below (this part of the proof in [19] also followed along the same lines as was used in the specific case of quad trees in [47]). Let $\bar{n} = (n_1, \ldots, n_b)$ denote the vector of cardinalities (subtree sizes in terms of number of balls) for the subtrees rooted at the children of the root. Then we have the recursive identity

$$\Psi(T^n) \overset{d}{=} \sum_{i=1}^{b} \Psi_i(T^{n_i}) + n - s_0, \tag{16}$$

where $\Psi_i(T^{n_i})$ are copies of $\Psi(T^{n_i})$ that are independent given the subtree sizes $(n_1, \ldots, n_b)$. Note that the recursive identity in (16) follows from that each subtree rooted in the children $i \in \{1, \ldots, b\}$ of the root is a split tree of cardinality n_i and given the sizes $(n_1, \ldots, n_b)$ they are also independent. By introducing the *normalized total path length*

$$X_n := \frac{\Psi(T^n) - \mathbf{E}[\Psi(T^n)]}{n},$$

we get that X_n is recursively given by the sum

$$X_n \overset{d}{=} \sum_{i=1}^{b} \frac{n_i}{n} X_{n_i} + C_n(\bar{n}),$$

where X_{n_i}, $i \in \{1, \ldots, b\}$ are independent given the sizes $(n_1, \ldots, n_b)$ of the subtrees rooted at the children of the root, and

$$C_n(\overline{n}) := 1 - \frac{s_0}{n} - \frac{\mathbf{E}[\Psi(T^n)]}{n} + \sum_{i=1}^{b} \frac{\mathbf{E}[\Psi(T^{n_i})]}{n}.$$

From the precise asymptotics of $\mathbf{E}[\Psi(T^n)]$ in Theorem 8 (i.e., [19, Theorem 3.1]), it follows that

$$\lim_{n \to \infty} C_n(\overline{n}) \overset{d}{=} 1 + \frac{1}{\mu} \sum_{i=1}^{b} V_i \ln V_i := C(\mathcal{V}),$$

where $\mathcal{V} = (V_1, \ldots, V_b)$ is the split vector of T^n. This distributional identity implies that if X_n converges in distribution to some limit X, then X should satisfy the following fixed point equation:

$$X \overset{d}{=} \sum_{i=1}^{b} V_i X^{(i)} + C(\mathcal{V}),$$

where $X^{(i)}$ are i.i.d. copies of X; also recalling that the subtree size $n_i \approx nV_i$. The point of the contraction method is then to show that the fixed point equation defined in (12) i.e.,

$$X \overset{d}{=} \sum_{i=1}^{b} V_i X^{(i)} + C(\mathcal{V}),$$

has a unique solution X and that $X_n \to X$ in distribution. This was done by proving that the recursive map defined by (12) is a so-called *contraction* in a suitable *complete metric space* of probability measures; hence the classical *Banach fixed point theorem* thereby implies this result.

Then it remained to actually show Theorem 8 (i.e., [19, Theorem 3.1]), which proof was rather involved and was based on the use of renewal theory to study split trees (recall from Sect. 3.2 the relation between renewal theory and the subtree sizes of split trees). Thus, to be able to apply renewal theory an important observation was that there is an equivalent definition of the total path length defined as the sum of all proper subtree sizes. By a *proper subtree size* we mean a subtree size that is strictly smaller than n (the size of the whole tree T^n). Thus, $\Psi(T^n) = \sum_{v \neq r} n_v$, where r is the root of the split tree T^n; see Fig. 9 which illustrates this equivalence.

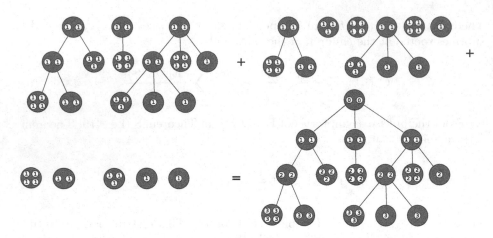

Fig. 9. This figure illustrates that the total path length which is defined as the sum of all the depths of the balls is also equal to the sum of all proper subtree sizes of the balls (i.e., summing the sizes of all subtrees except for the size n of the whole tree T^n).

6   Inversions and Permutations in Split Trees

Consider an unlabelled rooted tree T on node set V. Let r denote the root. Write $u < v$ if u is a *proper ancestor* of v, i.e., the unique path from r to v passes through u and $u \neq v$.

6.1   A Fixed-Point Equation for the Normalized Number of Inversions

Given a *bijection* $l : V \to \{1, \dots, |V|\}$ (i.e., a *node labelling*), define an *inversion* as a pair (u, v) with $u < v$, so that $l(u) > l(v)$; see Fig. 10. Below we define the *number of inversions* in a rooted tree T on node set V.

Definition 3. *Given a bijection* $l : V \to \{1, \dots, |V|\}$ *(a node labeling), define the number of* inversions

$$I(T) \overset{d}{=} \sum_{u < v} \mathbf{1}_{l(u) > l(v)},$$

where $\mathbf{1}_{l(u) > l(v)}$ *is the indicator function of the event* $A = \{l(u) > l(v)\}$, *i.e.,* $\mathbf{1}_A$ *is equal to 1 if A occurs and 0 otherwise.*

Note that if T is a path, then $I(T)$ is nothing but the number of inversions in a permutation. First we will explain how the number of inversions $I(T)$ is related to the total path length $\Psi(T)$ as defined in Definition 2 in Sect. 5.1. For any $u < v$ note that we have $\mathbf{P}(l(u) > l(v)) = 1/2$. Thus, it immediately follows that the expected number of inversions is equal to,

$$\mathbf{E}[I(T)] = \sum_{u < v} \mathbf{E}\left[\mathbf{1}_{l(u) > l(v)}\right] = \frac{1}{2}\Psi(T). \tag{17}$$

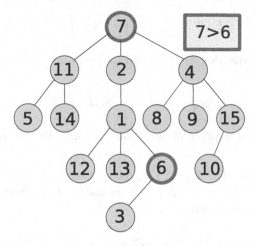

Fig. 10. This figure shows a node labelling of a rooted tree T on node set V i.e., a bijection $l : V \to \{1, \ldots, |V|\}$. In particular, the figure illustrates a pair (r, v) of nodes with r as the root, where $l(r) = 7$, $l(v) = 6$, hence $l(r) > l(v)$ so that (r, v) is an example of an inversion.

In [20] we analyzed the total number of inversions in two important classes of random trees, i.e., conditioned Galton-Watson trees as well as split trees. In the context of split trees we differentiated between $I(T_n)$ (the number of inversions on *nodes*), and $\hat{I}(T_n)$ (the number of inversions on *balls*). In the former case, the nodes are given labels, while in the latter case the individual balls are given labels. For balls β_1, β_2, write $\beta_1 < \beta_2$ if the node containing β_1 is a proper ancestor of the node containing β_2; if β_1, β_2 are contained in the same node we do not compare their labels. Below we state the result we obtained in [20, Theorem 1.12] for the number of inversions of balls (in fact [20, Theorem 1.12] also contained an alternative expression for the normalization that we will not present here). [20, Theorem 4.8] also described a similar convergence result for the number of inversions of the nodes, but we choose to not present that result in this survey.

We first derived an expression for the expected number of inversions. Any internal node contains s_0 balls, so any ball at height d has $s_0 d$ balls in its ancestors, recalling that each non-leaf (internal) node in a split tree holds s_0 balls. By a similar argument as we used to show (17) for the deterministic tree T it is now easy to show that for a general split tree T^n it holds that:

$$\mathbf{E}\left[\hat{I}(T_n)\right] = \frac{s_0}{2}\mathbf{E}\left[\Psi(T^n)\right]; \tag{18}$$

note that the expected value in the right handside of (18) occurs since split trees are random trees, while T is deterministic so that the total path length $\Psi(T)$ in (17) is not random. Thus, by [19, Theorem 3.1], see Theorem 8 in Sect. 5 we have that:

$$\mathbf{E}\left[\hat{I}(T_n)\right] = \frac{s_0}{2}\mathbf{E}\left[\Psi(T^n)\right] = \frac{s_0}{2}\left[\frac{1}{\mu}n\ln n + n\varpi(\ln n) + o(n)\right]. \qquad (19)$$

Let $C(\mathcal{V})$ be the function of $\mathcal{V}$ defined in Theorem 7 and let

$$\hat{X}_n = \frac{\hat{I}(T_n) - \mathbf{E}\left[\hat{I}(T_n)\right]}{n}; \qquad (20)$$

this is the normalized number of inversions of balls. In [20, Theorem 1.12] we showed that:

Theorem 9. *Assume that* $\mathbf{P}\left(\exists i : V_i = 1\right) < 1$ *and* $s_0 > 0$. *For the random variable* $\hat{X}_n$ *in (20) it holds that* $\hat{X}_n \to \hat{X}$ *in distribution, where* $\hat{X}$ *is the unique solution of the fixed point equation*

$$\hat{X} \stackrel{d}{=} \sum_{i=1}^{b} V_i \hat{X}^{(i)} + \sum_{j=1}^{s_0} U_j + (C(\mathcal{V}) - 1)\frac{s_0}{2}$$

Here $(V_1, \ldots, V_b), U_1, \ldots, U_{s_0}, \hat{X}^{(1)}, \ldots, \hat{X}^{(b)}$ *are independent, where* $U_1, \ldots, U_{s_0}$ *are distributed as a* $U(0, 1)$ *random variable, and* $\hat{X}^{(1)}, \ldots, \hat{X}^{(b)}$ *are distributed as* $\hat{X}$.

Theorem 9 was proved by again using the contraction method and applying the precise asymptotics for the expected value of $\hat{I}(T^n)$ in (19).

6.2   What About the Number of Longer Permutations in a Split Tree?

One can also extend the notion of inversions in labelled trees to *longer permutations*. For example, the *number of inverted triples* in a tree T with labelling l is the number of triples of nodes $u_1 < u_2 < u_3$ with labels such that $l(u_1) > l(u_2) > l(u_3)$. In general, we say a *permutation* α appears on the $|\alpha|$-tuple of nodes $u_1, \ldots, u_{|\alpha|}$, if $u_1 < \ldots < u_{|\alpha|}$ and the induced order $l(u) = (l(u_1), \ldots, l(u_{|\alpha|}))$ is α. Write $l(u) \approx \alpha$ to indicate the induced order is the same: for example $527 \approx 213$. The number of inverted triples in a fixed tree T is the random variable

$$R(321, T) = \sum_{u_1 < u_2 < u_3} \mathbf{1}[l(u_1) > l(u_2) > l(u_3)],$$

where the sum runs over all triples of nodes in T such that u_1 is an ancestor of u_2 and u_2 an ancestor of u_3. For a tree T and uniformly random node labelling define

$$R(\alpha, T) \stackrel{d}{=} \sum_{u_1 < \ldots < u_{|\alpha|}} \mathbf{1}[l(u) \approx \alpha],$$

so in particular $R(21, T)$ counts the number of inversions in a random labelling of T, which we earlier denoted by $I(T)$ (see Definition 3). For any $u_1 < \ldots < u_{|\alpha|}$ we have $\mathbb{P}[\pi(u) \approx \alpha] = 1/|\alpha|!$ and so it is easy to see that,

$$\mathbf{E}\left[R(\alpha,T)\right] = \sum_{u_1 < \ldots < u_{|\alpha|}} \mathbb{P}\left[\pi(u) \approx \alpha\right] = \frac{1}{|\alpha|!} \sum_{v} \binom{d(v)}{|\alpha| - 1}.$$

For inversions, $\mathbf{E}\left[R(21,T)\right] = \frac{1}{2}\Psi(T)$, where the tree parameter $\Psi(T)$ is the *total path length* of T as we also saw in (17). In [2] we considered the distribution of the number of appearances of a fixed permutation of random labellings of split trees i.e., the distribution of $R(\alpha,T^n)$. Similarly to the way one can describe a probability distribution by giving all finite moments, we may also describe a distribution via its *cumulant moments*. The cumulants, which we denote by $\varkappa_r = \varkappa_r(X)$, are the coefficients in the Taylor expansion of the log of the *moment generating function* of X about the origin (provided they exist)

$$\log \mathbb{E}(e^{\xi X}) = \sum_{r} \varkappa_r \xi^r / r!$$

thus $\varkappa_1(X) = \mathbf{E}\left[X\right]$ and $\varkappa_2(X) = \mathrm{Var}\,(X)$.

For general split trees T^n, we stated our results for cumulants of $R(\alpha,T^n)$ in terms of a tree parameter $\Psi_r^k(T^n)$ which generalises the notion $\Psi(T^n)$ of total path length. For the definition of $\Psi_r^k(T^n)$, see [2, Equation (1.2)]. Denote the random variable for the number of occurrences of α in a uniformly random ball labelling of an arbitrary split tree T^n by $R(\alpha,T^n)$. The next result is [2, Theorem 1.6].

Theorem 10. *Fix a permutation* $\alpha = \alpha_1 \ldots \alpha_k$ *of length* k. *Let* T^n *be a split tree with split vector* $V = (V_1,\ldots,V_b)$ *and* n *balls. Let* $\varkappa_r = \varkappa_r(R(\alpha,T_n))$ *be the* r-th *cumulant of* $R(\alpha,T^n)$. *With high probability the split tree* T^n *has the following property:*

$$\varkappa_r = D_{\alpha,r}\Psi_r^k(T^n) + o\big(\Psi_r^k(T^n)\big),$$

where $D_{\alpha,r}$ *is a constant defined in [2, Equation (6.1)].*

Theorem 10 says the following: Generate a random split tree T^n, with high probability it has the property that the random number of occurrences of any fixed subpermutation in a random ball labelling of T^n has variance and higher cumulant moments approximately a constant times a "simple" tree parameter of T^n.

7   Cuttings of Split Trees

Cuttings of trees were introduced by Meir and Moon [46] and has during the last decades been an active area, with many studies of cuttings for different types of random trees e.g., [1, 12, 21, 22, 30, 33, 37, 38, 41, 42].

7.1   Introduction to Cuttings and Records of Rooted Trees

Given a rooted tree T with n nodes. Make a random cut by choosing one node at random. Delete this node so that the tree separates into two parts and keep only

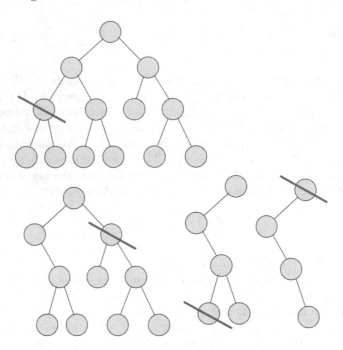

Fig. 11. This figure shows an example of cuttings of a rooted tree T. A random node is chosen and a cut is performed in that node so that the tree separates into two parts. One keeps only the part which contains the root. One continues recursively by performing a cut in a random node of the remaining tree (that gets smaller after each step of the cutting procedure), until the root is finally cut. Let $X(T)$ denote the random number of cuts until this happen. In this example $X(T) = 4$.

the part containing the root. Continue recursively until the root is cut. Then the total (random) *number of cuts* made is $X(T)$. See an illustration in Fig. 11 for an example with $X(T) = 4$. One may also consider the *number of records* in T. As shown by Janson [41, 42], this number is equivalent in distribution to the number of cuts needed to eliminate this type of tree, i.e., the number of records is also given by the random variable $X(T)$ (which is equal to the number of cuts). Given a rooted tree T with n nodes, let each node v have a random value λ_v attached to it, and assume that these values are i.i.d. with a continuous distribution. We say that the value λ_v is a *record* if it is the smallest value in the path from the root to v. Only the order relations of the λ_v's are important, so the distribution of λ_v does not matter, i.e., one can choose any continuous distribution for λ_v, or as another choice let the values λ_v be a random permutation of $\{1, 2, \dots, n\}$, see [41, 42]. Janson used the following natural coupling argument to show the equivalence between the number of cuts and the number of records: First generate the values λ_v in the tree T, for instance by using a node labelling such that the values $\{1, 2, \dots, n\}$ are distributed to the nodes uniformly at random (as was also used in Sect. 6). A node v is then cut at some time if and only if λ_v is a

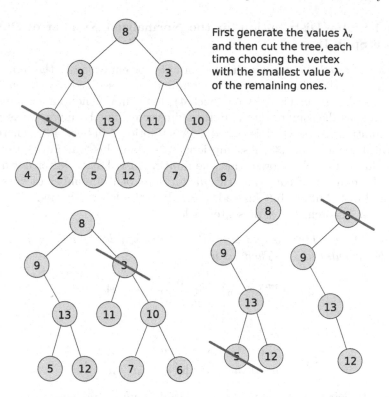

First generate the values λ_v and then cut the tree, each time choosing the vertex with the smallest value λ_v of the remaining ones.

Fig. 12. This figure shows an illustration of the equivalence between the number of cuts and the number of records in a rooted tree T. In this example, the number of cuts and the number of records is equal to $X(T) = 4$. A node v is cut at some time, if and only if, λ_v is a record.

record, and the order of the cuts is to always choose the node with the smallest λ_v value of the remaining ones. An illustration of the equivalence for the number of cuts respectively the number of records in a rooted tree is shown in Fig. 12, also compare with Fig. 11 which shows the same cutting procedure. Cuttings of trees have several different applications, for instance to *coalescent theory* (where one studies the physical phenomenon when several blocks merge into one block); see [33], where they described a representation of the Bolthausen-Sznitman coalescent in terms of cuttings of the random recursive tree. The model of cuttings can also be applied to *botnets*, i.e., malicious computer networks consisting of compromised machines which are often used in spamming or attacks; see [21]. The nodes in a tree (or another type of graph) represent the computers in a botnet, and the root represents the bot-master. The effectiveness of a botnet can be measured using the size of the component containing the root, which indicates the resources available to the bot-master [23]. To take down a botnet means to reduce the size of this root component as much as possible. The number of cuttings then measures how difficult it is to completely isolate the bot-master.

7.2   A 1-Stable Distribution for the Normalized Number of Cuts in a Split Tree

In [38] we analyzed the random number of cuts (or equivalently the number of records) in an arbitrary split tree. For simplicity we assumed [38, Assumption (A1)] (as we also did in Theorem 6 in Sect. 4) that $-\ln V$ is non-lattice, where V is distributed as the components V_i in the split vector. Furthermore, we assumed [38, Assumption (A3)], which is stated as Assumption 1 in Sect. 4.3. (There was also a third assumption [38, Assumption (A2)], but this was later proved to always hold in [19].) We showed that the limiting distribution (after normalization) of the number of cuts (i.e., the *fluctuations* of the number of cuts) is a non-normal distribution, but instead a weakly 1-stable distribution. The main theorem [38, Theorem 1.1] is presented below:

Theorem 11. *Let T^n be a split tree with n balls, and let $X(T^n)$ be the number of records (or cuts) in T^n. Then, as $n \to \infty$,*

$$(X(T^n) - C_n) \Big/ \frac{\alpha n}{\mu^{-2} \ln^2 n} \xrightarrow{d} -W, \tag{21}$$

where

$$C_n := \frac{\alpha n}{\mu^{-1} \ln n} + \frac{\alpha n \ln \ln n}{\mu^{-1} \ln^2 n} - \frac{\tau n}{\mu^{-1} \ln^2 n}, \tag{22}$$

and where W has a weakly 1-stable distribution, with characteristic function

$$\mathbf{E}\left(e^{itW}\right) = \exp\left(-\frac{\mu^{-1}}{2}\pi|t| + it\left(C - \mu^{-1}\ln|t|\right)\right), \tag{23}$$

where $C := -\mu^{-1}\ln\mu^{-1} + 2\mu^{-1} - \mu^{-2}\sigma^2 - \mu^{-1}\gamma - \frac{\sigma^2 - \mu^2}{2\mu^2}$, where μ and σ^2 are the constants in (9), where α is the constant in (6), and γ is the Euler constant and τ is another constant defined in [38, Theorem 1.1].

Remark 7. Note that the theorem tells in particular that the first order asymptotics of $X(T^n)$ is $\frac{\alpha n}{\mu^{-1}\ln n}$ and the second order asymptotics is $\frac{\alpha n \ln \ln n}{\mu^{-1}\ln^2 n}$. The first order asymptotics is easy to see directly as follows: It is easier to think of the number of records rather than the number of cuts. Note that the probability that $\lambda(v)$ in node v is a record is $\frac{1}{d(v)}$, where $d(v)$ is the depth of v. Recall from Sect. 4.1 that almost all nodes in a split tree are close to the depth $\mu^{-1}\ln n$. Thus, for almost each node v the probability that its $\lambda(v)$ value is a record is approximately $\frac{1}{\mu^{-1}\ln n}$. Thus, since the number of nodes in a split tree is (with high probability) $\alpha n + o(n)$ (see Theorem 6) it follows that the first order asymptotics of $X(T^n)$ is $\frac{\alpha n}{\mu^{-1}\ln n}$.

The proof of Theorem 11 used a classical limit theorem [44, Theorem 15.28] for convergence of sums of *triangular arrays* to *infinitely divisible distributions*. The class of *stable distributions* is included in the larger class of infinitely divisible distributions, and thus from the limit theorem we obtained that the random variable W in Theorem 11 is *weakly 1-stable*. Theorem 11 and its proof extended previous work by Janson [41] on the number of cuts (or records) in the (deterministic) complete binary tree and by Holmgren [37] for the binary search tree (which we recall is a specific example of a split tree).

7.3   Infinitely Divisible and Stable Distributions

In this section we recall the definitions of infinitely divisible and stable distributions. A triangular array is a sequence of random variables $\{(Z_{n,j},\ 1 \leq j \leq n), n \geq 1\}$, so that the variables in each row, n, are independent and identically distributed, typically the variables in different rows are not independent. A random variable Z has an infinitely divisible distribution, if and only if, for all n, there is a triangular array $\{(Z_{n,j},\ 1 \leq j \leq n), n \geq 1\}$, such that

$$Z \overset{d}{=} \sum_{j=1}^{n} Z_{n,j}.$$

Recall that the characteristic function of any real-valued random variable completely defines its probability distribution. The general formula for the characteristic function of a random variable Z, which has an infinitely divisible distribution, is equal to

$$\mathbf{E}\left(e^{itZ}\right) = \exp\left(itb - \frac{a^2 t}{2} + \int_{-\infty}^{\infty} (e^{itx} - 1 - itx\mathbf{1}[|x| < 1])d\nu(x)\right), \quad (24)$$

for constants $a \geq 0$, $b \in \mathbb{R}$ and ν is the so called Lévy measure. A random variable Z is defined as α-stable for $\alpha \in (0, 2]$, if for partial sums of independent and identically distributed random variables $Z, Z_1, Z_2 \ldots$, we have

$$\sum_{k=1}^{n} Z_k \overset{d}{=} n^{\frac{1}{\alpha}} Z + c_n \quad \forall n,$$

where $c_n \in \mathbb{R}$ are constants (and $\overset{d}{=}$ means equality in distribution). Hence, the normal distribution is one example of a stable distribution when $\alpha = 2$; moreover the stable distributions generalize the central limit theorem to random variables with infinite variances. If the Lévy measure ν in (24) satisfies $\frac{d\nu}{dx} = \frac{c_\pm}{|x|^{\alpha+1}}$ on $\mathbb{R}_\pm$, for $\alpha \in (0, 2)$ and constants $c_\pm$ the corresponding infinitely divisible distribution is weakly α-stable. The distribution is said to be strictly stable if for all n, $c_n = 0$, and weakly stable otherwise. If $\alpha < 2$, the moments $\mathbf{E}(Z^m)$ are only defined for $m > \alpha$. The characteristic function in (24) of an α-stable distribution, for $\alpha = 1$ can be simplified to

$$\mathbf{E}\left(e^{itZ}\right) = \exp\left(idt - c|t|\left(1 + i\beta\frac{2}{\pi}\text{sign}(t)\ln|t|\right)\right),$$

for constants $c > 0$, $\beta \in [-1, 1]$ and $d \in \mathbb{R}$. For further information about stable distributions, see e.g., [31, Section XVII.3].

Remark 8. The random variable W in Theorem 11 (which has a weakly 1-stable distribution) has support on $(-\infty, \infty)$, and has a *heavy tailed* distribution. Since $\alpha < 2$ the variance of W is infinite, and further, since $\alpha = 1$ not even the expected value of W is defined. The most well-known 1-stable distribution is the Cauchy distribution. However, in contrast to the distribution of W in Theorem 11, the Cauchy distribution is strictly 1-stable and symmetric.

7.4   An Extension of the Cutting Model

In [21], we have generalised the cutting model by introducing the so-called *k-cut model*, where each node has to be cut k times (instead of 1 time) before it is deleted. The obvious extension of the cutting number $X(T)$ is then: What is the total number of cuts needed until the root is cut exactly k times (i.e., the whole tree is destroyed)? This is called the *k-cut number* for a rooted tree T and is denoted by $\mathcal{K}(T)$. We also extended the notion of records in Sect. 7.1 to *r-records* for the k-cut model, where $r \le k$, and showed that the sum of all r-records is equal to the number of k-cuts.

In [21] we studied this model for different types of trees including split trees; however our main result was for a path. In [12] we analyzed the k-cut number in several different classes of random trees by calculating the moments, and showed that for split trees the k-cut number converges in probability to a constant after rescaling:

Proposition 2. *Let $\mathcal{K}(T^n)$ be the k-cut number of a split tree T^n, then it follows that*

$$\mu^{-1/k} n^{-1} (\ln n)^{1/k} \mathcal{K}(T^n) \xrightarrow{p} (k!)^{1/k} \Gamma(1 + 1/k),$$

where μ is the constant in (9), $\xrightarrow{p}$ denotes convergence in probability, and $\Gamma(\cdot)$ denotes the well-known gamma function.

In contrast to conditioned Galton-Watson trees (which were also studied in [12]), the moments are not enough to obtain the limiting distribution (i.e., the fluctuations) for the k-cut number of split trees. However, in [22] we generalized the work of Janson [41] for the cutting number of the complete binary tree to the general k-cut number, by again applying [44, Theorem 15.28] for convergence of sums of triangular arrays to infinitely divisible distributions. We are currently working on an extension of the approach in [22], to analyze the distribution of the k-cut number (after normalization) for general split trees and thereby also extend the work by Holmgren [38] for the ordinary cutting number of split trees.

8   Bond Percolation on Split Trees

Bond percolation was introduced in the mathematical literature by Broadbent and Hammersley [16] in 1957 (but had earlier been used in physics to study how liquid may travel through a porous material). It has now for decades, been an active area of research for deterministic and random graphs and trees; see e.g., the books by Bollobás [14] and Grimmet [34].

In bond percolation one considers a graph G or a tree T, and choose a value p, such that one keeps every edge in G (or T) with probability p and removes it with probability $1 - p$. Since we consider graphs and trees in the limit as their size n grows, the probability p is often a function of n, i.e., $p = p_n$. In [11] and in [13] we studied bond percolation on general split trees. For simplicity, in this survey we choose to only state the results under the assumption that $-lnV$ is non-lattice which holds true for most split trees. (However, in [11,13] we also treated the case when $-lnV$ is lattice.)

8.1   The Giant Cluster After Performing Bond Percolation in Split Trees

Consider a tree T_n of large but finite size $n \in \mathbb{N}$ and perform *Bernoulli bond-percolation* with *percolation parameter* $p_n \in [0,1]$ that depends on the size of the graph. This means that we remove each edge in T_n with probability $1 - p_n$, independently of the other edges, inducing a partition of the set of nodes into connected clusters. In particular, one is often interested in the *supercritical percolation regime*, in the sense that with high probability, there exists a *giant cluster*, that is of size comparable to that of the entire tree. Bertoin [6] established for several families of trees with n nodes that the supercritical regime corresponds to percolation parameters of the form $1 - p_n = c/\ell(n) + o(1/\ell(n))$ as $n \to \infty$, where $c > 0$ is fixed and $\ell(n)$ is an estimate of the depth of a typical node in the tree. The results of Bertoin [6] implied that under the previous regime the size Γ_n of the cluster converges in probability to $\lim_{n\to\infty} n^{-1}\Gamma_n = c_1$, where c_1 is a constant for important families of random trees with logarithmic height, such as random recursive trees, preferential attachment trees and binary search trees. In the supercritical regime, the size of the largest cluster (the giant cluster) also turns out to be the same as the cluster containing the root, which we refer to as the *root cluster*.

More recently, some authors have also analyzed the fluctuations (i.e., the limiting distribution after normalization) of the size of the largest percolation cluster as $n \to \infty$ for different families of trees with logarithmic height. Schweinsberg [50] and Bertoin [7] for random recursive trees, and Berzunza [10] for m-ary random increasing trees (these include binary search trees) and preferential attachment trees. They showed that for these examples of random trees the fluctuations of the size are described by an infinitely divisible distribution. This contrasts with analogous results on other random graphs, where the asymptotic normality for the size of the giant clusters on supercritical percolation is established; see e.g., Stepanov [52], Bollobás and Riordan [15] and Seierstad [51].

In [11] we analyzed analogously the case of arbitrary (random) split trees T^n; see Fig. 13 which shows an example of performing bond percolation in a split tree.

For split trees the size of the largest cluster (after bond percolation) could either be considered as the number of balls or as the number of nodes. In [11] we analyzed both cases. Loosely speaking, our main results showed that in the supercritical percolation regime the fluctuations of the "size" of the giant cluster has also non-normal fluctuations, where the "size" of the cluster can either be

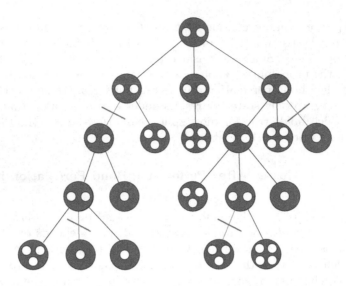

Fig. 13. This figure shows an example of performing bond percolation in a split tree T^n, such that for some value p, each edge in the tree is removed with probability $1 - p$. Since we consider trees T^n in the limit as n grows, the probability p is often a function of n, i.e., $p = p_n$.

defined as the number of nodes or the number of balls. We showed that the fluctuations of the "size" (either number of nodes or balls) of the giant cluster are described by an infinitely divisible distribution.

We consider the *supercritical regime* i.e., Regime:

$$1 - p_n = \frac{c}{\ln n} + o(\ln^{-1} n), \quad c > 0. \tag{25}$$

Definition 4. *Let T^n be an arbitrary split tree. We write G_{p_n} for the number of nodes in the root cluster, and similarly we write $\hat{G}_{p_n}$ for the number of balls in the root cluster, after performing bond percolation in T^n.*

Theorem 12 below, which is [11, Lemma 1 and Lemma 2], shows that also for general split trees there exist a unique giant cluster, which is indeed the root cluster. In Fig. 14, the different clusters after performing bond percolation in a split tree, according to the example in Fig. 13, are organized by size. We proved Theorem 12 below (i.e.,[11, Lemma 1 and Lemma 2]) by applying the general criterion for random trees in Bertoin [6] as we briefly described above together with Proposition 1 in Sect. 4.1.

Theorem 12. *Let μ be the constant in (9), c be the constant in (25), and α be the constant in (6). Then for $\hat{G}_{p_n}$ and G_{p_n} as defined in Definition 4, we have*

(i) $\dfrac{\hat{G}_{p_n}}{n} \xrightarrow[n \to \infty]{p} e^{-c/\mu}$ *and under Assumption 1,*

(ii) $\dfrac{G_{p_n}}{n} \xrightarrow[n \to \infty]{p} \alpha e^{-c/\mu},$

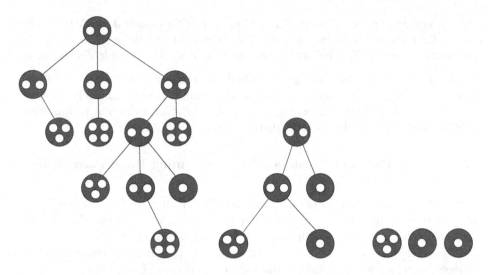

Fig. 14. Figure 13 showed an example of bond percolation in a split tree, where four edges were cut (each edge is cut with probability $1-p$). In this figure the clusters (after the edges were removed in that example) have been organized by size. The root cluster is the largest (this is always true for values of $p = p_n$ in the so-called supercritical phase, i.e., for a p_n tending to 1).

where α is the constant in Theorem 6. Moreover, the root cluster is for both cases the unique giant component.

In the next result which was one of the main results in [11] we analyzed the fluctuations of the giant cluster (where the size is now considered as the number of balls). In particular, we showed that the limiting distribution (after normalization) of $\hat{G}_{p_n}$ in Definition 4, obtains a weakly 1-stable distribution, the so-called *Luria-Delbrück distribution*. The theorem stated below is [11, Theorem 1], when we assume that $-lnV$ is non-lattice. (In [11, Theorem 1] we also stated the result when $-lnV$ is lattice.) The other main result of [11] was a similar theorem, where the size of the giant cluster instead was the number of nodes (i.e., we considered G_{p_n} in Definition 4); however we will not state this result here and instead refer to [11, Theorem 2].

Theorem 13. *Let $\hat{G}_{p_n}$ be defined as in Definition 4. As $n \to \infty$, there is the convergence in distribution*

$$\left(\frac{\hat{G}_{p_n}}{n} - e^{-\frac{c}{\mu}} \right) \ln n - \frac{c}{\mu} e^{-\frac{c}{\mu}} \ln \ln n$$

$$\xrightarrow{d} -\frac{c}{\mu} e^{-\frac{c}{\mu}} \left(Z + \ln \left(\frac{c}{\mu} \right) + \zeta \mu + \frac{(\mu^2 - \sigma^2)(c + \mu)}{2\mu^2} - \gamma + 1 \right),$$

where μ and σ^2 are the constants defined in (9), c is the constant in (25), ζ is the constant defined in Theorem 8, γ is the Euler constant and the variable Z has the continuous Luria-Delbrück distribution, which is a weakly 1-stable distribution.

The proof of Theorem 13 again applied the classical limit theorem in [44, Theorem 15.28] for convergence of sums of triangular arrays to infinitely divisible distributions. Recall that in Sect. 7.3 we gave a brief introduction to infinitely divisible and in particular 1-stable distributions.

8.2   Sizes of Different Clusters After Performing Bond Percolation in Split Trees

In Sect. 8.1 we described results for the giant cluster after performing bond percolation in random trees and in particular we stated a result about the limiting distribution for split trees in Theorem 13. A natural problem is also to estimate the sizes of the next largest clusters. (See again Fig. 14 which illustrates the cluster sizes in an example of bond percolation in a split tree.) Concerning trees with logarithmic height, Bertoin [8] proved that in the supercritical regime, the sizes of the next largest clusters of a uniform random recursive tree, normalized by a factor $\ln n/n$, converge to the atoms of some Poisson process; see also [5]. Thus, while the largest cluster has a size "close to" n, the next largest clusters have only a size "close to" $\frac{n}{\ln n}$. This result was extended by Bertoin and Bravo [9] to preferential attachment trees. In the next theorem [13, Theorem 1] we extended their results to general split trees (where we now assume that the sizes are in terms of the number of balls). (As for Theorem 13 in Sect. 8.1, we have also shown a similar result for the case when the sizes instead refer to the number of nodes; however, we refer to [13, Theorem 2] for that case.) We again consider the supercritical regime i.e., the regime in (25). Thus, [13, Theorem 1] states that:

Theorem 14. *Let $C_0 := \hat{G}_{p_n}$ be the number of balls of the root cluster and $C_1 \geq C_2 \geq \cdots$, the sequence of the number of balls of the remaining clusters ranked in the decreasing order. Then*

$$n^{-1}C_0 \xrightarrow{d} e^{-c/\mu}, \qquad as \ n \to \infty,$$

Further, for every fixed $i \in \mathbb{N}$, we have the convergence in distribution

$$\left(\frac{\ln n}{n}C_1, \ldots, \frac{\ln n}{n}C_i \right) \xrightarrow{d} (\mathrm{x}_1, \ldots, \mathrm{x}_i), \qquad as \ n \to \infty,$$

where μ is the constant in (9), c is the constant in (25), and where $\mathrm{x}_1 > \mathrm{x}_2 > \cdots$ denotes the sequence of the atoms of a Poisson random measure on $(0, \infty)$ with intensity $c\mu^{-1}e^{-c/\mu}x^{-2}\mathrm{d}x$.

Acknowledgements. I wish to thank all my collaborators on split trees at Uppsala University and internationally. A special thanks goes to Prof. Svante Janson who introduced me to the field and has remained an admired mentor and friend ever since. I also wish to gratefully acknowledge Johan Björklund for his helpful comments on the manuscript and for all his help with the figures.

References

1. Addario-Berry, L., Broutin, N., Holmgren, C.: Cutting down trees with a Markov chainsaw. Ann. Appl. Probab. **24**, 2297–2339 (2014)
2. Albert, M., Holmgren, C., Johansson, T., Skerman, F.: Embedding small digraphs and permutations in binary trees and split trees. Algorithmica **82**, 589–615 (2020)
3. Asmussen, S.: Applied Probability and Queues. Wiley, Chichester (1987)
4. Barabási, A.-L., Albert, R.: Emergence of scaling in random networks. Science **286**(5439), 509–512 (1999)
5. Baur, E.: Percolation on random recursive trees. Random. Struct. Alg. **48**, 655–680 (2016)
6. Bertoin, J.: Almost giant clusters for percolation on large trees with logarithmic heights. J. Appl. Probab. **50**(3), 603–611 (2013)
7. Bertoin, J.: On the non-Gaussian fluctuations of the giant cluster for percolation on random recursive trees. Electron. J. Probab. **19**(24), 1–15 (2014)
8. Bertoin, J.: Sizes of the largest clusters for supercritical percolation on random recursive trees. Random Struct. Algorithms **44**(1), 29–44 (2014)
9. Bertoin, J., Uribe Bravo, G.: Supercritical percolation on large scale-free random trees. Ann. Appl. Probab. **25**(1), 81–103 (2015)
10. Berzunza, G.: Yule processes with rare mutation and their applications to percolation on b-ary trees. Electron. J. Probab. **20**(43), 1–23 (2015)
11. Berzunza, G., Cai, X.S., Holmgren, C.: The asymptotic non-normality of the giant cluster for percolation on random split trees. https://arxiv.org/abs/1902.08109 (2019)
12. Berzunza, G., Cai, X.S., Holmgren, C.: The k-cut model in deterministic and random trees. Electron. J. Combin. **28**(1), 1–30 (2021)
13. Berzunza, G., Holmgren, C.: The asymptotic distribution of cluster sizes for supercritical percolation on random split trees. https://arxiv.org/abs/2003.12018 (2020)
14. Bollobás, B., Riordan, O.: Percolation. Cambridge University Press, Cambridge (2006)
15. Bollobás, B., Riordan, O.: Asymptotic normality of the size of the giant component in a random hypergraph. Random Struct. Algorithms **41**(4), 441–450 (2012)
16. Broadbent, S., Hammersley, J.: Percolation processes I. Crystals and mazes. Math. Proc. Cambridge Philos. Soc. **53**(3), 629–641 (1957)
17. Broutin, N., Devroye, L.: Large deviations for the weighted height of an extended class of trees. Algorithmica **46**(3–4), 271–297 (2006)
18. Broutin, N., Devroye, L., McLeish, E.: Weighted height of random trees. Acta Inform. **45**(4), 237–277 (2008)
19. Broutin, N., Holmgren, C.: The total path length of split trees. Ann. Appl. Probab. **22**(5), 1745–1777 (2012)
20. Cai, X.S., Holmgren, C., Janson, S., Johansson, T., Skerman, F.: Inversions in split trees and conditional Galton-Watson trees. Comb. Probab. Comput. **28**(3), 335–364 (2019)
21. Cai, X.S., Devroye, L., Holmgren, C., Skerman, F.: k-cut on paths and some trees. Electron. J. Probab. **24**(53), 1–22 (2019)
22. Cai, X.S., Holmgren, C.: Cutting resilient networks - complete binary trees. Electron. J. Combin. **26**(4), 1–28 (2019)
23. Dagon, D., Gu, G., Lee, C.P., Lee, W.: A taxonomy of botnet structures. In: Twenty-Third Annual Computer Security Applications Conference (ACSAC 2007), pp. 325–333 (2007)

24. Devroye, L.: A note on the height of binary search trees. J. Assoc. Comput. Mach. **33**, 489–498 (1986)
25. Devroye, L.: Applications of the theory of records in the study of random trees. Acta Inform. **26**, 123–130 (1988)
26. Devroye, L.: On the height of random m-ary search trees. Random Struct. Algorithms **1**(2), 191–203 (1990)
27. Devroye, L.: Branching processes and their applications in the analysis of tree structures and tree algorithms. In: Habib, M., McDiarmid, C., Ramirez-Alfonsin, J., Reed, B. (eds.) Probabilistic Methods for Algorithmic Discrete Mathematics, pp. 249–314. Springer, Berlin (1998). https://doi.org/10.1007/978-3-662-12788-9_7
28. Devroye, L.: Universal limit laws for depth in random trees. SIAM J. Comput. **28**, 409–432 (1998)
29. Drmota, M.: Random Trees. Springer, Vienna (2009)
30. Drmota, M., Iksanov, A., Moehle, M., Roesler, U.: A limiting distribution for the number of cuts needed to isolate the root of a random recursive tree. Random Struct. Alg. **34**, 319–336 (2009)
31. Feller, W.: An Introduction to Probability Theory and Its Applications. Volume II, 2nd edn. Wiley, New York (1971)
32. Friedkin, E.: Trie memory. Commun. ACM **3**, 490–500 (1960)
33. Goldschmidt, C., Martin, J.: Random recursive trees and the Bolthausen-Sznitman coalescent. Electron. J. Probab. **10**, 718–745 (2005)
34. Grimmett, G.: Percolation, 2nd edn. Springer, Heidelberg (1999)
35. Hoare, C.A.R.: Partition (Algorithm 63), Ouicksort (Algorithm 64), and Find (Algorithm 65). Comm. ACM **4**, 321–322 (1961)
36. Hoare, C.A.R.: Ouicksort. Comput. J. **5**, 10–15 (1962)
37. Holmgren, C.: Random records and cuttings in binary search trees. Comb. Probab. Comput. **19**, 391–424 (2010)
38. Holmgren, C.: A weakly 1-stable distribution for the number of random records and cuttings in split trees. Adv. in Appl. Probab. **43**, 151–177 (2011)
39. Holmgren, C.: Novel characteristics of split trees by use of renewal theory. Electron. J. Probab. **17**, 1–27 (2012)
40. Jacquet, P., Régnier, M.: Normal limiting distribution for the size and the external path length of tries. Technical report 827, INRIA-Rocquencourt (1988)
41. Janson, S.: Random records and cuttings in complete binary trees. In: Mathematics and Computer Science III Birkhäuser, Basel, pp. 241–253 (2004)
42. Janson, S.: Random cuttings and records in deterministic and random trees. Random Struct. Alg. **29**, 139–179 (2006)
43. Janson, S.: Random recursive trees and preferential attachment trees are random split trees. Comb. Probab. Comput. **28**(1), 81–99 (2019)
44. Kallenberg, O.: Foundations of Modern Probability, 2nd edn. Springer, New York (2002)
45. Knuth, D.E.: The Art of Computer Programming. Vol. 3: Sorting and Searching, 2nd edn. Addison-Wesley, Reading (1998)
46. Meir, A., Moon, J.W.: Cutting down random trees. J. Australian Math. Soc. **11**, 313–324 (1970)
47. Neininger, R., Rüschendorf, L.: On the internal pathlength of d-dimensional quad trees. Random Struct. Alg. **15**(1), 25–41 (1999)
48. Roesler, U.: A limit theorem for "Quicksort". RAIRO Inform. Théor. Appl. **25**, 85–100 (1991)
49. Roesler, U.: On the analysis of stochastic divide and conquer algorithms. Algorithmica **29**(1–2), 238–261 (2001)

50. Schweinsberg, J.: Dynamics of the evolving Bolthausen-Sznitman coalecent. Electron. J. Probab. **17**(91), 1–50 (2012)
51. Seierstad, T.G.: On the normality of giant components. Random Struct. Algorithms **43**(4), 452–485 (2013)
52. Stepanov, V.E.: Phase transitions in random graphs. Teor. Verojatnost. i Primenen. **15**, 200–216 (1970)

On Topological Analysis of Cells Organization in Biological Images

Maria-Jose Jimenez[✉][iD]

Departamento de Matematica Aplicada I, Universidad de Sevilla, Seville, Spain
majiro@us.es

Abstract. There are many problems inside the field of biomedical image analysis that can be dealt from a topological (and geometric) point of view. One of them refers to the way in which cells are self-organised inside a tissue, mainly motivated by the changes that may occur in such an organization in case of disease. This problem can be faced from different perspectives in terms, first, of how to model the cells and their 'connections' and second, how to computationally characterise their organization. We will discuss some topological approaches to this topic as well as future lines of research.

Keywords: Topological data analysis · Persistent homology · Epithelial organization

The present manuscript summarizes the main aspects treated in the invited talk, with the only purpose of giving some intuition on the main concepts, as well as a general overview.

1 Biological Setting

Epithelia are packed tissues formed by tightly assembled cells. Their apical surfaces describe regions that are similar to convex polygons forming a natural tessellation. This way, each cell can be identified with a polygon with as many sides as neighboring cells. Hence, epithelial organization has been analyzed in several works (see, for example [9,18,19]) with a common approach based on the study of the properties and the distribution of the polygons (cells). In [18], they show that geometrical and network characteristics of each cell are dependent on the number of sides. In [19], they focused on the polygons distribution to describe a range of values (obtained from mathematical tesselations) that can be taken as a reference in terms of cell arrangements. A different approach was developed in [21], were the authors provided an image analysis tool implemented in the open-access platform FIJI, to quantify epithelial organization based in computational

Project PID2019-107339GB-I00, funded by: Ministerio de Ciencia e Innovación – Agencia Estatal de Investigación/10.13039/501100011033.

J. Lindblad et al. (Eds.): DGMM 2021, LNCS 12708, pp. 58–63, 2021.
https://doi.org/10.1007/978-3-030-76657-3_3

geometry and graph theory concepts. More specifically, they looked *locally* for specific motifs represented by small graphs (graplets).

At this point, it seems interesting to introduce a measure that provides *global* topological information about the graph of cell contacts (not only about local configurations or number of neighbors), preserving part of the geometric information of the cells.

2   Topological Data Analysis

Topological Data Analysis (TDA) studies the shape of data from a topological viewpoint, having *persistent homology* [7,23] as its main tool. Persistent homology studies the evolution of *homology classes* and their life-times (persistence) in an increasing nested sequence of spaces (that is called a filtration). For a basic reference, see [8].

A *k-simplex* (or simplex of dimension k), for $k = 0, 1, 2, \ldots$ is given by the convex hull of $k + 1$ affinely independent points in $\mathbf{R}^d$, $k \leq d$. A 0-simplex is a vertex, a 1-simplex is an edge, a 2-simplex is a triangle and a k–simplex is the generalization to dimension k. A *simplicial complex* is a finite collection of simplices, K, such that for any simplex in K, all its faces are also simplices in K and the intersection of any two simplices is either empty or a simplex in K. The dimension of the simplicial complex is the maximum dimension of any of its simplices. In the case of the data dealt here, the topological spaces will be simplicial complexes of dimension 2 and homology classes will represent either connected components or holes.

For persistent homology computation on K, it is necessary to have a *filter function* that induces an increasing sequence of simplicial complexes (a filtration) $K_0 \subset K_1 \subset \cdots \subset K_n = K$, along which the lifespan of connected components and holes will be measured. The output of persistent homology computation can be codified under the form of a set of intervals, *barcode*, encoding the birth and death times of each homology class arising in the increasing sequence of spaces. However, the nature of the barcode makes it hard to combine it with usual algorithms of machine learning. There are several works in the literature to construct different types of topological summaries (out of the barcode) that are easier to compute and represent a global topological feature of the space. Some examples are persistence landscapes [4], persistence images [1] or tropical coordinates [13]. Another topological summary is *persistent entropy*, which is a parameter that can be computed from the lengths of the intervals in the barcode. The concept first arose in [6] and can be described as an adaptation of Shannon entropy to this context. In [2], the authors proved that persistent entropy is robust to small perturbations of the input data. There have been several successful applications of persistent entropy for classification or characterization purposes, as, for example, [15], for detecting the transition between the preictal and ictal states in EEG signals and [17], for classifying long-length noisy signals of DC electrical motors.

Lately, there has been a great proliferation of applications of TDA to many different areas, including the analysis of biomedical images (for example, [14,16]).

3 Topological Approaches to Describe the Organization of Cells

In a 2D image of a packed tissue, the cells are represented by different regions whose boundaries are always in contact with other cells. Once the image has been properly segmented, it becomes a map of regions tiling the whole image, as in Fig. 1.

A first approach that was used to model the structure of the tissue is by a simple abstract graph where each vertex represents a cell and each edge represents a cell-to-cell (physical) contact. Notice that the structure reflects only neighboring relations, but nothing about the geometry of the cells [18,19,21]. Apart from the already mentioned papers, there have been also several works in which the contact graph (a simplicial complex of dimension 1) has been provided with a filter function for the computation of persistent homology. In the case of [22], the filter was given by the number of contacts of each cell; in the case of [11] persistent entropy was first used as a topological summary to characterise topologically the cell arrangements.

Voronoi tessellations are commonly used for estimating the morphology of cells in epithelial tissues (see Fig. 2) by building the diagrams from the set of centroids of the cells. Indeed, in [12], the authors evaluate the suitability of approximating the cells by such a mathematical construction.

A natural construction, from the Voronoi diagram on the centroids of the cells, is the Delaunay triangulation that is dually associated to the Voronoi diagram. Hence, the Delaunay complex arise as a 2-dimensional simplicial complex that may represent the cells (vertices) and the neighboring relations between

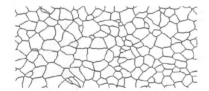
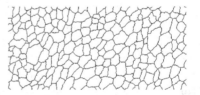

Fig. 1. Segmented images of tissues of Neuroepithelium from chicken embryos, on the left, and Wing imaginal disc from Drosophila, on the right.

Fig. 2. From left to right, a portion of segmented image of epithelial cells with centroids marked in red; Voronoi diagram of their centroids superimposed in blue; Voronoi diagrams and centroids of cells; Delaunay triangulation superimposed in purple; Delaunay triangulation representing the cells arrangement. For the color version we address the reader to the online publication. (Color figure online)

them (edges and triangles). Notice that this representation, unlike the previous one, keeps spatial information of the cells on the plane. Filter values to the edges and triangles will be assigned in terms of the distances between the centroids of the cells. This is, hence, a topological and geometrical model that is also a suitable input for persistent homology computation. A first work in this line was [3], also using persistent entropy as topological feature. However, other topological summaries can be explored in the search for new insights in the characterisation of different tissues as well as mathematical tessellations.

In the previous representation, there is, in fact an approximation of the cells by Voronoi regions, but sometimes there might be an important difference between the real and the Voronoi region. So it is natural to think about a way of keeping more information about the morphology of each cell. For this aim, the distance transform could be used to provide a filter function to all the pixels of each region in the segmented image, in which case, persistent homology could also be computed. This is an ongoing work that is closer to digital topology, since every pixel on the boundary (between cells) should be taken into account for the computation.

4 Future Work

A forward extension of the work introduced here is to deal with segmented 3D images of epithelia. In [10], the authors recently described the geometry of this tissues. The main drawback in this case is the huge amount of data that needs to be processed, what makes more difficult to get a database of segmented images to work with.

A different problem in the field is the analysis of images derived from clonogenic studies, aiming to investigate the formation of colonies in cell culture assays. A recent collaboration with CITRE, a Biotechnology research center focused on drug discovery, has provided a new context to learn features from those cell colonies in the segmented images. Persistent homology has been successfully applied to related problems such as clustering [5] and particles aggregation [20], pointing to a potential application of TDA methods to the study of colony forming assays.

Acknowledgments. The work introduced here has been (and/or is being) developed in collaboration with several colleagues: Luis M. Escudero and his team, from Instituto de Biomedicina de Sevilla (IBiS), Spain, who proposed the topological study of epithelia organization; Matteo Rucco, from United Technology Research Center, Italy; Nieves Atienza, Belen Medrano and Manuel Soriano-Trigueros (with a special mention to the latter, for his implication), from Universidad de Sevilla, Spain; and CITRE – A Bristol Myers Squibb Company.

References

1. Adams, H., et al.: Persistence images: a stable vector representation of persistent homology. J. Mach. Learn. Res. **18**, 1–35 (2017)

2. Atienza, N., Gonzalez-Diaz, R., Soriano-Trigueros, M.: On the stability of persistent entropy and new summary functions for topological data analysis. Pattern Recogn. **107**, 107509 (2020)
3. Atienza, N., Escudero, L.M., Jimenez, M.J., Soriano-Trigueros, M.: Characterising epithelial tissues using persistent entropy. In: Marfil, R., Calderón, M., Díaz del Río, F., Real, P., Bandera, A. (eds.) CTIC 2019. LNCS, vol. 11382, pp. 179–190. Springer, Cham (2019). https://doi.org/10.1007/978-3-030-10828-1_14
4. Bubenik, P.: Statistical topological data analysis using persistence landscapes. J. Mach. Learn. Res. **16**(1), 77–102 (2015)
5. Chazal, F., Guibas, L.J., Oudot, S.Y., Skraba, P.: Persistence-based clustering in Riemannian manifolds. J. ACM **60**(6), Article 41 (2013)
6. Chintakunta, H., Gentimis, T., Gonzalez-Diaz, R., Jimenez, M.J., Krim, H.: An entropy-based persistence barcod. Pattern Recogn. **48**(2), 391–401 (2015)
7. Edelsbrunner H., Letscher D., Zomorodian A.: Topological persistence and simplification. In: FOCS 2000, pp. 454–463. IEEE Computer Society (2000)
8. Edelsbrunner, H., Harer, J.L.: Computational Topology: An Introduction. American Mathematical Society (2010)
9. Gibson, M.C., Patel, A.B., Nagpal, R., Perrimon, N.: The emergence of geometric order in proliferating metazoan epithelia. Nature **442**(7106), 1038–1041 (2006)
10. Gomez-Galvez, P., Vicente-Munuera, P., Escudero, L.M., et al.: Scutoids are a geometrical solution to three-dimensional packing of epithelia. Nat. Commun. **9**(1), 2960 (2018)
11. Jimenez, M.J., Rucco, M., Vicente-Munuera, P., Gómez-Gálvez, P., Escudero, L.M.: Topological data analysis for self-organization of biological tissues. In: Brimkov, V.E., Barneva, R.P. (eds.) IWCIA 2017. LNCS, vol. 10256, pp. 229–242. Springer, Cham (2017). https://doi.org/10.1007/978-3-319-59108-7_18
12. Kaliman, S., Jayachandran, C., Rehfeldt, F., Smith, A.-S.: Limits of applicability of the voronoi tessellation determined by centers of cell nuclei to epithelium morphology. Front. Physio **7**, 551 (2016)
13. Kališnik, S.: Tropical coordinates on the space of persistence barcodes. Found. Comput. Math. **19**, 101–129 (2019)
14. Lawson, P., Sholl, A.B., Brown, J.Q., et al.: Persistent homology for the quantitative evaluation of architectural features in prostate cancer histology. Sci. Rep. **9**, 1139 (2019)
15. Merelli, E., Piangerelli, M., Rucco, M., Toller, D.: A topological approach for multivariate time series characterization: the epileptic brain. EAI Endorsed Trans. Self-Adaptive Syst. **16** (2016)
16. Qaiser, T., et al.: Fast and accurate tumor segmentation of histology images using persistent homology and deep convolutional features. Med. Image Anal. **55**, 1–14 (2019)
17. Rucco, M., et al.: A new topological entropy-based approach for measuring similarities among piecewise linear functions. Signal Process. **134**, 130–138 (2017)
18. Sánchez-Gutiérrez, D., Sáez, A., Pascual, A., Escudero, L.M.: Topological progression in proliferating epithelia is driven by a unique variation in polygon distribution. PLoS ONE **8**(11), e79227 (2013)
19. Sánchez-Gutiérrez, D., Tozluoglu, M., Barry, J.D., Pascual, A., Mao, Y., Escudero, L.M.: Fundamental physical cellular constraints drive self-organization of tissues. EMBO J. **35**(1), 77–88 (2016)
20. Topaz, C.M., Ziegelmeier, L., Halverson, T.: Topological data analysis of biological aggregation models. PLoS ONE **10**(5), e0126383e0126383 (2015)

21. Vicente-Munuera, P., et al.: EpiGraph: an open-source platform to quantify epithelial organization. Bioinformatics **36**(4), 1314–1316 (2019)
22. Villoutreix, P.: Randomness and variability in animal embryogenesis, a multi-scale approach. PhD dissertation. Université Sorbonne Paris Cité (2015)
23. Zomorodian, A., Carlsson, G.: Computing persistent homology. Discret. Comput. Geom. **33**(2), 249–274 (2005)

Applications in Image Processing, Computer Vision, and Pattern Recognition

A New Matching Algorithm Between Trees of Shapes and Its Application to Brain Tumor Segmentation

Nicolas Boutry$^{(\boxtimes)}$ and Thierry Géraud

EPITA Research and Development Laboratory (LRDE), EPITA,
Le Kremlin-Bicêtre, France
nicolas.boutry@lrde.epita.fr

Abstract. Many approaches exist to compute the distance between two trees in pattern recognition. These trees can be structures with or without values on their nodes or edges. However, none of these distances take into account the shapes possibly associated to the nodes of the tree. For this reason, we propose in this paper a new distance between two trees of shapes based on the Hausdorff distance. This distance allows us to make inexact tree matching and to compute what we call residual forests, representing where two trees differ. We will also see that thanks to these residual forests, we can obtain good preliminary results in matter of brain tumor segmentation. This segmentation not only provides a segmentation but also the tree of shapes corresponding to the segmentation and its depth map.

Keywords: Mathematical morphology · Tree of shape · Brain tumor segmentation

1 Introduction

The tree of shapes (ToS) is a hierarchical representation of the boundaries of the objects in an image (they are sometimes called *level-lines*). For sake of completeness, and because we think that many applications can be derived from it, we propose to introduce the first distance between two ToS based on the distance between their shapes. This distance makes us able to obtain a fast graph inexact matching algorithm, whose complexity is in $O(n_1 \times n_2 \times K + n_1^2 + n_2^2)$ where n_1 and n_2 are the numbers of nodes of the trees T_1 and T_2 respectively and where K is the number of operations needed to compute the distance between two shapes. Our methodology is related to the following topics.

Hausdorff Distance: The *Hausdorff distance* (HD) is a very powerful tool used in Pattern Recognition to compute the deformation needed to obtain a curve from

Electronic supplementary material The online version of this chapter (https://doi.org/10.1007/978-3-030-76657-3_4) contains supplementary material, which is available to authorized users.

another. It is much used in image matching [19]. Sometimes, we can prefer to use the ranked Hausdorff distance [18] (which is more robust), or the Gromov-Hausdorff Distance when we want to compute the distance between two metric trees [26].

Distance Between Graphs: Among the possible distances between trees, we can find the *tree-edit distances* [3]. When hierarchical structures contain cycles, they are graph and then specific distances can be used [5]. A co-spectral distance between graphs can also be found in [11], where in brief they compute the Laplacian of the adjacency matrix of two given graphs of same number of nodes; after having computed their respective eigenvalues, they compute the squared sum of the differences of the two spectra which leads to the desired distance. From the computational topology point of view, we can recall the distances between Reeb graphs [2].

Graph Matching (GM): The references presented here are not exhaustive since according to Conte *et al.* [12], more than 160 publications are related to GM. There exist several approaches: *exact matching methods* that require a strict correspondence among the two objects or among their subparts, and *inexact matching methods* where a matching can occur even if the two graphs being compared are structurally different. Exact ones can be based on tree search [4] or not [25]. Among them, several flavours exist. From the strongest to the weakest forms: the graph isomorphisms which are bijective, the subgraph isomorphisms, the monomorphisms, and the homomorphisms. An alternative approach is to compute maximal common subgraphs (MCS) [5]. These algorithms are NP-complete, and require exponential time in the worst case [12] except for special kinds of graphs. Concerning the inexact ones, they can be based on tree search [34], on continuous optimization [14], on spectral methods [35], or other techniques [20]. They are considered to be either optimal or approximate depending on the case. Usually, a matching cost is associated to these algorithms (like for the tree-edit distance [3]); the aim is then to find a mapping which minimizes this cost. As explained in [7], *relaxation labeling and probabilistic approaches* [6], *semidefinite relaxations* [32], *replicator equations* [29], and *graduated assignments* [17] can also be used to proceed to graph matching. GM algorithms can be based on *similarity functions* [?] to do for example face recognition. Finally, GM can be based on the tree of shapes (see [28]). However, as we will see later, this approach is not "differential" like ours, since it is deserved to locate patterns that are already known and not for patterns that are unknown.

The Tree of Shapes: the tree of shapes [10,16] is a hierarchical representation of the *shapes* in an image. Its origin can be found in [22,27], and its applications are numerous: grain filtering [9], object detection [13], object retrieval [28], texture analysis [36], image simplification and segmentation [37], and image classification [24]. It is mainly known as being the fusion of the min-tree and the max-tree [30].

The paper is organized as follows: Sect. 2 gives the mathematical background needed in this paper, Sect. 3 presents our proposition of distance between two trees, Sect. 4 introduces our tree-matching algorithm, Sect. 5 demonstrates that the provided tools can be used to do brain tumor segmentation, Sect. 6 concludes the paper.

2   Mathematical Background

The shapes of a real image defined in a finite rectangle Ω in $\mathbb{Z}^2$ are the saturations [10] of the connected components of its (upper and lower) threshold sets. A set T of shapes is then called *tree of shapes* [16] when any two shapes are either nested or disjoint. A *distance* d on a set E is a mapping from $E \times E$ to $\mathbb{R}^+$ which satisfies that for any two elements A, B of E, $d(A, B) = 0$ iff $A = B$, that it is symmetrical, and which satisfies the triangular inequality. Let us denote by μ the *cardinality operator* and by A, B two (finite) subsets of Ω. Then, the mapping d_μ from $E \times E$ to $\mathbb{R}^+$:

$$d_\mu(A, B) = \begin{cases} 0 & \text{if } A \text{ and } B \text{ are empty,} \\ 1 - \frac{\mu(A \cap B)}{\mu(A \cup B)} & \text{otherwise.} \end{cases}$$

is a distance [21] called the *Jaccard distance*. Let (E, d) be some metric space. The *Hausdorff distance* between two finite subsets E_1 and E_2 of E and based on a given distance d is defined as:

$$D_H(E_1, E_2) := \max \left\{ \max_{p_1 \in E_1} \min_{p_2 \in E_2} d(p_1, p_2), \max_{p_2 \in E_2} \min_{p_1 \in E_1} d(p_1, p_2) \right\}.$$

3   A Distance Between Two Trees of Shapes

Let I_1, I_2 be two images on Ω and T_1, T_2 their respective trees. We define the distance between a shape s_1 of T_1 and T_2 as $d_\mu(s_1, T_2) = \min_{s_2 \in T_2} d_\mu(s_1, s_2)$. Let T be the set of trees of shapes in Ω. We can define a mapping d_T from $T \times T$ to $\mathbb{R}^+$: $d_T(T_1, T_2) = \max_{s_1 \in T_1} d_\mu(s_1, T_2)$, which is not symmetrical. We finally define $D_T(T_1, T_2) = \max(d_T(T_1, T_2), d_T(T_2, T_1))$. Since Ω is supplied with the distance d_μ, it is metric, and then D_T is the Hausdorff distance based on the distance d_μ. Let us propose the following proof whose main steps are indicated at http://www.phys.ens.fr/~chevy/Tutorat/Hausdorff.pdf, that the mapping D_T is a distance.

Property 1. *For the Jaccard distance d_μ, the mapping D_T is a distance.*

Proof: let T_A, T_B, T_C be three elements of T. Then:

1. When $T_A = T_B$, for any $s_A \in T_A$, $\min_{s_B \in T_B} d_\mu(s_A, s_B) = 0$, then for any $s_A \in T_A$, we have $d_\mu(s_A, T_B) = 0$, and then $d_T(T_A, T_B) = 0$. A symmetrical reasoning shows that $d_T(T_B, T_A) = 0$, and then $D_T(T_A, T_B) = 0$. Conversely, $D_T(T_A, T_B) = 0$ implies that $d_T(T_A, T_B) = 0$ and $d_T(T_B, T_A) = 0$. Thanks

to $d_{\mathcal{T}}(T_A, T_B) = 0$, we know that for any $s_A \in T_A$, there exists some $s_B \in T_B$ with $d_\mu(s_A, s_B) = 0$ (and thus $s_A = s_B$). In other words, for any $s_A \in T_A$, $s_A \in T_B$, that is, $T_A \subseteq T_B$. Thanks to $d_{\mathcal{T}}(T_A, T_B) = 0$, we obtain $T_B \subseteq T_A$. We can conclude with $T_A = T_B$.

2. The symmetry is obtained by construction.

3. Triangular inequality: let us proceed in five steps:

   (a) For any $s_A \in T_A$ and any $s_B \in T_B$, let us prove that:

$$d_\mu(s_A, T_C) \leq d_\mu(s_A, s_B) + d_\mu(s_B, T_C).$$

   Since d_μ is a distance, for any $s_C \in T_C$:

$$d_\mu(s_A, s_C) \leq d_\mu(s_A, s_B) + d_\mu(s_B, s_C),$$

   which implies by applying the increasing min operator:

$$d_\mu(s_A, T_C) = \min_{s_C \in T_C} d_\mu(s_A, s_C)$$
$$\leq d_\mu(s_A, s_B) + \min_{s_C \in T_C} d_\mu(s_B, s_C)$$
$$\leq d_\mu(s_A, s_B) + d_\mu(s_B, T_C),$$

   which proves the inequality.

   (b) Now, let us prove that for any $s_A \in T_A$ and any $s_B \in T_B$:

$$d_\mu(s_A, s_B) + d_\mu(s_B, T_C) \leq d_\mu(s_A, s_B) + D_{\mathcal{T}}(T_B, T_C).$$

   This property is due to $d_\mu(s_B, T_C) \leq d_{\mathcal{T}}(T_B, T_C) \leq D_{\mathcal{T}}(T_B, T_C)$.

   (c) For any s_A in T_A, let us prove that:

$$d_\mu(s_A, T_C) \leq d_\mu(s_A, T_B) + D_{\mathcal{T}}(T_B, T_C).$$

   We already know that $d_\mu(s_A, T_C) \leq d_\mu(s_A, s_B) + D_{\mathcal{T}}(T_B, T_C)$, then thanks to the min operator, we obtain:

$$d_\mu(s_A, T_C) = \min_{s_B \in T_B} d_\mu(s_A, T_C),$$
$$\leq \min_{s_B \in T_B} d_\mu(s_A, s_B) + D_{\mathcal{T}}(T_B, T_C),$$
$$\leq d_\mu(s_A, T_B) + D_{\mathcal{T}}(T_B, T_C),$$

   which concludes this part of the proof.

   (d) Thus we obtain:

$$d_\mu(s_A, T_C) \leq d_\mu(s_A, T_B) + D_{\mathcal{T}}(T_B, T_C) \leq D_{\mathcal{T}}(T_A, T_B) + D_{\mathcal{T}}(T_B, T_C),$$

   which leads to:

$$d_{\mathcal{T}}(T_A, T_C) = \max_{s_A \in T_A} d_\mu(s_A, T_C)$$
$$\leq \max_{s_A \in T_A} \left(d_\mu(s_A, T_B) + D_{\mathcal{T}}(T_B, T_C) \right),$$
$$\leq D_{\mathcal{T}}(T_A, T_B) + D_{\mathcal{T}}(T_B, T_C),$$

then with a similar reasoning, we obtain that:

$$d_{\mathcal{T}}(T_C, T_A) \leq D_{\mathcal{T}}(T_A, T_B) + D_{\mathcal{T}}(T_B, T_C),$$

and then $D_{\mathcal{T}}(T_A, T_C) \leq D_{\mathcal{T}}(T_A, T_B) + D_{\mathcal{T}}(T_B, T_C)$, which concludes the proof. □

Property 2. *Let T_1 and T_2 be two trees of shapes defined on the same domain. Let us compute the subsets T_1' and T_2' with $\lambda \geq 0$ a given threshold:*

$$T_1' = \{s \in T_1 \mid d_\mu(s, T_2) \leq \lambda\},$$
$$T_2' = \{s \in T_2 \mid d_\mu(s, T_1) \leq \lambda\}.$$

Then the subtrees T_1' of T_1 and T_2' of T_2 satisfy $D_{\mathcal{T}}(T_1', T_2') \leq \lambda$.

Proof: Let us prove first that:

$$\forall s_1' \in T_1', \ \min_{\forall s_2' \in T_2'} d_\mu(s_1', s_2') \leq \lambda. \quad (P)$$

When (P) is false, there exists some $s_1' \in T_1'$ such that: $\min_{s_2' \in T_2'} d_\mu(s_1', s_2') > \lambda$, that is, for any $s_2' \in T_2'$, we have $d_\mu(s_1', s_2') > \lambda$. However, s_1' belongs to T_1', then $d_\mu(s_1', T_2) \leq \lambda$, then there exists $s_2 \in T_2 \setminus T_2'$ such that $d_\mu(s_1', s_2) \leq \lambda$. By symmetry of d_μ, we have that $d_\mu(s_2, s_1') \leq \lambda$, then $\min_{s_1 \in T_1} d_\mu(s_2, s_1) \leq \lambda$, then $s_2 \in T_2'$. We obtain a contradiction, then (P) is true. By symmetry, we obtain:

$$\forall s_2' \in T_2', \ \min_{s_1' \in T_1'} d_\mu(s_2', s_1') \leq \lambda,$$

thus for any $s_1' \in T_1'$ and for any $s_2' \in T_2'$, $d_\mu(s_1', T_2') \leq \lambda$ and $d_\mu(s_2', T_1') \leq \lambda$, which leads to $D_{\mathcal{T}}(T_1', T_2') \leq \lambda$. □

4 Tree-Matching and Residual Forests

In this section, we present our definition of tree-matching, we explain how we are able to ensure that the Hausdorff distance between two subtrees is lower than a given threshold, and then we introduce our *residual forests*.

4.1 Our Definition of Tree-Matching

In this paper, we consider that two trees T_1 and T_2 computed on the images I_1 and I_2 defined on Ω *match* relatively to a given $\lambda \in \mathbb{R}^+$ when their Hausdorff distance $D_{\mathcal{T}}(T_1, T_2)$ is lower than or equal to the threshold λ. A strong property is that when T_1 and T_2 match relatively to 0, they are identical sets of shapes, since it means that for any shape s_1 in T_1, there exists some shape s_2 in T_2 equal to s_1, and conversely (thanks to the symmetry of $D_{\mathcal{T}}$).

4.2 Subtrees Extraction

Now let us assume that we have two trees T_1 and T_2 corresponding to two images I_1 and I_2 respectively, both defined on Ω. We want to find two subtrees T_1' of T_1 and T_2' of T_2 satisfying: $D_{\mathcal{T}}(T_1', T_2') \leq \lambda$ for some $\lambda \in \mathbb{R}^+$. For this aim, it is sufficient to compute: $T_1' = \{s_1 \in T_1 ; d_\mu(s_1, T_2) \leq \lambda\}$ and $T_2' = \{s_2 \in T_2 ; d_\mu(s_2, T_1) \leq \lambda\}$. We are ensured that T_1' and T_2' are trees: they are both sets of shapes which are disjoint or nested and they both contain the maximal element Ω. Furthermore, by Property 2, we ensure that the Hausdorff distance between T_1' and T_2' satisfies: $D_{\mathcal{T}}(T_1', T_2') \leq \lambda$, and then we obtain subtrees of T_1 and T_2 which are as much similar as we want.

4.3 Residual Forests

Assuming we have computed T_1' and T_2' for a given $\lambda \in \mathbb{R}^+$, we can then remove from T_1 the elements of T_1' (we obtain the forest F_1) and from T_2 the elements of T_2' (we obtain the forest F_2). We call then F_1 and F_2 *residual forests* of T_1 (relatively to T_2) and of T_2 (relatively to T_1) respectively. The connected components of F_1 and F_2, called *residual trees* of I_1 and I_2 respectively, will then represent where I_1 and I_2 differ from each other. Obviously, the lower λ, the bigger the residual forests.

4.4 Complexity

Let us assume that some threshold λ is given. For two given trees of shapes T_1, T_2 of numbers of nodes n_1, n_2 respectively, we can compute a distance matrix M whose element $M_{i,j}$ is equal to $d_\mu(s_i^1, s_j^2)$ where $\{s_i^1\}_i$ are the shapes of T_1 and $\{s_j^2\}_j$ are the shapes of T_2. This is done in $O(n_1 \times n_2 \times K)$ where K is the time needed to compute the distance between two shapes. From this matrix, we can deduce in linear time each term $d_\mu(s_i^1, T_2)$ for each i and $d_\mu(s_j^2, T_1)$ for each j using the min operator. This decision step is then in $O(n_1 \times n_2)$. Now, for each shape that we want to remove from T_1 to finally obtain T_1', we have to remove the corresponding node in T_1 structure, which is linear time. At most we have to do this $(n_1 - 1)$ times, this part is then in $O(n_1^2)$. Obtaining T_2' from T_2 is then in $O(n_2^2)$. The total complexity of our algorithm is then in $O(n_1 \times n_2 \times K + n_1^2 + n_2^2)$.

5   An Application: Brain Tumor Segmentation

To show an application of our distance, we propose an algorithm able to do unsupervised brain tumor segmentation [31]. The key idea is the following: assuming that two brains look like each other, except that the first has a tumor and the second has not, the tumor should appear in one of the residual trees of the tumored brain since the residual trees encode the difference between two images. In this experiment, we use the MICCAI BraTS multi-modal[1] dataset [33] to obtain

---

[1] The BraTS dataset provides FLAIR, T1, T1CE, and T2 modalities for each brain but we will limit us to the FLAIR and T2 modalities.

Fig. 1. From left to right, the initial tumored slice where we want to locate the tumor, then the sane slices of similarities equal to 0.396337, 0.537613, 0.558739 and 0.604324 (relatively to the tumored slice). The last one is the best matching brain in the dataset.

tumored brains, and the OASIS-3 dataset [23] for brains with no tumor. We assume that the brains are aligned. Note that the following experiment is in 2D, but it can easily be extended to n-D [15,16].

Details of the Algorithm: For practical reasons, we normalize the FLAIR and T2 modalities using Gauss normalization (we remove the mean and divide by 5 times the standard deviation), we clip the values between -1 and 1, and we uniformly quantify so that the value space becomes $[\![0, 10]\!]$ for the two FLAIR modalities (tumored and sane brains) and $[\![0, 20]\!]$ for the T2 modality (tumored brain).

The Algorithm

1. We choose one of the 335 brains of size $240 \times 240 \times 155$ in the BraTS 2020 dataset and we extract the slice corresponding to $z = 77$ in the FLAIR modality file (see Fig. 1).
2. The similarity between two slices is computed this way:
   - We compute the cross-correlation between the intensities of the two slices (each one has been normalized by its L2 norm), that we name $\mathcal{I}_{\text{sim}}$.
   - We compute the norms of the gradient of both slices in a pixel-wise manner, we normalize by their L2 norms each of these images, and we deduce the cross-correlation $\mathcal{G}_{\text{sim}}$ between these two signals.
   - We compute on the FLAIR images the masks $\text{Fluids}_{\text{BraTS}}$[2] and $\text{Fluids}_{\text{OASIS}}$ (see Section A for the details). We deduce their cross-correlation $\mathcal{F}_{\text{sim}}$ (normalized by the L2 norm).
   - We finally compute the similarity as the weighted sum:

$$1/3 * \mathcal{I}_{\text{sim}} + 1/3 * \mathcal{G}_{\text{sim}} + 1/3 * \mathcal{F}_{\text{sim}}.$$

3. We choose in the database of 749 OASIS-3 FLAIR images (with no tumors) the slice of the brain which best matches with the slice coming from the BraTS database (see Fig. 1).
4. We compute the trees of shapes T^{sane} and T^{tum} of the sane brain and of the tumored brain respectively on quantified slices (to limit the number of components in the computed trees).

---

[2] We call *fluids* the cerebro-spinal fluids, which appears in dark gray in a FLAIR image and which are generally located at the center of the brain.

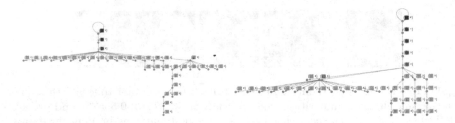

Fig. 2. Filtered trees of the slices of the ill brain and its best-matching sane brain (the background is in yellow, the shapes are in black).

Fig. 3. Extraction of the part of the tree of the tumored brain which best matches with the sane brain. The root of this tree is the only node which loops.

5. Using grain filtering, we keep in each computed ToS a maximal number of $n = 35$ nodes to obtain the most representative structures in the image. The grain filtering removes all the shapes in the two trees whose area is lower than the one of the nth greater component in each tree. This way we obtain $T^{sane}_{simp} = \{\mathcal{S}^{sane}_i\}_i$ and $T^{tum}_{simp} = \{\mathcal{S}^{tum}_i\}_i$ (see Fig. 2).

6. We fix a threshold $\lambda = 0.6$ (empirically chosen) which determines when two shapes will be considered as sufficiently similar.

7. We compute in a matrix M the distances between each shape of the first tree with each shape of the second tree: $M_{i,j} = d_\mu(\mathcal{S}^{tum}_i, \mathcal{S}^{sane}_j)$.

8. For T^{tum}_{simp}, we keep only the nodes whose corresponding shape has a distance lower than λ to the other tree T^{sane}_{simp}; we obtain then T^{tum}_{match} (see Fig. 3).

9. We compute the residual trees $\{T^{res}_i\}_i$ by removing to T^{tum}_{simp} the elements of T^{tum}_{match}: these residual trees correspond to the tumor(s) or to small differences between the two brains (see Fig. 4).

10. We set at zero the components of T^{res}_i whose amplitude is too low because low amplitudes are rarely tumors in FLAIR images, at whatever their position (see the black thumbnails in Fig. 4); we chose empirically the threshold $(\Xi + \sigma)$ of the BraTS FLAIR image (see Section A for the definitions of Ξ and σ).

11. Then we compute the tree of shapes T_{T2} of the T2 modality of the same tumored brain as before, we simplify it as usually using a grain filter keeping only the n greatest components, and we deduce the corresponding depth map $\text{depth}_{T_{T2}}$ (see Fig. 5) representing the minimal number of level-lines we have to cross to reach a pixel in an image.

12. Using the mask computed before, we deduce the image:

$$\text{depth}'_{T_{T2}} = (1 - \text{Fluids}_{\text{BraTS}}) * \text{depth}_{T_{T2}}$$

which represents the T2-weighted structures in the brain minus the fluids.

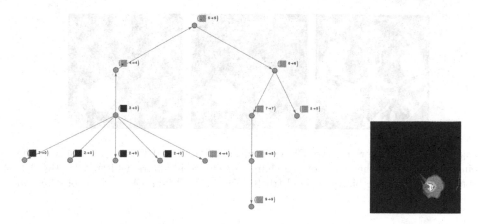

Fig. 4. From left to right, the most relevant residual tree extracted from T_{simp}^{tum} and its depth map. Other trees are filtered out because of their negligible corresponding area in the image.

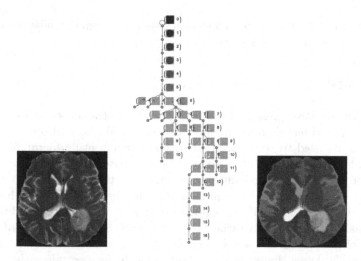

Fig. 5. From left to right, the T2-weighted slice of the tumored brain, its corresponding tree, and then its depth map.

13. By thresholding this depth map at $\beta = 0.35 * \max(\text{depth}'_{T_{T2}})^3$, we obtain the location(s) in the image where tumors should be (see Fig. 6 on its left side).
14. We apply this mask to the segmentation computed from the residual tree, we apply on it some morphological operator (here an opening of radius 2)

---

[3] We will see later that the value of β can be changed when the contrast around the tumor is very low.

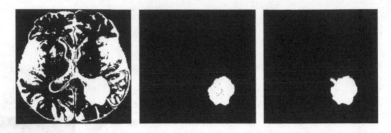

Fig. 6. From left to right, the T2-depth mask when $\beta = 0.35 * \max(\text{depth}'_{T_{T_2}})$, the binary prediction (the union of the shapes of the residual tree) filtered by the T2-depth mask, and the binary ground truth. We finally obtain a Dice score of 0.953 on this slice.

to obtain the final (smooth) result depicted in Fig. 6. Note that the morphological operator is optional and generally does not change much the final Dice score.

The set of parameters presented in this paper has been established based on many tests made on 50 test images.

6   Conclusion

In this paper, we have presented the first distance between two trees of shapes which is computed based on its shape-valued nodes. We have also seen that this distance can be used to compute residual trees representing hierarchies of the locations where two images differ. An application related to brain tumor segmentation is proposed with promising preliminary results, but furthermore it provides a tree of shapes of the segmentation and the corresponding depth map. In the future, we plan to study if our segmentation method can be applied to multimodal signals using the multi-variate tree of shapes [8], to optimize/automatize the choice of the parameters which is crucial in such a methodology, to see if we can increase the Dices thanks to some contrast enhancement algorithm [1], and more generally, we plan to find other applications of this distance. An exploration of tree-traversing methods will also be done to fasten our implementation.

Acknowledgements. Data were provided by OASIS-3 (Principal Investigators: T. Benzinger, D. Marcus, J. Morris; NIH P50AG00561, P30NS09857781, P01AG026276, P01AG003991, R01AG043434, UL1TR000448, R01EB009352. AV-45 doses were provided by Avid Radiopharmaceuticals, a wholly owned subsidiary of Eli Lilly).

References

1. Arici, T., Dikbas, S., Altunbasak, Y.: A histogram modification framework and its application for image contrast enhancement. IEEE Trans. Image Process. **18**(9), 1921–1935 (2009)

2. Bauer, U., Ge, X., Wang, Y.: Measuring distance between Reeb graphs. In: Proceedings of the Thirtieth Annual Symposium on Computational Geometry, pp. 464–473 (2014)
3. Bille, P.: A survey on tree edit distance and related problems. Theoret. Comput. Sci. **337**(1–3), 217–239 (2005)
4. Bron, C., Kerbosch, J.: Finding all cliques of an undirected graph (algorithm 457). Commun. ACM **16**(9), 575–576 (1973)
5. Bunke, H., Shearer, K.: A graph distance metric based on the maximal common subgraph. Pattern Recogn. Lett. **19**(3–4), 255–259 (1998)
6. Caetano, T.S., Caelli, T., Barone, D.A.C.: Graphical models for graph matching. In: Proceedings of the 2004 IEEE Computer Society Conference on Computer Vision and Pattern Recognition. IEEE (2004)
7. Caetano, T.S., McAuley, J.J., Cheng, L., Le, Q.V., Smola, A.J.: Learning graph matching. IEEE Trans. Pattern Anal. Mach. Intell. **31**(6), 1048–1058 (2009)
8. Carlinet, E., Géraud, T.: Getting a morphological tree of shapes for multivariate images: paths, traps, and pitfalls. In: Proceedings of the IEEE International Conference on Image Processing, Paris, France, October 2014
9. Caselles, V., Monasse, P.: Grain filters. J. Math. Imaging Vis. **17**(3), 249–270 (2002)
10. Caselles, V., Monasse, P.: Geometric Description of Images as Topographic Maps. Springer, Heidelberg (2009). https://doi.org/10.1007/978-3-642-04611-7
11. Chung, F.R.K., Graham, F.C.: Spectral graph theory. Number 92. American Mathematical Society (1997)
12. Conte, D., Foggia, P., Sansone, C., Vento, M.: Thirty years of graph matching in pattern recognition. Int. J. Pattern Recognit Artif Intell. **18**(03), 265–298 (2004)
13. Desolneux, A., Moisan, L., Morel, J.-M.: Edge detection by Helmholtz principle. J. Math. Imaging Vis. **14**(3), 271–284 (2001)
14. Fischler, M.A., Elschlager, R.A.: The representation and matching of pictorial structures. IEEE Trans. Comput. **100**(1), 67–92 (1973)
15. Géraud, T., Carlinet, E., Crozet, S.: Self-duality and digital topology: links between the morphological tree of shapes and well-composed gray-level images. In: Benediktsson, J.A., Chanussot, J., Najman, L., Talbot, H. (eds.) ISMM 2015. LNCS, vol. 9082, pp. 573–584. Springer, Cham (2015). https://doi.org/10.1007/978-3-319-18720-4_48
16. Géraud, T., Carlinet, E., Crozet, S., Najman, L.: A quasi-linear algorithm to compute the tree of shapes of nD images. In: Hendriks, C.L.L., Borgefors, G., Strand, R. (eds.) ISMM 2013. LNCS, vol. 7883, pp. 98–110. Springer, Heidelberg (2013). https://doi.org/10.1007/978-3-642-38294-9_9
17. Gold, S., Rangarajan, A.: A graduated assignment algorithm for graph matching. IEEE Trans. Pattern Anal. Mach. Intell. **18**(4), 377–388 (1996)
18. Huttenlocher, D.P., Klanderman, G.A., Rucklidge, W.J.: Comparing images using the Hausdorff distance. IEEE Trans. Pattern Anal. Mach. Intell. **15**(9), 850–863 (1993)
19. Huttenlocher, D.P., Leventon, M.E., Rucklidge, W.J.: Visually-guided navigation by comparing two-dimensional edge images. Cornell University, Department of Computer Science (1994)
20. Kitchen, L.: Discrete relaxation for matching relational structures. Technical report, Maryland Univ College Park Computer Science Center (1978)
21. Kosub, S.: A note on the triangle inequality for the Jaccard distance. Pattern Recogn. Lett. **120**, 36–38 (2019)

22. Kronrod, A.S.: On functions of two variables. Uspehi Math. Sci. **5**, 24–134 (1950). In Russian
23. LaMontagne, P.J., et al.: OASIS-3: longitudinal neuroimaging, clinical, and cognitive dataset for normal aging and Alzheimer disease. medRxiv (2019)
24. Luo, B., Zhang, L.: Robust autodual morphological profiles for the classification of high-resolution satellite images. IEEE Trans. Geosci. Remote Sens. **52**(2), 1451–1462 (2014)
25. McKay, B.D., et al.: Practical graph isomorphism. Department of Computer Science, Vanderbilt University Tennessee, USA (1981)
26. Memoli, F.: On the use of Gromov-Hausdorff distances for shape comparison. In: Botsch, M., Pajarola, R., Chen, B., Zwicker, M. (eds.) Eurographics Symposium on Point-Based Graphics. The Eurographics Association (2007)
27. Monasse, P., Guichard, F.: Fast computation of a contrast-invariant image representation. IEEE Trans. Image Process. **9**(5), 860–872 (2000)
28. Pan, Y., Birdwell, J.D., Djouadi, S.M.: Preferential image segmentation using trees of shapes. IEEE Trans. Image Process. **18**(4), 854–866 (2009)
29. Pelillo, M.: Replicator equations, maximal cliques, and graph isomorphism. Adv. Neural. Inf. Process. Syst. **11**, 1933–1955 (1999)
30. Salembier, P., Oliveras, A., Garrido, L.: Antiextensive connected operators for image and sequence processing. IEEE Trans. Image Process. **7**(4), 555–570 (1998)
31. Sauwen, N., et al.: Comparison of unsupervised classification methods for brain tumor segmentation using multi-parametric MRI. NeuroImage Clin. **12**, 753–764 (2016)
32. Schellewald, C.: Convex mathematical programs for relational matching of object views. Ph.D. thesis, Universität Mannheim (2004)
33. Bakas, S., et al.: Identifying the best machine learning algorithms for brain tumor segmentation, progression assessment, and overall survival prediction in the brats challenge. arXiv preprint arXiv:1811.02629 (2018)
34. Tsai, W.-H., King-Sun, F.: Error-correcting isomorphisms of attributed relational graphs for pattern analysis. IEEE Trans. Syst. Man Cybern. **9**(12), 757–768 (1979)
35. Umeyama, S.: An eigendecomposition approach to weighted graph matching problems. IEEE Trans. Pattern Anal. Mach. Intell. **10**(5), 695–703 (1988)
36. Xia, G.-S., Delon, J., Gousseau, Y.: Shape-based invariant texture indexing. Int. J. Comput. Vision **88**(3), 382–403 (2010)
37. Xu, Y., Géraud, T., Najman, L.: Salient level lines selection using the Mumford-Shah functional. In: Proceedings of the IEEE International Conference on Image Processing, pp. 1–5 (2013)

Combining Deep Learning and Mathematical Morphology for Historical Map Segmentation

Yizi Chen[1,2(✉)], Edwin Carlinet[1], Joseph Chazalon[1], Clément Mallet[2], Bertrand Duménieu[3], and Julien Perret[2,3]

[1] EPITA Research and Development Lab. (LRDE), EPITA,
Le Kremlin-Bicêtre, France
`yizi.chen@ign.fr`
[2] Univ. Gustave Eiffel, IGN-ENSG, LaSTIG, Saint-Mande, France
[3] LaDéHiS, CRH, EHESS, Paris, France

Abstract. The digitization of historical maps enables the study of ancient, fragile, unique, and hardly accessible information sources. Main map features can be retrieved and tracked through the time for subsequent thematic analysis. The goal of this work is the vectorization step, i.e., the extraction of vector shapes of the objects of interest from raster images of maps. We are particularly interested in closed shape detection such as buildings, building blocks, gardens, rivers, etc. in order to monitor their temporal evolution. Historical map images present significant pattern recognition challenges. The extraction of closed shapes by using traditional Mathematical Morphology (MM) is highly challenging due to the overlapping of multiple map features and texts. Moreover, state-of-the-art Convolutional Neural Networks (CNN) are perfectly designed for content image filtering but provide no guarantee about closed shape detection. Also, the lack of textural and color information of historical maps makes it hard for CNN to detect shapes that are represented by only their boundaries. Our contribution is a pipeline that combines the strengths of CNN (efficient edge detection and filtering) and MM (guaranteed extraction of closed shapes) in order to achieve such a task. The evaluation of our approach on a public dataset shows its effectiveness for extracting the closed boundaries of objects in historical maps.

Keywords: Deep learning · Convolutional neural networks · Mathematical morphology · Historical map segmentation · Object extraction

1 Introduction

The massive digitization of archival collections carried out by heritage institutions provides access to huge volumes of historical information encoded in the

Extra material for this paper (full-size figures, results, code, dataset) available at: https://github.com/soduco/paper-dgmm2021.

J. Lindblad et al. (Eds.): DGMM 2021, LNCS 12708, pp. 79–92, 2021.
https://doi.org/10.1007/978-3-030-76657-3_5

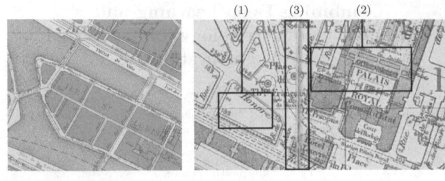

(a) Some geographical entities typically depicted in city maps: building blocks (orange), roads (green) and rivers (blue).

(b) Challenges in historical maps: (1) planimetric overlap, (2) text overlap, (3) paper folds.

Fig. 1. Contents of a 1925 urban topographic map along with an overview of their challenging properties for automatic feature extraction. (Color figure online)

available documents. Among them, maps are unfortunately still little exploited. Yet they are a gold mine of geographic data that allows to reconstruct and analyze the morphological and social evolution of a place over time [12,24]. In particular, topographic maps contain geographical features: their distribution in space, their topological relationships and various information encoded by the map legend or by text labels [8,17]. Transforming such graphical representations of geographic entities into discrete geographic data (or vector data) is a crucial step for numerous spatial and spatio-temporal analysis purposes. Such a transformation is most often manually retrieved by historians or with the help of crowdsourcing tools. This is extremely time-consuming, non-reproducible, and leads to heterogeneous data quality. Automating this tedious task is a key step towards building large volumes of reference geo-historical data.

Unfortunately, historical maps exhibit characteristics that hinder standard pattern recognition approaches and make them relatively inefficient at extracting data of good quality, i.e., that do not need to be manually post-processed. Unlike modern computer-generated maps which follow roughly the same semiotic rules, these maps vary in terms of legend, level of generalization, type of geographic features and text fonts [17]. They also usually lack texture information, which creates ambiguities in the detection of objects. For instance, building blocks and roads have very similar textures despite being of completely different nature (Fig. 1a). Popular semantic [3,18,28] and instance [4,7,27] image segmentation algorithms detect objects based on textures and are prone to fail in our context. Color is not a relevant cue either: the palette is usually highly restricted due to the technical limitations and financial constraints of their production. Objects in maps are often overlapping, some are thus partially hidden and hardly separable. Occlusion happens with overlaid textual and carto-geodetic information

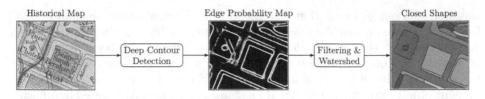

Fig. 2. Overview of the approach presented in the paper: we combine an efficient edge detection and filtering stage using a deep network with a fast closed shape extraction using mathematical morphology tools.

in particular (Fig. 1b, rectangles (1) and (2)). Last, preservation conditions of historical maps play a role as stains, folds or holes might cause gaps in the cartographic information. Such artifacts may lead to incorrect object detection (Fig. 1b, rectangle (3)).

Our contributions in this paper are as follows. After reviewing the limitations of the current approaches for segmenting maps in Sect. 2, we propose a simple pipeline (Fig. 2) that combines deep networks and mathematical morphology for object detection in maps. It takes benefit from their complementary strengths, namely image filtering and strong guarantees with respect to closed shapes. We derive edge probability maps using a multi-scale deep network approach depicted in Sect. 3 and then leverage mathematical morphology tools to extract closed shapes as explained in Sect. 4. Eventually, in Sect. 5, the second contribution lies in a thorough evaluation of the relevance of the mathematical morphology stage with novel visualizations and metrics to objectively assess our approach and better identify the strengths and weaknesses of each stage and of the workflow.

2   Approaches for Map Segmentation

We target to recover geometric structures from scans of historical maps. In literature, Angulo et al. [1] apply watershed in Mathematical morphology in color cartographic image to extract objects through color and geometrical features. Unfortunately, as mentioned above, due to the limited texture and color content of such data sources, standard semantic segmentation approaches of the literature would fail for most cases. Instead, we cast our problem as a vectorization challenge that can be turned into a region-based contour extraction task. Such a problem is traditionally solved through a two-step approach: the detection of edges or local primitives (lines, corners) followed by the retrieval of structures based on global constraints [34]. Recent works have shown the relevance of a coupled solution [13]. They remain tractable and efficient only for a limited number of structures. Region-based methods (e.g., based on PDEs [21]) may lead to oversimplified results and will not be further analyzed here.

The main issue of two-step solutions is the edge detection step. This low-level task is achieved by measuring locally pixel gradients. Due to the amount of noise (overlapping objects, map deformation), this would result in many tiny and spurious elements that any global solution would manage connecting. Instead, we focus on boundary detection, i.e., a middle-level image task that separates objects at the semantic level according to different geometric properties of images. This offers two main advantages: (i) a limited sensitivity to noise in maps and (ii) the provision of more salient and robust primitives for the subsequent object extraction step. We do not focus on a primitive-based approach since shapes on maps cannot be simply assumed.

Recently, among the vast amount of literature, convolution neural networks (CNN) have shown a high level of performance for boundary detection [15,33]. However, they only provide probability edge maps. Without topological constraints, image partitioning is not ensured. Conversely, watershed segmentation techniques in mathematical morphology can directly extract closed contours. They run fast for such a generation, but may lead to many false-positive results. Indeed, using only low-level image features such as image gradients, watershed techniques may not efficiently maintain useful boundary information [4]. Consequently, we propose here to merge the CNN-based and watershed image segmentation methods in order to benefit from the strengths of both strategies [32]. A supervised approach is conceivable since we both have access to reference vectorized maps and CNN architectures pre-trained with natural image.

3   Deep Edge Detection

We detail how we selected the network architecture used to detect and filter edges, with illustrations of the strengths of such approach, and describe the training procedure we followed to use the selected network (BDCN) on our dataset.

Network Architecture. Contour detection was first addressed with the design of handcrafted features based on brightness, color, textures [19]. Then, improvements lied in their efficient group through mono- or multi-scale attributes retrieving micro-structures: textons are a salient example [35]. Afterwards, main methods focused on combining all available cues, such as [2]. They used a global probability boundary by learning the weights of manually selected features (gradients and textons as features in several image scales) in order to detect contours and form better closed boundaries to represent the objects in images. Since CNNs have proved their relevance to extract and combine meaningful image features, a large amount of research has focused on detecting contours. The most famous one is the so-called Holistically-nested edge detector (HED) [33], which is an end-to-end multi-scale deep learning network. The novelty consisted in using skip-connections to merge different levels of features and learn different losses from intermediate layers of VGG-16 [30]. This allowed recovering multiscale representations of image features. Eventually, He et al. [15] proposed a so-called Bi-Directional Cascade Network (BDCN) by designing a scale enhancement module (SEM) on top of HED to enhance multiscale spatial contexts in images resulting in a better performance than humans in the BSDS500 dataset.

One advantage of BDCN is that the multiscale representatives combine semantically meaningful features to efficiently filter out the image textures and

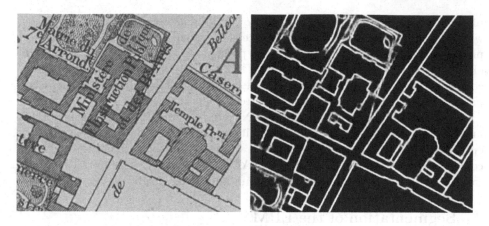

Fig. 3. BDCN produces an Edge Probability Map (*right*) with texts and textures removed from the input (*left*).

text information while maintaining useful contours and lines in the images. It is particularly suited for handling noise in our maps. Another advantage is that learnable dilated convolutions in SEM can learn fine-grained features with larger receptive fields that are beneficial when we want to accurately separate the texts with object contours. It is because building contours have much longer pixel continuity than text, resulting in higher activation after dilated convolution. After several iterations, the probability of text pixels will vanish, leading to their removal, similarly to texture, as shown in Fig. 3. However, the BDCN network works only at the pixel level and cannot guarantee the required topological properties in predicted edge probability maps without additional topological constraints [9], thus the current solution requires knowledge of the number of structures to be retrieved.

Training. Annotated historical maps are used to train a BDCN network. The final prediction which is a probability map where each pixel in the maps contain values in range $[0, 1]$ (zero means the pixel does not belong to a contour, one that it does). We train our network from scratch instead of using transfer learning on the edge weights learned from BSDS500 (dataset developed for image boundary detection and segmentation tasks): the features in natural images are very different from our historical map images. We need to filter out most of the texts in our maps, but the network trained on the BSDS dataset does not provide any useful features related to geometric filtering tasks. In order to handle data imbalance during training, we proceed as follows. We define our input image as $x \in \mathbb{R}^{H \cdot W}$ and ground truth label $y \in \{0, 1\}^{H \cdot W}$. The output of predicted image is $\hat{y} = f(x, w) \in [0, 1]^{H \cdot W}$ and every element of $\hat{y}$ is interpreted as the probability of pixel i having label 1: $\hat{y} \equiv p(Y_i = 1 | x, w)$. Since the edge detection is a binary classification task, binary cross entropy loss is used as loss function between predictions and ground truths. Due to highly imbalanced edge (97.5%) and non-edge (only 2.5%) classes, extra parameters α, β are used as weights to re-balance the binary cross entropy loss, as $\mathcal{L}_{BCE} = -\alpha \sum_{j \in Y_-} log(1 - \hat{y}_j) - \beta \sum_{j \in Y_+} log(\hat{y}_j)$ where Y_+ is the set of indices of edge pixels, Y_- is the set of indices of non-edge pixels, $\alpha = (\lambda \cdot |Y_-|/(|Y_+| + |Y_-|))$ is the percentage of edge pixels in each batch of historical map image and $\beta = (|Y_+|/(|Y_+| + |Y_-|)))$ is the percentage of non-edge pixels. An extra $\lambda = 1.1$ factor is used to enhance the percentage of edge pixels in order to give extra weights for edge responses.

We build our code based on the BDCN code repository to train our historical map dataset from scratch with a few modifications. We evaluate the loss for every epoch and also for choosing the best training weights. To make the network converge faster, we replace SGD with ADAM optimizer. The initial learning rate is set to 5×10^{-5} with 0.9 momentum and 0.002 weight decay.

4   Segmentation of the EPM

From the Edge Probability Map, we then need to extract boundaries of the objects. For natural images, Hanbury et al. [14] extract close shapes from learned gradient image similar to Edge Probability Map (EPM) by using watershed transform. In Mathematical Morphology, the Watershed Transform [20] is a *de facto* standard approach for image segmentation. It has been used in many applications and has been widely studied in terms of topological properties [11,26], in terms of algorithms and in terms on computation speed [10,26].

It has two well-known issues: the over-segmentation due to the high number of minima, and the gradient leakage that merges regions. There is a third general issue with the watershed that concerns the separation of overlapping or touching objects, but this is not a problem in our case since the map components do not overlap.

Solutions to the Over-Segmentation Problem. The first problem is generally solved by filtering the minima first. In [31], the h-minima characterize the importance of each local minimum through their *dynamic*. When flooding a basin, it actually refers to the water elevation required to merge with another basin. Attributes filter, filters by reconstruction [29] also allow to eliminate some minima based on their algebraic properties: size, shape, volume... Another efficient approach consists in first ordering the way the basins merge to create a hierarchy of partitions and then performing a cut in the hierarchy to get a segmentation with non-meaningful basins removed [5,6,23].

Solutions to the Early Leakage Problem. The second problem lies in the quality of the gradient. It has been noted [22], that (hierarchical) watersheds have better results on non-local supervised gradient estimators. The idea of combining the watershed with high performance contour detector dates back to [2].

The relevance of a simple closing by area and dynamic on the edge map produced by our deep-learning edge detector combined with the watershed for this application lies in three points.

First, the minimum size of the components is known. Indeed, the document represents a physical size, and regions whose area is below 100 m^2 are not represented in the map. Thus, we have a strong *a priori* knowledge we want to inject

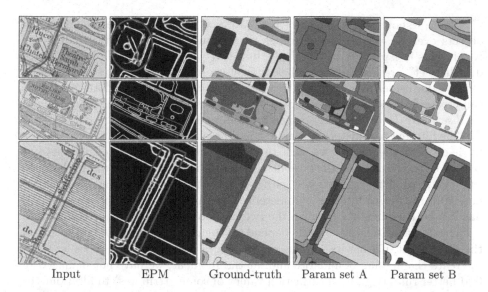

Input   EPM   Ground-truth   Param set A   Param set B

Fig. 4. Some *failures* and some *success stories* of the watershed segmentation. The parameter sets are A: $h = 3$, $\lambda = 250$, and B: $h = 7$, $\lambda = 400$. The first row shows the ability to recover weak boundaries. This sensitivity is not desirable in some cases as it leads the over-segmentation of the 2nd row. The third row suggests that the over-segmentation can be prevented by a stronger filtering but would also lead to a lower shape detection.

in the process, the minimum size of the regions (in pixels). This type of constrain is hard to infer in a deep-learning system and we cannot have such guarantees from its output. Having hard guaranties about the shapes and their size is at the foundation of the granulometries in Mathematical Morphology. Moreover, the connected (area) filter used for filtering the edge image ensure that we do not distort the signal at the boundaries of the meaningful regions.

Second, the watershed segmentation method does not rely on the strength of the gradients to select the regions. Even if the edge response is low (i.e., the gradient is weak), the watershed is able to consider this weak response and closes the contour of the region. We do not depend on the strength of the edge response from BDCN which is difficult to calibrate and normalize.

Last but not least, not only the watershed outputs a segmentation, but some implementations also produce watershed lines between regions. In our application, watershed lines are even more important than regions because we need to extract polygons for each shape. Event if we could extract boundaries from regions, it avoids an extra processing step. The watershed lines produced by the algorithm is one pixel-large and are located where the edges are the strongest, i.e., where the network has the strongest response on thick edges. The watershed lines form closed boundaries around regions which is a guarantee we cannot have from the output of a network.

Figure 4 shows the strength of the watershed to recover the boundaries of objects even on weak edge responses that would be lost by thresholding the EPM. This is especially visible in the first row where the boundaries of "Place du Châtelet" are leaking; nevertheless they are recovered in the segmentation. On the downside, this ability to recover weak edges is also a bottleneck that can create false-boundaries as shown in the middle row where the place around *"Eglise Notre-Dame"* is over-segmented because of some detection noise.

The filtering parameters (dynamic h and area λ) are important to control the trade-off between the fact we want to recover small/leaking regions (somewhat related to the *recall*) and the false-detection of boundaries (somewhat related to the *precision*). This is illustrated with two sets of parameters A and B where B has more restrictive filtering parameters. The third row of Fig. 4 shows that B has less over-segmentation but in the two first rows, it misses some boundaries.

The decision to merge objects depend on their context and not on the size of the component, neither its volume, nor its shape. The watershed "does its best" to create the missing boundaries and, at the moment, we have not managed to find better rules (e.g., with extinction values of some attributes) to filter out the basins of the watershed.

5   Evaluation

To assess the performance of the proposed approach, we conducted a series of experiments on a fully manually annotated map sheet. We report here details about this dataset we created and used, the experimental protocol as well as the calibration procedures we followed, the metrics we designed and used, and discuss some results.

Dataset. Among the multiple map sheets of the collection of Paris atlases, our work focuses on the particular sheet representing a central area of the city from year 1925 [25]. We encourage the reader to refer to the extra online material of this paper for a full-size view of this image. Indeed, such map sheets are large by nature and were digitized with high resolution, resulting in a 8500×6500 image for the area of interest.

We carefully annotated the original image by creating line vector information for each edge of each object of interest in the map. It should be noted that only a subset map strokes should be kept as many objects are not relevant for our current study: underground lines and railways, for instance, should be discarded. The resulting vector information was rasterized to produce: i) a reference edge map (a small dilation was applied, so the resulting edges have a thickness of 3 pixels); ii) a reference label map identifying each shape to be detected.

We divided the image into three disjoint subsets: a training set (rows 0 to 3999); a validation set (rows 4000 to 4999); and a test set (rows 5000 to 6500). These areas were divided into 228 disjoint tiles of 500×500 pixels.

Protocol. In the evaluation protocol we designed, our goal was to assess the impact of the watershed stage in our pipeline. We compared the performance of a baseline system, without watershed, with our proposed approach: the same baseline augmented by a watershed stage (see Fig. 2).

The baseline (without watershed) consists in a deep edge detection stage using the BDCN network presented in Sect. 3. This stage produces an edge probability map (EPM) as previously explained. The network was trained on the training set using the validation set as control set during training. To generate closed shapes, we simply thresholded the EPM and extracted the connected components. We selected the best performing threshold value (9) on the validation set for fair comparison.

The proposed approach (baseline plus watershed) consists in adding a joint filtering on area and dynamic of the EPM followed by a watershed. This approach produces a label map, i.e. a usable set of closed shapes, as detailed in Sect. 4. We selected the best performing values for area (λ) and dynamic h parameters on the validation set.

To avoid losing topological information during component labeling (baseline) or during watershed, these steps were performed on the full image (with training, validation and test sets merged) but the performance indicators were computed exclusively on the test set by masking other areas.

Metrics. While it is common in segmentation challenges to evaluate the quality of object detection by evaluating the precision and recall of edge detection at pixel, such an approach would only evaluate the process halfway to our target application: closed shapes detection. To evaluate shape detection, we need to identify pairs of matching shapes between a reference set (R) and a set of predictions (P). Because, in our particular case, shapes are disjoint among R and also among P (by construction), we can leverage the following property: as soon as the intersection over union (IoU) between $r_i \in R$ and $p_j \in P$ is strictly superior to 0.5, then we know that no other element $r_k \in R, i \neq k$ can have a higher IoU with $p_j \in R$ than $r_i \in R$, and reciprocally.

For each pair of shapes $(r_i, p_j) \in R \times P$ which verifies $\text{IoU}(r_i, p_j) = \text{area}(\frac{r_i \cap p_j}{r_i \cup p_j}) \geq T > 0.5$ we count a successful match under the threshold constraint T. We introduce this threshold value to consider all possible values between 0.5 (excluded) and 1 (included) and create a global indicator of the system under all potential quality requirements. This allows us to count the number of correctly detected shapes (*true positives* or TP), missed shapes ((*false negatives* or FN)), and wrongly predicted shapes ((*false positives* or FP)) for *every operating characteristics*. This is a very simple extension of the COCO Panoptic metric [16] which enables a finer evaluation of the system. We derive from this set of measures two analysis tools.

First a precision ($\frac{\text{TP}}{\text{TP+FP}}$), a recall ($\frac{\text{TP}}{\text{TP+FN}}$) and a F1 score ($\frac{2\text{TP}}{2\text{TP+FP+FN}}$) curves for all possible threshold values. They offer a condensed view of the behavior of a system under all possible operating characteristics. The area under the F1 score curve is equivalent, up to an offset, to the COCO PQ metric.

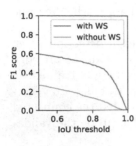

IoU	CC-labeling			Watersheding		
	Precision	Recall	F-score	Precision	Recall	F-score
0.50	0.20	0.39	0.27	0.74	0.50	0.59
0.80	0.10	0.19	0.13	0.60	0.40	0.48
0.90	0.04	0.09	0.06	0.45	0.30	0.36
0.95	0.01	0.02	0.02	0.25	0.16	0.20
COCO	PQ	RQ	RQ	PQ	SQ	RQ
(%)	0.21	0.77	0.27	0.52	0.88	0.59

Fig. 5. *Left*: comparison of the evolution of the shape detection F1-score across all possible IoU threshold with and without the watershed stage. *Right*: evaluation metrics with and without watershed. PQ ($SQ \times RQ$), SQ (segmentation) and RQ (retrieval) are COCO Panoptic [16] global metrics for each system.

The second tool is a pair of visualization maps: a precision map which associates for each predicted shape $p_j \in P$ the maximal IoU value b_{pj} such as $b_{pj} = \text{argmax}_{r_i \in R}(\text{IoU}(r_i, p_j))$, and a recall map which associates for each expected shape $r_i \in R$ the maximal IoU value b_{ri} such as $b_{ri} = \text{argmax}_{p_j \in P}(\text{IoU}(r_i, p_j))$. Each pixel of each shape is then assigned a color indicating the value of the maximal IoU: red to yellow for values between 0 and 0.5, and yellow to green for values between 0.5 and 1. The darker the green, the better the match (for both maps). The darker the red, the more serious the false positive (resp. negative) in precision (resp. recall) map.

Results and Discussion. We report here the results for the best calibrated variant of each of the two systems (baseline+connected component labeling vs baseline+watershed) under test. Figure 5 (left) compares the evolution of the F1 score indicator for both systems under each possible IoU threshold. Figure 5 (right) details the different indicators for several key values of IoU thresholds. We can see from those results that the watershed post-processing consistently and significantly improves the quality of the results. The precision and recall maps presented in Fig. 6 illustrate the benefits that the watershed post-processing bring to the deep edge segmentation: it adjusts the border of the shapes (improves precision and recall); it also removes small noise (improves precision); and it also efficiently recovers some weak boundaries (improves recall).

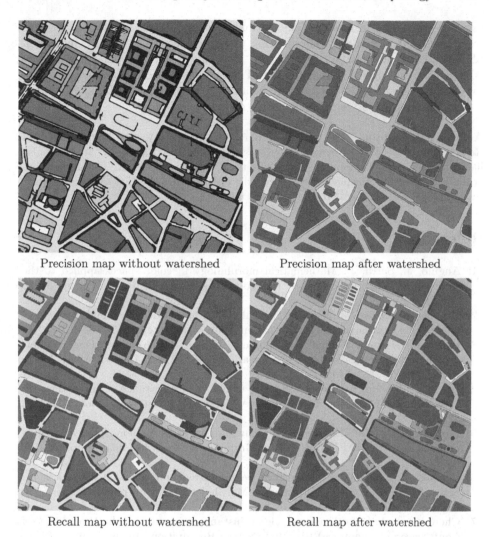

Precision map without watershed   Precision map after watershed

Recall map without watershed   Recall map after watershed

Fig. 6. Precision and recall maps without and with watershed. In all maps, the darker the green is, the better the match between predicted and reference shapes. Predicted shapes (precision map, top row) have thick and inaccurate borders which are effectively thinned by the watershed. In precision maps, red areas indicate false positives (over-, under-segmentations and noise). Reference shapes (recall map, bottom row) are better localized and sometimes recovered thanks to the restoration of weak boundaries by the watershed. In recall maps, red areas indicate false negatives (over-, under-segmentations and missed elements). (Color figure online)

6   Conclusion

In this paper, we propose an efficient combination of convolutional neural networks and mathematical morphology to address the problem of closed shapes extraction in historical maps. Convolutional neural networks (BDCN) allow us to efficiently detect edges while filtering unwanted features (text for instance). Mathematical morphology is applied to the edge probability map created by BDCN to create closed shapes reliably. The efficiency of our approach is shown by testing it on an open dataset. We believe such a method will make the digitization process of historical maps faster and more reliable.

Acknowledgements. This work was partially funded by the French National Research Agency (ANR): Project SoDuCo, grant ANR-18-CE38-0013. We would also like to thank the anonymous reviewers for their valuable feedback. The authors are grateful to the Bibliothèque de l'Hôtel de Ville (BHdV) and the City of Paris for their support for giving us access to the raw map images.

References

1. Angulo, J., Serra, J.: Mathematical morphology in color spaces applied to the analysis of cartographic images. In: Proceedings of GEOPRO, vol. 3, pp. 59–66 (2003)
2. Arbelaez, P., Maire, M., Fowlkes, C., Malik, J.: Contour detection and hierarchical image segmentation. IEEE Trans. Pattern Anal. Mach. Intel. **33**(5), 898–916 (2010)
3. Badrinarayanan, V., Kendall, A., Cipolla, R.: Segnet: a deep convolutional encoder-decoder architecture for image segmentation. IEEE Trans. Pattern Analy. Mach. Intell. **39**(12), 2481–2495 (2017)
4. Bai, M., Urtasun, R.: Deep watershed transform for instance segmentation. In: Proceedings of Conference on Computer Vision and Pattern Recognition, pp. 5221–5229 (2017)
5. Barcelos, I.B., et al.: Exploring hierarchy simplification for non-significant region removal. In: SIBGRAPI Conference on Graphics, Patterns and Images, pp. 100–107 (2019)
6. Beucher, S.: Watershed, hierarchical segmentation and waterfall algorithm. In: Serra, J., Soille, P. (eds.) Mathematical Morphology (ISMM), pp. 69–76. Springer, Dordrecht (1994)
7. Chen, K., et al.: Hybrid task cascade for instance segmentation. In: Proceedings of Conference on Computer Vision and Pattern Recognition, pp. 4974–4983 (2019)
8. Chiang, Y.-Y., Leyk, S., Knoblock, C.A.: Efficient and robust graphics recognition from historical maps. In: Kwon, Y.-B., Ogier, J.-M. (eds.) GREC 2011. LNCS, vol. 7423, pp. 25–35. Springer, Heidelberg (2013). https://doi.org/10.1007/978-3-642-36824-0_3
9. Clough, J.R., Oksuz, I., Byrne, N., Schnabel, J.A., King, A.P.: Explicit topological priors for deep-learning based image segmentation using persistent homology. In: Chung, A.C.S., Gee, J.C., Yushkevich, P.A., Bao, S. (eds.) IPMI 2019. LNCS, vol. 11492, pp. 16–28. Springer, Cham (2019). https://doi.org/10.1007/978-3-030-20351-1_2
10. Couprie, M., Najman, L., Bertrand, G.: Quasi-linear algorithms for the topological watershed. J. Math. Imaging Vis. **22**(2–3), 231–249 (2005)
11. Cousty, J., Bertrand, G., Najman, L., Couprie, M.: Watershed cuts: thinnings, shortest path forests, and topological watersheds. IEEE Trans. Pattern Anal. Mach. Intell. **32**(5), 925–939 (2009)

12. Dietzel, C., Herold, M., Hemphill, J.J., Clarke, K.C.: Spatio-temporal dynamics in California's central valley: empirical links to urban theory. Int. J. Geogr. Inf. Sci. (IJGIS) **19**(2), 175–195 (2005)
13. Favreau, J., Lafarge, F., Bousseau, A., Auvolat, A.: Extracting geometric structures in images with delaunay point processes. IEEE Trans. Pattern Anal. Mach. Intell. **42**(4), 837–850 (2020)
14. Hanbury, A., Marcotegui, B.: Morphological segmentation on learned boundaries. Image Vis. Comput. **27**(4), 480–488 (2009)
15. He, J., Zhang, S., Yang, M., Shan, Y., Huang, T.: BDCN: bi-directional cascade network for perceptual edge detection. IEEE Trans. Pattern Anal. Mach. Intell. (2020)
16. Kirillov, A., He, K., Girshick, R., Rother, C., Dollár, P.: Panoptic segmentation. In: Proceedings of Conference on Computer Vision and Pattern Recognition, pp. 9404–9413 (2019)
17. Leyk, S., Boesch, R., Weibel, R.: Saliency and semantic processing: extracting forest cover from historical topographic maps. Pattern Recogn. **39**(5), 953–968 (2006)
18. Long, J., Shelhamer, E., Darrell, T.: Fully convolutional networks for semantic segmentation. In: Proceedings of Conference on Computer Vision and Pattern Recognition, pp. 3431–3440 (2015)
19. Martin, D., Fowlkes, C., Tal, D., Malik, J.: A database of human segmented natural images and its application to evaluating segmentation algorithms and measuring ecological statistics. In: Proceedings of International Conference of Computer Vision (ICCV), vol. 2, pp. 416–423 (2001)
20. Meyer, F.: Topographic distance and watershed lines. Signal Process. **38**(1), 113–125 (1994)
21. Orzan, A., Bousseau, A., Winnemöller, H., Barla, P., Thollot, J., Salesin, D.: Diffusion curves: a vector representation for smooth-shaded images. ACM Trans. Graph. **27**(3), 1–8 (2008)
22. Perret, B., Cousty, J., Guimaraes, S.J.F., Maia, D.S.: Evaluation of hierarchical watersheds. IEEE Trans. Image Process. **27**(4), 1676–1688 (2017)
23. Perret, B., Cousty, J., Guimarães, S.J.F., Kenmochi, Y., Najman, L.: Removing non-significant regions in hierarchical clustering and segmentation. Pattern Recogn. Lett. **128**, 433–439 (2019)
24. Perret, J., Gribaudi, M., Barthelemy, M.: Roads and cities of 18th century France. Sci. Data **2**(1), 1–7 (2015)
25. Préfecture de la Seine, service du Plan: Atlas des vingt arrondissements de Paris [1 vol. (3 pl., 16 pl. doubles), 68 cm]. Paris. L. Wuhrer. ARK: 73873/pf0000935524 (1925), Bibliothèque de l'Hôtel de Ville, Ville de Paris, Paris
26. Roerdink, J.B., Meijster, A.: The watershed transform: definitions, algorithms and parallelization strategies. Fundam. Informaticae **41**(1, 2), 187–228 (2000)
27. Romera-Paredes, B., Torr, P.H.S.: Recurrent instance segmentation. In: Leibe, B., Matas, J., Sebe, N., Welling, M. (eds.) ECCV 2016. LNCS, vol. 9910, pp. 312–329. Springer, Cham (2016). https://doi.org/10.1007/978-3-319-46466-4_19
28. Ronneberger, O., Fischer, P., Brox, T.: U-Net: convolutional networks for biomedical image segmentation. In: Navab, N., Hornegger, J., Wells, W.M., Frangi, A.F. (eds.) MICCAI 2015. LNCS, vol. 9351, pp. 234–241. Springer, Cham (2015). https://doi.org/10.1007/978-3-319-24574-4_28
29. Salembier, P., Serra, J.: Flat zones filtering, connected operators, and filters by reconstruction. IEEE Trans. Image Process. **4**(8), 1153–1160 (1995)

30. Simonyan, K., Zisserman, A.: Very deep convolutional networks for large-scale image recognition. arXiv preprint arXiv:1409.1556 (2014)
31. Soille, P.: Morphological Image Analysis: Principles and Applications. Springer, Heidelberg (2013)
32. Xie, L., Qi, J., Pan, L., Wali, S.: Integrating deep convolutional neural networks with marker-controlled watershed for overlapping nuclei segmentation in histopathology images. Neurocomputing **376**, 166–179 (2020)
33. Xie, S., Tu, Z.: Holistically-nested edge detection. In: Proceedings of Conference on Computer Vision and Pattern Recognition, pp. 1395–1403 (2015)
34. Zhang, Z., et al.: Superedge grouping for object localization by combining appearance and shape informations. In: Proceedings of Conference on Computer Vision and Pattern Recognition, pp. 3266–3273 (2012)
35. Zhu, S.C., Guo, C.E., Wang, Y., Xu, Z.: What are textons? Intl. J. Comput. Vis. **62**(1–2), 121–143 (2005)

Automatic Forest Road Extraction from LiDAR Data of Mountainous Areas

Philippe Even$^{(\boxtimes)}$ and Phuc Ngo

Université de Lorraine, CNRS, LORIA, Nancy 54000, France
{philippe.even,hoai-diem-phuc.ngo}@loria.fr

Abstract. In this paper, a framework is proposed to extract forest roads from LiDAR (Light Detection and Ranging) data in mountainous areas. For that purpose, an efficient and simple solution based on discrete geometry and mathematical morphology tools is proposed. The framework is composed of two steps: (i) detecting road candidates in DTM (Digital Terrain Model) views using a mathematical morphology filter and a fast blurred segment detector in order to select a set of road seeds; (ii) extracting road sections from the obtained seeds using only the raw LiDAR points to cope with DTM approximations. For the second step, a previous tool for fast extraction of linear structures directly from ground points was adapted to automatically process each seed. It first performs a recognition of the road structure under the seed. In case of success, the structure is tracked and extended as far as possible on each side of the segment before post-processing validation and cleaning. Experiments on real data over a wide mountain area (about $78\,\mathrm{km}^2$) have been conducted to validate the proposed method.

Keywords: LiDAR data · Road detection · Point cloud processing · DTM image analysis

1 Introduction

Road location and characterization are important information used for various purposes in forest management, such as wood harvesting, road construction and maintenance, transport, ... In this context, airborne laser scanning, also called LiDAR for Light Detection And Ranging, is of great help to survey forested areas. It is a 3D acquisition technique based on the emission of a laser beam swept over the measured scene and on the detection of reflected signal from the surface. In forested areas, the received signal is composed of multiple echoes corresponding to the successive hit obstacles, from the forest canopy, down to lower vegetation levels and finally to the ground itself. A surface is interpolated through lower cloud points to classify them as ground and produce a *digital terrain model* (DTM). However, conifers are a strong obstacle. They impede the laser beam reaching the soil, and thus produce holes in the ground point distribution. Achieved point density is quite heterogeneous and this may cause large approximations in the delivered DTM.

© Springer Nature Switzerland AG 2021
J. Lindblad et al. (Eds.): DGMM 2021, LNCS 12708, pp. 93–106, 2021.
https://doi.org/10.1007/978-3-030-76657-3_6

In the literature, several methods have been developed for road extraction from LiDAR data [1,2,6,8,10,11]. Most of them rely on the shaded DTM analysis, using standard image processing tools. They show accurate enough for applications in urban and peri-urban areas in which road characteristics are quite regular and well contrasted. However, they are less efficient to detect forest roads in mountainous areas because of the strong geometric irregularities of these objects along their run. Their characteristics vary a lot depending on the terrain and also on the usage, from small narrow tracks up to very wide recent calibrated roads with the traffic line winding in the middle and vegetated sides for stocking timbers. The road surface is more or less rough depending on variable factors, such as erosion conditions or maintenance efforts. Lack of points due to LiDAR signal occlusions by a dense vegetation, especially nearby conifer plantations, drastically increases the detection difficulty. Some authors suggest to use raw data in complement to DTM analysis. Mostly signal intensity was used to recognize the road surface response [1,6], but this data depends strongly on local terrain features and parameters are difficult to set in practice. Raw altimetric information could help to better discriminate roads, but its processing is generally considered as too complicated.

To tackle this issue, we propose a framework based on efficient discrete geometry and mathematical morphology tools to automatically extract forest roads. First, the DTM is used to find relevant seeds, then only raw ground points are processed to extract road sections from each seed. By processing ground points, the new approach is more aware of point distribution in the raw data, and helps to overcome the limits of DTM interpolation. This second step relies on a recent approach for supervised extraction of linear structures from LiDAR raw data [4], that was designed for geomorphologists and archaeologists needs. In this approach, the human expertise is used to recognize roads in the DTM and provide correct initializations of the detection task. Fast responses ensure a good interaction level, that enables on-line corrections to quickly get long road sections with good accuracy. This is not any more possible in pure automatic context, so that adaptations were made to the initial step of this approach to accept less accurate road seeds. These adaptations and the seed selection procedure constitute the novelties put forward in this paper.

The method was tested on a LiDAR data set covering the Fossard mountain, in an area delimited by Remiremont, Docelles and Le Tholy (Vosges, France). Altitude ranges from $360\,$m up to $820\,$m. In such a wide region (about 78 km^2) with a dense network of forest roads, a manual extraction of all roads using the supervised approach would be too much time expensive and a tedious task.

The rest of the paper is organized as follows: Sect. 2 recalls the discrete geometry notions used in this work. Sect. 3 introduces the overall automatic extraction framework. It is composed of two steps, road seeds selection, then road extraction from each seed, that are described in Sect. 4 and Sect. 5. Held experiments are presented in Sect. 6, and Sect. 7 gives a conclusion and draws some perspectives.

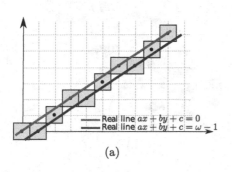
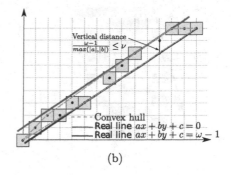

(a)   (b)

Fig. 1. (a) Naive digital straight line $\mathcal{L}(2,-3,0,3)$. (b) Blurred segment of assigned thickness $\varepsilon = 1.5$ and its optimal line $\mathcal{L}(3,-4,3,7)$.

2   Background Notions

We recall briefly in this section several notions of digital geometry [7] used in this work. We refer the reader to the given references for more details.

A **digital straight line** $\mathcal{L}(a,b,c,\nu)$ is the set of discrete points $p = (x,y) \in \mathbb{Z}^2$ satisfying the inequalities:

$$0 \le ax + by - c < \omega, \text{ with } (a,b,c,\omega) \in \mathbb{Z}^4 \qquad (1)$$

Hereafter, we note $\vec{V}(\mathcal{L}) = (a,b)$ the director vector of $\mathcal{L}$, $w(\mathcal{L}) = \nu$ its arithmetical width, $h(\mathcal{L}) = c$ its shift to origin, and $p(\mathcal{L}) = max(|a|,|b|)$ its period. When $\nu = p(\mathcal{L})$, then $\mathcal{L}$ is the narrowest 8-connected line and is called a *naive line* (see Fig. 1(a)). The thickness $\mu = \frac{\omega-1}{p(\mathcal{L})}$ of $\mathcal{L}$ is the minimum of the vertical and horizontal distances between lines $ax + by = c$ and $ax + by = c + \omega$.

A **digital straight segment** is a digital straight line restricted to $[x_{min}, x_{max}]$ interval if $|a| < |b|$, to $[y_{min}, y_{max}]$ interval otherwise.

A **blurred segment** [3] $\mathcal{B}$ of assigned thickness ε is a set of points in $\mathbb{Z}^2$ that are all covered by a digital straight line $\mathcal{L}$ of thickness $\mu \le \varepsilon$ (see Fig. 1(b)). The covering digital line with minimal thickness is called the *optimal line* of $\mathcal{B}$. Blurred segments can be detected in linear time by a recognition algorithm [3] based on an incremental growth of the convex hull of added points.

A **directional scan** is an ordered partition into scans S_i restricted to the grid domain $\mathcal{G} \subset \mathbb{Z}^2$ of a thick digital straight line $\mathcal{D}$, called *scan strip*. Each scan S_i is a segment of a naive line $\mathcal{N}_i$, called *scan line*, orthogonal to $\mathcal{D}$. The directional scan is defined as:

$$DS = \left\{ S_i = \mathcal{D} \cap \mathcal{N}_i \cap \mathcal{G} \; \middle| \; \begin{array}{l} \vec{V}(\mathcal{N}_i) \cdot \vec{V}(\mathcal{D}) = 0 \\ h(\mathcal{N}_i) = h(\mathcal{N}_{i-1}) + p(\mathcal{D}) \end{array} \right\} \qquad (2)$$

In this definition, $\vec{V}(\mathcal{N}_i) \cdot \vec{V}(\mathcal{D}) = 0$ expresses the orthogonality between the scan lines $\mathcal{N}_i$ and the scan strip $\mathcal{D}$. The shift $p(\mathcal{D})$ between successive scans $\mathcal{N}_{i-1}$ and

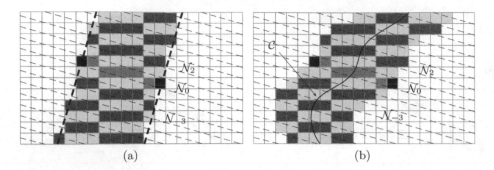

Fig. 2. (a) A directional scan in which the start scan S_0 in red, odd scans in green, even scans in blue, bounds of scan lines $\mathcal{N}_i$ with dashed lines and bounds of scan strip $\mathcal{D}$ with bold dashed lines. (b) An adaptive directional scan. The scan strip is dynamically fit to the curve position, and the scans are accordingly shifted to continuously cover the curve $\mathcal{C}$. (Color figure online)

$\mathcal{N}_i$ guarantees that all points of $\mathcal{D}$ are traversed only one time. The scans S_i can be iteratively parsed from a start scan S_0 to both ends (see Fig. 2(a)).

An **adaptive directional scan** [5] is a dynamical version of the directional scan with an on-line registration to a moving search direction. Compared to static directional scans where the scan strip remains fixed to the initial line $\mathcal{D}_0$, here the scan strip follows a curve $\mathcal{C}$ to track while scan lines remain fixed (see Fig. 2(b)). An adaptive directional scan is defined by:

$$ADS = \left\{ S_i = \mathcal{D}_i \cap \mathcal{N}_i \cap \mathcal{G} \left| \begin{array}{l} \vec{V}(\mathcal{N}_i) \cdot \vec{V}(\mathcal{D}_0) = 0 \\ h(\mathcal{N}_i) = h(\mathcal{N}_{i-1}) + p(\mathcal{D}_0) \\ \mathcal{D}_i = \mathcal{L}(\widehat{C}_i, w(\mathcal{D}_0)), i > 0 \end{array} \right. \right\} \tag{3}$$

where $\widehat{C}_i$ is a triplet composed of the director vector (a_i, b_i) and the shift to origin c_i of an estimate at position i of the tangent to the curve $\mathcal{C}$ to track. The obtained thick digital line is used to update the scan strip and lines. The last clause expresses the scan bounds update at iteration i.

3   Main Extraction Framework

The automatic road extraction process is composed of two steps: selection of road seeds, then extraction of road sections. Both steps are detailed in next sections. We give here an overview of the whole process, illustrated in Fig. 3.

The first step consists in selecting a complete set of *road seeds* in the shaded DTM –a 2D gray image. The quality of this seed selection step is very important and affects strongly the final result of full road extraction. These seeds are straight line segments that are assumed to cross the road to detect. To get this selection, a morphological operator is first applied to the slope-shaded DTM (see Fig. 3b) to enhance elongated structures such as roads (see Fig. 3c). The edges of these structures are then extracted using a fast blurred segment detector (see Fig. 3d). Orthogonal segments are distributed along them to provide the road seeds (see Fig. 3e).

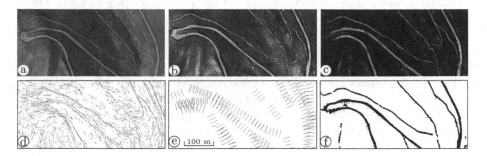

Fig. 3. Road extraction steps: (a) classical hill-shaded DTM, (b) slope-shaded DTM, (c) elongated structures enhancement, (d) edge detection, (e) road seeds selection, (f) road sections extraction.

In the second step, these seed segments are processed by a fast road extraction tool in raw LiDAR data –a 3D point cloud. For each seed, the road profile is detected under the input segment and then tracked on both sides of this segment to obtain a full road section (see Fig. 3f). This tool is based on a former approach [4] designed for a supervised context, that was adapted here to the automatic detection of forest roads using the selected seeds. In particular, the detector is modified to be more robust to possible road surface inclination and roughness using a blurred segment recognition algorithm, and shifted seed positions are tested if necessary to take into account possible inaccuracies in input data.

4   Road Seeds Selection from Shaded DTM

The DTM is encoded as a normal vector map obtained by derivation of the original height map. Hill-shading is a classical visualization technique based on controlable directional light sources (see Fig. 3a for instance). For detection purpose, we rather use slope-shading which can be seen as a lighting by a zenital source and is simply obtained by the normal vector z-component. As pointed out in [11], slope-shading ensures a good contrast between low gradient road surface and steep adjacent road cuts (see Fig. 3b).

A mathematical morphology filter, namely RORPO [9], is then applied to the slope-shaded map. Based on path operators, it contributes to enhance elongated structures without any blurring effect, that is important for the next step. Although RORPO also provides a local orientation measure, only the output intensity map is used here (see Fig. 3c). In order to speed up this step, we use a parallelized version based on OpenMP Application Program Interface[1].

---

[1] https://www.openmp.org.

Straight edges are then extracted using FBSD blurred segment detector [5]. Based on discrete geometry tools, this detector is fast, accurate and robust to noise. It can be run in pure automatic mode, with quite few parameter settings. In the present case, the only setting is a controlable thickness parameter fixed to 7 pixels in order to cope with slightly curved road edges (see Fig. 3d). Only longer edges (>80 pixels) are kept for the following.

Straight line segments are arranged across each output edges at regular interval along their run. They constitute seeds for the final road extraction step (see Fig. 3e). A seed length of 40 pixels (corresponding to 20 m in the tested data set) is selected to guarantee the crossing of wide roads with enough safety margin to cope with possible road curves. An interval value of 24 pixels is set as a good compromise between useless multiple detection of road parts and reduced risk to miss some road sections in occluded areas of the point cloud.

5   Road Sections Extraction from Raw Data

The input raw LiDAR data is a set of points, that are classified as ground. It is available in LAZ format files, organized in 500 m × 500 m tiles which cover the whole acquisition area. Ground points are extracted and projected on a high resolution grid $\mathcal{G}$ that covers all the tiles. A resolution of 0.1 m has been chosen in order to match the 0.5 m DTM grid resolution for result display.

The road detection from raw data is based on a previous work on supervised extraction of linear structures [4]. The adapted framework to automatic context is explained in this section. The road model used is first introduced, then the extraction algorithms are described.

5.1   Forest Road Model

In this work, the road is composed of a sequence of cross sections. A road cross section can be seen as a flat part, called *plateau*, bounded by steep cuts (see Fig. 4). The plateau thickness corresponds to road surface roughness and its length to road width. In order to cope with various forest road types, the plateau thickness is restricted to $\varepsilon = 0.25$m, and the length is set to a range interval from $L_{min} = 2$m up to $L_{max} = 6$m. Shorter plateaux are discarded, while longer ones are just considered as unreliable and only accepted as far as at least one of the cuts is clearly detected. As a result, the detected road sticks to its sharper cut. Compared to the model used in the former approach [4], a maximal tilt angle set to $\beta_{max} = 6°$ was added to cope with most road surface irregularities. A similar model was used in [6] for forestry management purposes, and showed well adapted to the context of mountainous relief. Even in the few flat areas found, bounds are often marked by surrounding ditches or by some hollowness because of terrain compaction. In valley or urban contexts, flat areas spread on each side of the road, and a denser point cloud is required to detect such subtle details as ditches or footpaths.

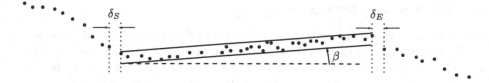

Fig. 4. Road model from an altimetric profile: the road plateau is a blurred segment with tilt angle β and bound accuracy δ_S and δ_E.

We assume here that the plateau point set is a blurred segment. Because of the discrete nature of the point cloud, each bound can not be exactly detected but just estimated within an interval lying between the end point of the blurred segment and next point in the scan. Bounds are considered as detected if this interval is less than $\delta_{max} = 0.5$ m. When both bounds are detected, the plateau is reliable and the road section position can be estimated. But very often, only one bound is detected because of the presence of trees nearby the road. In that case, the road position is interpolated between reliable plateaux.

5.2   Road Plateau Detection

The new initial plateau detection procedure described in Algorithm 1   consists in detecting a connected sub-sequence of nearly aligned points, namely a blurred segment, from a given start position, into a sequence of sorted 2D points along the scan direction.

The blurred segment is initialized with the nearest point to start position in the sequence. It is then extended by progressively inserting next points until outliers are met on each side. The blurred segment thickness is controlled by an initially assigned value ε. In order to optimize the output plateau thickness for further comparisons, a *pinching* procedure is performed after N_P points are inserted and an observation distance D_P is reached. The assigned thickness is set to a dilated value of the blurred segment thickness μ. The dilation ratio δ_P provides a security margin, that allows to cope with possible road roughness.

Moreover this pinching procedure helps to detects the plateau bounds with more accuracy. As long as the initial assigned thickness is set, no constraint is put on the segment orientation. It lets large freedom to incorporate spurious points: in our case, upper points on one side of the road, and lower points on the other side. The obtained blurred segment is thus thick and ill-oriented. To solve this drawback, the pinching procedure is essential.

5.3   Road Section Extraction

The automatic road extraction is performed on all the selected seed segments $s \in \mathcal{S}$. Algorithm 2 describes the road extraction procedure. At first, this segment is used to initialize an adaptive directional scan $\mathcal{A}$ in grid $\mathcal{G}$. An altimetric profile $\mathcal{V}$ is generated from $\mathcal{A}$ with the points in the start scan cells, sorted along the seed direction. In order to ensure a more regular repartition of the points along the profile, the grid is subdivided in smaller cells of factor N_R, and a set of N_R scans is processed instead of the initial grid scan. We select a resolution factor of $N_R = 5$, that leads to the final grid resolution of $0.1\,m$ for $\mathcal{G}$.

Then, a series of plateaux is detected from regularly spaced positions around the altimetric profile center. Too narrow or too tilted ones are discarded. Among remaining candidates, the thiner plateau is kept. Notice that this part of the algorithm differs from the former method described in [4]. In supervised context, the user is assumed to select a visually suitable start seed. At least, he can adjust his selection. Strict conditions are required on the found plateau (no tilt, accurate bounds) in order to get longer road sections. This can not be ensured in pure automatic mode, so that the default plateau detection must be run at several test positions instead. Expected road sections are thus shorter in automatic mode than in supervised mode.

The final part of the algorithm is identical to the former approach [4]. In case of success of the first plateau detection, the structure is tracked and extended as far as possible on each side of the seed. At each step, a template based on the last detected plateau is built. It contains the altitude of the plateau and the position of its bounds. In case of failure of a detection, these attributes are incremented with an estimate of the road drift in position and altitude.

The template is used to enforce spatial consistency between adjacent plateaux using tolerance thresholds for each attribute (isConsistent primitive in Algorithm 2). The adaptive directional scan is continuously centered on the last plateau template. The tracking stops when reaching grid bounds or after N_F (set to 5 here) successive detection failures. In case of sparse profiles (featuring less than 6 points), this counter is not incremented so that occluded areas can be crossed. When a denser profile is met, the extension goes on again if a reliable plateau is detected.

Algorithm 1: Detection of a road plateau profile (`detectPlateau`).

Input data : A sorted sequence of points $\mathcal{P}$, a start position c
Parameters: Initial assigned thickness ε, pinch size N_P, pinch length D_P and
               pinch ratio δ_P
Output   : The detected plateau $\mathcal{R}$

1 $\mathcal{R} \leftarrow \emptyset$;
2 $p \leftarrow$ `getNearestPoint` $(\mathcal{P}, c)$; // Select the nearest point to start position
3 $\mathcal{P} \leftarrow \mathcal{P}$ - p ;
4 $\mathcal{B} \leftarrow$ `initializeBlurredSegment` (p, ε) ; // Initialize $\mathcal{B}$ on point p
5 $pinched \leftarrow$ false ;
6 $right \leftarrow$ true ;
7 $left \leftarrow$ true ;
8 **while** $\mathcal{P} \neq \emptyset$ and (right or left) **do**
9 $\quad$ $p \leftarrow$ `getNearestPoint` $(\mathcal{P}, c)$; // Select next point
10 $\quad$ $\mathcal{P} \leftarrow \mathcal{P}$ - p ;
11 $\quad$ **if** $(p < c$ and left$)$ or $(p > c$ and right$)$ **then**
12 $\quad\quad$ $\mathcal{B}' \leftarrow \mathcal{B} + p$; // Extend $\mathcal{B}$ on both sides of start point
13 $\quad\quad$ **if** `isBlurredSegment` $(\mathcal{B}')$ **then**
14 $\quad\quad\quad$ $\mathcal{B} \leftarrow \mathcal{B}'$;
$\quad\quad\quad$ /* Blurred segment pinching   */
15 $\quad\quad\quad$ **if** (! pinched) and `size` $(\mathcal{B}) > N_P$ and `length` $(\mathcal{B}) > D_P$ **then**
16 $\quad\quad\quad\quad$ $\mathcal{B} \leftarrow$ `copyBlurredSegment` $(\mathcal{B}, \min (\varepsilon, \texttt{thickness} (\mathcal{B}) * \delta_P))$;
17 $\quad\quad\quad\quad$ $pinched \leftarrow$ true ;

18 $\quad\quad$ **else**
19 $\quad\quad\quad$ **if** $p < c$ **then**
20 $\quad\quad\quad\quad$ $left \leftarrow$ false ;
21 $\quad\quad\quad$ **else**
22 $\quad\quad\quad\quad$ $right \leftarrow$ false ;

23 **if** $pinched$ **then**
24 $\quad$ $\mathcal{R} \leftarrow$ `getPointsOfBlurredSegment` $(\mathcal{B})$;

Finally a post-processing validation is performed. The plateau density is checked in order to discard road sections with too much missing plateaux (less than 40% allowed). Moreover section tails (last sequence of adjacent plateaux) are also iteratively withdrawn if they are too short (10 plateaux at least). This last test contributes to remove most of wrong road ends.

Algorithm 2: Detection of a forest road from a seed line segment.

Input data : A seed line segment s and a grid of cloud points $\mathcal{G}$
Parameters: Number of initial trials N_I, minimal profile size N_P, maximum
number of successive fails N_F and grid subdivision factor N_R
Output   : The detected road section $\mathcal{R}$

1 $\mathcal{R} \leftarrow \emptyset$;
   /* First road detection at the segment s of $\mathcal{S}$ */
2 $\mathcal{A} \leftarrow$ getAdaptiveDirectionalScan (s); // See Sec. 2
3 $\mathcal{C} \leftarrow$ getFirstScans $(\mathcal{A}, N_R)$; // Get the first N_R scans of $\mathcal{A}$ around s
4 $\mathcal{V} \leftarrow$ getAltimetricProfile $(\mathcal{C}, \mathcal{G})$; // Get point cloud of $\mathcal{G}$ under $\mathcal{C}$
5 $\mathcal{P} \leftarrow \emptyset$;
6 $c \leftarrow$ getProfileCenter $(\mathcal{V})$; // Get the center of profile $\mathcal{V}$
7 **for** $\delta \leftarrow$ - $N_I/2$ **to** $N_I/2$ **do**
8      $\mathcal{P}' \leftarrow$ detectPlateau $(\mathcal{V}, \ c + \delta)$; // See Algo. 1 and Sec. 5.2
9   **if** isConsistent $(\mathcal{P})$ *and* thickness $(\mathcal{P}') <$ thickness $(\mathcal{P})$ **then**
10        $\mathcal{P} \leftarrow \mathcal{P}'$;
11     $\delta \leftarrow \delta + 1$;

12 **if** $\mathcal{P} \neq \emptyset$ **then**
13     $\mathcal{R} \leftarrow \mathcal{R} + \mathcal{P}$; // Add the first detection to result
14     $\mathcal{T}_0 \leftarrow$ getRoadParameters $(\mathcal{P})$; // Get parameters for the tracking step
       /* Road tracking on both sides of the segment s */
15   **for** side $\leftarrow$ *1* **to** *2* **do**
16        $\mathcal{T} \leftarrow \mathcal{T}_0$;
17        $N \leftarrow 0$;
18   **do**
19           $\mathcal{A} \leftarrow$ getCenteredADS $(\mathcal{A}, \mathcal{T})$; // Center ADS on template
20           $\mathcal{C} \leftarrow$ getNextScans $(\mathcal{A},$ side$, N_R)$; // Get the next N_R scans in $\mathcal{A}$
21           $\mathcal{V} \leftarrow$ getAltimetricProfile $(\mathcal{C}, \mathcal{G})$;
22   **if** size $(\mathcal{V}) > N_P$ **then**
23              $c \leftarrow$ getTemplateCenter $(\mathcal{T})$; // Get template center
24              $\mathcal{P} \leftarrow$ detectPlateau $(\mathcal{V}, c)$; // See Algo. 1 and Sec. 5.2
25   **if** $\mathcal{P} \neq \emptyset$ *and* isConsistent $(\mathcal{P}, \mathcal{T})$ **then**
26                 $\mathcal{R} \leftarrow \mathcal{R} + \mathcal{P}$; // Add the obtained tracking to result
27                 $\mathcal{T} \leftarrow$ getRoadParameters $(\mathcal{P})$; // Update road template
28                 $N \leftarrow 0$;
29   **else**
30                 $N \leftarrow N + 1$;
31                 $\mathcal{T} \leftarrow \mathcal{T} +$ getRoadDrift $(\mathcal{R},$ side$)$;
32   **else**
                /* N is not incremented in occluded areas   */
33              $\mathcal{T} \leftarrow \mathcal{T} +$ getRoadDrift $(\mathcal{R},$ side$)$;
34   **while** $\mathcal{C} \neq \emptyset$ *and* $N < N_F$;
35   side $\leftarrow$ side $+ 1$;
36     $\mathcal{R} \leftarrow$ cleanRoadSection $(\mathcal{R})$; // Final road section cleaning

---

6   Experiments

6.1   Usability Test on a Large-Scale LiDAR Set

At our knowledge, neither LiDAR test set, with raw data and forest road ground truth, nor concurrent approach code is publicly available for testing and comparing. This is certainly due to a large variability between different acquisition contexts and terrain configurations and maybe also to the huge data storage required. Nevertheless, the available data set used to validate the method is quite large, covering the whole Fossard mountain. Tiles corresponding to surrounding urbanized valleys were not considered as the road model used is not adapted to this context. The remaining part features 313 tiles (about 78 km^2) of forested mountain area, with about 755 millions of points for a mean density of 9.65 points/m^2. But the repartition is not homogeneous because of the presence of many tight conifer plantations.

Processing such a large amount of tiles exceeds the memory capacity of standard computers. Therefore, blocks of tiles are successively processed for each step, until the whole area is completely covered. For the seed selection step, resulting road seeds are classified into an array according to the DTM tile they belong to. In this step, the block must include a one-tile border to avoid the splitting of many blurred segments between two adjacent tiles during edge detection stage. For the road extraction step, buffers are used to collect point tiles. Tiles are processed one by one, but road sections may have a long extent, so that a two-tiles border is required. The whole data set is traversed using a specific path minimizing tile loading operations. This parcelling strategy allows the automatic processing of quite large data sets, at the price of an execution time increase because it performs more tile loading operations, and border tiles are processed multiple times in the first step.

In the present case, the whole Fossard processing takes **1747 s** on Intel i7 processor, 840 s for the seed selection step and 907 s for the final road extraction from raw data. Most output roads are visually correct (see a detail on Fig. 5). Because the simple road drift estimation procedure does not allow to predict rapid direction changes, many tight curves are left undetected. Most false detections occur in areas with low height gradient where the road model is not well adapted, or along some talwegs, old road relicts or cultivation terraces. More generally, output filtering should be added to deliver more accurate data such as road width or center line.

6.2   Performance Test on Smaller Size Areas

A finer evaluation was performed on four distinct 4×4 tile sets (4 km^2 each). Two of them include arranged areas for walking with some wood clearings and a large variety of tracks (*Saint-Mont* and *Tête des Cuveaux*) near the urbanized Mosel valley. The other ones (*Gris-Mouton* and *Grand-Rupt*) correspond to classical wood exploitation sectors, the last one featuring more standardized roads. During this long and tedious task, for each sector, most salient roads were carefully delineated in DTM views. Only the central line was manually extracted as the road width could not be easily estimated whatever the sector type. The achieved set of polylines constitutes the road ground truth.

Table 1. Road extraction performance measures

Sector	D (pts/m^2)	T (s)	R (%)	P (%)	F (%)
Saint-Mont	8.62	28.70	67.4616	74.8994	70.9862
Gris-Mouton	6.86	30.18	69.8697	88.1259	77.9431
Grand-Rupt	7.47	27.55	73.0137	78.3710	75.5976
Tête des Cuveaux	8.95	32.62	68.7293	80.4931	74.1475

For each extracted road section, missing plateaux were interpolated between detected ones. The set D of the covered pixels by these road sections was collected. It was compared to the set G_L of the covered pixels by the ground truth polylines to get a recall measure R, and to a 28 pixels dilation of G_L to get a precision measure P. This dilated set, called G_W, is assumed to enclose the real road set, including calibrated roads up to twelve meters wide, but also to take into account possible delineation errors. A F-measure F was also computed as the harmonic mean of R and P. All these measures are given by Eq. 4.

$$R = \frac{\overline{D \cap G_L}}{\overline{G_L}}, \quad P = \frac{\overline{D \cap G_W}}{\overline{D}}, \quad F = \frac{2 * R * P}{R + P} \tag{4}$$

Point density D, execution time T, recall R, precision P and F-measure F are reported in table Tab. 1, and Fig. 5 gives the results on one of the sectors.

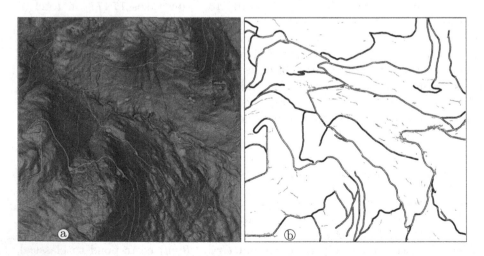

Fig. 5. Results on Grand-Rupt sector; (a) DTM map; (b) precision map: dilated ground truth in black, good detection $D \cap G_W$ in green, false detection in red.

Comparing the mean execution time for these tests $(7.44\,\mathrm{s/km^2})$ to the one observed for whole Fossard processing $(22.33\,\mathrm{s/km^2})$ gives an idea of the parcelling strategy overcost. Achieved accuracy results are consistent with local sector features, and quite correct if we consider the diversity of processed road types and most of all the variability of their geometric characteristics along their run. In addition, the supervised mode may be used to complete most of missing road sections.

7   Conclusion

This paper presents a new framework only based on mathematical morphology and discrete geometry tools to automatically extract forest roads in mountainous areas from LiDAR DTM and point cloud. In a first step, road seeds are selected in the shaded DTM using RORPO filter and FBSD blurred segment detector. Then a road extraction step is performed from provided seeds using only LiDAR raw data. It relies on the analysis of point profiles on each side of the seed position to detect road plateaux. Not only based on the interpolated DTM, this approach is more aware of the heterogeneous point repartition, giving better chance to cross occluded areas. The extraction framework was successfully tested on a large-scale LiDAR data set. Good execution time was obtained (**about 22 s/km²**). A comparison to selected ground truth in four smaller areas pointed out good accuracy measures (**around 70%**) regarding the quite simple road model used.

Achieved results showed the good potential of the used discrete geometry notions (blurred segment, adaptive directional scan) to process raw LiDAR data, and relevant seeds could be efficiently obtained using RORPO and FBSD. To get more effective results, it is left possible to combine this tools with more standard signal processing ones. In particular, a finer estimation of the extended road drift may help to detect rapid changes of direction, so that better aligned directional scans could be provided to extend the road section in tight curves. Some output filtering could also be added to deliver finer road width and position information. Moreover the integration of additional data such as low vegetation points could bring more clues to discriminate road sections from similar structures such as talwegs or ancient cultivation terraces. Also this new framework could probably be easily adapted to other application contexts, for instance urban structures detection and tracking from on-board sensors of smart vehicles.

Acknowledgements. This work was realized in the scope of SolHoM interdisciplinary project of Université de Lorraine. LiDAR data were acquired in the scope of PCR AGER project (*Projet collectif de recherche – Archéologie et GEoarchéologie du premier Remiremont et de ses abords*), and left available to SolHoM project.

References

1. Clode, S., Kootsookos, P., Rottensteiner, F.: The automatic extraction of roads from LiDAR data. Int. Archiv. Photogr. Remote Sens. Spatial Inf. Sci. **34**(B7) (2004)

2. David, N., Mallet, C., Pons, T., Chauve, A., Bretar, F.: Pathway detection and geometrical description from ALS data in forested montaneous areas. Int. Archiv. Photogr. Remote Sens. Spatial Inf. Sci. **38**(part 3/W8), 242–247 (2009)
3. Debled-Rennesson, I., Feschet, F., Rouyer-Degli, J.: Optimal blurred segments decomposition of noisy shapes in linear time. Comput. Graph. **30**(1), 30–36 (2006). https://doi.org/10.1016/j.cag.2005.10.007
4. Even, P., Ngo, P.: Live extraction of curvilinear structures from LiDAR raw data. ISPRS Annals Photogr. Remote Sens. Spatial Inf. Sci. **2**, 211–219 (XXIV ISPRS Congress 2020). https://doi.org/10.5194/isprs-annals-V-2-2020-211-2020
5. Even, P., Ngo, P., Kerautret, B.: Thick line segment detection with fast directional tracking. In: Ricci, E., Rota Bulò, S., Snoek, C., Lanz, O., Messelodi, S., Sebe, N. (eds.) ICIAP 2019. LNCS, vol. 11752, pp. 159–170. Springer, Cham (2019). https://doi.org/10.1007/978-3-030-30645-8_15
6. Ferraz, A., Mallet, C., Chehata, N.: Large-scale road detection in forested mountainous areas using airborne topographic Lidar data. ISPSR J. Photogr. Remote Sens. **112**, 23–36 (2016). https://doi.org/10.1016/j.isprsjprs.2015.12.002
7. Klette, R., Rosenfeld, A.: Digital Geometry: Geometric Methods for Digital Picture Analysis. Morgan Kaufmann, San Francisco (2004)
8. Liu, Q., Kampffmeyer, M., Jenssen, R., Salberg, A.B.: Road mapping in LiDAR images using a joint-task dense dilated convolutions merging network. In: IEEE International Geoscience and Remote Sensing Symposium (IGARSS 2019), pp. 5041–5044. Yokohama, Japan, 28 July – 2 August 2019. https://doi.org/10.1109/IGARSS.2019.8900082
9. Merveille, O., Naegel, B., Talbot, H., Najman, L., Passat, N.: 2D filtering of curvilinear structures by ranking the orientation responses of path operators (RORPO). Image Processing On Line **7**, 246–261 (2017). https://doi.org/10.5201/ipol.2017.207
10. Salberg, A.-B., Trier, Ø.D., Kampffmeyer, M.: Large-scale mapping of small roads in Lidar images using deep convolutional neural networks. In: Sharma, P., Bianchi, F.M. (eds.) SCIA 2017. LNCS, vol. 10270, pp. 193–204. Springer, Cham (2017). https://doi.org/10.1007/978-3-319-59129-2_17
11. White, R.A., Dietterick, B.C., Mastin, T., Strohman, R.: Forest roads mapped using LiDAR in steep forested terrain. Remote Sens. **2**(4), 1120–1141 (2010). https://doi.org/10.3390/rs2041120

Fast Pattern Spectra Using Tree Representation of the Image for Patch Retrieval

Behzad Mirmahboub[1]($\boxtimes$), Jérôme Moré[1], David Youssefi[2], Alain Giros[2], François Merciol[1], and Sébastien Lefèvre[1]

[1] IRISA - Université Bretagne Sud, UMR 6074, 56000 Vannes, France
{behzad.mirmahboub,jerome.more,francois.merciol,
sebastien.lefevre}@irisa.fr
[2] CNES - Centre National d'Etudes Spatiales, Toulouse, France
{david.youssefi,alain.giros}@cnes.fr
http://www.irisa.fr/obelix

Abstract. We extend the notion of content based image retrieval to patch retrieval where the goal is to find the similar patches to a query patch in a large image. Naive searching for similar patches by sequentially computing and comparing descriptors of sliding windows takes a lot of time in a large image. We propose a novel method to compute descriptors for all sliding windows independent from number of patches. We rely on tree representation of the image and exploit the histogram nature of pattern spectra to compute all the required descriptors in parallel. Computation time of the proposed method depends only on the number of tree nodes and is free from query selection. Experimental results show the effectiveness of the proposed method to reduce the computation time and its potential for object detection in large images.

Keywords: Content based image retrieval · Patch retrieval · Tree representation · Pattern spectra · Large satellite images

1 Introduction

Content based image retrieval (CBIR) is the problem of finding images in a database that are similar to a query image [10]. This is generally done by two main components. The first step is feature extraction that computes discriminating descriptors for the images. Therefore, each image is represented by a descriptor. The second step is to find the similarities between the images by calculating the distances between their descriptors. Then, the dataset images are ranked based on their distances to the query image and the most similar images are retrieved.

Current datasets for image retrieval consist of the cropped images with specific content for each image and the task of CBIR is to find the similar images. However, in some applications such as remote sensing we have a patch in a

J. Lindblad et al. (Eds.): DGMM 2021, LNCS 12708, pp. 107–119, 2021.
https://doi.org/10.1007/978-3-030-76657-3_7

large satellite image and we need to find similar patches in the same image or other images (Fig. 1). CBIR can be trivially adapted to this problem by dividing the large image into small patches and extracting a descriptor for each image patch. Then, distances between patches can be computed to find the most similar patches to the query patch. This approach is time consuming and is not applicable in real time retrieval system for very large images.

In order to address this problem, we propose in this paper a patch retrieval system to quickly compute descriptors for all the patches in an image. As far as we know, it is the first work that tries to find the similar patches in a large image. We rely on Pattern Spectra (PS) based on tree representation of the image [12]. It is a histogram-like morphological descriptor for images. We introduce fast-PS descriptor as an approximation of PS that is computed for all image patches in parallel. Therefore, the computation time will be independent from the number of patches which is important for processing large images.

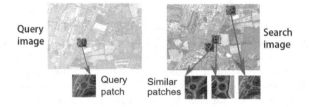

Fig. 1. Patch retrieval: User selects a query patch from an image and the goal is to find similar patches in the same or other images.

This paper is organized as follows. Section 2 reviews previous works on image retrieval, tree representation of the image and pattern spectra. We explain our proposed method in Sect. 4 and present experimental results in Sect. 5. Section 6 concludes the paper.

2   Related Works

2.1   Image Retrieval

CBIR is a challenging task to find the images in a database that are similar to a query image [10]. The general framework consists of two main steps as it is shown in Fig. 2. First, a discriminant feature vector is extracted from each image. Second, these descriptors are used to compute the distance of each dataset image to the query image. The dataset images are sorted according to their distances to the query image and we expect to see the similar images at the beginning of the ranked list.

Various types of feature vectors are proposed in literature based on color or texture of the image such as Gabor Filter [9], LBP (Local Binary Pattern) [8], SIFT (Scale-Invariant Feature Transform) [13], MSER (Maximally Stable

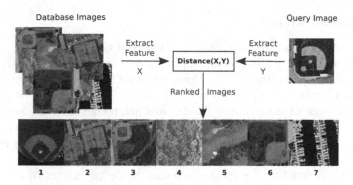

Fig. 2. General framework of content based image retrieval. The images in the database are sorted based on their distances to a query image. (The images are selected from Merced dataset [15] only for illustration.)

Extremal Regions) [16] and recently deep CNN features [4]. An ideal feature vector must be rotation/translation/scale-invariant and robust to changes in illumination. Descriptor computation is the bottleneck in image retrieval. Therefore, practical image retrieval systems pre-compute image descriptors and store them along the original images.

Distance between two feature vectors can be simply computed using their Euclidean distance. However, in many cases the feature vectors are high dimensional and different dimensions have different importance. An efficient solution is to learn a weighting matrix to emphasize the important dimensions [7]. On the other hand, some distance metrics are specially designed to compare histograms. Since pattern spectra is a histogram-like feature vector, we use the well-known χ^2 distance to compare them. The χ^2 distance of two histograms x and y tries to reduce the effect of large bins:

$$\mathcal{X}^2(x,y) = \sum_i \frac{(x_i - y_i)^2}{(x_i + y_i)} \tag{1}$$

Previous works on image retrieval mainly try to compare two whole images [5]. To the best of our knowledge, it is the first time that the concept of CBIR is extended to find similar patches in a large image as depicted in Fig. 1. We assume that the user selects a query patch with arbitrary size. Therefore, the query patch size is not known and the descriptors cannot be computed in advance. A closely related field of study exists in literature for patch matching [1]. However, the solutions to this problem cannot meet our requirements since patch matching is generally applied between two similar scenes such as stereo images. Therefore, those methods exploit similarity between images to restrict the possible locations of matching patches and expedite the search, while we do not impose that the images have similar content and therefore we cannot use such methods.

2.2   Evaluation Metric

When we calculate and sort the distances between the image patches and the query patch, ideally we expect to see all the relevant images (true positives or tp) at the beginning of the retrieved list. However, since the descriptors and distance metric are not perfect, we also see irrelevant images (false positives or fp) in the list and even some of the relevant images are not retrieved (false negatives or fn). Mean Average Precision (mAP) is a measure to evaluate the performance of a retrieval system [11]. It shows how well the relevant images are ranked in the retrieved list and it is calculated based on precision and recall.

Precision is defined as the ratio of relevant retrieved images to all retrieved images or $tp/(tp + fp)$ and it is computed for each rank position in the retrieved list. In the example of Fig. 2 if baseball field is the relevant class, the precisions for first seven ranks will be $\{\frac{1}{1}, \frac{1}{2}, \frac{2}{3}, \frac{2}{4}, \frac{2}{5}, \frac{3}{6}, \frac{3}{7}\}$. Recall is defined as the ratio of retrieved relevant images to all relevant images in the dataset or $tp/(tp + fn)$. Assume that there are four relevant baseball fields in Fig. 2 and that three of them are retrieved in the first seven ranks. Then, the recalls for these ranks will be $\{\frac{1}{4}, \frac{1}{4}, \frac{2}{4}, \frac{2}{4}, \frac{2}{4}, \frac{3}{4}, \frac{3}{4}\}$. Average Precision (AP) is the area under precision-recall curve. Practically, it is the average of precisions at the positions of relevant retrieved images (where the recall is changed). In the above example AP is calculated as $\frac{1}{3}(\frac{1}{1} + \frac{2}{3} + \frac{3}{6}) = \frac{13}{18}$. Finally, mAP is simply the mean value of all APs for different queries.

3   Theoretical Background

3.1   Tree Representation

Usual image representation by a matrix of pixels gray levels does not include spatial relations between image components. In contrast, hierarchical image representation shows an image with a tree structure that considers spatial relations [3]. Each node in this tree structure represents a connected component in the original image. Each pair of components in the image are either disjoint or one of them is a subset of the other one making a child-parent relationship. The leaves of the tree consist of the smallest components while the tree root represents the entire image. Such a hierarchical structure lets us quickly process the image in different scales. Depending on how to define the image components, various tree types can be built including max-tree, min-tree, Tree-of-Shapes, α-tree, ω-tree and Binary Partition Tree [3].

Figure 3 shows a simple gray scale image on the left with its max-tree representation in the middle. Max-tree is built using the connected components of upper level sets of the image [6] that is defined as:

$$\mathcal{H}_{max} = \{\mathcal{C}(\mathcal{L}^k)|v_{min} \leqslant k \leqslant v_{max}\} \tag{2}$$

where k is a threshold that is selected between minimum gray level v_{min} and maximum gray level v_{max}. Level set $\mathcal{L}^k$ is an image that is obtained by thresholding the original image with k. It only contains all the components with intensity values equal or greater than k and $\mathcal{C}(\mathcal{L}^k)$ is the set of those components. Therefore, $\mathcal{H}_{max}$ is the set of all connected components in different levels.

Each circle in Fig. 3 represents a node in the tree structure and the corresponding connected component in the image is shown beside it. Also the gray level k is written next to each tree node. The leaves of the max-tree consists of the brightest image components or local maxima while the tree root represents the whole image.

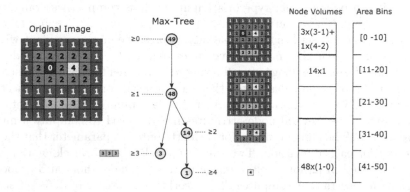

Fig. 3. Computation of pattern spectra using area attribute of a max-tree nodes. Tree nodes are shown by circles in the tree structure. The number inside each circle shows the node area and the number next to it represents the node altitude. Each tree node contributes to a histogram bin by its area multiplied by the gray level difference with its parent.

3.2   Pattern Spectra

Tree representation is an efficient structure to process the image at multliple scales. Pattern Spectra is a morphological image descriptor that can be computed using the tree representation [12]. It is a histogram-like feature vector that shows the distribution of components in an image. After building the tree structure, various geometrical or statistical attributes can be computed for each tree node and PS shows the frequencies of those attributes in the image. In the example of Fig. 3, area attribute for each node is written in the relevant circle in the tree structure and PS is shown on the right of the figure. In this example, PS consists of five bins for the area attribute and each tree node contributes to one relevant bin. Each node contributes by its volume which is defined as the node area multiplied by the difference between node gray level and its parent gray level. Contributions of all nodes to the relevant PS bin are summed and the final descriptor will be $[8, 14, 0, 0, 48]$. Generally, bin j of pattern spectra PS is computed based on tree representation of the image as:

$$PS_j = \sum_{\forall i, A(c_i) \in [b_{j-1}, b_j[} a(c_i)\big|k(c_i) - k\big(p(c_i)\big)\big| \tag{3}$$

where $a(c_i)$, $k(c_i)$ and $p(c_i)$ represent the area, the gray level and the parent of the component c_i respectively. Therefore, the area of each component is weighted by the gray level difference with its parent. $A(c_i)$ denotes an attribute of the component c_i. All the components that their $A(c_i)$ attributes are located in an interval between b_{j-1} and b_j contribute to bin j of the PS. The histogram bin edges $\{b_0, ..., b_{max}\}$ of the desired attribute can be divided linearly or logarithmic. In the above example, the attribute $A(c_i)$ is defined as the area of the component c_i and it is distributed linearly between 0 and 50 producing a size pattern spectra. However, it can be any other type of attribute such as compactness value that produces shape pattern spectra.

Pattern spectra describes the probability of presence of a component with a certain size or shape in the image. It gives an estimation for the amount of details that is removed from the image after filtering its components. PS is a translation and scale invariant image descriptor that can be used for image classification and retrieval. PS is also rotation invariant if it is computed based on rotation invariant attributes. Several node attributes may be used to compute a multi-dimensional PS. The number of bins for each attribute is a parameter that should be selected for each dimension. Two dimensional PS based on elongation and entropy was used in [2] for satellite image retrieval. The authors also proposed local pattern spectra that computes PS in overlapped patches in different scales of the image. The final image descriptor is the aggregation of all computed PS.

4   Proposed Method

In conventional image retrieval, similar images to a query image are selected according to the distances between their descriptors. In this paper, we aim to extend that concept to patch retrieval where we look for some patches in the dataset that are similar to a query patch. The similar patches can be found in the same image of the query patch or in other images as shown in the example of Fig. 1.

The naive method is to divide each image to a regular grid of patches with the same size of the query patch. Then, a sliding window is used and the similarity of each window with the query patch is computed. The computation time of this approach is proportional to the number of sliding windows that is too much for a large image. Notice that the size (and so the number) of the sliding windows depends on the user selected query patch. Therefore, it is not possible to compute and store the descriptors of the sliding windows in advance. In this section, we exploit the tree representation of the image and introduce the notion of fast pattern spectra descriptor to reduce the searching time. Finally the computation time is related to the number of tree nodes instead of number of the sliding windows.

4.1   Sub-pattern Spectra

As explained in Sect. 3.1, pattern spectra is a histogram-like image descriptor that represents the distribution of connected components in the image. Each image component contributes to one of the histogram bins and all image components build the whole histogram. Since a small patch in the image contains a subset of image components, we expect that it contributes to only a few number of histogram bins. This intuition is shown in Fig. 4 where all the image components make the full histogram and the components in the small patch make a sparse histogram called here "sub-pattern spectra". We also notice that each image patch consists of a subset of nodes in the tree structure of the whole image and this subset of nodes does not necessarily form a single integrated sub-tree structure.

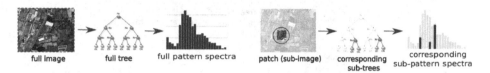

full image   full tree   full pattern spectra   patch (sub-image)   corresponding   corresponding
                                                                                     sub-trees   sub-pattern spectra

Fig. 4. (Left) Pattern spectra is a histogram-like distribution of image components. (Right) The components of each image patch contribute to a few numbers of histogram bins creating a sub-pattern spectra.

Having all related sub-pattern spectra computed, we can calculate the distance between the query patch and the sliding windows in the image to find the similar windows as depicted in Fig. 5. The straightforward approach to compute sub-pattern spectra for each image patch is to directly compute pattern spectra on each window that are called here "local pattern spectra". Two issues arise from this approach. First, the components in the patch are not the same components in the whole image, because some of the image components are partially located in the sliding window and create new local components (Fig. 6 Top). Second, there are many sliding windows in a large image and directly computing a pattern spectra for each of them will be time consuming. We address these problems by introducing "fast pattern spectra" in next section.

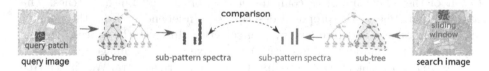

                                          comparison
query patch

query image   sub-tree   sub-pattern spectra   sub-pattern spectra   sub-tree   search image

Fig. 5. Similarity of a query patch with a sliding window is computed based on the distance between their sub-pattern spectra.

4.2   Fast Pattern Spectra

Our motivation in this paper is to avoid directly computing a descriptor for each image patch. For this purpose we rely on the histogram nature of pattern spectra and propose an indirect approach to compute the histogram in each patch. In this case, sub-pattern spectra consists of a subset of tree nodes in the whole image. Each node in the tree structure comes with some information such as center and bounding box of the corresponding component in the image. Therefore, when we select a patch from the image, we know which tree nodes are located inside it. We use these nodes to create our sub-pattern spectra and refer to it as fast pattern spectra (Fig. 6 Bottom).

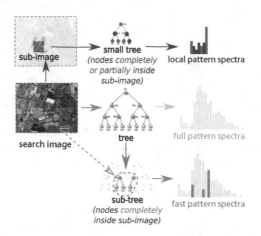

Fig. 6. Sub-pattern spectra of a given patch can be computed in two different ways. (Top) local pattern spectra: A small tree is built directly on the patch and its pattern spectra is computed. The nodes of this small tree may be different from the big tree of the whole image. (Bottom) fast pattern spectra: Tree representation of the whole image is constructed and only the nodes that are located inside the patch are selected to compute the histogram.

The proposed fast pattern spectra is an estimation of local pattern spectra. Instead of computing PS on the patch separately, we use the tree representation of the whole image to compute descriptors for all image patches. The computation time of the proposed descriptor is independent from the number of the patches that is determined by user selection. Instead, the processing time depends on the number of tree nodes that are fixed for each image.

Figure 7 shows the process of computing fast pattern spectra. We build a tree representation of a given image and compute the desired attribute for its nodes. According to Eq. 3, pattern spectra is the sum of components volumes which are computed by multiplying the nodes areas by their gray level differences with their parents $v(c_i) = a(c_i)|k(c_i) - k(p(c_i))|$. We calculate the volumes of all tree nodes and store them in a "tree table" along with nodes locations in the original image. Then, we compute a full pattern spectra and use it to assign a bin number to each tree node. The bin number is also stored in the tree table. When a user selects a query patch, we divide the image into sliding windows based on that selection and create an empty histogram for each sliding window.

Therefore, we have a three dimensional matrix $PS[x, y, b]$ spanned by image locations and histogram bins. Each tree node c_i contributes by its volume $v(c_i)$ to the histogram bin $b(c_i)$ in a specific location $(x(c_i), y(c_i))$.

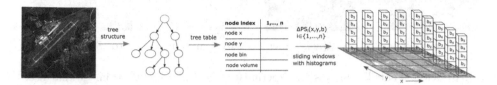

Fig. 7. Computation of fast pattern spectra consists of populating the local histograms for sliding windows. The algorithm traverses the tree structure and adds the contribution of each node to the corresponding histogram bin.

In order to compute fast pattern spectra, we traverse the tree table and simply add the volume of each node to the corresponding histogram bin. It is formulated in Eq. 4 that is computed with complexity $O(n)$ where n is number of the tree nodes. In this way, we get the sub-pattern spectra for all sliding windows of the image in one pass of the tree structure. We point out that the tree table is pre-computed and fixed. Every time that the user selects a new query patch, only Eq. 4 needs to be evaluated again.

$$PS[(x(c_i), y(c_i), b(c_i)] += v(c_i) \ , \ i = 1, ..., n \tag{4}$$

5   Experimental Results

In this section, we experimented our proposed fast pattern spectra for the application of patch retrieval on selected images and compare it with baseline local pattern spectra. All the experiments were done using Python with Higra package [14] in Windows 7 64-bit Operating System running on a laptop with Intel® Core™ i7 CPU @ 2 GHz and 8 GB RAM.

5.1   Dataset

The ideal dataset for patch retrieval, as we addressed in this paper, consists of a large aerial image with annotations for various landmarks on it. The available datasets for content based image retrieval such as "UC Merced Land Use Dataset" [15] are not suitable for this purpose since they consist of separate image patches with different classes. As far as we know, there is no available dataset with the desired annotations. Therefore, in order to conduct the reported experiments, we selected two satellite RGB images and manually annotated them with 6 classes including farm, forest, lake, airport, road and building as shown in Fig. 8. We used Fig. 8(a) as query image where the user selects its query patches. Three user selected patches with different sizes and contents are marked in this image. Figure 8(c) was used to search for the similar patches. Each image patch may contain various pixel labels. We assigned the majority of pixel labels inside each patch as its class.

(a) Query image   (b) Query labels   (c) Search image   (d) Search labels

Fig. 8. Selected satellite images for experiments. (a) 617×647 RGB image that user selects query patches from it. Three selected query patches are marked with red rectangles. (b) Ground truth labels for query image pixels showing 6 different classes. (c) 633×668 RGB image to search for similar patches (d) ground truth labels for search image pixels.

5.2   Experiment Settings

We compare the proposed fast pattern (fast-PS) spectra with the baseline local pattern spectra (local-PS). In both cases, the user selects an arbitrary patch from the query image. Then, the search image is divided to regular overlapped windows with the same size of the query patch and the stride of half size of the query patch horizontally and vertically. In the case of baseline local-PS, pattern spectra is computed independently on each search window. However, in the proposed fast-PS approach, a big tree is built on the whole search image and the pattern spectra is computed for all the search windows in parallel as detailed in Sect. 4.2.

In these experiments, pattern spectra is computed based on the max-tree structure that is built on the gray scale image. We use one dimensional pattern spectra based on $A(c_i) = compactness$ attribute. The compactness of a tree node is defined as its area divided by the square of its perimeter. We divided nodes compactness linearly into 30 bins resulting to 30-dimensional descriptors for image patches. The similarity between descriptors are computed using χ^2 distance.

5.3   Results and Discussions

We selected three query patches with different sizes and contents as marked in Fig. 8(a) and retrieved similar patches from Fig. 8(c). For each case, we computed the average precision as explained in Sect. 2.2 using baseline and proposed methods. The average precision and computation time are reported in Table 1. The first column of the table shows the size of the query patch. The second column of table shows the number of sliding windows in the search image that depends on the size of the selected query patch. As we expected, local-PS needs more computation time which depends on number of sliding windows. However, the computation time of fast-PS is almost constant since it depends only on the number of tree nodes. As we explained in Sect. 4.2, in proposed fast-PS we need to store a table of tree nodes attributes for search image. Last column of Table 1 reports the required time to compute this tree table. Notice that the tree table is independent from query patch. It is computed only one time for each search image and is stored along the image. The stored tree table can be used later for retrieval. Therefore, its computation time does not affect the retrieval time. However, it adds an extra cost for the storage. In our experiment the search image is 1.09 MB and the stored tree table is 4.42 MB.

Table 1. Average precision and computation time for retrieval of three selected query patches using local pattern spectra and fast pattern spectra.

Query size (pixels)	Total patches	Average precision (%)		Computation time (seconds)		
		Local-PS	Fast-PS	Local-PS	Fast-PS	Tree table
(a) 80 × 80	210	30.83	**38.89**	3.91	**1.97**	
(b) 70 × 90	221	41.03	**46.92**	4.12	**1.98**	2.86
(c) 90 × 70	234	30.54	**41.39**	4.38	**2.01**	

We basically proposed fast-PS (as its name suggests) as an approximation of local pattern spectra that is computed quickly over sliding windows in a large image. Table 1 shows that this novel descriptor also achieves higher precision than the original local-PS for the application of patch retrieval. A possible reason can be explained referring to Fig. 6. When we crop a patch from a large image, it contains additional components that do not exist in the original image and may be irrelevant to its contents. In this case, fast-PS helps to rely on relevant components from original image and achieves better results. Figure 9 shows the precision-recall curves for three selected query patches. We observe that fast-PS performs better than local-PS especially at first ranks.

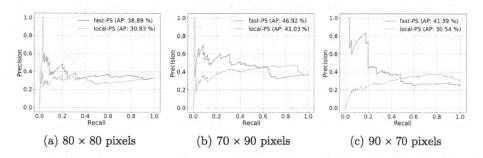

(a) 80 × 80 pixels   (b) 70 × 90 pixels   (c) 90 × 70 pixels

Fig. 9. Precision-recall curve and average precision (AP) for three selected query patches with different sizes using baseline local-PS and proposed fast-PS.

Our main goal in this work was to show the capability of fast-PS to offer interactive search in a large image when the size of the query patch is not known in advance. Although we found out that the average precision of fast-PS is higher than local-PS in our experiments, we recall that our objective is not to surpass previous descriptors in term of accuracy. Rather, we relied on the histogram nature of pattern spectra to propose a novel descriptor that is computed quickly on a large image while achieving acceptable accuracy.

We did not compare our proposed descriptor with state-of-the-art descriptors such as deep features and left it for future work. While other descriptors possibly lead to a higher accuracy, they cannot adapt to our framework for parallel computation since they have to be computed locally in each patch. We assume that the retrieval time is more important in a large-scale dataset where our proposed unsupervised method is more efficient.

6   Conclusion and Future Lines

We extended the problem of content-based image retrieval and introduced a new application that we call patch retrieval where a user selects a query patch with arbitrary size from a large image and we look for similar patches in the same or other image. Available solutions from conventional image retrieval need to divide the image into many sliding windows and compute an image descriptor on each window separately. This approach takes a lot of time especially for large images. We relied on pattern spectra that is an image descriptor based on tree representation of the image. We exploited the histogram nature of pattern spectra and proposed fast pattern spectra. It is computed for all sliding windows of the image in parallel and its processing time is independent from the number of the sliding windows. Interestingly, it is not only faster than local pattern spectra, but also achieves a higher average precision. However, the proposed method comes with a cost that needs to store an extra structure along with the original image. We showed that the fast pattern spectra is very promising on a simple dataset. More experiments on larger datasets are necessary to explore its full potential.

There is a lot of rooms for improvement of the proposed method. We used a max-tree structure and computed one dimensional pattern spectra based on the compactness attribute. However, our method is not restricted to the tree types and their attributes. We designed an efficient framework that can be realised with various types of pattern spectra to find the best combination of tree structures and attributes. Also, we simply calculated the distance between descriptors using χ^2. So, we did not rely on a learnt distance metric. We expect that learning a weighting matrix to compute the distances will lead to a higher accuracy.

Acknowledgment. This work was funded by DAJ-AR-NO-2018.0010814 project from CNES.

References

1. Barnes, C., Shechtman, E., Finkelstein, A., Goldman, D.B.: PatchMatch: a randomized correspondence algorithm for structural image editing. ACM Trans. Graph. (ToG) **28**, 24 (2009)

2. Bosilj, P., Aptoula, E., Lefèvre, S., Kijak, E.: Retrieval of remote sensing images with pattern spectra descriptors. ISPRS Int. J. Geo Inf. **5**(12), 228 (2016)
3. Bosilj, P., Kijak, E., Lefèvre, S.: Partition and inclusion hierarchies of images: a comprehensive survey. J. Imaging **4**(2), 33 (2018)
4. Chen, Y., Jiang, H., Li, C., Jia, X., Ghamisi, P.: Deep feature extraction and classification of hyperspectral images based on convolutional neural networks. IEEE Trans. Geosci. Remote Sens. **54**(10), 6232–6251 (2016)
5. Cheng, G., Xie, X., Han, J., Guo, L., Xia, G.S.: Remote sensing image scene classification meets deep learning: challenges, methods, benchmarks, and opportunities. IEEE J. Sel. Top. Appl. Earth Obser. Remote Sens. **13**, 3735–3756 (2020)
6. Jones, R.: Component trees for image filtering and segmentation. In: Coyle, E. (ed.) IEEE Workshop on Nonlinear Signal and Image Processing, Mackinac Island (1997)
7. Koestinger, M., Hirzer, M., Wohlhart, P., Roth, P.M., Bischof, H.: Large scale metric learning from equivalence constraints. In: 2012 IEEE Conference on Computer Vision and Pattern Recognition, pp. 2288–2295. IEEE (2012)
8. Li, W., Chen, C., Su, H., Du, Q.: Local binary patterns and extreme learning machine for hyperspectral imagery classification. IEEE Trans. Geosci. Remote Sens. **53**(7), 3681–3693 (2015)
9. Li, W., Du, Q.: Gabor-filtering-based nearest regularized subspace for hyperspectral image classification. IEEE J. Sel. Top. Appl. Earth Observ. Remote Sensing **7**(4), 1012–1022 (2014)
10. Liu, Y., Zhang, D., Lu, G., Ma, W.Y.: A survey of content-based image retrieval with high-level semantics. Pattern Recogn. **40**(1), 262–282 (2007)
11. Manning, C.D., Schütze, H., Raghavan, P.: Introduction to information retrieval. Cambridge University Press (2008)
12. Maragos, P.: Pattern spectrum and multiscale shape representation. IEEE Trans. Pattern Anal. Mach. Intell. **11**(7), 701–716 (1989)
13. Paul, S., Pati, U.C.: Remote sensing optical image registration using modified uniform robust sift. IEEE Geosci. Remote Sens. Lett. **13**(9), 1300–1304 (2016)
14. Perret, B., Chierchia, G., Cousty, J., Guimarães, S., Kenmochi, Y., Najman, L.: Higra: hierarchical graph analysis. SoftwareX **10**, (2019)
15. Yang, Y., Newsam, S.: Bag-of-visual-words and spatial extensions for land-use classification. In: Proceedings of the 18th SIGSPATIAL International Conference on Advances in Geographic Information Systems, pp. 270–279 (2010)
16. Yu, H.Y., Sun, J.G., Liu, L.N., Wang, Y.H., Wang, Y.D.: MSER based shadow detection in high resolution remote sensing image. In: 2010 International Conference on Machine Learning and Cybernetics, vol. 2, pp. 780–783. IEEE (2010)

Watershed-Based Attribute Profiles for Pixel Classification of Remote Sensing Data

Deise Santana Maia[✉], Minh-Tan Pham, and Sébastien Lefèvre

Univ. Bretagne Sud, UMR 6074, IRISA, 56000 Vannes, France
`deise.santana-maia@irisa.fr`

Abstract. We combine two well-established mathematical morphology notions: watershed segmentation and morphological attribute profile (AP), a multilevel feature extraction method commonly applied to the analysis of remote sensing images. To convey spatial-spectral features of remote sensing images, APs were initially defined as sequences of filtering operators on the max- and min-trees computed from the original data. Since its appearance, the notion of APs has been extended to other hierarchical representations including tree-of-shapes and partition trees such as α-tree and ω-tree. In this article, we propose a novel extension of APs to hierarchical watersheds. Furthermore, we extend the proposed approach to consider prior knowledge from training samples, leading to a more meaningful hierarchy. More precisely, in the construction of hierarchical watersheds, we combine the original data with the semantic knowledge provided by labeled training pixels. We illustrate the relevance of the proposed method with an application in land cover classification using optical remote sensing images, showing that the new profiles outperform various existing features.

1 Introduction

Mathematical morphology has a long history with the processing and analysis of remote sensing images, as attested by earlier surveys on this topic [20]. In particular, in the past decade, special attention has been given to a multi-level feature extraction method, known as Attribute Profile (AP) [8], which relies on hierarchical image representations to convey spatial-spectral features of remote sensing images.

In this article, we study the relevance of hierarchical watersheds for remote sensing applications. Our contributions are two-fold: (1) the introduction of the Watershed-AP, which is an extension of AP to hierarchical watersheds; and (2) an investigation on the use of prior knowledge in the construction of hierarchical watersheds for the classification of remote sensing images.

This article is organized as follows. In Sect. 2, we recall the definitions of graphs, hierarchical watersheds and AP, and we review the literature on prior knowledge for image processing. Section 3 introduces the Watershed-AP and our

© Springer Nature Switzerland AG 2021
J. Lindblad et al. (Eds.): DGMM 2021, LNCS 12708, pp. 120–133, 2021.
https://doi.org/10.1007/978-3-030-76657-3_8

method to integrate semantic knowledge in its construction. Finally, experiments with remote sensing images are given in Sect. 4.

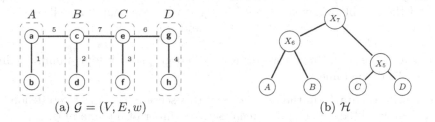

(a) $\mathcal{G} = (V, E, w)$ (b) $\mathcal{H}$

Fig. 1. (a): A weighted graph $\mathcal{G} = (V, E, w)$. (b): A tree representation of the hierarchical watershed $\mathcal{H}$ of $\mathcal{G}$ for the sequence (C, A, B, D) of minima of $\mathcal{G}$.

2   Background Notions

In this section, we first review graphs and hierarchical watersheds. Then, we review the literature on the use of prior knowledge and markers in the field of image processing. Finally, we recall the definition of Attribute Profile (AP).

2.1   Graphs and Hierarchical Watersheds

Watershed segmentation was proposed in the late 70's and, since then, this concept has been extended to several frameworks and implemented through a variety of algorithms. The intuition behind the various definitions of the watershed segmentation derive from the topographic definition of watersheds: dividing lines between catchment basins, which are, in their turn, areas where collected precipitation flows into the same regional minimum. These notions can be extended to gray-scale images and graphs, leading to different definitions of watershed segmentation. In this paper, we focus on watershed-cuts and hierarchical watersheds defined in the context of edge-weighted graphs, as formalized in [5,6]. In the remainder of this section, we present the notions of graphs, hierarchies of partitions and hierarchical watersheds.

A (edge) weighted graph is a triplet $\mathcal{G} = (V, E, w)$ where V is a finite set, E is a subset of $V \times V$, and w is a map from E into $\mathbb{R}$. The elements of V and E are called *vertices* and *edges* (of $\mathcal{G}$), respectively. Let $\mathcal{G} = (V, E, w)$ be a weighted graph and let $\mathcal{G}' = (V', E', w)$ be a graph such that $V' \subseteq V$ and $E' \subseteq E$. We say that $\mathcal{G}'$ *is a subgraph of* $\mathcal{G}$. A sequence $\pi = (x_0, \ldots, x_n)$ of vertices in V' is a *path (in G') from* x_0 *to* x_n if $\{x_{i-1}, x_i\}$ is an edge of $\mathcal{G}'$ for any $1 \leq i \leq n$. If $x_0 = x_n$ and if there are no repeated edges in π, we say that π *is a cycle* (in $\mathcal{G}'$). The subgraph $\mathcal{G}'$ of $\mathcal{G}$ is said to be *connected* if, for any x and x' in V', there exists a path from x to x'. Moreover, we say that $\mathcal{G}'$ *is a connected component of* $\mathcal{G}$ if

1. for any x and x' in V', if $\{x, x'\} \in E$ then $\{x, x'\} \in E'$; and
2. there is no edge $e = \{y, y'\} \in E$ such that $y \in V \setminus V'$ and $y' \in V'$.

Let $\mathcal{G} = (V, E, w)$ be a graph and let $\mathcal{G}' = (V', E', w)$ be a connected subgraph of $\mathcal{G}$. If the weight of any edge in E' is equal to a constant k and if $w(e) > k$ for any edge $e = \{x, y\}$ such that $x \in V'$ and $y \in V \setminus V'$, then G' is a (local) *minimum of G*.

For instance, Fig. 1(a) illustrates a weighted graph with four minima delimited by the dashed lines.

Important Remark: in the remainder of this section, $\mathcal{G} = (V, E, w)$ denotes a connected weighted graph and n denotes the number of minima of $\mathcal{G}$.

Let $\mathcal{G}' = (V', E', w)$ be a subgraph of G. A *Minimum Spanning Forest (MSF) of G rooted in G'* is a subgraph $\mathcal{G}'' = (V, E'', w)$ of $\mathcal{G}$ such that:

1. for every connected component X'' of $\mathcal{G}''$, there is exactly one connected component X' of $\mathcal{G}'$ such that X' is a subgraph of X'';
2. every cycle in $\mathcal{G}''$ is a cycle in $\mathcal{G}'$; and
3. $\sum_{e \in E''} w(e)$ is minimal among all graphs which satisfy conditions (1) and (2).

A *partition of V* is a set $\mathbf{P}$ of disjoint subsets of V such that the union of the elements in $\mathbf{P}$ is V. The *partition of V induced by a graph $\mathcal{G}'$* is the partition $\mathbf{P}$ such that every element of $\mathbf{P}$ is the set of vertices of a connected component of $\mathcal{G}'$. A *hierarchy of partitions of V* is a sequence $\mathcal{H} = (\mathbf{P}_0, \ldots, \mathbf{P}_n)$ of partitions of V such that $\mathbf{P}_n = \{V\}$ and such that, for any $0 < i \leq n$, every element of $\mathbf{P}_i$ is the union of elements of $\mathbf{P}_{i-1}$.

Any hierarchy of partitions $\mathcal{H}$ can be represented as a tree whose vertices correspond to the regions of $\mathcal{H}$ and whose edges link nested regions. For instance, Fig. 1(b) shows a tree representation of the hierarchy $\mathcal{H} = (\mathbf{P}_0, \mathbf{P}_1, \mathbf{P}_2, \mathbf{P}_3)$, where $\mathbf{P}_0 = \{\{a, b\}, \{c, d\}, \{e, f\}, \{g, h\}\}$, $\mathbf{P}_1 = \{\{a, b\}, \{c, d\}, \{e, f, g, h\}\}$, $\mathbf{P}_2 = \{\{a, b, c, d\}, \{e, f, g, h\}\}$ and $\mathbf{P}_3 = \{\{a, b, c, d, e, f, g, h\}\}$.

Let $\mathcal{S} = (\mathcal{M}_1, \ldots, \mathcal{M}_n)$ be a sequence of n distinct minima of $\mathcal{G}$ such that, for any $0 < i \leq n$, we have $\mathcal{M}_i = (V_i, E_i, w)$. The *hierarchy of Minimum Spanning Forests of $\mathcal{G}$ for $\mathcal{S}$*, also known as *hierarchical watershed of $\mathcal{G}$ for $\mathcal{S}$*, is a hierarchy $\mathcal{H} = (\mathbf{P}_1, \ldots, \mathbf{P}_n)$ of partitions of V such that each partition $\mathbf{P}_i$ is the partition induced by the MSF of G rooted in the graph $(\bigcup_{j \geq i} V_j, \bigcup_{j \geq i} E_j, w)$.

A hierarchical watershed of the graph $\mathcal{G}$ of Fig. 1(a) for the sequence (C, A, B, D) of minima of $\mathcal{G}$ is illustrated in Fig. 1(b).

2.2   Prior Knowledge for Image Processing

Unsupervised data pre-processing methods, such as watershed segmentation and image filtering (*e.g.* Sobel, Laplacian and morphological filters), are successfully employed in several computer vision tasks, including classification and detection problems. Moreover, when prior knowledge is provided, this can be used to provide further improvements.

In the context of image segmentation, a widespread method to introduce prior knowledge in the results is to consider user-defined markers, which are subsets of image pixels indicating the locations of objects of interest. Such markers guide the segmentation algorithm and assure that the objects of interest are segmented into distinct regions. The notion of markers has been especially explored in watershed segmentations, in which catchment basins are grown from input markers instead of the regional minima of an image (or graph) [2].

The use of markers in a watershed segmentation can go beyond the introduction of new regional minima. The values or spectral signatures of the marked pixels can provide further knowledge about the objects we aim to segment. For instance, in [10,11], the authors use the spectral signature of training samples in the construction of watershed segmentations of multi-spectral images. First, the spectral signature of training pixels is used to train a classifier, which is then applied on the whole image and used to obtain a probability map per class. Then, those maps are combined and used to obtain a single watershed segmentation. Another use of supervised classification for watersheds has been proposed in [15], where the watershed segmentation is computed from user-defined markers combined with probability maps computed for each targeted class. More precisely, catchment basins are grown from different markers, and the probability maps, combined with the original data, are used simultaneously in the process. In remote sensing, the later approach has been applied to the detection of buildings [1] and shorelines [17] in multi-spectral images. Finally, in [9], prior knowledge from markers is employed on several interactive image segmentation methods, including watersheds, in the framework of edge-weighted graphs. Edge weights are defined as a linear combination of the weights obtained from two sources: from the pixel values and from the classification probability maps computed from the markers that are incrementally provided by the users.

More generally, knowledge from markers can be used by other kinds of pre-processing methods beyond watershed segmentation. Namely, spectral signatures of training pixels have been used in [4] to optimize the data pre-processing with alternating sequential filters. Hence, training pixels are used for pre-processing the input data, as well as for the final pixel classification. A related approach is proposed in [23], where training pixels are used to optimize vector orderings for morphological operations applied to hyperspectral images.

In the context of hierarchical segmentation, prior knowledge can play a role in defining which regions should be highlighted at different levels of a hierarchy. In [21], a marker-based hierarchical segmentation is proposed for hyperspectral image classification. Labeled markers are derived from a probability classification map, which is obtained from training samples, as done in [10,11,15]. Then, those labeled markers guide the construction of a hierarchical segmentation by preventing regions of different classes to be merged, and by propagating the labeled markers to unlabeled regions. Another related approach, proposed in [12], uses prior knowledge to keep the regions of interest from being merged early in the hierarchy, *i.e.*, the details in the regions of interest are preserved at high levels of the hierarchy.

Finally, in [16], the authors propose a knowledge-based hierarchical representation for hyperspectral images. In their approach, a dissimilarity measure learned from training pixels is employed in the construction of α-trees.

2.3 Attribute Profiles

Attribute profile (AP) [8] is a multilevel feature extraction method commonly applied to the analysis of remote sensing images. To convey spatial-spectral features of remote sensing images, APs were initially defined as sequences of filtering operators on the max- and min-trees computed from the original data. Let $X : P \to \mathbb{Z}$, $P \subseteq \mathbb{Z}^2$ be a gray-scale image. The calculation of APs on X is achieved by applying a sequence of attribute filters based on a min-tree (i.e. attribute thickening operators $\{\phi_k^A\}_{k=1}^K$) and on a max-tree (i.e. attribute thinning operators $\{\gamma_k^A\}_{k=1}^K$) as follows:

$$(X) = \Big\{ \phi_K^A(X), \phi_{K-1}^A(X), \ldots, \phi_1^A(X), X,$$
$$\gamma_1^A(X), \ldots, \gamma_{K-1}^A(X), \gamma_K^A(X) \Big\}, \tag{1}$$

where ϕ_k^A and γ_k^A are respectively the thickening and thinning operators with respect to the attribute A and to the threshold k, and K is the number of selected thresholds. More precisely, the thickening $\phi_k^A(X)$ of X (resp. thinning $\gamma_k^A(X)$ of X) with respect to an attribute A and to a threshold k is obtained as follows: given the min-tree T (resp. max-tree T) of X, the A attribute values (e.g. area, circularity and contrast) of the nodes of T are computed. If the attribute A is increasing, the nodes whose attribute values are inferior to k are pruned from the tree T; otherwise other pruning strategies can be adopted [18]. Finally, the resulting image is reconstructed by projecting the gray levels of the remaining nodes of T into the pixels of X.

Since its appearance, the notion of APs has been extended to other hierarchical representations including tree-of-shapes [7] and partition trees such as α-tree and ω-tree [3]. To obtain a profile from a partition tree instead of a component tree, some adaptations have to be made to the original definition of APs, as discussed in [3]. For instance, the nodes of a partition tree are not naturally associated to gray-level values, as it is the case of component trees. The strategy adopted in [3] is to represent each node as its level in the tree or as the maximum, minimum, or average gray-level of the leaf nodes (pixels) of this node. For more details about APs' extensions, we invite readers to refer to a recent survey [18].

3   Watershed-Based Attribute Profiles

In this article, we extend the notion of AP to hierarchical watershed obtained in the framework of edge-weighted graphs.

As mentioned in Sect. 2.1, hierarchies of partitions, such as hierarchical watersheds, can be equally represented a (partition) tree. Hence, the filtering strategy of Watershed-APs is similar to the strategy described in [3] for the α- and ω-APs.

As discussed in [3], image reconstruction from partition trees is not straightforward as it is from component trees. For node representation, we adopt one of the solutions proposed in [3] and already mentioned in Sect. 2.3, in which a node is represented by the average gray-level of the pixels belonging to it. We highlight that, in the case of multiband images, the average grey level computed on each band might lead to spectral values not present in the input image. However, in the context of attribute profiles used for pixel classification, our aim is not image filtering. Hence, the fact that new spectral values (a.k.a. false colors) are created is not a problem as long as they allow us to distinguish between different semantic classes.

Note that hierarchical watersheds are usually constructed from a gradient of the original image, which contains more information about the contours between salient regions than about the spectral signature of those regions. Hence, we consider the original pixel values to obtain the nodes representation instead of the image gradient.

Formally, let $X : P \to \mathbb{Z}$ be a gray-scale image and let $\mathcal{G} = (V, E, w)$ be a weighted graph which represents a gradient of X, i.e., $V = P$ and, for every edge $e = \{x, y\}$ in E, the weight $w(e)$ represents the dissimilarity between x and y, e.g. $w(e) = |X(x) - X(y)|$. Let S be a sequence of minima of $\mathcal{G}$ ordered according to a given criterion C, and let $\mathcal{H}$ be the hierarchical watershed of $\mathcal{G}$ for the sequence S. Given the tree representation T of $\mathcal{H}$, a *Watershed-AP of X for the criterion C* is constructed as a sequence of image reconstructions from filtered versions of T.

As discussed in Sect. 2.2, user-defined markers and prior knowledge can boost the performance of watershed segmentations. In this article, we extend the use of prior knowledge to hierarchical watersheds. More specifically, we aim to *enforce* regional minima at the regions with high probability of belonging to any given ground-truth class. This is done through a combination of the methods proposed in [10,11,15]. Given a dataset I (*e.g.* a panchromatic or a RGB image) and its training set composed of c classes, we compute its hierarchical watershed using prior knowledge as follows:

1. Train a classifier using the training set of I and compute per-pixel classification probabilities $(p_1, \ldots, p_c)$ for all pixels of I;
2. Combine the classification probabilities into a single map $\mu : X \to [0, 1]$ such that, for any pixel x in X, $\mu(x) = 1 - \sqrt{p_1(x)^2 + \ldots p_c(x)^2}$
3. Compute a 4- or 8-connected weighted graph $\mathcal{G}_P = (V, E, w_P)$ from μ such that, for any edge $e = (x, y)$ in E, we have $w_P(e) = max(\mu(x), \mu(y))$;
4. Compute a 4- or 8-connected weighted graph $\mathcal{G}_G = (V, E, w_G)$ which represents a gradient of I. For instance, for any edge $e = (x, y)$ in E, we may have $w_G(e) = |I(x) - I(y)|$ or $w_G(e) = (I(x) - I(y))^2$;
5. Combine the weight maps w_P and w_G into a map w_{GP} such that, for any edge e in E, we have $w_{GP} = w_P(e) \times w_G(e)$; and
6. Finally, compute the hierarchical watershed of $\mathcal{G}_{GP} = (V, E, w_{GP})$ for a given sequence of minima of $\mathcal{G}$.

In the first step of our method, we are aware that: (1) there might be sample pixels of a given class whose spectral values are not represented in the training set, and (2) there might be pixels in the training set with very similar spectral signatures but which belong to distinct classes. In those cases, we expect the classifier to assign low classification probabilities to such pixels. This means that the watershed segmentation at those regions will be mostly guided by the original gray-levels of the image gradient. Then, in the second step of our method, we combine the classification probability maps into a single probability map μ. We expect this combination to provide flat zones of pixels with high probability of belonging to any given class, *i.e.*, subsets of pixels that should be merged *early* in the resulting hierarchical watershed. In the extreme case where the classifier assigns very high classification probabilities to all pixels of I, we would have a single flat zone and, consequently, a hierarchy with a single segmentation level. However, in that case, we might not need APs to improve the classification results on the image I. In the third step, a weighted graph (V, E, w_P) is obtained from the combined probability map μ. Our choice for computing edge weights as the maximum between the probability values of neighbouring pixels was actually heuristic and was based on a few experiments with the datasets described in the next section. In the steps 4 and 5, a gradient (V, E, w_G) of I is computed and then combined with (V, E, w_P) as a multiplication of edge weights, similarly to [15]. We note that the proposed method is related to ones introduced in [12,16], the main difference being the type of hierarchy under consideration and how the original data is combined with the prior knowledge.

In Fig. 2, we show that the proposed method can be effectively used to high-light objects of interest (e.g. cars) in hierarchical watersheds. Given the RGB image I of Fig. 2(a) and the set of labeled training samples for the car (in blue) and background (in red) classes of Fig. 2(b), we compute the probability map per class and combine them into the map μ of Fig. 2(c), in which dark regions are composed of pixels with high probability of belonging to any given class. Then, the weighted graph $\mathcal{G}_P = (V, E, w_P)$ is computed from μ as described in the third step of our method. Next, we compute the gradient $\mathcal{G}_G = (V, E, w_G)$ of the Red channel of I such that, for any edge $e = (x, y)$ in E, we have $w_G(e) = |I(x) - I(y)|$. At this step, the Green and Blue channels could have been chosen and other dis-similarity measures could be used as well. Then, the weight maps w_P and w_G are combined into w_{PG} as described previously. Finally, to compare the hierar-chies obtained with and without prior-knowledge, we computed the area-based hierarchical watersheds $\mathcal{H}$ and $\mathcal{H}'$ of $\mathcal{G}_G$ and $\mathcal{G}_{GP}$, respectively. In Figs. 2(d), (e) and (f), we show the reconstructions of the tree representation of $\mathcal{H}$ after filtering the nodes with area inferior to 100, 500 and 1000, respectively. Those reconstructions are represented with random colors. Similarly, Figs. 2(g), (h) and (i) show the reconstructions performed on the tree representation of $\mathcal{H}'$ for the same filterings. We observe that, by imposing regional minima at the location of the cars, we make those objects to appear earlier in the hierarchy $\mathcal{H}'$ when compared to $\mathcal{H}$ and, at the same time, to be merged later to their surrounding regions. For instance, the dark blue car on the top left appears as a single region in all reconstructions performed on the tree representation of $\mathcal{H}'$, which is not the case of $\mathcal{H}$.

4   Experimental Results

In this section, we evaluate the performance of Watershed-AP (computed with and without prior-knowledge) in the context of land-cover classification of remote sensing images. We first describe the panchromatic and RGB images considered in our study, as well as the experimental settings used for evaluation. Finally, we show that Watershed-AP outperforms AP and its variants including SDAP [7], α-AP [3] and ω-AP [3], on both datasets.

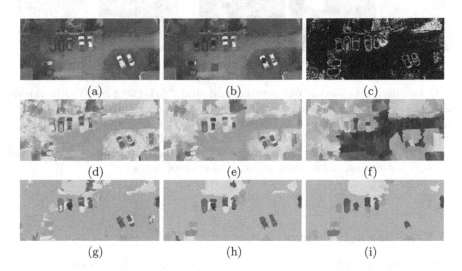

Fig. 2. (a): original image I. (b) training samples in red (background) and blue (cars). (c) classification probability map μ. (d), (e) and (f): image reconstructions obtained from a hierarchical watershed of I, by filtering the nodes with area inferior to 1000, 5000 and 10000, respectively. (g), (h) and (i): image reconstructions obtained from a hierarchical watershed of the combination of I and μ, by filtering the nodes with area inferior to 1000, 5000 and 10000, respectively. (Color figure online)

4.1   Datasets

We validate our approach on two remote sensing datasets: the panchromatic Reykjavik dataset and a RGB image of Zurich dataset [24].

The Reykjavik dataset is a panchromatic image of size 628×700 pixels acquired by the IKONOS Earth imaging satellite with 1-m resolution in Reykjavik, Iceland. This data consists of six thematic classes including residential, soil, shadow, commercial, highway and road. The image was provided with already-split training and test sets (22741 training samples and 98726 test samples). The input image together with its thematic ground truth map are shown in Fig. 3(a).

The Zurich Summer dataset [24] is a collection of 20 NIR+RGB images of various dimensions taken from a QuickBird acquisition of the city of Zurich, Switzerland, in August 2002. The ground-truth provided for each image consists of at most eight thematic classes: roads, buildings, trees, grass, bare soil, water, railways and swimming pools. In our experiments, we only consider the RGB channels of the first image *zh1.tif* of this dataset. The training set is composed of 1% of the labeled pixels randomly extracted for each class. The input image and its ground-truth are given in Fig. 3(b).

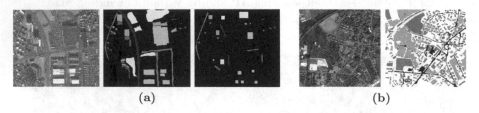

(a)   (b)

Fig. 3. Two data sets used in our experimental study. (a) The 628 × 700 Reykjavik data. Left to right: panchromatic image, ground truth and training set including six thematic classes: ■ residential, ░ shadow, ■ highway, ░ soil, ░ commercial, ■ road. (b) The 610 × 340 Zurich data. Left to right: RGB image and ground truth including seven thematic classes: ■ roads, ■ buildings, ■ trees, ░ grass, ■ bare soil, ░ railways, ■ swiming pools. (Color figure online)

4.2   Experimental Settings

AP and its variants were computed on the Reykjavik and Zurich datasets with the usual area and moment of inertia (MoI) attributes. The following ten area thresholds and four MoI thresholds were adopted for both datasets: $\lambda_{area} = \{25, 100, 500, 1000, 5000, 10000, 20000, 50000, 100000, 150000\}$ and $\lambda_{moi} = \{0.2, 0.3, 0.4, 0.5\}$. For the Zurich dataset, the APs and their extensions are computed independently on each of the RGB bands and are concatenated, leading to Extended Attribute Profiles (EAP).

For each dataset I, hierarchical watersheds were computed from two 4-connected edge-weighted graphs: from the graph $\mathcal{G}_G = (V, E, w_G)$ obtained from a gradient of the original data I (without any prior-knowledge from markers), and the second one computed from the combination of the graph $\mathcal{G}_G$ with the classification probability map obtained from the training set of I, as described in Sect. 3. For every edge $e = (x, y)$ in E, we define $w_G(e)$ as $|I(x) - I(y)|$.

Supervised pixel classification was performed twice, once for obtaining the classification probability map and then to provide the final land-cover pixel classification. Both were performed using a Random Forest classifier with 100 trees. The number of variables used for training was set to the square root of the feature vectors length. The different approaches are compared using the overall accuracy (OA), average accuracy per class (AA) and κ coefficient, as done in [8]. For each tested method, we report the average and standard deviation of the classification scores over ten runs.

As defined in Sect. 2.1, hierarchical watersheds can be computed for any given ordering on the minima of a weighted graph. In our experiments, such orderings are obtained from *extinction values* [14,22] based on the area, dynamics and volume attributes.

5   Results and Discussion

Tables 1, 2, 3 and 4 present the classification results of the Reykjavik and Zurich datasets. We compare the performance of the following methods: AP-maxT and AP-minT obtained by filtering the max- and min-tree, respectively; AP [8], obtained as a concatenation of AP-maxT and AP-minT; SDAP [7]; α-AP and ω-AP [3]; and the Watershed-AP computed with and without prior knowledge. To simplify the notations, Watershed-AP computed without and with prior knowledge are denoted respectively as A-WS-AP and A-PWS-AP, where A is the attribute used in the construction of the hierarchical watersheds, namely Area, Dynamics (Dyn) and Volume (Vol).

For the Reykjavik dataset, Watershed-AP constructed with the area and volume attributes outperform all other methods. As shown in Table 1, the best classification result, obtained with Area-WS-AP, outperforms AP by 1.39%, 3.08% and 1.76% in terms or OA, AA and κ, respectively. Moreover, considering only the APs obtained from partition trees (α-AP, ω-AP and Watershed-AP), the Watershed-AP outperform both α- and ω-APs by more than 10% with respect to OA, AA and κ. In terms of classification results per class (see Table 2), the highest scores are achieved by Watershed-AP and AP, except for the shadow class, for which none of those methods where able to outperform the classification based only on panchromatic pixel values.

Let us now analyse the influence of the use of prior knowledge in the performance of Watershed-AP. For the Watershed-AP computed with the dynamics attribute, the use of prior knowledge led to an improvement of 6.97%, 9.43% and 8.79% in terms of OA, AA and κ, respectively, and better accuracy scores for all six semantic classes. Whereas, the same did not happen for the Watershed-AP computed with area and volume, for which the overall classification results decreased by more than 1%.

On the Zurich dataset, the Watershed-APs outperform all other methods (see Table 3). The best method, Vol-PWS-AP, outperforms SDAP by 3.11%, 3.16% and 4.21% in terms of OA, AA and κ, respectively. Similar to the Reykjavik dataset, the Watershed-AP yielded the best results per class, except for the 'swimming pools' class (see Table 4). Moreover, on this dataset, the use of prior knowledge in the construction of hierarchical watersheds led to small improvements for all three Watershed-APs. More precisely, we observed an improvement of up to 0.25% in terms of OA, AA and κ for the Watershed-APs constructed with the area, dynamics, and volume attributes.

Table 1. Classification result of Reykjavik dataset obtained by different methods using the default 4-connectivity and 1-byte quantization.

Method	Dimension	Classification result		
		OA (%)	AA (%)	$\kappa \times 100$
Panchromatic	1	63.86 ± 0.01	53.35 ± 0.05	52.70 ± 0.00
AP-maxT [8]	16	76.01 ± 0.23	69.15 ± 0.10	68.89 ± 0.29
AP-minT [8]	16	72.90 ± 0.48	64.01 ± 0.56	64.97 ± 0.61
AP [8]	30	84.45 ± 0.43	78.22 ± 0.37	80.11 ± 0.53
SDAP [7]	16	81.51 ± 0.37	74.83 ± 0.34	76.45 ± 0.45
α-AP [3]	16	74.99 ± 0.14	66.48 ± 0.22	67.90 ± 0.19
ω-AP [3]	16	75.06 ± 0.22	66.69 ± 0.38	67.99 ± 0.30
Area-WS-AP	16	$\mathbf{85.84 \pm 0.33}$	$\mathbf{81.30 \pm 0.5}$	$\mathbf{81.87 \pm 0.41}$
Dyn-WS-AP	16	75.96 ± 0.13	66.86 ± 0.18	69.20 ± 0.14
Vol-WS-AP	16	85.62 ± 0.20	80.13 ± 0.31	81.54 ± 0.27
Area-PWS-AP	16	84.63 ± 0.09	79.05 ± 0.12	80.21 ± 0.07
Dyn-PWS-AP	16	82.93 ± 0.36	76.29 ± 0.55	77.99 ± 0.48
Vol-PWS-AP	16	83.86 ± 1.22	79.15 ± 1.25	79.41 ± 1.51

Table 2. Classification result of Reykjavik dataset obtained by different methods using the default 4-connectivity and 1-byte quantization.

Method	Dim.	Classification results per class					
		Residential	Soil	Shadow	Commercial	Highway	Road
Panchromatic	1	6.02 ± 0.36	73.81 ± 0.00	$\mathbf{90.94 \pm 0.01}$	77.86 ± 0.09	16.60 ± 0.00	54.87 ± 0.01
AP-maxT [8]	16	44.55 ± 0.42	79.64 ± 1.29	88.28 ± 0.01	88.41 ± 1.51	51.15 ± 0.15	62.88 ± 0.18
AP-minT [8]	16	30.79 ± 0.68	85.55 ± 0.22	81.04 ± 0.73	81.89 ± 0.16	39.93 ± 6.37	64.89 ± 3.42
AP [8]	30	62.11 ± 0.57	91.29 ± 1.52	76.58 ± 1.61	$\mathbf{92.79 \pm 0.36}$	74.45 ± 0.32	$\mathbf{72.08 \pm 0.27}$
SDAP [7]	16	56.73 ± 1.11	89.57 ± 1.17	78.50 ± 0.89	90.45 ± 0.12	62.77 ± 0.24	70.94 ± 1.20
α-AP [3]	16	45.68 ± 1.33	88.45 ± 0.39	81.40 ± 0.08	82.11 ± 0.05	69.34 ± 0.37	31.93 ± 0.58
ω-AP [3]	16	47.15 ± 1.24	88.60 ± 0.38	81.39 ± 0.06	82.02 ± 0.16	69.20 ± 0.31	31.75 ± 1.66
Area-WS-AP	16	$\mathbf{79.23 \pm 1.39}$	$\mathbf{94.98 \pm 0.28}$	87.38 ± 0.66	89.33 ± 0.08	87.99 ± 1.52	48.88 ± 1.65
Dyn-WS-AP	16	47.16 ± 1.08	89.74 ± 0.15	80.26 ± 0.16	84.47 ± 0.48	71.68 ± 0.19	27.85 ± 0.41
Vol-WS-AP	16	74.41 ± 0.82	95.98 ± 0.19	84.00 ± 0.70	89.85 ± 0.04	87.62 ± 1.33	48.90 ± 2.07
Area-PWS-AP	16	71.05 ± 0.11	91.08 ± 0.08	80.67 ± 0.50	91.90 ± 0.13	$\mathbf{88.09 \pm 0.00}$	51.54 ± 0.34
Dyn-PWS-AP	16	65.70 ± 1.17	94.87 ± 0.23	80.81 ± 0.67	88.55 ± 0.22	77.99 ± 0.19	49.84 ± 3.51
Vol-PWS-AP	16	75.05 ± 0.16	89.91 ± 3.24	79.50 ± 0.45	90.27 ± 0.16	84.89 ± 3.37	55.29 ± 5.31

Table 3. Classification result of Zurich dataset obtained by different methods using the default 4-connectivity and 1-byte quantization.

Method	Dimension	Classification result		
		OA (%)	AA (%)	$\kappa \times 100$
RGB	3	80.21 ± 0.03	69.32 ± 0.07	73.18 ± 0.04
AP-maxT [8]	48	88.44 ± 0.02	87.13 ± 0.04	84.37 ± 0.03
AP-minT [8]	48	87.90 ± 0.04	81.56 ± 0.07	83.69 ± 0.05
AP [8]	90	92.83 ± 0.04	92.43 ± 0.10	90.34 ± 0.05
SDAP [7]	48	93.78 ± 0.04	92.17 ± 0.10	91.61 ± 0.05
α-AP [3]	48	92.55 ± 0.05	86.10 ± 0.15	89.96 ± 0.07
ω-AP [3]	48	92.48 ± 0.03	85.79 ± 0.14	89.86 ± 0.04
Area-WS-AP	48	96.61 ± 0.03	95.30 ± 0.03	95.44 ± 0.04
Dyn-WS-AP	48	93.88 ± 0.03	88.70 ± 0.12	91.76 ± 0.04
Vol-WS-AP	48	96.79 ± 0.02	95.09 ± 0.11	95.69 ± 0.03
Area-PWS-AP	48	96.73 ± 0.04	95.46 ± 0.08	95.60 ± 0.05
Dyn-PWS-AP	48	94.07 ± 0.04	88.60 ± 0.16	92.01 ± 0.06
Vol-PWS-AP	48	$\mathbf{96.89 \pm 0.05}$	$\mathbf{95.33 \pm 0.17}$	$\mathbf{95.82 \pm 0.06}$

Table 4. Classification result of Zurich dataset obtained by different methods using the default 4-connectivity and 1-byte quantization.

Method	Dim.	Classification result per class						
		Roads	Buildings	Trees	Grass	Bare Soil	Railways	Swimming Pools
RGB	3	71.79 ± 0.15	79.14 ± 0.12	51.16 ± 0.19	91.76 ± 0.06	95.69 ± 0.10	5.87 ± 0.37	89.85 ± 0.43
AP-maxT [8]	48	85.88 ± 0.12	90.65 ± 0.09	56.41 ± 0.20	93.26 ± 0.07	99.30 ± 0.03	84.72 ± 0.14	**99.66 ± 0.26**
AP-minT [8]	48	78.62 ± 0.17	85.22 ± 0.13	77.31 ± 0.20	96.29 ± 0.04	97.64 ± 0.05	43.51 ± 0.44	92.37 ± 0.28
AP [8]	90	88.12 ± 0.19	93.49 ± 0.07	76.74 ± 0.20	96.53 ± 0.05	99.48 ± 0.02	93.75 ± 0.23	98.90 ± 0.43
SDAP [7]	48	90.94 ± 0.12	94.35 ± 0.07	79.42 ± 0.14	97.09 ± 0.04	99.56 ± 0.01	84.94 ± 0.54	98.92 ± 0.41
α-AP [3]	48	88.00 ± 0.11	95.25 ± 0.07	82.01 ± 0.26	95.30 ± 0.05	98.64 ± 0.03	58.07 ± 0.83	85.42 ± 0.46
ω-AP [3]	48	88.14 ± 0.10	95.12 ± 0.06	82.19 ± 0.34	95.21 ± 0.04	98.68 ± 0.04	55.65 ± 0.60	85.53 ± 0.70
Area-WS-AP	48	95.83 ± 0.08	97.32 ± 0.05	88.20 ± 0.11	97.88 ± 0.04	99.87 ± 0.01	94.55 ± 0.19	93.44 ± 0.24
Dyn-WS-AP	48	90.01 ± 0.10	96.24 ± 0.11	84.01 ± 0.14	96.18 ± 0.05	99.28 ± 0.03	67.94 ± 0.61	87.25 ± 0.50
Vol-WS-AP	48	**96.27 ± 0.06**	97.57 ± 0.05	88.74 ± 0.13	97.95 ± 0.03	**99.89 ± 0.01**	93.40 ± 0.34	91.79 ± 0.98
Area-PWS-AP	48	94.86 ± 0.09	97.71 ± 0.05	88.65 ± 0.16	98.14 ± 0.04	99.74 ± 0.06	**94.90 ± 0.40**	94.23 ± 0.40
Dyn-PWS-AP	48	90.23 ± 0.14	96.06 ± 0.07	85.25 ± 0.20	96.58 ± 0.04	99.22 ± 0.05	65.26 ± 0.63	87.64 ± 0.74
Vol-PWS-AP	48	95.11 ± 0.11	**97.90 ± 0.04**	**89.10 ± 0.21**	**98.31 ± 0.04**	99.82 ± 0.02	93.28 ± 0.67	93.77 ± 0.81

6   Conclusion

We proposed the Watershed-AP as an extension of AP to hierarchical watersheds computed from (edge) weighted graphs. Besides, we investigate the relevance of using semantic prior knowledge in the construction of such hierarchies. We validated our approach on the pixel classification of two remote sensing images, which showed the potential of hierarchical watersheds in this field. On both datasets, Watershed-APs, computed with and without prior-knowledge, presented the highest evaluation scores when compared to the standard APs.

As future work, we will to explore the versatility of hierarchical watersheds by considering other methods to obtain the gradient of remote sensing images, as well as different ways of including prior knowledge in the computation of those hierarchies. We are also interested in theoretical properties of the Watershed-AP, namely its link with other related methods such as Feature Profiles [19] and Extinction Profiles [13].

Acknowledgements. This work was partially supported by the ANR Multiscale project under the reference ANR-18-CE23-0022. The authors would like to thank Prof. Jon Atli Benediktsson for making available the Reykjavik image.

References

1. Aksoy, S., et al.: Performance evaluation of building detection and digital surface model extraction algorithms: outcomes of the PRRS 2008 algorithm performance contest. In: PRRS 2008, pp. 1–12. IEEE (2008)
2. Beucher, S., Meyer, F.: The morphological approach to segmentation: the watershed transformation. Math. Morphol. Image Process. **34**, 433–481 (1993)
3. Bosilj, P., Damodaran, B.B., Aptoula, E., Dalla Mura, M., Lefèvre, S.: Attribute profiles from partitioning trees. In: ISMM, pp. 381–392 (2017)
4. Courty, N., Aptoula, E., Lefèvre, S.: A classwise supervised ordering approach for morphology based hyperspectral image classification. In: ICPR 2012, pp. 1997–2000. IEEE (2012)
5. Cousty, J., Bertrand, G., Najman, L., Couprie, M.: Watershed cuts: minimum spanning forests and the drop of water principle. IEEE PAMI **31**(8), 1362–1374 (2008)

6. Cousty, J., Najman, L., Perret, B.: Constructive links between some morphological hierarchies on edge-weighted graphs. In: Hendriks, C.L.L., Borgefors, G., Strand, R. (eds.) ISMM 2013. LNCS, vol. 7883, pp. 86–97. Springer, Heidelberg (2013). https://doi.org/10.1007/978-3-642-38294-9_8

7. Dalla Mura, M., Benediktsson, J., Bruzzone, L.: Self-dual attribute profiles for the analysis of remote sensing images. In: ISMM, pp. 320–330 (2011)

8. Dalla Mura, M., Benediktsson, J.A., Waske, B., Bruzzone, L.: Morphological attribute profiles for the analysis of very high resolution images. IEEE TGRS 48(10), 3747–3762 (2010)

9. De Miranda, P.A., Falcão, A.X., Udupa, J.K.: Synergistic arc-weight estimation for interactive image segmentation using graphs. Comput. Vis. Image Underst. 114(1), 85–99 (2010)

10. Derivaux, S., Forestier, G., Wemmert, C., Lefèvre, S.: Supervised image segmentation using watershed transform, fuzzy classification and evolutionary computation. PRL 31(15), 2364–2374 (2010)

11. Derivaux, S., Lefevre, S., Wemmert, C., Korczak, J.: Watershed segmentation of remotely sensed images based on a supervised fuzzy pixel classification. In: IEEE IGARSS, pp. 3712–3715 (2006)

12. Fehri, A., Velasco-Forero, S., Meyer, F.: Prior-based hierarchical segmentation highlighting structures of interest. Math. Morphol. Theory Appl. 3(1), 29–44 (2019)

13. Ghamisi, P., Souza, R., Benediktsson, J.A., Zhu, X.X., Rittner, L., Lotufo, R.A.: Extinction profiles for the classification of remote sensing data. IEEE Trans. Geosci. Remote Sens. 54(10), 5631–5645 (2016)

14. Grimaud, M.: New measure of contrast: the dynamics. In: Image Algebra and Morphological Image Processing III,vol. 1769, pp. 292–305. International Society for Optics and Photonics (1992)

15. Lefèvre, S.: Knowledge from markers in watershed segmentation. In: Kropatsch, W.G., Kampel, M., Hanbury, A. (eds.) CAIP 2007. LNCS, vol. 4673, pp. 579–586. Springer, Heidelberg (2007). https://doi.org/10.1007/978-3-540-74272-2_72

16. Lefèvre, S., Chapel, L., Merciol, F.: Hyperspectral image classification from multiscale description with constrained connectivity and metric learning. In: 2014 WHISPERS, pp. 1–4. IEEE (2014)

17. Lefevre, S., Puissant, A., Levoy, F.: Weakly supervised image segmentation: application to mapping and monitoring of salt marsh vegetation in the mont-saint-michel bay from high resolution imagery. In: ESA-EUSC-JRC 2011, pp. 4-p (2011)

18. Maia, D.S., Pham, M.T., Aptoula, E., Guiotte, F., Lefèvre, S.: Classification of remote sensing data with morphological attributes profiles: a decade of advances. IEEE GRSM (2021)

19. Pham, M.T., Aptoula, E., Lefèvre, S.: Feature profiles from attribute filtering for classification of remote sensing images. IEEE J. Sel. Topics Appl. Earth Observ. Remote Sens. 11(1), 249–256 (2017)

20. Soille, P., Pesaresi, M.: Advances in mathematical morphology applied to geoscience and remote sensing. IEEE TGRS 40(9), 2042–2055 (2002)

21. Tarabalka, Y., Tilton, J.C., Benediktsson, J.A., Chanussot, J.: Marker-based hierarchical segmentation and classification approach for hyperspectral imagery. In: 2011 ICASSP, pp. 1089–1092. IEEE (2011)

22. Vachier, C., Meyer, F.: Extinction value: a new measurement of persistence. In: IEEE Workshop on Nonlinear Signal and Image Processing, vol. 1, pp. 254–257 (1995)

23. Velasco-Forero, S., Angulo, J.: Supervised ordering in $\mathbb{R}^p$: application to morphological processing of hyperspectral images. IEEE TIP **20**(11), 3301–3308 (2011)
24. Volpi, M., Ferrari, V.: Semantic segmentation of urban scenes by learning local class interactions. In: Proceedings of the IEEE Conference on Computer Vision and Pattern Recognition Workshops, pp. 1–9 (2015)

Discrete and Combinatorial Topology

Completions, Perforations and Fillings

Gilles Bertrand$^{(\boxtimes)}$

LIGM, Univ Gustave Eiffel, CNRS, ESIEE Paris, 77454 Marne-la-Vallée, France
g.bertrand@esiee.fr

Abstract. We introduce new local operations, *perforations* and *fillings*, which transform couples of objects (X, Y) such that $X \subseteq Y$. We complement these two operations by collapses and anti-collapses of both X and Y. This set of operations may be seen as a generalization of transformations based solely on collapses, thus a generalization of simple homotopy. Though these transformations do not, in general, preserve homotopy, we make a link between perforations, fillings, and simple homotopy. We give a full characterization of the collection of couples of objects that is closed under the above set of operations. Also, we provide a full characterization of the collection of couples of objects that is closed under the single operation of collapse.

Keywords: Combinatorial topology · Simple homotopy · Contractability · Collapse · Completions

1 Introduction

Simple homotopy, introduced by J. H. C. Whitehead in the early 1930's, may be seen as a refinement of the concept of homotopy [1]. Two complexes are simple homotopy equivalent if one of them may be obtained from the other by a sequence of elementary collapses and anti-collapses.

Simple homotopy plays a fundamental role in combinatorial topology [1–4]. Also, many notions relative to homotopy in the context of computer imagery rely on the collapse operation. In particular, this is the case for the notion of a simple point, which is crucial for all image transformations that preserve the topology of the objects [5–7], see also [8–10].

In this paper, we introduce new local operations, *perforations* and *fillings*, which transform couples of objects. Intuitively, a perforation of a couple (X, Y), with $X \subseteq Y$, consists of removing a part of Y which is also a part of X, a filling is the operation that reverses the effect of a perforation. We complement these two operations by collapses and anti-collapses of both X and Y, with the constraint that we keep the inclusion relation between X and Y. This set of operations may be seen as a generalization of transformations based solely on collapses, thus a generalization of simple homotopy. We formalize these notions by means of completions, which are inductive properties expressed in a declarative way [11].

J. Lindblad et al. (Eds.): DGMM 2021, LNCS 12708, pp. 137–151, 2021.
https://doi.org/10.1007/978-3-030-76657-3_9

Our main results include the following:

- Though these transformations do not, in general, preserve homotopy, we make a link between perforations, fillings, and simple homotopy.
- In particular, we show there is an equivalence between the collection of couples of objects that is closed under the above local operations and the collection made of all contractible objects.
- We give a full characterization of this collection. This characterization is given by four global properties, which are a subset of the five completions that describe acyclic pairs of objects.
- We also provide a full characterization of the collection of couples of objects that is closed under the single operation of collapse.

The paper is organized as follows. First, we give some basic definitions for simplicial complexes (Sect. 2) and simple homotopy (Sect. 3). Then, we recall some facts relative to completions (Sect. 4). We also recall the definitions of the completions that describe acyclic objects (dendrites) and acyclic pairs (dyads) in Sect. 5. Then, we present homotopic pairs and give preliminary results on contractibility (Sect. 6). In Sect. 7, we introduce our perforation and filling operations and give properties relative to these transformations. In Sect. 8, we present some results on collapsible pairs. Note that the paper is self contained.

2   Basic Definitions for Simplicial Complexes

Let X be a finite family composed of finite sets. The *simplicial closure of X* is the complex $X^- = \{y \subseteq x \mid x \in X\}$. The family X is a *(simplicial) complex* if $X = X^-$. We write $\mathbb{S}$ for the collection of all simplicial complexes.

Observe that $\emptyset \in \mathbb{S}$ and $\{\emptyset\} \in \mathbb{S}$. The complex $\emptyset$ is the *void complex*, and the complex $\{\emptyset\}$ is the *empty complex*.

Let $X \in \mathbb{S}$. An element of X is *a simplex of X* or *a face of X*. A *facet of X* is a simplex of X that is maximal for inclusion. For example, the family $X = \{\emptyset, \{a\}, \{b\}, \{a, b\}\}$ is a simplicial complex with four faces and one facet. Note that the empty set is necessarily a face of X whenever $X \neq \emptyset$.

A *simplicial subcomplex* of $X \in \mathbb{S}$ is any subset Y of X that is a simplicial complex. If Y is a subcomplex of X, we write $Y \preceq X$.

Let $X \in \mathbb{S}$. The *dimension* of $x \in X$, written $dim(x)$, is the number of its elements minus one. The *dimension of X*, written $dim(X)$, is the largest dimension of its simplices, the *dimension of $\emptyset$*, the void complex, being defined to be -1. Observe that the dimension of the empty complex $\{\emptyset\}$ is also -1.

A complex $A \in \mathbb{S}$ is *a cell* if $A = \emptyset$ or if A has precisely one non-empty facet x. We set $A^\circ = A \setminus \{x\}$ and $\emptyset^\circ = \emptyset$. We write $\mathbb{C}$ for the collection of all cells. A cell $\alpha \in \mathbb{C}$ is *a vertex* if $dim(\alpha) = 0$.

The *ground set* of $X \in \mathbb{S}$ is the set $\underline{X} = \cup\{x \in X \mid dim(x) = 0\}$. Thus, if $A \in \mathbb{C}$, with $A \neq \emptyset$, then $\underline{A}$ is precisely the unique facet of A. In particular, if α is a vertex, we have $\alpha = \{\emptyset, \underline{\alpha}\}$.

We say that $X \in \mathbb{S}$ and $Y \in \mathbb{S}$ are *disjoint*, or that X is *disjoint from Y*, if $\underline{X} \cap \underline{Y} = \emptyset$. Thus, X and Y are disjoint if and only if $X \cap Y = \emptyset$ or $X \cap Y = \{\emptyset\}$.

If $X \in \mathbb{S}$ and $Y \in \mathbb{S}$ are disjoint, the *join of X and Y* is the simplicial complex XY such that $XY = \{x \cup y \mid x \in X, y \in Y\}$. Thus $XY = \emptyset$ if $Y = \emptyset$, and $XY = X$ if $Y = \{\emptyset\}$. The join αX of a vertex α and a complex $X \in \mathbb{S}$ is a *cone*.

3   Collapse

Let us first recall some basic definitions related to the collapse operator [1].

Let $X \in \mathbb{S}$, and let x, y be two distinct faces of X. The couple (x, y) is a *free pair for X* if y is the only face of X that contains x. Thus, the face y is necessarily a facet of X. If (x, y) is a free pair for X, then $Y = X \setminus \{x, y\}$ is *an elementary collapse of X*, and X is *an elementary expansion of Y*. We say that X *collapses onto Y*, or that Y *expands onto X*, if there exists a sequence $\langle X_0, ..., X_k \rangle$ such that $X_0 = X$, $X_k = Y$, and X_i is an elementary collapse of X_{i-1}, $i \in [1, k]$. The complex X is *collapsible* if X collapses onto $\emptyset$. We say that X is *(simply) homotopic to Y*, or that X and Y are *(simply) homotopic*, if there exists a sequence $\langle X_0, ..., X_k \rangle$ such that $X_0 = X$, $X_k = Y$, and X_i is an elementary collapse or an elementary expansion of X_{i-1}, $i \in [1, k]$. The complex X is *(simply) contractible* if X is simply homotopic to $\emptyset$.

In the sequel, we will use the following facts.

1) Let $X, Y \in \mathbb{S}$, and let (x, y) be a free pair for Y. Thus $Y' = Y \setminus \{x, y\}$ is an elementary collapse of Y. If x and y are not in X, we observe that $X \cup Y'$ is an elementary collapse of $X \cup Y$. It follows that:

The complex Y collapses onto $X \cap Y$ if and only if $X \cup Y$ collapses onto X.

2) Let α be an arbitrary vertex. We have $\alpha = \{\emptyset, \underline{\alpha}\}$. The couple $(\emptyset, \underline{\alpha})$ is a free pair for α. Hence α collapses onto $\emptyset$, *i.e.*, the vertex α is collapsible.

3) Let $X \in \mathbb{S}$, $Y \preceq X$, and let α be a vertex disjoint from X. If $X = \{\emptyset\}$, we have $\alpha X = \alpha$, and either $Y = X$ or $Y = \emptyset$. Otherwise, let x be a facet of X that is not in Y. We observe that the couple $(x, \underline{\alpha} \cup x)$ is a free pair for the cone αX. Thus, the cone $\alpha X'$, with $X' = X \setminus \{x, \underline{\alpha} \cup x\}$, is an elementary collapse of the cone αX. By induction, we obtain:

If $Y \preceq X$, then αX collapses onto αY.

If $Y = \emptyset$, it means that αX collapses onto $\emptyset$. Thus:

Any cone is collapsible.

4) Let $X, Y \in \mathbb{S}$, with $X \preceq Y$, and let α be a vertex disjoint from Y. Let x and y be two faces of X. The couple (x, y) is a free pair for X if and only if $(\underline{\alpha} \cup x, \underline{\alpha} \cup y)$ is a free pair for $\alpha X \cup Y$. Thus, by induction, we have the following:

The complex X collapses onto Z if and only if $\alpha X \cup Y$ collapses onto $\alpha Z \cup Y$.

As a special case of this result, we obtain:

The complex X is collapsible if and only if $\alpha X \cup Y$ collapses onto Y.

5) Let $A \in \mathbb{C}$. If $A = \emptyset$, or if A is a vertex, then A is collapsible. Otherwise, there exists a vertex α and a cell B such that $A = \alpha B$, which is a cone. Thus, any cell is collapsible. Using the result 4), we see that the cell $A = \alpha B$ collapses onto the cell B. By induction, we obtain the following result:

If $A, B \in \mathbb{C}$, and if $B \preceq A$, then A collapses onto B.

6) Let $X, Y \in \mathbb{S}$, with $X \neq \emptyset$. We observe that X is an elementary collapse of Y if and only if there exists a cell D, $D \neq \emptyset$, such that $Y = X \cup \alpha D$ and $X \cap \alpha D = \alpha D°$, where α is a vertex disjoint from D.

Now, we recall the definition of the dual of a complex, and we give two related propositions which will be used for collapsibility.

Let $A \in \mathbb{C}$ and $X \preceq A$. The *dual of X for A* is the simplicial complex, written X_A^*, such that $X_A^* = \{x \in A \mid (\underline{A} \setminus x) \notin X\}$.

We have $\emptyset_A^* = A$ and $\{\emptyset\}_A^* = A°$, and, for any $A \in \mathbb{C}$, we have the following:

– If $X \preceq A$, then $(X_A^*)_A^* = X$.
– If $X \preceq A$, $Y \preceq A$, then $(X \cup Y)_A^* = X_A^* \cap Y_A^*$ and $(X \cap Y)_A^* = X_A^* \cup Y_A^*$.

Proposition 1 ([4,11]). *Let $A \in \mathbb{C}$. Let $X, Y \in \mathbb{S}$, with $X \preceq Y \preceq A$. The complex Y collapses onto X if and only if X_A^* collapses onto Y_A^*.*

Proposition 2 ([12]). *Let $A \in \mathbb{C}$. Let $X, Y \in \mathbb{S}$, with $X \preceq Y \preceq A$, and let α be a vertex disjoint from A. Then we have $(\alpha X \cup Y)_{\alpha A}^* = \alpha Y_A^* \cup X_A^*$.*

4   Completions

We give some basic definitions for completions. A completion may be seen as a rewriting rule that permits to derive collections of sets. See [11] for more details.

Let $\mathbf{S}$ be a given collection and let $\mathcal{K}$ be an arbitrary subcollection of $\mathbf{S}$. Thus, we have $\mathcal{K} \subseteq \mathbf{S}$. In the sequel of the paper, the symbol $\mathcal{K}$, with possible superscripts, will be a dedicated symbol (a kind of variable).

Let κ be a binary relation on $2^{\mathbf{S}}$, thus $\kappa \subseteq 2^{\mathbf{S}} \times 2^{\mathbf{S}}$. We say that κ is *finitary*, if $\mathbf{F}$ is finite whenever $(\mathbf{F}, \mathbf{G}) \in \kappa$.

Let $\langle \mathrm{K} \rangle$ be a property that depends on $\mathcal{K}$. We say that $\langle \mathrm{K} \rangle$ is a *completion (on $\mathbf{S}$)* if $\langle \mathrm{K} \rangle$ may be expressed as the following property:

$\rightarrow$ If $\mathbf{F} \subseteq \mathcal{K}$, then $\mathbf{G} \subseteq \mathcal{K}$ whenever $(\mathbf{F}, \mathbf{G}) \in \kappa$. $\langle \mathrm{K} \rangle$

where κ is a finitary binary relation on $2^{\mathbf{S}}$.

If $\langle \mathrm{K} \rangle$ is a property that depends on $\mathcal{K}$, we say that a given collection $\mathbf{X} \subseteq \mathbf{S}$ satisfies $\langle \mathrm{K} \rangle$ if the property $\langle \mathrm{K} \rangle$ is true for $\mathcal{K} = \mathbf{X}$.

Theorem 1 [11]. *Let* $\langle K \rangle$ *be a completion on* **S** *and let* $\mathbf{X} \subseteq \mathbf{S}$. *There exists, under the subset ordering, a unique minimal collection that contains* **X** *and that satisfies* $\langle K \rangle$.

If $\langle K \rangle$ is a completion on **S** and if $\mathbf{X} \subseteq \mathbf{S}$, we write $\langle \mathbf{X}; K \rangle$ for the unique minimal collection that contains **X** and that satisfies $\langle K \rangle$.

Let $\langle K \rangle$ be a completion expressed as the above property $\langle \kappa \rangle$. By a fixed point property, the collection $\langle \mathbf{X}; K \rangle$ may be obtained by starting from $\mathcal{K} = \mathbf{X}$, and by iteratively adding to $\mathcal{K}$ all the sets **G** such that $(\mathbf{F}, \mathbf{G}) \in \kappa$ and $\mathbf{F} \subseteq \mathcal{K}$ (see [11]). Thus, if $\mathbf{C} = \langle \mathbf{X}; K \rangle$, then $\langle \mathbf{X}; K \rangle$ may be seen as a dynamic structure that describes **C**; the completion $\langle K \rangle$ acts as a generator, which, from **X**, makes it possible to enumerate all elements in **C**. We will see now that $\langle K \rangle$ may in fact be composed of several completions.

Let $\langle K_1 \rangle, \langle K_2 \rangle, ..., \langle K_k \rangle$ be completions on **S**. We write $\wedge$ for the logical "and". It may be seen that $\langle K \rangle = \langle K_1 \rangle \wedge \langle K_2 \rangle ... \wedge \langle K_k \rangle$ is a completion. In the sequel, we write $\langle K_1, K_2, ..., K_k \rangle$ for $\langle K \rangle$. Thus, if $\mathbf{X} \subseteq \mathbf{S}$, the notation $\langle \mathbf{X}; K_1, K_2, ..., K_k \rangle$ stands for the smallest collection that contains **X** and that satisfies each of the properties $\langle K_1 \rangle, \langle K_2 \rangle, ..., \langle K_k \rangle$.

Let $\langle K \rangle, \langle Q \rangle$ be two completions on **S**. We say that $\langle K \rangle$ and $\langle Q \rangle$ are *equivalent* if, for any $\mathbf{X} \subseteq \mathbf{S}$, the collection **X** satisfies $\langle K \rangle$ if and only if **X** satisfies $\langle Q \rangle$.

5   Dyads and Dendrites

The notion of a dendrite was introduced in [11] as a way for defining a collection made of acyclic complexes. Let us consider the collection $\mathbf{S} = \mathbb{S}$, and let $\mathcal{K}$ denote an arbitrary collection of simplicial complexes.

We define the two completions $\langle D1 \rangle$ and $\langle D2 \rangle$ on $\mathbb{S}$: For any $S, T \in \mathbb{S}$,

   –> If $S, T \in \mathcal{K}$, then $S \cup T \in \mathcal{K}$ whenever $S \cap T \in \mathcal{K}$. $\langle D1 \rangle$

   –> If $S, T \in \mathcal{K}$, then $S \cap T \in \mathcal{K}$ whenever $S \cup T \in \mathcal{K}$. $\langle D2 \rangle$

Let $\mathbb{D} = \langle \mathbb{C}; D1, D2 \rangle$. Each element of $\mathbb{D}$ is a *dendrite* or an *acyclic complex*.

The collection $\mathbb{T}$ of all trees (*i.e.*, all connected acyclic graphs) provides an example of a collection of dendrites. It may be checked that $\mathbb{T}$ satisfies both $\langle D1 \rangle$ and $\langle D2 \rangle$, and that we have $\mathbb{T} \subseteq \mathbb{D}$. In fact, we have the general result [11]:

A complex is a dendrite if and only if it is acyclic in the sense of homology.

As a consequence, any contractible complex is a dendrite but there exist some dendrites that are not contractible. The punctured Poincaré homology sphere provides such a counter-example.

In the following, we will say that a dendrite is *twisted* if it is not contractible.

Intuitively, a dyad is a couple of complexes (X, Y), with $X \preceq Y$, such that the cycles of X are "at the right place with respect to the ones of Y".

We set $\ddot{\mathbb{S}} = \{(X, Y) \mid X, Y \in \mathbb{S}, X \preceq Y\}$, the notation $\ddot{\mathcal{K}}$ stands for an arbitrary subcollection of $\ddot{\mathbb{S}}$.

We define five completions on $\mathbb{\ddot{S}}$ (the symbols $\widetilde{T}$, $\widetilde{U}$, $\widetilde{L}$ stand respectively for "transitivity", "upper confluence", and "lower confluence"):

For any $S, T \in \mathbb{S}$,
$\rightarrow$ If $(S \cap T, T) \in \ddot{\mathcal{K}}$, then $(S, S \cup T) \in \ddot{\mathcal{K}}$. $\langle \widetilde{X} \rangle$
$\rightarrow$ If $(S, S \cup T) \in \ddot{\mathcal{K}}$, then $(S \cap T, T) \in \ddot{\mathcal{K}}$. $\langle \widetilde{Y} \rangle$
For any $(R, S), (S, T), (R, T) \in \mathbb{\ddot{S}}$,
$\rightarrow$ If $(R, S) \in \ddot{\mathcal{K}}$ and $(S, T) \in \ddot{\mathcal{K}}$, then $(R, T) \in \ddot{\mathcal{K}}$. $\langle \widetilde{T} \rangle$
$\rightarrow$ If $(R, S) \in \ddot{\mathcal{K}}$ and $(R, T) \in \ddot{\mathcal{K}}$, then $(S, T) \in \ddot{\mathcal{K}}$. $\langle \widetilde{U} \rangle$
$\rightarrow$ If $(R, T) \in \ddot{\mathcal{K}}$ and $(S, T) \in \ddot{\mathcal{K}}$, then $(R, S) \in \ddot{\mathcal{K}}$. $\langle \widetilde{L} \rangle$

We set $\ddot{\mathbb{C}} = \{(A, B) \in \mathbb{\ddot{S}} \mid A, B \in \mathbb{C}\}$ and $\ddot{\mathbb{X}} = \langle \ddot{\mathbb{C}}; \widetilde{X}, \widetilde{Y}, \widetilde{T}, \widetilde{U}, \widetilde{L} \rangle$.
Each couple of $\ddot{\mathbb{X}}$ is a *dyad* or an *acyclic pair*.

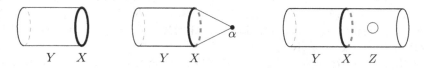

Y X Y X Y X Z

Fig. 1. Left: A circle X in a cylindrical surface Y, the couple (X, Y) is a dyad. Middle: The complex $\alpha X \cup Y$ is a dendrite. Thus, $\alpha X \cup Y$ is acyclic. Right, illustration of the completion $\langle \widetilde{X} \rangle$: If a complex Z is such that $Z \cap Y = X$, then $(Z, Z \cup Y)$ is a dyad. Here Z is a punctured cylindrical surface. Again, the complex $\alpha Z \cup (Z \cup Y)$ is acyclic.

In [12], the following fundamental relation between dyads and dendrites was given. Intuitively, Theorem 2 indicates that (X, Y) is a dyad if and only if we cancel out all cycles of Y (*i.e.*, we obtain an acyclic complex), whenever we cancel out those of X (by the way of a cone). See Fig. 1, which also gives an illustration of the completion $\langle \widetilde{X} \rangle$.

Theorem 2. *Let $(X, Y) \in \mathbb{\ddot{S}}$ and let α be a vertex disjoint from Y.*
   The couple (X, Y) is a dyad if and only if $\alpha X \cup Y$ is a dendrite.

In particular, setting $X = \emptyset$ in the above theorem, we see that a complex $D \in \mathbb{S}$ is a dendrite if and only if $(\emptyset, D)$ is a dyad. Therefore, we have the following corollary of Theorem 2.

Corollary 1. *Let $(X, Y) \in \mathbb{\ddot{S}}$ and let α be a vertex disjoint from Y.*
   The couple (X, Y) is a dyad if and only if $(\emptyset, \alpha X \cup Y)$ is a dyad.

If $(X, Y) \in \mathbb{\ddot{S}}$ and α is a vertex disjoint from Y, we say that $\alpha X \cup Y$ is the Δ-*form of* (X, Y) *(with vertex α)*. Such forms play a decisive role in our framework. We see that Theorem 2 may be formulated in this way: *A couple $(X, Y) \in \mathbb{\ddot{S}}$ is a dyad if and only if its Δ-form is a dendrite.*

6   Homotopy and Contractibiliy

We now recall the completions introduced in [13] for simple homotopy.

If $(I, O) \in \ddot{\mathbb{S}}$, the basic idea is to continuously deform both the inner complex I and the outer complex O, these deformations keeping I inside O.

If $X, Y \in \mathbb{S}$, we write $X \overset{E}{\longmapsto} Y$, whenever Y is an elementary expansion of X. We define four completions on $\ddot{\mathbb{S}}$: For any (R, S), (R, T), (S, T) in $\ddot{\mathbb{S}}$,

> $\rightarrow$ If $(R, S) \in \ddot{\mathcal{K}}$ and $S \overset{E}{\longmapsto} T$, then $(R, T) \in \ddot{\mathcal{K}}$. $\qquad\qquad\qquad\qquad$ $\langle \mathrm{O^+} \rangle$

> $\rightarrow$ If $(R, T) \in \ddot{\mathcal{K}}$ and $S \overset{E}{\longmapsto} T$, then $(R, S) \in \ddot{\mathcal{K}}$. $\qquad\qquad\qquad\qquad$ $\langle \mathrm{O^-} \rangle$

> $\rightarrow$ If $(R, T) \in \ddot{\mathcal{K}}$ and $R \overset{E}{\longmapsto} S$, then $(S, T) \in \ddot{\mathcal{K}}$. $\qquad\qquad\qquad\qquad$ $\langle \mathrm{I^+} \rangle$

> $\rightarrow$ If $(S, T) \in \ddot{\mathcal{K}}$ and $R \overset{E}{\longmapsto} S$, then $(R, T) \in \ddot{\mathcal{K}}$. $\qquad\qquad\qquad\qquad$ $\langle \mathrm{I^-} \rangle$

We set $\ddot{\mathbb{I}} = \{(X, X) \mid X \in \mathbb{S}\}$ and $\ddot{\mathbb{H}} = \langle \ddot{\mathbb{I}}; \mathrm{O^+}, \mathrm{O^-}, \mathrm{I^+}, \mathrm{I^-} \rangle$.

Each element of $\ddot{\mathbb{H}}$ is a *homotopic pair*.

Observe that X is homotopic to Y whenever $(X, Y) \in \ddot{\mathbb{H}}$. The following is a variant of a result given in [13] which gives a basic characterization of the collection $\ddot{\mathbb{H}}$.

Theorem 3. *Let* $(X, Y) \in \ddot{\mathbb{S}}$. *We have* $(X, Y) \in \ddot{\mathbb{H}}$ *if and only if there exists a complex* Z *such that* Z *collapses onto* X *and* Z *collapses onto* Y.

The following theorem is the important global result given in [13]. It shows that four of the five completions that describe dyads allow for a characterization of the collection made of all homotopic pairs. Thus, in this framework, we have a unified presentation of both acyclicity and homotopy.

Theorem 4. *We have* $\ddot{\mathbb{H}} = \langle \ddot{\mathbb{C}}; \tilde{\mathrm{X}}, \tilde{\mathrm{T}}, \tilde{\mathrm{U}}, \tilde{\mathrm{L}} \rangle$.

We now present some facts relative to contractibility. This special case of homotopy will play a crucial role in the sequel of this paper. The following is a direct consequence of the definition of contractibility and that of the collection $\ddot{\mathbb{H}}$.

Proposition 3. *A complex* $X \in \mathbb{S}$ *is contractible if and only if* $(\emptyset, X) \in \ddot{\mathbb{H}}$.

The next proposition is a basic fact which will be useful in the sequel. It indicates that, if a subcomplex X of Y is contractible, then we can add or remove a cone on X without changing the homotopy of Y. Observe that it is not easy to prove this fact by using direct arguments, since the steps involved in the contractibility of X may "extend X outside Y". With the global properties given by Theorem 4, only few lines are needed for establishing this result. See [17], Proposition 0.17, for a counterpart of this property in the framework of general homotopy.

Proposition 4. *Let* $(X, Y) \in \ddot{\mathbb{S}}$ *and let* α *be a vertex disjoint from* Y. *Suppose the complex* X *is contractible, i.e., we have* $(\emptyset, X) \in \ddot{\mathbb{H}}$. *Then we have* $(Y, \alpha X \cup Y) \in \ddot{\mathbb{H}}$. *In particular, we have* $(\emptyset, Y) \in \ddot{\mathbb{H}}$ *if and only if* $(\emptyset, \alpha X \cup Y) \in \ddot{\mathbb{H}}$.

Proof. Since any cone is collapsible, we have $(\emptyset, \alpha X) \in \ddot{\mathbb{H}}$. By Theorem 4, the collection $\ddot{\mathbb{H}}$ satisfies $\langle \widetilde{U} \rangle$. Thus $(X, \alpha X) \in \ddot{\mathbb{H}}$. We may write $(Y \cap \alpha X, \alpha X) \in \ddot{\mathbb{H}}$. By $\langle \widetilde{X} \rangle$, we obtain $(Y, \alpha X \cup Y) \in \ddot{\mathbb{H}}$. Since $\ddot{\mathbb{H}}$ satisfies also $\langle \widetilde{T} \rangle$ and $\langle \widetilde{L} \rangle$, we see that $(\emptyset, Y) \in \ddot{\mathbb{H}}$ if and only if $(\emptyset, \alpha X \cup Y) \in \ddot{\mathbb{H}}$. $\qquad\square$

Proposition 5. *Let* $(X, Y) \in \ddot{\mathbb{S}}$ *and let* α *be a vertex disjoint from* Y. *If* $(X, Y) \in \ddot{\mathbb{H}}$, *then* $\alpha X \cup Y$ *is contractible, i.e., we have* $(\emptyset, \alpha X \cup Y) \in \ddot{\mathbb{H}}$.

Proof. Let $(X, Y) \in \ddot{\mathbb{H}}$. We have $\alpha X \cap Y = X$, thus $(\alpha X \cap Y, Y) \in \ddot{\mathbb{H}}$. By $\langle \widetilde{X} \rangle$, we obtain $(\alpha X, \alpha X \cup Y) \in \ddot{\mathbb{H}}$. Since any cone is collapsible, we have $(\emptyset, \alpha X) \in \ddot{\mathbb{H}}$. The result follows from $\langle \widetilde{T} \rangle$. $\qquad\square$

The converse of Proposition 5 is not true. Let X be a twisted dendrite, *i.e.*, a non contractible dendrite. If A is a cell, with $X \preceq A$, then X is not homotopic to A, but $\alpha X \cup A$ is contractible. This fact is well known in algebraic topology [17].

Let $\ddot{\mathbb{K}}$ be a collection of couples such that $(\emptyset, \emptyset)$ is in $\ddot{\mathbb{K}}$. We define the *kernel* *of* $\ddot{\mathbb{K}}$ as the collection of all complexes $X \in \mathbb{S}$ such that $(\emptyset, X)$ is in $\ddot{\mathbb{K}}$.

Let us consider the two following properties, where α is a vertex disjoint from Y:

If $(X, Y) \in \ddot{\mathbb{K}}$, then the Δ-form $\alpha X \cup Y$ is in the kernel of $\ddot{\mathbb{K}}$. $\qquad$ (K1)

If $(X, Y) \in \ddot{\mathbb{S}}$ and the Δ-form $\alpha X \cup Y$ is in the kernel of $\ddot{\mathbb{K}}$, then $(X, Y) \in \ddot{\mathbb{K}}$. $\qquad$ (K2)

By Corollary 1, the collection $\ddot{\mathbb{X}}$ of dyads satisfies both (K1) and (K2). In particular, it means that $\ddot{\mathbb{X}}$ may be recovered from its kernel. By Proposition 5 and the above remark, the collection of all homotopic pairs $\ddot{\mathbb{H}}$ satisfies (K1) but not (K2).

In the next section, we will investigate the collection $\ddot{\mathbb{W}}$ of all pairs (X, Y) such that the Δ-form $\alpha X \cup Y$ is contractible. Thus, $\ddot{\mathbb{W}}$ will satisfy both (K1) and (K2) and, by the preceding remark, $\ddot{\mathbb{W}}$ will strictly contain the collection $\ddot{\mathbb{H}}$. Furthermore, we will see that $\ddot{\mathbb{W}}$ has a remarkable global characterization.

7   Filling and Perforation

Let $(X, Y) \in \ddot{\mathbb{S}}$ and $(X', Y') \in \ddot{\mathbb{S}}$. We say that (X, Y) is *an elementary perforation* *of* (X', Y'), if there exists a face f of both X' and Y' such that $X = X' \setminus \{f\}$ and $Y = Y' \setminus \{f\}$. Thus f is necessarily a facet of both X' and Y'.

We say that (X', Y') is *an elementary filling of* (X, Y), if (X, Y) is an elementary perforation of (X', Y').

Let $(X, Y) \in \ddot{\mathbb{S}}$. We write $(X, Y) \overset{F}{\longmapsto} (X', Y')$ if (X', Y') is an elementary filling of (X, Y). We define two completions on $\ddot{\mathbb{S}}$: For any (S, T), (S', T') in $\ddot{\mathbb{S}}$,

$\rightarrow$ If $(S, T) \in \ddot{\mathbb{K}}$ and $(S, T) \overset{F}{\longmapsto} (S', T')$, then $(S', T') \in \ddot{\mathbb{K}}$. $\qquad$ $\langle \text{F} \rangle$

$\rightarrow$ If $(S, T) \in \ddot{\mathbb{K}}$ and $(S', T') \overset{F}{\longmapsto} (S, T)$, then $(S', T') \in \ddot{\mathbb{K}}$. $\qquad$ $\langle \text{P} \rangle$

We set $\ddot{\emptyset} = \{(\emptyset, \emptyset)\}$ and $\ddot{W} = \langle \ddot{\emptyset}; O^+, O^-, I^+, I^-, F, P \rangle$.

Each couple of $\ddot{W}$ is a *contractible pair*.

We first observe that any couple $(X, X) \in \ddot{I}$ may be obtained from $(\emptyset, \emptyset)$ by iteratively applying the completion $\langle F \rangle$. In fact we see that $\ddot{I} = \langle \ddot{\emptyset}; F \rangle$.

Thus, we have $\ddot{W} = \langle \ddot{I}; O^+, O^-, I^+, I^-, F, P \rangle$.

See Fig. 2 which provides an illustration of four contractible pairs.

We have $\ddot{H} = \langle \ddot{I}; O^+, O^-, I^+, I^- \rangle$. Therefore, we immediately note that $\ddot{H} \subseteq \ddot{W}$ since the two additional completions $\langle F \rangle$ and $\langle P \rangle$ allow to generate more couples in $\ddot{W}$. Also, we observe that, if we apply the completions $\langle F \rangle$ or $\langle P \rangle$ on a couple $(X, Y) \in \ddot{S}$, then the homotopy of both X and Y is changed. Nevertheless, we will see that all couples $(X, Y) \in \ddot{W}$ satisfy an homotopy condition.

It can be checked that each transformation of a couple $(X, Y) \in \ddot{S}$, that is made according to the six completions which constitute the definition of $\ddot{W}$, may be seen as a collapse or an expansion of a complex of the form $\alpha X \cup Y$.

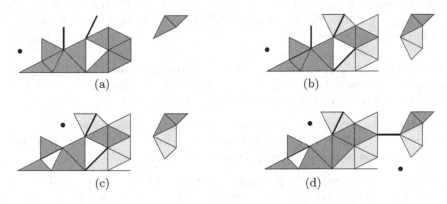

(a)

(b)

(c)

(d)

Fig. 2. Four contractible pairs. For each pair (X, Y), the subcomplex X of Y is highlighted by its facets (triangles, segments, or vertices). (a) A pair $(X_0, Y_0) \in \ddot{I}$, thus $X_0 = Y_0$. (b) A pair (X_1, Y_1): the complex Y_1 is obtained from Y_0 with a sequence of expansions, and X_1 is obtained from X_0 with a sequence of collapses. (c) A pair (X_2, Y_2) obtained from (X_1, Y_1) with three perforations. (d) A pair (X_3, Y_3) obtained from (X_2, Y_2) with three fillings.

Proposition 6. *Let* $(X, Y) \in \ddot{S}$, *let* $X' \preceq X$, $Y' \preceq Y$, *and let* α *be a vertex disjoint from* Y. *The complex* $\alpha X' \cup Y'$ *is an elementary collapse of* $\alpha X \cup Y$ *if and only if the couples* (X', Y') *and* (X, Y) *are such that:*

- *we have* $Y' = Y$ *and the complex* X' *is an elementary collapse of* X, *or*
- *we have* $X' = X$ *and the complex* Y' *is an elementary collapse of* Y, *or*
- *the couple* (X', Y') *is an elementary perforation of* (X, Y).

As a consequence, we have the following result which makes the connection between the collection $\ddot{W}$ and contractible complexes.

Proposition 7. *Let $(X,Y) \in \ddot{\mathbb{S}}$ and let α be a vertex disjoint from Y.
We have $(X,Y) \in \ddot{\mathbb{W}}$ if and only if $\alpha X \cup Y$ is contractible, i.e., if and only if
$(\emptyset, \alpha X \cup Y) \in \ddot{\mathbb{H}}$.*

By Theorem 3, we have $(\emptyset, \alpha X \cup Y) \in \ddot{\mathbb{H}}$ if and only if there exists a complex
Z such that Z collapses onto $\emptyset$, and Z collapses onto $\alpha X \cup Y$. Thus, the complex
Z is collapsible. It follows that Proposition 7 and Theorem 3 permits to establish
a link between the collection $\ddot{\mathbb{W}}$ and collapsibility. The following proposition will
allow us to refine this result.

Proposition 8. *Let $X \in \mathbb{S}$, $A \in \mathbb{C}$, with $X \preceq A$, and let α be a vertex disjoint
from A. If X collapses onto $\emptyset$, then αA collapses onto X.*

Proof. Suppose X collapses onto $\emptyset$, and let $A \in \mathbb{C}$ with $X \preceq A$. By Proposition 1, since $\emptyset^*_A = A$, the complex A collapses onto X^*_A. Let α be a vertex
disjoint from A, and let $Z = \alpha X^*_A \cup A$. Since A collapses onto X^*_A, the complex Z collapses onto αX^*_A. Thus Z collapses onto $\emptyset$. By Proposition 2, we have
$Z^*_{\alpha A} = \alpha A^*_A \cup (X^*_A)^*_A$. We obtain $Z^*_{\alpha A} = X$. By Proposition 1, we deduce that
αA collapses onto X. $\qquad\square$

By Proposition 7, Theorem 3, and Proposition 8, we deduce that $(X,Y) \in \ddot{\mathbb{W}}$
if and only if there exists a cell $B \in \mathbb{C}$ such that B collapses onto $\alpha X \cup Y$.
Clearly, we can assume that B is of the form αC, with $C \in \mathbb{C}$.

Proposition 9. *Let $(X,Y) \in \ddot{\mathbb{S}}$ and let α be a vertex disjoint from Y.
We have $(X,Y) \in \ddot{\mathbb{W}}$ if and only if there exists a cell $C \in \mathbb{C}$, which is disjoint
from α, such that the cell αC collapses onto $\alpha X \cup Y$.*

Under the conditions of Proposition 9, we may write $\alpha C = \alpha C \cup C$. Hence, we
have $(X,Y) \in \ddot{\mathbb{W}}$ if and only if there exists a cell $C \in \mathbb{C}$ such that (X,Y) may be
obtained from (C,C) by using the three transformations given in Proposition 6.
Since $(C,C) \in \ddot{\mathbb{I}}$, we obtain the following result.

Proposition 10. *We have $\ddot{\mathbb{W}} = \langle \ddot{\mathbb{I}}; O^-, I^-, P \rangle$.*

A direct consequence of Theorem 3 is that we have $\ddot{\mathbb{H}} = \langle \ddot{\mathbb{I}}; O^-, I^- \rangle$ (see
Proposition 7 of [13]). Thus, the previous result may be seen as a counterpart
of this characterization of homotopic pairs. We now point out the following
equivalence.

Proposition 11. *The two completions $\langle F \rangle$ and $\langle \widetilde{X} \rangle$ are equivalent and the two
completions $\langle P \rangle$ and $\langle \widetilde{Y} \rangle$ are equivalent.*

Proof. Let $\ddot{\mathbb{K}}$ be an arbitrary subcollection of $\ddot{\mathbb{S}}$.

i) Suppose $\ddot{\mathbb{K}}$ satisfies property $\langle P \rangle$ and let $(S, S \cup T) \in \ddot{\mathbb{K}}$. Let x be a facet
of S, with $x \notin T$. If no such a facet exists, we are done. Otherwise, the face
x is also a facet of $S \cup T$. Thus the couple $(S \setminus \{x\}, (S \setminus \{x\}) \cup T)$ is an
elementary perforation of $(S, S \cup T)$. By induction on the number of such
facets, $\ddot{\mathbb{K}}$ satisfies $\langle \widetilde{Y} \rangle$.

ii) Suppose $\ddot{\mathbb{K}}$ satisfies property $\langle \widetilde{Y} \rangle$. Let $(S,T) \in \ddot{\mathbb{K}}$ and let (S',T') be an elementary perforation of (S,T). We have $(S,T) = (S, S \cup T')$, thus by $\langle \widetilde{Y} \rangle$, we have $(S \cap T', T') = (S', T') \in \ddot{\mathbb{K}}$. We deduce that $\ddot{\mathbb{K}}$ satisfies $\langle P \rangle$.

It follows that $\langle P \rangle$ and $\langle \widetilde{Y} \rangle$ are equivalent. By reversing the above operations, we get the equivalence between $\langle F \rangle$ and $\langle \widetilde{X} \rangle$.   □

Since the collection $\ddot{\mathbb{H}}$ satisfies the property $\langle \widetilde{X} \rangle$, a consequence of the above equivalence is that we can add another local transformation to the four completions which define homotopic pairs.

Corollary 2. *We have* $\ddot{\mathbb{H}} = \langle \ddot{\mathbb{I}}; O^+, O^-, I^+, I^-, F \rangle$.

The following is one of the main results of this paper. It shows that the collection of contractible pairs is uniquely identified by four global properties.

Theorem 5. *We have* $\ddot{\mathbb{W}} = \langle \ddot{\mathbb{C}}; \widetilde{X}, \widetilde{Y}, \widetilde{T}, \widetilde{U} \rangle$.

Proof. We set $\ddot{\mathbb{W}}' = \langle \ddot{\mathbb{C}}; \widetilde{X}, \widetilde{Y}, \widetilde{T}, \widetilde{U} \rangle$.

1) In this part, we show that we have $\ddot{\mathbb{W}} \subseteq \ddot{\mathbb{W}}'$.

   For that purpose we first establish the following result (R).

   Let $X, Y \in \mathbb{S}$. *If* $(\emptyset, Y) \in \ddot{\mathbb{W}}'$ *and if* Y *collapses onto* X, *then* $(\emptyset, X) \in \ddot{\mathbb{W}}'$.   (R)

   Suppose $(\emptyset, Y) \in \ddot{\mathbb{W}}'$ and X is an elementary collapse of Y. If $X = \emptyset$, we are done. Otherwise, there exists a cell D, $D \neq \emptyset$, such that $Y = X \cup \alpha D$ and $X \cap \alpha D = \alpha D^\circ$, where α is a vertex disjoint from D. The complex αD being a cone, we have $(\emptyset, \alpha D) \in \ddot{\mathbb{W}}'$ (see Lemma 1 in [13]). By $\langle \widetilde{U} \rangle$, we get $(\alpha D, Y) = (\alpha D, X \cup \alpha D) \in \ddot{\mathbb{W}}'$. By $\langle \widetilde{Y} \rangle$, we derive $(\alpha D^\circ, X) \in \ddot{\mathbb{W}}'$. Since αD° is a cone, we have $(\emptyset, \alpha D^\circ) \in \ddot{\mathbb{W}}'$, and we obtain $(\emptyset, X) \in \ddot{\mathbb{W}}'$ (by $\langle \widetilde{T} \rangle$). By induction, we derive the result (R).

   Now, let $(X,Y) \in \ddot{\mathbb{S}}$ and let α be a vertex disjoint from Y. Suppose $(X,Y) \in \ddot{\mathbb{W}}$. Then, there exists a cell $\alpha C \in \mathbb{C}$ such that αC collapses onto $\alpha X \cup Y$ (Proposition 9). We have $(\emptyset, \alpha C) \in \ddot{\mathbb{W}}'$. By (R) we obtain $(\emptyset, \alpha X \cup Y) \in \ddot{\mathbb{W}}'$. We also have $(\emptyset, \alpha X) \in \ddot{\mathbb{W}}'$. Hence, by $\langle \widetilde{U} \rangle$, we have $(\alpha X, \alpha X \cup Y) \in \ddot{\mathbb{W}}'$. By $\langle \widetilde{Y} \rangle$, we deduce $(\alpha X \cap Y, Y) \in \ddot{\mathbb{W}}'$. Therefore $(X,Y) \in \ddot{\mathbb{W}}'$. Thus, we proved that $\ddot{\mathbb{W}} \subseteq \ddot{\mathbb{W}}'$.

2) By Proposition 11, $\ddot{\mathbb{W}}$ satisfies $\langle \widetilde{X} \rangle$ and $\langle \widetilde{Y} \rangle$. We also have $\ddot{\mathbb{C}} \subseteq \ddot{\mathbb{W}}$. So, we only have to show that $\ddot{\mathbb{W}}$ satisfies $\langle \widetilde{U} \rangle$ and $\langle \widetilde{T} \rangle$ in order to prove the inclusion $\ddot{\mathbb{W}}' \subseteq \ddot{\mathbb{W}}$.

   Let $(R,S), (S,T), (R,T) \in \ddot{\mathbb{S}}$. Suppose $(R,S) \in \ddot{\mathbb{W}}$, and let α, β be two distinct vertices disjoint from T. Let us consider the complex $U = \beta(\alpha R \cup S) \cup (\alpha R \cup T)$. By Proposition 7, we have $(\emptyset, \alpha R \cup S) \in \ddot{\mathbb{H}}$. Hence, by Proposition 4, we have $(\alpha R \cup T, U) \in \ddot{\mathbb{H}}$. But we can write $U = \alpha(\beta R) \cup (\beta S \cup T)$. Since $(\emptyset, \beta R) \in \ddot{\mathbb{H}}$, again by Proposition 4, we have $(\beta S \cup T, U) \in \ddot{\mathbb{H}}$.

   i) Suppose $(S,T) \in \ddot{\mathbb{W}}$. We have $(\emptyset, \beta S \cup T) \in \ddot{\mathbb{H}}$ (Proposition 7) and $(\beta S \cup T, U) \in \ddot{\mathbb{H}}$. Thus $(\emptyset, U) \in \ddot{\mathbb{H}}$ (since $\ddot{\mathbb{H}}$ satisfies $\langle \widetilde{T} \rangle$). We have $(\alpha R \cup T, U) \in \ddot{\mathbb{H}}$ and $(\emptyset, U) \in \ddot{\mathbb{H}}$. Thus $(\emptyset, \alpha R \cup T) \in \ddot{\mathbb{H}}$ (since $\ddot{\mathbb{H}}$ satisfies $\langle \widetilde{L} \rangle$). We obtain $(R,T) \in \ddot{\mathbb{W}}$ (Proposition 7). It follows that the collection $\ddot{\mathbb{W}}$ satisfies $\langle \widetilde{T} \rangle$.

ii) Suppose $(R, T) \in \ddot{\mathbb{W}}$. We have $(\emptyset, \alpha R \cup T) \in \ddot{\mathbb{H}}$ (Proposition 7) and $(\alpha R \cup T, U) \in \ddot{\mathbb{H}}$. Thus $(\emptyset, U) \in \ddot{\mathbb{H}}$ (since $\ddot{\mathbb{H}}$ satisfies $\langle \widetilde{T} \rangle$). We have $(\beta S \cup T, U) \in \ddot{\mathbb{H}}$ and $(\emptyset, U) \in \ddot{\mathbb{H}}$. Thus $(\emptyset, \beta S \cup T) \in \ddot{\mathbb{H}}$ (since $\ddot{\mathbb{H}}$ satisfies $\langle \widetilde{L} \rangle$). We obtain $(S, T) \in \ddot{\mathbb{W}}$ (Proposition 7). Consequently, the collection $\ddot{\mathbb{W}}$ satisfies $\langle \widetilde{U} \rangle$.   □

Therefore, the completion $\langle \widetilde{L} \rangle$ is the unique property that makes the distinction between the collections $\ddot{\mathbb{W}}$ and $\ddot{\mathbb{X}}$. Observe also the differences between $\ddot{\mathbb{W}}$ and $\ddot{\mathbb{H}}$: the collection $\ddot{\mathbb{W}}$ satisfies $\langle \widetilde{Y} \rangle$, but it no longer satisfies $\langle \widetilde{L} \rangle$. Nevertheless, as discussed before, the collection $\ddot{\mathbb{W}}$ strictly contains the collection $\ddot{\mathbb{H}}$.

Let us consider now the three following completions on $\ddot{\mathbb{S}}$.

For any $(R, S) \in \ddot{\mathbb{S}}$, $T \in \mathbb{S}$,

$\rightarrow$ If $(R, S) \in \ddot{\mathcal{K}}$ and $(S \cap T, T) \in \ddot{\mathcal{K}}$, then $(R, S \cup T) \in \ddot{\mathcal{K}}$. $\langle \widetilde{Z}1 \rangle$

$\rightarrow$ If $(R, S) \in \ddot{\mathcal{K}}$ and $(R, S \cup T) \in \ddot{\mathcal{K}}$, then $(S \cap T, T) \in \ddot{\mathcal{K}}$. $\langle \widetilde{Z}2 \rangle$

$\rightarrow$ If $(R, S \cup T) \in \ddot{\mathcal{K}}$ and $(S \cap T, T) \in \ddot{\mathcal{K}}$, then $(R, S) \in \ddot{\mathcal{K}}$. $\langle \widetilde{Z}3 \rangle$

We set $\ddot{\mathbb{Z}} = \langle \ddot{\mathbb{C}}; \widetilde{Z}1, \widetilde{Z}2, \widetilde{Z}3 \rangle$.

It has been shown that this set of completions allows for a characterization of dyads [12]. More precisely, it has been proved that we have $\ddot{\mathbb{X}} = \ddot{\mathbb{Z}}$.

The following theorem may be derived from Theorem 5 and from arguments used in four proofs given in [12] (Proofs of Prop. 1, 2, 3 and Th. 2).

Theorem 6. *We have* $\ddot{\mathbb{W}} = \langle \ddot{\mathbb{C}}; \widetilde{Z}1, \widetilde{Z}2 \rangle$.

Thus, Theorem 6 provides another characterization of contractibility by means of only two global properties.

8   Collapsible Pairs

In this section, we show that the three completions $\langle \widetilde{X} \rangle$, $\langle \widetilde{Y} \rangle$, $\langle \widetilde{T} \rangle$ allow us to describe collapsibility.

Let $(X, Y) \in \ddot{\mathbb{S}}$. If α is a vertex disjoint from Y, we say that $(\alpha X, \alpha Y)$ is a *cone pair*. We denote by $\ddot{\triangle}$ the collection composed of all cone pairs.

Observe that we have $\ddot{\mathbb{C}} \subseteq \ddot{\triangle}$. In [11], it was shown that the collection $\ddot{\triangle}$ is closed under duality, but not the collection $\ddot{\mathbb{C}}$.

In the previous sections, we have often considered $\ddot{\mathbb{C}}$ as the starting collection of our completion systems. In the following, we will replace $\ddot{\mathbb{C}}$ by $\ddot{\triangle}$ in order to obtain an exact characterization of collapsibility.

We set $\ddot{\mathbb{E}} = \langle \ddot{\triangle}; \widetilde{X}, \widetilde{T} \rangle$. Each couple of $\ddot{\mathbb{E}}$ is a *collapsible pair*.

Proposition 12. *We have* $(X, Y) \in \ddot{\mathbb{E}}$ *if and only if* Y *collapses onto* X.

Proof

i) Suppose X is an elementary collapse of Y. If $X = \emptyset$, we are done. Otherwise, there exists a cell D, $D \neq \emptyset$, such that $Y = X \cup \alpha D$ and $X \cap \alpha D = \alpha D°$, where α is a vertex disjoint from D. The couple $(\alpha D°, \alpha D)$ being a cone pair, we have $(\alpha D°, \alpha D) \in \ddot{\mathbb{E}}$. By $\langle \widetilde{X} \rangle$, since $X \cap \alpha D = \alpha D°$, we obtain $(X, X \cup \alpha D) \in \ddot{\mathbb{E}}$. Thus, $(X, Y) \in \ddot{\mathbb{E}}$. By induction, we have $(X, Y) \in \ddot{\mathbb{E}}$ whenever Y collapses onto X.

ii) If $(\alpha X, \alpha Y) \in \ddot{\triangle}$, then αY collapses onto αX. Furthermore:
   - If T collapses onto $S \cap T$, then $S \cup T$ collapses onto S.
   - If T collapses onto S, and S collapses onto R, then T collapses onto R.
   By induction on $\langle \widetilde{X} \rangle$ and $\langle \widetilde{T} \rangle$, the complex Y collapses onto X if $(X, Y) \in \ddot{\mathbb{E}}$. $\qquad\square$

Thus, we have $(\emptyset, X) \in \ddot{\mathbb{E}}$ if and only if X collapses onto $\emptyset$. In other words, the kernel of $\ddot{\mathbb{E}}$ is precisely made of all collapsible complexes.

The following corollary is a direct consequence of Proposition 12.

Corollary 3. *We have* $\ddot{\mathbb{E}} = \langle \ddot{\mathbb{I}}; I^- \rangle = \langle \ddot{\mathbb{I}}; O^+ \rangle = \langle \ddot{\mathbb{I}}; O^+, I^- \rangle$.

In addition, the following proposition shows that $\ddot{\mathbb{E}}$ satisfies the property $\langle \widetilde{Y} \rangle$.

Proposition 13. *We have* $\ddot{\mathbb{E}} = \langle \ddot{\triangle}; \widetilde{X}, \widetilde{Y}, \widetilde{T} \rangle$.

Proof. Let $(S, S \cup T) \in \ddot{\mathbb{E}}$. The complex $S \cup T$ collapses onto S (Proposition 12). Thus, T collapses onto $S \cap T$. Therefore $(S \cap T, T) \in \ddot{\mathbb{E}}$ (Again by Proposition 12). It means that the collection $\ddot{\mathbb{E}}$ satisfies the property $\langle \widetilde{Y} \rangle$. Thus $\ddot{\mathbb{E}} = \langle \ddot{\triangle}; \widetilde{X}, \widetilde{Y}, \widetilde{T} \rangle$. $\qquad\square$

The dunce hat [16] provides an example of a complex X that is contractible, but not collapsible. Since X is contractible, there exists a complex Y, which is collapsible, and which collapses onto X (a consequence of Proposition 3 and Theorem 3). Thus, we have $(X, Y) \in \ddot{\mathbb{E}}$ and $(\emptyset, Y) \in \ddot{\mathbb{E}}$, but $(\emptyset, X) \notin \ddot{\mathbb{E}}$. This shows that the collection $\ddot{\mathbb{E}}$ does not satisfy the property $\langle \widetilde{L} \rangle$. By considering the dual of the dunce hat in a cell, we obtain an example which shows that the collection $\ddot{\mathbb{E}}$ does not satisfy the property $\langle \widetilde{U} \rangle$.

Thanks to Proposition 12, we easily derive the following.

Proposition 14. *We have* $\ddot{\mathbb{H}} = \langle \ddot{\triangle}; \widetilde{X}, \widetilde{T}, \widetilde{L} \rangle$.

Proof. Let $\ddot{\mathbb{K}} = \langle \ddot{\triangle}; \widetilde{X}, \widetilde{T}, \widetilde{L} \rangle$. Clearly, we have $\ddot{\triangle} \subseteq \ddot{\mathbb{H}}$. Furthermore, we have $\ddot{\mathbb{H}} = \langle \ddot{\mathbb{C}}; \widetilde{X}, \widetilde{T}, \widetilde{U}, \widetilde{L} \rangle$ (Theorem 4). Thus $\ddot{\mathbb{K}} \subseteq \ddot{\mathbb{H}}$. Now, let $(X, Y) \in \ddot{\mathbb{H}}$. By Theorem 3, there exists a complex Z such that Z collapses onto X and Z collapses onto Y. The collection $\ddot{\mathbb{K}}$ contains the collection $\langle \ddot{\triangle}; \widetilde{X}, \widetilde{T} \rangle$. Thus, by Proposition 12, we have $(X, Z) \in \ddot{\mathbb{K}}$ and $(Y, Z) \in \ddot{\mathbb{K}}$. By $\langle \widetilde{L} \rangle$, it means that $(X, Y) \in \ddot{\mathbb{K}}$. $\qquad\square$

9   Conclusion

In a previous work, it was shown that only five completions $\langle \widetilde{X} \rangle$, $\langle \widetilde{Y} \rangle$, $\langle \widetilde{T} \rangle$, $\langle \widetilde{U} \rangle$, and $\langle \widetilde{L} \rangle$, were sufficient to describe the whole collection $\ddot{X}$ made of all acyclic pairs of complexes. Furthermore, it has been found that a subset of these completions permits to characterize the collection $\ddot{H}$ made of all homotopic pairs.

In this paper, we show that another subset of these completions corresponds to a collection $\ddot{W}$, the couples of which are linked by a contractibility relation. These couples, we called contractible pairs, may be generated by two local transforms, perforations and fillings, completed by collapses. We also show that a certain collection $\ddot{E}$ provides an exact characterization of all collapsible pairs.

Thus, these five completions allow us to describe, in a unified framework, the four remarkable nested collections composed of all collapsible pairs, homotopic pairs, contractible pairs, and acyclic pairs. Not only these completions may act as generators of these collections, but also they provide structural and global properties of them.

References

1. Whitehead, J.H.C.: Simplicial spaces, nuclei, and m-groups. Proc. London Math. Soc. **2**(45), 243–327 (1939)
2. Björner, A.: Topological methods. In: Graham, R., Grötschel, M., Lovász, L. (eds.) Handbook of Combinatorics, North-Holland, Amsterdam, pp. 1819–1872 (1995)
3. Jonsson, J.: Simplicial Complexes of Graphs. Springer, Hidelberg (2008). https://doi.org/10.1007/978-3-540-75859-4
4. Kalai, G.: Enumeration of Q-acyclic simplicial complexes. Israel J. Math. **45**(4), 337–351 (1983)
5. Yung Kong, T.: Topology-preserving deletion of 1's from 2-, 3- and 4-dimensional binary images. In: Ahronovitz, E., Fiorio, C. (eds.) DGCI 1997. LNCS, vol. 1347, pp. 3–18. Springer, Heidelberg (1997). https://doi.org/10.1007/BFb0024826
6. Couprie, M., Bertrand, G.: New characterizations of simple points in 2D, 3D and 4D discrete spaces. IEEE Trans. Pattern Anal. Mach. Intell. **31**(4), 637–648 (2009)
7. Bertrand, G.: On critical kernels. Comptes Rendus de l'Académie des Sciences, Série Math. **345**, 363–367 (2007)
8. Rosenfeld, A.: Digital topology. Amer. Math. Monthly **621–630** (1979)
9. Kovalevsky, V.: Finite topology as applied to image analysis. Comput. Vis. Graph. Image Proc. **46**, 141–161 (1989)
10. Kong, T.Y., Rosenfeld, A.: Digital topology: introduction and survey. Comput. Vis. Graph. Image Proc. **48**, 357–393 (1989)
11. Bertrand, G.: Completions and simplicial complexes, HAL-00761162 (2012)
12. Bertrand, G.: New structures based on completions. In: Gonzalez-Diaz, R., Jimenez, M.-J., Medrano, B. (eds.) DGCI 2013. LNCS, vol. 7749, pp. 83–94. Springer, Heidelberg (2013). https://doi.org/10.1007/978-3-642-37067-0_8
13. Bertrand, G.: Completions and simple homotopy. In: Barcucci, E., Frosini, A., Rinaldi, S. (eds.) DGCI 2014. LNCS, vol. 8668, pp. 63–74. Springer, Cham (2014). https://doi.org/10.1007/978-3-319-09955-2_6
14. Giblin, P.: Graphs, Surfaces and Homology. Chapman and Hall, Boca Raton (1981)

15. Bing, R.H.: Some aspects of the topology of 3-manifolds related to the Poincaré Conjecture, Lectures on Modern Mathematics II. Wiley **93–128** (1964)
16. Zeeman, E.C.: On the dunce hat. Topology **2**, 341–358 (1964)
17. Hatcher, A.: Algebraic Topology. Cambridge University Press, Cambridge (2001)

Body Centered Cubic Grid - Coordinate System and Discrete Analytical Plane Definition

Lidija Čomić[1], Rita Zrour[2(✉)], Gaëlle Largeteau-Skapin[2], Ranita Biswas[3], and Eric Andres[2]

[1] Faculty of Technical Sciences, University of Novi Sad, Novi Sad, Serbia
comic@uns.ac.rs
[2] University of Poitiers, Laboratory XLIM, ASALI, UMR CNRS 7252, BP 30179, 86962 Futuroscope Chasseneuil, France
{rita.zrour,gaelle.largeteau.skapin,eric.andres}@univ-poitiers.fr
[3] Institute of Science and Technology (IST) Austria, 3400 Klosterneuburg, Austria
ranita.biswas@ist.ac.at

Abstract. We define a new compact coordinate system in which each integer triplet addresses a voxel in the BCC grid, and we investigate some of its properties. We propose a characterization of 3D discrete analytical planes with their topological features (in the Cartesian and in the new coordinate system) such as the interrelation between the thickness of the plane and the separability constraint we aim to obtain.

Keywords: Discrete geometry · BCC grid · 3D coordinate system · Discrete analytical plane

1 Introduction

The cubic grid is the only regular grid in the 3D space, with the alternative body centered cubic (BCC), face-centered cubic (FCC) and diamond cubic grids receiving an increasing amount of research attention. Due to their properties, such as better packing densities [9] (with applications e.g. in the design of error-correcting codes) and larger number of face directions (in comparison with the usual cubic grid), the FCC and BCC grids have several advantages in various fields including discretization, ray tracing and ray casting, volume rendering and repairing [5–7,10,14–16], to name just a few.

The retrieval of adjacency relations between grid voxels can be achieved through various data structures proposed in the literature, or through a definition of suitable coordinate systems. Both 3- and 4-valued coordinate systems have been developed for the centers of voxels in the BCC and FCC grids [4,12,13,17,18,20]. In the Cartesian coordinates, each voxel in the BCC grid has three even or three odd coordinates, while the sum of coordinates is even for voxels in the FCC grid.

© Springer Nature Switzerland AG 2021
J. Lindblad et al. (Eds.): DGMM 2021, LNCS 12708, pp. 152–163, 2021.
https://doi.org/10.1007/978-3-030-76657-3_10

We propose a new non-orthogonal 3-valued coordinate system for $\mathbb{R}^3$. When restricted to points in $\mathbb{Z}^3$ with all three coordinates of the same parity, it defines a compact integer-valued 3-coordinate system in which each voxel in the BCC grid is represented through three integer coordinates and each integer triplet addresses some BCC voxel. This new coordinate system may be viewed as a bijective mapping between the BCC and the cubic grid, which will facilitate the work on the BCC grid, e.g. by eliminating the need to check the input coordinate triplet for consistency (whether it corresponds to a BCC voxel or not).We define discrete analytical planes in the BCC grid, both in the Cartesian and in the new coordinate system.

2 Preliminaries

In this section, we recall some basic definitions relevant to our problem.

2.1 Discrete Analytical Planes in the Cubic Grid

The cubic grid is the set $\mathbb{Z}^3$ of points with three integer coordinates. With each point we can associate its Voronoi region, which is a cube.

A discrete analytical object in the cubic grid is a set of cubes. The object is l-connected, $l \in \{0, 1, 2\}$, if there is an l-path (a sequence of cubes where two consecutive cubes are l-adjacent) consisting of object cubes, connecting any two object cubes. The 0, 1 and 2-connectivities correspond to the usual 26, 18 and 6-connectivities, respectively. A discrete analytical object is l-separating if its complement is not l-connected.

Definition 1 (Discrete Analytical Plane [19]). *The discrete analytical plane* P_ρ *of thickness* $\rho \in \mathbb{R}^+$, *corresponding to the Euclidean plane* $P : ax + by + cz + d = 0$, *is the set of cubes centered at* $(x, y, z) \in \mathbb{Z}^3$ *such that*

$$-\frac{\rho}{2} \le ax + by + cz + d < \frac{\rho}{2},$$

Different values of ρ in this definition lead to different digitization models with respect to connectedness and separability.

Definition 2 (Discrete Analytical Standard Plane [1–3,11]). *The discrete analytical standard plane is a discrete analytical plane with*

$$\rho = |a| + |b| + |c|.$$

Equivalently, the discrete analytical standard plane can be defined as the set of the cubes which have a non-empty intersection with the Euclidean plane P (which do not have all the vertices in one of the two open halfspaces defined by P), with the exception of the cubes which have vertices either on P or in the halfspace defined by P into which the normal vector (a, b, c) points (to account for one of the two inequalities being strict). It is 2-connected and 0-separating.

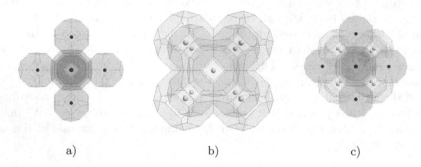

a)   b)   c)

Fig. 1. A truncated octahedron with its a) six quad-neighbors, b) eight hex-neighbors and c) both hex- and quad-neighbors.

Definition 3 (Discrete Analytical Naive Plane [1,3,8]**).** *The discrete analytical naive plane is a discrete analytical plane with*

$$\rho = \max\{|a|, |b|, |c|\}.$$

Equivalently, the discrete analytical naive plane can be defined as the set of the cubes which do not have the centers of all the faces in one of the two open halfspaces defined by P, with the exception of those cubes which have the face centers either on P or in the halfspace defined by P into which the normal vector (a, b, c) points. It is 0-connected and 2-separating.

2.2 The BCC Grid

The grid points in the BCC grid are either at the center or at a vertex of a cube in the cubic grid. The grid points partition the 3D space into Voronoi regions, which are truncated octahedra, with eight regular hexagonal faces and six quadrangular (square) faces, 36 edges and 24 vertices. Each edge is shared by three octahedra, and each vertex is shared by four octahedra. Two octahedra in the BCC grid are either disjoint or they share an entire (hexagonal or quadrangular) face. Each octahedron in the BCC grid has eight hex-neighbors (one across each of its hexagonal faces) and six quad-neighbors (one across each of its quadrangular faces), as shown in Fig. 1.

After rescaling by factor two, (the centers of) all octahedra have integer Cartesian coordinates: the (even) octahedra centered at cube centers have three even coordinates and the (odd) octahedra centered at cube vertices have three odd coordinates.

3   The New Coordinate System

We propose a new non-orthogonal coordinate system for $\mathbb{R}^3$, which is integer-valued on the octahedra in the BCC grid, providing a bijective correspondence

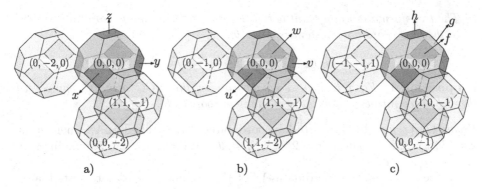

Fig. 2. A neighborhood of the truncated octahedron $(0,0,0)$, with octahedra labeled by their a) Cartesian b) new and c) coordinates in the system by He et al.

between triplets of integers and truncated octahedra. We review the alternative coordinate system proposed by He et al. [12], we show how the neighbors of each octahedron can be retrieved from the octahedron coordinates through simple integer operations and we derive the equation of the Euclidean plane in the new coordinates.

3.1   Definition

We make a rescaling in the x and y directions by the factor 2, we slant the z axis to pass through one of the vertices of the cube centered at the origin and we make a rescaling in the z direction by the factor $\sqrt{3}$ (see Fig. 2 (b)). Thus, the linear transformation between the Cartesian and the new coordinate system is a composition of non-uniform scaling and shear. We denote the Cartesian coordinates by (x, y, z), and the new ones by (u, v, w). We will write the coordinates both in the row and in the column form.

The transformation matrix from the new to the Cartesian coordinate system is

$$M_{\mathcal{NC}} = \begin{bmatrix} 2 & 0 & 1 \\ 0 & 2 & 1 \\ 0 & 0 & 1 \end{bmatrix},$$

and from the Cartesian to the new coordinate system is its inverse

$$M_{\mathcal{CN}} = M_{\mathcal{NC}}^{-1} = 1/2 \begin{bmatrix} 1 & 0 & -1 \\ 0 & 1 & -1 \\ 0 & 0 & 2 \end{bmatrix}.$$

Definition 1.   – *The transformation of a point $p(u, v, w)$ from the new coordinate system to the Cartesian coordinate system is given by*

$$\mathcal{C}(p) = (2u + w, 2v + w, w).$$

– *The transformation of a point $p(x, y, z)$ from the Cartesian coordinate system to the new coordinate system is given by:*

$$\mathcal{N}(p) = ((x - z)/2, (y - z)/2, z).$$

Proposition 1. *Each truncated octahedron in the BCC grid is represented through three integer coordinates in the new coordinate system.*

Proof. Let (x, y, z) be the Cartesian coordinates of an octahedron V. Then either $x = 2k$, $y = 2l$, $z = 2m$ or $x = 2k + 1$, $y = 2l + 1$, $z = 2m + 1$ for some integers k, l and m.

In the first case, the coordinates V' of V in the new coordinate system are

$$V' = M_{\mathcal{CN}}V = 1/2 \begin{bmatrix} 1 & 0 & -1 \\ 0 & 1 & -1 \\ 0 & 0 & 2 \end{bmatrix} \begin{bmatrix} 2k \\ 2l \\ 2m \end{bmatrix} = \begin{bmatrix} k - m \\ l - m \\ 2m \end{bmatrix}.$$

In the second case,

$$V' = M_{\mathcal{CN}}V = 1/2 \begin{bmatrix} 1 & 0 & -1 \\ 0 & 1 & -1 \\ 0 & 0 & 2 \end{bmatrix} \begin{bmatrix} 2k + 1 \\ 2l + 1 \\ 2m + 1 \end{bmatrix} = \begin{bmatrix} k + 1/2 - m - 1/2 \\ l + 1/2 - m - 1/2 \\ 2m + 1 \end{bmatrix} = \begin{bmatrix} k - m \\ l - m \\ 2m + 1 \end{bmatrix}.$$

In both cases, V' has integer coordinates. As a byproduct, we see that we can distinguish between the even and odd octahedra in the new coordinate system through the last coordinate: it is even for even octahedra and odd for odd ones.

Proposition 2. *Each integer triplet represents an octahedron in the BCC grid in the new coordinate system.*

Proof. Let $V' = (u, v, w)$ be a triplet of integers. Then

$$V = M_{\mathcal{NC}}V' = \begin{bmatrix} 2 & 0 & 1 \\ 0 & 2 & 1 \\ 0 & 0 & 1 \end{bmatrix} \begin{bmatrix} u \\ v \\ w \end{bmatrix} = \begin{bmatrix} 2u + w \\ 2v + w \\ w \end{bmatrix}.$$

If w is even, then all three coordinates of V are even; if w is odd, all three coordinates are odd. Thus, V represents a BCC octahedron in the Cartesian coordinates.

3.2   Comparison with the Coordinate System by He et al.

Another non-orthogonal coordinate system has been proposed by He et al. [12], with the aim of adapting the Bresenham line drawing algorithm to the BCC grid. In it, the basis vectors are $(1, 1, 1)$, $(-1, 1, 1)$ and $(0, 1, 2)$, and the transformation matrix from the coordinate system by He et al. to the Cartesian coordinate system is

$$M_{\mathcal{HC}} = \begin{bmatrix} 1 & -1 & 0 \\ 1 & 1 & 0 \\ 1 & 1 & 2 \end{bmatrix}.$$

Thus, the coordinate axes f, g and h pass through the centers of two adjacent hex-faces of the voxel centered at the origin, and the center of a quad-face adjacent to both, respectively. The rescaling factor for the axes f and g is $\sqrt{3}$, and for h it is 2.

The transformation matrix from the Cartesian to the coordinate system by He et al. is

$$M_{\mathcal{CH}} = 1/2 \begin{bmatrix} 1 & 1 & 0 \\ -1 & 1 & 0 \\ 0 & -1 & 1 \end{bmatrix}.$$

Similarly to our coordinate system, it provides a bijection between integer triplets and truncated octahedra but it is less regular than ours (as no two of its axes are orthogonal) and it requires more arithmetical operations to pass to and from the Cartesian coordinates (as the number of non-zeroes is greater in the corresponding transformation matrices).

In Fig. 2, we illustrate a neighborhood of the octahedron centered at $(0,0,0)$, with octahedra labeled by their coordinate values in the Cartesian, the new, and the coordinate system by He et al..

3.3 Adjacency Relation

Each BCC octahedron has fourteen neighbors: six across its quad-faces, and eight across its hex-faces.

Proposition 3. *The six quad-neighbors of an octahedron $V = (u, v, w)$ are $(u \pm 1, v, w)$, $(u, v \pm 1, w)$, $(u - 1, v - 1, w + 2)$ and $(u + 1, v + 1, w - 2)$.*

Proof. In the Cartesian coordinates, the six neighbors of an (even or odd) octahedron (x, y, z) are $(x \pm 2, y, z)$, $(x, y \pm 2, z)$ and $(x, y, z \pm 2)$, i.e., the translation vectors from (x, y, z) to its quad-neighbors are $(\pm 2, 0, 0)$, $(0, \pm 2, 0)$ and $(0, 0, \pm 2)$. In the new coordinate system, the translation vectors are

$$\begin{bmatrix} \frac{1}{2} & 0 & -\frac{1}{2} \\ 0 & \frac{1}{2} & -\frac{1}{2} \\ 0 & 0 & 1 \end{bmatrix} \begin{bmatrix} \pm 2 \\ 0 \\ 0 \end{bmatrix} = \begin{bmatrix} \pm 1 \\ 0 \\ 0 \end{bmatrix},$$

$$\begin{bmatrix} \frac{1}{2} & 0 & -\frac{1}{2} \\ 0 & \frac{1}{2} & -\frac{1}{2} \\ 0 & 0 & 1 \end{bmatrix} \begin{bmatrix} 0 \\ \pm 2 \\ 0 \end{bmatrix} = \begin{bmatrix} 0 \\ \pm 1 \\ 0 \end{bmatrix},$$

$$\begin{bmatrix} \frac{1}{2} & 0 & -\frac{1}{2} \\ 0 & \frac{1}{2} & -\frac{1}{2} \\ 0 & 0 & 1 \end{bmatrix} \begin{bmatrix} 0 \\ 0 \\ \pm 2 \end{bmatrix} = \begin{bmatrix} \mp 1 \\ \mp 1 \\ \pm 2 \end{bmatrix}.$$

Proposition 4. *The eight hex-neighbors of an octahedron* $V = (u, v, w)$ *are* $(u, v, w\pm 1)$, $(u+1, v+1, w-1)$, $(u-1, v-1, w+1)$, $(u+1, v, w-1)$, $(u-1, v, w+1)$, $(u, v+1, w-1)$ *and* $(u, v-1, w+1)$.

Proof. The eight translation vectors for hex-neighbors in the Cartesian coordinates are $(t_1, t_2, t_3) \in \{-1, 1\}^3$. As for quad-neighbors, the transformed coordinates of these vectors are translation vectors in the new coordinate system.

3.4 Distance

Observation 1 (Euclidean Distance). *The Euclidean distance* $d_2(A, B)$ *between two points* $A(a_u, a_v, a_w)_N$ *and* $B(b_u, b_v, b_w)_N$ *in the new coordinate system is equal to*

$$d_2(A, B) = \sqrt{(2(a_u - b_u) + (a_w - b_w))^2 + (2(a_v - b_v) + (a_w - b_w))^2 + (a_w - b_w)^2}.$$

The Cartesian coordinates of the two points are $A(2a_u + a_w, 2a_v + a_w, a_w)_C$ and $B(2b_u + b_w, 2b_v + b_w, b_w)_C$, and the Euclidean distance between them is equal to $d_2(A, B)$.

3.5 Euclidean Plane Equation

We give the formulas for converting Euclidean plane equations between the new and the Cartesian coordinate system.

Observation 2 (New to Cartesian Plane Equation). *Let* $\alpha u + \beta v + \gamma w + \delta = 0$ *be a plane equation in the new coordinate system. The equation of the same plane in the Cartesian coordinate system is given by:*

$$\frac{\alpha}{2}x + \frac{\beta}{2}y + (-\frac{\alpha}{2} - \frac{\beta}{2} + \gamma)z + \delta = 0. \tag{1}$$

Observation 3 (Cartesian to New Plane Equation). *Let* $ax + by + cz + d = 0$ *be a plane equation in the Cartesian coordinate system. The equation of the same plane in the new coordinate system is given by:*

$$2au + 2bv + (a + b + c)w + d = 0. \tag{2}$$

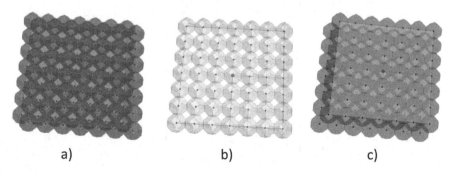

Fig. 3. The discrete analytical a) standard plane b) hex-plane and c) quad-plane $z = 0$ (in the Cartesian coordinates), i.e., $w = 0$ (in the new coordinates).

4   Discrete Analytical Plane in the BCC Grid

We define the discrete analytical plane in the BCC grid, both in the Cartesian and in the new coordinate system, using the classical definition (to our knowledge there is no previous work defining the plane and its topology control for the BCC grid).

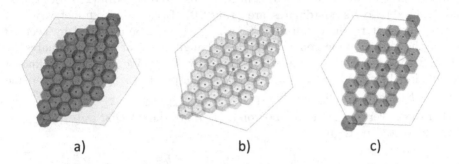

Fig. 4. The discrete analytical a) standard plane b) hex-plane and c) quad-plane $x + y + z = 0$ (in Cartesian coordinates).

4.1   Discrete Analytical Plane in the Cartesian Coordinates

Definition 4 (Discrete Analytical Plane). *Let $ax + by + cz + d = 0$ be a plane equation in the Cartesian coordinate system. The discrete analytical plane P_ρ of thickness ρ in the BCC grid is the set of octahedra with centers (x, y, z) satisfying the following inequality:*

$$-\frac{\rho}{2} \le ax + by + cz + d < \frac{\rho}{2}. \tag{3}$$

We can control the topology of the digital plane by changing the value of ρ.

Definition 5. *The discrete analytical standard plane corresponding to the Euclidean plane $P : ax + by + cz + d = 0$ is the set of octahedra that do not have all the vertices in one of the two open halfspaces defined by P, with the exception of octahedra which have all vertices either on P or in the halfspace defined by P into which the normal vector (a, b, c) points.*

Proposition 1. *The discrete analytical standard plane corresponding to the Euclidean plane $ax + by + cz + d = 0$ is obtained using:*

$$\rho_{St} = 2 \max\{|b| + \frac{|c|}{2}, \frac{|b|}{2} + |c|, |a| + \frac{|b|}{2}, \frac{|a|}{2} + |b|, |a| + \frac{|c|}{2}, \frac{|a|}{2} + |c|\}. \tag{4}$$

Proof. Let $ax + by + cz + d = 0$ be a Euclidean plane in the Cartesian coordinate system. We shall find the signed distances between all 24 vertices of the octahedron and the plane.

All 24 octahedron vertices are at the same distance from the octahedron center. Each vertex is incident to exactly four octahedra, and is equidistant from the four octahedra centers. Thus, we can find the vertex coordinates as the arithmetic mean of the four centers. For example, one of the 24 vertices incident to the octahedron $(0,0,0)$ is incident also to octahedra $(2,0,0)$, $(1,1,1)$ and $(1,1,-1)$, so its coordinates are $(1,1/2,0)$. Because of symmetry, all 24 vertices incident to the octahedron $(x,y,z) \in \mathbb{Z}^3$ can be obtained by keeping one of the three octahedron coordinates, changing one of the coordinates by ± 1, and changing the remaining coordinate by $\pm 1/2$. We use the well-known formula for the signed distance $d(p,P)$ between a point $p(p_x, p_y, p_z)$ and a plane $P : ax + by + cz + d = 0$, namely $d(p,P) = \frac{ap_x + bp_y + cp_z + d}{\sqrt{a^2 + b^2 + c^2}}$.

For the 24 vertices of the octahedron (x_0, y_0, z_0), the set $Dist = \{d_i\}_{i=1,2,\ldots,24}$ of the 24 distances is given by:

$$
\begin{aligned}
Dist = \{ & d_0 + \frac{-b - \frac{c}{2}}{r}, d_0 + \frac{-b + \frac{c}{2}}{r}, d_0 + \frac{b - \frac{c}{2}}{r}, d_0 + \frac{b + \frac{c}{2}}{r}, \\
& d_0 + \frac{-\frac{b}{2} - c}{r}, d_0 + \frac{-\frac{b}{2} + c}{r}, d_0 + \frac{\frac{b}{2} - c}{r}, d_0 + \frac{+\frac{b}{2} + c}{r}, \\
& d_0 + \frac{-a - \frac{b}{2}}{r}, d_0 + \frac{-a + \frac{b}{2}}{r}, d_0 + \frac{a - \frac{b}{2}}{r}, d_0 + \frac{a + \frac{b}{2}}{r}, \\
& d_0 + \frac{-\frac{a}{2} - b}{r}, d_0 + \frac{-\frac{a}{2} + b}{r}, d_0 + \frac{\frac{a}{2} - b}{r}, d_0 + \frac{\frac{a}{2} + b}{r}, \\
& d_0 + \frac{-a - \frac{c}{2}}{r}, d_0 + \frac{-a + \frac{c}{2}}{r}, d_0 + \frac{a - \frac{c}{2}}{r}, d_0 + \frac{a + \frac{c}{2}}{r}, \\
& d_0 + \frac{-\frac{a}{2} - c}{r}, d_0 + \frac{-\frac{a}{2} + c}{r}, d_0 + \frac{\frac{a}{2} - c}{r}, d_0 + \frac{\frac{a}{2} + c}{r} \}
\end{aligned}
$$

where $d_0 = \frac{ax_0 + by_0 + cz_0 + d}{\sqrt{a^2 + b^2 + c^2}}$ and $r = \sqrt{a^2 + b^2 + c^2}$. If the octahedron and the plane cross, then

$$\min\{d_i\} < 0 \leq \max\{d_i\}, \qquad i = 1, 2, \ldots, 24. \tag{5}$$

We have that

$$\min\{d_i\} = d_0 - \frac{1}{r}\max\{|b| + \frac{|c|}{2}, \frac{|b|}{2} + |c|, |a| + \frac{|b|}{2}, \frac{|a|}{2} + |b|, |a| + \frac{|c|}{2}, \frac{|a|}{2} + |c|\} = d_0 - \frac{\rho_{St}}{2r},$$

$$\max\{d_i\} = d_0 + \frac{1}{r}\max\{|b| + \frac{|c|}{2}, \frac{|b|}{2} + |c|, |a| + \frac{|b|}{2}, \frac{|a|}{2} + |b|, |a| + \frac{|c|}{2}, \frac{|a|}{2} + |c|\} = d_0 + \frac{\rho_{St}}{2r}.$$

Therefore, Eq. 5 becomes

$$-\frac{\rho_{St}}{2r} \leq d_0 < \frac{\rho_{St}}{2r}, \tag{6}$$

i.e.,

$$-\frac{\rho_{St}}{2} \leq ax_0 + by_0 + cz_0 + d < \frac{\rho_{St}}{2}. \tag{7}$$

Definition 6. *The discrete analytical (naive) hex-plane corresponding to the Euclidean plane $P : ax + by + cz + d = 0$ is the set of octahedra that do not have the centers of all hex-faces in one of the two open halfspaces defined by P, with the exception of octahedra which have the hex-face centers either on P or in the halfspace defined by P into which the normal vector (a, b, c) points.*

Proposition 2. *The discrete analytical hex-plane corresponding to the Euclidean plane $ax + by + cz + d = 0$ in the Cartesian coordinate system is obtained with:*

$$\rho_{hex} = |a| + |b| + |c| . \tag{8}$$

Proof. The proof follows the same reasoning as the proof of Proposition 1, considering the signed distance from the plane to the centers of the hex-faces. For an octahedron (x_0, y_0, z_0) in the Cartesian coordinates, the centers of hex-faces are $(x_0 + \frac{1}{2}, y_0 - \frac{1}{2}, z_0 \pm \frac{1}{2})$, $(x_0 + \frac{1}{2}, y_0 + \frac{1}{2}, z_0 \pm \frac{1}{2})$, $(x_0 - \frac{1}{2}, y_0 - \frac{1}{2}, z_0 \pm \frac{1}{2})$, $(x_0 - \frac{1}{2}, y_0 + \frac{1}{2}, z_0 \pm \frac{1}{2})$. The discrete analytical hex-plane $ax + by + cz + d = 0$ in the Cartesian coordinates is the collection of octahedra (x_0, y_0, z_0) satisfying

$$-\frac{\rho_{hex}}{2} \leq ax_0 + by_0 + cz_0 + d < \frac{\rho_{hex}}{2}. \tag{9}$$

Definition 7. *The discrete analytical (naive) quad-plane corresponding to the Euclidean plane $P : ax + by + cz + d = 0$ is the set of octahedra that do not have the centers of all quad-faces in one of the two open halfspaces defined by P, with the exception of octahedra which have the quad-face centers either on P or in the halfspace defined by P into which the normal vector (a, b, c) points.*

Proposition 3. *The discrete analytical quad-plane $ax + by + cz + d = 0$ in the Cartesian coordinates is obtained with:*

$$\rho_{quad} = 2 \max(|a| , |b| , |c|). \tag{10}$$

Proof. The signed distance from the plane to the centers of the quad-faces is computed. For an octahedron (x_0, y_0, z_0) in the Cartesian coordinates, the centers of quad-faces are $(x_0 \pm 1, y_0, z_0)$, $(x_0, y_0 \pm 1, z_0)$ and $(x_0, y_0, z_0 \mp 1)$.

The discrete analytical quad-plane $ax + by + cz + d = 0$ in the Cartesian coordinates is the collection of octahedra (x_0, y_0, z_0) satisfying

$$-\frac{\rho_{quad}}{2} \leq ax_0 + by_0 + cz_0 + d < \frac{\rho_{quad}}{2}. \tag{11}$$

4.2 Discrete Analytical Plane in the New Coordinate System

Following the same reasoning as in the previous section, we define the plane in the new coordinate system. Due to lack of space, we only give the final equations that can be proven directly in the new coordinate system or deduced from the equations in the Cartesian system.

Proposition 4. *The discrete analytical standard plane corresponding to the Euclidean plane $\alpha u + \beta v + \gamma w + \delta = 0$ in the new coordinate system is obtained with:*

$$\rho_{St_N} = 2\max\{\left|\frac{\beta}{2}\right| + \frac{\left|-\frac{\alpha}{2} - \frac{\beta}{2} + \gamma\right|}{2}, \left|\frac{\beta}{4}\right| + \left|-\frac{\alpha}{2} - \frac{\beta}{2} + \gamma\right|, \left|\frac{\alpha}{2}\right| + \left|\frac{\beta}{4}\right|,$$

$$\left|\frac{\alpha}{4}\right| + \left|\frac{\beta}{2}\right|, \left|\frac{\alpha}{2}\right| + \frac{\left|-\frac{\alpha}{2} - \frac{\beta}{2} + \gamma\right|}{2}, \left|\frac{\alpha}{4}\right| + \left|-\frac{\alpha}{2} - \frac{\beta}{2} + \gamma\right|\}.$$

Proposition 5. *The discrete analytical hex-plane corresponding to the Euclidean plane $\alpha x + \beta y + \gamma z + \delta = 0$ in the new coordinate system is obtained with:*

$$\rho_{hex_N} = \left|\frac{\alpha}{2}\right| + \left|\frac{\beta}{2}\right| + \left|-\frac{\alpha}{2} - \frac{\beta}{2} + \gamma\right|. \tag{12}$$

Proposition 6. *The discrete analytical quad-plane corresponding to the Euclidean plane $\alpha x + \beta y + \gamma z + \delta = 0$ in the new coordinate system is obtained with:*

$$\rho_{quad_N} = 2\max\{\left|\frac{\alpha}{2}\right|, \left|\frac{\beta}{2}\right|, \left|-\frac{\alpha}{2} - \frac{\beta}{2} + \gamma\right|\}. \tag{13}$$

5   Summary and Future Work

We have defined a new non-orthogonal coordinate system for $\mathbb{R}^3$, which induces a bijection between the centers of BCC voxels and integer triplets. We have defined discrete analytical planes in the BCC grid with standard, hex and quad thickness. In future work, we want to define discrete analytical spheres and to study further the topological properties of these discrete analytical objects. We also plan to study some graphical transforms in the BCC grid, such as rotation, and to implement them using the proposed coordinate system.

Acknowledgement. This work has been partially supported by the Ministry of Education, Science and Technological Development of the Republic of Serbia through the project no. 451-03-68/2020-14/200156: "Innovative scientific and artistic research from the FTS (activity) domain" (LČ), the European Research Council (ERC) under the European Union's Horizon 2020 research and innovation programme, grant no. 788183 (RB), and the DFG Collaborative Research Center TRR 109, 'Discretization in Geometry and Dynamics', Austrian Science Fund (FWF), grant no. I 02979-N35 (RB).

References

1. Andres, E.: Le plan discret. Actes du 3eme Colloque Geometrie discrete en imagerie: fondements et applications, Strasbourg, France (1993)
2. Andres, E.: Discrete linear objects in dimension n: the standard model. Graph. Model. **65**(1–3), 92–111 (2003)

3. Andres, E., Acharya, R., Sibata, C.: Discrete analytical hyperplanes. Graph. Models Image Process. **59**(5), 302–309 (1997)
4. Biswas, R., Largeteau-Skapin, G., Zrour, R., Andres, E.: Rhombic dodecahedron grid—coordinate system and 3D digital object definitions. In: Couprie, M., Cousty, J., Kenmochi, Y., Mustafa, N. (eds.) DGCI 2019. LNCS, vol. 11414, pp. 27–37. Springer, Cham (2019). https://doi.org/10.1007/978-3-030-14085-4_3
5. Čomić, L., Magillo, P.: Repairing 3D binary images using the BCC grid with a 4-valued combinatorial coordinate system. Inf. Sci. **499**, 47–61 (2019)
6. Čomić, L., Magillo, P.: Repairing 3D Binary Images Using the FCC Grid. J. Math. Imaging Vis. **61**(9), 1301–1321 (2019). https://doi.org/10.1007/s10851-019-00904-0
7. Csébfalvi, B.: An evaluation of prefiltered b-spline reconstruction for quasi- interpolation on the body-centered cubic lattice. IEEE Trans. Visual Comput. Graphics **16**(3), 499–512 (2010)
8. Debled-Rennesson, I., Reveilles, J.: New approach to digital planes. Vision Geometry III, vol. 2356, pp. 12–21. International Society for Optics and Photonics, SPIE (1995)
9. Edelsbrunner, H., Iglesias-Ham, M., Kurlin, V.: Relaxed disk packing. In: CCCG 2015, pp. 128–135 (2015)
10. Finkbeiner, B., Entezari, A., Van De Ville, D., Möller, T.: Efficient volume rendering on the body centered cubic lattice using box splines. Comput. Graph. **34**(4), 409–423 (2010)
11. Françon, J.: Discrete combinatorial surfaces. Graph. Models Image Process. **57**(1), 20–26 (1995)
12. He, L., Liu, S., Yun, J., Liu, Y.: A line generation algorithm over 3D body-centered cubic lattice. J. Multimed. **8**(1), 40–47 (2013)
13. Her, I.: Description of the F.C.C. lattice geometry through a four-dimensional hypercube. Acta Crystallographica Section A, **51**(5), 659–662 (1995)
14. Ibáñez, L., Hamitouche, C., Roux, C.: Ray-tracing and 3D Objects Representation in the BCC and FCC Grids. In: 7th International Workshop on Discrete Geometry for Computer Imagery (DGCI), pp. 235–242 (1997)
15. Ibáñez, L., Hamitouche, C., Roux, C.: Ray casting in the BCC grid applied to 3D medical image visualization. In: IEEE Engineering in Medicine and Biology Society, vol. 2, pp. 548–551 (1998)
16. Meng, T., et al.: On visual quality of optimal 3d sampling and reconstruction. In: Graphics Interface, pp. 265–272 (2007)
17. Nagy, B., Strand, R.: Distances based on neighbourhood sequences in non-standard three-dimensional grids. Discret. Appl. Math. **155**(4), 548–557 (2007)
18. Nagy, B., Strand, R.: Non-traditional grids embedded in Z^n. Int. J. Shape Model. **14**(2), 209–228 (2008)
19. Reveillès, J.-P.: Géométrie discrète, calcul en nombres entiers et algorithmique. State thesis, University Louis Paster, December 1991
20. Strand, R., Nagy, B., Borgefors, G.: Digital distance functions on three-dimensional grids. Theor. Comput. Sci. **412**, 1350–1363 (2011)

Digital Convex + Unimodular Mapping = 8-Connected (All Points but One 4-Connected)

Loïc Crombez[(✉)]

Université Clermont Auvergne and LIMOS, Clermont-Ferrand, France
lcrombez@uca.fr

Abstract. In two dimensional digital geometry, two lattice points are *4-connected* (resp. *8-connected*) if their Euclidean distance is at most one (resp. $\sqrt{2}$). A set $S \subset \mathbb{Z}^2$ is *4-connected* (resp. *8-connected*) if for all pair of points p_1, p_2 in S there is a path connecting p_1 to p_2 such that every edge consists of a 4-connected (resp. 8-connected) pair of points. The original definition of digital convexity which states that a set $S \subset \mathbb{Z}^d$ is *digital convex* if $\mathrm{conv}(S) \cap \mathbb{Z}^d = S$, where $\mathrm{conv}(S)$ denotes the convex hull of S does not guarantee connectivity. However, multiple algorithms assume connectivity. In this paper, we show that in two dimensional space, any digital convex set S of n points is unimodularly equivalent to a 8-connected digital convex set C. In fact, the resulting digital convex set C is 4-connected except for at most one point which is 8-connected to the rest of the set. The matrix of $SL_2(\mathbb{Z})$ defining the affine isomorphism of $\mathbb{Z}^2$ between the two unimodularly equivalent lattice polytopes S and C can be computed in roughly $O(n)$ time. We also show that no similar result is possible in higher dimension.

Keywords: Digital geometry · Convex · Unimodular affine transformation · 4-connected · 8-connected · Lattice diameter

1 Introduction

Digital Geometry studies the geometry of *lattice points*, those are the points with integer coordinates [1]. Convexity, a fundamental concept in continuous geometry [2], is naturally also fundamental in digital geometry. However, unlike in any linear space where convexity is clearly defined, several definitions have been investigated for convexity in digital geometry [3–7]. Just in two dimensions, we encounter several definitions such as triangle line [3], HV convexity [8], Q convexity [9].

Some of those definitions where created in order to guarantee that under those definitions a convex object is connected (in terms of the induced grid subgraph). No such guarantee is given by the original following definition of digital convexity which is the one that will be used throughout this paper: A set S of n lattice

© Springer Nature Switzerland AG 2021
J. Lindblad et al. (Eds.): DGMM 2021, LNCS 12708, pp. 164–176, 2021.
https://doi.org/10.1007/978-3-030-76657-3_11

points is said to be *digital convex* if $\text{conv}(S) \cap \mathbb{Z}^d = S$, where $\text{conv}(S)$ is the convex hull of S. This definition is equivalent to saying that there exist a convex polyhedron P in $\mathcal{R}^d$ such that $P \cap \mathbb{Z}^d = S$. However, this definition of digital convexity provides a lot of mathematical properties such as being preserved under unimodular affine transformations $SL_2(\mathbb{Z})$. These transformations are the lattice preserving mappings that also preserves parallel lines and area [10, 11].

In this paper we prove that any digital convex sets S is unimodularly equivalent to an *almost 4-connected* set C. We say that a set C is *almost 4-connected* if C is 4-connected except for at most one point which is 8-connected to the rest of the set. We also propose an algorithm that computes such a set C in roughly $O(n)$ time.

The demonstration of existence of such a set, and the algorithm are both based on the same technique which consists in mapping a *lattice diameter* [12] of S to a horizontal line. A *lattice diameter* of digital convex set S is the longest string of integer points on any line in the Euclidean space that is contained in $\text{conv}(S)$. The lattice diameter is invariant under the group of unimodular affine transformations $SL_2(\mathbb{Z})$.

The construction consists in computing a lattice diameter d of S, applying an affine isomorphism $\mathbb{Z}^2$ that maps S to a horizontal line, and finally adjusts through *horizontal shear mapping* in order to obtain an almost 4-connected set. A horizontal shear mapping is a unimodular affine transformation that preserves the y-coordinates. The matrix of defining those transformations are of the form $\begin{bmatrix} 1 & k \\ 0 & 1 \end{bmatrix}$.

Note that in 3 dimension the tetrahedron $((0,0,0),(0,1,0),(1,0,0),(1,1,k))$ has no lattice points in its interior and has a volume of $\frac{k}{6}$. As a consequence, since unimodular affine transformations preserve volume, similar results as the one presented in this paper for unimodular affine transformations are impossible in dimension higher than 2.

2   Unimodularly Equivalence to Connected Set

The purpose of this section is to provide a constructive proof of the following theorem.

Theorem 1. *For any digital convex set S there is a unimodular affine transformations that maps S to an almost 4-connected set C.*

In order to prove theorem 1 we describe a multiple step construction that results in an almost 4-connected set. The steps of this constructions are the following:

- First, we find a lattice diameter d of S. We define k as the number of lattice points located on d.
- We then map S to S_1 using a unimodular affine transformation that maps d to d_1 such that d_1 starts from the point $(0, 0)$ and end at the point $(k - 1, 0)$. Note that d_1 is a lattice diameter of S_1.

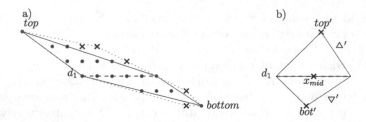

Fig. 1. a) The lattice diameter d_1 of S_1 is represented by the dashed red line segment. The subset $\diamond$ that is the union of an upper and lower triangle with the diameter as a common horizontal edge is represented by the blue dots. The set S_1 is the union of the blue dots, and the black crosses. b) Representation of $\diamond'$ after horizontal shearing. x_{mid} is the middle of d_1 the lattice diameter of $\diamond$. $\triangle$ and $\triangledown$ are the triangles on each side of d_1. The points top' and bot' are the two points with the most extreme y-coordinates. (Color figure online)

- In Sect. 2.1 we reduce the problem to the study of a subset $\diamond \in S_1$ such that the convex hull of $\diamond$ is the convex hull of: d_1, the topmost point in S_1, and the bottommost point in S_1. From here on in, for simplicity, we assume that the lattice point $p\diamond$ that is the furthest from the line $y = 0$ is the topmost point in S_1.
- In Sect. 2.2 we only consider $\triangle \in \diamond$, the subset of lattice points above d_1. We show that there is a horizontal shear mapping $\mathcal{M}_\infty$ (a unimodular affine that does not affect y-coordinates) such that $\mathcal{M}_\infty(\triangle)$ is 4-connected, to the exception of one special case that is treated in Sect. 2.3.
- Sect. 2.4 focuses $\triangledown \in \diamond$, the subset of lattice points below d_1 and show that there is a shear mapping $\mathcal{M}_\in$ such that $\mathcal{M}_\in(\diamond)$ is almost 4-connected.
- Finally, in Sect. 2.5 we go back to the general case and explain why our construction not only almost 4-connects $\diamond$, but also S.

2.1   Reduction to a Quadrilateral

For simplicity, in most of this proof, we will consider the subset $\diamond \in S_1$ that is the intersection of the lattice grid $\mathbb{Z}^2$ with the convex hull of: d_1, top, and $bottom$, where top and $bottom$ are respectively the topmost and bottommost points of S_1 (See Fig. 1 a). This reduction is mostly possible thanks to Lemma 1, even though we will have to take additional precautions detailed in Sect. 2.5 when $\diamond$ is almost 4-connected but not 4-connected.

Lemma 1. *For any digital convex set S_1 such that top (resp. bottom) are one of the topmost (resp. bottommost points) in S_1, and such that $\diamond \in S_1$ is the intersection of $\mathbb{Z}^2$ with the convex hull of the horizontal diameter d_1 and the two points top, bottom. For any horizontal shear mapping $\mathcal{M}$, if $\mathcal{M}(\diamond)$ is 4-connected, then $\mathcal{M}(S_1)$ is also 4-connected.*

Proof. We denote y_t and y_b the y-coordinates of the topmost and bottommost point in $\Diamond$. As $\mathcal{M}$ is a horizontal shear mapping, it preserves y-coordinates, and as $\Diamond_c$ is 4-connected, for each integer y_i such that $y_t \geq y_i \geq y_b$ there is a point in $\Diamond_c$ whose y-coordinate is equal to y_i. Now, we consider a point $p(x_p, y_p)$ in $S_c \notin \Diamond_c$. We have $y_t \geq y_p \geq y_b$, and hence there is a point $p_q \in \Diamond_c$ whose y-coordinate is y_p. As S_c is digital convex, all lattice points on the horizontal line segment from p to p_q are in S_c, hence p is 4-connected to $\Diamond_c$.   □

We denote k the number of lattice points in d, the lattice diameter of S_1, which we mapped to a horizontal line segment d_1. The leftmost and rightmost lattice points in d_1 are p_0 and p_{k-1}. For simplicity, we choose the coordinates such that p_0 is the origin. Hence the coordinates of p_0 and p_{k-1} are $p_0(0,0)$ and $p_{k-1}(k-1, 0)$. For simplicity, from now on, we assume that the y-coordinate of *top* is larger or equal than the absolute value of the y-coordinate of *bottom*. The opposite case being symmetrically equivalent.

2.2   Connecting the Top

In this section, we will only consider $\triangle$, the top part of $\Diamond$, consisting of all the points above d_1, d_1 included. We denote $x_{mid} = \frac{k-1}{2}$. We now apply to $\triangle$ the horizontal shear mapping $\mathcal{M}_h$ such that all lattice points on d_1 maps to itself and such that $top(x_t, y_t)$ is mapped as close as possible to (x_{mid}, y_t). The image of top by $\mathcal{M}_h$ is $top'(top'_x, y_t)$. For simplicity we assume that $top'_x \leq x_{mid}$. The case when $top'_x \leq x_{mid}$ is symmetrically equivalent. We define $\triangle'$ as the set to which $\mathcal{M}_h$ maps $\triangle$ (See Fig. 1 b). We now assume that $\triangle'$ is not 4-connected and study the possible location of *top'*.

We know that:

i $top'_x < 0$ or $top'_x > k-1$. Otherwise $\triangle'$ would be 4-connected since the vertical segment going from *top'* to d_1 is in $\triangle'$ (See Fig. 2 a).

ii *top'* cannot be located both left of the line $x=0$ and above the line $y = k - \frac{k}{k-1}x$. Otherwise the $k+1$ points: $(0,0), (0,1), ..., (0,k)$ would be in $\triangle'$ and d_1 would not be a lattice diameter of $\triangle'$ (See Fig. 2 b).

iii *top'* is located above or on the line $y = \frac{k-1}{2} - 2x$. Otherwise there would be a horizontal mapping that maps *top* closer to the point (x_{mid}, y_t) (See Fig. 2 c).

iv *top'* is not located both above the line $y = d - 1 - x$ and below the line $y = -dx$. Otherwise the $k+1$ points: $(k-1,0), (k-2,1), ...(0, k-1), (-1, k)$ would be in $\triangle'$ and d_1 would not be a lattice diameter (See Fig. 2 d).

Hence, the only possible location for *top'* is strictly inside the quadrilateral *Quad* whose vertices are: $(0, k-1), (0, k), (-1, k), (-\frac{k-1}{k-2}, k\frac{k-1}{k-2})$. As $x = -1$ is the only vertical line of integer coordinates intersecting the inside of *Quad*, all lattice points strictly inside *Quad* are located on the line $x=-1$. The line $x=-1$ intersects the edges of *Quad* at the points: $(-1, k)$ and $(-1, k + \frac{k}{k-1})$. Since $k \geq 2$, we have $k/(k-1) \leq 2$. Hence there is exactly one lattice point located strictly inside *Quad*; $(-1, k+1)$.(See Fig. 2 e).

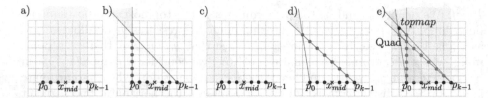

Fig. 2. a): *top'* cannot lie within the red region. Otherwise △' is 4-connected. b): *top'* cannot lie within the red region. Otherwise the red line which contains more lattice point than the diameter would be in △'. c): *top'* cannot lie within the red region. Otherwise there would be a horizontal shear mapping such that *top'* is closer to x_{mid} than it currently is. d): *top'* cannot lie within the red region. Otherwise the red line which contains more lattice point than the diameter would be in △'. e): Superposition of the four previous figures. There is only one potential location for *top'*. (Color figure online)

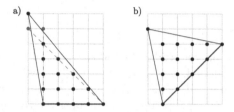

Fig. 3. a) The only possible location for *top'* in which △' is not 4-connected. As d_1 (shown in blue) is a lattice diameter neither of the red points can be in △'. As adding any points above d would imply adding one of the red points, the set to which S maps does not contain any other points above d than those in △' shown here in black. b) Representation of the same set as in a) after the vertical shear mapping making it almost 4-connected (Color figure online)

2.3   Special Case Study

In this section, we consider the situation where *top'* is located at $(-1, k+1)$. Notice that in this situation no lattice points can be added to △' above d_1 without adding either the point $(-1, k)$ or the point $(0, k)$. Adding either of those points to △' is impossible as d_1 would no longer be the lattice diameter (See Fig. 3 a). This means that at this step, △' is not 4-connected if and only if △' is equal to the intersection of $\mathbb{Z}^2$ with the triangle $(0, 0), (k-1, 0), (-1, k+1)$. This also implies that the part of S_1 above d is equal to △'.

We also notice that in this situation △' is not only not 4-connected, but not even 8-connected. However it is possible to apply to △' a vertical shearing that maps d_1 to the line $y = x$. The resulting set that we call a *pompom tri-angle* is almost 4-connected (See Fig. 3 b). Note that the pompom triangle is not unimodularly equivalent to any 4-connected set. The pompom triangle for $k = 2$ consisting of only 4 points is the smallest digital convex set that is not unimodularly equivalent to any 4-connected set.

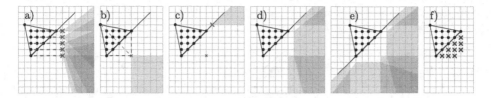

Fig. 4. a–e) Visual representation of the different location in which bot' cannot be located when the top part is a pompom triangle. f) The remaining possible location for bot' are all within a position that makes $\triangledown'$ 4-connected.

We now consider S', the image of S after all the mappings previously described in the case where the top part of the set is equal to the not 4-connected pompom triangle. The position of $bottom$ after the mappings is denoted bot'. We consider the potential location of bot'.

- S' cannot contain any of the points $(k,0),(k,1),...(k,k-1)$. Otherwise d_1 would not be a lattice diameter. This defines $k-1$ cones in which bot' cannot be located (See Fig. 4 a).
- S' cannot contain the point $(0,k-1)$. Otherwise d_1 would not be a lattice diameter. This defines a cone in which bot' cannot be located (See Fig. 4 b).
- S' does not contain the point (k,k). This defines a cone in which bot' cannot be located (See Fig. 4 c).
- The same constraints can also by applied using symmetry by reflection on the line $y = k - 1 - x$ (See Fig. 4 e).

All those location constraint results in the x-coordinate of bot' being between 0 and $k-1$, which implies the bottom part of S' located below d is 4-connected, and hence S' is almost 4-connected.

2.4 Connecting the Bottom

We now consider the case where $\triangle'$, the top part of the set $\lozenge'$, which is the intersection of $\mathbb{Z}^2$ with the triangle $p_0(0,0), p_{k-1}(k-1,0), top'(a,b)$ is 4-connected, and we will focus on finding a horizontal shear mapping that almost 4-connects $\lozenge'$.

As we moved top' as close as possible to the middle of p_0p_{k-1} in Sect. 2.2, we have the following inequality: $|\frac{k-1}{2} - a| \leq \frac{b}{2}$. We also showed that $0 \leq a \leq k-1$.

We say that a point $p(x,y)$ is *directly above* (resp. *directly below*) the line segment $d_1((0,0),(k-1,0))$ if $0 \leq x \leq k-1$ and $y \geq 0$ (resp. $y \leq 0$).

We now consider the possible locations for the bottommost point bot'. If bot', is directly below d_1, $\lozenge'$ is trivially 4-connected, so we only consider the situation where bot' is not directly below d_1. We forget the assumption made in Sect. 2.2 about the position of top' relative to x_{mid}. However, we still keep the assumption that states that top' is furthest or equally furthest than $bottom'$ from the line $y = 0$. For simplicity, we assume that $bot'(x_b, y_b)$ is located to the right of d_1, that

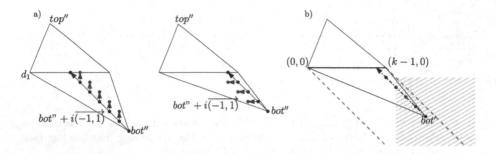

Fig. 5. a) The final set $\diamond''$ that we will obtain always contains the lattice points on the diagonal $bot'' + i(-1, 1)$ shown by the black arrow. $\diamond''$ will also contains, either the lattice points to the left or above the diagonal, shown in blue here. b) bot' is contained in the green tiled region. As long as bot' is located left of the line $y = k - 1 - x$, ∇' contains the upper left diagonal from bot', making ∇' 8-connected. (Color figure online)

is when the x-coordinate of bot' is larger than $k - 1$. The situation where bot' is located to the left of d_1 is symmetrically identical. We will not consider it.

As we are not in the pompom case, we know that top' is located directly above d_1, hence its y-coordinate is at most $k - 1$ as otherwise d_1 would not be a lattice diameter. In order to prove that $\diamond'$ is 4-connected, we will now prove that there is a horizontal shear mapping that maps $\diamond'$ to $\diamond''$ such that

- top'' remains directly above d_1, which guarantees the connectivity of $\triangle''$
- The lattice points $p_i(x, y) = bot'' + i\overrightarrow{(-1, 1)}$ such that $i > 0, y < 0$ are all within ∇''
- For either $v = \overrightarrow{(-1, 0)}$ or $v = \overrightarrow{(0, 1)}$, the following is true for all $p_i(x, y)$ such that $i \neq 0, y \neq 0$: $p_i + v \in \diamond$ (See Fig. 5 a).

In order for ∇' to contain the lattice points $p_i(x, y)$ located on the diagonal $bot'' + i\overrightarrow{(-1, 1)}$ such that $i > 0$ and $y < 0$, $bot'(x_b, y_b)$ has to be located in between the two lines $y = -x$ and $y = k - 1 - x$. As we know that $x_b > k - x$, and $y_b \leq k - 1$, this is equivalent to bot' being located left of the line $y = k - 1 - x$ (See Fig. 5 b).

First, we consider the situation where $bot'(x_b, y_b) = (k + \lambda, y_b)$ is in the region to the right of the line $y = k - 1 - x$, and we will show that there is an horizontal shear mapping that maps $bottom'$ to the left of the line $y = k - 1 - x$, and also maps top' directly above d_1. As $(k, 0)$ is not in $\diamond'$, we know that top' lies below the line supported by $(k, 0)$ and bot', that is the line $y = -\frac{x}{\lambda} + \frac{yk}{\lambda}$. As bot' is a lattice point right of the line $y = k - 1 - x$, bot' is to the right of, or on, the line $y = k - x$. Hence, $y_b \geq \lambda$. We call m the integer such that $m < \frac{|y_b|}{\lambda} \leq m + 1$. The point bot' is located in the wedge above $y = -\frac{1}{m}x + \frac{k}{m}$ and below $y = -\frac{1}{m+1}x + \frac{k}{m+1}$ (See Fig. 6). We now apply to $\diamond'$ the horizontal shear mapping that maps $y = -\frac{1}{m}x + \frac{k}{m}$ to the vertical line $x = k$. This mapping, maps $y = -\frac{1}{m}x + \frac{k}{m}$ to the diagonal line $y = k - x$. Hence, after mapping bot' is now located left of the line $y = k - x$. As top' is located left of the line $y = k - x$, after this mapping top' still strictly lies

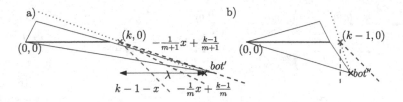

Fig. 6. a) *bot'* is located in between the blue and red dashed lines, and *top'* is located left to the red dotted line. b) Once the horizontal shear mapping is applied, *bot''* is located left to the line $y = k - 1 - x$. As *top''* is located left to the red dotted line, *top''* is located left to the vertical line $x = k - 1$. Hence $\triangle''$ is 4-connected. (Color figure online)

to the left of the line $x = k$, and hence the triangle $(0,0), (k-1,0)top'$ is still 4-connected.

We will now consider the last case remaining, that is when *bot'* is located left of, or on, the line $y = k - 1 - x$. To show that, in this situation, $\triangledown'$ is almost 4-connected we will use two lemmas. Lemma 2 that explicits the location in which $\triangledown'$ contains the lattice points on the line $bot' + (-2, 1) + i\overrightarrow{(-1,1)}$ (See, Fig. 7 b)), and Lemma 3 that explicits the location in which $\triangledown'$ contains the lattice points on the line $bot' + (-1, 2) + i\overrightarrow{(-1,1)}$ (See, Fig. 7 a).

Lemma 2. *Any triangle* $\triangledown : (0,0), (l,0), p(x,y)$, *with* $l > 0$, $y < 0$ *and* $x > l > 1$ *is almost 4-connected when:* $\frac{-x}{2} \leq y \leq l - x$,

Proof. We first consider the case where the point $p(l - y, y)$ is located on the line $y = l - x$. The right edge of $\triangledown$ is supported by the line $y = l - x$, and hence contains a lattice point for each integer y-coordinate the edge crosses. The horizontal width of $\triangledown$ at the y-coordinate $y + 1$ is equal to $\frac{l}{y} \geq 1$. Hence both the points $(l - y - 1, y + 1)$ and $(l - y - 2, y + 1)$ are in $\triangledown$ and, the width of the triangle increasing with y-coordinate, the same reasoning can be done with the other points on the diagonal $p + k\overrightarrow{(-1,1)}$.

We now study, how far to the left of the line $y = l - x$ can $p(x, y)$ move, such that $\triangledown$ still contains $(x-1, y+1)$ and $(x-2, y+1)$. We rewrite the coordinates of p in the following manner: $p(l - y - \lambda, y)$. The left edge of the triangle is supported by the line $y = \frac{y}{l - y - \lambda}$. At height $y + 1$ the left edge x-coordinate x_l is equal to:
$$x_l = \frac{(y+1)(l-y-\lambda)}{y} = l - \lambda - 1 - y + \frac{l-\lambda}{y}.$$
As we want, $(l - y - \lambda - 2, y + 1)$ to be inside $\triangledown$, we need $x_l \leq l - y - \lambda - 2$. That is
$$l - \lambda - 1 - y + \frac{l-\lambda}{y} < l - y - \lambda - 2$$
$$\lambda \leq y + l$$

As $\triangledown$ contains all points on the diagonal $p + k\overrightarrow{(-1,1)}$, $\triangledown$ is almost 4-connected when $0 \leq \lambda \leq y + l$. That means, considering $p(x, y)$, $\triangledown$ is connected if $-2y \leq x \leq l - y$. Which is equivalent to $\frac{-x}{2} \leq y \leq l - x$.   □

Lemma 3. *The triangle* $\triangledown : (0,0), (l,0), p(x,y)$, *with* $l > 0$, $-l < y < 0$ *and* $x > l > 1$ *is almost 4-connected when:* $y \leq 2l - 2x$.

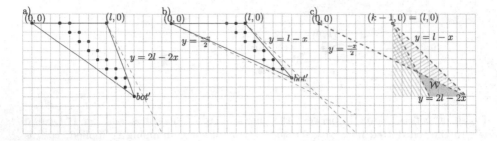

Fig. 7. a) If *bot′* is located left of the dashed red line, then ▽′ contains the black diagonal, and the blue one just above it. b) If *bot′* is located in the top wedge in between the two dashed red line, then ▽′ contains the black diagonal, and the blue one just left of it. c) If *bot′* is located in one of the green tiled surfaces, then ▽′ is almost 4-connected. That is not true when *bot′* is located in the blue wedge $\mathcal{W}$ (Color figure online)

Proof. We first consider the case where the point $p(l, y)$ is located on the line $x = l$. In this situation, it is clear that $(l - 1, y + 2)$ is in ▽. Now we want to know, how far to the right of the line $x = l$ can $p(x, y)$ be located such that $(x - 1, y + 2)$ is in ▽. We rewrite the coordinates of p in the following manner: $p(l + \lambda, y)$. The right edge of ▽ is located on the line $y = \frac{y}{\lambda}x - \frac{yl}{\lambda}$. As we want $(x - 1, y + 2)$ in ▽, we want the right edge to intersect the line $y = l + \lambda - 1$ above $y + 2$. That is

$$\frac{y}{\lambda}(l + \lambda - 1) - \frac{yl}{\lambda} \geq y + 2$$

$$\frac{y(\lambda - 1)}{\lambda} \geq y + 2$$

$$\lambda \leq \frac{-y}{2}$$

That means, considering $p(x, y)$, ▽ is connected when $x \leq l - \frac{y}{2}$. Which is the same as $y \leq 2l - 2x$.   □

We now consider the last surface in which *bot′* can be located such that ▽′ is not almost 4-connected. That is the wedge $\mathcal{W}$ located below $y = -\frac{x}{2}$ and above $y = 2k - 2 - 2x$ (See Fig. 7 c).

To study the situation where *bot′* is located inside $\mathcal{W}$ we have to consider multiple possible location for *top′*. If *top′* is located below the line $y = k - 1 - x$, then we can apply a horizontal shearing that maps $\mathcal{W}$ to a surface that is directly below d_1, and maps △′ to a 4 connected set. If *top′* is located above the line $y = k - 1 - x$ and above the line $y = x$, then the lattice points inside the triangle $\triangle_{top}(0, 0), (k - 1, 0), (\frac{k-1}{2}, \frac{k-1}{2})$ are in ◇′, and since $\triangle_{top} + \overrightarrow{k(1, -1)}$ covers $\mathcal{W}$, *bot′* cannot be located inside $\mathcal{W}$ with d_1 being a lattice diameter. Finally, *top′* cannot be located below the line $y = x$ as the fact that *top′* is the furthest point from the line $y = 0$ and that the point $(k, 0)$ is not in ◇′ makes it impossible for *top′* to be located below the line $y = x$. With $\mathcal{W}$ covered, we now have considered all possibilities for the location of bot' and found in each case a mapping that maps $\diamond'$ to an almost 4-connected set. Furthermore, in the event where $\diamond'$ is not mapped to 4-connected, the point that is not 4-connected to the set is bot'' that image of bot'.

2.5   Back to the General Case

We now show how to obtain a mapping that maps S to an almost 4-connected set from the mapping $\mathcal{M}$ we previously described that maps $\diamond$ to $\diamond''$, an almost 4-connected set. Using Lemma 1 we can discard the case where $\diamond''$ is 4-connected, and using the same arguments as in the proof of Lemma 1 we can conclude that the only points that might not be 4-connected in S'' are the one that have the same y-coordinate as bot''. As bot'' is not 4-connected, it means that bot'' is located right of the line $x = k-1$, in this situation we showed that $\diamond''$ contains the lattice point $bot'' + (-1, 1)$. As a consequence, adding a point to the left of bot'' would make S'' 4-connected. Now, if we add the point $p_a = bot'' + (1, 0)$, either p_a is below or on the line $y = k - 1 - x$, and hence S'' contains $p_a + (-1, 1)$, which is the point above bot'' which makes S'' 4-connected, or p_a is located exactly on the line $y = k - x$. As a consequence, since S'' does not contain $(k, 0)$, top'' is located left to the line $y = k - x$ which means we can apply to S'' the horizontal shearing that maps $y = k - x$ to $x = k$ in order to obtain a 4-connected set.

3   Algorithm

In this section we study the algorithmic complexity of finding an almost 4-connected unimodularly equivalent set to a digital convex set S of n points and of diameter r. We propose an algorithm that runs in $O(n + h \log r) = O(n + n^{\frac{1}{3}} \log r)$ time, where h is the number of vertices on the convex hull of S. This algorithm mimics the construction done in Sect. 2, and hence relies on the computation of a lattice diameter of S.

3.1   Computing the Lattice Diameter

We present here an algorithm to compute a lattice diameter of a digital convex set S in $O(n + h \log r)$ time, or $O(h\sqrt{n} + h \log r)$ when the convex hull of S is known. This algorithm relies on the fact that at least one of the vertices of the convex hull of S is located on a lattice diameter [11,12]. Hence, we only have to consider vertices of the convex hull to find a lattice diameter. We consider v, a vertex of $\mathrm{conv}(S)$ such that v is on a lattice diameter d of S. We now consider the fan triangulation $\mathcal{T}$ rooted on v of $\mathrm{conv}(S)$. That is the triangulation defined by all the diagonals going from the vertex v to all of the other vertices of $\mathrm{conv}(S)$. There is a triangle t_1 in $\mathcal{T}$ that contains d (See Fig. 8 a). The base of t_1 is an edge of $\mathrm{conv}(S)$, and its opposite vertex is v. Hence, in order to find a lattice diameter of S we can test each triangle of each of the h fan triangulations. In each of those triangles $t_1(v_1, v_2, v_3)$, we want to compute a line ℓ such that v_1

is located on ℓ, and ℓ maximizes $|\ell \cap t_1 \cap \mathbb{Z}^2|$. In order to compute such a line, we apply to t_1 a unimodular affine transformation that maps the line supported by $v_2 v_3$ to a horizontal line (See Fig. 8 b). The three vertices of t_1: v_1, v_2 and v_3 are now mapped to $v_1'(x_1', y_1')$, $v_2'(0,0)$ and $v_3'(x_3', 0)$, the three vertices of t_1'. For simplicity, we assume that v_1' is above v_2' and v_3'. We now consider any line ℓ that goes through v_1' and intersects the segment $v_2' v_3'$, more specifically we consider the line segment supported by ℓ that is located inside t_1' $\ell_s = \ell \cap t_1'$. As we mapped $v_2' v_3'$ to an horizontal line, the number of horizontal lines $y = i, i \in \mathbb{Z}$ intersected by ℓ_s is always equal to the constant number $y_1' - y_3' + 1$. As lattice points on a line are evenly separated, and since v_1' is a lattice point, maximizing the number of lattice points on ℓ_s is equivalent to finding $l_t(x_t, y_t) \neq v_1'$, the top most lattice point in t_1'. The number of lattice points in t_1' on the line $v_1' l_t$ is equal to $\left\lfloor \frac{y_1'}{y_1' - y_t} \right\rfloor + 1$ (See Fig. 8 b).

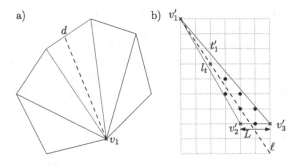

Fig. 8. a) The lattice diameter d that goes through v_1 is in one of the triangles of the fan triangulation of S from v_1. This triangle is shown in blue. b) The triangle is mapped to a triangle where $v_2' v_3'$ is a horizontal line. In this triangle, finding ℓ (the longest digital segment in t_1' going through v_1') is equivalent to finding l_t the topmost lattice point in $t_1' \cap \mathbb{Z}^2 \setminus v_1'$. (Color figure online)

Naively looking for l_t by testing the intersection of all horizontal lines with t_1' starting from y_1' to 0 leads to at most $y_1' + 1$ tests. We now show that $y_1' \leq 2n$. Indeed, the area of t_1' is equal to $\frac{(y_1')x_3'}{2}$, and we now that $x_3' \geq 1$. Using Pick's formula [13], that states that in a lattice triangle $\triangle$ we have the following equality: $A = i + \frac{b}{2} - 1$ where A is the area of $\triangle$, i is the number of lattice points strictly inside $\triangle$ and b is the number of lattice points on the edges of $\triangle$, we can deduce that the number of lattice points inside t_1' $n_t = t_1' \cap \mathbb{Z}^2$ is at least equal to $\frac{y_1'}{2}$. This means that l_t can be found in $O(n_t)$ time. However, using a lower bound λ on n_d the number of lattice points on a lattice diameter of S, we can stop computation in t_1' before finding l_t after $\frac{y_1'}{\lambda - 1}$ steps. Indeed, any below that would be too far from v_1' to result in a lattice diameter of S as the segment would contain less than $\frac{y_1'}{\frac{y_1'}{\lambda - 1}} + 1 = \lambda$ lattice points. We use a very rough lower bound to n_d that can be directly deduced from the construction we made in order to prove Theorem 1 by fattening the bounding box of $\diamond''$ by n_d both to the left and to the right. The resulting bound is the following: $\frac{1}{8}\sqrt{n} \leq n_d$.

We now consider the total number of operations needed to compute a lattice diameter of S. For a given vertex v of $\mathrm{conv}(S)$, we denote $(n_1, n_2, ..., n_i, ...n_h)$ the number of lattice points inside each of the h triangles of the fan triangulation rooted on v. As v is in all the h triangles, and any other lattice point can only be in at most 2 triangles, $\sum_i n_i \leq 2n + h \leq 3n$. In addition to the computation of all the h unimodular affine transformations required in order to map the edges of $\mathrm{conv}(S)$ to horizontal lines, computing a potential lattice diameter that goes through a given vertex v takes at most, $\sum_i \dfrac{2n_i}{\frac{1}{8}\sqrt{n}-1} \leq \dfrac{6n}{\frac{1}{8}\sqrt{n}-1} = O(\sqrt{n})$ time. Repeating this process for all the h vertices of $\mathrm{conv}(S)$ takes at most $O(h\sqrt{n})$ time. Now, adding to this the computing time required to compute the h unimodular affine transformations that map each edge of $\mathrm{conv}(S)$ to a horizontal line, we obtain a time complexity of $O(h\sqrt{n} + h\log r)$ to compute the lattice diameter of S, given its convex hull.

Computing the convex hull of a digital convex set can be done in linear time using the quickhull algorithm [14], and since there is at most $O(n^{1/3})$ vertices on $\mathrm{conv}(S)$ [15,16] the total time complexity of the algorithm in order to compute a lattice diameter of a digital convex set is $O(n + h\log r) = O(n + n^{\frac{1}{3}}\log r)$.

3.2 Computing a Unimodularly Equivalent Almost 4-Connected Set

Once the lattice diameter computed, the computation of an affine isomorphism of $\mathbb{Z}^2$ resulting in an almost 4-connected set requires:

- The computation of the affine isomorphism mapping d to a horizontal line in $O(\log r)$ time, where r is the diameter of S.
- Applying the mapping to $conv(S)$ in $O(h)$ time.
- Computing $\diamond$ in $O(h)$ time.
- Computing the horizontal shear mapping positioning the topmost point in $O(1)$ time.
- Computing a horizontal shear mapping from the positions of the top most and bottom most points in order to make $\diamond'$ 4-connected in $O(1)$ time.
- applying all the mappings to $conv(S)$ in $O(h)$ time.
- Eventually computing one last shear mapping in the event where S does not map to an almost 4-connected set in $O(h)$.

Hence, the total time complexity sums to $O(n)$. Adding the time complexity of the lattice diameter algorithm results to a time complexity of $O(n + h\log r)$ time in order to compute a unimodularly equivalent almost 4-connected set.

4   Perspective

In this paper, in 2 dimension, we proved for every digital convex set the existence of a unimodularly equivalent almost 4-connected set. While it is proven that an infinite amount of digital convex sets cannot be mapped to unimodularly equivalent 4-connected sets, the algorithm proposed in this paper does not ensure to provide a 4-connected unimodularly equivalent set when such a set exists.

References

1. Klette, R., Rosenfeld, A.: Digital Geometry: Geometric Methods for Digital Picture Analysis. Elsevier, Boston (2004)
2. Ronse, C.: A bibliography on digital and computational convexity (1961–1988). IEEE Trans. Pattern Anal. Mach. Intell. **11**(2), 181–190 (1989)
3. Kim, C.E., Rosenfeld, A.: Digital straight lines and convexity of digital regions. IEEE Trans. Pattern Anal. Mach. Intell. **4**(2), 149–153 (1982)
4. Kim, C.E., Rosenfeld, A.: Convex digital solids. IEEE Trans. Pattern Anal. Mach. Intell. **4**(6), 612–618 (1982)
5. Chassery, J.-M.: Discrete convexity: definition, parametrization, and compatibility with continuous convexity. Comput. Vis. Graph. Image Proces. **21**(3), 326–344 (1983)
6. Kishimoto, K.: Characterizing digital convexity and straightness in terms of length and total absolute curvature. Comput. Vis. Image Underst. **63**(2), 326–333 (1996)
7. Chaudhuri, B.B., Rosenfeld, A.: On the computation of the digital convex hull and circular hull of a digital region. Pattern Recogn. **31**(12), 2007–2016 (1998)
8. Barcucci, E., Del Lungo, A., Nivat, M., Pinzani, R.: Reconstructing convex polyominoes from horizontal and vertical projections. Theoret. Comput. Sci. **155**(2), 321–347 (1996)
9. Daurat, A.: Salient points of q-convex sets. Int. J. Pattern Recogn. Artif. Intell. **15**(7), 1023–1030 (2001)
10. Haase, C., Nill, B., Paffenholz, A.: Lecture notes on lattice polytopes. Fall School on Polyhedral Combinatorics (2012)
11. Bárány, I., Pach, J.: On the number of convex lattice polygons. Comb. Probab. Comput. **1**(4), 295–302 (1992)
12. Bárány, I., Füredi, Z.: On the lattice diameter of a convex polygon. Discret. Math. **241**(1), 41–50 (2001)
13. Pick, G.: Geometrisches zur zahlenlehre. Sitzungsberichte des Deutschen Naturwissenschaftlich-Medicinischen Vereines für Böhmen "Lotos" in Prag., vol. 47–48 1899–1900 (1899)
14. Crombez, L., da Fonseca, G.D., Gérard, Y.: Efficient algorithms to test digital convexity. In: Couprie, M., Cousty, J., Kenmochi, Y., Mustafa, N. (eds.) DGCI 2019. LNCS, vol. 11414, pp. 409–419. Springer, Cham (2019). https://doi.org/10.1007/978-3-030-14085-4_32
15. Žunić, J.: Notes on optimal convex lattice polygons. Bull. Lond. Math. Soc. **30**(4), 377–385 (1998)
16. Bárány, I.: Extremal problems for convex lattice polytopes: a survey. Contemp. Math. **453**, 87–104 (2008)

Distance-Oriented Surface Skeletonization on the Face-Centered Cubic Grid

Gábor Karai[✉]

Department of Image Processing and Computer Graphics, University of Szeged,
Szeged, Hungary
karai@inf.u-szeged.hu

Abstract. Strand proposed a distance-based sequential thinning algorithm for producing surface skeletons of binary objects sampled on the face-centered cubic (FCC) grid. In this paper, we present two modified versions of his algorithm, which are faster and - according to our experiments - less sensitive to the visiting order of border points in the sequential thinning phase.

Keywords: FCC grid · Distance transform · Skeletonization · Thinning

1 Introduction

Skeletons are region-based shape descriptors which summarize the general form of (segmented) digital binary objects [13]. In 3D, *surface skeletons* should contain 2D thin surface patches. Various examples for the application of surface skeletons are reviewed in [12]. Distance-based skeletonization techniques focus on the detection of ridges or local maxima in the *distance map* by using a properly selected distance and a computationally efficient algorithm for distance transform [1,2]. Another strategy for skeletonization, called as *thinning*, is an iterative object reduction in a topology preserving way [10]. Distance-based methods are often combined with thinning [4,14,19].

Most 3D skeletonization algorithms work on digital pictures sampled on the conventional cubic grid. An alternative structure is the *face-centered cubic* (FCC) grid that tessellates the 3D Euclidean space into rhombic dodecahedra [7]. The motivation behind choosing this structure lies in its beneficial topological and geometrical properties [6,17]. Three adjacency relations are generally considered on the conventional cubic grid, since two voxels (i.e., unit cubes) can share a face, an edge, or a vertex. In contrast, if two voxels (i.e., rhombic dodecahedra) share edges on the FCC grid, they also share a face. Hence, we can take only two adjacencies into consideration.

Strand proposed a skeletonization algorithm for binary objects sampled on the FCC grid [15]. His algorithm focuses only on the d_{12} distance. In this paper, we present two faster versions of Strand's method that are tested on Chamfer and Euclidean distance maps [3].

© Springer Nature Switzerland AG 2021
J. Lindblad et al. (Eds.): DGMM 2021, LNCS 12708, pp. 177–188, 2021.
https://doi.org/10.1007/978-3-030-76657-3_12

The rest of the paper is organized as follows. Section 2 reviews the basic notions of 3D digital topology. In Sect. 3, Strand's algorithm is reported, then in Sect. 4, we propose the modified versions of that algorithm for producing surface skeletons. Some experimental results are shown in Sect. 5. Finally, we round off this work with some concluding remarks in Sect. 6.

2 Basic Notions and Definitions

In this section, we review the basic concepts of digital topology and distance transform.

The FCC grid is denoted by $\mathbb{F}$, whose elements are called *points*. The FCC grid is defined as the following subset of $\mathbb{Z}^3$:

$$\mathbb{F} = \{(x, y, z) \in \mathbb{Z}^3 \mid x + y + z \equiv 0 \ (\text{mod } 2)\}. \tag{1}$$

We make a distinction among the following three types of neighborhood of a point $p = (p_x, p_y, p_z) \in \mathbb{F}$, see Fig. 1.

$$N_{12}^*(p) = \{(q_x, q_y, q_z) \in \mathbb{F} \mid (p_x - q_x)^2 + (p_y - q_y)^2 + (p_z - q_z)^2 = 2\}$$
$$N_6^*(p) = \{(q_x, q_y, q_z) \in \mathbb{F} \mid (p_x - q_x)^2 + (p_y - q_y)^2 + (p_z - q_z)^2 = 4\} \tag{2}$$
$$N_{18}^*(p) = N_{12}^*(p) \cup N_6^*(p)$$

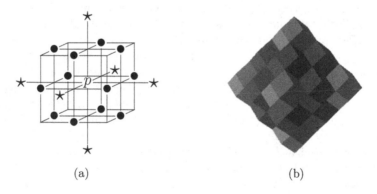

(a)   (b)

Fig. 1. Adjacency relations studied on $\mathbb{F}$ (a) and their voxel representation (b). $N_{12}^*(p)$ contains the points marked "•" (red voxels in (b)), $N_6^*(p)$ contains the points marked "⋆" (green voxels in (b)). (Color figure online)

Let $N_i(p) = N_i^*(p) \cup \{p\}$ $(i \in \{6, 12, 18\})$. Two points $p, q \in \mathbb{F}$ are *i-adjacent* if $q \in N_i(p)$. Furthermore, two points $q, r \in N_{12}^*(p)$ are *opposite*, if $q - p = p - r$.

The sequence of distinct points $\langle x_0, x_1, \ldots, x_s \rangle$ in a non-empty set of points X is called an *i-path* from x_0 to x_s in X if x_k is *i*-adjacent to x_{k-1}, $(k = 1, \ldots, s)$. Two points $p, q \in \mathbb{F}$ are said to be *i-connected* in X if there is an *i*-path from p

to q in X. The set X is *i-connected* in the set of points $Y \supseteq X$ if any two points in X are i-connected in Y.

Next, we apply the fundamental concepts of digital topology as reviewed by Kong and Rosenfeld [9]. An (m, n) *binary digital picture* on the FCC grid is a quadruple $\mathcal{P} = (\mathbb{F}, m, n, B)$, where each point in $B \subseteq \mathbb{F}$ is called a *black point* and has a value of 1 assigned to it, and each point in $\mathbb{F} \setminus B$ is called a *white point* with a value of 0. A *black component* or an *object* is a maximal m-connected set of points in B, while a *white component* is a maximal n-connected set of points in $\mathbb{F} \setminus B$. A black point p is called a *border point* if $N_n(p) \setminus B \neq \emptyset$, else it is an *interior point*. Picture $\mathcal{P}$ is *finite* if B contains finitely many black points. We refer to the total number of picture points as $|\mathcal{P}|$. In a finite picture, there is a unique white component that is called the *background*. A finite white component is called a *cavity*. In this paper, our attention is focused on $(12, 12)$ pictures as well as Strand did [15].

A *reduction* transforms a binary picture only by changing some black points to white ones (which is referred to as the *deletion* of 1's). A 3D reduction does *not* preserve topology [8] if

- any object in the input picture is split or is completely deleted,
- any cavity in the input picture is merged with the background or another cavity,
- a cavity is created where there was none in the input picture, or
- a *hole* (that e.g. donuts have) is eliminated, merged with other holes or created.

A *simple point* is a black point whose deletion is a topology preserving reduction [9]. Sequential thinning algorithms traverse the border points of a picture, and focus on the actually visited single point for possible deletion, hence for such algorithms, the deletion of only simple points ensures topology preservation. The following theorem states that simpleness is a local property in $\mathbb{F}$ which can be verified by investigating the 18-neighborhood of points. Figure 2 shows examples for simple and non-simple points.

Theorem 1. [6] *Let p be a black point in a $(\mathbb{F}, 12, 12, B)$ picture. Then p is a simple point if and only if the following conditions hold:*

1. *Point p is 12-adjacent to exactly one 12-component of $N_{18}^*(p) \cap B$.*
2. *Point p is 12-adjacent to exactly one 12-component of $N_{18}(p) \setminus B$.*

The *distance transform* DT assigns the distance of the closest white point to each point in the input binary picture [4]. The most frequently used distance functions for two points p and q are *neighborhood distances*, where $d_i(p, q)$ denotes the length of the shortest i-path between p and q [11], and the Euclidean distance, $d_e(p, q) = \sqrt{(p_x - q_x)^2 + (p_y - q_y)^2 + (p_z - q_z)^2}$. In $\mathbb{F}$, neighborhood distances $d_{12}(p, q)$ and $d_{18}(p, q)$ are taken into consideration. For better approximations to the exact Euclidean distance, the length of the moves from a point to its neighbors can be weighted according to some criteria, and the path of

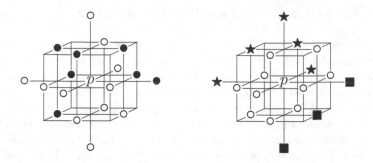

Fig. 2. Example for a simple (left) and a non-simple point (right) on the FCC grid. The distinct black 12-components are labeled with different symbols.

the minimal sum of weights called as *Chamfer distances* can be used [3]. Let $\langle a, b, c \rangle$ denote the general *Chamfer mask* for the FCC grid, where a and b are the weights assigned to all points in $N_{12}^*(p)$ and $N_6^*(p)$, respectively, and weight c is assigned to all q points such that $d_e(p, q) = \sqrt{6}$. If c is not used, denotation $\langle a, b \rangle$ is applied. Note that distances d_{12} and d_{18} are equivalent to the $\langle 1, 2 \rangle$ and $\langle 1, 1 \rangle$ Chamfer distances, respectively. Some examples are presented in [5,18].

Point $p \in B$ is called a *local maximum* if for any point $q \in N_i^*(p)$, $DT(p) \geq DT(q)$, where $i = 18$ for d_{18} distance, otherwise $i = 12$. Point $p \in B$ is a *center of maximal ball (CMB)* [4] if $DT(q) < DT(p) + w$ for any point q concerned in the Chamfer mask, where w is the weight that corresponds to q in the Chamfer mask. Note that local maxima coincide with CMB's on d_{12} and d_{18} distance maps.

3   Strand's Algorithm

The sequential thinning method proposed by Strand [15] assumes d_{12} distance, considers CMB's as safe skeletal (anchor) points, and it can produce surface skeletons by preserving non-simple points and surface edge points. A point $p \in B$ is called a *surface edge point*, if the following conditions hold:

- there are two opposite black points $q, r \in N_{12}^*(p) \cap B$, and
- there is no point $s \in N_{12}^*(p) \cap B$ such that $N_{12}(s) \cap N_{18}(p) \subseteq B$.

The thinning process consists of two phases. Forward thinning reduces the input objects to 2–3 voxel thick surface patches, which are peeled further in the backward thinning phase. The unusual in this process is that during backward thinning, object points are visited in descending order of their distance value. The sequential thinning is rather sensitive to the visiting order of border points. As a result the final skeleton usually has some unwanted branches or surface segments. This situation is illustrated in Fig. 3, and an example for its occurrence can be found in Fig. 4.

Algorithm 1: Strand's distance-based thinning - SDT

Input: picture $(\mathbb{F}, 12, 12, X)$ with the initial objects in it
Output: picture $(\mathbb{F}, 12, 12, X)$ containing the surface skeleton
`// Distance mapping and identifying CMB's`
1 $DT \leftarrow$ `computeDT`(X, d_{12})
2 $A \leftarrow$ `CMB`(DT, d_{12})
`// Forward thinning`
3 **for** $k \leftarrow 1$ **to** $\max(DT)$ **do**
4 $\quad$ **foreach** $p \in X$ **do**
5 $\quad\quad$ **if** $DT(p) = k$, p is *simple* and $N_{12}(p) \cap A = \emptyset$ **then**
6 $\quad\quad\quad$ $X \leftarrow X \setminus \{p\}$

$\quad$ `// Backward thinning`
7 **repeat**
8 $\quad$ $changed1 \leftarrow false$
9 $\quad$ **for** $k \leftarrow \max(DT)$ **downto** 1 **do**
10 $\quad\quad$ **repeat**
11 $\quad\quad\quad$ $changed2 \leftarrow false$
12 $\quad\quad\quad$ $D \leftarrow \{p \in X \setminus A \mid DT(p) = k$, and p is *simple*$\}$
13 $\quad\quad\quad$ **foreach** $p \in D$ **do**
14 $\quad\quad\quad\quad$ **if** p is *simple* and not a *surface edge point* **then**
15 $\quad\quad\quad\quad\quad$ $X \leftarrow X \setminus \{p\}$
16 $\quad\quad\quad\quad\quad$ $D \leftarrow D \setminus \{p\}$
17 $\quad\quad\quad\quad\quad$ $changed1 \leftarrow true$
18 $\quad\quad\quad\quad\quad$ $changed2 \leftarrow true$

19 $\quad\quad$ **until** $changed2 = false$
20 **until** $changed1 = false$

Theorem 2. *The runtime complexity of Strand's algorithm (see Algorithm 1) is $O(|\mathcal{P}|^{4/3})$.*

Proof. The first phase of the algorithm is linear since computing the d_{12} distance transform requires two raster scans [18], and the sequential detection of CMB's takes one extra scan. Forward thinning is linear too, because the object point's local neighborhood is examined exactly once. In worst case, border points of the original object may appear as surface edge or line endpoints, which means the algorithm may classify black points incorrectly in the first iteration. As a consequence, the length of these segments is equal to half of the object's thickness, which we refer to as T_{obj}.

During the *for* loop in Line 9, all remaining object points are visited, but only a one-unit layer from each segment can be peeled due to the descending distance-based visiting order, since only simple points can be deletable. Hence, each object point will be visited at most T_{obj} times until the algorithm terminates, so the runtime complexity is $O(T_{obj} \cdot |\mathcal{P}|)$. T_{obj} is maximal, if the input picture has a cubic shape, i.e. it coincides with its bounding box. In this case, $T_{obj} = \sqrt[3]{|\mathcal{P}|} / 2$, if all points are black in the input picture. By inserting it into the previous formula, we get $O(\sqrt[3]{|\mathcal{P}|}/2 \cdot |\mathcal{P}|) = O(|\mathcal{P}|^{4/3})$, which is not just the third phase's, but also the whole algorithm's runtime complexity. $\square$

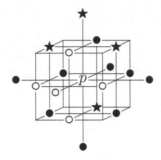

Fig. 3. Example for an unfavorable visiting order during sequential thinning. Border point p becomes non-simple after deletion of its neighbors marked "★".

Original object	Result of forward	Final skeleton
(84 911)	thinning	(6 847)

Fig. 4. Result of Strand's algorithm on an A-shaped object. In the middle figure the longest branch is depicted in red, whose endpoint was a border point in the initial picture. In the right figure CMB's are gray and further skeleton voxels are red. Numbers in parentheses show the number of black points. (Color figure online)

4   Two Modified Versions

In order to construct linear time algorithms, we merge the thinning phases in Algorithm 1 and simplify the organization of the thinning iterations. Our deletion rule also preserves non-simple points and surface edge points. Each sequential thinning iteration consists of two substeps in our modified algorithms. First, we collect the deletable points from the actual picture into set D. Then we individually delete each point that is still simple when visited. In order to lessen the occurrence of false skeleton points, elements of D are visited twice.

The first improved variant, called APT, is an anchor-based thinning method that considers local maxima instead of CMB's to be safe skeletal points because many CMB's are proved to be "false" skeleton points on weighted distance maps [4]. At the end of each iteration, all visited but non-deleted points are insterted into set S of skeleton points. This operation guarantees that not any already visited border point will be examined during the further iterations.

The second improved variant, called DOT, is a distance-ordered thinning method that omits the detection of anchor points, i.e. the set A is not used in this version. The border points are visited in ascending order of their distance value during the thinning phase. In case of d_{18} distance, collection of deletable points must be repeated twice, since some points with corresponding distance value may be interior points during the first examination.

Algorithm 2: Anchor-preserving thinning - APT

Input: picture $(\mathbb{F}, 12, 12, X)$ with the initial objects in it and distance d
Output: picture $(\mathbb{F}, 12, 12, S)$ containing the surface skeleton
`// Distance mapping and identifying local maxima`
1 $DT \leftarrow \texttt{computeDT}(X, d)$
2 $A \leftarrow \texttt{LocalMaxima}(DT, d)$
3 $S \leftarrow \emptyset$
`// Thinning`
4 **repeat**
5      $\quad L \leftarrow \{p \in X \setminus (S \cup A) \mid p \text{ is a } border\ point\}$
6      $\quad D \leftarrow \{p \in L \mid p \text{ is } simple \text{ for } X \text{ and not a } surface\ edge\ point\}$
7      $\quad t \leftarrow 0$
8      $\quad$**repeat**
9          $\quad\quad t \leftarrow t + 1$
10         $\quad\quad$**foreach** $p \in D$ **do**
11            $\quad\quad\quad$**if** p is $simple$ for X **then**
12               $\quad\quad\quad\quad X \leftarrow X \setminus \{p\}$
13               $\quad\quad\quad\quad L \leftarrow L \setminus \{p\}$
14               $\quad\quad\quad\quad D \leftarrow D \setminus \{p\}$
15      $\quad$**until** $t = 2$ or no points are deleted
16      $\quad S \leftarrow S \cup L$
17 **until** $D = \emptyset$

Theorem 3. *The runtime complexity of Algorithm 2 and Algorithm 3 is linear.*

Proof. The distance map is computable in linear time for the d_{12}, d_{18}, and any Chamfer distance [5,18]. Moreover, there are also linear time adaptations of the (non-errorfree) Euclidean distance transform for the FCC grid [16].

Algorithm 3: Distance-ordered thinning - DOT

Input: picture $(\mathbb{F}, 12, 12, X)$ with the initial objects in it and distance d
Output: picture $(\mathbb{F}, 12, 12, X)$ containing the surface skeleton
`// Distance mapping`
1 $DT \leftarrow$ `computeDT`(X, d)
`// Thinning`
2 **if** $d = d_{18}$ **then**
3 $\quad$ $it \leftarrow 2$
4 **else**
5 $\quad$ $it \leftarrow 1$
6 **for** $k \leftarrow 1$ **to** $\max(DT)$ **do**
7 $\quad$ **for** $l \leftarrow 1$ **to** it **do**
8 $\quad\quad$ $D \leftarrow \{p \in X \mid DT(p) = k, p$ is *simple* and not a *surface edge point*$\}$
9 $\quad\quad$ $t \leftarrow 0$
10 $\quad\quad$ **repeat**
11 $\quad\quad\quad$ $t \leftarrow t + 1$
12 $\quad\quad\quad$ **foreach** $p \in D$ **do**
13 $\quad\quad\quad\quad$ **if** p is *simple* **then**
14 $\quad\quad\quad\quad\quad$ $X \leftarrow X \setminus \{p\}$
15 $\quad\quad\quad\quad\quad$ $D \leftarrow D \setminus \{p\}$
16 $\quad\quad$ **until** $t = 2$ or no points are deleted

During the thinning phase, each object point is visited in exactly one iteration. In Algorithm 2, this is ensured by collecting the previously investigated but not deleted, i.e. skeleton points into set S. In Algorithm 3, the distance-ordered strategy guarantees this property. It is easy to see that each border point is visited up to k times in each thinning iteration, where $k = 4$ for Algorithm 3 with d_{18} distance, otherwise $k = 2$. As a consequence, all object points are visited maximum two or four times during the thinning phase. Hence, the computational cost is $O(k \cdot |\mathcal{P}|)$, $k \in \{2, 4\}$. $\qquad\square$

5   Results

The reported algorithms were tested on numerous objects of different shapes. Figures 5, 6 and 7 show the produced surface skeletons from different distance maps. In the case of methods SDT and APT the gray and red points indicate the anchor points and further skeletal points, respectively. Numbers in parentheses show the number of black points. In order to visualise the sensitivity of the sequential thinning to the visiting order of border points, we also extracted the surface skeleton of two rotated versions of the amphora-shaped object (see Fig. 6).

We can observe that the anchor-preserving thinning (i.e., APT) method generates less insignificant surface patches than Strand's (i.e., SDT) algorithm. Most of the unwanted branches are not generated by the purely distance-ordered algorithm. However, it may overshrink the object, especially in d_{12} distance map. This phenomenon is successfully handled by choosing weighted distance suggested in [5].

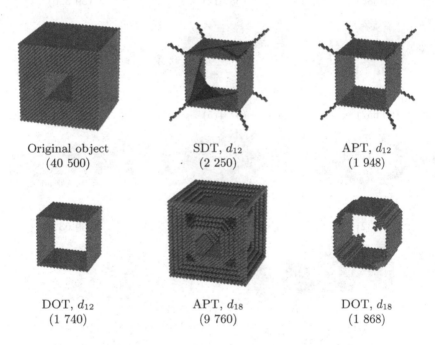

| Original object | SDT, d_{12} | APT, d_{12} |
| (40 500) | (2 250) | (1 948) |

| DOT, d_{12} | APT, d_{18} | DOT, d_{18} |
| (1 740) | (9 760) | (1 868) |

Fig. 5. Produced surface skeletons of a holey cube with various parameters.

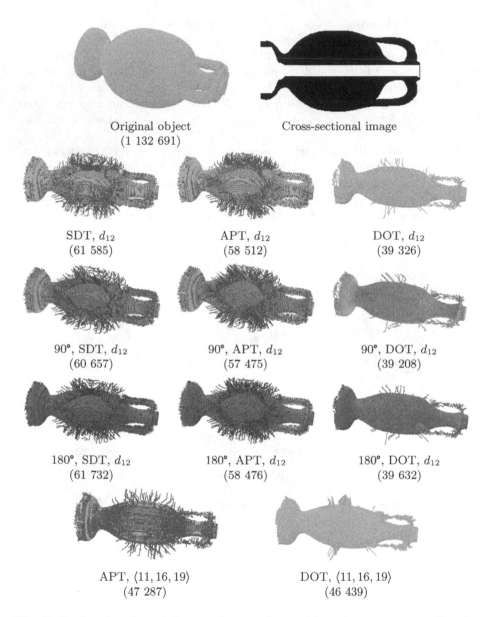

Original object
(1 132 691)

Cross-sectional image

SDT, d_{12}
(61 585)

APT, d_{12}
(58 512)

DOT, d_{12}
(39 326)

90°, SDT, d_{12}
(60 657)

90°, APT, d_{12}
(57 475)

90°, DOT, d_{12}
(39 208)

180°, SDT, d_{12}
(61 732)

180°, APT, d_{12}
(58 476)

180°, DOT, d_{12}
(39 632)

APT, $\langle 11, 16, 19 \rangle$
(47 287)

DOT, $\langle 11, 16, 19 \rangle$
(46 439)

Fig. 6. Produced surface skeletons of an amphora with various parameters. Results from the rotated objects can be found in third and fourth rows. Note that all objects are depicted in the same orientation. The rotation was performed around the vertical axis.

| Original object (38 640) | SDT, d_{12} (7 491) | APT, d_{12} (6 631) |
| DOT, d_{12} (3 842) | APT, $\langle 2,3 \rangle$ (5 488) | DOT, $\langle 2,3 \rangle$ (4 570) |

Fig. 7. Produced surface skeletons of a P-shaped object with various parameters.

6 Conclusion and Future Work

Both the proposed algorithms APT and DOT have linear runtime complexity. All examined algorithms preserve topology due to the fact that only a single simple point is deleted at a time. Furthermore, unlike Strand's method, algorithms APT and DOT are not constrained to a given distance.

Future research will be directed to constructing similar distance-based algorithms combined with parallel thinning strategies and adapting the presented results to $(18, 12)$ and $(12, 18)$ pictures. Moreover, quantitative comparison of these algorithms will be presented as well.

Acknowledgments. This research was supported by the project "Integrated program for training new generation of scientists in the fields of computer science", no EFOP-3.6.3-VEKOP-16-2017-00002. The project has been supported by the European Union and co-funded by the European Social Fund.

References

1. Arcelli, C., Sanniti di Baja, G.: Finding local maxima in a pseudo-Euclidean distance transform. Comput. Vision, Graphics Im. Proc. **43**(3), 361–367 (1988)
2. Arcelli, C., Sanniti di Baja, G.: Ridge points in Euclidean distance maps. Pattern Recogn. Lett. **13**(4), 237–243 (1992)
3. Borgefors, G.: Distance transformations in arbitrary dimensions. Comput. Vision, Graphics, Im. Proc. **27**(3), 321–345 (1984)

4. Borgefors, G., Nyström, I., Sanniti di Baja, G.: Discrete Skeletons from Distance Transforms in 2D and 3D. In: Siddiqi, K., Pizer, S.M. (eds.) Medial Representations. Computational Imaging and Vision, vol. 37. Springer, Dordrecht (2008). https://doi.org/10.1007/978-1-4020-8658-8_5
5. Fouard, C., Strand, R., Borgefors, G.: Weighted distance transforms generalized to modules and their computation on point lattices. Patt. Rec. **40**(9), 2453–2474 (2007)
6. Gau, C.J., Kong, T.Y.: Minimal nonsimple sets of voxels in binary pictures on a face-centered cubic grid. Int. J. Patt. Rec. Artif. Intell. **13**(4), 485–502 (1999)
7. Kittel, C.: Crystal structures. In: Kittel, C. (ed.) Introduction to Solid State Physics, 8th edn. Wiley, New York (2004)
8. Kong, T.Y.: On topology preservation in 2-D and 3-D thinning. Int. J. Patt. Rec. Artif. Intell. **9**(5), 813–844 (1995)
9. Kong, T.Y., Rosenfeld, A.: Digital topology: introduction and survey. Comput. Vision, Graphics, Im. Proc. **48**(3), 357–393 (1989)
10. Lam, L., Lee, S.-W., Suen, C.Y.: Thinning methodologies - a comprehensive survey. IEEE Trans. Pattern Anal. Mach. Intell. **14**(9), 869–885 (1992)
11. Marchand-Maillet, S., Sharaiha Y.M.: Binary Digital Picture Processing: A Discrete Approach. Academic Press (2000)
12. Saha, P.K., Borgefors, G., Sanniti di Baja, G.: A survey on skeletonization algorithms and their applications. Patt. Rec. Lett. **76**(1), 3–12 (2016)
13. Saha, P.K., Borgefors, G., Sanniti di Baja, G.: Skeletonization: Theory, Methods and Applications. 1st edn. Academic Press (2017)
14. Saito, T., Toriwaki, J.: A sequential thinning algorithm for three dimensional digital pictures using the Euclidean distance transformation. In: Proceedings of the 9th Scandinavian Conference on Picture Analysis, pp. 507–516, Uppsala, Sweden (1995)
15. Strand, R.: Surface skeletons in grids with non-cubic voxels. In: Proceedings of the 17th International Conference on Pattern Recognition, 2004, ICPR 2004, vol. 1, pp. 548–551. IEEE, Cambridge (2004)
16. Strand, R.: The Euclidean distance transform applied to the FCC and BCC grids. In: Marques, J.S., Pérez de la Blanca, N., Pina, P. (eds.) IbPRIA 2005. LNCS, vol. 3522, pp. 243–250. Springer, Heidelberg (2005). https://doi.org/10.1007/11492429_30
17. Strand, R.: The face-centered cubic grid and the body-centered cubic grid: a literature survey. Technical Report 35, Centre for Image Analysis, Uppsala University, Uppsala, Sweden (2005)
18. Strand, R., Borgefors, G.: Distance transforms for three-dimensional grids with non-cubic voxels. Comput. Vision Im. Underst. **100**(3), 294–311 (2005)
19. Svensson, S.: Reversible surface skeletons of 3D objects by iterative thinning of distance transforms. In: Bertrand, G., Imiya, A., Klette, R. (eds.) Digital and Image Geometry. LNCS, vol. 2243, pp. 400–411. Springer, Heidelberg (2001). https://doi.org/10.1007/3-540-45576-0_24

Homotopic Digital Rigid Motion: An Optimization Approach on Cellular Complexes

Nicolas Passat[1]([envelope]) [iD], Phuc Ngo[2] [iD], and Yukiko Kenmochi[3] [iD]

[1] Université de Reims Champagne Ardenne, CReSTIC EA 3804, 51097 Reims, France
nicolas.passat@univ-reims.fr
[2] Université de Lorraine, LORIA, UMR, 7503 Villers-lès-Nancy, France
[3] LIGM, Univ Gustave Eiffel, CNRS, Marne-la-Vallée, France

Abstract. Topology preservation is a property of rigid motions in $\mathbb{R}^2$, but not in $\mathbb{Z}^2$. In this article, given a binary object $X \subset \mathbb{Z}^2$ and a rational rigid motion $\mathcal{R}$, we propose a method for building a binary object $X_{\mathcal{R}} \subset \mathbb{Z}^2$ resulting from the application of $\mathcal{R}$ on a binary object X. Our purpose is to preserve the homotopy type between X and $X_{\mathcal{R}}$. To this end, we formulate the construction of $X_{\mathcal{R}}$ from X as an optimization problem in the space of cellular complexes with the notion of collapse on complexes. More precisely, we define a cellular space $\mathbb{H}$ by superimposition of two cubical spaces $\mathbb{F}$ and $\mathbb{G}$ corresponding to the canonical Cartesian grid of $\mathbb{Z}^2$ where X is defined, and the Cartesian grid induced by the rigid motion $\mathcal{R}$, respectively. The object $X_{\mathcal{R}}$ is then computed by building a homotopic transformation within the space $\mathbb{H}$, starting from the cubical complex in $\mathbb{G}$ resulting from the rigid motion of X with respect to $\mathcal{R}$ and ending at a complex fitting $X_{\mathcal{R}}$ in $\mathbb{F}$ that can be embedded back into $\mathbb{Z}^2$.

Keywords: Rigid motions · Cartesian grid · Homotopy type · Binary images · Cubical complexes · Cellular complexes

1 Introduction

Rigid motions built by composition of rotations and translations are isometric transformations in the Euclidean spaces $\mathbb{R}^n$ ($n \geq 2$). In particular, they are bijective and they preserve geometric and topological properties between an object and its image. This is no longer the case when rigid motions are considered in the Cartesian grids $\mathbb{Z}^n$.

Translations [4,14], rotations [1,2,5,9,19,20,23,25] and more generally rigid motions [16–18,21,24] in the Cartesian grids have been studied with various purposes: describing the combinatorial structure of these transformations with respect to $\mathbb{R}^n$ vs. $\mathbb{Z}^n$ [4,5,14,22,26], guaranteeing their bijectivity [1,2,9,19,23–25] or transitivity [20] in $\mathbb{Z}^n$, preserving geometrical properties [17] and, less frequently, ensuring their topological invariance [16,18] in $\mathbb{Z}^n$. These are non-trivial questions, and their difficulty increases with the dimension of the Cartesian grid [21]. Indeed, most of these works deal with $\mathbb{Z}^2$ [1,2,4,5,9,14,16,18–20,24,25]; fewer with $\mathbb{Z}^3$ [17,23,26].

This work was supported by the French *Agence Nationale de la Recherche* (Grants ANR-15-CE23-0009 and ANR-18-CE23-0025).

© Springer Nature Switzerland AG 2021
J. Lindblad et al. (Eds.): DGMM 2021, LNCS 12708, pp. 189–201, 2021.
https://doi.org/10.1007/978-3-030-76657-3_13

In this preliminary study we investigate how it may be possible to preserve the topological properties of a digital object defined in the Cartesian grid when applying a rigid motion. In [18] a specific family of digital objects in $\mathbb{Z}^2$, called "regular", was proved to preserve their topology under any rigid motion. But all the digital objects in $\mathbb{Z}^2$ are not regular, and the required modifications for generating a regular object from a non-regular one induce asymmetric operations between the object and its background. In [16] the putative topology preservation between an object and its image in $\mathbb{Z}^2$ by a rigid motion was checked by searching a path in the combinatorial space of digital rigid motions that corresponds to a point-by-point homotopic transformation between both. But this process allows to assess the topological invariance, not to ensure it.

We propose a new, alternative way of tackling the problem of digital rigid motion under the constraint of topological invariance. As in [16,18], we consider the case of digital objects in $\mathbb{Z}^2$. Since a digital object X and its usual digital image by a rigid motion $\mathcal{R}$ are not guaranteed to present the same topology, our purpose is to compute a digital object $\mathsf{X}_\mathcal{R}$ that (1) has the same topology as X and (2) is "as similar as possible" to the usual digital image of X by $\mathcal{R}$. To reach that goal, we embed our digital objects in the Euclidean space and we process them in the (continuous but discrete) space of cellular complexes. This allows us to model/manipulate these objects in a way compliant with both their digital nature and their continuous interpretation (in particular from a topological point of view), but also to carry out basic transformations at a scale finer than that of $\mathbb{Z}^2$. The definition of $\mathsf{X}_\mathcal{R}$ from X and $\mathcal{R}$ is then formulated as an optimization problem, which presents similarities with the topology-preserving paradigms developed in the framework of deformable models.

2 Problem Statement

Let $\mathsf{X} \subset \mathbb{Z}^2$ be a digital object. Let $X \subset \mathbb{R}^2$ be the continuous analogue of X, defined as $X = \mathsf{X} \oplus \square$ where $\oplus$ is the usual dilation operator and $\square$ is the structuring element $[\frac{1}{2}, \frac{1}{2}]^2 \subset \mathbb{R}^2$. In other words, X is the union of the pixels (i.e. closed, unit squares) centered at the points of X. We note $\square : 2^{\mathbb{Z}^2} \to 2^{\mathbb{R}^2}$ the function that defines this continuous analogue, i.e. such that $\square(\mathsf{X}) = \mathsf{X} \oplus \square = X$.

Let $\mathcal{R} : \mathbb{R}^2 \to \mathbb{R}^2$ be a rigid motion, defined as the composition of a rotation and a translation. Usually, the image of the digital object $\mathsf{X} \subset \mathbb{Z}^2$ by the rigid motion $\mathcal{R}$, noted $\mathsf{X}_\mathcal{R}$ is a digital object of $\mathbb{Z}^2$ defined as $\mathsf{X}_\mathcal{R} = X_\mathcal{R} \cap \mathbb{Z}^2$, with $X_\mathcal{R} = \mathcal{R}(X) = \{\mathcal{R}(\mathbf{x}) \mid \mathbf{x} \in X\} \subset \mathbb{R}^2$. In other words, $\mathsf{X}_\mathcal{R}$ is defined as the Gauss digitization of the continuous object $X_\mathcal{R}$. We note $\square : 2^{\mathbb{R}^2} \to 2^{\mathbb{Z}^2}$ the function that defines the Gauss digitization of a continuous object, i.e. such that $\square(Y) = Y \cap \mathbb{Z}^2$. The usual overall process is exemplified in Fig. 1.

Our purpose is that $\mathsf{X}_\mathcal{R}$ be as similar as possible to X, up to the rigid motion $\mathcal{R}$. Reaching the best similarity can be formalized as solving the following optimization problem:

$$\mathsf{X}_\mathcal{R} = \arg_{Y \in 2^{\mathbb{Z}^2}} \min \mathcal{D}_{\mathcal{R},\mathsf{X}}(Y) \qquad (1)$$

where $\mathcal{D}_{\mathcal{R},\mathsf{X}} : 2^{\mathbb{Z}^2} \to \mathbb{R}_+$ is an error measure (parameterized by $\mathcal{R}$ and X) that allows us to estimate the (dis)similarity between two digital objects. For instance, when considering the Gauss digitization we set $\mathcal{D}_{\mathcal{R},\mathsf{X}}^{\square}(Y) = |\square(\mathcal{R}(\square(\mathsf{X}))) \setminus Y| + |Y \setminus \square(\mathcal{R}(\square(\mathsf{X})))|$ and the unique solution $\mathsf{X}_\mathcal{R}$ is reached when $\mathcal{D}_{\mathcal{R},\mathsf{X}}^{\square}(\mathsf{X}_\mathcal{R}) = 0$.

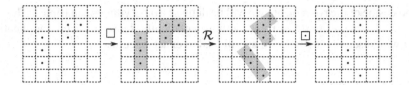

Fig. 1. Digitized rigid motion (here, by Gauss digitization). From left to right: $X \subset \mathbb{Z}^2$, $X = \square(X) \subset \mathbb{R}^2$, $\mathcal{R}(\square(X)) \subset \mathbb{R}^2$ and the result $\square(\mathcal{R}(\square(X))) = X_{\mathcal{R}} \subset \mathbb{Z}^2$. (Dots: points of $\mathbb{Z}^2$; grey zones: parts of $\mathbb{R}^2$). This transformation does not preserve the topology between X and $X_{\mathcal{R}}$.

However, in this work, we also want to guarantee that $X_{\mathcal{R}}$ has the same topology as X. In other words, we now want to solve the optimization problem (1) under an additional constraint that excludes the candidates $Y \subset \mathbb{Z}^2$ that have a different topology from X. Still considering the Gauss digitization policy, a solution $X_{\mathcal{R}}$ may then be reached for $\mathcal{D}^{\square}_{\mathcal{R},X}(X_{\mathcal{R}}) > 0$, i.e. without fully satisfying the minimality requirements on the error measure. Our purpose is to solve this constrained optimization problem, i.e. to develop a method for computing the homotopic images of digital objects under rigid motions.

3   Hypotheses

Digital Topology, Adjacency – The digital objects of $\mathbb{Z}^2$ are considered in the usual framework of digital topology. In this framework, an object X has to be considered with the 8- (resp. 4-) adjacency, whereas its background $\mathbb{Z}^2 \setminus X$ is considered with the dual 4- (resp. 8-) adjacency, in order to avoid topological paradoxes related to the Jordan theorem. Without loss of generality, we choose to consider $X \subset \mathbb{Z}^2$ with the 8-adjacency (otherwise, it is sufficient to consider the complementary of X instead of X as the object).

Cellular/Cubical Complexes – In order to handle the digital-continuous analogy between the objects of $\mathbb{Z}^2$ and those of $\mathbb{R}^2$, we consider the (intermediate) framework of cellular complexes, formalized in [11] in the case of cubical complexes induced by the Cartesian grid and proved compliant with both digital and continuous topologies [12,15]. The cellular complexes can be generalized, without loss of generality to non-cubic partitions (see e.g. [6]), and in particular to partitions of $\mathbb{R}^2$ made of convex polygons.

Homotopy Type, Simple Points/Cells – By "same topology", we mean that the objects we manipulate should have the same homotopy type. This choice is relevant for two reasons. First, in dimension 2, the homotopy type is equivalent to most of the other usual topological invariants. Second, there exist efficient topological tools that allow one to modify an object whereas preserving its homotopy type. In particular, we will rely on the notion of simple points/simple cells that are defined in the framework of digital topology and cubical complexes [8], and which can be extended without difficulty to any cellular complex thanks to the atomic notion of collapse.

Rational Rigid Motions – We define our rigid motions such that their parameters have rational values. In particular, the translation vectors will be defined on $\mathbb{Q}^2$ whereas the sine and cosine of the rotation angles will be defined from Pythagorean triples. This will allow us to handle a family of rigid motions sufficiently dense for actual applications [3], but with discrete parameters that will lead to exact calculus.

4   Rigid Motions

In the sequel, a point of $\mathbb{R}^2$ is noted in bold ($\mathbf{p}$); its coordinates are noted with subscripts ($\mathbf{p} = (p_x, p_y)^t$). The transpose symbol is omitted by abuse of notation ($\mathbf{p} = (p_x, p_y)$).

4.1   Basics on Rigid Motions

Let $\theta \in [0, 2\pi)$. Let $\mathbf{t} \in \mathbb{R}^2$. The rigid motion $\mathcal{R}_{(\theta,\mathbf{t})} : \mathbb{R}^2 \to \mathbb{R}^2$ is defined, for any $\mathbf{p} \in \mathbb{R}^2$ as:

$$\mathcal{R}_{(\theta,\mathbf{t})}(\mathbf{p}) = R(\theta) \cdot \mathbf{p} + \mathbf{t} \quad \text{where} \quad R(\theta) = \begin{bmatrix} \cos\theta & -\sin\theta \\ \sin\theta & \cos\theta \end{bmatrix} \tag{2}$$

is the rotation matrix of angle θ and $\mathbf{t}$ is the translation vector.

As stated in Sect. 3, we only consider rotation angles within the subset of $[0, 2\pi)$ that contains values built from Pythagorean triples [3], called rational rotations. More precisely, for any such θ, there exists a triple $(a, b, c) \in \mathbb{Z}^3$ such that $a^2 + b^2 = c^2$, that satisfies $\cos\theta = a/c$ and $\sin\theta = b/c$. In other words, we have the guarantee that $\cos\theta$ and $\sin\theta$ are rationals. In addition, we will also assume that $\mathbf{t} \in \mathbb{Q}^2$.

From now on, we will set $\alpha = \cos\theta = a/c$ and $\beta = \sin\theta = b/c \in [-1, 1] \cap \mathbb{Q}$ and the rotation matrix of Eq. (2) is rewritten as:

$$R(\theta) = R(\alpha, \beta) = \begin{bmatrix} \alpha & -\beta \\ \beta & \alpha \end{bmatrix} = \frac{1}{c} \begin{bmatrix} a & -b \\ b & a \end{bmatrix} \tag{3}$$

The rigid motion $\mathcal{R}_{(\theta,\mathbf{t})}$ of Eq. (2), simply noted $\mathcal{R}$ from now on, can then be expressed from $(\alpha, \beta, t_x, t_y) \in \mathbb{Q}^4$, with $\alpha^2 + \beta^2 = 1$, and is called rational rigid motion. In particular, for any $\mathbf{p} \in \mathbb{Q}^2$, we have:

$$\mathcal{R}(\mathbf{p}) = \begin{pmatrix} \alpha p_x - \beta p_y + t_x \\ \beta p_x + \alpha p_y + t_y \end{pmatrix} \in \mathbb{Q}^2 \tag{4}$$

4.2   Rigid Motion of a Digital Object

Let $\mathsf{X} \subset \mathbb{Z}^2$ be a digital object. Let $\mathcal{R} : \mathbb{Q}^2 \to \mathbb{Q}^2$ be a rational rigid motion such as defined by Eq. (4). Our purpose is to compute a digital object $\mathsf{X}_\mathcal{R} \subset \mathbb{Z}^2$ that corresponds to the image of X by $\mathcal{R}$, with regards to our two constraints: the preservation of the homotopy type between X and $\mathsf{X}_\mathcal{R}$; and the optimality of $\mathsf{X}_\mathcal{R}$ with respect to the optimization problem (1).

In general, the object $\mathcal{R}(\mathsf{X}) = \{\mathcal{R}(\mathbf{x}) \mid \mathbf{x} \in \mathsf{X}\}$ does not fulfill the required properties. Indeed, by definition, we have $\mathcal{R}(\mathsf{X}) \subset \mathbb{Q}^2$, but in general we do not have $\mathcal{R}(\mathsf{X}) \subset \mathbb{Z}^2$. A usual solution consists of applying the rigid motion $\mathcal{R}$ on a continuous analogue of X. This continuous analogue is often chosen as $X = \square(\mathsf{X})$, i.e. by associating to each

$\mathbf{x} \in X$ the pixel centered on $\mathbf{x}$. We then obtain a continuous object $X \subset \mathbb{R}^2$, and we can relevantly build $X_{\mathcal{R}} = \mathcal{R}(X) = \{\mathcal{R}(\mathbf{x}) \mid \mathbf{x} \in X\}$. This object $X_{\mathcal{R}}$ has the same topology as X and thus as X [12, 15] but it is not defined in $\mathbb{Z}^2$. To define a digital object $\mathsf{X}_{\mathcal{R}}$ from $X_{\mathcal{R}}$, we generally rely on a digitization. But then, we can no longer guarantee that $\mathsf{X}_{\mathcal{R}}$ has the same topology as $X_{\mathcal{R}}$, X and X.

To tackle this issue, once $X_{\mathcal{R}} = \mathcal{R}(X) = \mathcal{R}(\square(\mathsf{X})) \subset \mathbb{R}^2$ has been built, we propose to transform it into another continuous object $Y \subset \mathbb{R}^2$, with three constraints: (1) the transformation between $X_{\mathcal{R}}$ and Y has to be homotopic; (2) Y may be the continuous analogue of a digital object of $\mathbb{Z}^2$, i.e. $Y = \square(\boxdot(Y))$; and (3) the digital object $\mathsf{Y} = \boxdot(Y) \subset \mathbb{Z}^2$ associated to Y may satisfy the optimality in Eq. (1) for the chosen measure $\mathcal{D}_{\mathcal{R},\mathsf{X}}$.

To reach that goal, we propose to work in the space of cellular complexes, that allows to model the continuous space $\mathbb{R}^2$ in a discrete way, but also to carry out homotopic transformations.

5   Cellular Complexes

5.1   Basics on Cellular Complexes

Let $P \subset \mathbb{R}^2$ be a closed, convex polygon. Let $\mathring{P}$ be the interior of P and $\partial P = P \setminus \mathring{P}$ the boundary of P. We note $\mathcal{P}(P) = \{\mathring{P}\}$. Let $E \subset \partial P$ be a maximal, closed line segment of ∂P. Let $\mathring{E}$ be the interior (i.e. the open line segment) of E, and $\partial E = E \setminus \mathring{E}$ be the boundary of E. The open line segment $\mathring{E}$ is called an edge of P. We note $\mathcal{E}(P)$ the set of all the edges of P. Let $\mathbf{v} \in \partial E$ be a point of ∂E; the singleton set $V = \{\mathbf{v}\}$ is called a vertex of P. We note $\mathcal{V}(P)$ the set of all the vertices of P. The set $\mathcal{F}(P) = \mathcal{P}(P) \cup \mathcal{E}(P) \cup \mathcal{V}(P)$ is a partition of P.

Let $\Omega \subset \mathbb{R}^2$ be a closed, convex polygon. Let $\mathcal{K}$ be a set of closed, convex polygons such that $\Omega = \bigcup \mathcal{K}$ and for any two distinct polygons $P_1, P_2 \in \mathcal{K}$, we have $\mathring{P}_1 \cap \mathring{P}_2 = \emptyset$. We set $\mathbb{K}(\Omega) = \bigcup_{P \in \mathcal{K}} \mathcal{F}(P)$. It is plain that $\mathbb{K}(\Omega)$ is a partition of Ω. We call $\mathbb{K}(\Omega)$, or simply $\mathbb{K}$, a cellular space (associated to Ω).

Each element $\mathfrak{f}_2$ (resp. $\mathfrak{f}_1$, resp. $\mathfrak{f}_0$) of $\mathbb{K}$ which is the interior (resp. an edge, resp. a vertex) of a polygon $P \in \mathcal{K}$ is called a 2-face (resp. 1-face, resp. 0-face). We set $\mathbb{K}_d$ ($0 \leq d \leq 2$, $d \in \mathbb{Z}$) the set of all the d-faces of $\mathbb{K}$. More generally, each element of $\mathbb{K}$ is called a face.

Let $\mathfrak{f} \in \mathbb{K}$ be a face. The cell $C(\mathfrak{f})$ induced by $\mathfrak{f}$ is the subset of faces of $\mathbb{K}$ such that $\bigcup C(\mathfrak{f})$ is the smallest closed set that includes $\mathfrak{f}$. If $\mathfrak{f}_0$ is a 0-face, then $C(\mathfrak{f}_0) = \{\mathfrak{f}_0\}$. If $\mathfrak{f}_1$ is a 1-face, then $C(\mathfrak{f}_1) = \{\mathfrak{f}_1, \mathfrak{f}_0^1, \mathfrak{f}_0^2\}$ with $\mathfrak{f}_0^1, \mathfrak{f}_0^2$ the two vertices bounding $\mathfrak{f}_1$, such that $\bigcup C(\mathfrak{f}_1)$ is a closed line segment. If $\mathfrak{f}_2$ is a 2-face, then $C(\mathfrak{f}_2) = \{\mathfrak{f}_2, \mathfrak{f}_1^1, \ldots, \mathfrak{f}_1^k, \mathfrak{f}_0^1, \ldots, \mathfrak{f}_0^k\}$ ($k \geq 3$) and $\bigcup C(\mathfrak{f}_2)$ is the closed polygon of interior $\mathfrak{f}_2$ with k edges $\mathfrak{f}_1^\star$ and k vertices $\mathfrak{f}_0^\star$. For any cell $C(\mathfrak{f})$, the face $\mathfrak{f}$ is called the principal face of $C(\mathfrak{f})$, and $C(\mathfrak{f})$ is also called the closure of $\mathfrak{f}$. The star $S(\mathfrak{f})$ of a face $\mathfrak{f}$ is the set of all the faces $\mathfrak{f}'$ such that $\mathfrak{f} \in C(\mathfrak{f}')$.

Remark. A face $\mathfrak{f}$ and its induced cell $C(\mathfrak{f})$ are characterized by the list of the 0-faces in $C(\mathfrak{f})$. By abuse of notation, we will sometimes assimilate $\mathfrak{f}$ and $C(\mathfrak{f})$ to the sorted (e.g. clockwise) series of the k points $\mathbf{v}_i$ ($1 \leq i \leq k$) that correspond to these 0-faces $\{\mathbf{v}_i\}$.

A complex of $\mathbb{K}$ is a subset $K \subset \mathbb{K}$ defined as a union of cells of $\mathbb{K}$. The embedding of K into $\mathbb{R}^2$ is the set noted $\Pi_{\mathbb{R}^2}(K) \subset \mathbb{R}^2$ defined by $\Pi_{\mathbb{R}^2}(K) = \bigcup K$. Let $X \subset \mathbb{R}^2$. If there exists a complex $K \subset \mathbb{K}$ such that $X = \Pi_{\mathbb{R}^2}(K)$, then we say that K is the embedding of X into $\mathbb{K}$ and we note $K = \Pi_{\mathbb{K}}(X)$.

5.2 The Initial Cubical Space $\mathbb{F}$

The initial digital object X is defined in $\mathbb{Z}^2$, and so is the final digital object $\mathsf{X}_{\mathcal{R}}$ that we aim to build. Both have a continuous analogue in $\mathbb{R}^2$. The continuous analogue X of X is defined as $X = \square(\mathsf{X})$. The continuous analogue Y of $\mathsf{X}_{\mathcal{R}}$ is characterized by $Y = \square(\mathsf{X}_{\mathcal{R}})$ (see Sect. 4.2). In other words, both are defined as unions of unit, closed squares (i.e. pixels) centered on the points of X and $\mathsf{X}_{\mathcal{R}}$, respectively. In order to model/manipulate these two continuous objects X and Y of $\mathbb{R}^2$ as complexes, we build the cellular (actually, cubical) complex space $\mathbb{F}$ as follows.

Let $\Delta = \mathbb{Z} + \frac{1}{2} = \{k + \frac{1}{2} \mid k \in \mathbb{Z}\}$. Let $\delta \in \Delta$. We define the vertical line $V_\delta \subset \mathbb{R}^2$ and the horizon line $H_\delta \subset \mathbb{R}^2$ by the following equations, respectively:

$$(V_\delta) \quad x - \delta = 0 \tag{5}$$

$$(H_\delta) \quad y - \delta = 0 \tag{6}$$

We set $\mathcal{V}_\Delta = \{V_\delta \mid \delta \in \Delta\}$, $\mathcal{H}_\Delta = \{H_\delta \mid \delta \in \Delta\}$ and $\mathcal{G}_\Delta = \mathcal{V}_\Delta \cup \mathcal{H}_\Delta$. This set $\mathcal{G}_\Delta$ is the square grid that subdivides $\mathbb{R}^2$ into unit squares centered on the points of $\mathbb{Z}^2$. In other words, $\mathcal{G}_\Delta$ generates the Voronoi diagram of $\mathbb{Z}^2$ in $\mathbb{R}^2$.

The induced cellular complex space $\mathbb{F}(\mathbb{R}^2)$, simply noted $\mathbb{F}$, is then composed of:

- the set of 0-faces $\mathbb{F}_0 = \{\{\mathbf{d}\} \mid \mathbf{d} \in \Delta^2\}$;
- the set of 1-faces $\mathbb{F}_1 = \{]\mathbf{d}, \mathbf{d} + \mathbf{e}_x[\mid \mathbf{d} \in \Delta^2\} \cup \{]\mathbf{d}, \mathbf{d} + \mathbf{e}_y[\mid \mathbf{d} \in \Delta^2\}$; and
- the set of 2-faces $\mathbb{F}_2 = \{]\mathbf{d}, \mathbf{d} + \mathbf{e}_x[\times]\mathbf{d}, \mathbf{d} + \mathbf{e}_y[\mid \mathbf{d} \in \Delta^2\}$;

where $\mathbf{e}_x = (1, 0)$ and $\mathbf{e}_y = (0, 1)$. In particular, we have $\bigcup \mathbb{F}_0 = \mathcal{V}_\Delta \cap \mathcal{H}_\Delta$, $\bigcup \mathbb{F}_1 = \mathcal{G}_\Delta \setminus (\mathcal{V}_\Delta \cap \mathcal{H}_\Delta)$ and $\bigcup \mathbb{F}_2 = \mathbb{R}^2 \setminus \mathcal{G}_\Delta$.

For a digital object $\mathsf{X} \subset \mathbb{Z}^2$ and its continuous analogue $X = \square(\mathsf{X})$, we define the associated complex $F = \Pi_{\mathbb{F}}(X)$ as:

$$F = \bigcup_{\mathbf{x} \in \mathsf{X}} C(\blacksquare(\mathbf{x})) = \{\mathfrak{f} \in \mathbb{F} \mid \mathfrak{f} \subset X\} \tag{7}$$

where $\blacksquare : \mathbb{Z}^2 \to \mathbb{F}_2$ is the bijective function that maps each $\mathbf{p} \in \mathbb{Z}^2$ to the unit, open square (i.e. 2-face) $\blacksquare(\mathbf{p}) = \mathbf{p} \oplus] - \frac{1}{2}, \frac{1}{2}[^2$. We set $\mathbb{F}_d(F)$ $(0 \leq d \leq 2)$ the set of all the d-faces of F. In particular, we have:

$$X = \square(\mathsf{X}) = \bigcup \Pi_{\mathbb{F}}(X) = \Pi_{\mathbb{R}^2}(F) \tag{8}$$

$$\mathsf{X} = \boxdot(X) = \blacksquare^{-1}(\mathbb{F}_2(F)) \tag{9}$$

5.3   The Cubical Space $\mathbb{G}$ Induced by the Rigid Motion $\mathcal{R}$

The rigid motion $\mathcal{R}$ is applied on the continuous analogue $X \subset \mathbb{R}^2$ of X. The new continuous object $X_{\mathcal{R}} \subset \mathbb{R}^2$ is defined as $X_{\mathcal{R}} = \mathcal{R}(X) = \{\mathcal{R}(\mathbf{x}) \mid \mathbf{x} \in X\}$ (see Eq. (4)).

Similarly to X, that can be modeled by a complex F in the cubical space $\mathbb{F}$ defined in Sect. 5.2, the object $X_{\mathcal{R}}$ can also be modeled by a complex G in a cubical space $\mathbb{G}$. This second cubical space $\mathbb{G}$ is the image of $\mathbb{F}$ by the rigid motion $\mathcal{R}$. In particular, $\mathcal{R}$ trivially induces an isomorphism between these two cubical spaces.

More precisely, $\mathbb{G}$ derives from the square grid $\mathcal{R}(\mathscr{G}_\Delta)$ which subdivides $\mathbb{R}^2$ into unit squares centered on the points of $\mathcal{R}(\mathbb{Z}^2)$. We have $\mathcal{R}(\mathscr{G}_\Delta) = \mathcal{R}(\mathscr{V}_\Delta) \cup \mathcal{R}(\mathscr{H}_\Delta)$, with $\mathcal{R}(\mathscr{V}_\Delta) = \{\mathcal{R}(V_\delta) \mid \delta \in \Delta\}$ and $\mathcal{R}(\mathscr{H}_\Delta) = \{\mathcal{R}(H_\delta) \mid \delta \in \Delta\}$. For each $\delta \in \Delta$, the lines $\mathcal{R}(V_\delta)$ and $\mathcal{R}(H_\delta)$ are defined by the following equations, respectively:

$$(\mathcal{R}(V_\delta)) \qquad \alpha x + \beta y - \alpha t_x - \beta t_y - \delta = 0 \qquad\qquad (10)$$

$$(\mathcal{R}(H_\delta)) \qquad -\beta x + \alpha y + \beta t_x - \alpha t_y - \delta = 0 \qquad\qquad (11)$$

The induced cubical space $\mathbb{G}$ is then composed of the three sets of d-faces $\mathbb{G}_d = \mathcal{R}(\mathbb{F}_d) = \{\mathcal{R}(\mathfrak{f}) \mid \mathfrak{f} \in \mathbb{F}_d\}$ $(0 \le d \le 2)$.

The continuous object $X_{\mathcal{R}} \subset \mathbb{R}^2$ is then modeled by the complex $G = \Pi_{\mathbb{G}}(X_{\mathcal{R}}) \subset \mathbb{G}$:

$$G = \mathcal{R}(F) = \mathcal{R}(\Pi_{\mathbb{F}}(X)) = \{\mathcal{R}(\mathfrak{f}) \mid \mathfrak{f} \in \Pi_{\mathbb{F}}(X)\} \qquad\qquad (12)$$

We set $\mathbb{G}_d(G)$ $(0 \le d \le 2)$ the set of all the d-faces of G.

5.4   The Cellular Space $\mathbb{H}$ Refining the Cubical Spaces $\mathbb{F}$ and $\mathbb{G}$

Although $X_{\mathcal{R}}$ presents good topological properties with respect to X, it cannot be directly used for building the final digital object $\mathsf{X}_{\mathcal{R}}$. Indeed, $X_{\mathcal{R}}$ is the continuous analogue of a digital object defined on $\mathcal{R}(\mathbb{Z}^2)$ but not $\mathbb{Z}^2$. In other words, the complex G that models $X_{\mathcal{R}}$ is defined on $\mathbb{G}$ and not on $\mathbb{F}$.

At this stage, our purpose is to build from the complex G in $\mathbb{G}$, a new cubical complex H in $\mathbb{F}$, that will be used to finally define the resulting digital object $\mathsf{X}_{\mathcal{R}}$. In order to guarantee the preservation of the homotopy type between X and $\mathsf{X}_{\mathcal{R}}$, it is indeed necessary that G and H also have the same homotopy type, i.e. we have to build H from G via a homotopic transformation. This requires that both of these complexes be defined in the same cellular space.

Then, we build a new cellular space $\mathbb{H}$ that refines both $\mathbb{F}$ and $\mathbb{G}$. This space $\mathbb{H}$ is not cubical; its 2-faces are convex polygons (with 3 to 8 edges). Practically, $\mathbb{H}$ is built from the subdivision of the Euclidean plane $\mathbb{R}^2$ by the union of the two square grids $\mathscr{G}_\Delta$ and $\mathcal{R}(\mathscr{G}_\Delta)$. In particular, for each 2-face $\mathfrak{h}_2$ of $\mathbb{H}$, there exists exactly one 2-face $\mathfrak{f}_2$ of $\mathbb{F}$ and one 2-face $\mathfrak{g}_2$ of $\mathbb{G}$ such that $\mathfrak{h}_2 = \mathfrak{f}_2 \cap \mathfrak{g}_2$. Based on this property, we define the two functions $\phi : \mathbb{H}_2 \to \mathbb{F}_2$ and $\gamma : \mathbb{H}_2 \to \mathbb{G}_2$, such that $\phi(\mathfrak{h}_2) = \mathfrak{f}_2$ and $\gamma(\mathfrak{h}_2) = \mathfrak{g}_2$. Reversely, we build the two functions $\Phi : \mathbb{F}_2 \to 2^{\mathbb{H}_2}$ and $\Gamma : \mathbb{G}_2 \to 2^{\mathbb{H}_2}$ such that for any $\mathfrak{f}_2 \in \mathbb{F}_2$ and $\mathfrak{g}_2 \in \mathbb{G}_2$, we have $\Phi(\mathfrak{f}_2) = \phi^{-1}(\{\mathfrak{f}_2\}) = \{\mathfrak{h}_2 \in \mathbb{H}_2 \mid \phi(\mathfrak{h}_2) = \mathfrak{f}_2\}$ and $\Gamma(\mathfrak{g}_2) = \gamma^{-1}(\{\mathfrak{g}_2\}) = \{\mathfrak{h}_2 \in \mathbb{H}_2 \mid \gamma(\mathfrak{h}_2) = \mathfrak{g}_2\}$.

Due to space limitations, we do not present here the (exact calculus) algorithmic process for building $\mathbb{H}$ from $\mathbb{F}$ and $\mathbb{G}$.

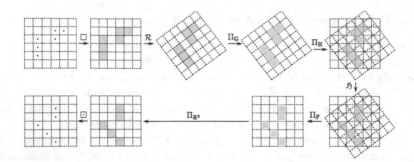

Fig. 2. Proposed framework for homotopy type preserving rigid motion. Following the flowchart: $X \subset \mathbb{Z}^2, \square(X) = X \subset \mathbb{R}^2, \mathcal{R}(X) = X_\mathcal{R} \subset \mathbb{R}^2, \Pi_\mathbb{G}(X_\mathcal{R}) = G \subset \mathbb{G}, \Pi_\mathbb{H}(G) = H \subset \mathbb{H}, \mathfrak{H}(H) = \widehat{H} \subset \mathbb{H},$ $\Pi_\mathbb{F}(\widehat{H}) = \widehat{F} \subset \mathbb{F}, \Pi_{\mathbb{R}^2}(\widehat{F}) = Y \subset \mathbb{R}^2$ and $\square(Y) = X_\mathcal{R} \subset \mathbb{Z}^2$.

Based on the above functions, each complex F on $\mathbb{F}$ (resp. G of $\mathbb{G}$) can be embedded into $\mathbb{H}$ be defining a complex $\Pi_\mathbb{H}(F)$ (resp. $\Pi_\mathbb{H}(G)$) as $\Pi_\mathbb{H}(F) = \bigcup_{f_2 \in \mathbb{F}_2(F)} \bigcup_{\mathfrak{h}_2 \in \Phi(f_2)} C(\mathfrak{h}_2)$ (resp. $\Pi_\mathbb{H}(G) = \bigcup_{\mathfrak{g}_2 \in \mathbb{G}_2(G)} \bigcup_{\mathfrak{h}_2 \in \Gamma(\mathfrak{g}_2)} C(\mathfrak{h}_2)$), and we say that $\Pi_\mathbb{H}(F)$ (resp. $\Pi_\mathbb{H}(G)$) is the embedding of F (resp. G) in $\mathbb{H}$. For any complex H on $\mathbb{H}$, if there exists a complex F on $\mathbb{F}$ (resp. G on $\mathbb{G}$) such that $H = \Pi_\mathbb{H}(F)$ (resp. $H = \Pi_\mathbb{H}(G)$), then we write $F = \Pi_\mathbb{F}(H)$ (resp. $G = \Pi_\mathbb{G}(H)$) and we say that F (resp. G) is the embedding of H in $\mathbb{F}$ (resp. $\mathbb{G}$). In such case, we have in particular $\Pi_\mathbb{F}(H) = \bigcup_{\mathfrak{h}_2 \in \mathbb{H}_2(H)} C(\phi(\mathfrak{h}_2))$ (resp. $\Pi_\mathbb{G}(H) = \bigcup_{\mathfrak{h}_2 \in \mathbb{H}_2(H)} C(\gamma(\mathfrak{h}_2))$).

6   Optimization-Based Rigid Motion

By contrast to the process depicted in Fig. 1 that does not handle topological constraints, our approach (Fig. 2) aims to guarantee that X and $X_\mathcal{R}$ will have the same topology.

The first four steps of this process (from X to H) and the last three ones (from $\widehat{H}$ to $X_\mathcal{R}$) can be dealt with by considering Sects. 4 and 5 (keep in mind that all these steps are topology-preserving). The only part that remains to be described is the construction of the transformation $\mathfrak{H}$ from H to $\widehat{H}$. In particular, it is mandatory that:

- $\mathfrak{H}$ be a homotopic transformation (to preserve the topology between X and $X_\mathcal{R}$);
- $\widehat{H}$ can be embedded into $\mathbb{F}$, i.e. $\widehat{F} = \Pi_\mathbb{F}(\widehat{H})$ exists; and
- the digital analogue $\square(\Pi_{\mathbb{R}^2}(\widehat{H})) \subset \mathbb{Z}^2$ of $\widehat{H}$ be (as close as possible to) the exact solution of the optimization problem (1).

The space $\mathbf{C}_\mathbb{H}$ of all the complexes H of $\mathbb{H}$ has a size $2^{|\mathbb{H}_2|}$. Some of these complexes H are such that $\Pi_\mathbb{F}(H)$ exists, i.e. they can be embedded as complexes F of $\mathbb{F}$, and then in $\mathbb{Z}^2$. These complexes form a supbspace $\mathbf{S}_\mathbb{H}$ of $\mathbf{C}_\mathbb{H}$ of size $2^{|\mathbb{F}_2|}$. We can endow $\mathbf{C}_\mathbb{H}$ with a graph structure, by defining the following adjacency relation $\frown$: for any distinct complexes H_1 and H_2 of $\mathbb{H}$, we have $H_1 \frown H_2$ iff $H_1 = H_2 \cup C(\mathfrak{f})$ or $H_2 = H_1 \cup C(\mathfrak{f})$ for a 2-face $\mathfrak{f} \in \mathbb{H}_2$, and if the associated cell $C(\mathfrak{f})$ is simple for H_1 and H_2. The graph $(\mathbf{C}_\mathbb{H}, \frown)$ is composed of connected components, each one corresponding to a family of complexes that have the same homotopy type. In particular, the connected component

$C_{\mathbb{H}}^{\star}$ that contains H is the set of all the complexes of $\mathbb{H}$ that can be obtained from H by a homotopic transformation $\mathfrak{H}$. This subspace $C_{\mathbb{H}}^{\star}$ can be built with a time complexity $\theta(|C_{\mathbb{H}}^{\star}|)$. The solution of Eq. (1) lies in $C_{\mathbb{H}}^{\star}$, and more precisely in the subspace $S_{\mathbb{H}}^{\star} = C_{\mathbb{H}}^{\star} \cap S_{\mathbb{H}}$. In theory, it is then possible to solve Eq. (1) by building $C_{\mathbb{H}}^{\star}$ and by finding in $S_{\mathbb{H}}^{\star}$ the complex that optimizes the chosen error measure. Although this process ends in finite time, it is generally not tractable in practice, since the time $\theta(|C_{\mathbb{H}}^{\star}|)$ may be exponentially high.

In the next sections, we deal with two points. On the one hand, we show how to guarantee that we search the solution within $C_{\mathbb{H}}^{\star}$, i.e. that the transformation $\mathfrak{H}$ we build is indeed homotopic. On the other hand, we discuss heuristic strategies for computing a solution fairly close to the true optimum relatively to Eq. (1) whereas avoiding to carry out an exhaustive search within $C_{\mathbb{H}}^{\star}$ which would require an exponential time cost.

6.1 Homotopic Transformations and Simple Cells in the Cellular Space

To guarantee that $\mathfrak{H}$ is a homotopic transformation, it is built as a sequence of additions/removals of simple cells. This notion of simple cell is directly derived from that considered in [8] which relies on the notion of collapse in complexes.

Let K be a complex defined in a cellular space $\mathbb{K}$ on $\mathbb{R}^2$. Let $\mathfrak{f}_2$ be a 2-face of K. Let $D_0(\mathfrak{f}_2)$ (resp. $D_1(\mathfrak{f}_2)$) be the subset of $C(\mathfrak{f}_2)$ composed by the 0- (resp. 1-) faces $\mathfrak{f}$ the star of which intersects K only within $C(\mathfrak{f}_2)$, i.e. $S(\mathfrak{f}) \cap K = S(\mathfrak{f}) \cap C(\mathfrak{f}_2)$. We say that $C(\mathfrak{f}_2)$ is a simple 2-cell (for K) if $|D_1(\mathfrak{f}_2)| = |D_0(\mathfrak{f}_2)| + 1$ (which is equivalent to say that the intersection of the border of $C(\mathfrak{f}_2)$ and K is connected and with a Euler characteristics of 1). In such case, the detachment of this 2-cell $C(\mathfrak{f}_2)$ from K, i.e. the operation that transforms K into $K \otimes C(\mathfrak{f}_2) = K \setminus (\{\mathfrak{f}_2\} \cup D_1(\mathfrak{f}_2) \cup D_0(\mathfrak{f}_2))$ corresponds to a collapse operation from K to $K \otimes C(\mathfrak{f}_2)$, and both complexes have the same homotopy type. Reversely, if $\mathfrak{f}_2$ is a 2-face of $\mathbb{K} \setminus K$, and if $C(\mathfrak{f}_2)$ is a simple 2-cell for the complex $K \cup C(\mathfrak{f}_2)$, then the operation of attachment that transforms K into $K \cup C(\mathfrak{f}_2)$ corresponds to the inverse collapse operation from K into $K \cup C(\mathfrak{f}_2)$, and both complexes also have the same homotopy type.

6.2 Optimization Problem: Heuristics

Even if we consider a finite part of $\mathbb{Z}^2$ (which is the case in digital imaging), the induced finite space of the solutions of the optimization problem (1) is huge, and the topological constraints induced by the homotopy type equivalence between X and X_R are not sufficient to reduce this space to a tractable size allowing for an exhaustive investigation. Thus, we do not aim at solving exactly the optimization problem (1) (although we will sometimes succeed), but to find a solution reasonably close to the optimum. In particular, we only explore a part of the space of solutions. Our purpose is then to make this exploration as relevant as possible. We briefly discuss hereafter a non-exhaustive list of ideas that can be relevant to reach that goal.

Border Processing – In general, the complex H (resp. its complement) cannot be directly embedded into $\mathbb{F}$. However, some parts of H (resp. its complement) already correspond to 2-cells of $\mathbb{F}$. In most cases, these parts that constitute the "internal" (resp. "external") part of H will not be modified during the optimization process. More formally, this means that in most application cases, the addition/removal of simple 2-cells to/from H will occur for 2-faces $\mathfrak{f}_2$ such that $\Phi(\mathfrak{f}_2)$ intersects—but is not included in (resp. excluded from)—H. In other words, it is generally sufficient to work on the "border" of H to build $\widehat{H}$ and thus $\widehat{F}$.

Measure Separability and Gradient Climbing – Most error measures aim to emulate the behaviour of usual digitization policies. For instance the two ones considered in our experiments (Sect. 7) correspond to the Gaussian (Eq. (13)) and the majority vote (Eq. (14)) digitizations. From these very definitions, it is plain that it is possible to process the 2-cells of $\mathbb{F}$ (in particular the "border" ones) one after another, either by addition or removal of 2-cells of $\mathbb{H}$. In this context, it may be relevant to process them by giving the highest priority to the cells that induce the lowest increase of the error measure, following a (reverse) gradient climbing paradigm.

Homotopic Transformations in $\mathbb{F}$ – Once a first candidate complex $\widehat{H}$ belonging to $\mathbf{S}_{\mathbb{H}}^{\star}$ has been built, it can be associated to a complex $\widehat{F} = \Pi_{\mathbb{F}}(\widehat{H})$ of $\mathbb{F}$. Then, some next candidate complexes can be sought in $\mathbf{S}_{\mathbb{H}}^{\star}$ by "starting" from $\widehat{H}$, or more precisely from $\widehat{F}$ by considering the search space $\mathbf{S}_{\mathbb{F}}^{\star}$ of $\mathbb{F}$ (defined the same way as $\mathbf{S}_{\mathbb{H}}^{\star}$) instead of $\mathbf{S}_{\mathbb{H}}^{\star}$. This is motivated by the fact that $\mathbb{F}$ is much smaller that $\mathbb{H}$, whereas the candidate complexes have to be defined in $\mathbb{F}$.

Termination Issues – For most objects, the optimization process based on the above heuristics will converge directly towards the true optimum for Eq. (1). For other objects, in particular those presenting complex details, it may be required to explore the search space in a less straightforward way, and in particular to go backward in the putative path built from H in the graph $(\mathbf{S}_{\mathbb{H}}^{\star}, \frown)$. Such forward–backward steps may potentially lead to non-termination issues of the optimization process. To deal with this difficulty, a solution may consist of storing the different complexes already explored, in order to guarantee that they will not be processed many times.

Non-existence of Solutions – For complex objects, a solution (with a reasonably low error measure) may not exist. This may be caused by the non-existence of a solution, under the topological constraints, in the context of a finite support image. For instance, this may happen for a checkerboard configuration with 1-pixel-sized squares. This problem may be tackled by multigrid paradigm, for instance by considering $(\frac{1}{2}\mathbb{Z})^2$ instead of $\mathbb{Z}^2$ as output space.

7 Experiments

We implemented a first algorithm (Algorithm 1) that builds upon some of the heuristics discussed above. This algorithm, although very simple, actually works in most cases. We process only the 2-faces at the border of the complex ($\mathbb{B}_2$, line 2). We sequentially deal with these faces in order to fully include or fully exclude them from the final

---

Algorithm 1: Definition of $\widehat{H}$ by construction of $\mathfrak{H}$.

Input: $H \subset \mathbb{H}$, $\mathcal{D}_{\mathcal{R},X} : 2^{\mathbb{Z}^2} \to \mathbb{R}_+$

Output: $\widehat{H} \subset \mathbb{H}$

1   $\widehat{H} \leftarrow H$

2   $\mathbb{B}_2 \leftarrow \{f_2 \in \mathbb{F}_2 \mid \Phi(f_2) \not\subseteq H_2(H) \wedge \Phi(f_2) \cap H_2(H) \neq \emptyset\}$

3   **while** $\mathbb{B}_2 \neq \emptyset$ **do**

4   choose $f_2 \in \mathbb{B}_2$ wrt $\mathcal{D}_{\mathcal{R},X}$

5      $\mathbb{B}_2 \leftarrow \mathbb{B}_2 \setminus \{f_2\}$

6      $(I_2, O_2) \leftarrow (\Phi(f_2) \cap H_2(\widehat{H}), \Phi(f_2) \setminus H_2(\widehat{H}))$ (the roles of O_2 and I_2 may be reversed, depending on the priority of either removing or adding the face f_2.)

7   **while** $\exists b_2 \in I_2$ *s.t.* $C(b_2)$ *is simple for* $\widehat{H}$ **do**

8         $\widehat{H} \leftarrow \widehat{H} \oslash C(b_2)$

9         $(I_2, O_2) \leftarrow (I_2 \setminus \{b_2\}, O_2 \cup \{b_2\})$

10   **if** $I_2 \neq \emptyset$ **then**

11   **while** $\exists b_2 \in O_2$ *s.t.* $C(b_2)$ *is simple for* $\widehat{H}$ **do**

12            $\widehat{H} \leftarrow \widehat{H} \cup C(b_2)$

13            $(I_2, O_2) \leftarrow (I_2 \cup \{b_2\}, O_2 \setminus \{b_2\})$

14   **if** $O_2 \neq \emptyset$ **then** Failure of the process

---

complex. Depending on the induced increasing of the considered error metric $\mathcal{D}_{\mathcal{R},X}$, we aim at removing (or adding) all the 2-faces in $\mathbb{H}$ that compose the current 2-face of $\mathbb{B}_2$ (while loop, line 7). If this attempt fails (line 10), we alternatively aim at adding (or removing) all the 2-faces in $\mathbb{H}$ that compose the current 2-face of $\mathbb{B}_2$ (while loop, line 11). If this second attempt also fails, the algorithms is not able to provide a solution, and it stops. The process ends when $\mathbb{B}_2$ is empty, i.e. when all the border 2-faces have been fully included in/excluded from $\widehat{X}$. Of course, many other—more sophisticated—algorithms may be proposed, but such study is beyond the scope of this article.

We consider the two following error measures:

$$\mathcal{D}_{\mathcal{R},X}^{\boxdot}(Y) = |\boxdot (\mathcal{R}(\square(X))) \setminus Y| + |Y \setminus \boxdot(\mathcal{R}(\square(X)))| \tag{13}$$

$$\mathcal{D}_{\mathcal{R},X}^{\square}(Y) = |\mathcal{R}(\square(X)) \setminus \square(Y)| + |\square(Y) \setminus \mathcal{R}(\square(X))| \tag{14}$$

where $|\cdot|$ is the cardinal for discrete sets (Eq. (13)), and the area for continuous objects (Eq. (14)). The first (resp. the second) corresponds to the Gauss (resp. majority vote) digitization. This will allow us to compare the results obtained by our method with these two usual digitization policies.

Results are illustrated in Fig. 3. They are proposed for small, yet complex objects. Indeed, we focus on objects that present details which are the most likely to be topologically altered by a rigid motion, namely small connected components and thin structures.

The first image (ellipse) illustrates the fact that in the most simple cases (here, no complex details and a globally smooth border), our method provides the same results as usual transformations-by-digitization approaches. Indeed, when such methods do not alter the topology, our method has the same behaviour. Without surprise, we also observe that the results with Eq. (14) have smoother boundaries than with Eq. (13).

In the other three examples (head, circles and DGMM logo), the transformations-by-digitization (second and fourth columns) fail to preserve the topology, leading to broken or merged connected components. By contrast, our method (third and fifth

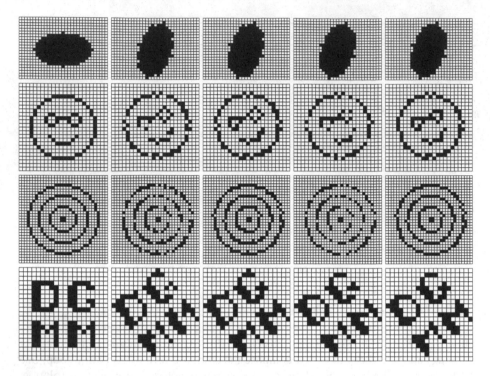

Fig. 3. From left to right: input image $X \subset \mathbb{Z}^2$; Gaussian digitization of $\mathcal{R}(\square(X))$ and its analogue version with our method (Eq. (13)); majority vote digitization of $\mathcal{R}(\square(X))$ and its analogue version with our method (Eq. (14)). From top to bottom, the used rigid motion paramaters $(\alpha, \beta, t_x, t_y) \in \mathbb{Q}^4$ (Eq. (3)) are: $(\frac{22}{25}, \frac{7}{25}, 0, 0)$, $(\frac{5}{13}, \frac{12}{13}, \frac{1}{5}, \frac{2}{3})$, $(\frac{3}{5}, \frac{4}{5}, \frac{1}{3}, \frac{1}{3})$, $(\frac{3}{5}, \frac{4}{5}, \frac{1}{5}, \frac{1}{4})$.

columns) succeed in preserving the topology, whereas leading to results with as few as possible differences with the transformations-by-digitization results.

8   Conclusion

The proposed approach of digital rigid motion allows us to ensure topological invariance between the initial object and its image. It relies on an optimization strategy under topological constraints. Since the definition of the final object is obtained by a constructive process, these topological constraints may lead to a non-convergence of the method when the structure of the object is too close to the resolution of the grid. A short term perspective will consist of considering multigrid strategies to handle such cases.

As mid-term perspectives, we will also investigate our approach with other kinds of topological models (e.g. the well-composed sets), but also with non-binary images. Longer-term perspectives will consist of investigating transformations in higher dimensions and/or for richer families of transformations [7, 10, 13]. It would be also interesting to combine topological and geometric constraints, such as perimeter or curvature minimization, convexity preservation, etc.

References

1. Andres, É.: The quasi-shear rotation. In: DGCI, pp. 307–314 (1996)
2. Andres, É., Dutt, M., Biswas, A., Largeteau-Skapin, G., Zrour, R.: Digital two-dimensional bijective reflection and associated rotation. In: DGCI, pp. 3–14 (2019)
3. Anglin, W.S.: Using Pythagorean triangles to approximate angles. Am. Math. Monthly **95**, 540–541 (1988)
4. Baudrier, É., Mazo, L.: Combinatorics of the Gauss digitization under translation in 2D. J. Math. Imaging Vision **61**, 224–236 (2019)
5. Berthé, V., Nouvel, B.: Discrete rotations and symbolic dynamics. Theoret. Comput. Sci. **380**, 276–285 (2007)
6. Bloch, I., Pescatore, J., Garnero, L.: A new characterization of simple elements in a tetrahedral mesh. Graphical Models **67**, 260–284 (2005)
7. Blot, V., Coeurjolly, D.: Quasi-affine transformation in higher dimension. In: DGCI, pp. 493–504 (2009)
8. Couprie, M., Bertrand, G.: New characterizations of simple points in 2D, 3D, and 4D discrete spaces. IEEE Trans. Pattern Anal. Mach. Intell. **31**, 637–648 (2009)
9. Jacob, M.A., Andres, É.: On discrete rotations. In: DGCI, pp. 161–174 (1995)
10. Jacob-Da Col, M., Mazo, L.: nD quasi-affine transformations. In: DGCI, pp. 337–348 (2016)
11. Kovalevsky, V.A.: Finite topology as applied to image analysis. Comput Vision Graphics Image Process. **46**, 141–161 (1989)
12. Mazo, L., Passat, N., Couprie, M., Ronse, C.: Paths, homotopy and reduction in digital images. Acta Applicandae Mathematicae **113**, 167–193 (2011)
13. Mazo, L.: Multi-scale arithmetization of linear transformations. J. Math. Imag. Vision **61**, 432–442 (2019)
14. Mazo, L., Baudrier, É.: Object digitization up to a translation. J. Comput. Syst. Sci. **95**, 193–203 (2018)
15. Mazo, L., Passat, N., Couprie, M., Ronse, C.: Digital imaging: a unified topological framework. J. Math. Imaging Vision **44**, 19–37 (2012)
16. Ngo, P., Kenmochi, Y., Passat, N., Talbot, H.: Topology-preserving conditions for 2D digital images under rigid transformations. J. Math. Imaging Vision **49**, 418–433 (2014)
17. Ngo, P., Passat, N., Kenmochi, Y., Debled-Rennesson, I.: Convexity invariance of voxel objects under rigid motions. In: ICPR, pp. 1157–1162 (2018)
18. Ngo, P., Passat, N., Kenmochi, Y., Talbot, H.: Topology-preserving rigid transformation of 2D digital images. IEEE Trans. Image Process. **23**, 885–897 (2014)
19. Nouvel, B., Rémila, E.: Characterization of bijective discretized rotations. In: IWCIA, pp. 248–259 (2004)
20. Nouvel, B., Rémila, E.: Incremental and transitive discrete rotations. In: IWCIA, pp. 199–213 (2006)
21. Passat, N., Kenmochi, Y., Ngo, P., Pluta, K.: Rigid motions in the cubic grid: a discussion on topological issues. In: DGCI, pp. 127–140 (2019)
22. Pluta, K., Moroz, G., Kenmochi, Y., Romon, P.: Quadric arrangement in classifying rigid motions of a 3D digital image. In: CASC, pp. 426–443 (2016)
23. Pluta, K., Romon, P., Kenmochi, Y., Passat, N.: Bijectivity certification of 3D digitized rotations. In: CTIC, pp. 30–41 (2016)
24. Pluta, K., Romon, P., Kenmochi, Y., Passat, N.: Bijective digitized rigid motions on subsets of the plane. J. Math. Imaging Vision **59**, 84–105 (2017)
25. Roussillon, T., Coeurjolly, D.: Characterization of bijective discretized rotations by Gaussian integers. Technical report (2016). https://hal.archives-ouvertes.fr/hal-01259826
26. Thibault, Y., Sugimoto, A., Kenmochi, Y.: 3D discrete rotations using hinge angles. Theoret. Comput. Sci. **412**, 1378–1391 (2011)

Locally Turn-Bounded Curves
Are Quasi-Regular

Étienne Le Quentrec[(✉)], Loïc Mazo, Étienne Baudrier, and Mohamed Tajine

ICube-UMR 7357, 300 Bd Sébastien Brant - CS 10413, 67412 Illkirch Cedex, France
elequentrec@unistra.fr

Abstract. The characteristics of a digitization of a Euclidean planar shape depends on the digitization process but also on the shape border regularity. The notion of Local Turn Boundedness (LTB) was introduced by the authors in *Le Quentrec, É. et al.: Local Turn-Boundedness: A curvature control for a good digitization, DGCI 2019* so as to have multigrid convergent perimeter estimation on Euclidean shapes. If it was proved that the par-regular curves are locally turn bounded, the relation with the quasi-regularity introduced in *Ngo, P. et al.: Convexity-Preserving Rigid Motions of 2D Digital Objects, DGCI 2017* had not yet been explored. Our paper is dedicated to prove that for planar shapes, local turn-boundedness implies quasi-regularity.

1 Introduction

A loss of information is inherent to any digitization of a continuous shape. The control of the shape border can allow the digitization to inherit of continuous shape properties. Thus the notion of local-turn boundedness (LTB) introduced in [6] by the authors makes it possible to preserve the shape connectivity and well-composedness for a Gauss digitization under a condition on the grid step. The class of LTB curves is not the first attempt to control the shape border for digitization. One can cite the par-regularity [12] and its generalizations including shapes with spikes: half-regularity [13], r-stability [9], quasi(r)-regularity [10] and the μ-reach [2]. There are links among the existing notions and also with the LTB notion. The following equivalences have already been shown: in [3], the equivalence between the class $C^{1,1}$ (curves with Lipschitz unit tangents) and the par-regular class; in [4], the equivalence between par-regular class and the class of curves with a positive reach; in [7,8], the equivalence between the class of curves with a positive reach and LTB curves with Lipschitz turn.

This paper is dedicated to show that LTB implies the quasi-regularity. The proof is composed of several but necessary steps. The key point of the proof is the connectivity of the eroded of a LTB shape. It consists in showing that close points, or on the contrary distant points, in the eroded set can be joined by a path inside the shape. The main difficulty is to define precisely the terms "close" and "far" to cover all the point distances in the eroded set while making proof possible.

© Springer Nature Switzerland AG 2021
J. Lindblad et al. (Eds.): DGMM 2021, LNCS 12708, pp. 202–214, 2021.
https://doi.org/10.1007/978-3-030-76657-3_14

In Sect. 2, the main notions and some useful properties –some of them revisited– are recalled. Section 3 gives all the steps to prove the implication. The conclusion and some perspectives are given in Sect. 4.

2 Definitions

Notations. The complementary of a subset S of $\mathbb{R}^2$ is noted S^c. We write $\bar{A}$ for the topological closure of a set A and ∂A for its topological boundary. We note $B(c, r)$ the open disk centered in c and of radius r. The notation $[x_i]_{i=0}^N$ designates the polygonal line whose ordered sequence of vertices is $(x_i)_{i=0}^N$. When $x_0 = x_N$, the polygonal line is actually a polygon. The geometric angle between two vectors $\vec{u}$ and $\vec{v}$, or between two directed straight lines oriented by $\vec{u}$ and $\vec{v}$, is denoted by $\angle(\vec{u}, \vec{v})$. It is the absolute value of the reference angle taken in $(-\pi, \pi]$ between the two vectors. Given three points x, y, z, we also write $\widehat{xyz}$ for the geometric angle between the vectors $x - y$ and $z - y$. We write $\mathcal{C}_{a,b}$ for an arc of a curve $\mathcal{C}$ between the points a and b.

The two following definitions introduce the notion of *local turn boundedness*.

Definition 1 (Turn, [1]).

- The turn $\kappa(L)$ of a polygonal line $L = [x_i]_{i=0}^N$ is defined by:

$$\kappa(L) := \sum_{i=1}^{N-1} \angle(x_i - x_{i-1}, x_{i+1} - x_i).$$

- The turn $\kappa(P)$ of a polygon $P = [x_i]_{i=0}^N$ (where $x_N = x_0$ and $x_{N+1} = x_1$) is defined by:

$$\kappa(P) := \sum_{i=1}^{N} \angle(x_i - x_{i-1}, x_{i+1} - x_i).$$

- The turn $\kappa(\mathcal{C})$ of a simple curve $\mathcal{C}$ (respectively of a Jordan curve) is the supremum of the turn of its inscribed polygonal lines (respectively of its inscribed polygons).

At each point c of a curve whose turn is finite, there exists a left-hand and a right-hand tangent vectors, denoted by $e_l(c)$ and $e_r(c)$ [1].

Property 1 (Fenchel's Theorem, [1] Theorem 5.1.5). The turn of a Jordan curve is greater than or equal to 2π. The equality case occurs if and only if the interior of $\mathcal{C}$ is convex.

Definition 2 (Proposition 2 [7]). *A Jordan curve $\mathcal{C}$ is (θ, δ)-LTB if for any two points a and b in $\mathcal{C}$ such that $d(a, b) < \delta$, the turn of one of the arcs of the curve $\mathcal{C}$ delimited by a and b is less than or equal to θ.*

As the (θ, δ)-LTB-curve set is growing with θ, the properties established for $\theta = \theta_0$ are also available for $\theta \leq \theta_0$. In the rest of the paper, θ is fixed to $\pi/2$ and we write δ-LTB instead of $(\pi/2, \delta)$-LTB.

Notice that two distinct points of a Jordan curve delimit two arcs of the curve. The notion of *straightest arc* introduced in [6] makes it possible to distinguish these two arcs.

Property 2 *([7], Definition 6 and Proposition 4).* Let a, b be two distinct points of a δ-LTB curve $\mathcal{C}$. If $\mathrm{d}(a, b) < \delta$, then there exists a unique arc of $\mathcal{C}$ between a and b whose turn is less than or equal to $\pi/2$. This arc, denoted by $\mathcal{C}|_a^b$, is included in the closed disk with diameter $[a, b]$ and is called *the straightest arc between a and b.*

Let us quote a recent result which makes easier the use of straightest arcs.

Property 3 *([8], Lemma 1).* Let a and b two points of a δ-LTB curve such that $\mathrm{d}(a, b) < \delta$. Let $\mathcal{C}|_a^b$ be the straightest arc between a and b. Then,

$$\angle(e_l(a), e_r(a)) + \kappa(\mathcal{C}|_a^b) + \angle(e_l(b), e_r(b)) \leq \frac{\pi}{2}.$$

The following proposition is a very slight quantitative improvement of [7, Proposition 5]. Nevertheless, this improvement is absolutely necessary to get the main result of this paper.

Proposition 1. *Let $\mathcal{C}$ be a (θ, δ)-LTB curve and $a \in \mathcal{C}$. Then, for any $\epsilon < \delta$, the intersection of $\mathcal{C}$ with the closed disk $\bar{B}(a, \epsilon)$ is path-connected and is therefore an arc of $\mathcal{C}$. Furthermore, the turn of this arc is less than or equal to 2θ.*

Proof. The proof is exactly the same as the one given in [7] except that, taking into account Property 3, we can omit the term $\angle(e_l(a), e_r(a))$ so as to upper bound the curvature of the arc $\mathcal{C} \cap B(a, \epsilon)$ by 2θ instead of 3θ.   □

We recall here the notions of par-regularity and quasi-regularity.

Definition 3 (par(r)-regularity, [5]). *Let $\mathcal{C}$ be a Jordan curve of interior K.*

- *A closed disk $\bar{B}(c_i, r)$ is an inside osculating disk of radius r to $\mathcal{C}$ at point $a \in \mathcal{C}$ if $\mathcal{C} \cap \bar{B}(c_i, r) = \{a\}$ and $\bar{B}(c_i, r) \subset K \cup \{a\}$.*
- *A closed disk $\bar{B}(c_e, r)$ is an outside osculating disk of radius r to $\mathcal{C}$ at point $a \in \mathcal{C}$ if $\mathcal{C} \cap \bar{B}(c_e, r) = \{a\}$ and $\bar{B}(c_e, r) \subset \mathbb{R}^2 \setminus (\mathcal{C} \cup K) \cup \{a\}$.*
- *A curve $\mathcal{C}$ or a set K is par(r)-regular if there exist inside and outside osculating disks of radius r at each $a \in \mathcal{C}$.*

As noticed in [11], for a bounded simply connected set S, the above definition can be rephrased in the following way [1]:

The set S is par(r)-regular if and only if

---

[1] Actually, the equivalence does not perfectly hold as seen taking $S = \bar{B}(0, r)$.

- $S \ominus \bar{B}(0, r)$ is non-empty and connected,
- $S^c \ominus \bar{B}(0, r)$ is connected,
- $S = (S \ominus \bar{B}(0, r)) \oplus \bar{B}(0, r)$,
- $S^c = (S^c \ominus \bar{B}(0, r)) \oplus \bar{B}(0, r)$,

with $\oplus$, $\ominus$ the standard dilation and erosion operators.

In order to consider shapes with angles, the two last items of par-regularity are relaxed in Definition 4 allowing the border of the shape to oscillate in a margin around its opening.

Definition 4 (Quasi-regularity [10,11] **).** *Let $S \subset \mathbb{R}^n$ ($n = 2,3$) be a bounded, simply connected set. We say that S is quasi-r-regular with margin $r' - r$ (with $0 < r \leq r'$) if it satisfies the following four properties*

- $S \ominus \bar{B}(0, r)$ *is non-empty and connected,*
- $S^c \ominus \bar{B}(0, r)$ *is connected,*
- $S \subset S \ominus \bar{B}(0, r) \oplus \bar{B}(0, r')$,
- $S^c \subset S^c \ominus \bar{B}(0, r) \oplus \bar{B}(0, r')$.

3   Main Result

For sake of readability of the article, we state the main result—Theorem 1— before the propositions and lemmas needed for the proof.

Theorem 1. *Let S be a compact subset of the plane $\mathbb{R}^2$ whose boundary is δ-LTB. Then S is quasi-r-regular with margin $(\sqrt{2}-1)r$ for any $r < \delta/\sqrt{10 + 4\sqrt{2}}$.*

Our first intermediate result is an improvement of a proposition about turn originally stated in [7] that roughly asserts that avoiding a convex obstacle bounds from below the turn of a curve. It was stated for convex polygonal obstacles in [7, Lemma 3] and for convex obstacles in a particular configuration in [8, Lemma 10]. The new version presented below is valid in a more general configuration (see Fig. 1).

Proposition 2. *Let $\mathcal{C}$ be a simple curve with endpoints a, b. Let $H_{a,b}$ be a half-plane having a and b in its boundary and S be a closed set included in the closure of a bounded connected component of $\mathbb{R}^2 \setminus (\mathcal{C} \cup [a, b])$ and whose intersection with the half-plan $H_{a,b}$ is not included in the line passing through a and b. Then,*

$$\kappa(\mathcal{C}) \geq \kappa(\partial \operatorname{conv}(H_{a,b} \cap S \cup [a, b]) \setminus (a, b)),$$

where $\partial \operatorname{conv}(\cdot)$ stands for the boundary of the convex hull.

Proof. We begin the proof by stating and proving some facts about convex polygons. So, let $n \geq 3$ and $P = [a_i]_{i=0}^n$ with $a_n = a_0$ be a convex polygon. Then,

Claim 1. *Let σ be a permutation of $[1, n]$ such that $\sigma(1) = 1$ and $\sigma(n) = n$. Then, $\kappa([a_{\sigma(i)}]_{i=1}^n) \geq \kappa([a_i]_{i=1}^n)$.*

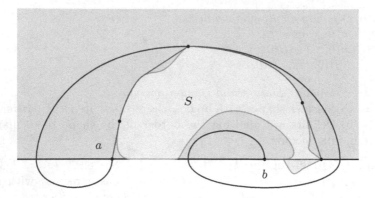

Fig. 1. Proposition 2 states that the blue curve with endpoints a and b has a turn greater than the red curve $\partial \operatorname{conv}(H_{a,b} \cap S \cup [a,b]) \setminus (a,b)$ where $H_{a,b}$ is the grey half-plane.

Claim 2. Let $[a,b,c,d,a]$ be a convex polygon of $\mathbb{R}^2$. For any point b' on the half-line $\overrightarrow{c,b} \setminus [c,b)$, for any point c' on the half-line $\overrightarrow{a,c} \setminus [a,c)$ and for any point b'' on the half-line $\overrightarrow{d,b} \setminus [d,b)$, the polygons $[a,b',c,a]$ and $[a,b'',c',d,a]$ are convex and the turn of the polygonal lines $[a,b',c]$ and $[a,b'',c',d]$ are respectively greater than or equal to the turn of the polygonal lines $[a,b,c]$ and $[a,b,c,d]$. Claim 2 is also valid in the degenerate case where b, a, d, c are aligned in this order.

Proof of Claim 1: The turn of the polygonal line $Q = [a_{\sigma(i)}]_{i=1}^n$ is the sum of the turns at each vertex $a_{\sigma(i)}$, $2 \le i \le n-1$. Since P is convex, the turn of Q at $a_{\sigma(i)}$ is bounded from below by the turn of the polyline $[a_{\sigma(i)-1}, a_{\sigma(i)}, a_{\sigma(i)+1}]$, that is by the turn at $a_{\sigma(i)}$ for the polyline $[a_i]_{i=1}^n$.

Proof of Claim 2: The triangle $[a,b',c,a]$ is obviously convex. The interior angle of $[a,b'',c',d,a]$ at b'' is maximum when $b'' = b$ and c' at infinity. Thereby, it is never a reflex angle. Alike, the angle at c' is never reflex and the quadrilateral $[a,b'',c',d,a]$ is convex. Furthermore, the turn of the polygons $[a,b',c,a]$, $[a,b'',c',d,a]$, $[a,b,c]$ and $[a,b,c,d]$ are equal to 2π by Property 1 and, by definition of b', c' and b'', the interior angles at a and c for $[a,b',c,a]$ (resp. a and d for $[a,b'',c',d,a]$) are greater than or equal to those for $[a,b,c,a]$ (resp. for $[a,b,c,d,a]$). We derive that $\kappa([a,b',c]) \ge \kappa([a,b,c])$ and $\kappa([a,b'',c',d]) \ge \kappa([a,b,c,d])$. In the degenerate case, the reader can check that the turns of the polygons $[a,b,c]$, $[a,b,c,d,a]$, $[a,b',c,a]$ and $[a,b'',c',d,a]$ are equal to 2π.

Let us go back to the main proof. Put $T = \partial \operatorname{conv}(H_{a,b} \cap S \cup [a,b])$. Observe that T includes the straight segment $[a,b]$ and has a non-empty interior (for $H_{a,b} \cap S$ is not included in the line passing through a and b). Thus, $P = T \setminus (a,b)$ is a curve with endpoints a and b. Firstly, we assume that T is a polygon. Consequently, P is a polygonal line and we denote by c, resp. d the vertex of the edge of P whose other end is a, resp b. Notice that the turns of T and of the quadrilateral (or triangle) $[a, c, d, b, a]$ are equal to 2π (Property 1). Also, observe that if $[a, c, d, b, a]$ is degenerate, then c, a, b and d are aligned in this order. Thereby, the turn of P and the polyline $[a, c, d, b]$ are equal.

Since the component of $\mathbb{R}^2 \setminus (C \cup [a, b])$ whose closing includes S is bounded and $c, d \in S$, any half-line with initial point c or d cuts the curve $C \cup [a, b]$. Then, thanks to Claim 2, we define the points c' and d' on C such that $[a, c', d', b, a]$ is a convex quadrilateral with vertices on C. Observe that either the polygonal line $[a, c', d', b]$ or the polygonal line $[a, d', c', b]$ is inscribed in C. We denote by Q the one which is inscribed in C and we set $Q' = [a, c', d', b]$.

We now end the proof in the same manner as in [7, Lemma 3]. We have $\kappa(C) \geq \kappa(Q)$ by definition of $\kappa(C)$. Moreover, $\kappa(Q) \geq \kappa(Q')$ by Claim 1. Besides, $\kappa(Q') \geq \kappa([a, c, d, b])$ by Claim 2. Since $\kappa([a, c, d, b]) = \kappa(P)$, we get $\kappa(C) \geq \kappa(P)$. In the general case where P is not polygonal, it suffices to observe that the result is valid for any polygonal line inscribed in P. Then, taking the supremum of the turns of all such polygonal lines, we obtain the desired result.   □

The statement of the following proposition should be compared with the definition of par-regularity. Indeed, from Definition 3, it is possible to derive that any point lying in the closure S of the interior, or the exterior, of a Jordan par(r)-regular curve is contained in a circle with radius r included in S.

Proposition 3. *Let C be a δ-LTB curve and S be the closure of either the interior or the exterior of C. For each point $p \in S$, there exists a square of diameter δ included in S and containing p.*

Proof. The proof is divided in three parts. In the first one, we prove the statement for the points of C. In the second part, we prove the statement for points in $S \setminus (S \ominus \bar{B}(0, \delta/2))$. The last part treats the obvious case of points in $S \ominus \bar{B}(0, \delta/2)$ and concludes the proof.

1. This part of the proof is illustrated by Fig. 2 (center). In a first step, given a point $p \in C$, we prove that there exists a circular sector of center p and radius $\frac{\delta}{\sqrt{2}}$ included in S. In the second step, we prove that this circular sector extends to a square in S. We set $\bar{B}_p = \bar{B}(p, \delta/\sqrt{2})$.

   (a) Let $p \in C$. By Proposition 1, the intersection of the curve C with any disk included in $\bar{B}_p$ is an arc of C (it is connected) and its turn is less than or equal to π. We set $C_{a,b} = C \cap \bar{B}_p$. Then, the arc $C_{a,b}$ splits the disk $\bar{B}_p$ into three connected components: the arc $C_{a,b}$ itself, included in C, another one included in the interior of S called I and the last one included in the exterior of S. Let A be the circular sector of $\bar{B}_p$ delimited by $[a, p, b]$ including $I \cap \partial \bar{B}_p$. Let $C_{a,p}$, resp. $C_{b,p}$, be the subarc of $C_{a,b}$ from a, resp. b, included to p excluded.

   Let $c_a \in C_{a,p}$ and $c_b \in C_{b,p}$ and, by contradiction, assume that $\widehat{c_a p c_b} < \frac{\pi}{2}$. Then, the distance between c_a and c_b is less than δ. We derive that $C|_{c_a}^{c_b}$ exists. Nevertheless, on the one hand, the turn of the arc from c_a to c_b passing through p is greater than or equal to $\kappa([c_a, p, c_b])$ which is greater than $\pi/2$ (by the contradiction assumption). On the other hand, the turn of the arc from c_a to c_b not passing through p is greater than $\kappa(\mathcal{C} \setminus \mathcal{C}_{a,b})$ which is greater than π (by Fenchel's Theorem and the additivity of turns). We get an absurdity. Hence, $\widehat{c_a p c_b} \geq \frac{\pi}{2}$ for any $c_a \in \mathcal{C}_{a,p}$, $c_b \in \mathcal{C}_{b,p}$. Furthermore, a basic calculation of angles shows that the radius $[p, m]$ which is the angle bisector of $\widehat{apb}$ is not intersected by the arc $\mathcal{C}_{a,b}$ (see Fig. 2-left).

Let A be the smallest angular sector of $\partial \bar{B}_p$ containing I. Notice that $A \neq \partial \bar{B}_p$. Let D be the subset of the circular sectors of $\partial \bar{B}_p$ delimited by radii intersecting $\mathcal{C}_{a,p}$ and $\mathcal{C}_{b,p}$ and included in A. Since the radius $[p, m]$ is included in all the sectors of D and since any intersection of angular sectors is a sector or empty, the set $\bigcap\{d \mid d \in D\}$ is a circular sector. Put $\Omega = \bigcap\{d \mid d \in D\} \cap \partial \bar{B}_p$.

Let x_0 and x_1 be the ends of the arc of circle Ω. For any $\epsilon > 0$, there exists $x_0', x_1' \in \partial B(p, \frac{\delta}{\sqrt{2}}) \setminus \Omega$ such that $\widehat{x_0 p x_0'} < \epsilon$ and $\widehat{x_1 p x_1'} < \epsilon$. One of the segments $(p x_0']$ and $(p x_1']$ intersects $\mathcal{C}_{a,p}$ at a point c_a', and the other intersects $\mathcal{C}_{b,p}$ at a point c_b'. Moreover $\widehat{c_a' p c_b'} \geq \frac{\pi}{2}$, then for any $\epsilon > 0$, $\widehat{x_0 p x_1} \geq \widehat{x_0' p x_1'} - 2\epsilon$. Then the sector $\bigcap\{d \mid d \in D\}$ is included in $I \cap \bar{B}_p$ and has its angle greater than or equal to $\frac{\pi}{2}$.

(b) Let Q be a square of side length $\delta/\sqrt{2}$ having p as vertex and two edges included in A'. By contradiction, let c be a point of $\mathcal{C}$ lying in the interior of Q. Then, $\mathrm{d}(c, p) < \delta$. Thus there exists a straightest arc $\mathcal{C}|_c^p$ between c and p. This straightest arc is included in the disk D with diameter $[c, p]$ by Property 2. Besides, an elementary geometric reasoning shows that D intersects the circle $\partial \bar{B}_p$ in two points that lie in Q. Thus, $\mathcal{C}|_c^p$ does not contain neither a nor b, which is absurd.

2. This part of the proof is illustrated by Fig. 2 (right). Let us consider a point q in $S \setminus (S \ominus \bar{B}(0, \delta/2))$ and $q \notin \mathcal{C}$ (if $q \in \mathcal{C}$, we are done by Part 1). Then, there exists a point $p \in \mathcal{C}$ such that $\mathrm{d}(p, q) = \mathrm{d}(q, \mathcal{C}) < \delta/2$. Thereby, the open disk $B(q, \mathrm{d}(p, q))$ is included in the interior of S. With the notations of Part 1, we consider the sector A of $\bar{B}_p$ containing q and delimited by $[a, p, b]$ where a and b are the endpoints of $\mathcal{C} \cap \bar{B}_p$. Let $R = [p, c]$ be the radius of $\bar{B}_p$ passing through q. If the arc $\mathcal{C}_{a,p}$, resp. $\mathcal{C}_{b,p}$, cuts the radius R, thanks to Proposition 2, we have that the turn of $\mathcal{C}_{a,p}$, resp. $\mathcal{C}_{b,p}$, is greater than the turn of a quarter of circle and, as in Part 1, we derive a contradiction with Fenchel's Theorem (Property 1). Thus, $\mathcal{C}_{a,b}$ does not intersect the radius R. We end the proof as in Part 1, just noticing that the angular sector $\bigcap\{d \mid d \in D\}$ does contain the radius R which ensures the existence of a square in S including R, and thus containing q.

3. When the considered point lies in $S \ominus \bar{B}(0, \delta/2)$, the result follows from the very definition of the operator $\ominus$.

Eventually, we partitioned the set S in three subsets and in each of these subsets we proved that any point is contained in a square with diameter δ included in S. Hence, the result holds. $\square$

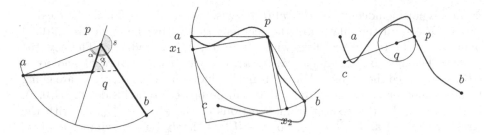

Fig. 2. Left: the turn of the polyline [a,q,p,b], $\gamma + \delta$, is greater than $\alpha + \delta = \pi$ (the half-line passing through p and q is the bisector of the angle $\widehat{apb}$). Center: in blue the arc $C_{a,b} = C \cap \bar{B}(p, \frac{\delta}{\sqrt{2}})$. In red, another arc of C whose end c is inside a square having p for vertex and included in the shape bounded by C. Right: in blue, the arc $C_{a,b} = C \cap \bar{B}(p, \frac{\delta}{\sqrt{2}})$. The disk $\bar{B}(p, \mathrm{d}(p,q))$ is included in the shape bounded by C. Since the arc $C_{a,b}$ cuts the radius $[p,c]$ passing through q, its turn is greater than that of the quarter of circle in red.

Thanks to the previous proposition, we get in Corollary 1 that a shape S having a δ-LTB curve for boundary with $r \leq \delta/(2\sqrt{2})$ verifies the two last items of Definition 4 with $r' = \sqrt{2}r$ and that $S \ominus \bar{B}(0,r)$ is non-empty.

Corollary 1. *Let S be closed shape having a δ-LTB curve C for boundary. Let $r \leq \delta/(2\sqrt{2})$. Then,*

- *$S \ominus \bar{B}(0,r)$ is non-empty,*
- *$S \subset S \ominus \bar{B}(0,r) \oplus \bar{B}(0, \sqrt{2}r)$,*
- *$S^c \subset S^c \ominus \bar{B}(0,r) \oplus \bar{B}(0, \sqrt{2}r)$,*

Proof. Let $p \in S$. By Proposition 3, there exists a square Q of edge length $2r$ containing p and included in S. Then, the center q of Q belongs to $S \ominus \bar{B}(0,r)$ which is therefore non-empty. Furthermore, $Q \subseteq S \ominus \bar{B}(0,r) \oplus \bar{B}(0, \sqrt{2}r)$. Then, $S \subseteq S \ominus \bar{B}(0,r) \oplus \bar{B}(0, \sqrt{2}r)$. Alike, applying Proposition 3 to $\bar{S}^c$, we get $S^c \subset S^c \ominus \bar{B}(0,r) \oplus \bar{B}(0, \sqrt{2}r)$.   □

It remains to prove that the erosion by a disk of radius r of a connected component of the plan deprived of a LTB curve is path connected (for well chosen values of r/δ). The rest of the proof is made by contradiction: we assume that $S \ominus B(0,r)$ has at least two distinct connected components and we consider the infimum distance d_0 between two connected components. The reasoning is split into two cases: $d_0 \leq 2\sqrt{2}r$ (Lemma 1) and $2\sqrt{2}r < d_0$ (Lemma 2).

Lemma 1. *Let C be a δ-LTB curve and A be a connected component of $\mathbb{R}^2 \setminus C$. Let $r < \delta/\sqrt{10 + 4\sqrt{2}}$. Two points of $\partial(A \ominus \bar{B}(0,r))$ at distance less than or equal to $2\sqrt{2}r$ are path-connected in $A \ominus \bar{B}(0,r)$.*

Proof. Let x_0 and x_1 two points of $A \ominus \bar{B}(0,r)$ at distance less than or equal to $2\sqrt{2}r$ from each other (Fig. 4 illustrates the proof). Assume that the segment

$[x_0, x_1]$ is not included in $A \ominus \bar{B}(0, r)$ (otherwise, we are done). Then, there exists a point $a \in \mathcal{C}$ in the dilation of the segment $[x_0, x_1]$ by the open disk $B(0, r)$ deprived of the two closed disks with center x_0, x_1 and radius r. Thereby, the point a belongs to a rectangle $[x_0, u_0, u_1, x_1]$ with $u_0 \in \partial B(x_0, r)$ and $u_1 \in \partial B(x_1, r)$. Since the segment $[x_0, x_1]$ is not included in $A \ominus \bar{B}(0, r)$, we are going to build another arc from x_0 to x_1 that will be proved to lie inside $A \ominus \bar{B}(0, r)$. Let i_0 and i_1 be the respective intersections of the segment $[x_0, x_1]$ with the circles $\partial \bar{B}(x_0, r)$ and $\partial \bar{B}(x_1, r)$. For $k \in \{0, 1\}$, let l_k be the symmetric of u_k with respect to x_k. Let p_0 and p_1 be the intersection points of the segment $[l_0, l_1]$ and the circles $\partial B(i_0, r)$ and $\partial B(i_1, r)$. Let P be the simple arc $\overparen{x_0 i_0 p_0} \cup [p_0, p_1] \cup \overparen{p_1 i_1 x_1}$ where $\overparen{xcy}$ denotes the quarter of circle with center c linking the points x and y. We claim that P is included in $A \ominus \bar{B}(0, r)$. By contradiction, assume that $P \not\subseteq A \ominus \bar{B}(0, r)$. Then, there exists a point $b \in \mathcal{C} \cap (P \oplus B(0, r))$ deprived of $B(x_0, r)$ and $B(x_1, r)$. The distance between a and b is upper-bounded by the distance between o_0 and u_1 (o_0 is the symmetric of i_0 with respect to x_0), that is by $\sqrt{10 + 4\sqrt{2}} r$. Thus, $\mathrm{d}(a, b) < \delta$. Therefore, there exists a straightest arc $\mathcal{C}|_a^b$ between a and b. We set $X = \mathbb{R}^2 \setminus (B(x_0, r) \cup B(x_1, r))$. For $k \in \{0, 1\}$, let H_k be the set of simple arcs $\mathcal{D}_{a,b}$ in X between a and b such that any arc between a and b homotopic to $\mathcal{D}_{a,b}$ in X intersects the quarter of plane Q_k delimited by the rays $\overrightarrow{x_k o_k}$ and $\overrightarrow{x_k u_k}$. Alike, let I be the set of simple arcs $\mathcal{D}_{a,b}$ in X between a and b such that any arc between a and b homotopic to $\mathcal{D}_{a,b}$ in X intersects the segment $[x_0, x_1]$ (see Fig. 3). Notice that I is empty if $\mathrm{d}(x_0, x_1) \leq 2r$. Observe that any arc between a and b in X that does not intersect the segment $[x_0, x_1]$ belongs to $H_0 \cup H_1$. Furthermore, by definition of H_0 and H_1, any arc between a and b in X homotopic to an arc not in $H_0 \cup H_1$ is not in $H_0 \cup H_1$ and thereby intersects $[x_0, x_1]$. Then, any arc between a and b in X that is not in $H_0 \cup H_1$ is in I.

In other words, we split the set of simple arcs between a and b in X in two classes: those passing in between the disks $B(x_0, r)$ and $B(x_1, r)$ and the others that turn around $B(x_0, r)$ or $B(x_1, r)$. Be aware that actually this splitting is not a partition for we make no restriction about the turn of the arcs in both sets. Hence, these arcs can do several turns around any of the two disks. The problem is that the only tool to link homotopy and turn to our knowledge is Proposition 2 and it is not sufficient to easily constrain the behavior of the arcs.

- Firstly, assume that the arc $\mathcal{C}|_a^b$ belongs to I. Let z be a point of intersection of the arc $\mathcal{C}|_a^b$ and the segment $[i_0, i_1]$.

  • Let t_0 and t_1 be the tangents from z to the quarters of circle $\overparen{i_0 x_0 u_0}$ and $\overparen{i_1 x_1 u_1}$ at points q_0 and q_1. Put $\alpha_k := \overparen{i_k x_k q_k}$ for $k \in \{0, 1\}$. Since $\mathrm{d}(x_0, x_1) \leq 2\sqrt{2}r$ and the secant function is increasing and strictly convex, we derive

$$\sec\left(\frac{\alpha_0 + \alpha_1}{2}\right) \leq \frac{\sec(\alpha_0) + \sec(\alpha_1)}{2} \leq \sqrt{2} \leq \sec\left(\frac{\pi}{4}\right), \qquad (1)$$

that is $\alpha_0 + \alpha_1 \leq \pi/2$ and the equality occurs only if $\alpha_0 = \alpha_1 = \frac{\pi}{4}$. Since $\widehat{q_0 z q_1} = \alpha_0 + \alpha_1$, we get $\widehat{q_0 z q_1} \leq \frac{\pi}{2}$.

- According to Proposition 3, there exists a square S with edge length $\delta/\sqrt{2}$, having z for vertex and whose interior is included in the exterior of A. Observe that, since $d(x_0, x_1) \leq 2\sqrt{2}r$, the distances $d(q_0, z)$ and $d(q_1, z)$ are upper-bounded by $d(i_0, u_1) = \sqrt{10 - 4\sqrt{2}}r$ which is less than the edge length of S. Thus, the square S has to be included in the sector delimited by the tangents t_0 and t_1 and not containing x_0 and x_1. Then, $\widehat{q_0 z q_1} \geq \frac{\pi}{2}$, and by Eq. 1, $\alpha_0 = \alpha_1 = \frac{\pi}{4}$, that is z is the middle of $[i_0, i_1]$. Therefore, z is the unique point of $\mathcal{C}$ lying on $[i_0, i_1]$.

- Noting that the three points q_0, q_1 and z are at distance less than $\delta/2$ from each other, we derive from [8, Lemma 8.a] that one of the three subarcs of $\mathcal{C}$ delimited by the three points z, q_0, q_1 has a turn greater than $\pi/2$. Since $\kappa([q_0, q_1, z]) > \pi/2$ and $\kappa([q_1, q_0, z]) > \pi/2$, the arc of $\mathcal{C}$ between q_0 and z not containing q_1 and the arc of $\mathcal{C}$ between q_1 and z not containing q_0 have a turn bounded from above by $\pi/2$. Thus, the third arc delimited by the three points z, q_0, q_1, which is the arc between q_0 and q_1 not containing z has a turn greater than $\pi/2$. Hence, the turn of the arc $\mathcal{C}_{q_0, q_1}$ between q_0 and q_1 containing z is less than or equal to $\pi/2$. As the $\kappa(\mathcal{C}_{q_0,q_1}) \geq \kappa(\widehat{q_0, z, q_1}) = \pi/2$ by the definition of the turn, we derive that $\kappa(\mathcal{C}_{q_0,q_1}) = \pi/2$. Thus, $\mathcal{C}_{q_0,q_1}$ is the polyline $[q_0, z, q_1]$. Then, the arc $\mathcal{C}|_a^b$ is the disjoint union of two or three arcs, an arc $\mathcal{C}_k$ between a and a point q_k, $k \in \{0, 1\}$, the open polyline (q_0, z, q_1) and an arc $\mathcal{C}_{1-k}$ between q_{1-k} and b if $a \notin [q_0, z, q_1]$, or the polyline $[a, z, q_{1-k}]$ and the arc $\mathcal{C}_{1-k}$ if $a \in [q_k, z)$. Thus, $\mathcal{C}|_a^b$ is homotopic in X to $\mathcal{C}_k \sqcup (q_0, q_1) \sqcup \mathcal{C}_{1-k}$, or to $[a, q_{1-k}) \sqcup \mathcal{C}_{1-k}$ which do not intersect $[x_0, x_1]$ (for z is the unique point of $\mathcal{C}$ on $[i_0, i_1]$ and $\mathcal{C}$ is simple). Contradiction!

– Secondly, assume that for some $k \in \{0, 1\}$, $\mathcal{C}|_a^b \in H_k$.

We denote by O_k, $k \in \{0, 1\}$, the convex hull of the quarter of the circle $\partial B(x_k, r)$ delimited by u_k and o_k.

O_k is included in a bounded component of $\mathbb{R}^2 \setminus (\mathcal{C}|_a^b \cup [a, b])$. Then, according to the definition of $\mathcal{C}|_a^b$ and Proposition 2,

$$\pi/2 \geq \kappa(\mathcal{C}|_a^b) \geq \kappa(\partial \operatorname{conv}(O_k \cup [a, b]) \setminus (a, b)) > \pi/2,$$

which is absurd (the last inequality comes from the fact that a, resp. b, cannot lie on the tangent at u_k, resp. o_k, to the circle $\partial B(x_k, r)$).

Finally, in each studied case, the assumption that the path P is not included in the eroded set $A \ominus \bar{B}(0, r)$ leads to a contradiction. We conclude that the points x_0 and x_1 are path-connected in $A \ominus \bar{B}(0, r)$. $\qquad \square$

Lemma 2. *Let S be a closed subset of $\mathbb{R}^2$ whose boundary is a δ-LTB curve. Let $r < \frac{\sqrt{2}}{2}\delta$. The minimal distance d_0 between two connected components of the eroded shape $S \ominus \bar{B}(0, r)$ (respectively $S^c \ominus \bar{B}(0, r)$) is upper-bounded by $2\sqrt{2}r$.*

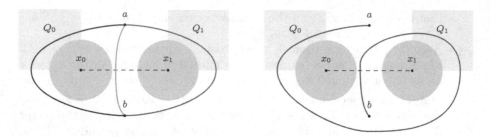

Fig. 3. On the left the blue arc belongs to the set H_0, the green arc to I and the red arc to H_1. On the right, the purple arc belongs to H_0, H_1 and I. (Color figure online)

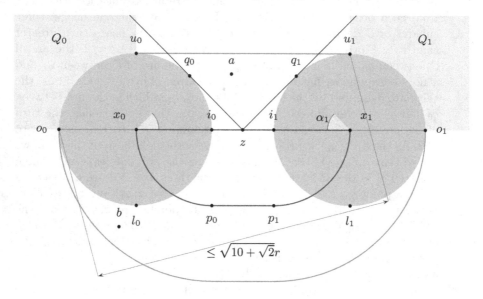

Fig. 4. The figure illustrates the notations used in the proof of Lemma 1. The proof consists in showing that one of the two red paths joining the points x_0 x_1 is included in $A \ominus \bar{B}(0, r)$.

Proof. Assume by contradiction that $d_0 > 2\sqrt{2}r$. Let A be $S \ominus \bar{B}(0, r)$ (the case $A = S^c \ominus \bar{B}(0, r)$ is similar). If A has two, or more, connected components, then it is the same with $A \oplus \bar{B}(0, \sqrt{2}r)$ for the dilation of a path connected set by a path connected structural element containing the origin is path connected and the radius of the dilation is less than the half of d_0. Therefore, S, which is connected and included in $A \oplus \bar{B}(0, \sqrt{2}r)$ is included in just one connected component of $A \oplus \bar{B}(0, \sqrt{2}r)$. Therefore, the others components do not contain any point of S. Hence, there are at least one non empty component of $S \ominus \bar{B}(0, r)$ which does not contain any point of S which is absurd. □

Proof (Theorem 1).
By Corollary 1:

- $S \ominus \bar{B}(0, r)$ is non-empty,
- $S \subset S \ominus \bar{B}(0, r) \oplus \bar{B}(0, \sqrt{2}r)$,
- $S^c \subset S^c \ominus \bar{B}(0, r) \oplus \bar{B}(0, \sqrt{2}r)$.

Assume by contradiction that $S \ominus \bar{B}(0, r)$ or $S^c \ominus \bar{B}(0, r)$ is not path-connected. Since S is a compact set, S can be covered by a finite number of disks of radius $\frac{r}{2}$, then $S \ominus \bar{B}(0, r)$ is also covered by a finite number of disks of radius $\frac{r}{2}$. Moreover, by Lemma 1, in each disk of radius $\frac{r}{2}$ there is at most one connected component of $S \ominus \bar{B}(0, r)$. Then $S \ominus \bar{B}(0, r)$ has a finite number of connected components. Since S is compact, $S^c \ominus \bar{B}(0, r)$ has just one unbounded component, say S_0^c, and $(S^c \ominus \bar{B}(0, r)) \setminus S_0^c$ is bounded. Thereby, by the same reasoning as for $S \ominus \bar{B}(0, r)$, we have that $S^c \ominus \bar{B}(0, r)$ has a finite number of connected components. Then the minimal distance d_0 between two connected components is well-defined for both $S \ominus \bar{B}(0, r)$ and $S^c \ominus \bar{B}(0, r)$. More precisely, d_0 is defined by:

$$d_0 := \min \left\{ \inf_{x_0 \in A_0, x_1 \in A_1} \mathrm{d}(x_0, x_1) | A_0, A_1 \text{ distinct connected components of } A \right\},$$

where A is $S \ominus \bar{B}(0, r)$ or $S^c \ominus \bar{B}(0, r)$. But by Lemmas 1, 2, $d_0 \notin [0, 2\sqrt{2}r] \cup (2\sqrt{2}r, +\infty)$. Contradiction ! $\square$

4   Conclusion

This paper establishes that the Local Turn Boundedness implies the quasi-regularity in 2D. Therefore the set of quasi(r)-regular curves is larger than the set of LTB curves for a $r < \delta/\sqrt{10 + 4\sqrt{2}}$. On the one hand, quasi-regularity allows the corresponding shape digitization to keep its convexity and its topological properties under rigid motion [10, 11]. On the other hand, Local Turn Boundedness has been introduced to map ordered samplings of the digital boundary to close ordered samplings of the continuous curve in order to compare the lengths of the continuous curve and of its digitization.

It is possible to build a quasi(r)-regular curve having arbitrary numerous small (against r) oscillations leading to an arbitrary large length. Thus, the results obtained in [8] on the length estimation of LTB curves cannot be extended to quasi-regular curves. Nevertheless, the link between Local Turn Boundedness and quasi-regularity can be useful for the generalization of Local Turn Boundedness to higher dimension and this is the perspective of our work.

References

1. Alexandrov, A.D., Reshetnyak, Y.G.: General Theory of Irregular Curves, Mathematics and Its Applications, vol. 29. Springer, Dordrecht (1989). https://doi.org/10.1007/978-94-009-2591-5

2. Chazal, F., Cohen-Steiner, D., Lieutier, A.: A sampling theory for compact sets in Euclidean space. Discr. Comput. Geom. **41**(3), 461–479 (2009). https://doi.org/10.1007/s00454-009-9144-8
3. Federer, H.: Curvature measures. Trans. Am. Math. Soc. **93**(3), 418–491 (1959). http://www.jstor.org/stable/1993504
4. Lachaud, J., Thibert, B.: Properties of Gauss digitized shapes and digital surface integration. J. Math Imaging Vis. **54**(2), 162–180 (2016). https://doi.org/10.1007/s10851-015-0595-7
5. Latecki, L., Conrad, C., Gross, A.: Preserving topology by a digitization process. J. Math. Imaging Vision **8**, 131–159 (1998). https://doi.org/10.1023/A:1008273227913
6. Le Quentrec, É., Mazo, L., Baudrier, É., Tajine, M.: Local Turn-Boundedness: A curvature control for a good digitization. In: Couprie, M., Cousty, J., Kenmochi, Y., Mustafa, N. (eds.) Discrete Geometry for Computer Imagery, pp. 51–61. Springer, Cham (2019). https://rd.springer.com/chapter/10.1007/978-3-030-14085-4_5
7. Le Quentrec, É., Mazo, L., Baudrier, É., Tajine, M.: Local Turn-Boundedness: A Curvature Control for Continuous Curves with Application to Digitization. J. Math. Imaging Vis. **62**(5), 673–692 (2020). https://doi.org/10.1007/s10851-020-00952-x
8. Le Quentrec, É., Mazo, L., Baudrier, É., Tajine, M.: Monotonic sampling of a continuous closed curve from its Gauss digitization. Application to length estimation. Technical report, Icube Laboratory, University of Strasbourg, CNRS (2020). https://hal.archives-ouvertes.fr/hal-02987858, submitted
9. Meine, H., Köthe, U., Stelldinger, P.: A topological sampling theorem for robust boundary reconstruction and image segmentation. Discrete Appl. Math. **157**(3), 524–541 (2009). https://doi.org/10.1016/j.dam.2008.05.031, http://www.sciencedirect.com/science/article/pii/S0166218X08002643
10. Ngo, P., Kenmochi, Y., Debled-Rennesson, I., Passat, N.: Convexity-preserving rigid motions of 2D digital objects. In: Kropatsch, W.G., Artner, N.M., Janusch, I. (eds.) DGCI 2017. LNCS, vol. 10502, pp. 69–81. Springer, Cham (2017). https://doi.org/10.1007/978-3-319-66272-5_7
11. Ngo, P., Passat, N., Kenmochi, Y., Debled-Rennesson, I.: Geometric Preservation of 2D Digital Objects Under Rigid Motions. J. Math. Imaging Vis. **61**(2), 204–223 (2019). https://doi.org/10.1007/s10851-018-0842-9, http://link.springer.com/10.1007/s10851-018-0842-9
12. Pavlidis, T.: Algorithms for Graphics and Image Processing. Springer, Heidelberg (1982)
13. Stelldinger, P., Terzic, K.: Digitization of non-regular shapes in arbitrary dimensions. Image Vis. Comput. **26**(10), 1338–1346 (2008). https://doi.org/10.1016/j.imavis.2007.07.013, http://www.sciencedirect.com/science/article/pii/S0262885607001370

Discrete Geometry - Models, Transforms, Visualization

Shear Based Bijective Digital Rotation in Hexagonal Grids

Eric Andres[✉], Gaëlle Largeteau-Skapin, and Rita Zrour

University of Poitiers, Laboratory XLIM, ASALI, UMR CNRS 7252, BP 30179,
86962 Futuroscope Chasseneuil, France
{eric.andres,gaelle.largeteau.skapin,rita.zrour}@univ-poitiers.fr

Abstract. In this paper, a new *bijective* digital rotation algorithm for the hexagonal grid is proposed. The method is based on an decomposition of rotations into shear transforms. It works for any angle with an hexagonal centroid as rotation center and is easily invertible. The algorithm achieves an average distance between the digital rotated point and the continuous rotated point of about 0.42 (for 1.0 the distance between two neighboring hexagon centroids).

Keywords: Hexagonal grid · Digital rotation · Bijective digital rotation · Shear transform

1 Introduction

Our motivation for lossless rotations in hexagonal grids came from recent work by K. Pluta et al. [7,8,13] on digitized bijective rotations in the hexagonal grid, and more generally from the work of P. Ngo et al. on bijective rigid motions [10] and the possibilities it opens to approximate general digital transforms as a sequence of such transforms. There are also some more direct and practical perspectives such 3D bijective rotations of honeycomb structures for 3D printing [3].

Transformations in hexagonal grids are still a largely open problem with few references (see [4,5] for a discussion on transforms on hexagonal grids). K. Pluta et al. showed that, as for the square grid [6,11], there are angles for which the digitized rotation is bijective in the hexagonal grid [7,8]. The problem is that these angles represent only a subset of all angles. This is the limitation we propose to overcome in this paper.

We propose to revisit an idea already applied to bijective rigid motions in the classical square grid: decomposing a rotation into a sequence of shear transforms [2] developed out of an original idea by [14]. A. W. Paeth used such a decomposition in order to propose a fast anti-aliasing method for image rotations [12]. The idea proposed by J.-P. Reveilles [14] is to use the same decomposition to push rows of pixels by an integer number of pixels, ensuring bijectivity and trivial reversibility. In the square grid, a rotation is defined by the composition of three shear transforms with the first and the last equal. For the hexagonal grid,

© Springer Nature Switzerland AG 2021
J. Lindblad et al. (Eds.): DGMM 2021, LNCS 12708, pp. 217–228, 2021.
https://doi.org/10.1007/978-3-030-76657-3_15

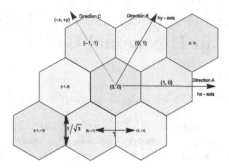

Fig. 1. Coordinate system, hexagon size and A, B and C directions

three different shear transforms are considered, one for each hexagonal symmetry direction (see [5] for some additional insight in such shear transforms in the hexagonal grid). This leads to a rotation decomposition. The idea behind the shear transforms is to push whole rows of hexagons successively in those three directions by integer numbers of hexagons, ensuring *bijectivity and reversibility*. As for the shear based rotation proposed by A. W. Paeth, we have angles where the formula diverges. There are however simple ways to overcome these problems. This leads to a digital rotation algorithm for all angles in the hexagonal grid with a grid point (the centroid of an hexagon) as center. A future work could consist in lifting this limitation and proposing a rigid motion transform (rotation with an arbitrary center).

The organization of the paper is as follows: in section two, we present the preliminaries. In particular, we present shear transforms and how shear transforms have been used in the classical square grid to define digital rotations. An error criteria based on the distance between the continuous and the digital rotated points is presented. In section three, we introduce our method of digital bijective rotation for the hexagonal grids. We conclude and present perspectives in the last section.

2   Preliminaries

2.1   Hexagonal Grid

We are considering an hexagonal cell (centered on the grid point) with "pointy top" hexagons (two sides of the hexagons are parallel to the ordinate axis of the classical Euclidean coordinate system). It should be easy to transpose this work for other hexagon orientations or hexagon sizes. The hexagons are regular with a side length of $\frac{1}{\sqrt{3}}$. This means that between the centroids of two neighboring hexagons, there is a distance of 1. There are various ways of creating a coordinate system for hexagonal grids. We chose a simple 2D coordinate system where the first coordinate, the hexagonal hx-axis, is given by the Cartesian vector $(1, 0)$ and the second coordinate, the hexagonal hy-axis, is defined by the Cartesian vector $\left(\frac{1}{2}, \frac{\sqrt{3}}{2}\right)$. This corresponds to the coordinate system proposed by W. E.

Snyder [16]. The coordinate transforms from hexagonal grid to the Cartesian grid and vice versa are given by the following transforms [16]:

$$Cart2Hex : (x, y) \mapsto (x_h, y_h) = \left(x - \frac{y}{\sqrt{3}}, \frac{2y}{\sqrt{3}} \right)$$

$$Hex2Cart : (x_h, y_h) \mapsto (x, y) = \left(x_h + \frac{y_h}{2}, \frac{y_h\sqrt{3}}{2} \right)$$

In the chosen coordinate system, we consider three directions: direction A defined by vector $(1, 0)$, direction B defined by vector $(0, 1)$ and direction C defined by vector $(-1, 1)$. These vectors are given in the hexagonal coordinate system (see Fig. 1). A A-row is a set of hexagonal grid points generated by hexagonal coordinate vector $(1, 0)$ and a grid point. A B-row is a set of hexagonal grid points generated by hexagonal coordinates vector $(0, 1)$ and a grid point. A C-row is a set of hexagonal grid points with hexagonal coordinates $(x, y) \in \mathbb{Z}^2$ such that they have all the same $x + y$ value, generated by hexagonal coordinates vector $(-1, 1)$ and a grid point.

2.2 Shear Transforms

The idea behind the method proposed in this paper is to decompose a rotation transform into a sequence of shear transforms in the hexagonal coordinate system. A *shear transform* is a linear mapping that translates a point in a given direction by a vector proportional (by a shear factor) to the signed distance to a line parallel to that direction. A typical shear transform uses the axis-lines as shear lines. Shear transforms preserve areas. This is why it is quite natural to decompose isometries into sequences of shear transforms (as atomic transforms). As an example, the decomposition of a rotation into three shear transforms with the Cartesian axes as shear lines (an 'ULU' decomposition with '1's on the diagonals) used by A. W. Paeth [12]:

$$\begin{pmatrix} \cos\theta & -\sin\theta \\ \sin\theta & \cos\theta \end{pmatrix} = \begin{pmatrix} 1 & -\tan\frac{\theta}{2} \\ 0 & 1 \end{pmatrix} \begin{pmatrix} 1 & 0 \\ \sin\theta & 1 \end{pmatrix} \begin{pmatrix} 1 & -\tan\frac{\theta}{2} \\ 0 & 1 \end{pmatrix}$$

This rotation decomposition leads to a fast and simple antialised rotation that is still in use in some image libraries such as ImageMagick.

2.3 Bijective Digital Rotations in the Square Grid

In this paper we are interested in bijective digital rotations on hexagonal grids. Let us first see how this problem has been tackled in the classical square grid. Let us consider a digitized rotation $DR(\theta)$:

$$DR(\theta) : \begin{pmatrix} x \\ y \end{pmatrix} \mapsto \begin{pmatrix} \lfloor x\cos\theta + y\sin\theta + 0.5 \rfloor \\ \lfloor -x\sin\theta + y\cos\theta + 0.5 \rfloor \end{pmatrix}$$

with $\lfloor u \rfloor$ the biggest integer smaller or equal to u (i.e. floor function).

A digitized rotation is not, in general, surjective or injective, except for some angles where the transform is actually bijective. For the interested reader, please refer to [6,11,15]. The problem with the angles for which the digitized rotation is bijective is that they do not cover all angles. J. P. Reveilles [14] took the same decomposition than A. W. Paeth to propose a bijective digital rotation that works for all angles [14]. E. Andres improved on this idea and gave formulas for an improved bijective digital rotation with lower errors and a bijective rigid motion in the classical square grid [2]. The idea that leads to a bijective digital rotation is the following: the shear coefficients multiplied by x and y respectively, $-x\tan\frac{\theta}{2}$ and $y\sin\theta$, are approximated by their closest integer $\lfloor -x\tan\frac{\theta}{2} + 0.5 \rfloor$ and $\lfloor y\sin\theta + 0.5 \rfloor$. Moving a grid point in the x-axis direction and y-axis direction by an integer displacement is obviously a reversible operation. One could think that this leads to a very coarse approximation of the continuous rotation, but this is not the case. The way to measure this "approximation" [1,2] is presented in the next subsection.

2.4   Error Measure

Each grid point has one and only one image through a bijective digital rotation but that does not mean that the digital rotation is a good approximation of the continuous one. To measure how "wrong" we are by choosing the digital rotation over the continuous one, we are considering two distance criteria [1,2]. Let us first denote $\mathcal{G} = \{ax_g + by_g | (a,b) \in \mathbb{Z}^2\} \subset \mathbb{R}^2$ a grid defined by the point $(0,0)$ and two vectors x_g and y_g. Let us denote $R_\theta(p)$ the continuous rotation of center $(0,0)$ and angle θ of a grid point $p \in \mathcal{G}$ and $\mathcal{R}_\theta(p)$ its digital rotation of center $(0,0)$ ($\mathcal{R}_\theta(p)$ is a grid point in this case).

Computing $\max_{p\in\mathcal{G}}(d(R_\theta(p), \mathcal{R}_\theta(p))$ defines the *Maximum Distance error criteria (MD)*, and $avg_{p\in\mathcal{G}}(d(R_\theta(p), \mathcal{R}_\theta(p))$ the *Average Distance error criteria (AD)*, where $avg_{p\in\mathcal{G}}$ is the average distance over the grid.

3   Bijective Digital Rotation in the Hexagonal Grid

3.1   Rotation Decomposition into Shears in the Hexagonal Grid

K. Pluta showed that there exists a subset of angles for which the digitized rotation of an hexagonal grid is bijective [7,8]. Our idea is to use shear transforms to define a digital rotation in the hexagonal grid that works for all angles by *pushing* grid points into specific directions. Since we are working in an hexagonal grid, we have three privileged directions to push grid points: direction A defined by vector $(1,0)$, direction B defined by vector $(0,1)$ and direction C defined by vector $(-1,1)$ (vectors expressed in the hexagonal coordinate system, see Fig. 1). Note that I. Her presents the same shear transforms along these three directions but with a different coordinate system [5]. The shear transforms for the three directions, expressed in the hexagonal grid coordinate system, correspond to the following matrices:

$$matA_{hex} = \begin{pmatrix} 1 & a \\ 0 & 1 \end{pmatrix} ; matB_{hex} = \begin{pmatrix} 1 & 0 \\ b & 1 \end{pmatrix}$$

$$matC_{hex} = \begin{pmatrix} 1-c & -c \\ c & 1+c \end{pmatrix}$$

$matA_{hex}$ and $matB_{hex}$ are classical shear transforms with the axis as shear lines. For the third matrix, it is the same idea: We have a direction C defined by the vector $(-1, 1)$ with an orthogonal direction $(1, 1)$ (in the hexagonal coordinate system). This leads to matrix $matC_{hex}$. One can easily see that this matrix preserves the sum of coordinates $x + y$ over a C-row.

Since the rotation matrix is simplest expressed in the classical Cartesian coordinate system, let us switch back to the Cartesian coordinate system, which leads to:

$$matA_{cart} = \begin{pmatrix} 1 & \frac{2a}{\sqrt{3}} \\ 0 & 1 \end{pmatrix}$$

$$matB_{cart} = \begin{pmatrix} 1+\frac{b}{2} & -\frac{b}{2\sqrt{3}} \\ \frac{b\sqrt{3}}{2} & 1-\frac{b}{2} \end{pmatrix}$$

$$matC_{cart} = \begin{pmatrix} 1-\frac{c}{2} & -\frac{c}{2\sqrt{3}} \\ \frac{c\sqrt{3}}{2} & 1+\frac{c}{2} \end{pmatrix}$$

The idea of our method is to push grid points in those three directions in order to approximate a rotation. Therefore, we solved the following equation:

$$matA_{cart}.matB_{cart}.matC_{cart} = \begin{pmatrix} \cos\theta & -\sin\theta \\ \sin\theta & \cos\theta \end{pmatrix}$$

which has a unique solution for a, b, c:

$$\begin{pmatrix} a \\ b \\ c \end{pmatrix} = \begin{pmatrix} -1 + \frac{\sqrt{3}-2\sin\theta}{\sqrt{3}\cos\theta-\sin\theta} \\ 1 - \cos\theta + \frac{\sin\theta}{\sqrt{3}} \\ 1 - \frac{\sqrt{3}-2\sin\theta}{\sqrt{3}\cos\theta-\sin\theta} \end{pmatrix}$$

For this solution, $a = -c$. Let us note that from here on, we are only going to work in the hexagonal coordinate system with matrices $matA_{hex}, matB_{hex}$ and $matC_{hex}$.

3.2   Dealing with the Divergence and the Inverse

The proposed solution for (a, b, c) is not universal: the denominator for a (and c), $\sqrt{3}\cos\theta - \sin\theta$, is equal to 0 for θ equal to $\pi/3$ or $4\pi/3$ (with angles between 0 and 2π). As one can see in Fig. 2, the zeros for $\pi/3$ and $4\pi/3$ are not of the same nature: $lim_{t \to \frac{\pi}{3}} a = 0.5$ and $lim_{t \to \frac{4\pi}{3}} a = \pm\infty$. Angle $\pi/3$ is a point singularity that can be easily dealt with since it is trivial to rotate an hexagonal image by an angle $\frac{\pi}{3}$ bijectively and thus avoid this singularity: $R_{\frac{\pi}{3}}(x, y) = (-y, x + y)$ is

---

Algorithm 1: $RotCBA(x, y, \theta)$: POINT ROT. CBA OF CENTER $(0,0)$ AND
ANGLE θ

$\quad$ **Input** $\;:\; (x, y) \in \mathbb{Z}^2, 0 \leq \theta < 2\pi/3$
$\quad$ **Output:** $(x_3, y_3) \in \mathbb{Z}^2$

1 $\quad$ If $\theta = \frac{\pi}{3}$ Then $\begin{pmatrix} x_3 \\ y_3 \end{pmatrix} \leftarrow \begin{pmatrix} -y \\ x + y \end{pmatrix}$

2 $\quad$ Else $\begin{pmatrix} a \\ b \end{pmatrix} \leftarrow \begin{pmatrix} -1 + \frac{\sqrt{3} - 2\sin\theta}{\sqrt{3}\cos\theta - \sin\theta} \\ 1 - \cos\theta + \frac{\sin\theta}{\sqrt{3}} \end{pmatrix}$

3 $\quad \begin{pmatrix} x_1 \\ y_1 \end{pmatrix} \leftarrow \begin{pmatrix} \lfloor x * (1+a) + y * a + 0.5 \rfloor \\ \lfloor -x * a + y * (1-a) + 0.5 \rfloor \end{pmatrix}$

4 $\quad \begin{pmatrix} x_2 \\ y_2 \end{pmatrix} \leftarrow \begin{pmatrix} x_1 \\ y_1 + \lfloor x_1 * b + 0.5 \rfloor \end{pmatrix}$

5 $\quad \begin{pmatrix} x_3 \\ y_3 \end{pmatrix} \leftarrow \begin{pmatrix} x_2 + \lfloor y_2 * a + 0.5 \rfloor \\ y_2 \end{pmatrix}$

6 $\quad$ **return** (x_3, y_3)

Algorithm 2: $RotCBANeg(x, y, \theta)$: INVERSE ROT. CBA OF CENTER
$(0,0)$ AND ANGLE θ

$\quad$ **Input** $\;:\; (x, y) \in \mathbb{Z}^2, 0 \leq \theta < 2\pi/3$
$\quad$ **Output:** $(x_3, y_3) \in \mathbb{Z}^2$

1 $\quad$ If $\theta = \frac{\pi}{3}$ Then $\begin{pmatrix} x_3 \\ y_3 \end{pmatrix} \leftarrow \begin{pmatrix} x + y \\ -x \end{pmatrix}$ – Rotation $-\pi/3$

2 $\quad$ Else $\begin{pmatrix} a \\ b \end{pmatrix} \leftarrow \begin{pmatrix} 1 - \frac{\sqrt{3} - 2\sin\theta}{\sqrt{3}\cos\theta - \sin\theta} \\ -1 + \cos\theta - \frac{\sin\theta}{\sqrt{3}} \end{pmatrix}$

3 $\quad \begin{pmatrix} x_1 \\ y_1 \end{pmatrix} \leftarrow \begin{pmatrix} x + \lfloor y * a + 0.5 \rfloor \\ y \end{pmatrix}$

4 $\quad \begin{pmatrix} x_2 \\ y_2 \end{pmatrix} \leftarrow \begin{pmatrix} x_1 \\ y_1 + \lfloor x_1 * b + 0.5 \rfloor \end{pmatrix}$

5 $\quad \begin{pmatrix} x_3 \\ y_3 \end{pmatrix} \leftarrow \begin{pmatrix} \lfloor x_2 * (1+a) + y_2 * a + 0.5 \rfloor \\ \lfloor -x_2 * a + y_2 * (1-a) + 0.5 \rfloor \end{pmatrix}$

6 $\quad$ **return** (x_3, y_3)

---

still an hexagonal grid point if (x, y) is one. This leads to the digital Rotation
RotCBA (Algorithm 1) that rotates an hexagonal grid point for angles from 0
to $2\pi/3$.

Let us see how we can handle the problems around $4\pi/3$. In Fig. 2, we can see
that all the values around angle $4\pi/3$ are affected by the divergence. Actually,
starting at $2\pi/3$ the average (AD) and maximal (MD) distance error values start
to increase. The idea here is to consider only rotations for angles between 0 and
$2\pi/3$ and map all the other angles to this interval. This is not that difficult
since, in the hexagonal grid, rotations by angles $k\pi/3$ (with a grid point center)
are naturally bijective so that for an angle θ, we can always decompose it as
$\theta = k\pi/3 + \theta'$ where $0 \leq \theta' \leq 2\pi/3$.

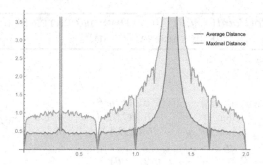

Fig. 2. Average (AD) and Maximal (MD) Distance error criteria for Rotation CBA for angles 0 to 2π

There is however another problem that needs to be dealt with if we want an easy to use digital rotation algorithm: *invertibility*. Indeed, consider Algorithm 1, $RotCBA(x, y, \theta)^{-1}$ is not equal to $RotCBA(x, y, -\theta)$. Having a bijective transform does not automatically mean that we have an easy way of performing the inverse transform. The approach we adopted to be sure that our rotation for angle θ is easily invertible is to perform all hexagon push operations in the reverse order for an angle $-\theta$ (or more precisely $2\pi - \theta$). For this, we introduce Algorithm 2 which is literally the inverse mapping of the mapping of Algorithm 1, with shear factors $-(a, b)$ and reversed shear transform applications order. With all this, we can now propose a rotation for an hexagonal point. The rotation of a point (x, y) by an angle $0 \leq \theta < 2\pi$ is performed in the following way (see Algorithm 3):

1. For an angle $0 \leq \theta 2 \leq 2\pi/3$: apply Algorithm 1 to the point (x, y) with angle θ;
2. For an angle $2\pi/3 < \theta \leq \pi$: apply Algorithm 1 to the point $(-x - y, x)$ with angle $\theta - 2\pi/3$ (the point $(-x - y, x)$ is the rotation of (x, y) by $2\pi/3$);
3. For an angle $\pi < \theta < 4\pi/3$: apply Algorithm 2 to the point (x, y) with angle $4\pi/3 - \theta$ and then perform a rotation of angle $-2\pi/3$ on the resulting point (x', y') which maps the point (x', y') to $(y', -x' - y')$ (This ensures that operations are exactly performed in the reverse order of point 2 (angle $2\pi/3 < \theta \leq \pi$));
4. For an angle $4\pi/3 \leq \theta < 2\pi$: apply Algorithm 2 to the point (x, y) with an angle $2\pi - \theta$. This ensures that operations are exactly performed in the reverse order of point 1 (angle $0 \leq \theta \leq 2\pi/3$).

This leads to point rotation *RotPoint* (Algorithm 3) and to the final bijective rotation *RotHexa* (Algorithm 4). We'll explain why Rotation Algorithm 4 is bijective in the next subsection.

Algorithm 3: $RotPoint(x, y, \theta)$: ROTATION OF CENTER $(0,0)$ AND ANGLE θ OF POINT (X,Y) IN THE HEXAGONAL GRID.

    Input   : $(x, y) \in \mathbb{Z}^2, \theta \in \mathbb{R}^2$
    Output: $(x'', y'') \in \mathbb{Z}^2$
1 $\theta = \theta \bmod 2\pi$
2 If $0 \leq \theta \leq \frac{2\pi}{3}$ Then $(x'', y'') \leftarrow RotCBA(x, y, \theta)$
3   Elsif $\frac{2\pi}{3} < \theta \leq \pi$ Then $(x'', y'') \leftarrow RotCBA\left(-x - y, x, \theta - \frac{2\pi}{3}\right)$
4   Elsif $\pi < \theta < \frac{4\pi}{3}$ Then $(x', y') \leftarrow RotCBANeg\left(x, y, \frac{4\pi}{3} - \theta\right)$;
    $(x'', y'') \leftarrow (y', -x' - y')$
5   Else $(x'', y'') \leftarrow RotCBANeg(x, y, 2\pi - \theta)$
6 Return(x'', y'')

Algorithm 4: $RotHexa(ImageIn, \theta)$: BIJECTIVE ROTATION FOR THE HEXAGONAL GRID OF CENTER $(0,0)$ AND ANGLE θ

    Input   : ImageIn, θ
    Output: ImageOut
1 For all (x, y) in ImageIn
2    $(x', y') \leftarrow$ RotPoint(x, y, θ)
3   ImageOut $(x', y') \leftarrow$ ImageIn(x, y)
4 **return** *ImageOut*

Let us quickly comment Fig. 3 which illustrates the distance error criteria for the bijective digital rotation algorithm and for the (non bijective) digitized rotation. In order to create the figure, and since both error measures cannot be computed on an infinite sized image, we used an rotation angle step of $\pi/300$ with all the points of domain $[-1000, 1000]^2$ that are then multiplied by random integers between 1 and 1000. That means that we considered a sample of 2001^2 points in a window of $[-10^6, 10^6]^2$. The idea is to have a large sample of points in a large enough domain in order to check if the error values diverge when the distances to the rotation center increase. As we can see on the figure, the digitized rotation has a very stable average error (AD error) value of ≈ 0.38 (except of course for angles $k\pi/3$ where the digitized rotation is trivially bijective with an error of 0) compared to our rotation that has an average error value of ≈ 0.43. The maximum distance error value (MD Error) is ≈ 0.707 for the digitized rotation and varies between 0.96 and 1.09 (with an average maximum distance error value of ≈ 1.04). The distance between two hexagon centroids being 1. This means basically that an hexagon *lands* usually in the immediate grid neighborhood of the spot where he should be. Note that the bijective angles of K. Pluta [7,8,13] are particular angles for which the digitized rotation is bijective but otherwise the error measures are similar than for the other angles of the digitized rotation. Let us note also that we obtain slightly better error values than for the similar type of algorithm in the square grid [2].

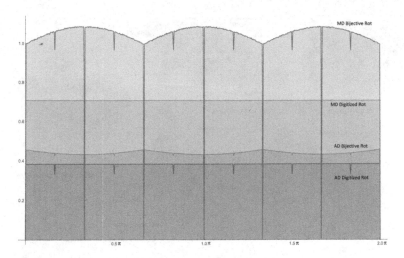

Fig. 3. Average (AD) and Maximal (MD) Distance error criteria for the hexagonal grid Rotation Algorithm and for the digitized rotation for angles 0 to 2π.

3.3   Bijectivity

All has been done in order to have a bijective transform $RotHexa$. Let us show that formally by examining the mappings of Algorithm 1. Let us define the mapping $\mathcal{A}(a) : (x,y) \mapsto (x + \lfloor ay + 0.5 \rfloor, y)$, $a \in \mathbb{R}^2$ (line 5 in Algorithm 1). For a A-row, we have $\mathcal{A}(a)(S) = S$ (y does not change). All hexagons on a given A-row are *pushed* in direction A by the same integer number of grid points.

In the same way, for the mapping $\mathcal{B}(b) : (x,y) \mapsto (x, y + \lfloor bx + 0.5 \rfloor, y)$, $b \in \mathbb{R}^2$ (line 4 in Algorithm 1), if S is a B-row, we have $\mathcal{B}(b)(S) = S$ (x does not change). All hexagons on a given B-row are pushed in direction B by the same integer number of grid points.

Let us now consider the mapping $\mathcal{C}(a) : (x,y) \mapsto (\lfloor x(1 + a) + ay + 0.5 \rfloor,$ $\lfloor -ax + y(1 - a) + 0.5 \rfloor)$, $a \in \mathbb{R}^2$ (line 3 in Algorithm 1). Let us show that for a C-row S, we have $\mathcal{C}(a)(S) = S$ (i.e. that $x + y$ does not change). Let us consider a point $(x,y) \in S$ and $(x',y') = \mathcal{C}(a)(x,y)$. We have $x' + y' = \lfloor x(1 + a) + ay + 0.5 \rfloor + \lfloor -ax + y(1 - a) + 0.5 \rfloor = x + y + \lfloor a(x + y) + 0.5 \rfloor + \lfloor -a(x + y) + 0.5 \rfloor$. Now $\lfloor x + 0.5 \rfloor$ is the rounding function and it is easy to see that $\lfloor -u + 0.5 \rfloor = -\lfloor u + 0.5 \rfloor$ which proves that $x' + y' = x + y$ and thus that $\mathcal{C}(a)(x,y) \in S$. All hexagons on a given C-row are pushed in direction C by the same integer number of grid points. All this is of course verified as well for Algorithm 2. All the other mappings (rotations by $k\pi/3$ angles) are trivially bijective.

This proves that the transform presented in Algorithm 4 is bijective (as sequence of bijective mappings). By design, we have made sure that the transform is also easily invertible such that, for $ImageIn$ an image in the hexagonal grid, we have $RotHexa(ImageIn, \theta)^{-1} = RotHexa(ImageIn, \theta)$.

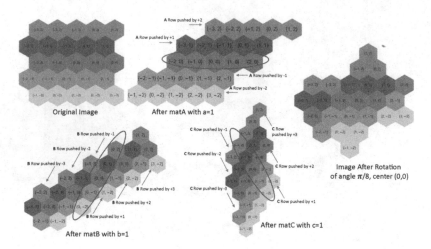

Fig. 4. Shears in directions A, B, C with $a = b = c = 1$. The rows containing $(0,0)$ (marked by an ellipse) do not move. On the right, rotation by $\pi/8$ and center $(0,0)$.

3.4 Illustrations

Figure 4 shows how each matrix acts on the hexagons. In order to illustrate the action of each shear transform, we set $a = b = c = 1$. The rows containing the rotation center $(0,0)$ (marked by an ellipse) do not move. On the right of Fig. 4, we show an actual rotation of the proposed image by an angle of $\pi/8$. Lastly, Fig. 5 shows the result of the bijective rotation on two type of images. On the left, we took a 'Lena' image of size 512×512 mapped on an hexagonal grid and applied our bijective rotation with an angle $\pi/8$. 128×128 for different angles. The original image can be seen below the rotated image. On the left, an artificial image made of some figures rotated by an angle of $\pi/6$. As can be seen, there are deconnections since, although the bijectivity is guaranteed, the topology is not necessarily preserved. The rotation modifies neighborhoods and therefore, locally, one cannot expect that an hexagon keeps all of its neighbors which means topological changes may occur. Characterizing objects that preserve topology under rotation in the hexagonal grid would certainly be an interesting question to explore [9, 10].

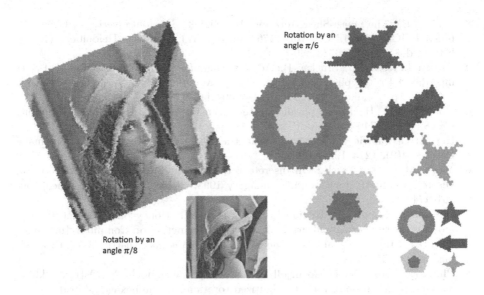

Fig. 5. Example with a Lena Image of size 512×512 and with an artificial image of 128×128 mapped on hexagonal grids.

4   Conclusions

In this paper, we have proposed a bijective digital rotation method for hexagonal grids that works for all angles. For this rotation, the average distance, on an image, between a point rotated by a continuous rotation and the point rotated by the digital rotation is about 0.42 (with a distance of 1.0 between two neighboring hexagons), while the maximal distance is bounded by 1.1. This algorithm extends the direct mapping proposed by K. Pluta in the sense that it works for all angles [7,8]. However, contrary to his work, we have only an algorithm that works for hexagonal grid points as center. Extending our method to arbitrary centers would be a very interesting extension. There are many other questions that are raised by this work. Using shears for directions that are not the usual axis directions can of course be applied to the regular square grid. How does that type of rotation measure up to the previously published bijective rotations [2]? the triangular grid is a dual grid of the hexagonal grid however so may be there is something that can be done there. What about bijective rotations in arbitrary grids? One aspect that is of special interest for us is the extension to higher dimensions especially for honeycomb type grids with possible applications in 3D printing.

References

1. Andres, E.: Cercles Discrets et Rotations Discretes. Ph.D. thesis, Université Louis Pasteur, Strasbourg, France (1992)

2. Andres, E.: The Quasi-Shear rotation. In: Miguet, S., Montanvert, A., Ubéda, S. (eds.) DGCI 1996. LNCS, vol. 1176, pp. 307–314. Springer, Heidelberg (1996). https://doi.org/10.1007/3-540-62005-2_26
3. Gibson, I., Rosen, D., Stucker, B.: Additive Manufacturing Technologies: 3D Printing, Rapid Prototyping, and Direct Digital Manufacturing. Springer, New York (2014). https://doi.org/10.1007/978-1-4939-2113-3
4. Golay, M.J.E.: Hexagonal parallel pattern transformations. IEEE Trans. Comput. **C−18**(8), 733–740 (1969)
5. Her, I.: Geometric transformations on the hexagonal grid. IEEE Trans. Image Process. **4**(9), 1213–1221 (1995)
6. Jacob, M.A., Andres, E.: On discrete rotations. In: International Workshop on Discrete Geometry for Computer Imagery 1995, Clermont-Ferrand (France), pp. 161–174 (1995)
7. Pluta, K., Romon, P., Kenmochi, Y., Passat, N.: Honeycomb geometry: rigid motions on the hexagonal grid. In: Discrete Geometry for Computer Imagery - 20th IAPR International Conference, DGCI 2017, Vienna, Austria, 2017, Proceedings, pp. 33–45 (2017)
8. Pluta, K., Roussillon,T., Coeurjolly, D., Romon, P., Kenmochi, Y., Ostromoukhov, V.: Characterization of bijective digitized rotations on the hexagonal grid
9. Ngo, P., Kenmochi, Y., Passat, N., Talbot, H.: Topology-preserving conditions for 2D digital images under rigid transformations. J. Math. Imaging Vision **49**(2), 418–433 (2013). https://doi.org/10.1007/s10851-013-0474-z
10. Ngo, P., Passat, N., Kenmochi, Y., Debled-Rennesson, I.: Geometric preservation of 2D digital objects under rigid motions. J. Math. Imaging Vision **61**(2), 204–223 (2018). https://doi.org/10.1007/s10851-018-0842-9
11. Nouvel, B., Rémila, E.: Characterization of bijective discretized rotations. In: Klette, R., Žunić, J. (eds.) IWCIA 2004. LNCS, vol. 3322, pp. 248–259. Springer, Heidelberg (2004). https://doi.org/10.1007/978-3-540-30503-3_19
12. Paeth, A.W.: A fast algorithm for general raster rotation. In: Graphic Interface 86 (reprinted with Corrections in Graphic Gems (Glassner Ed.) Academic 1990, pp. 179–195), pp. 77–81 (1986)
13. Kacper Pluta. Rigid motions on discrete spaces. PhD thesis, Université Paris Est, Paris, France
14. Reveillès, J.-P.: Calcul en Nombres Entiers et Algorithmique. Ph.D thesis, Université Louis Pasteur, Strasbourg, France (1991)
15. Roussillon, T., Coeurjolly, D.: Characterization of bijective discretized rotations by Gaussian integers. Research report, LIRIS UMR CNRS 5205, January 2016
16. Snyder, W.E., Qi, H., Sander, W.A.: Coordinate system for hexagonal pixels. In: Medical Imaging 1999: Image Processing, vol. 3661, pp. 716–728. International Society for Optics and Photonics (1999)

An Isometry Classification of Periodic Point Sets

Olga Anosova and Vitaliy Kurlin[(⊠)][iD]

University of Liverpool, Liverpool L69 3BX, UK
vkurlin@liv.ac.uk
http://kurlin.org

Abstract. We develop discrete geometry methods to resolve the data ambiguity challenge for periodic point sets to accelerate materials discovery. In any high-dimensional Euclidean space, a periodic point set is obtained from a finite set (motif) of points in a parallelepiped (unit cell) by periodic translations of the motif along basis vectors of the cell.

An important equivalence of periodic sets is a rigid motion or an isometry that preserves interpoint distances. This equivalence is motivated by solid crystals whose periodic structures are determined in a rigid form.

Crystals are still compared by descriptors that are either not isometry invariants or depend on manually chosen tolerances or cut-off parameters. All discrete invariants including symmetry groups can easily break down under atomic vibrations, which are always present in real crystals.

We introduce a complete isometry invariant for all periodic sets of points, which can additionally carry labels such as chemical elements. The main classification theorem says that any two periodic sets are isometric if and only if their proposed complete invariants (called isosets) are equal.

A potential equality between isosets can be checked by an algorithm, whose computational complexity is polynomial in the number of motif points. The key advantage of isosets is continuity under perturbations, which allows us to quantify similarities between any periodic point sets.

Keywords: Lattice · Periodic set · Isometry invariant · Classification

1 Introduction: Motivations and Problem Statement

One well-known challenge in applications is the *curse of dimensionality* meaning that any dataset seems sparse in a high-dimensional space. This paper studies the more basic *ambiguity challenge* in data representations meaning that equivalent real-life object can often be represented in infinitely many different ways.

Data ambiguity makes any comparison unreliable. For example, humans should be not be compared or identified by the average color of their clothes, though such colors are easily accessible in photos. Justified comparisons should use only *invariant* features that are independent of an object representation.

Supported by the EPSRC grant Application-driven Topological Data Analysis.

Our objects are periodic point sets, which model all solid crystalline materials (crystals). Solid crystal structures are determined in a rigid form with well-defined atomic positions. Atoms form strong bonds only in molecules, while all inter-molecular bonds are much weaker and have universally agreed definitions. Points at atomic centers can be labeled by chemical elements or any other properties, e.g. radii. Later we explain how to easily incorporate labels into our invariants. We start with the most fundamental model of a periodic point set.

The simplest example is a *lattice* Λ, a discrete set of points that are integer linear combinations of any linear (not necessarily orthogonal) basis in $\mathbb{R}^n$, see Fig. 1. More generally, a *periodic point set* is obtained from a finite collection (*motif*) of points by periodic translations along all vectors of a lattice Λ.

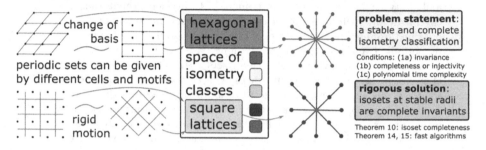

Fig. 1. Left: periodic sets represented by different cells are organized in isometry classes, which form a continuous space. **Right**: the new *isoset* resolves the ambiguity.

The same periodic point set S can be obtained from infinitely many different Minkowski sums $\Lambda + M$. For example, one can change a linear basis of Λ and get a new motif of points with different coordinates in the new basis.

The above ambiguity with respect to a basis is compounded by infinitely many rigid motions or isometries that preserve inter-point distances, hence produce equivalent crystal structures. Shifting all points by a fixed vector changes all point coordinates in a fixed basis, but not the isometry class of the set.

The curse of ambiguity for periodic sets can be resolved only by a complete isometry invariant as follows: two periodic sets given by any decompositions $\Lambda + M$ into a lattice and a motif should be isometric if and only if their complete invariants coincide. Such a complete invariant should have easily comparable values from which we could explicitly reconstruct an original crystal structure.

The final requirement for a complete invariant is its continuity under perturbations, which was largely ignored in the past despite all atoms vibrate above the absolute zero temperature. All discrete invariants including symmetry groups are discontinuous under perturbation of points. A similarity between crystals should be quantified in a continuous way to filter out nearly identical crystals obtained as approximations to local energy minima in Crystal Structure Prediction [17].

Problem 1 formalizes the above curse of ambiguity for crystal structures.

Problem 1 (complete isometry classification of periodic point sets). Find a function I on the space of all periodic point sets in $\mathbb{R}^n$ such that

(1a) *invariance*: if any periodic sets S, Q are isometric, then $I(S) = I(Q)$;
(1b) *continuity*: $I(S)$ continuously changes under perturbations of points;
(1c) *computability*: $I(S) = I(Q)$ is checked in a polynomial time in a motif size;
(1d) *completeness*: if $I(S) = I(Q)$, then the periodic sets S, Q are isometric. ∎

The main contribution is the new invariant *isoset* in Definition 9 whose completeness is proved in Theorem 10. Conditions (1cd) are proved in the recent work [3] introducing the new research area of Periodic Geometry and Topology.

2   A Review of the Relevant Work on Periodic Crystals

Despite any lattice can be defined by infinitely many primitive cells, there is a unique Niggli's reduced cell, which can be theoretically used for comparing periodic sets [11, section 9.2]. Niggli's and other reduced cells are discontinuous under perturbations in the sense that a reduced cell of a perturbed lattice can have a basis that substantially differs from that of a non-perturbed lattice [2].

Continuity condition (1b) fails not only for Niggli's reduced cell, but also for all discretely-valued invariants including symmetry groups. The 230 crystallographic groups in $\mathbb{R}^3$ cut the continuous space of isometry classes into disjoint pieces. This stratification shows many nearly identical crystals as distant.

The first step towards a complete isometry classification of crystals has recently been done in [14] by introducing two proper distances between arbitrary lattices that satisfy the metric axioms and are also continuous under perturbations. Also [14, section 3] reviews many past tools to compare crystals.

The world's largest Cambridge Structural Database (CSD) has more than 1M crystals. Each crystal is represented by one of infinitely many choices of a unit cell and a motif M in the form of a Crystallographic Information File (CIF). The CSD is a super-long list of CIFs with limited search tools, mainly by chemical compositions, and without any organization by geometric similarity.

Quantifying crystal similarities is even more important for Crystal Structure Prediction (CSP). A typical CSP software starts from a given chemical composition and outputs thousands of predicted crystals as approximations to local minima of a complicated energy function. Any iterative optimization produces many approximations to the same local minimum. These nearly identical crystals are currently impossible to automatically identify in a reliable way [17].

Crystals are often compared by the Radial Distribution Function (RDF) that measures the probability of finding one atom at a distance of r from a reference atom, which is computed up to a manually chosen cut-off radius.

The new concept of a stable radius in Definition 8 gives exact conditions for a required radius depending on a complexity of a periodic set. The crystals indistinguishable by their RDF or diffraction patterns are known as homometric [15]. The most recent survey [16, Fig. S4] has highlighted pairs of finite atomic arrangements that cannot be distinguished by any known crystal descriptors.

On a positive side, the mathematical approach in [10] has solved the already non-trivial 1-dimensional case for sets whose points have only integer (or rational) coordinates. Briefly, any given points $c_0, \ldots, c_{m-1}$ on the unit circle $S^1 \subset \mathbb{C}$ are converted into the Fourier coefficients $d(k) = \sum_{j=0}^{m-1} c_j \exp \frac{2\pi ijk}{m}$, $k = 0, \ldots, m - 1$. Then all point sets in the unit circle can be distinguished up to circular rotations by the n-th order invariants up to $n = 6$, which are all products of the form $d(k_1) \cdots d(k_n)$ with $k_1 + \cdots + k_n \equiv 0 \pmod{m}$.

The more recent advances in Problem 1 are Density Functions [9] and Average Minimum Distances [18]. The k-density function $\psi_k[S]$ of a periodic point set $S \subset \mathbb{R}^n$ measures the fractional area of the region within a unit cell U covered by exactly k closed balls with centers $a \in S$ and a radius $t \geq 0$. The density functions satisfy conditions (1abc) and completeness (1d) in general position.

However, the density functions do not distinguish the following 1-dimensional sets $S_{15} = \{0, 1, 3, 4, 5, 7, 9, 10, 12\} + 15\mathbb{Z}$ and $Q_{15} = \{0, 1, 3, 4, 6, 8, 9, 12, 14\} + 15\mathbb{Z}$ with period 15, see [3, Example 11]. The sets S_{15}, Q_{15} were introduced at the beginning of section 5 in [9] as $U \pm V + 15\mathbb{Z}$ for $U = \{0, 4, 9\}$ and $V = \{0, 1, 3\}$.

The above sets S_{15}, Q_{15} are distinguished by the faster Average Minimum Distances (AMD), see [3, Example 6]. For any integer $k \geq 1$, $\mathrm{AMD}_k(S)$ is the distance from a point $p \in S$ to its k-th nearest neighbor, averaged over all points p in a motif of S. For $k \to +\infty$, $\mathrm{AMD}_k(S)$ behaves as $\sqrt[n]{k}$, see [18, Theorem 14].

3   Necessary Concepts from Computational Geometry

In the Euclidean space $\mathbb{R}^n$, any point $p \in \mathbb{R}^n$ is represented by the vector $\boldsymbol{p}$ from the origin of $\mathbb{R}^n$ to p. The *Euclidean* distance between points $p, q \in \mathbb{R}^n$ is denoted by $|pq| = |\boldsymbol{p} - \boldsymbol{q}|$. For a standard orthonormal basis $\boldsymbol{e}_1, \ldots, \boldsymbol{e}_n$, the integer lattice $\mathbb{Z}^n \subset \mathbb{R}^n$ consists of all points with integer coordinates.

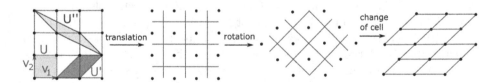

Fig. 2. Left: three primitive cells U, U', U'' of the square lattice S. Other pictures show different periodic sets $\varLambda + M$, which are all isometric to the square lattice S.

Definition 2 (a lattice $\varLambda$, a unit cell U, a motif M, a periodic set $S = \varLambda + M$). For any linear basis $\boldsymbol{v}_1, \ldots, \boldsymbol{v}_n$ in $\mathbb{R}^n$, a *lattice* is $\varLambda = \{ \sum_{i=1}^n \lambda_i \boldsymbol{v}_i : \lambda_i \in \mathbb{Z} \}$. The *unit cell* $U(\boldsymbol{v}_1, \ldots, \boldsymbol{v}_n) = \left\{ \sum_{i=1}^n \lambda_i \boldsymbol{v}_i : \lambda_i \in [0, 1) \right\}$ is the parallelepiped spanned by the basis. A motif M is any finite set of points $p_1, \ldots, p_m \in U$. A *periodic point set* is the Minkowski sum $S = \Lambda + M = \{u + v : u \in \Lambda, v \in M\}$. A unit cell U of a periodic set $S = \Lambda + M$ is *primitive* if any vector v that translates S to itself is an integer linear combination of the basis of the cell U, i.e. $v \in \Lambda$. ∎

A primitive unit cell U of any lattice has a motif of one point (the origin). If U is defined as the closed parallelepiped in $\mathbb{R}^n$, hence includes 2^n vertices, one could count every vertex with weight 2^{-n} so that the sum is 1. All closed unit cells in Fig. 2 are primitive, because four corners are counted as one point in U.

The first picture in Fig. 3 shows a small perturbation of a square lattice. The new periodic set has a twice larger primitive unit cell with two points in a motif instead of one. All invariants based on a fixed primitive unit cell such as Niggli's reduced cell [11, section 9.2] fail continuity condition (1b) in Problem 1.

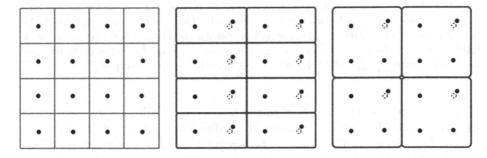

Fig. 3. A continuous invariant should take close values on these nearly identical periodic sets, though their symmetry groups and primitive cells substantially differ.

The auxiliary concepts in Definitions 3, 4, 5 follow Dolbilin's papers [5,7].

Definition 3 (bridge distance $\beta(S)$). For a periodic point set $S \subset \mathbb{R}^n$, the *bridge* distance is a minimum $\beta(S) > 0$ such that any two points $a, b \in S$ can be connected by a finite sequence $a_0 = a, a_1, \ldots, a_m = b$ such that any two successive points a_i, a_{i+1} are close, i.e. the Euclidean distance $|a_{i-1} - a_i| \leq \beta(S)$ for $i = 1, \ldots, m$. Figure 4 shows periodic sets with different bridge distances. ∎

Definition 4 (m-regularity of a periodic set). For any point a in a periodic set $S \subset \mathbb{R}^n$, the *global cluster* $C(S, a)$ is the infinite set of vectors $b - a$ for all points $b \in S$. Points $a, b \in S$ are called *isometrically equivalent* if there is an isometry $f : C(S, a) \to C(S, b)$ such that $f(a) = b$. A periodic set $S \subset \mathbb{R}^n$ is called *regular* if all points $a, b \in S$ are isometrically equivalent. A periodic set S is m-*regular* if all global clusters of S form exactly $m \geq 1$ isometry classes. ∎

For any point $a \in S$, its global cluster is a view of S from the position of a, e.g. how we view all astronomical stars in the universe S from our planet Earth. Any lattice is 1-regular, because all its global clusters are related by translations.

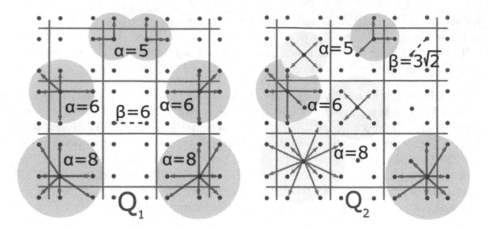

Fig. 4. Left: the periodic point set Q_1 has the four points $(\pm 2, \pm 2)$ in the square unit cell $[0, 10]^2$, so Q_1 isn't a lattice, but is 1-regular by Definition 4, also $\beta(Q_1) = 6$. All local α-clusters are isometric, shown by red arrows for radii $\alpha = 5, 6, 8$, see Definition 5. **Right**: the periodic point set Q_2 has the extra point $(5, 5)$ in the center of $[0, 10]^2$ and is 2-regular with $\beta(Q_2) = 3\sqrt{2}$. Local clusters have two isometry types.

Though the global clusters $C(S, a)$ and $C(S, b)$ at any different points $a, b \in S$ seem to contain the same set S, they can be different even modulo translations.

The first picture in Fig. 4 shows the 1-regular set $Q_1 \subset \mathbb{R}^2$, where all points have isometric global clusters related by translations and rotations through $\frac{\pi}{2}, \pi, \frac{3\pi}{2}$, so Q_1 is not a lattice. The global clusters are infinite, hence distinguishing them up to isometry is not easier than distinguishing the original sets. However, m-regularity can be checked in terms of local clusters defined below.

Definition 5 (local α-clusters $C(S, a; \alpha)$ and symmetry groups $\mathrm{Sym}(S, a; \alpha)$). For a point a in a crystal $S \subset \mathbb{R}^n$ and any radius $\alpha \geq 0$, the *local cluster* $C(S, a; \alpha)$ is the set of vectors $\boldsymbol{b} - \boldsymbol{a}$ of lengths $|\boldsymbol{b} - \boldsymbol{a}| \leq \alpha$ for $b \in S$. An isometry $f \in \mathrm{Iso}(\mathbb{R}^n)$ between clusters should match their centers. The *symmetry* group $\mathrm{Sym}(S, a; \alpha)$ consists of *self-isometries* of $C(S, a; \alpha)$ that fix the center a. ∎

If $\alpha > 0$ is smaller than the minimum distance between any points, then every cluster $C(S, a; \alpha)$ is the single-point set $\{a\}$ and its symmetry group $\mathrm{O}(\mathbb{R}^n)$ consists of all isometries fixing the center a. When the radius α is increasing, the α-clusters $C(S, a; \alpha)$ become larger and can have fewer self-isometries, so the symmetry group $\mathrm{Sym}(S, a; \alpha)$ becomes smaller and eventually stabilizes.

The 1-regular set Q_1 in Fig. 4 for any point $a \in Q_1$ has the symmetry group $\mathrm{Sym}(Q_1, a; \alpha) = \mathrm{O}(\mathbb{R}^2)$ for $\alpha \in [0, 4)$. The group $\mathrm{Sym}(Q_1, a; \alpha)$ stabilizes as $\mathbb{Z}_2$ for $\alpha \geq 4$ as soon as the local α-cluster $C(Q_1, a; \alpha)$ includes one more point.

4   The Isotree of Isometry Classes and a Stable Radius

This section introduces the isotree and a stable radius in Definitions 6 and 8 by comparing local clusters at radii $\alpha - \beta$ and β, where β is the bridge distance.

Any isometry $A \rightarrow B$ between local clusters should map the center of A to the center of B. The *isotree* in Definition 6 is inspired by a dendrogram of hierarchical clustering, though points are partitioned according to isometry classes of local α-clusters at different radii α, not by a distance threshold.

Definition 6 (isotree $IT(S)$ of α-partitions). Fix a periodic set $S \subset \mathbb{R}^n$ and $\alpha \geq 0$. Points $a, b \in S$ are called α-*equivalent* if their α-clusters $C(S, a; \alpha)$ and $C(S, b; \alpha)$ are isometric. The α-*equivalence class* $[C(S, a; \alpha)]$ consists of all α-clusters isometric to $C(S, a; \alpha)$. The α-*partition* $P(S; \alpha)$ is the splitting of S into α-equivalence classes of points. The number of α-equivalence classes of α-clusters is the *cluster count* $|P(S; \alpha)|$. When the radius α is increasing, the α-partition can be refined by subdividing α-equivalence classes of points of S into subclasses. If we represent each α-equivalence class by an abstract point, the resulting points form the *isotree* $IT(S)$ of all α-partitions, see Fig. 5, 6. ∎

The α-equivalence and isoset in Definition 9 can be refined by labels of points such as chemical elements. Theorem 10 will remain valid for labelled points. Recall that isometries include reflections, however an orientation sign can be easily added to α-clusters, hence we focus on the basic case of all isometries.

When a radius α is increasing, α-clusters $C(S, a; \alpha)$ include more points, hence are less likely to be isometric, so $|P(S; \alpha)|$ is a non-increasing function of α. Figure 5, 6 show α-clusters and isotrees of non-isometric 1D periodic sets S, Q [15, p. 197, Fig. 2], which have identical 1D analogs of diffraction patterns.

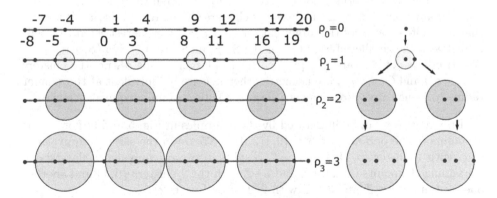

Fig. 5. Left: $S = \{0, 1, 3, 4\} + 8\mathbb{Z}$ has $t = 4$ and is 2-regular by Definition 4. **Right**: Local clusters with radii $\alpha = 0, 1, 2, 3$ represent vertices of the isotree $IT(S)$ in Definition 6. All α-clusters are isometric for $\alpha < 2$, form two isometry classes for $\alpha \geq 2$.

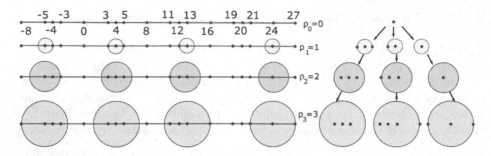

Fig. 6. Left: $Q = \{0, 3, 4, 5\} + 8\mathbb{Z}$ has $t = 3$ and is 3-regular by Definition 4. **Right**: Local clusters with radii $\alpha = 0, 1, 2, 3$ represent vertices of the isotree $\mathrm{IT}(Q)$ in Definition 6. All α-clusters are isometric for $\alpha < 1$, form three isometry classes for $\alpha \geq 1$.

Any α-equivalence class from $P(S; \alpha)$ may split into two or more classes, which will not merge at any larger radius α'. Lemma 7 justifies that the isotree $\mathrm{IT}(S)$ can be visualized as a merge tree of α-equivalence classes of clusters.

Lemma 7 (isotree properties). *The isotree $\mathrm{IT}(S)$ has the following properties:*

(7a) *for $\alpha = 0$, the α-partition $P(S; 0)$ consists of one class;*
(7b) *if $\alpha < \alpha'$, then $\mathrm{Sym}(S, a; \alpha') \subseteq \mathrm{Sym}(S, a; \alpha)$ for $a \in S$;*
(7c) *if $\alpha < \alpha'$, the α'-partition $P(S; \alpha')$ refines $P(S; \alpha)$, i.e. any set from the α'-partition $P(S; \alpha')$ is included into a set from the α-partition $P(S; \alpha)$.*   ∎

Proof. (7a) If $\alpha \geq 0$ is smaller than the minimum distance r between point of S, every cluster $C(S, a; \alpha)$ is the single-point set $\{a\}$. All these single-point clusters are isometric to each other. So $|P(S; \alpha)| = 1$ for all small radii $\alpha < r$.
(7b) For any point $a \in S$, the inclusion of clusters $C(S, a; \alpha) \subseteq C(S, a; \alpha')$ implies that any self-isometry of the larger cluster $C(S, a; \alpha')$ can be restricted to a self-isometry of the smaller cluster $C(S, a; \alpha)$. So $\mathrm{Sym}(S, a; \alpha') \subseteq \mathrm{Sym}(S, a; \alpha)$.
(7c) If points $a, b \in S$ are α'-equivalent at the larger radius α', i.e. the clusters $C(S, a; \alpha')$ and $C(S, b; \alpha')$ are isometric, then a, b are α-equivalent at the smaller radius α. Hence any α'-equivalence class is a subset of an α-equivalence class.

Property (7c) can be illustrated by the examples in Fig. 5 and 6. For $\alpha = 1$, all points of the periodic set $S = \{0, 1, 3, 4\} + 8\mathbb{Z}$ are in the same α-equivalence class with 1-cluster $\{0, 1\}$. For $\alpha' = 2$, S splits in two α'-equivalence classes: one containing the points from $0 + \mathbb{Z}$ and $4 + \mathbb{Z}$ with the 2-clusters $\{0, 1\}$ and another one containing $1 + \mathbb{Z}$ and $3 + \mathbb{Z}$ with 2-clusters $\{-1, 0, 2\}$.

If a point set S is periodic, the α-partitions of S stabilize in the sense below.

Definition 8 (a stable radius). Let a periodic point set $S \subset \mathbb{R}^n$ and β be an upper bound of its bridge distance $\beta(S)$ from Definition 3. A radius $\alpha \geq \beta$ is called *stable* if both conditions below hold:

(8a) the α-partition $P(S; \alpha)$ coincides with the $(\alpha - \beta)$-partition $P(S; \alpha - \beta)$;

(8b) the symmetry groups stabilize: $\text{Sym}(S, a; \alpha) = \text{Sym}(S, a; \alpha - \beta)$ for all points $a \in S$, which is enough to check for points only from a finite motif of S.   ∎

A minimum radius α satisfying the above conditions for the bridge distance $\beta(S)$ from Definition 3 can be called *the minimum stable radius* and denoted by $\alpha(S)$. Upper bounds of $\alpha(S)$ and $\beta(S)$ will be enough for all results below.

Due to Lemma (7bc), conditions (8ab) imply that the α'-partitions $P(S; \alpha')$ and the symmetry groups $\text{Sym}(S, a; \alpha')$ remain the same for all $\alpha' \in [\alpha - \beta, \alpha]$.

Condition (8b) doesn't follow from condition (8a) due to the following example. Let Λ be the 2D lattice with the basis $(1, 0)$ and $(0, \beta)$ for $\beta > 1$. Then β is the bridge distance of Λ. Condition (8a) is satisfied for any $\alpha \geq 0$, because all points of any lattice are equivalent up to translations. However, condition (8b) fails for any $\alpha < \beta + 1$. Indeed, the α-cluster of the origin $(0, 0)$ contains five points $(0, 0), (\pm 1, 0), (0, \pm \beta)$, whose symmetries are generated by the two reflections in the axes x, y, but the $(\alpha - \beta)$-cluster of the origin consists of only $(0, 0)$ and has the symmetry group $O(2)$.

Condition (8b) might imply condition (8a), but in practice it makes sense to verify (8b) only after checking much simpler condition (8a). Both conditions are essentially used in the proofs of Isometry Classification Theorem 10.

For the set $S = \{0, 1, 3, 4\} + 8\mathbb{Z}$ in Fig. 5 with the bridge distance $\beta(S) = 4$, any $\alpha \geq 6$ is a stable radius, because the partition $P(S; \alpha - 4)$ splits S into the same two classes for any $\alpha \geq 6$. For the periodic set $Q = \{0, 3, 4, 5\} + 8\mathbb{Z}$ in Fig. 6 with the bridge distance $\beta(Q) = 3$, any $\alpha \geq 4$ is a stable radius.

Any periodic set $S \subset \mathbb{R}^n$ with m motif points has at most m α-equivalence classes, because any point of S can be translated to a motif point. Hence it suffices to check condition (8a) about α-partitions only for the m motif points. Condition (8b) can be practically checked by testing if the inclusion $\text{Sym}(S, a; \alpha') \subset \text{Sym}(S, a; \alpha)$ from (7b) is surjective, which is needed only for one representative cluster from at most m isometry classes (exactly m is S is m-regular).

A stable radius in [5] was defined by using the notations ρ and $\rho + t$. This pair changed to $\alpha - \beta$ and α, because subsequent Theorem 10 is more conveniently stated for the larger radius α. Any 1-regular set in $\mathbb{R}^3$ with a bridge distance β has a stable radius $\alpha = 7\beta$ or $\rho = 6t$ in the past notations of [7].

5   Isosets Completely Classify Periodic Sets up to Isometry

A criterion of m-regular sets [8, Theorem 1.3] has inspired us to introduce the new invariant isoset in Definition 9, whose completeness (injectivity) in the isometry classification of periodic sets will be proved in main Theorem 10.

Definition 9 (isoset $I(S; \alpha)$ of a periodic point set S at a radius α). Let a periodic point set $S \subset \mathbb{R}^n$ have a motif M of m points. Split all points $a \in M$ into α-equivalence classes. Then each α-equivalence class consisting of (say) k

points in M can be associated with the *isometry class* of $\sigma = [C(S,a;\alpha)]$ of an α-cluster centered at one of these k points $a \in M$. The *weight* of the class σ is defined as $w = k/m$. Then the *isoset* $I(S;\alpha)$ is defined as the unordered set of all isometry classes with weights $(\sigma; w)$ over all points $a \in M$.   ∎

All points a of a lattice $\Lambda \subset \mathbb{R}^n$ are α-equivalent for any $\alpha \geq 0$, because all α-clusters $C(\Lambda, a; \alpha)$ are isometrically equivalent to each other by translations. Hence the isoset $I(\Lambda; \alpha)$ is one isometry class of weight 1 for any α.

All isometry classes $\sigma \in I(S; \alpha)$ are in a 1-1 correspondence with all α-equivalence classes in the α-partition $P(S; \alpha)$ from Definition 6. So $I(S; \alpha)$ without weights is a set of points in the isotree $\mathrm{IT}(S)$ at the radius α. The size of the isoset $I(S; \alpha)$ equals the cluster count $|P(S; \alpha)|$. Formally, $I(S; \alpha)$ depends on α, because α-clusters grow in α. To distinguish periodic point sets S, Q up to isometry, we will compare their isosets at a common stable radius α.

An equality $\sigma = \xi$ between isometry classes of clusters means that there is an isometry f from a cluster $C(S, a; \alpha)$ representing σ to a cluster $C(Q, b; \alpha)$ representing ξ such that $f(a) = b$, i.e. f respects the centers of the clusters.

The set $S = \{0, 1, 3, 4\} + 8\mathbb{Z}$ in Fig. 5 has the isoset $I(S; 6)$ of two isometry classes of 6-clusters represented by $\{-4, -3, -1, 0, 1, 4, 5\}$ and $\{-3, -2, 0, 1, 5, 6\}$ centered at 0. The set $Q = \{0, 3, 4, 5\} + 8\mathbb{Z}$ in Fig. 6 has the isoset $I(Q; 4)$ of three isometry classes of 4-clusters represented by $\{-4, -3, 0, 3, 4\}$, $\{-3, 0, 1, 2\}$, $\{-4, -1, 0, 1, 4\}$. To conclude that S, Q are not isometric, Theorem 10 will require us to compare their isosets at a common stable radius $\alpha \geq 6$. In the above case it suffices to say that the stabilized cluster counts differ: $2 \neq 3$.

An equality $\sigma = \xi$ between isometry classes means that there is an isometry f from a cluster in σ to a cluster in ξ so that f respects the centers of the clusters. This equality is checked in time $O(k^{n-2} \log k)$ for any dimension $n \geq 3$ by [1, Theorem 1(a)], where k is the maximum number of points in the clusters.

Theorem 10 (complete isometry classification of periodic point sets). For any periodic point sets $S, Q \subset \mathbb{R}^n$, let α be a common stable radius satisfying Definition 8 for an upper bound β of $\beta(S), \beta(Q)$. Then S, Q are isometric if and only if there is a bijection between their isosets respecting weights: $I(S; \alpha) = I(Q; \alpha)$ means that any isometry class $(\sigma; w) \in I(S; \alpha)$ of a weight w coincides with a class $(\xi; w) \in I(Q; \alpha)$ of the same weight w and vice versa.   ∎

Theoretically a complete invariant of S should include isosets $I(S; \alpha)$ for all sufficiently large radii α. However, when comparing two sets S, Q up to isometry, it suffices to build their isosets only at a common stable radius α.

The α-equivalence and isoset in Definition 9 can be refined by labels of points such as chemical elements, which keeps Theorem 10 valid for labeled points.

Recall that isometries include reflections, however an orientation sign can be easily added to α-clusters, hence we focus on the basic case of all isometries.

The proposed complete invariant for classification Problem 1 is the function $S \mapsto I(S; \alpha)$ from any periodic point set S to its isoset at a stable radius α, which doesn't need to be minimal. All points a of a lattice $\Lambda \subset \mathbb{R}^n$ are α-equivalent to each other for $\alpha \geq 0$, because all α-clusters $C(\Lambda, a; \alpha)$ are related by translations, hence the isoset $I(\Lambda; \alpha)$ of any lattice is a single isometry class for any α.

Lemmas 11 and 12 help to extend an isometry between local clusters to full periodic sets to prove the complete isometry classification in Theorem 10.

Lemma 11 (local extension). Let periodic sets $S, Q \subset \mathbb{R}^n$ have bridge distances at most β and a common stable radius α such that α-clusters $C(S, a; \alpha)$ and $C(Q, b; \alpha)$ are isometric for some $a \in S$, $b \in Q$. Then any isometry $f : C(S, a; \alpha - \beta) \to C(Q, b; \alpha - \beta)$ extends to an isometry $C(S, a; \alpha) \to C(Q, b; \alpha)$.   ∎

Proof. Let $g : C(S, a; \alpha) \to C(Q, b; \alpha)$ be any isometry, which may not coincide with f on the $(\alpha - \beta)$-subcluster $C(S, a; \alpha - \beta)$. The composition $f^{-1} \circ g$ isometrically maps $C(S, a; \alpha - \beta)$ to itself. Hence $f^{-1} \circ g = h \in \mathrm{Sym}(S, a; \alpha - \beta)$ is a self-isometry. Since the symmetry groups stabilize by condition (8b), the isometry h maps the larger cluster $C(S, a; \alpha)$ to itself. Then the initial isometry f extends to the isometry $g \circ h^{-1} : C(S, a; \alpha) \to C(Q, b; \alpha)$ as required.   □

Lemma 12 (global extension). For any periodic point sets $S, Q \subset \mathbb{R}^n$, let α be a common stable radius satisfying Definition 8 for an upper bound β of both $\beta(S), \beta(Q)$. Assume that $I(S; \alpha) = I(Q; \alpha)$. Fix a point $a \in S$. Then any local isometry $f : C(S, a; \alpha) \to C(Q, f(a); \alpha)$ extends to a global isometry $S \to Q$.   ∎

Proof. We shall prove that the image $f(b)$ of any point $a' \in S$ belongs to Q, hence $f(S) \subset Q$. Swapping the roles of S and Q will prove that $f^{-1}(Q) \subset S$, i.e. f is a global isometry $S \to Q$. By Definition 3 the above points $a, a' \in S$ are connected by a sequence of points $a = a_0, a_1, \ldots, a_m = a' \in S$ such that $|a_{i-1} - a_i| \leq \beta$, $i = 1, \ldots, m$, where β is an upper bound of both $\beta(S), \beta(Q)$.

The cluster $C(S, a; \alpha)$ is the intersection $S \cap B(a; \alpha)$. The ball $B(a; \alpha)$ contains the smaller ball $B(a_1; \alpha - \beta)$ around the closely located center a_1. Indeed, since $|a - a_1| \leq \beta$, the triangle inequality for the Euclidean distance implies that any $c \in B(a_1; \alpha)$ with $|a_1 - c| \leq \alpha - \beta$ satisfies $|a - c| \leq |a - a_1| + |a_1 - c| \leq \alpha$.

Due to $I(S; \alpha) = I(Q; \alpha)$ the isometry class of $C(S, a_1; \alpha)$ coincides with an isometry class of $C(Q, b; \alpha)$ for some $b \in Q$, i.e. $C(S, a_1; \alpha)$ is isometric to $C(Q, b; \alpha)$. Then the clusters $C(S, a_1; \alpha - \beta)$ and $C(Q, b; \alpha - \beta)$ are isometric.

By condition (8a), the splitting of Q into α-equivalence classes coincides with the splitting into $(\alpha - \beta)$-equivalence classes. Take the $(\alpha - \beta)$-equivalence class $[C(Q, b; \alpha - \beta)]$ containing b. This class includes the point $f(a_1) \in Q$, because f restricts to the isometry $f : C(S, a_1; \alpha - \beta) \to C(Q, f(a_1); \alpha - \beta)$ and $C(S, a_1; \alpha - \beta)$ was shown to be isometric to $C(Q, b; \alpha - \beta)$.

The α-equivalence class $[C(Q, b; \alpha)]$ includes both b and $f(a_1)$. The isometry class $[C(Q, b; \alpha)] = [C(S, a_1; \alpha)]$ can be represented by the cluster $C(Q, f(a_1); \alpha)$, which is now proved to be isometric to $C(S, a_1; \alpha)$.

We apply Lemma 11 for f restricted to $C(S, a_1; \alpha - \beta) \to C(Q, f(a_1), \alpha - \beta)$ and conclude that f extends to an isometry $C(S, a_1; \alpha) \to C(Q, f(a_1); \alpha)$.

Continue applying Lemma 11 to the clusters around the next center a_2 and so on until we conclude that the initial isometry f maps the α-cluster centered at $a_m = a' \in S$ to an isometric cluster within Q, so $f(a') \in Q$ as required.

Lemma 13 (all stable radii of a periodic set). If α is a stable radius of a periodic point set $S \subset \mathbb{R}^n$, then so is any larger radius $\alpha' > \alpha$. Then all stable radii form the interval $[\alpha(S), +\infty)$, where $\alpha(S)$ is the minimum stable radius of S.   ∎

Proof. Due to Lemma (7bc), conditions (8ab) imply that the α'-partition $P(S; \alpha')$ and the symmetry groups $\mathrm{Sym}(S, a; \alpha')$ remain the same for all $\alpha' \in [\alpha - \beta, \alpha]$. We need to show that they remain the same for any larger $\alpha' > \alpha$.

Below we will apply Lemma 12 for the same set $S = Q$ and $\beta = \beta(S)$. Let points $a, b \in S$ be α-equivalent, i.e. there is an isometry $f : C(S, a; \alpha) \to C(S, b; \alpha)$. By Lemma 12 the local isometry f extends to a global self-isometry $S \to S$ such that $f(a) = b$. Then all larger α'-clusters of a, b are isometric, i.e. a, b are α'-equivalent and $P(S; \alpha) = P(S, \alpha')$. Similarly, any self-isometry of $C(S, a; \alpha)$ extends to a global self-isometry, i.e. the symmetry group $\mathrm{Sym}(S, a; \alpha')$ for any $\alpha' > \alpha$ is isomorphic to $\mathrm{Sym}(S, a; \alpha')$.   □

Proof of Theorem 10. The part *only if* $\Rightarrow$ follows by restricting any given global isometry $f : S \to Q$ between the infinite sets of points to the local α-clusters $C(S, a; \alpha) \to C(Q, f(a); \alpha)$ for any point a in a motif M of S.

Hence the isometry class $[C(S, a; \alpha)]$ is considered equivalent to the class $[C(Q, f(a); \alpha)]$, which can be represented by the α-cluster $C(Q, b; \alpha)$ centered at a point b in a motif of Q. Since f is a bijection and the point $a \in M$ was arbitrary, we get a bijection between isometry classes with weights in $I(S; \alpha) = I(Q; \alpha)$.

The part *if* $\Leftarrow$. Fix a point $a \in S$. The α-cluster $C(S, a; \alpha)$ represents a class with a weight $(\sigma, w) \in I(S; \alpha)$. Due to $I(S; \alpha) = I(Q; \alpha)$, there is an isometry $f : C(S, a; \alpha) \to C(Q, f(a); \alpha)$ to a cluster from an equal class $(\sigma, w) \in I(Q; \alpha)$. By Lemma 12 the local isometry f extends to a global isometry $S \to Q$.   □

6   A Discussion of Further Properties of Isosets

This paper has resolved the ambiguity challenge for crystal representations, which is common for many data objects [4]. Crystal descriptors [13] are often based on ambiguous unit cells or computed up to a manual cut-off radii. Representations of 2-periodic textiles [6] should be similarly studied up to periodic isotopies [3, section 10] without fixing a unit cell. Definition 8 gives conditions for a stable radius so that larger clusters will not bring any new information.

The recent survey of atomic structure representations [16] confirmed that there was no complete invariant that distinguishes all crystals up to isometry.

Theorem 10 provides a complete invariant for the first time. The follow-up paper [3] discusses computations and continuity of the new invariant isoset. Isosets consisting of different numbers of isometry classes will be compared by the Earth Mover's Distance [12]. We thank all reviewers for their time and suggestions.

References

1. Alt, H., Mehlhorn, K., Wagener, H., Welzl, E.: Congruence, similarity, and symmetries of geometric objects. Discret. Comput. Geom. **3**, 237–256 (1988)
2. Andrews, L., Bernstein, H., Pelletier, G.: A perturbation stable cell comparison technique. Acta Crystallogr. A **36**(2), 248–252 (1980)
3. Anosova, O., Kurlin, V.: Introduction to periodic geometry and topology. arXiv:2103.02749 (2021)
4. Bright, M., Anosova, O., Kurlin, V.: A proof of the invariant-based formula for the linking number and its asymptotic behaviour. In: Proceedings of NumGrid (2020)
5. Bouniaev, M., Dolbilin, N.: Regular and multi-regular t-bonded systems. J. Inf. Process. **25**, 735–740 (2017)
6. Bright, M., Kurlin, V.: Encoding and topological computation on textile structures. Comput. Graph. **90**, 51–61 (2020)
7. Dolbilin, N., Bouniaev, M.: Regular t-bonded systems in R^3. Eur. J. Comb. **80**, 89–101 (2019)
8. Dolbilin, N., Lagarias, J., Senechal, M.: Multiregular point systems. Discret. Comput. Geom. **20**(4), 477–498 (1998)
9. Edelsbrunner, H., Heiss, T., Kurlin, V., Smith, P., Wintraecken, M.: The density fingerprint of a periodic point set. In: Proceedings of SoCG (2021)
10. Grünbaum, F., Moore, C.: The use of higher-order invariants in the determination of generalized Patterson cyclotomic sets. Acta Crystallogr. A **51**, 310–323 (1995)
11. Hahn, T., Shmueli, U., Arthur, J.: Intern. Tables crystallogr. **1**, 750 (1983)
12. Hargreaves, C.J., Dyer, M.S., Gaultois, M.W., Kurlin, V.A., Rosseinsky, M.J.: The earth mover's distance as a metric for the space of inorganic compositions. Chem. Mater. **32**, 10610–10620 (2020)
13. Himanen, L., et al.: Dscribe: library of descriptors for machine learning in materials science. Comput. Phys. Commun. **247**, 106949 (2020)
14. Mosca, M., Kurlin, V.: Voronoi-based similarity distances between arbitrary crystal lattices. Cryst. Res. Technol. **55**(5), 1900197 (2020)
15. Patterson, A.: Ambiguities in the x-ray analysis of crystal structures. Phys. Rev. **65**, 195 (1944)
16. Pozdnyakov, S., Willatt, M., Bartók, A., Ortner, C., Csányi, G., Ceriotti, M.: Incompleteness of atomic structure representations. Phys. Rev. Lett. **125**, 166001 (2020). http://arxiv.org/abs/2001.11696
17. Pulido, A., et al.: Functional materials discovery using energy-structure-function maps. Nature **543**, 657–664 (2017)
18. Widdowson, D., Mosca, M., Pulido, A., Kurlin, V., Cooper, A.: Average minimum distances of periodic point sets. arXiv:2009.02488 (2020)

Visiting Bijective Digitized Reflections and Rotations Using Geometric Algebra

Stéphane Breuils[1]([✉])[iD], Yukiko Kenmochi[2][iD], and Akihiro Sugimoto[1][iD]

[1] National Institute of Informatics, Tokyo, Japan
{breuils,sugimoto}@nii.ac.jp
[2] LIGM, Univ Gustave Eiffel, CNRS, Marne-la-Vallée, France
yukiko.kenmochi@esiee.fr

Abstract. Geometric algebra has become popularly used in applications dealing with geometry. This framework allows us to reformulate and redefine problems involving geometric transformations in a more intuitive and general way. In this paper, we focus on 2D bijective digitized reflections and rotations. After defining the digitization through geometric algebra, we characterize the set of bijective digitized reflections in the plane. We derive new bijective digitized rotations as compositions of bijective digitized reflections since any rotation is represented as the composition of two reflections. We also compare them with those obtained through geometric transformations by computing their distributions.

1 Introduction

Bijectivity of digitized rotations in two and three dimensions has been studied. Its characterization was initiated by the work on two-dimensional rotations followed by a digitized operator in the square grid [8,9]. It was then shown in [12] that an arithmetic proof of the characterization is provided through Gaussian integers. Similar arithmetic characterization on the hexagonal grid was also shown using the Eisenstein integers [11]. Concerning digitized rotations in the space, using the Lipschitz quaternions [7] allows to verify the bijectivity of a given digitized rotation [10]. A bijective reflection algorithm over the plane, on the other hand, was proposed in [3] where the line of reflection is digitized.

These arithmetic approaches using algebraic numbers are intuitive and convenient for providing the proofs of the characterization/certification of the bijective digitized rotations. However, each algebraic number provides different operations and definitions, so that they cause the lack of generality and extensibility. In contrast, geometric algebra is designed to retain generality and offer operators that are capable of computing considered geometric transformations for any geometric objects of the algebra in any dimension. This algebra was defined thanks to the work of Clifford [4] to unify and generalize Grassmann algebra and Hamilton's quaternion into a whole algebra. Geometric algebra is a framework that encompasses both quaternion algebra and complex numbers and extends rigid

© Springer Nature Switzerland AG 2021
J. Lindblad et al. (Eds.): DGMM 2021, LNCS 12708, pp. 242–254, 2021.
https://doi.org/10.1007/978-3-030-76657-3_17

transformations of geometric objects to higher dimensions by expressing them as composition of reflections. We therefore consider that geometric algebra is a natural tool for reasoning with both digitized reflections and rotations.

In order to exploit the generality and extensibility of geometric algebra for tackling digital geometry problems, we first formulate digitization of reflections and rotations in n dimensions using geometric algebra. We then focus on two dimensions and study bijectivity of digitized reflections. We then show that composition of bijective digitized reflections results in new bijective digitized rotations, allowing us to approximate any digitized rotation by bijective digitized rotations.

2   Reflections and Rotations via Geometric Algebra

Geometric algebra of a vector space is an algebra over a field such that its multiplication called geometric product is defined on a space of elements, i.e., multivectors [6]. Geometric algebra is an intuitive and geometric object-oriented algebra that allows to define geometric transformations in an efficient way. Definitions and compositions of geometric transformations are given through geometric products which are invertible. Let us briefly review geometric product rules.

2.1   Geometric Product

Given two vectors $\mathbf{m}, \mathbf{n}$, the geometric product is defined as

$$\mathbf{mn} = \mathbf{m} \cdot \mathbf{n} + \mathbf{m} \wedge \mathbf{n}, \tag{1}$$

where $\mathbf{m} \cdot \mathbf{n} = \|\mathbf{m}\|\|\mathbf{n}\| \cos(\alpha)$ and $\mathbf{m} \wedge \mathbf{n} = \|\mathbf{m}\|\|\mathbf{n}\| \sin(\alpha)\mathbf{I}$ with angle α between $\mathbf{m}$ and $\mathbf{n}$, and $\mathbf{I}$ as the bivector basis spanned by $\mathbf{m}$ and $\mathbf{n}$. Briefly, a bivector (or 2-vector) is an element of the algebra different from a scalar and a vector such that it geometrically represents an oriented area spanned by two vectors. Here the bivector $\mathbf{I}$ represents the unit oriented area element of the plane spanned by the vectors $\mathbf{m}$ and $\mathbf{n}$.

Letting d be the dimension of the vector space, the geometric product acts on the basis vectors $\mathbf{e}_i, \mathbf{e}_j$ and basis bivectors $\mathbf{e}_{ij}$ $(i, j \in [1, d])$ as follows:

$$\mathbf{e}_i\mathbf{e}_j = \begin{cases} 1 & \text{if } i = j \\ -\mathbf{e}_{ji} & \text{otherwise} \end{cases} \quad \text{and} \quad \mathbf{e}_{ij}\mathbf{e}_k = \begin{cases} \mathbf{e}_{ijk} & \text{if } j \neq k, i \neq k \\ \mathbf{e}_i & \text{if } j = k \\ -\mathbf{e}_j & \text{if } i = k \end{cases}. \tag{2}$$

If we permute $\mathbf{e}_k$ and $\mathbf{e}_{ij}$, the above multiplication becomes

$$\mathbf{e}_k\mathbf{e}_{ij} = \begin{cases} \mathbf{e}_{kij} & \text{if } j \neq k, i \neq k \\ -\mathbf{e}_i & \text{if } j = k \\ \mathbf{e}_j & \text{if } i - k \end{cases}. \tag{3}$$

Given a vector $\mathbf{x}$ defined as the weighted sum of components over the basis vectors, namely, $\mathbf{x} = \sum_{i=1,\dots,d} u_i \mathbf{e}_i$ the norm of $\mathbf{x}$ is defined as $\|\mathbf{x}\| = \sqrt{\mathbf{x} \cdot \mathbf{x}} = \sqrt{\sum_{i=1,\dots,d} u_i^2}$. Then, by definition, the inverse of $\mathbf{x}$ is defined as $\mathbf{x}^{-1} = \frac{1}{\mathbf{x}} = \frac{\mathbf{x}}{\|\mathbf{x}\|^2} = \frac{\mathbf{x}}{\mathbf{x}\mathbf{x}}$. The geometric product is invertible. In addition, the geometric product is associative and distributive over the addition but not commutative. The inner product results in a scalar. Namely, for given $\mathbf{e}_i, \mathbf{e}_j$

$$\mathbf{e}_i \cdot \mathbf{e}_j = \begin{cases} 1 & \text{if } i = j \\ 0 & \text{otherwise.} \end{cases} \tag{4}$$

2.2   Reflections

A reflection is the isometric mapping from $\mathbb{R}^n$ to itself with a hyperplane as a set of fixed (invariant) points. It is defined as follows with geometric algebra when the hyperplane goes through the origin.

Definition 1. *Given a hyperplane passing through the origin, with its normal vector* $\mathbf{m} \in \mathbb{R}^d$, *denoted by* $H(\mathbf{m})$, *the reflection of point* $\mathbf{x} \in \mathbb{R}^n$ *with respect to* $H(\mathbf{m})$ *is defined as*

$$\left| \begin{array}{l} \mathcal{U}^{\mathbf{m}} : \mathbb{R}^d \to \qquad\qquad\qquad \mathbb{R}^d \\ \quad \mathbf{x} \mapsto -\mathbf{m}\mathbf{x}\mathbf{m}^{-1} = -\frac{1}{\|\mathbf{m}\|^2}\mathbf{m}\mathbf{x}\mathbf{m}. \end{array} \right.$$

Reflections $\mathcal{U}^{\mathbf{m}}$ are said rational if all the components of $\mathbf{m}$ are rational. Note that any rational reflection $\mathcal{U}^{\mathbf{m}}$ can be represented by $\mathbf{m} = \sum_{i=1\dots d} u_i \mathbf{e}_i$ such that $u_i \in \mathbb{Z}$ and $\gcd(u_1, \cdots, u_d) = 1$.

2.3   Rotations

Any rotation is expressed as the composition of two reflections with geometric algebra. If a first reflection w.r.t. $H(\mathbf{m})$ followed by a second reflection w.r.t. $H(\mathbf{n})$, is applied to point $\mathbf{x} \in \mathbb{R}^d$, we have point $\mathbf{x}'$ such that

$$\mathbf{x}' = -\mathbf{n}(-\mathbf{m}\mathbf{x}\mathbf{m}^{-1})\mathbf{n}^{-1} = (\mathbf{n}\mathbf{m})\mathbf{x}(\mathbf{n}\mathbf{m})^{-1}. \tag{5}$$

In other words, $\mathbf{x}'$ is the rotation of $\mathbf{x}$ around the intersection of $\mathbf{m}$ and $\mathbf{n}$. Indeed, assuming $\mathbf{n}$ and $\mathbf{m}$ are both normalized, we have

$$\mathbf{x}' = (\cos\phi + \sin\phi\,\mathbf{I})\mathbf{x}(\cos\phi - \sin\phi\,\mathbf{I}), \tag{6}$$

where ϕ is the angle between $\mathbf{n}$ and $\mathbf{m}$ in the rotation plane whose bivector is $\mathbf{I}$ (cf. Eq. (1)). Note that the angle of this rotation corresponds to 2ϕ.

More generally, the algebraic entity representing the rotation of angle θ in the rotation plane whose bivector is $\mathbf{I}$ is defined as

$$Q = \cos\frac{\theta}{2} + \sin\frac{\theta}{2}\frac{\mathbf{I}}{\|\mathbf{I}\|}. \tag{7}$$

Then, a point $\mathbf{x}$ is rotated to $\mathbf{x}'$ as follows:

$$\mathbf{x}' = Q\mathbf{x}Q^\dagger, \tag{8}$$

where $Q^\dagger = \cos(\frac{\theta}{2}) - \sin\frac{\theta}{2}\frac{\mathbf{I}}{\|\mathbf{I}\|}$. Note that $\mathbf{I} = \mathbf{e}_{12}$ in the 2D case.

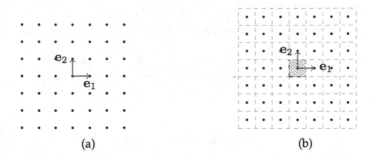

(a)   (b)

Fig. 1. (a) shows the set of points (denoted with small black point) in the digital square grid $\mathbb{Z}\mathbf{e}_1 \oplus \mathbb{Z}\mathbf{e}_2$ (also denoted by $\mathbb{Z}^2$). Such points are obtained as linear combinations of the basis vectors $\mathbf{e}_1$ and $\mathbf{e}_2$ illustrated with red arrow. (b) shows digitization cells associated to the square grid using gray dashed square. The red hatched square denote the digitization cell associated with the origin or $\mathcal{C}(\mathbf{0})$; the border of $\mathcal{C}(\mathbf{0})$ is denoted with red circle. (Color figure online)

3   Digitized Reflections

In order to define digitized reflection, we need to define the grid to digitize points.

3.1   Cubic Grids

In a similar way as the state-of-the-art, we denote the set of vectors of geometric algebra in the space with real coordinates as

$$\mathbb{R}^d = \Big\{\mathbf{x} = \sum_{i=1,\dots,d} u_i\mathbf{e}_i \mid u_i \in \mathbb{R}\Big\}.$$

Those with integer coordinates are called the cubic grid or the integer lattice in $\mathbf{R}^d$, defined as

$$\mathbb{Z}^d = \Big\{\mathbf{x} = \sum_{i=1,\dots,d} a_i\mathbf{e}_i \mid a_i \in \mathbb{Z}\Big\}.$$

This cubic grid is also written as $\mathbb{Z}^d = \oplus_{i=1,\dots,d}\mathbb{Z}\mathbf{e}_i$. An illustration of such a cubic grid in the plane is given in Fig. 1a.

By extension, any square grid generated by two orthogonal vectors, $u\mathbf{e}_1 + v\mathbf{e}_2$ and $-v\mathbf{e}_1 + u\mathbf{e}_2$, in the plane is defined as:

$$\mathbb{Z}(u\mathbf{e}_1 + v\mathbf{e}_2) \oplus \mathbb{Z}(-v\mathbf{e}_1 + u\mathbf{e}_2). \tag{9}$$

3.2   Digitization of Reflections

A digitized reflection is a reflection followed by a digitization. Let us consider reflections of points in $\mathbb{Z}^d$.

As $\mathcal{U}^{\mathbf{m}}(\mathbb{Z}^d) \not\subseteq \mathbb{Z}^d$ in general, we need to define the digitization operator again on the cubic grid in order to obtain digitized reflection points. To this end, we first define digitization cells of the cubic grid $\mathbb{Z}^d$.

Definition 2. *For any $\kappa \in \mathbb{Z}^d$, we define the digitization cell of κ as*

$$\mathcal{C}(\kappa) := \big\{ \mathbf{x} \in \mathbb{R}^d \mid \forall i \in [1, d] \ \|\mathbf{x} - \kappa\| \leq \|\mathbf{x} - \kappa + \mathbf{e}_i\|$$
$$\text{and} \ \ \|\mathbf{x} - \kappa\| < \|\mathbf{x} - \kappa - \mathbf{e}_i\| \big\}.$$

This can be rewritten as:

$$\mathcal{C}(\kappa) := \Big\{ \mathbf{x} = \textstyle\sum_{i=1,\dots,d} x_i \mathbf{e}_i \in \mathbb{R}^d \mid \forall i \in [1, d] \ a_i - \tfrac{1}{2} \leq x_i < a_i + \tfrac{1}{2} \Big\},$$

where $\kappa = \sum_{i=1,\dots,d} a_i \mathbf{e}_i$. An example of the set of digitization cells obtained from the square grid of Fig. 1a is shown in Fig. 1b.

We also define the digitization cell associated to a transformation such as reflection, rotation, and scaling.

Definition 3. *Given a transformation $\mathcal{T}$ such that any basis vector $\mathbf{e}_i$ is transformed to $\mathcal{T} \mathbf{e}_i \mathcal{T}^\dagger$, the digitization cell of $\kappa \in \mathbb{Z}^d$ transformed by $\mathcal{T}$ is defined as*

$$\mathcal{C}_{\mathcal{T}}(\kappa) := \big\{ \mathbf{x} \in \mathbb{R}^d \mid \forall i \in [1, d] \ \|\mathbf{x} - \kappa\| \leq \|\mathbf{x} - \kappa + \mathcal{T} \mathbf{e}_i \mathcal{T}^\dagger\|$$
$$\text{and} \ \ \|\mathbf{x} - \kappa\| < \|\mathbf{x} - \kappa - \mathcal{T} \mathbf{e}_i \mathcal{T}^\dagger\| \big\}.$$

Note that Definition 3 covers:

- a reflected digital cell, if $\mathcal{T} = \sum_{i=1,\dots,d} u_i \mathbf{e}_i$ with $\sum_{i=1,\dots,d} u_i^2 = 1$;
- a rotated digital cell, if $\mathcal{T} = u + v\mathbf{I}$ with $u^2 + v^2 = 1$;
- a scaled digital cell, if $\mathcal{T} = u$ with $u \in \mathbb{R}$ and the digital cell is scaled by a factor u^2.

We also note that the non-transformed digitization cell centered in κ is identical with the digitization cell defined in Definition 2: $\mathcal{C}_1(\kappa) = \mathcal{C}(\kappa)$. This comes simply from the fact that the multiplication of the scalar 1 and the basis vector $\mathbf{e}_i$ is $1\mathbf{e}_i = 1\mathbf{e}_i = \mathbf{e}_i$.

Similarly to [11], we define the digitization operator as follows:

Definition 4. *The digitization operator on a cubic grid is defined as*

$$\left|\begin{array}{ccc} \mathcal{D}: & \mathbb{R}^d & \to & \mathbb{Z}^d \\ & \sum_{i=1,\dots,d} u_i \mathbf{e}_i & \mapsto & \sum_{i=1,\dots,d} \lfloor u_i + \tfrac{1}{2} \rfloor \mathbf{e}_i \end{array}\right.$$

where $\lfloor u \rfloor$ $(u \in \mathbb{R})$ denotes the greatest integer not greater than u.

Now we define the digitized reflection as the composition of the reflection and the digitization.

Definition 5. *Given a hyperplane $H(\mathbf{m})$, a digitized reflection with respect to $H(\mathbf{m})$ is defined as*

$$\left| \begin{aligned} \mathcal{R}^{\mathbf{m}} : \mathbb{Z}^d &\to \quad\quad \mathbb{Z}^d \\ \mathbf{x} &\mapsto \mathcal{D} \circ \mathcal{U}^{\mathbf{m}}(\mathbf{x}). \end{aligned} \right.$$

Hereafter we focus on the case of $d = 2$.

4   Bijective Digitized Reflections

In order to describe the bijectivity of digitized reflections $\mathcal{R}^{\mathbf{m}}$ in the plane, we need to explore the structure of the square grid after reflection $\mathcal{U}^{\mathbf{m}}$. We start by the reflection of the basis vectors. Let us denote the reflection of $\mathbf{e}_1$ and $\mathbf{e}_2$ with respect to $H(\mathbf{m})$ by ϕ and ψ, respectively. Applying Definition 1 to $\mathbf{e}_1$ and $\mathbf{e}_2$ results in

$$\phi = \mathcal{U}^{\mathbf{m}}(\mathbf{e_1}) = -\mathbf{m}\mathbf{e}_1\mathbf{m}^{-1} = \frac{v^2 - u^2}{u^2 + v^2}\mathbf{e}_1 + \frac{2uv}{u^2 + v^2}\mathbf{e}_2, \tag{10}$$

$$\psi = \mathcal{U}^{\mathbf{m}}(\mathbf{e_2}) = -\mathbf{m}\mathbf{e}_2\mathbf{m}^{-1} = \frac{2uv}{u^2 + v^2}\mathbf{e}_1 + \frac{u^2 - v^2}{u^2 + v^2}\mathbf{e}_2. \tag{11}$$

Reflection of any point $\mathbf{y} \in \mathbb{Z}^2$ is expressed as a linear combination of the reflected unit vectors ϕ and ψ. Namely, the reflected points of $\mathbb{Z}\mathbf{e}_1 + \mathbb{Z}\mathbf{e}_2$ are the points of the grid $\mathbb{Z}\phi + \mathbb{Z}\psi$. An example of the transformed grid is shown in Fig. 2.

4.1   Set of Remainders

In a similar way as in [12], let us first consider the set of remainders to give the definition of bijective reflections.

Definition 6. *Given a reflection $\mathcal{U}^{\mathbf{m}}$, the set of remainders $\mathcal{S}^{\mathbf{m}}$ is defined as*

$$\left| \begin{aligned} \mathcal{S}^{\mathbf{m}} : \mathbb{Z}^2 \times \mathbb{Z}^2 &\to \quad\quad \mathbb{R}^2 \\ (\mathbf{x}, \mathbf{y}) &\mapsto \mathcal{U}^{\mathbf{m}}(\mathbf{x}) - \mathbf{y}. \end{aligned} \right.$$

Definition 7. *A digitized reflection $\mathcal{R}^{\mathbf{m}} = \mathcal{D} \circ \mathcal{U}^{\mathbf{m}}$ is bijective if and only if*

$$\forall \mathbf{y} \in \mathbb{Z}^2, \exists! \mathbf{x} \in \mathbb{Z}^2, \mathcal{S}^{\mathbf{m}}(\mathbf{x}, \mathbf{y}) \in \mathcal{C}_1(\mathbf{0}), \tag{12}$$

where $\mathbf{0}$ corresponds to the null vector.

Note that this definition can be divided into two parts like [12]:

$$\begin{cases} \forall \mathbf{y} \in \mathbb{Z}^2, \exists \mathbf{x} \in \mathbb{Z}^2, \mathcal{S}^{\mathbf{m}}(\mathbf{x}, \mathbf{y}) \in \mathcal{C}_1(\mathbf{0}) \\ \forall \mathbf{x} \in \mathbb{Z}^2, \exists \mathbf{y} \in \mathbb{Z}^2, \mathcal{S}^{\mathbf{m}}(\mathbf{x}, \mathbf{y}) \in \mathcal{C}_{\frac{\mathbf{m}}{\|\mathbf{m}\|}}(\mathbf{0}) \end{cases} \tag{13}$$

provided $\mathcal{S}^{\mathbf{m}}(\mathbb{Z}^2, \mathbb{Z}^2) \cap \mathcal{C}_1(\mathbf{0}) = \mathcal{S}^{\mathbf{m}}(\mathbb{Z}^2, \mathbb{Z}^2) \cap \mathcal{C}_{\frac{\mathbf{m}}{\|\mathbf{m}\|}}(\mathbf{0})$, that is to say:

$$\mathcal{I} = \mathcal{S}^{\mathbf{m}}(\mathbb{Z}^2, \mathbb{Z}^2) \cap \left(\mathcal{C}_1(\mathbf{0}) \cup \mathcal{C}_{\frac{\mathbf{m}}{\|\mathbf{m}\|}}(\mathbf{0}) \right) \setminus \left(\mathcal{C}_1(\mathbf{0}) \cap \mathcal{C}_{\frac{\mathbf{m}}{\|\mathbf{m}\|}}(\mathbf{0}) \right) = \emptyset. \tag{14}$$

As an illustration, Fig. 2b verifies the above condition whereas Fig. 2c does not.

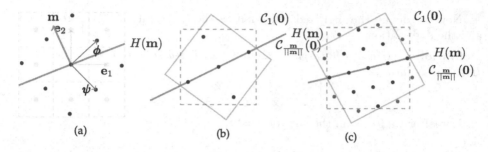

Fig. 2. (a) Discrete square grid $\mathbb{Z}\mathbf{e}_1 \oplus \mathbb{Z}\mathbf{e}_2$ illustrated by gray dots, and their associated digitization cells illustrated by gray dashed squares, using the geometric algebra implementation ganja.js [5]. The reflected points with respect to the line $H(\mathbf{m})$, i.e., $\mathbb{Z}\phi + \mathbb{Z}\psi$, illustrated by blue dots. (b) and (c) Set of remainders illustrated by blue (and red) dots; in (b) the set of remainders satisfies the bijectivity condition while (c) does not (see the red dots for non-bijective points). (Color figure online)

4.2   Non-rational Reflection

We first show that non-rational digitized reflections are not bijective. To achieve this, we study the structure of the set of remainders $\mathcal{S}^{\mathbf{m}}(\mathbb{Z}^2, \mathbb{Z}^2)$ with respect to the parameters of the digitized reflection.

$\mathcal{G}$ denotes the set composed of the lattice $\mathbb{Z}\mathbf{e}_1 \oplus \mathbb{Z}\mathbf{e}_2$ and $\mathbb{Z}\phi \oplus \mathbb{Z}\psi$:

$$\mathcal{G} = \mathbb{Z}\mathbf{e}_1 \oplus \mathbb{Z}\mathbf{e}_2 \oplus \mathbb{Z}\phi \oplus \mathbb{Z}\psi. \tag{15}$$

Proposition 1. *If the digitized reflection $\mathcal{U}^{\mathbf{m}}$ is non-rational, the set $\mathcal{G}$ is dense and infinite.*

Non-rational digitized reflection means the reflected components computed in Eq. (10) and Eq. (11), are not integers (non-Pythagorean primitive triples). From [9], the set obtained with non-Pythagorean is dense and infinite. In such a case the digitized reflection is not bijective since the two vectors $\mathbf{a}, \mathbf{b} \in \mathcal{S}^{\mathbf{m}}(\mathbb{Z}^2, \mathbb{Z}^2) \cap \mathcal{C}_1(\mathbf{0})$ exist such that $\mathcal{S}^{\mathbf{m}}(\mathbf{x}, \mathbf{y}) = \mathbf{a}$ and $\mathcal{S}^{\mathbf{m}}(\mathbf{x} + \mathbf{e}_1, \mathbf{y}) = \mathbf{b}$ (consequence of the Bolzano-Weirstrass theorem). This violates the bijectivity condition because both $\mathcal{R}^{\mathbf{m}}(\mathbf{x}) \in \mathcal{C}_1(\mathbf{y})$ and $\mathcal{R}^{\mathbf{m}}(\mathbf{x} + \mathbf{e}_1) \in \mathcal{C}_1(\mathbf{y})$.

4.3   Bijectivity of Digitized Reflections

In order to characterize normal vectors leading to bijective digitized reflections, we use the bijectivity condition of digitized rotations. Let us first recall the bijectivity condition defined in [9] with primitive Pythagorean triples. A digitized rotation whose rotation angle is θ is bijective if and only if

$$\{\cos(\theta), \sin(\theta)\} = \left\{ \tfrac{2k+1}{2k^2+2k+1}, \tfrac{2k(k+1)}{2k^2+2k+1} \right\}, k \in \mathbb{N}. \tag{16}$$

Using the half angle formula and $\theta \in [0, \frac{\pi}{2}]$, we have

$$\cos(\tfrac{\theta}{2}) = \sqrt{\tfrac{1}{2} + \tfrac{2k+1}{2(2k^2+2k+1)}}, \quad \sin(\tfrac{\theta}{2}) = \sqrt{\tfrac{1}{2} - \tfrac{2k+1}{2(2k^2+2k+1)}}.$$
$$= \tfrac{(k+1)}{\sqrt{2k^2+2k+1}}, \qquad\qquad = \tfrac{k}{\sqrt{2k^2+2k+1}}$$

With Eq. (7), this can be rewritten using the digitized rotation given by the entity $\mathbf{Q}$ as

$$\mathbf{Q} = k + 1 + k\mathbf{e}_{12}, \quad k \in \mathbb{N}. \tag{17}$$

Note that the rotation operator does not change by any scale.

Conversely, from Eq. (17) we have

$$\mathbf{QxQ}^{-1} = \left(((k+1)x + ky)\mathbf{e}_1 + ((k+1)y - kx)\mathbf{e}_2 \right)\left(\tfrac{k+1-k\mathbf{e}_{12}}{2k^2+2k+1} \right)$$
$$= \tfrac{1}{2k^2+2k+1}\left((2k+1)x - 2k(k+1)y \right)\mathbf{e}_1 + \left((2k+1)y + 2k(k+1)x \right)\mathbf{e}_2, \tag{18}$$

which leads to Eq. (16). This indicates that Eq. (17) and Eq. (16) are equivalent with each other. Equation (17) is thus the bijectivity condition of digitized rotations with geometric algebra. We remark that Gaussian integers defined in [12] gives us similar argument.

Proposition 2. *Given a rational reflection line $H(\widetilde{\mathbf{m}})$ such that*

$$\widetilde{\mathbf{m}} = -k\mathbf{e}_1 + (k+s)\mathbf{e}_2, \quad k \in \mathbb{N}, s \in \mathbb{N},$$

the rational digitized reflection $\mathcal{R}^{\widetilde{\mathbf{m}}}$ is bijective if and only if $s = 1$.

Proof. The idea is simply to express the set of remainders of digital reflections $\mathcal{S}^{\widetilde{\mathbf{m}}}$ by the set of remainders of digitized rotations $\mathcal{S}^Q$ where Q is a digitized rotation entity. This is performed through the fact that a composition of any digitized reflection and the digitized reflection with respect to $H(\mathbf{e}_2$ does not induce any change in the set of remainders, namely

$$\mathcal{S}^{\mathbf{e}_2\mathbf{m}}(\mathbb{Z}^2, \mathbb{Z}^2) = \mathcal{S}^{\mathbf{m}}(\mathbb{Z}^2, \mathbb{Z}^2).$$

Besides,

$$\mathcal{C}_{\frac{\widetilde{\mathbf{m}}}{\|\mathbf{m}\|}}(\mathbf{0}) = \mathcal{C}_{\mathbf{e}_2\frac{\widetilde{\mathbf{m}}}{\|\mathbf{m}\|}}(\mathbf{0}).$$

Algebraically $\mathbf{e}_2\mathbf{m} = \mathbf{e}_2(-k\mathbf{e}_1 + (k+s)\mathbf{e}_2) = (k+s) + k\mathbf{e}_{12}$. The resulting entity is homogeneous to a rotation. From Eq. (17), this entity is bijective if and only if $s = 1$. $\qquad\square$

4.4   Finding the Closest Bijective Digitized Reflections

Let us consider the set of all bijective digitized reflections such that the reflection lines have slant angles $\theta \in [0, \frac{\pi}{4}[$:

$$\mathbf{B} = \{\mathcal{U}^{\widetilde{\mathbf{m}}} \mid \widetilde{\mathbf{m}} = -k\mathbf{e}_1 + (k+1)\mathbf{e}_2, \, k \in \mathbb{N}\}.$$

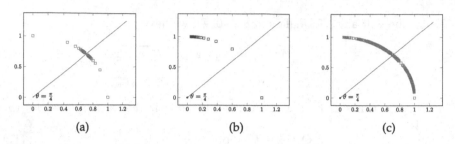

Fig. 3. Distribution of (reflection and rotation) angles within $[0, \frac{\pi}{2}]$ that make the digitized transformations bijective: (a) digitized reflections, (b) digitized rotations, (c) digitized rotation approximations through pairs of bijective digitized reflections. The same integer parameter $k = 20$ was used for the three figures. Note that the distribution of angles in $[\frac{\pi}{4}, \frac{\pi}{2}]$ is obtained by the reflection of that in $[0, \frac{\pi}{4}]$ with respect to the line $x = y$.

We show in Fig. 3a that the slant angles of such reflection lines defined by $\widetilde{\mathbf{m}}$ are sparse around $\theta = 0$ while dense around $\theta = \frac{\pi}{4}$. We remark that in practice, we have to limit the maximum value $k_{\max}$ of k because of the image size. Hereafter, instead of $\mathbf{B}$, we use $\mathbf{B}_{k_{\max}}$ with the condition $k \leq k_{\max}$.

The sparsity of $\mathbf{B}_{k_{\max}}$ motivates us to approximate a given reflection $\mathcal{R}^{\mathbf{m}}$ with the closest digitized reflection $\mathcal{R}^{\widetilde{\mathbf{m}}}$ such that $\mathcal{U}^{\widetilde{\mathbf{m}}} \in \mathbf{B}_{k_{max}}$. More precisely, given $\mathbf{m}$ with slant angle θ of its reflection line, we seek for $\widetilde{\mathbf{m}}$ with reflection line having the slant angle $\widetilde{\theta}$ that minimizes the absolute difference between the angles:

$$\underset{\mathcal{U}^{\widetilde{\mathbf{m}}} \in \mathbf{B}_{k_{\max}}}{\arg \min} \left| \widetilde{\theta} - \theta \right|.$$

Since tan monotonically increases in $[-\frac{\pi}{4}, \frac{\pi}{4}]$, we can consider $|\tan(\widetilde{\theta} - \theta)|$ instead of $|\widetilde{\theta} - \theta|$. This minimization is thus equivalent with

$$\underset{\mathcal{U}^{\widetilde{\mathbf{m}}} \in \mathbf{B}_{k_{\max}}}{\arg \min} \left| \frac{\tan(\widetilde{\theta}) - \tan(\theta)}{1 + \tan(\widetilde{\theta}) \tan(\theta)} \right| = \underset{\widetilde{k} \in \mathbb{N}}{\arg \min} \left| \frac{\widetilde{k}x - (\widetilde{k}+1)y}{(\widetilde{k}+1)x + \widetilde{k}y} \right|,$$

where (x, y) are the components of $\mathbf{m}$, i.e., $\tan \theta = y/x$ $(x > y)$. Assuming $\widetilde{k} \in \mathbb{Q}_+$, we find that $\widetilde{k} = \frac{y}{x-y}$ achieves the minimum of the objective function by making its numerator 0 because the denominator is always positive. As the function $f(k) = (kx - (k+1)y)/((k+1)x + ky)$ is increasing for all $k \geq 0$, $f(k) \leq 0$ when $0 \leq k \leq y/(x-y)$ and $f(k) > 0$ otherwise, we can find $\widetilde{k} \in \mathbb{N}$ such as

$$\widetilde{k} = \underset{\widetilde{k} \in \{\lfloor \frac{y}{x-y} \rfloor, \lceil \frac{y}{x-y} \rceil\}}{\arg \min} \left| \frac{\widetilde{k}x - (\widetilde{k}+1)y}{(\widetilde{k}+1)x + \widetilde{k}y} \right|.$$

Note that we consider the case where $x > y$. If $x = y$, we have $\widetilde{\mathbf{m}} = -\mathbf{e}_1 + \mathbf{e}_2$.

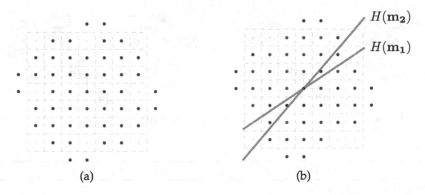

(a)   (b)

Fig. 4. (a) digitized rotation with $\theta = \frac{\pi}{6}$ which yields holes and double points. (b) same digitized rotation approximation with composition of two bijective reflections.

5   Bijective Digitized Rotations via Bijective Digitized Reflections

From the set of bijective digitized reflections, we can obtain the set of bijective digitized rotations since any rotation can be expressed as the composition of two reflections. We see such rotation angles are distributed sparsely as shown in Fig. 3b, thus most digitized rotations are likely to be non-bijective and yield holes and/or double points (see Fig. 4a).

5.1   Composition of Bijective Digitized Reflections Without Error

As any rotation is composed of a pair of reflections, it is easy to see that if the first digitized reflection $\mathcal{R}^{\mathbf{m}} = \mathcal{D} \circ \mathcal{U}^{\mathbf{m}}$ induces no digitization error and the second digitized reflection $\mathcal{R}^{\mathbf{n}}$ is bijective, then $\mathcal{R}^{\mathbf{n}} \circ \mathcal{R}^{\mathbf{m}}$ is a bijective digitized rotation. Such cases occur when $\mathbf{m} = \mathbf{e}_1$, $\mathbf{e}_2$ or $\mathbf{e}_1 + \mathbf{e}_2$. However, it is also easy to see that such composed bijective digitized rotations have the same rotation angle distribution as that of direct (non-composed) bijective digitized rotations, which is illustrated in Fig. 3b. This leads us to investigate the composition of two general bijective digitized reflections.

5.2   Approximating Digitized Rotations with Bijective Digitized Reflections

Any composition of bijective digitized reflections is also bijective, that is $\mathcal{R}^{\mathbf{n}} \circ \mathcal{R}^{\mathbf{m}}$ is bijective if $\mathcal{R}^{\mathbf{n}}$ and $\mathcal{R}^{\mathbf{m}}$ are both bijective. Based on this fact, given a rotation angle θ, the aim here is to find the best approximated rotation composed of a pair of bijective digitized reflections.

The idea of our algorithm for this is simple (see Algorithm 1). Given a maximum possible integer $k_{\max}$ that defines the set of bijective digitized reflections $\mathbf{B}_{k_{\max}}$ (see Sect. 4.4), we first loop over all possible bijective digitized reflections

Fig. 5. Original image and its digitized rotations by bijective digitized reflections for angles $\frac{\pi}{8}, \frac{\pi}{4}, \frac{7\pi}{16}$ from left to right.

Algorithm 1: Digitized rotation by bijective digitized reflections

---

1 **Function** `approxRotation`
  Input: $k_{\max}$, rotation angle θ
  Output: mutivector R for bijective approximated digitized rotation
2   $\tilde{R} \leftarrow$ geometric algebra rotation ; $\tilde{\theta} = 2\pi$
3   **for** $\mathcal{U}^{\mathbf{m}_1} \in \mathbf{B}_{k_{\max}}$ **do**
4     $x = \mathbf{m}_1 \cdot \mathbf{e}_2 \cos(\frac{\theta}{2}) + \mathbf{m}_1 \cdot \mathbf{e}_1 \sin(\frac{\theta}{2}), \quad y = \mathbf{m}_1 \cdot \mathbf{e}_2 \sin(\frac{\theta}{2}) - \mathbf{m}_1 \cdot \mathbf{e}_1 \cos(\frac{\theta}{2})$
      // look for the best approximation of digitized reflection
5   **if** $x = y$ **then**
6       $\mathbf{m}_2 = -\mathbf{e}_1 + \mathbf{e}_2$
7   **else**
8       $\tilde{k} = \displaystyle\arg\min_{\tilde{k} \in \{\lfloor \frac{y}{x-y} \rfloor, \lceil \frac{y}{x-y} \rceil\}} \left| \frac{\tilde{k}x - (\tilde{k}+1)y}{(\tilde{k}+1)x + \tilde{k}y} \right|$
9       $\mathbf{m}_2 = -\tilde{k}\mathbf{e}_1 + (\tilde{k}+1)\mathbf{e}_2$
10    $\theta_{12} \leftarrow \cos^{-1}(\mathbf{m}_1 \cdot \mathbf{m}_2)$
11   **if** $|\theta_{12}| < \tilde{\theta}$ **then**
12      $R = \mathbf{m}_2 \mathbf{m}_1, \quad \tilde{\theta} = \theta_{12}$

13 **return** R

$\mathcal{U}^{\mathbf{m}_1} \in \mathbf{B}_{k_{\max}}$. For each $\mathcal{U}^{\mathbf{m}_1} \in \mathbf{B}_{k_{\max}}$, the second bijective digitized reflection $\mathcal{U}^{\mathbf{m}_2} \in \mathbf{B}_{k_{\max}}$ is then selected such that $\arccos \mathbf{m}_1 \cdot \mathbf{m}_2$ is closest to $\frac{\theta}{2}$. For that, we use the approximation method proposed in Sect. 4.4. We remark that since the computations in the loop run in constant time, the overall complexity of Algorithm 1 is linear with respect to $\text{card}(\mathbf{B}_{k_{\max}})$.

The proposed algorithm was implemented in C++ and also with the library DGtal [1] for the digital geometry part. The code is available online[1]. Figure 5 shows some results on a single image with different rotation angles.

5.3 Distributions of Bijective Digitized Reflections and Rotations

With a similar idea to Algorithm 1, we can compute from $\mathbf{B}_{k_{\max}}$, all the rotation angles of such bijective approximations of digitized rotations. The distribution

---

[1] https://github.com/sbreuils/GADigitizedTransformations.git.

of such angles is shown in Fig. 3c; it is even less sparse compared to those of
Fig. 3a and 3b.

In order to compare the angle distribution between Fig. 3a and 3b, we define
the angle sparsity as the maximum angle between two successive bijective digitized transformations. The angle sparsity for bijective digitized reflections is 0.46
rad whereas it is 0.93 rad for bijective digitized rotations. Note that they do not
depend on the value of $k_{\max}$.

We can also evaluate the angle fineness (denoted by $\Delta\theta_{\min}$) in terms of the
minimum angle between two successive bijective digitized transformations. We
have from a given $k_{\max}(>1)$, $\Delta\theta_{\min} = \arctan\left(\frac{1}{2k_{\max}^2}\right)$ for bijective digitized
reflections while $\Delta\theta_{\min} = \arctan\left(\frac{4k_{\max}^2}{4k_{\max}^4-1}\right)$ for bijective digitized rotations. Note
that if $k_{\max} = 1$, $\Delta\theta_{\min} = \arctan(\frac{1}{2k_{\max}+1}) = \arctan(\frac{1}{3}) \approx 0.32$ rad for bijective
digitized reflections while it is $\Delta\theta_{\min} = \arccos\left(\frac{2k_{\max}(k_{\max}+1)}{2k_{\max}^2+2k_{\max}+1}\right) = \arccos\left(\frac{4}{5}\right) \approx$
0.64 rad for bijective digitized rotations.

We easily check that $\forall k_{\max} \in \mathbb{N}^*$, $\Delta\theta_{\min}$ of the bijective digitized reflection
is lower than that of bijective digitized rotation. For example, with $k_{\max} = 20$
in Fig. 3, $\Delta\theta_{\min}$ of bijective digitized reflections is 1.25×10^{-3} rad whereas that
of bijective digitized rotations is 2.5×10^{-3} rad.

6   Conclusion

We visited reflections, rotations, and their digitization using geometric algebra.
The geometric algebra framework allows us to characterize the bijective digitized reflections. We first showed that compositions of bijective digitized reflections result in new bijective digitized rotations using geometric algebra. We then
demonstrated that any digitized rotation is approximated by one of these new
bijective digitized rotations.

There are other approximation methods that preserve bijectivity for rotations or reflections on $\mathbb{Z}^2$, such as quasi-shear rotations [2] and digital bijective
reflections [3]. Naturally, a comparative study of our approach with them is
expected as a perspective of this article. We are also interested in adapting the
presented algorithm to the case where the number of considered points of $\mathbb{Z}^2$ is
finite; there would be more bijective digitized reflections. Finally, an extension
of the concept to higher dimensions is also our interest.

References

1. DGtal: Digital geometry tools and algorithms library. https://dgtal.org/
2. Andres, E.: The quasi-shear rotation. In: Miguet, S., Montanvert, A., Ubéda, S.
   (eds.) DGCI 1996. LNCS, vol. 1176, pp. 307–314. Springer, Heidelberg (1996).
   https://doi.org/10.1007/3-540-62005-2_26
3. Andres, E., Dutt, M., Biswas, A., Largeteau-Skapin, G., Zrour, R.: Digital two-
   dimensional bijective reflection and associated rotation. In: Couprie, M., Cousty,
   J., Kenmochi, Y., Mustafa, N. (eds.) DGCI 2019. LNCS, vol. 11414, pp. 3–14.
   Springer, Cham (2019). https://doi.org/10.1007/978-3-030-14085-4_1

4. Clifford, W.K.: Applications of Grassmann's extensive algebra. Am. J. Math. **1**(4), 350–358 (1878)
5. De Keninck, S.: ganja.js (2020). https://doi.org/10.5281/ZENODO.3635774, https://zenodo.org/record/3635774
6. Dorst, L., Fontijne, D., Mann, S.: Geometric algebra for computer science. An Object-Oriented Approach to Geometry. Morgan Kaufmann (2007)
7. Hamilton, W.R.: On quaternions; or on a new system of imaginaries in algebra. Philos. Mag. **25**(3), 489–495 (1844)
8. Jacob, M.A., Andres, E.: On discrete rotations. In: 5th International Workshop on Discrete Geometry for Computer Imagery, Clermont-Ferrand (France), pp. 161–174. Université de Clermont-Ferrand I (September 1995)
9. Nouvel, B., Rémila, E.: Characterization of bijective discretized rotations. In: Klette, R., Žunić, J. (eds.) IWCIA 2004. LNCS, vol. 3322, pp. 248–259. Springer, Heidelberg (2004). https://doi.org/10.1007/978-3-540-30503-3_19
10. Pluta, K., Romon, P., Kenmochi, Y., Passat, N.: Bijectivity certification of 3D digitized rotations. In: Bac, A., Mari, J.-L. (eds.) CTIC 2016. LNCS, vol. 9667, pp. 30–41. Springer, Cham (2016). https://doi.org/10.1007/978-3-319-39441-1_4
11. Pluta, K., Roussillon, T., Cœurjolly, D., Romon, P., Kenmochi, Y., Ostromoukhov, V.: Characterization of bijective digitized rotations on the hexagonal grid. J. Math. Imaging Vis. **60**(5), 707–716 (2018)
12. Roussillon, T., Coeurjolly, D.: Characterization of bijective discretized rotations by Gaussian integers. Research report, LIRIS UMR CNRS 5205 (2016)

Digital Straight Segment Filter
for Geometric Description

Rémi Decelle[1(✉)], Phuc Ngo[1], Isabelle Debled-Rennesson[1], Frédéric Mothe[2],
and Fleur Longuetaud[2]

[1] CNRS, LORIA, UMR 7503, Université de Lorraine, 54506 Nancy,
Vandoeuvre-lès-Nancy, France
`remi.decelle@loria.fr`
[2] AgroParisTech, INRAE, SILVA, Université de Lorraine, 54000 Nancy, France

Abstract. In this paper, an algorithmic scheme is proposed to estimate
different local characteristics of image structures using discrete geometry tools. The arithmetic properties of Digital Straight Lines and their
link with the Farey sequences allow the introduction of a new directional
filter. In an incremental process, it provides local geometric information
at each point in an image, such as the length, orientation and thickness of the longest Digital Straight Segment passing through that point.
Experiments on binary and grayscale images are proposed and show the
interest of this tool. Comparisons to a well-known morphological filter
for grayscale images are also presented.

Keywords: Farey sequences · Stern-Brocot tree · Digital straight
segment · Directional filter · Orientation field · Thickness

1 Introduction

Image processing applications often require computing, analyzing and studying
the characteristics of objects contained in digital images. In the field of digital geometry [4], new mathematical definitions of basic geometric objects are
introduced to better fit these discrete data, in particular, the notion of digital
straight segments (DSS) [4,12]. DSS has been used in many different contexts
to study and analyze the geometrical characteristics of digital curves extracted
from contour of objects in digital image [2,6,7,10,11], such as detection of the
convex and concave parts of the shape, the discrete geometric estimators: length,
tangent, curvature estimators, dominant point detection, ...

In this paper we are interested in computing the local geometric characteristics at every pixels in image, not only object contours, using the discrete tools.
More precisely, based on the arithmetic properties of DSS and their link with
the Farey sequences, we develop a new directional filter from which different

This research was made possible by support from the French National Research Agency,
in the framework of the project TreeTrace, ANR-17-CE10-0016.

J. Lindblad et al. (Eds.): DGMM 2021, LNCS 12708, pp. 255–268, 2021.
https://doi.org/10.1007/978-3-030-76657-3_18

geometric features at each image point can be estimated, such as the length and directional features, the thickness map and the distance to object boundary.

In the context of directional filters using discrete geometry and mathematical morphology tools, several works have been proposed [8,9,13,14,17]. Van Herk has presented in [17] an efficient algorithm for local min/max filters using linear structuring element (SE). Soille *et al.* [13] proposed later a generalized implementation of this method for erosion and dilation along discrete line at arbitrary angles. The method was promising, however the obtained results are not translation invariant (TI) to the image frame. Indeed, depending on the position of SE along the line, the obtained results can slightly vary from one position to another. To tackle this issue, Soille and Talbot [14] proposed a TI method allowing to efficiently compute the different morphological operations of grayscale images. However, no implementation is given to reproduce the proposed method. Still in [14], an application of orientation fields is presented. For this, a segment of fixed length is used for the local orientation estimation. The obtained result depends strongly on this length parameter. Using the path operator from mathematical morphology, Merveille *et al.* have introduced in [8,9] a non-linear operator, called Ranking the Orientation Responses of Path Operators (RORPO), for 2D and 3D curvilinear structure analysis. RORPO is well adapted to extract the characteristic of objects with intrinsic anisotropy, and has been applied as a curvilinear filter in segmentation framework of blood vessels in medical imaging.

Inspired by Soille and Talbot work [14] and to overcome the limitations of their method, we introduce a new directional filter using DSS. More precisely, we consider the DSS of adaptive lengths to exploit all possible directional spaces around a pixel and report the longest segments passing through it. Moreover the thickness of the selected DSS is also considered in our proposed method. These length and thickness of the segments are used to extract the local geometric features of pixels as they contain relevant information for the pixel description. This study leads to the proposal of an exact numerical and incremental algorithm allowing to efficiently compute these local features. Applications to the grayscale images and comparisons to a morphological filter are also detailed in the paper.

In the next section we recall definitions and results useful in our study. The Sect. 3 presents the proposed method on binary images and the deduced local geometric features are illustrated. The approach for grayscale images is presented in Sect. 4. Applications and results are shown in Sect. 5.

2   Digital Straight Lines and Stern-Brocot Tree

2.1   Digital Straight Lines [12]

Definition 1. *A digital straight line (DSL), denoted by $\mathcal{D}(a, b, \mu, \omega)$, with a, b, μ and ω integer numbers and $\gcd(a, b) = 1$, is the set of points $(x, y) \in \mathbb{Z}^2$ verifying $\mu \leq ax - by < \mu + \omega$.*

ω is called the *thickness* of the DSL. If $\omega = \max(|a|, |b|)$, the DSL is 8-connected and is named *naive DSL*, and denoted $\mathcal{D}(a, b, \mu)$. If $\omega > |a| + |b|$, the

DSL is named *thick DSL*. A (naive) *digital straight segment* (DSS) is a finite connected part of a (naive) DSL.

A naive DSS is called *symmetric* for a point (x, y) iff (x, y) belongs to it and if the DSS contains the same number of elements on both sides of (x, y).

It can be noticed that for given values of a, b and μ, the set of naive DSL $\mathcal{D}(a, b, \mu + i \max(|a|, |b|))$, with $i \in \mathbb{Z}$, tessellates $\mathbb{Z}^2$ (see Fig. 1).

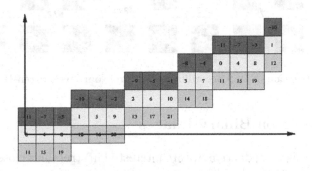

Fig. 1. Naive segments of 15 pixels for $x \in [0, 14]$ of $\mathcal{D}(4, 11, 0)$(white), $\mathcal{D}(4, 11, 11)$(light gray), $\mathcal{D}(4, 11, -11)$(dark gray). The thick DSS $TDSS_{1,1}(4, 11)$ contains all the pixels of the figure for $x \in [0, 14]$. The value $4x - 11y$ labels each pixel (x, y).

In the rest of this paper, we study the neighborhood of naive segments and we extend a DSS by stacking its closest naive DSS to obtain a thick DSS. In this way, we define on $[0, l] \times [0, L]$ in $\mathbb{Z}^2$ a *thick DSS of indices k, j and characteristics (a, b)*, noted $TDSS_{k,j}(a, b)$, as the set of points $(x, y) \in [0, l] \times [0, L]$ belonging to $\mathcal{D}(a, b, -k \max(|a|, |b|), (1 + k + j) \max(|a|, |b|))$ with k and j in $\mathbb{N}$. k and j are the number of naive DSS of characteristics (a, b) that could be stack on one side or on the opposite side (according to current octant) of the naive DSS of $\mathcal{D}(a, b, 0)$. This naive DSS is named the *seed* of the $TDSS$.

2.2   Stern-Brocot Tree and Farey Sequence [3, 15]

Constructing the set of reduced positives fractions $\frac{a}{b}$ can be done iteratively. From two successive fractions $\frac{a}{b}$ and $\frac{a'}{b'}$, we insert a new fraction $\frac{a+a'}{b+b'}$. The *Stern-Brocot tree* is created by starting with $0 = \frac{0}{1}$ and $1 = \frac{1}{1}$ (see Fig. 2).

Definition 2. *The Farey sequence of order $n \in \mathbb{N}^*$, denoted by $\mathcal{F}_n$ is the set of reduced fractions between 0 and 1 for which denominators are lower or equal to n.*

Some examples of Farey sequences: $\mathcal{F}_2 = \{\frac{0}{1}, \frac{1}{2}, \frac{1}{1}\}$, $\mathcal{F}_3 = \{\frac{0}{1}, \frac{1}{3}, \frac{1}{2}, \frac{2}{3}, \frac{1}{1}\}$. The number of DSS in the first octant in a grid of size $N \times N$ is exactly the number of elements of Farey sequence of order N (see Berenstein et al. [1] and Koplowitz et al. [5]). The Farey sequence of order N is a subtree of Stern-Brocot tree.

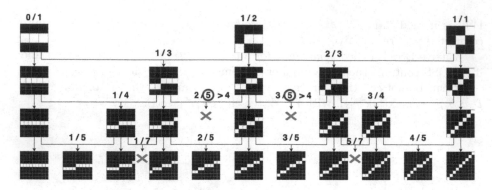

Fig. 2. Construction of DSS according to the first four levels Stren-Brocot tree.

3   DSS Filter on Binary Images

As described in the introduction, we are interested in structures close to DSS but with different lengths and widths. By using Stern-Brocot tree, we can take the advantage of constructing DSS of variable lengths and widths to retrieve more precise features. We will first describe our approach in binary images, where objects under consideration are white and the background is black.

3.1   Proposed Method

The method is described in Algorithm 1 and hereafter we summarize it. Let be p the pixel under consideration of a binary image I, the aim of the presented method is to obtain the longest naive DSS, centered in p, with its pixels in the same color than p. We can, without loss of generality, consider the first octant, i.e. the Stern-Brocot tree goes from $0/1$ up to $1/1$.

The method is as follows from the level n to the level $n + 1$. The considered set of fractions issued from Stern-Brocot tree at the level n is denoted by T_n (T in Algorithm 1). The segments which can be extended are stored in $\mathcal{S}$. At the start of each iteration, $\mathcal{S}$ is set to T_n (all segments may be extended). For each a/b in T_n, we create the DSS of $\mathcal{D}(a, b, \mu)$ with $2n + 1$ pixels (all μ are considered). If the segment cannot be extended, i.e. one of its points belongs to the background, then (a, b, μ) is removed from $\mathcal{S}$. Once all elements of T_n have been checked, if $\mathcal{S} = \emptyset$, meaning that no segment can be extended, the algorithm stops. Otherwise, T_{n+1} is built according to T_n and $\mathcal{S}$, and then we proceed to the iteration $n + 1$.

At each iteration, for a given n, the set T_n contains the fractions of the Farey sequence of order $n + 1$ for which the corresponding DSS could be extended. The incremental process to construct the considered fractions level $n + 1$ from the level n of Stern-Brocot tree is described in Algorithm 2. The slopes of DSS, kept in $\mathcal{S}$ at the previous iteration, permit to add in T_{n+1} only the useful fractions for which DSS could be extended at iteration $n + 1$.

Algorithm 1: Computation of the longest naive DSS at a given point (first octant)

Input: An image I and a pixel $p(x, y)$ of I
Output: The set S of all longest naive DSS centered at p
1 $T \leftarrow \{^0/_1, ^1/_1\}$ // Stern-Brocot sequence represented by fractions $^a/_b$
2 $n \leftarrow 0$
3 **repeat**
4    $S \leftarrow \emptyset$
5    $T' \leftarrow increment(T, S)$ // See Algo. 2
6    $n \leftarrow n + 1$
7    $cut \leftarrow 0$
8   **foreach** $^a/_b \in T'$ **do** // Compute for each node of T'
9   **for** $\mu \leftarrow -b + 1$ to 0 **do** // This loop is applied for small n
10          $S_n \leftarrow (a, b, \mu, n)$ // Characteristics of a DSS of $\mathcal{D}(a, b, \mu)$
11   **if** $b < n + 2$ **then** // Case $(a, b, \mu, n - 1) \in S$ at the level $n - 1$
12             $L_n \leftarrow \{(n, \frac{a*n-\mu}{b}), (-n, -\frac{a*n-\mu}{b})\}$ // End-points of the DSS
13   **else** // Case $b = n - 2$ $(a, b, \mu, n - 1) \notin S$ at the level $n - 1$
14             $L_n \leftarrow$ all the $2n + 1$ pixels of the DSS centered in p
15          $S \leftarrow S \cup S_n$
16   **foreach** $(i, j) \in L_n$ **do**
17             $q \leftarrow (x + i, y + j)$
18   **if** $q \notin I$ or $I(p) \neq I(q)$ **then**
19                $cut \leftarrow cut + 1$
20                $S \leftarrow S \setminus S_n$
21   **break**

22    $T \leftarrow T'$
  until $cut \neq size(T)$
24 **Return** S

Algorithm 2: Stern-Brocot incrementation (first octant)

Input: A Stern-Brocot sequence T at level n and the set S of extensible DSS
Output: The Stern-Brocot sequence T' at level $n + 1$
1 $T' \leftarrow \emptyset$
2 **for** $i \leftarrow 0$ to $size(T) - 1$ **do**
3    $node \leftarrow ^{a_i}/_{b_i}$ // The current node in T
4    $prev \leftarrow ^{a_{i-1}}/_{b_{i-1}[size(T)]}$ // The previous node of $node$ in T
5    $next \leftarrow ^{a_{i+1}}/_{b_{i+1}[size(T)]}$ // the next node of $node$ in T
6   **if** *(prev or next or node is the slope of a DSS in S)* **then**
7   Insert $^{a_i}/_{b_i}$ in T' // Keep the current node in the next level T'
8    $p \leftarrow a_i + a_{i+1}$
9    $q \leftarrow b_i + b_{i+1}$
10   **if** $(q <= n + 2)$ **then**
11   Insert $^p/_q$ in T' // Insert the node to the next level T'

12 **Return** T'

Figure 2 shows the constructed DSS at the first four levels of the tree (with $\mu = 0$) according to the incremental Algorithm 2. At the level three, we only consider the fraction of the Farey sequence of order four then both $^2/_5$ and $^3/_5$ are not inserted at this level but at the level four.

Figure 3 illustrates some steps of the method for one pixel. On the left side of the four subfigures, the red color shows the covered area by DSS at different levels. At each step, extensible DSS are kept. On the right side, we show on a circle the fractions of Stern-Brocot tree. Green lines indicate extendable DSS with the corresponding fractions. Red lines indicate new DSS which cannot be extended. And blue lines are DSS required for building new DSS at the next level. The tree starts at the level 1, with 8 fractions: $^0/_1, ^1/_2, \ldots, ^{-1}/_1$. All DSS can be extended. We move to the level 2. At this level, some DSS are not extendable, we removed them and move to the level 3. The algorithm stops at the level 6 (meaning DSS of length 13 pixels) where only the DSS $\mathcal{D}(5, 4, -4)$ is still extensible. This last DSS is shown in Fig. 4.

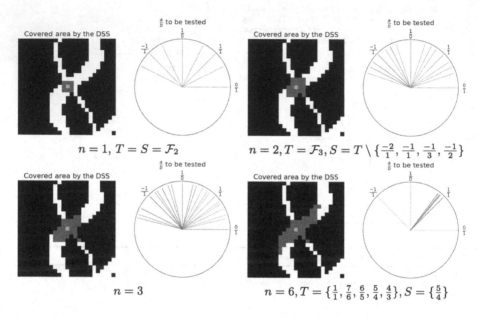

Fig. 3. Step-by-step method for binary image. In green the extendable fractions. In red the rejected ones. In blue those that are required to build new fractions. (Color figure online)

3.2 Geometric Description

At the end of the Algorithm 1, for a given point p in I, we get a value of n and the set $\mathcal{S}_n$ of all the DSS with $2n + 1$ pixels which fit the object. From this set, the following features can be retrieved.

Local Elongation. The elongation L at p is the length of the longest final DSS. We use the simple estimator, $length(DSS) = N_e + \sqrt{2}N_o$, with N_e et N_o the number of even and odd codes in the freeman code of the DSS. This feature indicates if locally the object is elongated. A high value indicates a straight and long shape.

Local Orientation and Direction. Another feature is the local orientation θ of the shape at p and the local vector direction Δ. It is computed by the average of the local orientations obtained in $\mathcal{S}_n$ at the final step of the method.

Orthogonal Span. By constructing, pixel by pixel in both directions, the orthogonal DSS of $\mathcal{D}(-b, a, \mu)$, we can extract the number of pixels ND or the length D of this DSS which reach object's border.

Directional Thickness. The directional thickness of the local shape at the point p. To obtain this feature, we thicken each DSS in $\mathcal{S}_n$ according to our definition of $TDSS_{k,j}(a, b)$ (see Sect. 2). The directional thickness DT is then defined as the sum of the maximum of j and k for which the $TDSS$ is included in I.

Those geometric features are visible in Fig. 4 for DSS of $\mathcal{D}(5, 4, -4)$ with 13 pixels previously extracted (see Fig. 3). The length is $L = 15.7$, the orientation is $\theta = \tan^{-1}(\frac{5}{4}) \approx 51°$, the orthogonal span is $ND = 6$, $D = 6.65$ and finally the directional thickness is $DT = 0 + 1 = 1$.

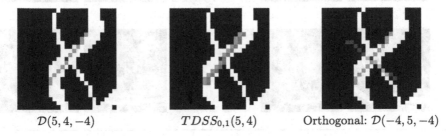

$\mathcal{D}(5, 4, -4)$ $TDSS_{0,1}(5, 4)$ Orthogonal: $\mathcal{D}(-4, 5, -4)$

Fig. 4. On the left, $\mathcal{D}(5, 4, -4)$ from last step of the algorithm (see Fig. 3). The blue pixel is the one under consideration. Green pixels are those of the DSS which are valid. On the right, red pixels are outside of the object, orange are inside of it. (Color figure online)

The Table 1 shows these geometric features for some images. The first image is an annulus, the second one is two circles with different center points and the last one represents vessels. At first, it can be observed that the length map is uniform for the annulus as expected. The length is higher close to the centre. It is the same for the two circles. For vessel image, elongated vessels have high value. For orientation map, the black/white transition corresponds to the change between π and 0. Note that the angle θ is considered for the orientation; e.g., the naive DSL $\mathcal{D}(1, 3, 0)$ and $\mathcal{D}(-1, -3, 0)$ have the same orientation (which is around 18 degree). For this, we use arctan of a/b. This explains the discontinuity between 0 (black pixel) and π (white pixel).

There are some artefacts due to the mean of orientation. The obtained orientation is close to the theoretical one. For the directional thickness we show that it is higher outside the annulus. It is due to small length of DSS at the end, which lead to being able to thicken easily. Small number of pixels reduces the probability to be outside the object when thickening. At last, the orthogonal span has to be high when the object is width or when it is in an intersection as it can be shown in vessel image.

Table 1. Extracted geometric features from DSS filter

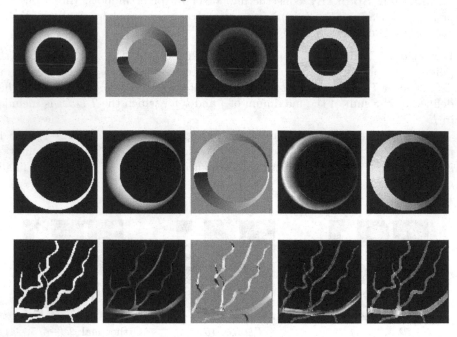

4   DSS Filter on Grayscale Images

In this section, we focus on a similar approach as the one developed by Soille and Talbot [14] for orientation field. Therefore, we are interested in local orientation of image structure with different lengths and widths. Similar approach to the binary case to take account different lengths and widths is used.

As the binary case, we start with first level of the Stern-Brocot tree and with small segments. We compute the erosion by it of the image. We thicken the segment if possible, and compute again the erosion by it. Then, we move to the next level of the tree. We stop until we reach a certain level n. We sum up the results of each erosion. Algorithm 3 sums up the method. Erosions should be used for brighter image structure than background. For darker image structure than the background, the invert image $J = 255 - I$ combined with erosions should be used. For each a/b, we denoted $B_{i,j,k}$ the $TDSS_{k,j}(a, b)$ for which the seed DSS contains $2i + 1$ points, we compute the image $Y_{a,b}$ equal to:

$$Y_{a,b} = \frac{1}{n - \alpha + 1} \sum_{i=\alpha}^{n} \left[\sum_{k=0}^{i-1} \varepsilon_{B_{i,k,0}}(I) + \sum_{j=1}^{i-1} \varepsilon_{B_{i,0,j}}(I) \right] \tag{1}$$

where ε is the morphological erosion and $\alpha = \max(max(|a|, |b|) - 1, 1)$. We introduce a coefficient of normalization in order to take account that the number

---

Algorithm 3: Grayscale method (first octant)

Input: An image I, n the maximum number of points of the DSS
Output: An array of images Y

1   $Y \leftarrow$ matrix of size $(2n+1, 2n+1)$ // The element at position (a,b) is the image $Y_{a,b}$
2   $A \leftarrow (0, 1, 1)^T$
3   **for** $i \leftarrow 0$ **to** n **do** // Iterate over points number up to n
4   **foreach** $a/b \in \mathcal{F}_{i+2}$ **do** // Iterate over Farey sequence of order $i+2$
5         $B \leftarrow TDSS_{0,0}(a,b)$ with $2i+1$ points
6         $E \leftarrow \varepsilon_B(I)$ // Compute erosion of I by the DSS $B_{i,0}$
7         $E^+ \leftarrow E$
8         $E^- \leftarrow E$
9         $Y_{a,b} \leftarrow Y_{a,b} + E$
10   **for** $k \leftarrow 1$ **to** i **do** // Iterate over the segment width
11           $E^+ \leftarrow \varepsilon_A(E^+)$ // Erosion by $TDSS_{0,k}(a,b)$
12           $E^- \leftarrow \varepsilon_{\check{A}}(E^-)$ // Erosion by $TDSS_{k,0}(a,b)$
13           $Y_{a,b} \leftarrow Y_{a,b} + E^+$
14           $Y_{a,b} \leftarrow Y_{a,b} + E^-$

15   Return Y

---

of DSS depends on a/b. Indeed, if a/b is defined at a level p, it will still be there for all the next levels. For instances, $1/2$ will be considered n times (DSS with $3, 5, 7, \ldots, 2n+1$ points), $1/3$ will be considered $n-1$ times (DSS with $5, 7, \ldots, 2n+1$ points), and so on.

Figure 5 presents, for a pixel in an image of tree rings (last image, bottom right, in the figure), different steps of the proposed process if we only consider length variations. The pixel under consideration is the one located at the center of the image (green cross).

The ten first images of the Fig. 5 highlight the results of each erosion, i.e. erosion of I by $B_{i,0,0}$ with i varying from 1 to 10. Axes are the parameters a and b from DSS. For each couple (a, b) from Farey sequences, we plot the gray value resulting from the erosion of I by $B_{i,0,0}$. In each image, the pixel with the highest value is framed with red color.

The image before the last one shows for each couple (a, b) the sum of each erosion for the different lengths. The orientation is assumed to be the one with the highest gray value (red pixel).

We can see that the a/b for which the maximum is reached changes several times depending on the value of i, i.e. the number of points in the segment. By choosing $i = 9$ (i.e. 19 points), the orientation seems correct, but by choosing $i = 10$ (i.e. 21 points) points, the direction is not the expected one (the corresponding segment is represented in red in the last image). However, the direction resulting from the sum is correct (the corresponding segment is represented in green in the last image).

4.1   Fast Implementation

For the first octant, with $A = (1 \ 1 \ 0)^T$, we have the relation $B_{i,j,k+1} = B_{i,j,k} \oplus A$, where $\oplus$ is the morphological dilatation. Knowing the property $\varepsilon_{A \oplus B}(I) =$

$\varepsilon_A(\varepsilon_B(I))$, the relation becomes $\varepsilon_{B_{i,j,k+1}}(I) = \varepsilon_A(\varepsilon_{B_{i,j,k}}(I))$. The relation holds for $B_{i,j,k} \subset B_{i,j+1,k}$ by taking the symmetric of A, i.e. $\check{A} = (0\ 1\ 1)^T$. This relation is shown by the loop line 10 in Algorithm 3. A is adjusted according to the octant. Moreover, for elements $B_{i,0,0}$ we can take the advantage of the algorithm developed by [14] (line 6 in Algorithm 3).

4.2   Orientation Feature

We can consider three variants of the proposed method. First, we consider all possible lengths and thicknesses (see Eq. 1). Second variant, we consider all thicknesses but the length is fix to n (in Algorithm 2 line 3 is removed and $i = n$). Third variant, we consider all lengths but the thickness is fix to 0 (loop line 10 is removed). The Eq. 1 is then respectively for the second and third variants:

$$Y_{a,b} = \sum_{k=0}^{n-1} \varepsilon_{B_{n,k,0}}(I) + \sum_{j=1}^{n-1} \varepsilon_{B_{n,0,j}}(I) \qquad Y_{a,b} = \frac{1}{n-\alpha+1} \sum_{i=\alpha}^{n} \varepsilon_{B_{i,0,0}}(I)$$

In all three cases, we are back to estimate only the dominant orientation for each point x since we have images rely only on a/b. Inspired by [14], we define at each point x the bright orientation as:

$$Dir^+(x) = \{\tan^{-1}(\frac{a_i}{b_i}), Y_{a_i,b_i}(x) \geq Y_{a_j,b_j}(x) \forall (a_i, b_i) \neq (a_j, b_j)\}$$

We introduce an additional quantity to allow us to determine which orientations to choose.

$$G^+(x) = \max_{a,b}(Y_{a,b}(x)) - \min_{a,b}(Y_{a,b}(x))$$

$G^+(x)$ can be interpreted as the strength of the bright structures at the point x. The orientation Dir^- for dark structures and their strength G^- are computed on the same principle as G^+ but using $J = 255 - I$ instead of I. Finally the orientation is computed as follows:

$$\theta(x) = \begin{cases} Dir^+(x)\,, \text{ if } G^+(x) \geq G^-(x) \\ Dir^-(x)\,, \qquad \text{otherwise} \end{cases}$$

It can be noticed that during the process of orientation, we can extract filtered images. These are images resulting from the calculation of $\max_{a,b}(Y_{a,b}(x))$ for both I and J (see Fig. 7).

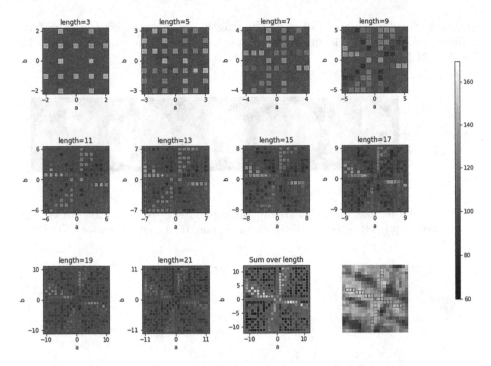

Fig. 5. Illustration of the approach with length variations. (Color figure online)

5   Results and Discussion

A deep analysis of grayscale method is out of range of this article. Nonetheless, we provide a comparison with the proposed method by Soille and Talbot [14]. To do this, we first consider circular objects (see Fig. 6) since the ground truth can easily be determined. This information is useful to demonstrate the efficiency of the proposed method. For the comparison, we computed the deviation between a ground truth and output using the formulae proposed by Turroni et al. [16]:

$$RMSD = \sqrt{\frac{\sum_x d^2(\theta_x, \theta_x^G)}{\text{Number of pixels}}}, \ d(\theta_1, \theta_2) = \begin{cases} \theta_1 - \theta_2 & if \ -\frac{\pi}{2} \leq \theta_1 - \theta_2 < \frac{\pi}{2} \\ \pi + \theta_1 - \theta_2 & if \quad \theta_1 - \theta_2 < -\frac{\pi}{2} \\ \pi - \theta_1 + \theta_2 & if \quad \theta_1 - \theta_2 \geq \frac{\pi}{2} \end{cases}$$

with θ_x^G theoretical orientation at x, θ_x its estimated orientation. The proposed method by [14] provides a RMSD of 0.2809. For our methods, we have respectively for thick variations a value of 0.3111, length variations a value of 0.2436 and both variations a value of 0.2317. For each methods, we set $n = 10$ (i.e. $\lambda = 21$ for [14]). Including thickness does not improve orientation, but including length (with or without thickness) improved the orientation.

Figure 7(a) shows the circles in Fig. 6 corrupted by the additive Gaussian noise of mean 0 and standard deviation 0.1, on top right a zoom on the central part of the image). Figure 7(b) and (c) are results filtered by both [14] and our approach. Since Soille and Talbot fix the length parameter, their approach fails to filter areas where this length is not suitable. As the orientation, we fix $n = 10$ (i.e. $\lambda = 21$ for [14]).

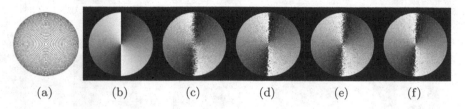

Fig. 6. Results for our method and [14]. (a) Input image of concentric circles. (b) Ground truth. (c) [14] method. (d) Thick variations. (e) Length variations. (f) Both variations.

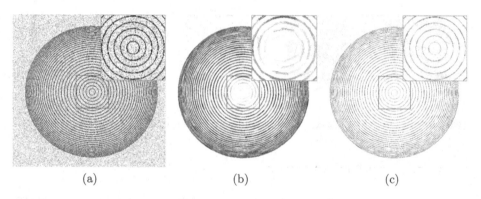

Fig. 7. (a) Noisy image, (b) filtering with [14] of image (a), (c) filtering with our method (length variations) of image (a). On the top right a zoom on the central part of the image.

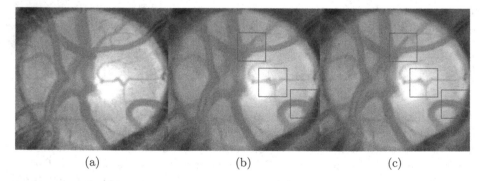

Fig. 8. (a) Original image, (b) filtering with [14], (c) filtering with our method (length variations).

Figure 8 shows an other example. From left to right, we can see the original image, the resulting images filtered by [14] and our approach We can see that our method preserves better the circular structures and is smoother at intersections.

6 Conclusion

In this paper, we firstly presented a method able to estimate geometric features on binary images such as the local elongation, local orientation, directional thickness and orthogonal distance. These features are based on digital straight segments (DSS) according to Farey sequences. By taking the advantage of widths and lengths of DSS, we secondly introduced an adaptation of the previous method to grayscale images. Comparisons to the approach proposed in [14] shows its interest. In particular, using the orientation estimation as a directional filtering, our method allows to reduce noise along line segments. It should be noted that such filter is well adapted for images with structures of different lengths such as tree ring and fingerprint images.

Applications of the extracted features for image analysis framework are currently under study. As short-term perspective, we would like to use the method for wood-log-section image analysis. In this context, several problems could be addressed such as pith detection, tree-ring delineation, ...

In other perspectives, we plan to investigate improvements for selecting optimal orientation among line segments. We are also looking to retrieve pertinent thickness information in grayscale images.

Acknowledgment. This research was made possible by support from the French National Research Agency, in the framework of the project TreeTrace, ANR-17-CE10-0016.

References

1. Berenstein, C.A., Lavine, D.: On the number of digital straight line segments. IEEE Trans. Pattern Anal. Mach. Intell. **10**(6), 880–887 (1988)
2. Dorksen-Reiter, H., Debled-Rennesson, I.: Convex and concave parts of digital curves. In: Klette, R., Kozera, R., Noakes, L., Weickert, J. (eds.) Geometric Properties for Incomplete data. Springer, Dordrecht (2006). https://doi.org/10.1007/1-4020-3858-8_8
3. Hardy, E.W.G.: An Introduction to the Theory of Numbers (1989)
4. Klette, R., Rosenfeld, A.: Digital Geometry - Geometric Methods for Digital Picture Analysis. Morgan Kaufmann, Burlington (2004)
5. Koplowitz, J., Lindenbaum, M., Bruckstein, A.: The number of digital straight lines on an n*n grid. IEEE Trans. Inf. Theory **36**(1), 192–197 (1990)
6. Lachaud, J.-O.: Digital shape analysis with maximal segments. In: Köthe, U., Montanvert, A., Soille, P. (eds.) WADGMM 2010. LNCS, vol. 7346, pp. 14–27. Springer, Heidelberg (2012). https://doi.org/10.1007/978-3-642-32313-3_2
7. Lachaud, J.-O., Vialard, A., de Vieilleville, F.: Fast, accurate and convergent tangent estimation on digital contours. Image Vis. Comput. **25**(10), 1572–1587 (2007)

8. Merveille, O., Naegel, B., Talbot, H., Najman, L., Passat, N.: 2D filtering of curvilinear structures by ranking the orientation responses of path operators (RORPO). Image Process. On Line **7**, 246–261 (2017)
9. Merveille, O., Talbot, H., Najman, L., Passat, N.: Curvilinear structure analysis by ranking the orientation responses of path operators. IEEE Trans. Pattern Anal. Mach. Intell. **40**(2), 304–317 (2018)
10. Nasser, H., Ngo, P., Debled-Rennesson, I.: Dominant point detection based on discrete curve structure and applications. J. Comput. Syst. Sci. **95**, 177–192 (2018)
11. Provençal, X., Lachaud, J.-O.: Two linear-time algorithms for computing the minimum length polygon of a digital contour. In: Brlek, S., Reutenauer, C., Provençal, X. (eds.) DGCI 2009. LNCS, vol. 5810, pp. 104–117. Springer, Heidelberg (2009). https://doi.org/10.1007/978-3-642-04397-0_10
12. Reveillès, J.-P.: Géométrie discrete, calcul en nombres entiers et algorithmique (1991)
13. Soille, P., Breen, E.J., Jones, R.: Recursive implementation of erosions and dilations along discrete lines at arbitrary angles. IEEE Trans. Pattern Anal. Mach. Intell. **18**(5), 562–567 (1996)
14. Soille, P., Talbot, H.: Directional morphological filtering. IEEE Trans. Pattern Anal. Mach. Intell. **23**(11), 1313–29 (2001)
15. Stern, M.: Uber eine verallgemeinerung der kreistheilung. J. fur die reine und angewandte Mathematik **55**, 193–220 (1858)
16. Turroni, F., Maltoni, D., Cappelli, R., Maio, D.: Improving fingerprint orientation extraction. IEEE Trans. Inf. Forensics Secur. **6**(3), 1002–1013 (2011)
17. van Herk, M.: A fast algorithm for local minimum and maximum filters on rectangular and octagonal kernels. Pattern Recognit. Lett. **13**(7), 517–521 (1992)

An Alternative Definition for Digital Convexity

Jacques-Olivier Lachaud$^{(\boxtimes)}$ iD

LAMA, Université Savoie Mont Blanc, Le Bourget-du-lac, France
jacques-olivier.lachaud@univ-smb.fr

Abstract. This paper proposes *full convexity* as an alternative definition of digital convexity, which is valid in arbitrary dimension. It solves many problems related to its usual definitions, like possible non connectedness or non simple connectedness, while encompassing its desirable features. Fully convex sets are digitally convex, but are connected and simply connected. They have a morphological characterisation, which induces a simple convexity test algorithm. Arithmetic planes are fully convex too. We obtain a natural definition of tangent subsets to a digital surface, which gives rise to the tangential cover in 2D, and to its extensions in arbitrary dimension. Finally it leads to a simple algorithm for building a polygonal mesh from a set of digital points.

Keywords: Digital geometry · Digital convexity · Simple connectedness · Arithmetic planes · Tangential cover · Digital surface reconstruction

1 Introduction

A subset $X \subset \mathbb{Z}^d$ is generally said to be *digitally convex* whenever

$$X = \text{cvxh}(X) \cap \mathbb{Z}^d, \tag{1}$$

where $\text{cvxh}(\cdot)$ denotes the convex hull (so called H-convexity [Eck01]). In contrast with continuous convexity, this definition does not imply digital connectedness of X starting from dimension $d \geq 2$ (see Fig. 1abcd). Therefore, especially in 2D, many works add a connectedness constraint or propose a definition that implies it (e.g. [KR82b] or see overviews of [Ron89, Eck01]). As already foreseen in [KR82a], 2D definitions do not extend well to 3D. Their own 3D digital convexity definition relies on the triangle chordal property plus connectedness, and induces a quite burdensome convexity check algorithm. But for $d \geq 3$ a connectedness constraint is not enough to build meaningful digital convex sets. For instance, when cut by a slice, they may lose connectedness (see Fig. 1e). Other convexity definitions rely on progressive intersections with half-planes [Soi04].

This work has been partly funded by CoMeDiC ANR-15-CE40-0006 research grant.

J. Lindblad et al. (Eds.): DGMM 2021, LNCS 12708, pp. 269–282, 2021.
https://doi.org/10.1007/978-3-030-76657-3_19

Connectedness is preserved in the first steps at the price of a coarse approximation of convexity, and at the limit this definition is equivalent to H-convexity.

We present here a more consistent definition of digital convexity, which naturally entails connectedness as well as simple connectedness, and that is valid in arbitrary dimension. This new definition, called *full convexity*, encompasses digital arithmetic planes or digitizations of thick enough convex shapes. We give a morphological characterisation of full convexity, which shares—but not originates from—the thickening idea present in [CdFG20] for connecting 2D digital convex sets. This induces a practical full convexity check algorithm. Finally full convexity nicely addresses classical digital geometry problems, like a tangential cover in arbitrary dimension or piecewise affine reconstruction of digital shapes.

2   Full Convexity

Let $\mathbb{Z}^d$ be the d-dimensional digital space, $d > 0$. Let $\mathscr{C}^d$ be the *(cubical) cell complex* induced by the lattice $\mathbb{Z}^d$: its 0-cells are the points of $\mathbb{Z}^d$, its 1-cells are the open unit segments joining two 0-cells at distance 1, its 2-cells are the open unit squares, etc., and its d-cells are the d-dimensional open unit hypercubes with vertices in $\mathbb{Z}^d$. We denote $\mathscr{C}_k^d$ the set of its k-cells. In the following, a *cell* will always designate an element of $\mathscr{C}^d$, and the term *subcomplex* always designates a subset of $\mathscr{C}^d$. A cell σ is a *face* of another cell τ whenever σ is a subset of the topological closure $\bar{\tau}$ of τ, and we write $\sigma \preccurlyeq \tau$. Given any subcomplex K of $\mathscr{C}^d$, the *closure* Cl (K) of K is the complex $\{\tau \in \mathscr{C}^d, \text{ s.t. } \exists \sigma \in K, \tau \preccurlyeq \sigma\}$ and the *star* Star (K) of K is $\{\tau \in \mathscr{C}^d, \text{ s.t. } \exists \sigma \in K, \sigma \preccurlyeq \tau\}$.

In combinatorial topology, a subcomplex K with Star $(K) = K$ is *open*, while being *closed* when Cl $(K) = K$. The *body* of a subcomplex K, i.e. the union of its cells in $\mathbb{R}^d$, is written $\|K\|$. Finally, if Y is any subset of the Euclidean space $\mathbb{R}^d$, we denote by $\mathscr{C}_k^d[Y]$ the set of k-cells whose topological closure has a non-empty intersection with Y, i.e. $\mathscr{C}_k^d[Y] := \{c \in \mathscr{C}_k^d, \bar{c} \cap Y \neq \emptyset\}$. The complex made of all k-cells having a non-empty intersection with Y, $0 \leqslant k \leqslant d$ is called the *intersection (cubical) complex of Y* and denoted by $\mathscr{C}^d[Y]$.

Lemma 1. *The intersection complex of a set Y is open and its body covers Y.*

Proof. If Y is the empty set, $\bar{\mathscr{C}}^d[Y]$ is empty and is open. If Y is not empty, let σ be any cell of $\bar{\mathscr{C}}^d[Y]$. Let τ be any cell of $\mathscr{C}^d$ with $\sigma \preccurlyeq \tau$. Thus $\sigma \subset \bar{\tau} \Rightarrow \bar{\sigma} \subset \bar{\tau}$ (since topological closure is increasing and idempotent). It follows that $\sigma \in \bar{\mathscr{C}}^d[Y] \Leftrightarrow \bar{\sigma} \cap Y \neq \emptyset \Rightarrow \bar{\tau} \cap Y \neq \emptyset \Leftrightarrow \tau \in \bar{\mathscr{C}}^d[Y]$. We have just proved that Star $(\sigma) \subset \bar{\mathscr{C}}^d[Y]$, hence Star $(\bar{\mathscr{C}}^d[Y]) \subset \bar{\mathscr{C}}^d[Y]$. The converse inclusion being obvious, $\bar{\mathscr{C}}^d[Y]$ is open. The fact that $Y \subset \|\bar{\mathscr{C}}^d[Y]\|$ is straightforward.   □

Definition 1 (Full convexity). A non empty subset $X \subset \mathbb{Z}^d$ is *digitally k-convex* for $0 \leqslant k \leqslant d$ whenever

$$\mathscr{C}_k^d[X] = \mathscr{C}_k^d[\text{cvxh}(X)]. \tag{2}$$

Subset X is *fully (digitally) convex* if it is digitally k-convex for all $k, 0 \leqslant k \leqslant d$.

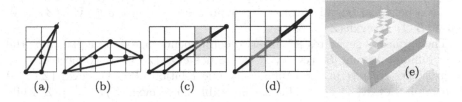

Fig. 1. (abcd) Digital triangles that are not fully convex: digital points are depicted as black disks, missing 1-cells for digital 1-convexity as blue lines, missing 2-cells for digital 2-convexity as green squares. (e) Usual digital convexity (1) plus 3D connectivity does not imply connectedness on the upper slice; it is also not fully convex. (Color figure online)

Equivalently, the intersection complex of a fully convex set Z covers the convex hull of Z. We can already make the following observation:

Lemma 2. *Common digital convexity is the digital 0-convexity.*

Proof. Remark (1): for any $Y \subset \mathbb{R}^d$, we have $\bar{\mathscr{C}}_0^d[Y] = \{c \in \mathscr{C}_0^d, c \cap Y \neq \emptyset\} = \{c \in \mathbb{Z}^d, c \cap Y \neq \emptyset\} = Y \cap \mathbb{Z}^d$. Now, from (1), $X \subset \mathbb{Z}^d$ is digitally convex iff $X = \mathrm{cvxh}(X) \cap \mathbb{Z}^d$, otherwise said $X \cap \mathbb{Z}^d = \mathrm{cvxh}(X) \cap \mathbb{Z}^d$. With remark (1), it is equivalent to $\bar{\mathscr{C}}_0^d[X] = \bar{\mathscr{C}}_0^d[\mathrm{cvxh}(X)]$, which is exactly (2) for $k = 0$. □

Figure 1 shows several digitally 0-convex sets, but which are not full convex. Clearly full convexity forbids too thin convex sets, which are typically the ones that are not connected or simply connected in the digital sense. Denoting by $\#(X)$ the cardinal of a finite set X, the straightforward lemma below shows that is suffices to count intersected cells to check for full convexity.

Lemma 3. *A finite non-empty subset $X \subset \mathbb{Z}^d$ is digitally k-convex for $0 \leqslant k \leqslant d$ iff $\#(\bar{\mathscr{C}}_k^d[X]) \geqslant \#(\bar{\mathscr{C}}_k^d[\mathrm{cvxh}(X)])$.*

For example, the tetrahedra $T(l) = \{(0,0,0), (1,0,0), (0,1,0), (1,1,l)\}$, for positive integer l, is digitally 0-convex. However $\mathrm{cvxh}(T(l))$ intersects as many 2-cells and 3-cells as wanted above the unit square with vertices $(0,0,0)$, $(1,0,0)$, $(1,1,0)$, $(1,0,0)$, just by increasing l. Meanwhile, $T(l)$ only intersects the same finite number of 2-cells and 3-cells. Hence, for $l \geqslant 2$, $T(l)$ is not fully convex.

It is not necessary to check digital d-convexity to verify if a digital set is fully convex, and this property is useful to speed up algorithms to check for full convexity. You can observe the contraposition of this lemma on Fig. 1, left, where non digitally 2-convex sets in 2D are not digitally 1-convex too.

Lemma 4. *If $Z \subset \mathbb{Z}^d$ is digitally k-convex for $0 \leqslant k < d$, it is also digitally d-convex, hence fully convex.*

Proof. Let Z be such set, so Z not empty, and let $\sigma \in \bar{\mathscr{C}}_d^d[\mathrm{cvxh}\,(Z)]$. Let $B = \partial\bar{\sigma}$ be the topological boundary of σ. By hypothesis, we have $\bar{\sigma} \cap \mathrm{cvxh}\,(Z) \neq \emptyset$, hence $(B \cup \sigma) \cap \mathrm{cvxh}\,(Z) \neq \emptyset$.

The surface B separates $\mathbb{R}^d$ into two components, one finite equal to σ, the other infinite. Assume $B \cap \mathrm{cvxh}\,(Z) = \emptyset$. Since $\mathrm{cvxh}\,(Z)$ is arc-connected, then $\mathrm{cvxh}\,(Z)$ lies entirely in one component. The relation $(B \cup \sigma) \cap \mathrm{cvxh}\,(Z) \neq \emptyset$ then implies $\mathrm{cvxh}\,(Z) \subset \sigma$. This is impossible since $\mathrm{cvxh}\,(Z) \cap \mathbb{Z}^d = Z$ while $\sigma \cap \mathbb{Z}^d = \emptyset$.

It follows that $B \cap \mathrm{cvxh}\,(Z) \neq \emptyset$. But B is a union of k-cells $(b_i)_{i=0...m}$ of $\mathscr{C}^d$, with $0 \leqslant k < d$. There exists at least one k-cell b_j with $b_j \cap \mathrm{cvxh}\,(Z) \neq \emptyset$. Thus $b_j \in \bar{\mathscr{C}}_k^d[\mathrm{cvxh}\,(Z)]$. But Z is digitally k-convex for $0 \leqslant k < d$, so $b_j \in \bar{\mathscr{C}}_k^d[Z]$. To conclude, $\sigma \in \mathrm{Star}\,(b_j)$ and $\mathscr{C}^d[Z]$ is open, so σ belongs also to $\mathscr{C}^d[Z]$. We have just shown that every d-cell of $\bar{\mathscr{C}}^d[\mathrm{cvxh}\,(Z)]$ are in $\mathscr{C}^d[Z]$, so Z is digitally d-convex and hence fully convex. $\qquad\square$

Other implications of digital k-convexities over digital l-convexities are unlikely. For instance in 3D, some digital sets are digitally 0-convex, 1-convex, 3-convex but are not 2-convex, like $\{(0,0,0),(1,1,2),(1,1,3),(1,2,3),(2,1,3)\}$.

3   Topological Properties of Fully Convex Digital Sets

We give below the main topological properties of fully convex digital sets.

Theorem 1. *If the digital set $Z \subset \mathbb{Z}^d$ is fully convex, then the body of its intersection cubical complex is connected.*

Proof. Let x, x' be two points of $\|\bar{\mathscr{C}}^d[X]\|$. Since $\bar{\mathscr{C}}^d$ is a partition of $\mathbb{R}^d$, there are two cells c, c' of $\bar{\mathscr{C}}^d[X]$ such that $x \in c$, $x' \in c'$. Since Z is fully convex, then $\bar{\mathscr{C}}^d[X] = \bar{\mathscr{C}}^d[\mathrm{cvxh}\,(X)]$. Hence there exist $y \in \bar{c} \cap \mathrm{cvxh}\,(X)$ and $y' \in \bar{c} \cap \mathrm{cvxh}\,(X)$. By convexity of cells, the segment $[x, y[$ lies entirely in c hence in $\|\bar{\mathscr{C}}^d[X]\|$. Similarly, the segment $]y', x']$ lies entirely in c' hence in $\|\bar{\mathscr{C}}^d[X]\|$.

Now by definition of convexity, the segment $[y, y']$ lies in $\mathrm{cvxh}\,(X)$. But $\mathrm{cvxh}\,(X) \subset \|\bar{\mathscr{C}}^d[\mathrm{cvxh}\,(X)]\| = \|\bar{\mathscr{C}}^d[X]\|$ by cell convexity. We have just built an arc from x to x' which lies entirely in $\|\bar{\mathscr{C}}^d[X]\|$. We conclude since arc-connectedness implies connectedness. $\qquad\square$

Two elements x, y of $\mathbb{Z}^d$ are *d-adjacent* if $\|x-y\|_\infty \leqslant 1$. The transitive closure of this relation defines the *d-connectedness* relation. Historically, it was called 8-connectivity in 2D, and 26-connectivity in 3D.

Theorem 2. *If the digital set $Z \subset \mathbb{Z}^d$ is fully convex, then Z is d-connected.*

Proof. We show first that 0-cells of $\bar{\mathscr{C}}^d[Z]$ are face-connected, i.e. for any points $z, z' \in Z = \bar{\mathscr{C}}_0^d[Z]$, there is a path of cells $(c_i)_{i=0..m}$ of $\bar{\mathscr{C}}^d[Z]$, such that $c_0 = \sigma$, $c_m = \tau$, and for all $i \in \mathbb{Z}, 0 \leqslant i < m$, either $c_i \preccurlyeq c_{i+1}$ or $c_{i+1} \preccurlyeq c_i$.

The straight segment $[z, z']$ is included in $\mathrm{cvxh}\,(Z)$, hence any one of its point belongs to a cell of $\bar{\mathscr{C}}^d[\mathrm{cvxh}\,(Z)]$ so a cell of $\bar{\mathscr{C}}^d[Z]$ by full convexity.

Let $p(t) = (1-t)z + tz'$ for $0 \leqslant t \leqslant 1$ be a parameterization of segment $[z, z']$. The above remark implies that, for any $t \in [0, 1]$, the point $p(t)$ belongs to a cell $c(t)$ of $\bar{\mathscr{C}}^d[Z]$. The sequence of intersected cells from $t = 0$ to $t = 1$ is obviously finite, and we denote it by $(c_0, c_1, \ldots, c_m)$ with $c_0 = p(0) = z$ and $c_m = p(1) = z'$. Since it corresponds to an infinitesimal change of t, two consecutive cells of this sequence are necessary in the closure of one of them, hence $c_i \preccurlyeq c_{i+1}$ or $c_{i+1} \preccurlyeq c_i$.

We use Lemma 5 given just after. We associate to each cell c_i one of its Z-corner, denoted z_i. We obtain a sequence of digital points $z = z_0, z_1, \ldots, z_m = z'$. Now any two incident faces (like c_i and c_{i+1}) belong to the closure of a d-cell σ. It follows that both corner z_i and z_{i+1} are vertices of $\bar{\sigma}$, a unit hypercube. Obviously $\|z_i - z_{i+1}\|_\infty \leqslant 1$ and these two points are d-adjacent. We have just built a sequence of d-adjacent points in Z, which concludes.   □

Lemma 5. *Let $Z \subset \mathbb{Z}^d$. If σ is a cell of $\bar{\mathscr{C}}^d[Z]$, there exists $z \in \bar{\mathscr{C}}_0^d[Z] = Z$ such that $z \preccurlyeq \sigma$. We call such digital point a Z-corner for σ. If σ is a 0-cell, its only Z-corner is itself.*

Proof. By definition of $\bar{\mathscr{C}}^d[Z]$ we have $\bar{\sigma} \cap Z \neq \emptyset$. It follows that $\exists z \in Z$ such that $z \in \bar{\sigma}$. So $z \preccurlyeq \sigma$ and also $z \in Z = \bar{\mathscr{C}}_0^d[Z]$.   □

We can show an even stronger result on fully convex sets: they present no topological holes. Indeed, we have:

Theorem 3. *If the digital set $Z \subset \mathbb{Z}^d$ is fully convex, then the body of its intersection cubical complex is simply connected.*

Proof. Let $\mathscr{A} := \{x(t), t \in [0, 1]\}$ be a closed curve in $\|\bar{\mathscr{C}}^d[X]\|$, i.e. $x(0) = x(1)$ and $x(t) \in \|\bar{\mathscr{C}}^d[X]\|$. We must show that there is a homotopy from $\mathscr{A}$ to a point $a \in \|\bar{\mathscr{C}}^d[X]\|$.

The curve $x(t)$ visits cells of $\bar{\mathscr{C}}^d[X]$. Let $c(t)$ be these cells. By finiteness of $\mathscr{A}$, $c(t)$ defines a finite sequence of cells $c_0, c_1, \ldots, c_m$ from $t = 0$ to $t = 1$, with $c_m = c_0$. We can also associate a sequence of parameters $t_0, t_1, \ldots, t_m$, such that $x(t_i) \in c_i = c(t_i)$. As in the proof of Theorem 2, two consecutive cells of this sequence are necessary in the closure of one of them. Let us set d_i to c_i or c_{i+1} such that both are in $\bar{d}_i$.

The path $x([t_i, t_{i+1}])$ lies in $c_i \cup c_{i+1}$. For each cell c_i we pick one of its Z-corner z_i. Clearly z_i and z_{i+1} belong to $\bar{d}_i$. By convexity of $\bar{d}_i$, it is in particular simply-connected and there is a homotopy in $\bar{d}_i$ between $x([t_i, t_{i+1}])$ and the segment $[z_i, z_{i+1}]$. Since $\bar{\mathscr{C}}^d[Z]$ is open and both points are in $\|\bar{\mathscr{C}}^d[Z]\|$, $[z_i, z_{i+1}] \subset \|\bar{\mathscr{C}}^d[Z]\|$ as well as the whole homotopy. Gathering all these local homotopies for every i, $0 \leqslant i < m$, we have defined a homotopy between $\mathscr{A}$ and the polyline $[z_i]_{i=0\ldots m}$.

By full convexity, every $z_i \in Z$ is also in cvxh(Z). It follows that the vertices of the polyline $[z_i]_{i=0\ldots m}$ belong to cvxh(Z). By convexity of cvxh(Z), the whole polyline is a subset of cvxh(Z). Being a closed curve in a simply connected set, the polyline $[z_i]_{i=0\ldots m}$ is continuously deformable to a point of this set, say z_0, by some homotopy. Composing the two homotopies finishes the argument.   □

Finally we can determine a relation between the numbers of k-cells of the intersection complex of a fully convex set. For K a subcomplex, let $\#_k(K)$ be its number of k-cells. The *Euler characteristic* of a subcomplex K is $\chi(K) := \sum_{k=0}^{d}(-1)^k\#_k(K)$. Its proof is omitted for space reasons.

Theorem 4. *The Euler characteristic of the intersection cubical complex of a fully convex set is $(-1)^d$.*

4   Morphological Properties and Recognition Algorithm

We provide first a morphological characterization of full convexity that will help us to design a practical algorithm for checking this property.

Morphological Characterisation. Let $I^d := \{1, \ldots, d\}$. The set of subsets of cardinal k of I^d is denoted by I_k^d, for $0 < k \leqslant d$. For $i \in I^d$, let $\mathcal{S}_i := \{t\mathbf{e}_i, t \in [0,1]\}$ be the unit segments aligned with axis vectors $\mathbf{e}_i$. For any point x of $\mathbb{R}^d$, we write its d coordinates with superscripts: $x^1, \ldots, x^d$. Let us also denote the Minkowski sum of two sets A and B by $A \oplus B$. We further build axis-aligned unit squares, cubes, etc., by suming up the unit segments: for any $\alpha \in I_k^d$, $\mathcal{S}_\alpha := \bigoplus_{i \in \alpha} \mathcal{S}_i$. For instance, in 3D, the three unit segments are $\mathcal{S}_1, \mathcal{S}_2, \mathcal{S}_3$ (or equiv. $\mathcal{S}_{\{1\}}, \mathcal{S}_{\{2\}}, \mathcal{S}_{\{3\}}$), the three unit squares are $\mathcal{S}_{\{1,2\}}, \mathcal{S}_{\{1,3\}}, \mathcal{S}_{\{2,3\}}$, the unit cube is $\mathcal{S}_{\{1,2,3\}}$. To treat the 0-dimensional case uniformly, we set $I_0^d = \{0\}$ and $\mathcal{S}_{\{0\}} = \{\vec{0}\}$.

We can partition the k-cells of $\mathscr{C}_k^d$ into $\#(I_k^d)$ subsets such that, for any $\alpha \in I_k^d$, each subset denoted by $\mathscr{C}_\alpha^d$ contains all the k-cells parallel to $\mathcal{S}_\alpha$. For instance, $\mathscr{C}_{\{1\}}^d$ and $\mathscr{C}_{\{2\}}^d$ partition the set $\mathscr{C}_1^d$ in dimension $d = 2$. Now let us define the mapping $\mathcal{Z} : \mathscr{C}^d \to \mathbb{Z}^d$ which associates to any cell σ, the digital vertex of $\bar{\sigma}$ with highest coordinates. Its restriction to $\mathscr{C}_\alpha^d$ is denoted by $\mathcal{Z}_\alpha$.

Lemma 6. *For any $\alpha \in I_k^d$, the mapping $\mathcal{Z}_\alpha$ is a bijection.*

Proof. Clearly every digital point of $\mathbb{Z}^d$ forms the highest vertex of all possible kind of cells, so $\mathcal{Z}_\alpha$ is a surjection. Now no two cells of $\mathscr{C}_\alpha^d$ can have the same highest vertex, since all cells of $\mathscr{C}_\alpha^d$ are distinct translations of the same set. $\square$

The intersection subcomplex of some set Y restricted to cells of $\mathscr{C}_\alpha^d$ is naturally denoted by $\bar{\mathscr{C}}_\alpha^d[Y]$. We relate k-cells intersected by set Y to digital points included in the set Y dilated in some directions, as illustrated in Fig. 2.

Lemma 7. *For any $Y \subset \mathbb{R}^d$, for any $\alpha \in I_k^d$, $\mathcal{Z}_\alpha(\bar{\mathscr{C}}_\alpha^d[Y]) = \bar{\mathscr{C}}_0^d[Y \oplus \mathcal{S}_\alpha]$.*

Proof. We proceed by equivalences (the logical "and", symbol $\wedge$, has more priority than "if and only if", symbol $\Leftrightarrow$, but less than any other operations):

$$z \in \mathcal{Z}_\alpha(\bar{\mathscr{C}}_\alpha^d[Y]) \Leftrightarrow \sigma \in \bar{\mathscr{C}}_\alpha^d[Y] \wedge \sigma = \mathcal{Z}_\alpha^{-1}(z) \qquad (\mathcal{Z}_\alpha \text{ is a bijection, Lemma 6})$$
$$\Leftrightarrow \exists y \in Y, y \in \bar{\sigma} \wedge \sigma = \mathcal{Z}_\alpha^{-1}(z)$$
$$\Leftrightarrow \exists y \in Y, (\forall i \in \alpha, z^i - 1 \leqslant y^i \leqslant z^i \wedge \forall j \in I^d \setminus \alpha, z^j = y^j)$$

$$\Leftrightarrow \exists y \in Y, (\forall i \in \alpha, 0 \leqslant x^i \leqslant 1 \wedge \forall j \in I^d \setminus \alpha, x^j = 0) \wedge x = z - y$$

$$\Leftrightarrow \exists y \in Y, x \in \mathcal{S}_\alpha \wedge z = x + y \in \mathbb{Z}^d$$

$$\Leftrightarrow z \in (Y \oplus \mathcal{S}_\alpha) \cap \mathbb{Z}^d.$$

We conclude since $(Y \oplus \mathcal{S}_\alpha) \cap \mathbb{Z}^d = \bar{\mathscr{C}}_0^d[Y \oplus \mathcal{S}_\alpha].$ $\qquad\qquad\square$

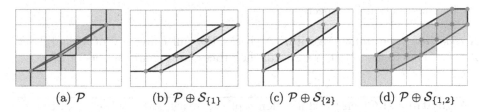

(a) $\mathcal{P}$ (b) $\mathcal{P} \oplus \mathcal{S}_{\{1\}}$ (c) $\mathcal{P} \oplus \mathcal{S}_{\{2\}}$ (d) $\mathcal{P} \oplus \mathcal{S}_{\{1,2\}}$

Fig. 2. Let $\mathcal{P} = \mathrm{cvxh}(\{(0,0),(2,1),(5,3)\})$. (a) $\bar{\mathscr{C}}^d[\mathcal{P}]$. (bcd) We can see that $\#(\bar{\mathscr{C}}_{\{1\}}^d[\mathcal{P}]) = \#(\bar{\mathscr{C}}_0^d[\mathcal{P} \oplus \mathcal{S}_{\{1\}}]) = 7$, $\#(\bar{\mathscr{C}}_{\{2\}}^d[\mathcal{P}]) = \#(\bar{\mathscr{C}}_0^d[\mathcal{P} \oplus \mathcal{S}_{\{2\}}]) = 9$, and $\#(\bar{\mathscr{C}}_{\{1,2\}}^d[\mathcal{P}]) = \#(\bar{\mathscr{C}}_0^d[\mathcal{P} \oplus \mathcal{S}_{\{1,2\}}]) = 14$. The bijections $\mathcal{Z}_{\{1\}}, \mathcal{Z}_{\{2\}}, \mathcal{Z}_{\{1,2\}}$ are made clear in (b), (c), (d) respectively.

We arrive to our morphological characterization of full convexity: full convexity can thus be checked with common algorithms for checking digital convexity. We denote by $\mathbf{x}(Z)$ the set Z translated by some lattice vector $\mathbf{x}$. Let $U_\emptyset(Z) := Z$, and, for $\alpha \subset I^d$ and $i \in \alpha$, we define recursively $U_\alpha(Z) := U_{\alpha \setminus i}(Z) \cup \mathbf{e}_i(U_{\alpha \setminus i}(Z))$. The previous definition is consistent since it does not depend on the order of the sequence $i \in \alpha$.

Theorem 5. *A non empty subset $X \subset \mathbb{Z}^d$ is digitally k-convex for $0 \leqslant k \leqslant d$ iff*

$$\forall \alpha \in I_k^d, (X \oplus \mathcal{S}_\alpha) \cap \mathbb{Z}^d = (\mathrm{cvxh}(X) \oplus \mathcal{S}_\alpha) \cap \mathbb{Z}^d, \tag{3}$$

$$or \ (X \oplus \mathcal{S}_\alpha) \cap \mathbb{Z}^d = (\mathrm{cvxh}(X \oplus \mathcal{S}_\alpha)) \cap \mathbb{Z}^d, \tag{4}$$

$$or \ U_\alpha(X) = \mathrm{cvxh}(U_\alpha(X)) \cap \mathbb{Z}^d. \tag{5}$$

It is thus fully convex if the previous relations holds for all $k, 0 \leqslant k \leqslant d$.

Proof. We proceed by equivalence for (3):

$$\bar{\mathscr{C}}_k^d[X] = \bar{\mathscr{C}}_k^d[\mathrm{cvxh}(X)]$$

$$\Leftrightarrow \forall \alpha \in I_k^d, \bar{\mathscr{C}}_\alpha^d[X] = \bar{\mathscr{C}}_\alpha^d[\mathrm{cvxh}(X)]$$

$$\Leftrightarrow \forall \alpha \in I_k^d, \mathcal{Z}_\alpha(\bar{\mathscr{C}}_\alpha^d[X]) = \mathcal{Z}_\alpha(\bar{\mathscr{C}}_\alpha^d[\text{cvxh}(X)]) \qquad (\mathcal{Z}_\alpha \text{ is a bijection})$$

$$\Leftrightarrow \forall \alpha \in I_k^d, \bar{\mathscr{C}}_0^d[X \oplus \mathcal{S}_\alpha] = \bar{\mathscr{C}}_0^d[\text{cvxh}(X) \oplus \mathcal{S}_\alpha] \qquad (\text{Lemma 7})$$

$$\Leftrightarrow \forall \alpha \in I_k^d, (X \oplus \mathcal{S}_\alpha) \cap \mathbb{Z}^d = (\text{cvxh}(X) \oplus \mathcal{S}_\alpha) \cap \mathbb{Z}^d.$$

(4) follows since convex hull operation commutes with Minkowski sum. Using Lemma 9 on left handside and Lemma 10 on right handside implies (5). □

Finally we can remark that, if $X \subset \mathbb{Z}^d$ is d-connected, then necessarily all $U_\alpha(X)$ are by construction d-connected.

Recognition Algorithm. Algorithm 1 checks the full convexity of a digital set $Z \subset \mathbb{Z}^d$. Due to the bijections $\mathcal{Z}_\alpha$, all the processed sets are subsets of $\mathbb{Z}^d$.

Algorithm 1: Given the dimension d of the space and a subset Z of the digital space $\mathbb{Z}^d$, IsFullyConvex returns true iff Z is fully convex.

    Function IsConvex(**In** S: subset of $\mathbb{Z}^d$) : boolean;
    begin
1 | Polytope $\mathcal{P} \leftarrow$ ConvexHull(S);
2 | **return** Cardinal(S) = CountLatticePoint($\mathcal{P}$);

    Function IsFullyConvex(**In** d : integer, **In** Z: subset of $\mathbb{Z}^d$) : boolean;
    Var C : array[0 ... d-1] of lists of subsets of I^d;
    Var X : map associating subsets of $I^d \to$ sets of digital points ;
    begin
3 | **if** Cardinal(Z) = 0 *or* ¬IsDConnected(Z, d) **then return** *false*;
      | $C[0] \leftarrow (\{0\})$; $X[0] \leftarrow Z$;
4 | **if** ¬IsConvex(Z) **then return** *false*;
5 | **for** $k \leftarrow 1$ **to** $d - 1$ **do**
        | $C[k] \leftarrow \emptyset$;
        | **foreach** $\beta \in C[k - 1]$ **do**
            | **for** $j \leftarrow 1$ **to** d **do**
                | $\alpha \leftarrow$ Append(β, j) ;
6 | **if** IsStrictlyIncreasing(α) **then**
                    | $C[k] \leftarrow$ Append($C[k], \alpha$);
7 | $X[\alpha] \leftarrow$ Union($X[\beta], \mathbf{e}_j(X[\beta])$) ;
8 | **if** ¬IsConvex($X[\alpha]$) **then return** *false*;

    └ **return** *true*

Theorem 6. *Algorithm 1 correctly checks if a digital set Z is fully convex.*

Proof. First of all, IsConvex checks the classical digital convexity of any digital set S by counting lattice points within cvxh(S) (Lemma 3).

    Looking now at IsFullyConvex, line 3 checks the d-connectedness of Z and outputs false if Z is not connected (valid since Lemma 2).

Line 4 checks for digital 0-convexity (i.e. usual digital convexity). The loop starting at line 5 builds, for each dimension k from 1 to $d-1$ the possible $\alpha \in I_k^d$, and stores them in $C[k]$. Line 6 guarantees that each possible subsets of I_k^d is used exactly once.

Line 7 builds $X[\alpha] = U_\alpha(Z)$. Indeed, by induction assume $X[\beta] = U_\beta(Z)$. Then $X[\alpha] = X[\beta] \cup \mathbf{e}_j(X[\beta])$ which is exactly the definition of $U_\alpha(Z)$.

Finally line 8 verifies $U_\alpha(Z) = \mathrm{cvxh}(U_\alpha(Z)) \cap \mathbb{Z}^d$. Since it does this check for every $\alpha \in I_k^d$, it checks digital k-convexity according to Theorem 5, (5). Now Lemma 4 tells that it is not necessary to check digital d-convexity if all the other digital k-convexities are satisfied. This establishes the correctness. $\qquad \square$

First, letting $n = \#(Z)$, function IsDCONNECTED takes $O(n)$ operations by depth first algorithm and bounded number of adjacent neighbors. There are less than $d2^d$ calls to IsSTRICTLYINCREASING, which takes $O(d)$ time complexity. However the total number of calls to IsCONVEX is exactly $2^d - 1$, and its time complexity dominates (by far) the previous $d^2 2^d$ in practical uses. The overall complexity $T(n)$ of this algorithm is thus governed by (1) the complexity $T_1(n)$ for computing the convex hull and (2) the complexity $T_2(n)$ for counting the lattice points within. For $T_1(n)$, it is: $O(n)$ for $d = 2$ (since we know S is d-connected, it can be done in linear time [HKV81] or [BLPR09] for a fast practical algorithm), $O(n \log n)$ in 3D [Cha96] and $n^{\lfloor d/2 \rfloor}/\lfloor d/2 \rfloor!$ in dD [Cha93, BDH96]. For $T_2(n)$, it is $O(n)$ in 2D using Pick's formula, and in general dimension it is related to Ehrhart's theory [Ehr62], where best algorithms run in $O(n^{O(d)})$ [Bar94].

In practice, when the number of facets of the convex hulls is low, counting lattice points within the hull is done efficiently by visiting all lattice points within the bounding box, which is not too big since all digital sets are connected.

5 Digital Planarity, Tangency and Linear Reconstruction

We sketch here a few nice geometric applications of full convexity.

Thick Enough Arithmetic Planes are Fully Convex. An arithmetic plane of intercept $\mu \in \mathbb{Z}$, positive thickness $\omega \in \mathbb{Z}, \omega > 0$, and irreducible normal vector $N \in \mathbb{Z}^d$ is defined as the digital set $P(\mu, N, \omega) := \{x \in \mathbb{Z}^d, \mu \leqslant x \cdot N < \mu + \omega\}$.

Theorem 7. *Arithmetic planes are digitally 0-convex for arbitrary thickness, and fully convex for thickness $\omega \geqslant \|N\|_\infty$.*

Proof. Let $Q = P(\mu, N, \omega)$ be some arithmetic plane. Let $Y^- := \{x \in \mathbb{R}^d, \mu \leqslant x \cdot N\}$, $Y^+ := \{x \in \mathbb{R}^d, x \cdot N < \mu + \omega\}$, and $Y = Y^- \cap Y^+$. We have $Q = Y \cap \mathbb{Z}^d$, with Y convex, so Q is digitally 0-convex (non emptyness comes from $\omega > 0$).

Now let c be any k-cell of $\bar{\mathscr{C}}^d[\mathrm{cvxh}(Q)]$, $1 \leqslant k \leqslant d$. Let us show that at least one vertex of c is in $\bar{\mathscr{C}}^d[Q]$. There exists $x \in c$ with $x \in \mathrm{cvxh}(Q) = \mathrm{cvxh}(Y \cap \mathbb{Z}^d) \subset \mathrm{cvxh}(Y) = Y$. It follows that $\mu \leqslant x \cdot N < \mu + \omega$.

Let $(z_i)_{i=1...2^k}$ be the vertices of $\bar{c}$ ordered from lowest to highest scalar product with N. There exists $j \in \{1, 2^k\}$ such that $\forall i \in \{1, j\}, z_i \cdot N \leqslant x \cdot N$ and $\forall i \in \{j+1, 2^k\}, x \cdot N < z_i \cdot N$. Should no z_i belong to Y, since x belongs to Y, we have :

$$\forall i \in \{1, j\}, z_i \cdot N < \mu, \quad \text{and} \quad \forall i \in \{j+1, 2^k\}, \mu + \omega \leqslant z_i.$$

It follows that $(z_{j+1} - z_j) \cdot N > \omega$. Now the (z_i) are vertices of a hypercube of side one, and ordered according to their projection along vector N. It is easy to see that any $(z_{i+1} - z_i) \cdot N \leqslant \max_{i=1...d}(|N_i|)$ (for instance by constructing a subsequence moving along axes in order, it achieves the bound, and then the actual sequence (z_i) is much finer than this one), so it holds in particular for $i = j$. To sum up, should no z_i belongs to Y, then $\omega < \max_{i=1...d}(|N_i|)$.

Otherwise, for $\omega \geqslant \max_{i=1...d}(|N_i|)$, either z_j or z_{j+1} or both belong to Y, and thus to Q. It follows that $c \in \mathscr{C}_k^d[Q]$, which concludes. $\qquad \square$

The classic 2D and 3D machinery of digital straight lines and planes, very rich in results and applications, thus belongs to the fully convex framework.

Tangent Subsets. Tangency as defined below induces strong geometric properties. In $\mathbb{Z}^d$, two cotangent points of a digital surface X delineates a straight segment that stays "in" the surface, i.e. a tangent vector. A simplex made of d cotangent points to X lies "in" the surface, so defines a (local) tangent plane.

Definition 2. The digital set $A \subset X \subset \mathbb{Z}^d$ is said to be *k-tangent to* X for $0 \leqslant k \leqslant d$ whenever $\bar{\mathscr{C}}_k^d[\text{cvxh}(A)] \subset \bar{\mathscr{C}}_k^d[X]$. It is *tangent to* X if the relation holds for all such k. Elements of A are called *cotangent*.

As always in digital geometry, objects that are too local are not precise enough. We are thus more interested in "big" tangent subsets: a set A, tangent to X, that is not included in any other tangent set to X is said *maximal in X*. In 2D, they give rise to the classical tangential cover of a contour [FT99]:

Theorem 8. *When $d = 2$, if C is a simple 2-connected digital contour (i.e. 8-connected in Rosenfeld's terminology), then the fully convex subsets of C that are maximal and tangent are the classical maximal naive digital straight segments.*

Proof. Let M be a fully convex subset of C, both maximal and tangent. Full convexity implies that M is 2-connected (Theorem 2). Full convexity implies H-convexity, and connected convex subsets of simple 2-connected contours are digital straight segments. Maximality implies that they are inextensible. The converse is obvious from Theorem 7. $\qquad \square$

Maximal fully convex tangent subsets to X seem a good candidate for a sound definition of maximal digital plane segments. Our definition avoids the classical problem of 3D planes that are not tangent to the surface (as noted in [CL11]) as well as the many heuristics to cope with this issue [SDC04, PDR09].

An Elementary Linear Reconstruction Algorithm for Digital Sets. Let X be a finite subset of $\mathbb{Z}^d$ and let $\mathrm{Del}\,(X)$ be its Delaunay complex.

Definition 3. *The* tangent Delaunay complex $\mathrm{Del}_T(X)$ *to* X *is the complex made of the cells* τ *of* $\mathrm{Del}\,(X)$ *such that the vertices of* τ *are tangent to* X.

The boundary of tangent Delaunay complexes is a linear reconstruction of voxel shapes, and is the boundary of the convex hull for fully convex shapes:

Lemma 8. *If* X *fully convex,* $\mathrm{Del}_T(X) = \mathrm{Del}\,(X)$. *So* $\partial\,\|\mathrm{Del}_T(X)\| = \partial\mathrm{cvxh}\,(X)$.

Proof. Let $\tau \in \mathrm{Del}\,(X)$. Let $A \subset X$ be the vertices of cell τ (i.e. $\tau = \mathrm{cvxh}\,(A)$). For all k, $0 \leqslant k \leqslant d$, $\bar{\mathscr{C}}_k^d[\mathrm{cvxh}\,(A)] \subset \bar{\mathscr{C}}_k^d[\mathrm{cvxh}\,(X)] = \bar{\mathscr{C}}_k^d[X]$ (by full convexity). This shows that A is tangent to X. $\qquad\square$

From Theorem 7, the tangent Delaunay complex of an arithmetic plane is the boundary of its convex hull, hence its facets have exactly the same normal as the arithmetic plane. Furthermore, since the tangent Delaunay complex is built with local geometric considerations, it is able to capture the geometry of local pieces of planes on digital objects, and nicely reconstructs convex and concave parts (see Fig. 3). It is also a tight and reversible reconstruction of X:

Theorem 9. *The body of* $\mathrm{Del}_T(X)$ *is at Hausdorff* L_∞-*distance 1 to* X. $\mathrm{Del}_T(X)$ *is a reversible polyhedrization, i.e.* $\|\mathrm{Del}_T(X)\| \cap \mathbb{Z}^d = X$.

Proof. First the distance of any point of X to $\|\mathrm{Del}_T(X)\|$ is zero, since any point of X is tangent to X and is also a 0-cell of $\mathrm{Del}\,(X)$. Second, any point

  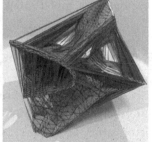

Input digital shape X Our reconstruction $\mathrm{Del}_T(X)$ Bad simplices of $\mathrm{Del}\,(X)$

Fig. 3. The tangent Delaunay complex $\mathrm{Del}_T(X)$ (middle) is a piecewise linear reconstruction of the input digital surface X (left). On (right), we display in red simplices of $\mathrm{Del}\,(X)$ which avoid lattice points of $\mathbb{Z}^3 \backslash X$ but are not tangent to X. Tangency thus eliminates the "sliver" simplices of $\mathrm{Del}\,(X)$ that are not geometrically informative. ($\#(X) = 6013$, $\mathrm{Del}\,(X)$ has 36361 tetrahedra, $\mathrm{Del}_T(X)$ has 25745 tetrahedra, computing $\mathrm{Del}\,(X)$ takes 70 ms, computing $\mathrm{Del}_T(X)$ takes 773 ms.) (Color figure online)

y of $\|\mathrm{Del}_T(X)\|$ belongs to a simplex $\tau \in \mathrm{Del}_T(X)$. Let A be the vertices of τ. By tangency, for any k, $0 \leqslant k \leqslant d$, $\bar{\mathscr{C}}_k^d[\tau] = \bar{\mathscr{C}}_k^d[\mathrm{cvxh}(A)] \subset \bar{\mathscr{C}}_k^d[X]$. Hence the point y belongs to some cell σ of $\bar{\mathscr{C}}_k^d[X]$ and is at most at L_∞-distance 1 of any one of its X-corner. Finally by tangency, $\|\mathrm{Del}_T(X)\| \subset \bar{\mathscr{C}}^d[X]$, so $\|\mathrm{Del}_T(X)\| \cap \mathbb{Z}^d \subset \bar{\mathscr{C}}^d[X] \cap \mathbb{Z}^d = X$. $X \subset \|\mathrm{Del}_T(X)\| \cap \mathbb{Z}^d$ is obvious. □

6 Conclusion and Perspectives

We have proposed an original definition for digital convexity in arbitrary dimension, called full convexity, which possesses topological and geometric properties that are more akin to continuous convexity. We exhibited an algorithm to check full convexity, which relies on standard algorithms. We illustrated the potential of full convexity for addressing classic discrete geometry problems like building a tangential cover or reconstructing a reversible first-order polygonal surface approximation. We believe that full convexity opens the path to d-dimensional digital shape geometry analysis. This work opens many perspectives. On a fundamental level, we work on a variant of full convexity that keeps the intersection property of continuous convexity. We also wish to improve the convexity check algorithm, especially in 3D, for instance when cvxh(X) has few facets.[1] Finally we wish to explore the properties of a tangential cover made of the maximal fully convex tangent subsets that are included in some arithmetic plane.

A Proofs of Some Properties

Lemma 9. For any $X \subset \mathbb{Z}^d$, for any $\alpha \subset I^d$, $(X \oplus \mathcal{S}_\alpha) \cap \mathbb{Z}^d = U_\alpha(X)$.

Proof. Let $\alpha = \{i_1, \ldots, i_k\} \subset I^d$, non empty. Note that $\mathcal{S}_\alpha = \bigoplus_{j=1}^k \mathcal{S}_j$ and that it does not depend on the chosen order. For conciseness, we write $X^\gamma := X \oplus \mathcal{S}_\gamma$ for any subset γ of I^d. Let $\beta = \{i_1, \ldots, i_{k-1}\}$ and let us first show that:

$$(X^\alpha) \cap \mathbb{Z}^d = (X^\beta \cap \mathbb{Z}^d) \cup \mathbf{e}_{i_k}(X^\beta \cap \mathbb{Z}^d). \tag{6}$$

⊃ $(X^\beta \cap \mathbb{Z}^d) \cup \mathbf{e}_{i_k}(X^\beta \cap \mathbb{Z}^d) = (X^\beta \cup \mathbf{e}_{i_k}(X^\beta)) \cap \mathbb{Z}^d \subset (X^\beta \oplus \mathcal{S}_{i_k}) \cap \mathbb{Z}^d = (X^\alpha) \cap \mathbb{Z}^d$.

⊂ Let $z \in X^\alpha \cap \mathbb{Z}^d$. We can write z as $z = x + t_{i_1}\mathbf{e}_{i_1} + \cdots + t_{i_k}\mathbf{e}_{i_k}$, with $x \in X$ and every $t_{i_j} \in [0,1]$. More precisely, since $z \in \mathbb{Z}^d$ and $x \in \mathbb{Z}^d$ and $\mathbf{e}_{i_j}$ is a unit vector, every $t_{i_j} \in \{0,1\}$. Clearly $z' = z + \sum_{j=1}^{k-1} t_{i_j}\mathbf{e}_{i_j}$ belongs to $X^\beta \cap \mathbb{Z}^d$. If $t_{i_k} = 0$ then $z' = z$ and we are done. Otherwise, $t_{i_k} = 1$ then $z' = z + \mathbf{e}_{i_k}$, which belongs to $\mathbf{e}_{i_k}(X^\beta \cap \mathbb{Z}^d)$.

We prove the lemma by induction on the cardinal of α. For $\alpha = \emptyset$, $(X \oplus \mathcal{S}_\emptyset) \cap \mathbb{Z}^d = X = U_\emptyset(X)$. Assume the lemma is true for any β of cardinal $k - 1 \geq 0$, and let us show it for $\alpha = \beta \cup \{i\}$.

$$(X^\alpha) \cap \mathbb{Z}^d = (X^\beta \cap \mathbb{Z}^d) \cup \mathbf{e}_i(X^\beta \cap \mathbb{Z}^d) \qquad \text{(Using (6))}$$

---

[1] See for instance the full convexity implementation in DGTAL.

$$= U_\beta(X) \cup \mathbf{e}_i(U_\beta(X)) \qquad \text{(Induction)}$$
$$= U_\alpha(X) \qquad \text{(Definition)}.$$

$\square$

Lemma 10. *For any $X \subset \mathbb{Z}^d$, for any $\alpha \subset I^d$, $\mathrm{cvxh}\,(X) \oplus \mathcal{S}_\alpha = \mathrm{cvxh}\,(U_\alpha(X))$.*

Proof. We prove it by induction on the cardinal k of α. It holds obviously for $k = 0$. Otherwise let $\alpha = \beta \cup \{i\}$ of cardinal k.

$\boxed{\supset}$ We have $U_\alpha(X) = U_\beta(X) \cup \mathbf{e}_i(U_\beta(X)) \subset U_\beta(X) \oplus \mathcal{S}_i$. Since convex hull is increasing, $\mathrm{cvxh}\,(U_\alpha(X)) \subset \mathrm{cvxh}\,(U_\beta(X) \oplus \mathcal{S}_i)$ holds. But convex hull commutes with Minkowski sum, so $\mathrm{cvxh}\,(U_\beta(X) \oplus \mathcal{S}_i) = \mathrm{cvxh}\,(U_\beta(X)) \oplus \mathcal{S}_i = \mathrm{cvxh}\,(X) \oplus \mathcal{S}_\beta \oplus \mathcal{S}_i$ by induction hypothesis. We conclude with $\mathcal{S}_\beta \oplus \mathcal{S}_i = \mathcal{S}_\alpha$.

$\boxed{\subset}$ Let $y \in \mathrm{cvxh}\,(X) \oplus \mathcal{S}_\alpha = \mathrm{cvxh}\,(X) \oplus \mathcal{S}_\beta \oplus \mathcal{S}_i = \mathrm{cvxh}\,(U_\beta(X)) \oplus \mathcal{S}_i$. Denoting by z_j, $j \in B$ the points of $U_\beta(X)$, it follows that y can be written as a convex linear combination of these points plus a point of $\mathcal{S}_i$, i.e. $y = (\sum_{j \in B} \mu_j z_j) + t\mathbf{e}_i$, $\sum_{j \in B} \mu_j = 1$, $\forall j \in B, \mu_j \geq 0$ and $t \in [0, 1]$. All following sums are taken over $j \in B$. Since $\sum \mu_j = 1$, we rewrite y as

$$y = \left(\sum \mu_j z_j\right) + t\left(\sum \mu_j\right)\mathbf{e}_i$$
$$= \sum (1 - t)\mu_j z_j + t\mu_j z_j + t\mu_j \mathbf{e}_i$$
$$= \left(\sum (1 - t)\mu_j z_j\right) + \left(\sum t\mu_j (z_j + \mathbf{e}_i)\right)$$
$$\in \mathrm{cvxh}\,(U_\beta(X) \cup \mathbf{e}_i(U_\beta(X))),$$

which shows that $y \in \mathrm{cvxh}\,(U_\alpha(X))$. $\square$

References

Bar94. Barvinok, A.I.: Computing the Ehrhart polynomial of a convex lattice polytope. Discret. Comput. Geom. **12**(1), 35–48 (1994). https://doi.org/10.1007/BF02574364

BDH96. Barber, C.B., Dobkin, D.P., Huhdanpaa, H.: The quickhull algorithm for convex hulls. ACM Trans. Math. Softw. **22**(4), 469–483 (1996)

BLPR09. Brlek, S., Lachaud, J.-O., Provençal, X., Reutenauer, C.: Lyndon + Christoffel = digitally convex. Pattern Recognit. **42**(10), 2239–2246 (2009)

CdFG20. Crombez, L., da Fonseca, G.D., Gérard, Y.: Efficiently testing digital convexity and recognizing digital convex polygons. J. Math. Imaging Vis. **62**, 693–703 (2020)

Cha93. Chazelle, B.: An optimal convex hull algorithm in any fixed dimension. Discret. Comput. Geom. **10**(4), 377–409 (1993). https://doi.org/10.1007/BF02573985

Cha96. Chan, T.M.: Optimal output-sensitive convex hull algorithms in two and three dimensions. Discret. Comput. Geom. **16**(4), 361–368 (1996). https://doi.org/10.1007/BF02712873

282   J.-O. Lachaud

CL11.   Charrier, E., Lachaud, J.-O.: Maximal planes and multiscale tangential cover
        of 3D digital objects. In: Aggarwal, J.K., Barneva, R.P., Brimkov, V.E.,
        Koroutchev, K.N., Korutcheva, E.R. (eds.) IWCIA 2011. LNCS, vol. 6636,
        pp. 132–143. Springer, Heidelberg (2011). https://doi.org/10.1007/978-3-
        642-21073-0_14

Eck01. Eckhardt, U.: Digital lines and digital convexity. In: Bertrand, G., Imiya, A.,
        Klette, R. (eds.) Digital and Image Geometry. LNCS, vol. 2243, pp. 209–228.
        Springer, Heidelberg (2001). https://doi.org/10.1007/3-540-45576-0_13

Ehr62. Eugène, E.: Sur les polyèdres rationnels homothétiques à n dimensions. C.R.
        Acad. Sci. **254**, 616–618 (1962)

FT99.   Feschet, F., Tougne, L.: Optimal time computation of the tangent of a dis-
        crete curve: application to the curvature. In: Bertrand, G., Couprie, M.,
        Perroton, L. (eds.) DGCI 1999. LNCS, vol. 1568, pp. 31–40. Springer, Hei-
        delberg (1999). https://doi.org/10.1007/3-540-49126-0_3

HKV81. Hübler, A., Klette, R., Voss, K.: Determination of the convex hull of a
        finite set of planar points within linear time. Elektron. Informationsverarb.
        Kybern. **17**(2–3), 121–139 (1981)

KR82a. Kim, C.E., Rosenfeld, A.: Convex digital solids. IEEE Trans. Pattern Anal.
        Mach. Intell. **6**, 612–618 (1982)

KR82b. Kim, C.E., Rosenfeld, A.: Digital straight lines and convexity of digital
        regions. IEEE Trans. Pattern Anal. Mach. Intell. **2**, 149–153 (1982)

PDR09. Provot, L., Debled-Rennesson, I.: 3D noisy discrete objects: segmentation
        and application to smoothing. Pattern Recognit. **42**(8), 1626–1636 (2009)

Ron89. Ronse, C.: A bibliography on digital and computational convexity (1961–
        1988). IEEE Trans. Pattern Anal. Mach. Intell. **11**(2), 181–190 (1989)

SDC04. Sivignon, I., Dupont, F., Chassery, J.-M.: Decomposition of a three-
        dimensional discrete object surface into discrete plane pieces. Algorithmica
        38(1), 25–43 (2004)

Soi04. Soille, P.: Morphological Image Analysis: Principles and Applications,
        2nd edn. Springer, Heidelberg (2004). https://doi.org/10.1007/978-3-662-
        05088-0

Digital Geometry on the Dual of Some Semi-regular Tessellations

Mohammadreza Saadat and Benedek Nagy[(⊠)]

Department of Mathematics, Faculty of Arts and Sciences, Eastern Mediterranean University, Mersin-10, 99450 Famagusta, North Cyprus, Turkey
mohammed.saadet@emu.edu.tr, nbenedek.inf@gmail.com

Abstract. There are various tessellations of the plane. There are three regular ones, each of them using a sole regular tile. The square grid is self-dual, and the two others, the hexagonal and triangular grids are duals of each other. There are eight semi-regular tessellations, they are based on more than one type of tiles. In this paper, we are interested to their dual tessellations. We show a general method to obtain coordinate system to address the tiles of these tessellations. The properties of the coordinate systems used to address the tiles are playing crucial roles. For some of those grids, including the tetrille tiling D(6, 4, 3, 4) (also called deltoidal trihexagonal tiling and it is the dual of the rhombihexadeltille, T(6, 4, 3, 4) tiling), the rhombille tiling, D(6, 3, 6, 3) (that is the dual of the hexadeltille T(6, 3, 6, 3), also known as trihexagonal tiling) and the kisquadrille tiling D(8, 8, 4) (it is also called tetrakis square tiling and it is the dual of the truncated quadrille tiling T(8, 8, 4) which is also known as Khalimsky grid) we give detailed descriptions. Moreover, we are also presenting formulae to compute the digital, i.e., path-based distance based on the length of a/the shortest path(s) through neighbor tiles for these specific grids.

Keywords: Semi-regular grids · Nontraditional grids · Digital distance · Dual tessellations · Path-based distance · Coordinate system

1 Introduction

Since our world is generally not rectangular, various non-traditional grids and tilings have significant importance in various disciplines [18]. A tiling or a tessellation of the plane is based on one or more geometric shapes, referred as tiles, without overlaps and gaps. The tiling is periodic if it has a repeating pattern, i.e., the tessellation is resistant to some transitions of the plane, in the sense that translating the whole plane the tessellation pattern is exactly the same. There are three regular tessellations; they are named after the tile is used: the square, the triangular and the hexagonal tiling, each uses exactly one type of tile, equilateral triangles, squares, and regular hexagons, respectively. Thus, all three of these tilings (also called grids) are monohedral. They are also isogonal: all corners of tiles, the gridpoints, i.e., vertices, crossing points of the grid, are identical. Actually, with isometric geometric transformations every vertex can be mapped to any

© Springer Nature Switzerland AG 2021
J. Lindblad et al. (Eds.): DGMM 2021, LNCS 12708, pp. 283–295, 2021.
https://doi.org/10.1007/978-3-030-76657-3_20

other vertex. The square grid is self-dual, its pixels and gridpoints have exactly the same neighborhood structure. The triangular and the hexagonal grids are dual grids of each other. Based on this fact, there is mixture of their terminology in various disciplines depending on whether the tiles (e.g., as pixels in image processing or computer graphics) play importance, or the connections between vertices (e.g., communication networks). On the one hand, both the square and hexagonal tilings are point lattices: translations with any grid vectors (here grid vectors are defined as vectors connecting the midpoints of pixels, these vectors are actually the directed variants of the edges of the dual grid) map the grid into itself. On the other hand, the triangular grid does not have this property, i.e., it is not a point lattice, more precisely, it is not a discrete subgroup of the Euclidean space. There are grid vectors, e.g., the vectors connecting the center of two neighbor tiles, such that the translations of the grid with these vectors do not map the grid into itself. This can also be seen on the dual, hexagonal grid: the vertices of the grid are seemingly two different types, i.e., no translation can map them to each other. In fact, one needs to use either rotation or mirroring as isometric transformation to show that they are identical. Correspondingly, there are two different orientations of the tiles in the triangular grid.

In each of the eight semi-regular (also called Archimedean) tessellations there are more than one type tiles, but they are still only be regular polygons with the same side length, and the tiling is isogonal [3]. One of the naming conventions of these tessellations (or grids) is based on their vertex configuration: numbers in a list representing the number of sides of the regular polygons joined at that vertex in a given order [18]. Using this method T(6, 4, 3, 4) represents the grid in which at every vertex, a hexagon, a square, a triangle and another square meet. None of these grids is self-dual, their dual tessellations can be named, based on the semi-regular grid name showing that the dual grid is subjected, e.g., the dual of the aforementioned grid is named as D(6, 4, 3, 4), meaning that it is the dual tessellation of T(6, 4, 3, 4). Another possible naming of these sixteen grids is provided in chapter 21 of the book "The symmetries of things" by Conway et al. [2]. None of the semi-regular grids and their dual grids are point lattices. In each of those grids it is easy to find grid vectors that do not transform the grid into itself. While each semi-regular tiling uses more than one type of tiles, but the vertices of the grid are identical, in each of the dual tilings of them, there is exactly one type of tile which is on one hand, not regular, and on the other hand, used in various orientations. Moreover, every of the mentioned grids has the edge-to-edge property which means that if two tiles share more than one point on their boundary, then they share exactly one of their full edges (sides).

Coordinate systems are used to address the elements, e.g., tiles, of the space by fixed number of coordinates, usually, by numbers, to uniquely determine the position of them. Here we are interested to have coordinate systems for the above mentioned tessellations. There are various coordinate systems known for the hexagonal and triangular grids, e.g., discrete: addressing only the tiles [4, 11, 14], topological: addressing also the edges and vertices of the grid [10, 13], continuous: addressing the whole plane [5, 12]. Here, we will show a general method with some examples to address the tiles (we call them block coordinate systems), and also we provide more specific coordinate systems for some of the grids that reflect better the symmetries and the structure of the grid (we call them symmetric coordinate systems). In this paper, we concentrate on the dual of three of the semi-regular tilings. We give a description of the neighborhood structures of the grids D(6,4,3,4), D(6, 3, 6, 3) and D(8, 8, 4) based on their symmetric coordinate systems.

Digital distances are usually based on the number of steps in shortest paths, where a path is a finite sequence of tiles such that any two consecutive tiles are neighbors. In this paper we use the concept of neighbor for those pairs of tiles that share a side. Digital distances (even more general ones with a large variety) are frequently used in image processing and they are already defined and computed on various grids, including the hexagonal [11], the triangular [14, 16], T(8 ,8, 4) [8], T(6, 3, 6, 3) [9] and T(4, 3, 4, 3, 3) grids. This latter one is studied in [1] and its well applicability in image acquisition and in displaying images with various practical algorithms were shown. Its dual is also known as Cairo pattern and digital geometry on it was recently investigated [7]. To show the applicability of our newly investigated coordinate systems, and also as the first results on digital geometry of D(6, 4, 3, 4), D(6, 3, 6, 3) and D(8, 8, 4) grids, we give formulae to compute digital distances based on the number of steps in a shortest path between the tiles. Our results can be applied in networking, these grids support specific architectures, which could be beneficial, e.g., where the number of neighbors and the structure of neighborhood match. Also, it can be used in image processing, in cases, where the symmetry or the structure of the image, the image grid or the capturing device fits better to the studied grids than to the classical regular grids. For instance, the tiling D(6, 3, 6, 3) can be seen as an oblique mesh of the cubic grid, where the square faces on the surface give exactly the same structure as the tiles of this grid. In this way, it is natural to consider images on such surfaces and the underlying geometry is exactly the one we propose here.

2   A General Method for Addressing Tiles: Block Coordinate Systems

Whenever, a periodic tiling is given, one may consider the smallest connected regions of the grid as unit cells (or block of tiles) with the property that the tiling, the grid, is built up by translating this unit cell by the unit vectors. For all regular and semi-regular grids, and for their duals the unit cell is always contain a relatively small finite number of tiles and we may use two of the shortest independent unit vectors to generate the whole tessellation by translations. One unit cell can be chosen to be addressed by the nullvector (0, 0). Every other unit cell will be addressed by the coefficients of the unit vectors in the vector connecting the (center) of the unit cell (0,0) to the (center) of the given unit cell. The usual description $\mathbb{Z}^2$ of the square grid is based also on this approach, by using unit cells of the size of a tile and two orthogonal unit vectors. For the other point lattice grid, the hexagonal grid, the oblique coordinate system having axes with 60° or 120° angle between them, reflects also this method [11]. For the other grids, the unit cell contains more than one tile, thus one needs to identify the elements of the unit cell inside the cell as well. In general, we may use various letters to refer to the tiles inside a unit cell. This method gives us a coordinate system, that we call block coordinate system for the given grid. Each tile is addressed by two integer numbers (that show the position of the block in the plain) and one letter (that shows the type and the position of tile in the unit cell). In

Fig. 1 and 2 we show two examples for such systems, the coordinate axes are also shown, the unit vectors can also easily be identified from the figures. Figure 1 shows a block coordinate system for the T(4,3,4,3,3) semi-regular grid (also known as snub quadrille tiling). Table 1 gives also the neighborhood criteria for each type of tile. The coordinate system that was used in [1] is different from this, it uses the idea to have somewhat rows of the tiles and a kind of other, let us say, column coordinate, was assigned to each element of a row of tiles. Figure 2 shows a block coordinate system for the dual semi-regular grid D(6,3,3,3,3) that is also called 6-fold pentille tiling. Table 2 presents the neighborhood criteria based on this coordinate system. Block coordinate systems can effectively be used in programming, in graphical algorithms etc., since usually the conversion to the usual Cartesian coordinates (that is needed, e.g., if one uses traditional monitor with square pixels) is very simple and the neighborhood can be described by relatively small tables, similarly to those we have shown.

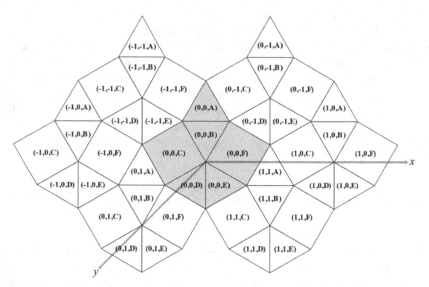

Fig. 1. A block coordinate system for the snub quadrille tiling, T(4, 3, 4, 3, 3).

Table 1. Neighborhood sets of tiles of snub quadrille tiling

Tile type	Tile coordinate	Neighborhood set
A	(i, j, A)	$\{(i, j, B), (i, j-1, C), (i-1, j-1, F)\}$
B	(i, j, B)	$\{(i, j, A), (i, j, C), (i, j, F)\}$
C	(i, j, C)	$\{(i, j, B), (i, j, D), (i, j+1, A), (i-1, j-1, E)\}$
D	(i, j, D)	$\{(i, j, C), (i, j, E), (i, j+1, F)\}$
E	(i, j, E)	$\{(i, j, D), (i, j, F), (i+1, j+1, C)\}$
F	(i, j, F)	$\{(i, j, E), (i, j, B), (i, j-1, D), (i+1, j+1, A)\}$

Table 2. Neighborhood sets of the 6-fold pentille tiling

Tile type	Tile coordinate	Neighborhood set
A	(i, j, A)	$\{(i, j, F), (i + 1, j, C), (i, j - 1, E), (i, j - 1, D), (i, j, B)\}$
B	(i, j, B)	$\{(i, j, A), (i, j - 1, D), (i - 1, j - 1, F), (i - 1, j - 1, E), (i, j, C)\}$
C	(i, j, C)	$\{(i, j, B), (i - 1, j - 1, E), (i - 1, j, A), (i - 1, j, F), (i, j, D)\}$
D	(i, j, D)	$\{(i, j, C), (i - 1, j, F), (i, j + 1, B), (i, j + 1, A), (i, j, E)\}$
E	(i, j, E)	$\{(i, j, D), (i, j + 1, A), (i + 1, j + 1, C), (i + 1, j + 1, B), (i, j, F)\}$
F	(i, j, F)	$\{(i, j, E), (i + 1, j + 1, B), (i + 1, j, D), (i + 1, j, C), (i, j, A)\}$

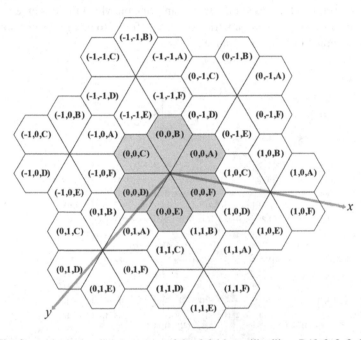

Fig. 2. A block coordinate system of the 6-fold pentille tiling, D(6, 3, 3, 3, 3).

In many cases, maybe based on the previously described block coordinate system, one can have a coordinate system that reflects better the symmetry of the grid. Symmetric coordinate systems for the hexagonal and triangular grids use integer triplets to address the tiles, zero-sum triplets are used for the hexagons [4], while zero-sum and one-sum triplets for the triangles oriented in two different ways, respectively [14]. Symmetric coordinate system with two coordinates was used for T(8,8,4) [8], while three integer coordinates were used to address the tiles of T(6,3,6,3) [9]. The symmetric coordinate system is based on three integer coordinates, especially on those grids, where the lengths of the unit vectors are the same in three different directions with angles 120° of them. In this paper, we continue the research on this field and in the next sections we provide symmetric coordinate systems and formulae for digital distances for the dual grids of three of the semi-regular grids.

3   Tetrille Tiling

In this section we deal with the dual D(6, 4, 3, 4) of the semi-regular rhombihexadeltille grid T(6, 4, 3, 4). There are various names for D(6, 4, 3, 4), it is called deltoidal tri-hexagonal tiling [17] and tetrille tiling [2]. The edges of this tiling can be formed by the intersection overlay of two regular grids: a regular triangular grid and a hexagonal grid (see Fig. 3 left, a unit cell is shown also on the right). Each deltoid (also called kite face) of the tiling has angles 120°, 90°, 60° and 90°, and there are six different orientations of them. Moreover, this tiling is symmetric to any line on which there is an edge [6]. In the next subsections we give a symmetric coordinate frame to this grid and also we give a formula to compute the digital distance.

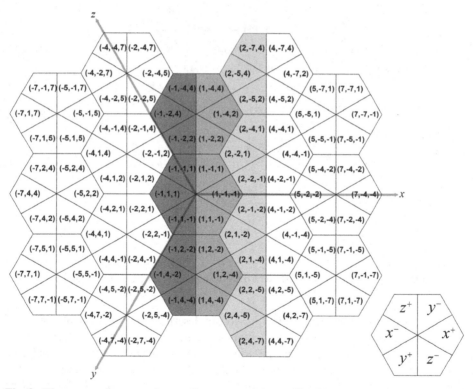

Fig. 3. The proposed symmetric coordinate system for tetrille tiling with three coordinate axes (left) and the types of the tiles (right). (Color figure online)

3.1 Symmetric Coordinate System

Figure 3 (left) shows a symmetric coordinate system for D(6, 4, 3, 4). Since the grid has similar symmetries as the triangular and hexagonal grids (as we have described it by intersection of those grids), it is worthwhile to use coordinate triplets to address the deltoid tiles. There is a chain of hexagons perpendicular to a given axis, e.g., the chain is vertical if axis x is considered. Each of such chains are divided to two chains of tiles with a gridline perpendicular to the axis, and these chains of tiles built up by repeating three types of tiles (i.e., tiles with three different orientations). The corresponding coordinate is fixed for the tiles of these chains of tiles in such a way that their difference is 2 inside a chain of hexagons, e.g., $x = 1$ (yellow highlights) and $x = -1$ (orange highlights) in the chain of hexagons in the middle of Fig. 3. The corresponding coordinate difference is 1 for neighbor chains of tiles having tiles of the same triangles, see, e.g., the tiles having $x = 1$ (yellow) and $x = 2$ (pink highlights). The three coordinate axes are also shown in the figure. They meet in a gridpoint which is in the middle of a hexagon (and corner point of six triangles, and also six tiles, one from each type). For the chain of tiles going through on some of these six tiles we fix the corresponding coordinate value by ± 1. Since the chains of tiles in the three directions altogether uniquely determines the tile, the coordinate system is correct, each tile is uniquely addressed. Of course, the three coordinate values are not independent of each other, the coordinate sum is always ± 1 (reflecting the idea of the coordinate system for the triangular grid). This also implies that there is at least 1 odd coordinate value for each tile. We name the six different orientations of the tiles based on their directions in a hexagon (see Fig. 3, right): we have tiles of type x^+, y^+, z^+ and x^-, y^-, z^-. The coordinate sum is -1 for the former tiles and $+1$ for the latter ones. As each tile is a deltoid, each has exactly four neighbor tiles. Let $i \in \{x, y, z\}$ be one of the coordinate letters and $j, k \in \{x, y, z\}$ be the other two, further let $s \in \{+, -\}$ and let z denote the opposite sign. Then a tile i^s has two neighbors, type j^s and k^s, respectively, through the shorter sides of the deltoid. From the i^s-type tile coordinate i is modified by $s1$ (i.e., by $+1$ or -1, according to the type of the tile), and one of the other coordinates, j or k, is modified by $z1$, to reach the neighbor of type j^s or k^s, respectively. By stepping through on a longer side of an i^s-type kite shape, one of the coordinates j and k is modified by $s2$ (i.e., by $+2$ or -2) to reach the neighbor of type j^z or k^z, respectively. Remember that in this paper only side-neighbors are used, thus we have already discussed all of them. We can see that at each step to a neighbor either two of the coordinates are changing, one of them with $+1$, the other with -1; or exactly one coordinate is changed by ± 2. The coordinate system reflects nicely the symmetry, the structure of the tessellation.

Proposition 1. *Each $+1$-sum and -1-sum integer triplet not containing any number that is divisible by 3 is addressing a tile.*

Proof. Let us fix two of the coordinates, let us say x and y such that none of them is divisible by 3. Under this condition they determine two chains of tiles which have either one or two common tiles as intersection. If the two fixed number has same remainder (mod 3), then the third value must have a different remainder (otherwise the sum of the coordinates will be divisible by 3 contradicting to the condition that the sum is either $+1$ or -1). Thus, if the two fixed values are $3i + 1$ and $3j + 1$, then the third value must be $-3(i+j) - 1$ (and the sum of the coordinates is $+1$). Similarly, if the two fixed values are $3i + 2$ and $3j + 2$, then the third value must be $-3(i+j) - 5$ (and the sum of the coordinates is -1). Whenever, the two fixed value has different remainder (mod 3), the third value could also have either 1 or 2 (mod 3): with two fixed values $3i + 1$ and $3j + 2$, the third value could either be $-3(i+j) - 2$ (having sum $+1$) and $-3(i+j) - 4$ (having sum -1). The proof is finished.   □

3.2 Digital Distance

In this subsection we address the shortest path problem, we prove a formula which gives the digital distance, i.e., the minimal number of steps between any two tiles, when the path is built up by steps between neighbor tiles.

Theorem 1. *Let $p = (p_1, p_2, p_3)$ and $q = (q_1, q_2, q_3)$ be two tiles of the tetrille tiling. Their digital distance is denoted and given by*

$$D(p, q) = \sum_{i=1}^{3} \frac{|p_i - q_i|}{2}.$$

Proof. If two tiles share a common coordinate (e.g., their x coordinate is the same, i.e., $p_1 = q_1$), then the shortest path contains the tiles between them in their common chain of tiles (including also the start and end tiles, p and q). The two other coordinates are changing in every step, either one of them by 2, or both of them by 1 to the direction to reach the value in the end tile. Thus, the sum of the absolute coordinate differences of the tiles, $\sum_{i=1}^{3} |p_i - q_i|$ gives the double of the number of steps, i.e., the double of the distance. On the other hand, since in any step to a neighbor tile, the absolute sum of the coordinate change is always 2, there is no way to connect the tiles in a shorter path than the one we have studied.

If there is no shared coordinate between the tiles p and q, then we may connect them by two chains of tiles that have 120° between them, based on the coordinate differences with the same sign (e.g., for tiles $p = (-5, 5, -1)$ and $q = (2, -7, 4)$, the two coordinate differences with the same sign are in coordinates x and z, consequently a chain of tiles with fixed $z = -1$ and another chain $x = 2$ can be used). In such chains, in every step either two coordinates are changed by $+1$ and -1, respectively, or only one coordinate is changed by either $+2$ or -2, in each step to the direction of the endpoint. Consequently, the number of steps in the path will be the half of the sum of the absolute coordinate differences for the two tiles. Here we can apply the same argument as in the first case of the proof to ensure that shorter paths than the one we have just studied do not exist.   □

In the rest of the section, to support readers' understanding, we give some examples.

Example 1. Let us consider the tiles $p = (-5, 5, -1)$, $q = (2, -7, 4)$, $r = (4, -4, 1)$, and $s = (2, 4, -5)$. Then,

$$D(p, q) = (7 + 12 + 5)/2 = 12, \quad D(p, r) = (9 + 9 + 2)/2 = 10,$$
$$D(p, s) = (7 + 1 + 4)/2 = 6, \quad D(q, r) = (2 + 3 + 3)/2 = 4,$$
$$D(q, s) = (0 + 11 + 9)/2 = 10, \quad D(r, s) = (2 + 8 + 6)/2 = 8.$$

In each case it is easy to check the result also by finding a shortest path on Fig. 3.

4 Rhombille Tiling

In this section we work with D(6,3,6,3). This, also called rhombille tiling, is a tessellation of the plane by identical rhombuses (Fig. 4, left). Each tile has two 60° and two 120° angles. Three rhombuses share a common gridpoint at their 120° angles and six tiles share a gridpoint at their 60° angles. The tiles appear in three different orientations (we refer to them as types). A block coordinate system for this grid may use the hexagons built up by one of each type of tiles.

4.1 Symmetric Coordinate System

This grid is mentioned also in [15], as the dual of the 3-plane triangular grid, the tri-hexagonal grid, but with a coordinate system not showing the structure of the grid as smoothly as our new coordinate system. Here we present a new coordinate system on Fig. 4 using integer triplets to address the tiles, reflecting the symmetry of the grid. The three axes can also be seen. They meet in a gridpoint where six rhombuses have their 60° angles. Let us consider chains of tiles perpendicular to an axis, there are two types of chains, and they are following each other alternately. In Fig. 4 (left) we show horizontal chains which are perpendicular to axis y, correspondingly the second coordinate of the tiles in such chains is fixed: $y = 0$ (see the yellow highlighted chain) and $y = -1$ (orange). Whenever a fixed coordinate has an even value, the chain is built up by exactly one type of rhombuses. We name the type of tiles based on that, e.g., the tiles are type y in our earlier example, with $y = 0$. Thus, the shapes, i.e., more precisely, the orientations of the tiles are shown in Fig. 4 (right). Other chains contain exactly two types of tiles alternately. Every tile has a 0-sum triplet, moreover there is exactly one even and two odd coordinates in a triplet. The even coordinate specifies also the type of the tile. Every two chains of tiles have either exactly one common tile (intersection, where the third coordinate value is uniquely determined by the 0-sum property) or they do not share any tiles (exactly in the cases when both fixed coordinates are even). Thus, one can see that the coordinate system is correct, every tile is addressed in a unique way.

Each tile has four neighbors, each of different type than the original tile. From a tile to any of its neighbors one coordinate is changing by $+1$ and one by -1, such that the even coordinate must change, in this way all the four neighbors are described.

4.2 Digital Distance

Theorem 2. *Let* $p = (p_1, p_2, p_3)$ *and* $q = (q_1, q_2, q_3)$ *be two tiles of the rhombille tiling. Their digital distance is denoted and given by*

$$
D(p, q) = \begin{cases} \sum_{i=1}^{3} \frac{|p_i - q_i|}{2} + 1, & \text{if } p \text{ and } q \text{ share an even coordinate value,} \\ \sum_{i=1}^{3} \frac{|p_i - q_i|}{2}, & \text{otherwise.} \end{cases}
$$

Proof. We start the proof by the cases where the second formula can be used, and in the end of the proof we prove the case of the first formula.

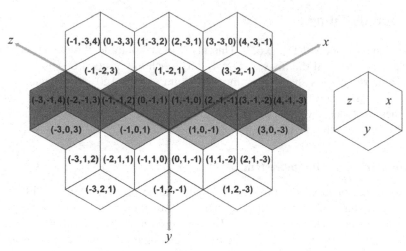

Fig. 4. The proposed symmetric coordinate system for the rhombille tiling with some highlighted chains perpendicular to axis y (left) and the types of the tiles (right). (Color figure online)

If p and q share an odd coordinate, i.e., they are on a common chain of tiles (with a fixed odd coordinate value), then the shortest path is to connect these tiles through the tiles between them in the chain. Since in every step, the absolute sum of the coordinate change is 2, the number of steps is the half of the sum of the coordinate change. Whenever, the two tiles do not share any coordinate, we show that we can still connect them with a path such that in every step the changes of the coordinates are in the aimed direction, which provides a shortest path with the length given by our formula. In this case, two of the coordinate differences $q_1 - p_1, q_2 - p_2, q_3 - p_3$ have the same sign (i.e., either two of them are positive or two of them are negative), let us call the corresponding coordinates i and j, while the third coordinate is referred as k. Let us start from tile p. (At least) one of its i and j coordinate is odd, let us fix this coordinate (any of those if both are odd) and make steps on the chain to the direction of q (i.e., by increasing the coordinate different from i and having positive coordinate difference, and at the same time, decreasing the coordinate different from i and having negative coordinate difference). If coordinate j has odd value in q, we may go on the previous chain by reaching that value and then we can connect this intermediate tile with q on a chain with fixed odd j coordinate. However, if q has an even value at coordinate j, then we should stop one step before reaching this value in the chain with fixed coordinate i and we should use a chain with fixed odd coordinate j to reach a neighbor of tile q. In such a path, in every step, coordinate k is changed (and one of the other coordinates) to the direction of q, thus the formula of the theorem is proven for this case.

Finally, the case of the first formula is investigated. When the tiles on the same chain, but with a fixed even coordinate value, since there is no neighbor tile without changing it, the first step must change this even value, as well, to reach one of the neighbor chains. Then from this tile, by the previously proven case, the path could go to a neighbor of q to reach q in the last step. Formally, let coordinate i with positive coordinate difference of q and p, and j with negative (while coordinate k is shared by the tiles, i, j, k are pairwise different elements of $\{x, y, z\}$). The first step could be by increasing coordinate i by 1 and decreasing coordinate k by 1. In this way, the first step does not decrease (nor increase) the sum of coordinate differences, but keeps it. Thus, the first case of our formula has also been proven.   □

Now, for the clarity of the presentation, we provide some examples.

Example 2. Let the tiles $p = (-3, 0, 3)$, $q = (3, 0, -3)$, $r = (4, -1, -3)$, $s = (-3, -1, 4)$ be considered. Then $D(p,q) = ((6 + 0 + 6)/2) + 1 = 7$, which can also be verified by a shortest path $(-3, 0, 3)$, $(-2, -1, 3)$, $(-1, -1, 2)$, $(0, -1, 1)$, $(1, -1, 0)$, $(2, -1, -1)$, $(3, -1, -1)$, $(3, 0, -3)$ from p to q, where p and q share an even coordinate. Further, $D(r, s) = (7 + 0 + 7)/2 = 7$, since these tiles have a common odd coordinate. $D(p, r) = (7 + 1 + 6)/2 = 7$ and $D(q, s) = (6 + 1 + 7)/2 = 7$, since nor p and r, neither q and s share a coordinate value. Finally, $D(p, s) = (0 + 1 + 1)/2 = 1$ and $D(q, r) = (1 + 1 + 0)/2 = 1$. The cases can also be followed on Fig. 4.

5   Kisquadrille Tiling

The grid D(8,8,4) is also called kisquadrille and tetrakis square tiling (see Fig. 5, left) [2]. It is obtained from the square grid by dividing each square into four isosceles right-angled triangles from the center point. It is also referred as the Union Jack lattice because of the resemblance to the UK flag of the triangles surrounding its degree-8 vertices [19]. It is the dual of the truncated quadrille tiling which was also called Khalimsky grid [8].

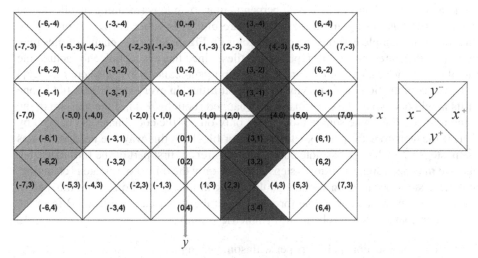

Fig. 5. The proposed symmetric coordinate system for the kisquadrille tiling (left) and the tile types (right). (Color figure online)

5.1 A Symmetric Coordinate Frame

The block coordinate system for $D(8, 8, 4)$ is based on the square grid. Now, instead of adding a letter, e.g., A, B, C, D, or maybe by the directions, N, W, S, E, we may modify the original pairs based on the direction of the tile from the center of the unit square. In this way, and by rescaling with a factor 3, to get integer coordinates, we have obtained the system shown in Fig. 5 (left). Each tile has three neighbors: A tile of type j^s ($j \in \{x,y\}$, $s \in \{+, -\}$) has a neighbor j^z by changing its coordinate j by $s1$ and has two neighbors k^+ and k^- changing its coordinate j by $z1$ and coordinate k by ± 1, respectively ($k \in \{x, y\}, k \neq j, z \in \{+, -\}, z \neq s$).

5.2 Digital Distance

Theorem 3. *Let $p = (p_1, p_2)$ and $q = (q_1, q_2)$ be tiles of the tetrakis square tiling. The digital distance of p and q is denoted and computed by*

$$D(p, q) = \begin{cases} |p_1 - q_1| + \left| \left\lfloor \frac{p_2+1}{3} \right\rfloor - \left\lfloor \frac{q_2+1}{3} \right\rfloor \right|, & \text{if } |p_1 - q_1| \geq |p_2 - q_2|, \\ |p_2 - q_2| + \left| \left\lfloor \frac{p_1+1}{3} \right\rfloor - \left\lfloor \frac{q_1+1}{3} \right\rfloor \right|, & \text{if } |p_1 - q_1| < |p_2 - q_2|. \end{cases}$$

Proof. (Because lack of space, the idea of the proof is given, the formal details are left for the reader.) We may use chains of tiles which have 45° or 135° (from any of the axis), in such chains, the four type of tiles are following each other periodically (see the yellow chain on Fig. 5, left). Any two tiles can be connected by one such chain (if they are in a common chain) or by a path perpendicular to an axis (meaning parallel to the other axis in this gird) going through on square unit cells either horizontal or vertical direction (an example is shown by orange color on Fig. 5, left, such chains of tiles may built up either three or all four types of the tiles, since there is a choice to include the tile of type j^+ or j^- at any unit square in a chain perpendicular to axis j, where $j \in \{x,y\}$) or by the combination of the previous two possibilities. In every step from a tile to one of its neighbors two of the coordinates are changed (both by ± 1), if the tiles are in the same unit cell (i.e. same square block); and only one of the coordinates is changed by ± 1, if the step goes from a block to a neighbor block. In this way, we need at least as many steps as the largest absolute coordinate difference (first part of the formulae), but also we may need some additional steps in which only the other coordinate is changing, and these steps occur when we change the block in this, second direction: this part is reflected in the second part of the formula (remember we used scale factor 3 from the block coordinate system to obtain the symmetric frame). $\qquad \square$

We close the section and the paper with some examples.

Example 3. Let $p = (-7, -3), q = (4, -3), r = (4,0)$ and $s = (2,3)$ be the tiles. The tiles p and s are x^- tiles, while the type of the other two tiles is x^+. Then the distances of these tiles are computed as follows. The upper formula applies for every distance between p and another tile and the lower formula between other pairs of tiles. The tiles, and the shortest paths between them can be checked on Fig. 5.

$$D(p,q) = 11 + |-1 - (-1)| = 11 \quad D(p,r) = 11 + |-1 - 0| = 12,$$
$$D(p,s) = 9 + |-1 - 1| = 11, \quad\quad\quad D(q,r) = 3 + |1 - 1| = 3,$$
$$D(q,s) = 6 + |1 - 1| = 6, \quad\quad\quad\quad D(r,s) = 3 + |1 - 1| = 3.$$

References

1. Borgefors, G.: A semiregular image grid. J. Vis. Commun. Image Repres. **1**(2), 127–136 (1990)
2. Conway, J.H., Burgiel, H., Goodman-Strass, C.: The symmetries of things. AK Peters, (2008)
3. Grünbaum, B., Shephard, G.C.: Tilings by regular polygons. Math. Mag. **50**(5), 227–247 (1977)
4. Her, I.: A symmetrical coordinate frame on the hexagonal grid for computer graphics and vision. ASME J. Mech. Des. **115**(3), 447–449 (1993)
5. Her, I.: Geometric transformations on the hexagonal grid. IEEE Trans. Image Process. **4**(9), 1213–1222 (1995)
6. Kirby, M., Umble, R.: Edge tessellations and stamp folding puzzles. Math. Mag. **84**(4), 283–289 (2011)
7. Kovács, G., Nagy, B., Turgay, N.D.: Distance on the Cairo pattern. Pattern Recogn. Lett. **145**, 141–146 (2021)
8. Kovács, G., Nagy, B., Vizvári, B.: On weighted distances on the Khalimsky grid. In: Normand, N., Guédon, J., Autrusseau, F. (eds.) DGCI 2016. LNCS, vol. 9647, pp. 372–384. Springer, Cham (2016). https://doi.org/10.1007/978-3-319-32360-2_29
9. Kovács, G., Nagy, B., Vizvári, B.: Weighted distances on the trihexagonal grid. In: Kropatsch, W.G., Artner, N.M., Janusch, I. (eds.) DGCI 2017. LNCS, vol. 10502, pp. 82–93. Springer, Cham (2017). https://doi.org/10.1007/978-3-319-66272-5_8
10. Kovalevsky, V.A.: Geometry of locally finite spaces: Computer agreeable topology and algorithms for computer imagery. Bärbel Kovalevski, Berlin (2008)
11. Luczak, E., Rosenfeld, A.: Distance on a hexagonal grid. IEEE Trans. Comput. **5**, 532–533 (1976)
12. Nagy, B., Abuhmaidan, K.: A continuous coordinate system for the plane by triangular symmetry. Symmetry **11**(2), paper 191 (2019)
13. Nagy, B.: Cellular topology and topological coordinate systems on the hexagonal and on the triangular grids. Annals Math. Artif. Intell. **75**(1–2), 117–134 (2014). https://doi.org/10.1007/s10472-014-9404-z
14. Nagy, B.: Finding shortest path with neighbourhood sequences in triangular grids. In: International Symposium on Image and Signal Processing and Analysis conjunction with 23rd International Conference on Information Technology Interfaces 2001, pp. 55–60. IEEE, Pula (2001)
15. Nagy, B.: Generalized triangular grids in digital geometry. Acta Mathematica Academiae Paedagogicae Nyíregyháziensis **20**, 63–78 (2004)
16. Nagy, B.: Weighted distances on a triangular grid. In: Barneva, R., Brimkov, V.E., Šlapal, J. (eds.) IWCIA 2014. LNCS, vol. 8466, pp. 37–50. Springer, Cham (2014). https://doi.org/10.1007/978-3-319-07148-0_5
17. Räbinä, J., Kettunen, L., Mönkölä, S., Rossi, T.: Generalized wave propagation problems and discrete exterior calculus. ESAIM Math. Modell. Numer. Anal. **52**(3), 1195–1218 (2018)
18. Radványi, A.G.: On the rectangular grid representation of general CNN networks. Int. J. Circuit Theory Appl. **30**(2–3), 181–193 (2002)
19. Stephenson, J.: Ising model with antiferromagnetic next-nearest-neighbor coupling: spin correlations and disorder points. Phys. Rev. B **1**(11), 4405–4409 (1970)

Discrete Tomography and Inverse Problems

On Some Geometric Aspects of the Class of hv-Convex Switching Components

Paolo Dulio[1] and Andrea Frosini[2（✉）]

[1] Dipartimento di Matematica, Politecnico di Milano, Piazza Leonardo da Vinci 32,
20133 Milano, Italy
paolo.dulio@polimi.it
[2] Dipartimento di Matematica e Informatica "U. Dini", Università di Firenze,
viale Morgagni 65, 50134 Firenze, Italy
andrea.frosini@unifi.it

Abstract. In the usual aim of discrete tomography, the reconstruction of an unknown discrete set is considered, by means of projection data collected along a set U of discrete directions. Possible ambiguous reconstructions can arise if and only if switching components occur, namely, if and only if non-empty images exist having null projections along all the directions in U. In order to lower the number of allowed reconstructions, one tries to incorporate possible extra geometric constraints in the tomographic problem, such as the request for connectedness, or some reconstruction satisfying special convexity constraints. In particular, the class $\mathbb{P}$ of horizontally and vertically convex connected sets (briefly, hv-convex polyominoes) has been largely considered.

In this paper we introduce the class of hv-convex switching components, and prove some preliminary results on their geometric structure. The class includes all switching components arising when the tomographic problem is considered in $\mathbb{P}$, which highly motivates the investigation of such configurations, also in view of possible uniqueness results for hv-convex polyominoes.

It turns out that the considered class can be partitioned into two disjointed subclasses of closed patterns, called windows and curls, respectively, according to whether the pattern can be travelled by turning always clockwise (or always counterclockwise), or whether points with different turning directions exist. It follows that all windows have a unique representation, while curls consist of interlaced sequences of subpatterns, called Z-paths, which leads to the problem of understanding the combinatorial structure of such sequences.

We provide explicit constructions of families of curls associated to some special sequences, and also give additional details on further allowed or forbidden configurations by means of a number of illustrative examples.

Keywords: Curl · Discrete tomography · hv-convex set · Polyomino · Projection · Switching-component · Window · X-ray

Math subject classification: 52A30 · 68R01 · 52C30 · 52C45

© Springer Nature Switzerland AG 2021
J. Lindblad et al. (Eds.): DGMM 2021, LNCS 12708, pp. 299–311, 2021.
https://doi.org/10.1007/978-3-030-76657-3_21

1 Introduction

Discrete Tomography is a part of the wider area of Computerized Tomography, which relates to a huge number of applications where image reconstruction from X-ray collected data is required. While Computerized Tomography involves analytical techniques and continuous mathematics (see, for instance [18,19]), Discrete Tomography is mainly concerned with discrete and combinatorial structures, it works with a small number of density values, in particular with homogeneous objects, and usually allows very few X-rays directions to be considered (see [16,17] for a general introduction to the main problems of Discrete Tomography).

The reconstruction problem is usually *ill-posed*, meaning that *ambiguous and unstable reconstructions* are expected ([16,17]). To limit the number of allowed configurations, further information is usually incorporated in the tomographic problem, which sometimes leads to a unique solution (see for instance [13]) in the case of convex reconstructions), or to the enumeration of the allowed solutions (an example with two projections is [7]).

In case different discrete sets Y_1 and Y_2 are *tomographically equivalent* with respect to a set U of directions, namely Y_1, Y_2 can be reconstructed by means of the same X-rays with respect to U, then there exist specific patterns, called *switching components* which turn Y_1 into Y_2. Understanding the combinatorial and the geometric structure of the switching components is a main issue in discrete tomography (see, for instance [2–4,6,9,11–15,20]).

A largely investigated case concerns the class $\mathbb{P}$ of hv-convex polyominoes, i.e., finite connected subsets of $\mathbb{Z}^2$ that are horizontally and vertically convex. Early results for two projections can be found in [3], where a uniqueness conjecture has been also stated, later disproved in [8]. On this regard, a main role is played by switching components with respect to the horizontal and to the vertical directions, respectively denoted by $\overrightarrow{h}$ and $\overrightarrow{v}$. In [14], such switching components have been studied from an enumerative and an algorithmic point of view, which provided a very interesting and illustrative presentation of their connection with the complexity of the reconstruction problem. In this paper we also focus on such switching components, but we follow a different approach, based on a special geometrical condition (see Definition 2), which defines a class of patterns called hv-*convex switching components*. It includes the classes of *regular* and of *irregular* switching components considered in [14], that we redefine in terms of hv-convex *windows* and hv-convex *curls*, respectively.

The geometric condition in Definition 2 is always satisfied when the switching component is determined by a pair of sets Y_1, Y_2 both internal to the class $\mathbb{P}$. This motivates a deep investigation of the structure of hv-convex switching components, in view of possible uniqueness results for hv-convex polyominoes.

We give a geometric characterization of hv-convex windows (Theorem 1), and a necessary condition for a curl to be a hv-convex switching component (Theorem 2). In general, the condition is not sufficient, but it provides a basic information concerning the geometric structure of hv-convex curls, which leads to the problem of understanding their geometric and combinatorial structure.

2   Notations and Preliminaries

We first introduce some notations and basic definitions. As usual, $\mathbb{R}^2$ denotes the Euclidean two-dimensional space, and $\mathbb{Z}^2 \subset \mathbb{R}^2$ is the lattice of points having integer coordinates. If A is a subset of $\mathbb{R}^2$, we denote by $int(A)$ and by $conv(A)$ the *interior* and the *convex hull* of A, respectively. If A consists of two distinct points v and w, then $conv(A)$ is a *segment*, denoted by $s(v, w)$. If A is a finite set of $\mathbb{Z}^2$, then A is said to be a *lattice set*, and $|A|$ denotes the number of elements of A. A *convex lattice set* is a lattice set $A \subset \mathbb{Z}^2$ such that $A = (conv(A)) \cap \mathbb{Z}^2$.

By $\overrightarrow{h}, \overrightarrow{v}$ we mean the horizontal and the vertical directions, respectively. For any point $v \in \mathbb{R}^2$, we indicate by $L_h(v)$ and $L_v(v)$ the horizontal and the vertical line passing through v, respectively.

Finally, we define *horizontal* (resp. *vertical*) *projection* of a finite set $A \subset \mathbb{Z}^2$ to be the integer vector $H(A)$ (resp. $V(A)$) counting the number of points of A that lie on each horizontal (resp. vertical) line passing through it. We underline that such a notion of projection can be defined for a generic set of discrete lines parallel to a given (discrete) direction.

In the literature, the word *polyomino* indicates a connected finite discrete set of points. In particular, a polyomino is *hv-convex* if each one of its rows and columns is connected. As it is commonly assumed, a polyomino is composed by rows and columns due to the habit of representing it by a binary matrix whose dimensions are those of its minimal bounding rectangle. The class of all *hv*-convex polyominoes is denoted by $\mathbb{P}$.

Given a point $v = (i, j) \in \mathbb{Z}^2$, the four following closed regions are defined (with the same notations as in [5, 10]):

$$Z_0(v) = \{(i', j') \in \mathbb{R}^2 : i' \leq i, j' \leq j\}, \quad Z_1(v) = \{(i', j') \in \mathbb{R}^2 : i' \geq i, j' \leq j\},$$
$$Z_2(v) = \{(i', j') \in \mathbb{R}^2 : i' \geq i, j' \geq j\}, \quad Z_3(v) = \{(i', j') \in \mathbb{R}^2 : i' \leq i, j' \geq j\}.$$

A set of points A is said to be *Q-convex* (quadrant convex) along the horizontal and vertical directions if $Z_l(v) \cap A \neq \emptyset$ for all $l = 0, 1, 2, 3$ implies $v \in A$.

Lemma 1. *Let P be a hv-convex polyomino, and consider a point $v \in \mathbb{Z}^2$. If $w_1, w_2, w_3 \in P$ exist such that $Z_i(v) \cap \{w_1, w_2, w_3\} \neq \emptyset$ for all $i = 0, 1, 2, 3$, then $v \in P$.*

Proof. By [5, Proposition 2.3], a *hv*-convex set is also *Q*-convex with respect to the horizontal and to the vertical directions. The statement follows immediately by the hv-convex property of P. □

2.1   Switching Components and the Uniqueness Problem

Definition 1. *A pair $S = (S^0, S^1)$ of sets of points is a hv-switching if:*

- *$S^0 \cap S^1 = \emptyset$ and $|S^0| = |S^1|$;*
- *$H(S^0) = H(S^1)$ and $V(S^0) = V(S^1)$, i.e., S^0 and S^1 have the same horizontal and vertical projections.*

Each set S^0 and S^1 is indicated as hv-switching component. We underline that also the notion of switching can be extended to the projections along a generic set of discrete directions (again refer to [16,17] for these definitions and the related main results).

A discrete set A contains a hv-switching component if $S^0 \subseteq A$ and $S^1 \cap A = \emptyset$. In this case, we consider $A = Y \cup S^0$, with Y being a (possibly void) discrete set; we define the set $A' = Y \cup S^1$ as the *dual* of A, and we say that the switching S is *associated* to A and A'.

2.2 hv-convex Switching

A classical result in [20] states that if A_1 and A_2 are two discrete sets sharing the same horizontal and vertical projections, then A_2 is the dual of A_1 with respect to a hv-switching. So, for any point $v \in S^0$ (resp. $v \in S^1$), there exist points $w_1, w_2 \in S^1$ (resp. $w_1, w_2 \in S^0$) such that $w_1 \in L_h(v)$ and $w_2 \in L_v(v)$.

If the sets A_1 and A_2 are hv-convex polyominoes, then, due to Lemma 1, for any $x \in S$ there exists one and only one $i \in \{0, 1, 2, 3\}$ such that $Z_i(x) \cap S$ consists of points all belonging to the same component of S as x. The quadrant $Z_i(x)$ is said to be the *free region* of x, or the S-free region of x in case we wish to emphasize that the free region relates to the switching S. We denote by $F(x)$ (or by $F_S(x)$) the free region of $x \in S$. Also, $F_i(S)$ denotes the subset of S consisting of all points having free region $Z_i(x)$, namely $F_i(S) = \{x \in S, F_S(x) = Z_i(x)\}$, $i \in \{0, 1, 2, 3\}$.

We have the following

Lemma 2. *Let $S = (S^0, S^1)$ be a hv-switching. Then, the following conditions are equivalent*

(1) $\bigcup_{i=0}^{3} F_i(S) = S$;
(2) $v, w \in F_i(S), i \in \{0, 1, 2, 3\}, v \in S^0, w \in S^1$ *implies* $v \notin Z_j(w)$, $w \notin Z_j(v)$ *with $j = i + 2 \,(mod\, 4)$.*

Proof. Let $v, w \in S$ such that $v, w \in F_i(S)$, with $v \in S^0, w \in S^1$. Suppose that $v \in Z_j(w)$, with $j = i + 2 \,(mod\, 4)$. Then $w \in Z_i(v)$, a contradiction. Analogously, if $w \in Z_j(v)$, with $j = i + 2 \,(mod\, 4)$, then $v \in Z_i(w)$, a contradiction. Therefore, (2) holds. Conversely, assume that (2) holds. Let $v \in S$, and suppose $v \in S^0$. Since S is a hv-switching, then there exist three values of $k \in \{0, 1, 2, 3\}$ such that $Z_k(v) \cap S^1 \neq \emptyset$. Suppose that $w \in S^1$ exists such that $w \in Z_i(v)$ for $i \neq k$. Then $v \in Z_j(w)$, where $j = i + 2 \,(mod\, 4)$, which contradicts (2). Therefore, v has a free region, namely $F(v) = Z_i(v)$. With the same argument we get that any $w \in S^1$ has a free region. Therefore, each point of S has a nonempty free region, and (1) follows.   □

Definition 2. *Let $S = (S^0, S^1)$ be a hv-switching. Then, S is said to be a hv-convex switching if one of the equivalent conditions of Lemma 2 holds.*

Remark 1. *By the above discussion, if* $S = (S^0, S^1)$ *is a hv-switching associated to a pair of hv-convex polyominoes, then* (1) *holds, so* S *is a hv-convex switching. However, the converse is not necessarily true, namely it could exist two polyominoes* P_1 *and* P_2, *where one is the dual of the other with respect to* S, *and such that one or both of them are not hv-convex polyominoes. An interesting case is Figure 23 in [14], or Fig. 2 below.*

2.3   Squared Spirals

A closed polygonal curve K in $\mathbb{R}^2$ is said to be a *squared spiral* if K consists of segments having, alternatively, horizontal and vertical direction. Their endpoints form the *set of vertices* of the polygonal, denoted by $V(K)$. Two squared spirals are said to *intersect* in case some of (possibly all) their segments intersect. Assume to travel K according to a prescribed orientation. A vertex v of K is said to be a *counterclockwise point* if, crossing v, implies a counterclockwise change of direction. Differently, v is a *clockwise point.* Of course, by reversing the travelling orientation, clockwise and counterclockwise vertices mutually exchange. The *bounding rectangle* of K is the smallest rectangle R_K containing K.

2.4   Windows and Curls

We now introduce two classes of special squared spirals that provide a geometric reformulation of the notions of *regularity* and of *irregularity* discussed in [14], which, in addition, constitute the main focus of our study. A squared spiral W is said to be a *window* if it can be traveled by turning always clockwise, or always counterclockwise. Differently, the squared spiral is said to be a *curl.* Therefore, travelling a curl needs changes of turning direction.

Obviously a rectangle is a particular case of window that coincides with its bounding rectangle.

Remark 2. *Each window and each curl form a hv-switching* $S = (S^0, S^1)$ *by considering the corresponding vertices alternatively belonging to* S^0 *and* S^1.

2.5 Z-paths

A Z-path is a staircase shaped pattern consisting of a monotone sequence of horizontal and vertical segments, whose vertices alternate between clockwise and counterclockwise points. Note that staircase like patterns play a similar role in [1,21]. We say that the Z-path is of *type* SE-NW, or SW-NE, according as it can be travelled moving from South-East to North-West (or conversely), or from South-West to North-East (or conversely), respectively. A *simple,* or *one-level,* Z-path consists of just three segments, horizontal-vertical-horizontal, or vertical-horizontal-vertical, referred to as *hvh,* or *vhv* Z-path, respectively. Excluding its endpoints, a simple Z-path exhibits a pair of vertices having a specified orientation, clockwise-counterclockwise, or clockwise-counterclockwise, according to the considered type, and moving from south to north along the pattern. In general, for $q > 0$, we have a q-*level* Z-path if, excluding its endpoints, it consists of $q+1$ vertices having alternating orientations. Therefore, if q is odd, we have q horizontal and $q-1$ vertical segments, or conversely, and we refer to the corresponding Z-path with the notation $h(vh)_{q-1}$ and $v(hv)_{q-1}$, respectively. If q is even, then the Z-path consists of q horizontal and of q vertical segments, and we adopt the notation $(hv)_q$, or $(vh)_q$, according as the first segment is horizontal or vertical (see Fig. 1). Any Z-path is a hv-convex set. In a SE-NW Z-path, any vertex v, different from an endpoint, has free region $Z_0(v)$ or $Z_2(v)$, while, in a SW-NE Z-path, the free region is $Z_1(v)$ or $Z_3(v)$. In any case, the elements of the sets of free regions $\{Z_0(v), Z_2(v)\}$, or $\{Z_1(v), Z_3(v)\}$ alternate along the Z-path. Since the vertices of a Z-path are, alternatively, clockwise and counterclockwise oriented, then no q-level Z-path, with $q > 0$, can be found in a window, while any curl surely includes some Z-paths. Differently, if $q = 0$, we have an L-shaped path, consisting of an horizontal and a vertical segment, with just one intermediate point. We refer to such a path as a *degenerate Z-path*. Note that a window can be considered as a consecutive sequence of degenerate Z-paths, while, in a curl, different Z-paths (possibly degenerate) can appear. In what follows, we provide a precise characterization of how these paths can be combined together.

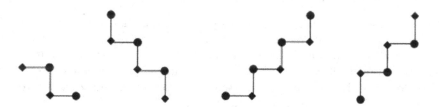

Fig. 1. Different types of Z-paths. From left to right: a simple hvh SE-NW Z-path, a $v(hv)_2$ SE-NW Z-path, a $(hv)_3$ SW-NE Z-path, and a $v(hv)_2$ SW-NE Z-path.

3   Characterization of hv-convex Windows

We give a necessary and sufficient condition for a window to be hv-convex. This leads to a geometric characterization of the hv-convex switchings that have the structure of a hv-convex window.

Theorem 1. *Let W be a window of size $n \geq 1$ and $\{w_1, w_2, ..., w_{4n}\}$ be the set of its vertices. Then W is a hv-convex switching if and only if a point $x \in \mathbb{R}^2$ exists such that $w_i \in Z_0(x) \cup Z_2(x)$ for all the odd indices, and $w_i \in Z_1(x) \cup Z_3(x)$ for all the even indices.*

Proof. Assume that a point $x \in \mathbb{R}^2$ exists such that $w_i \in Z_0(x) \cup Z_2(x)$ for all the odd indices, and $w_i \in Z_1(x) \cup Z_3(x)$ for all the even indices. Then, by definition of window, W has the same number of vertices in each $Z_i(x)$, $i = 0, 1, 2, 3$, namely, $w_i \in Z_0(x)$, for $i = 1 \, (mod \, 4)$, $w_i \in Z_1(x)$, for $i = 2 \, (mod \, 4)$, $w_i \in Z_2(x)$, for $i = 3 \, (mod \, 4)$, and $w_i \in Z_3(x)$, for $i = 0 \, (mod \, 4)$. Therefore, if W^0 and W^1 are, respectively, the set of the even and of the odd labeled vertices of W, then each point of W^0 has a horizontal and a vertical corresponding in W^1 and conversely. This implies that the free regions of all points in W^0 are contained in $Z_1(x)$ or in $Z_3(x)$, and the free regions of all points in W^1 are contained in $Z_0(x)$ or in $Z_2(x)$. Therefore, W is hv-convex.

Conversely, suppose that W is a hv-convex switching. Without loss of generality we can assume that W is traveled counterclockwise, starting from w_1. Also, up to a rotation (which does not change the argument) we can always assume that the free region of w_1 is $Z_0(w_1)$. Then the free region of w_i is $Z_j(w_i)$ where $i - j = 1 \, (mod \, 4)$. For $j = 0, 1, 2, 3$, let H_j be the set $H_j = \bigcup_{i=j+1 \, (mod \, 4)} Z_j(w_i)$.

Due to the hv-convexity of W, the sets H_i are mutually disjointed. Consider the strip bounded by the two horizontal lines supporting $H_0 \cup H_1$ and $H_2 \cup H_3$, and the strip bounded by the two vertical lines supporting $H_0 \cup H_3$ and $H_1 \cup H_2$ (see Fig. 2).

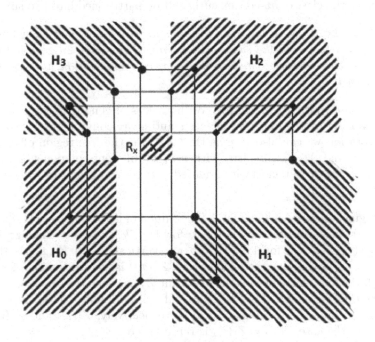

Fig. 2. The four regions H_i, $i \in \{0, 1, 2, 3\}$ related to a hv-convex window. The rectangle R_x contains all the points having the property stated in Theorem 1.

The intersection of such strips forms a rectangle R, having horizontal and vertical sides, and with no points of W belonging to the internal $int(R)$ of R. Let x be any point such that $x \in int(R)$. Then $H_j \subseteq Z_j(x)$ for all $j = 0, 1, 2, 3$, and the statement follows.   □

Remark 3. *The property stated in Theorem 1 is not restricted to a single point, but it involves all the points belonging to the interior of the rectangle R.*

For any hv-convex window W, and point $x \in \mathbb{R}^2$ as in Theorem 1, all quadrants $Z_i(x)$, $i \in \{0, 1, 2, 3\}$ contain the same number of points of W, which is said the *size* of the window. Note that a window can be a switching component with respect to the horizontal and vertical directions without being hv-convex.

4   Characterization of hv-convex Curls

Moving to curls, a deeper analysis is required, as it has been pointed out in [14] in terms of irregular switching components. Here we push the study a step ahead, by investigating the geometric nature and the main features of those curls that form hv-convex switching. As a first result, we prove a necessary condition for a curl to be a hv-convex switching, say hv-*convex curl*. In general, the given condition is not sufficient, but it spreads light on the geometric structure of the hv-convex curls, and leads to their characterization in terms of Z-paths. As a consequence, the class of hv-convex curls will be partitioned into two subclasses.

Theorem 2. *Let C be a curl that forms a hv-switching, and let v and w be two points in $V(C)$ with the same orientation. If precisely $2n > 0$ consecutive vertices between v and w exist, with their opposite orientation, then C is not a hv-convex curl.*

Proof. Suppose that C is hv-convex. Without loss of generality, we can assume that travelling C from v to w the vertices v and w are counterclockwise oriented. Up to a rotation we can also assume that $Z_0(v)$ is the free region of v, so that a vertex $v_1 \in Z_2(v) \cap Z_3(v)$ exists, with v, v_1 in different components of C. Let $x_1, ..., x_{2n}$ be the $2n > 0$ clockwise oriented vertices of C that are crossed when moving from v to w.

The segment $s(v, x_1)$ is horizontal. The same holds for the segment $s(x_{2n}, w)$, and also for all segments $s(x_{2k}, x_{2k+1})$, for $1 \le k \le n - 1$. Analogously, all segments $s(x_{2k-1}, x_{2k})$, for $1 \le k \le n$ are vertical. Then $Z_2(x_1)$ is the free region of x_1, $Z_1(x_2)$ is the free region of x_2, $Z_0(x_3)$ is the free region of x_3, and, in general, the free region of x_i is the quadrant $Z_j(x_i)$ such that $i+j = 3 \,(mod\,4)$. Therefore, the free region of x_{2n} is $Z_1(x_{2n})$ if n is odd, and $Z_3(x_{2n})$ if n is even, which implies that the free region of w is, respectively, $Z_3(w)$ and $Z_1(w)$ (see Fig. 3, where the case $F(w) = Z_1(w)$ is represented).

Now, all the vertices x_h, with h odd, belong to a component different from that v, so they do not lie in $F(v) = Z_0(v)$. Since $x_1 \in Z_1(v) \cap Z_2(v)$, then $x_2 \in int(Z_1(v))$. If $n = 1$, then $w \in int(Z_1(v))$, and $F(w) = Z_3(w)$, so that $v \in F(w)$, a contradiction, since v and w belong to different components of C. So, the statement follows for $n = 1$. If $n > 1$, then $x_3 \in int(Z_1(v))$, so $x_4 \in Z_1(v) \cup Z_2(v)$. However, $x_4 \notin Z_1(v)$, since, otherwise $v_1 \in F(x_4) = Z_3(x_4)$, a contradiction, being v and v_1 in different components of C. By iterating the

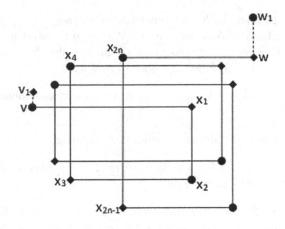

Fig. 3. Positions of consecutive vertices having a same orientation in a curl.

argument, we get that all the vertices of the form x_{2k}, with $k \leq n$ and k even must belong to $Z_2(v)$. Analogously, all the vertices of the form x_{2k}, with $k \leq n$ and k odd must belong to $Z_1(v)$, since, differently, x_{2k} would belong to $F(x_1) = Z_3(x_1)$, or conversely, $x_1 \in F(x_{2k}) = Z_1(x_{2k})$, a contradiction, since x_1 and x_{2k} belong to different components of C. This implies that the vertices x_{2k-1} with $2 \leq k \leq n$ and k even belong to $Z_1(v)$, while the vertices x_{2k-1} with $1 < k \leq n$ and k odd belong to $Z_2(v)$. Consequently, also $w \in Z_1(v) \cup Z_2(v)$.

Suppose that $w \in Z_1(v)$. Since the segment $s(x_{2n}, w)$ is horizontal, then x_{2n} also belongs to $Z_1(v)$. As shown above, this implies that n is odd, so $F(w) = Z_3(w)$, and consequently $v \in F(w)$, a contradiction.

Hence $w \in Z_2(v)$, then x_{2n} also belongs to $Z_2(v)$, which implies that n is even, and consequently $F(x_{2n}) = Z_3(x_{2n})$, and $F(w) = Z_1(w)$. Therefore, w must belong to $Z_2(x_2)$, otherwise $w \in Z_3(x_2)$, and consequently $x_2 \in F(w) = Z_1(w)$, a contradiction. From $w \in Z_2(x_2)$, and $w \in Z_2(v)$, it follows that $w \in Z_2(x_1)$. Since C is a switching with respect to the vertical direction, then there exists a vertex $w_1 \in Z_2(w) \cap Z_3(w)$, with w and w_1 in different components, and consequently also x_1 and w_1 belong to different components of C (see Fig. 3). Since $w \in Z_2(x_1)$, then also $w_1 \in F(x_1) = Z_2(x_1)$, a contradiction.

Consequently, the assumption that C is hv-convex always leads to a contradiction, and the statement follows. □

5   On Some Sequences Associated to hv-convex Switchings

Let S be a squared spiral. We associate to S an integer sequence $(k_1, k_2, ..., k_n)$, say hv-sequence, where each k_i represents the i-th maximal sequence of k_i vertices that can be travelled clockwise or counterclockwise, with $i = 1, 2, ..., n$. The starting vertex is not indicated, so the sequence can be considered up to circular shifts. If the sequence $(k_1, k_2, ..., k_n)$ is periodic, then we adopt the notation

$(k_1, ..., k_{n'})_h$, to represent the h time repetition of the sequence $(k_1, ..., k_{n'})$, with $n = n' \cdot h$; if $h = 1$, we choose to omit it. We are interested in characterizing the hv-sequences that admit a hv-convex switching, called hv-*convex sequences*. Therefore, we are led to the following general problem.

Problem 1. For which $k_1, ..., k_n, h \in \mathbb{N}$ does a hv-convex sequence $(k_1, ..., k_n)_h$ exist?

Concerning windows, Problem 1 has an easy solution.

Theorem 3. *For each $n > 0$, $(4n)$ is a hv-convex sequence if and only if the associated squared spiral is a window.*

Proof. By Theorem 2, the hv-sequence associated to a curl is of the form $(k_1, ..., k_{n'})_h$, where $k_1, ..., k_n$ are odd. Therefore, $(4n)$ cannot be the hv-sequence associated to a curl. Let W be a window. By Theorem 1, $x \in \mathbb{R}^2$ exists such that all quadrants $Z_i(x)$, $i \in \{0, 1, 2, 3\}$ contain the same number of points of W. Then, W has $4n$ vertices, for some $n > 0$, which implies that the hv-sequence associated to W is $(4n)$.   □

Differently, Problem 1 seems to require a deeper investigation of the geometrical and combinatorial structure of the set of vertices of a curl. In this view, we give here some preliminary remarks. First of all, note that, having a $(k_1, ..., k_n)_h$ curl, is in general not sufficient to get hv-convexity, since the conditions in Lemma 2 do not automatically hold.

A vertex v of a curl whose turning direction differs from that of the previous encountered point, is said to be a *changing point*. In order to improve our knowledge on the hv-sequences allowed for curls, it is worth focusing on the possible Z-paths that can be included in a hv-convex curl, which reflects in the understanding of how changing points can occur. As already observed, the simplest hv-convex curl is the $(3, 3)_1$ curl shown in Fig. 4 (a). Its vertices consists of six points $x_1, ..., x_6$, where $F(x_1) = Z_1(x_1)$, $F(x_2) = Z_0(x_2)$, $F(x_3) = Z_3(x_3)$, $F(x_4) = Z_1(x_4)$, $F(x_5) = Z_2(x_5)$, $F(x_6) = Z_3(x_6)$. The $(3, 3)_1$ curl can also be considered as the union of two simple hvh and vhv SW-NE Z-paths (see Sect. 2.5) having x_2 and x_5 in common. This means that x_1, x_4 are changing points (or x_3, x_6, depending on the starting choice for the walking direction). Figure 4 (b) shows a $(3, 3)_2$ hv-convex curl, consisting of two different pairs of intersecting simple SW-NE Z-paths (analogous constructions can be performed by using SE-NW Z-paths). Analogously, for any integer number $h \geq 1$, a curl C can be constructed having h pairs of intersecting SW-NE Z-paths. These can be consecutively arranged, or, differently, connected by means of L-shaped paths, as described above. See Fig. 4 (c) for an example where $h = 4$.

Of course, curl containing Z-paths of higher level can be also constructed.

However, different Z-paths of a same curl are not necessarily consecutive. For instance, Fig. 5 shows how to insert degenerate Z-paths (L-shaped paths) between the bottom-left endpoint of a Z-path and the upper-right endpoint of a different Z-path, so transforming a $(3, 3)_1$ curl into a $(5, 5)_1$ curl.

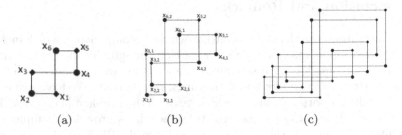

Fig. 4. (a) A $(3,3)_1$ hv-convex curl C, corresponding to a SW-NE vertex-gluing of two rectangles. (b) A $(3,3)_2$ hv-convex curl with two pairs of intersecting simple Z-paths. (c) Example of curl associated to the hv-sequence $(3,3)_4$.

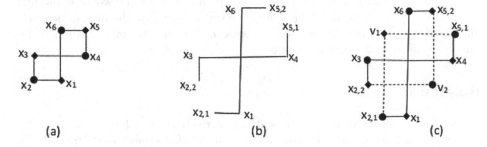

Fig. 5. Including L-shaped paths in a given curl. (a) The starting curl. (b) The split of the two constituent simple Z-paths. (c) The connection of the two simple Z-paths by joining their extremal vertices with two degenerate Z-paths.

Further constructions also exist having associated hv-convex sequence of type $(k_1, k_2)_h$, with $k_1 \neq k_2$. Figure 6 shows a curl associated to the hv-convex sequence $(3,5)_2$.

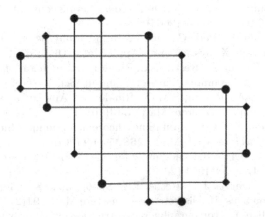

Fig. 6. The hv-convex curl associated to the hv-convex sequence $(3,5)_2$.

6 Conclusion and Remarks

We have introduced the class of hv-convex switching components, which includes all switching components associated to a pair of tomographically equivalent hv-convex polyominoes. We have separated the class in two disjointed subclasses of closed patterns, the windows and the curls, respectively. We have given geometrical results on both subclasses, which leads to the problem of characterizing them in terms of hv-convex sequences. While windows provide a complete and easy solution, deeper investigation is required for curls. We have discussed a few preliminary allowed or forbidden hv-sequences, which provide partial answers to Problem 1 in the case of curls. For a complete solution to Problem 1 it becomes relevant to understand how, in general, different Z-paths can be connected between them in a same curl. In particular, it would be worth exploring possible connections between the allowed levels of the Z-paths in a same curl, and the degree of convexity of L-convex sets [7]. We wish to investigate in these directions in separated further works

References

1. Alpers, A., Brunetti, S.: On the stability of reconstructing lattice sets from X-rays along two directions. In: Andres, E., Damiand, G., Lienhardt, P. (eds.) DGCI 2005. LNCS, vol. 3429, pp. 92–103. Springer, Heidelberg (2005). https://doi.org/10.1007/978-3-540-31965-8_9
2. Alpers, A., Tijdeman, R.: The two-dimensional Prouhet-Tarry-Escott problem. J. Number Theory **123**(2), 403–412 (2007)
3. Barcucci, E., Del Lungo, A., Nivat, M., Pinzani, R.: X-rays characterizing some classes of discrete sets. Linear Algebra Appl. **339**, 3–21 (2001)
4. Barcucci, E., Dulio, P., Frosini, A., Rinaldi, S.: Ambiguity results in the characterization of hv-convex polyominoes from projections. In: Kropatsch, W.G., Artner, N.M., Janusch, I. (eds.) DGCI 2017. LNCS, vol. 10502, pp. 147–158. Springer, Cham (2017). https://doi.org/10.1007/978-3-319-66272-5_13
5. Brunetti, S., Daurat, A.: An algorithm reconstructing convex lattice sets. Theoret. Comput. Sci. **304**(1–3), 35–57 (2003)
6. Brunetti, S., Dulio, P., Peri, C.: Discrete tomography determination of bounded lattice sets from four X-rays. Discrete Appl. Math. **161**(15), 2281–2292 (2013)
7. Castiglione, G., Frosini, A., Restivo, A., Rinaldi, S.: Enumeration of L-convex polyominoes by rows and columns. Theoret. Comput. Sci. **347**(1–2), 336–352 (2005)
8. Dulio, P., Frosini, A., Pagani, S.M.C., Rinaldi, S.: Ambiguous reconstructions of hv-convex polyominoes. Discrete Math. **343**(10), 111998 (2020)
9. Dulio, P., Pagani, S.M.C.: A rounding theorem for unique binary tomographic reconstruction. Discrete Appl. Math. **268**, 54–69 (2019)
10. Dulio, P., Peri, C.: Discrete tomography for inscribable lattice sets. Discrete Appl. Math. **161**(13–14), 1959–1974 (2013)
11. Fishburn, P.C., Lagarias, J.C., Reeds, J.A., Shepp, L.A.: Sets uniquely determined by projections on axes. II. Discrete case. Discrete Math. **91**(2), 149–159 (1991)
12. Frosini, A., Vuillon, L.: Tomographic reconstruction of 2-convex polyominoes using dual horn clauses. Theor. Comput. Sci. **777**, 329–337 (2019)

13. Gardner, R.J., Gritzmann, P.: Discrete tomography: determination of finite sets by X-rays. Trans. Amer. Math. Soc. **349**(6), 2271–2295 (1997)
14. Gerard, Y.: Regular switching components. Theor. Comput. Sci. **777**, 338–355 (2019)
15. Hajdu, L., Tijdeman, R.: Algebraic aspects of discrete tomography. J. Reine Angew. Math. **534**, 119–128 (2001)
16. Herman, G.T., Kuba, A.: Discrete Tomography: Foundations, Algorithms, and Applications. Birkhäuser, Boston (1999)
17. Herman, G.T., Kuba, A.: Advances in Discrete Tomography and its Applications. Birkhäuser, Boston (2006)
18. Kak, A.C., Slaney, M.: Principles of Computerized Tomographic Imaging. Society for Industrial and Applied Mathematics, Philadelphia (2001)
19. Natterer, F.: The Mathematics of Computerized Tomography. Teubner, Stuttgart (1986)
20. Ryser, H.J.: Combinatorial properties of matrices of zeros and ones. Canad. J. Math. **9**, 371–377 (1957)
21. van Dalen, B.: Stability results for uniquely determined sets from two directions in discrete tomography. Discrete Math. **309**(12), 3905–3916 (2009)

Properties of Unique Degree Sequences of 3-Uniform Hypergraphs

Michela Ascolese[1], Andrea Frosini[1]([✉]), William Lawrence Kocay[2], and Lama Tarsissi[3,4]

[1] Dipartimento di Sistemi e Informatica, Universitàdi Firenze, Firenze, Italy
andrea.frosini@unifi.it
[2] Department of Computer Science and St. Pauls College, University of Manitoba, Winnipeg, MB, Canada
[3] LIGM, University Gustave Eiffel, CNRS, ESIEE, F-77454 Marne-la -Vallée, Paris, France
[4] Department of Sciences and Engineering, Sorbonne University Abu Dhabi, Abu Dhabi, United Arab Emirates

Abstract. In 2018 Deza et al. proved the NP-completeness of deciding wether there exists a 3-uniform hypergraph compatible with a given degree sequence. A well known result of Erdös and Gallai (1960) shows that the same problem related to graphs can be solved in polynomial time. So, it becomes relevant to detect classes of uniform hypergraphs that are reconstructible in polynomial time. In particular, our study concerns 3-uniform hypergraphs that are defined in the NP-completeness proof of Deza et al. Those hypergraphs are constructed starting from a non-increasing sequence s of integers and have very interesting properties. In particular, they are unique, i.e., there do not exist two non isomorphic 3-uniform hypergraphs having the same degree sequence d_s. This property makes us conjecture that the reconstruction of these hypergraphs from their degree sequences can be done in polynomial time. So, we first generalize the computation of the d_s degree sequences by Deza et al., and we show their uniqueness. We proceed by defining the equivalence classes of the integer sequences determining the same d_s and we define a (minimal) representative. Then, we find the asymptotic growth rate of the maximal element of the representatives in terms of the length of the sequence, with the aim of generating and then reconstructing them. Finally, we show an example of a unique 3-uniform hypergraph similar to those defined by Deza et al. that does not admit a generating integer sequence s. The existence of this hypergraph makes us conjecture an extended generating algorithm for the sequences of Deza et al. to include a much wider class of unique 3-uniform hypergraphs. Further studies could also include strategies for the identification and reconstruction of those new sequences and hypergraphs.

Keywords: Hypergraph · Graphic sequence · Uniqueness problem · Analysis of algorithms

AMS classification: 05C65 · 05C60 · 05C99

© Springer Nature Switzerland AG 2021
J. Lindblad et al. (Eds.): DGMM 2021, LNCS 12708, pp. 312–324, 2021.
https://doi.org/10.1007/978-3-030-76657-3_22

1   Introduction

The notion of hypergraph naturally extends that of graphs, where each edge is defined to be a subset of the vertices (see [2] for basic definitions and results on hypergraphs).

In this paper, we consider *simple* hypergraphs, i.e., hypergraphs that are loopless and with distinct edges. Here a loop is considered to be an edge consisting of just one vertex. A hypergraph is *h-uniform* if every edge is a subset of cardinality h.

The degree sequence of a simple hypergraph is the sequence of the degrees of its vertices arranged in non-increasing order. A degree sequence usually does not characterize the hypergraph it is related to, but however it reveals interesting properties. For this reason, degree sequences are widely studied in (hyper)graph theory. We say that a *reconstruction* of a degree sequence d is a (hyper)graph whose degree sequence is d. One of the most challenging problems related to hypergraph degree sequences is the reconstruction of a compatible (hyper)graph, if any [4]. For graphs, this problem has been solved in a milestone paper by Erdös and Gallai [7], in 1960, providing an efficiently computable characterization of them. And an algorithm to construct a graph with a given degree sequence was given by Havel [11] and Hakimi [10]. On the other hand, the same problem for hypergraphs remained unsolved till 2018, when Deza et al. in [6] proved its NP-completeness, even for the simplest case of 3-uniform hypergraphs.

As a consequence, the study of wide classes of uniform hypergraphs whose reconstruction from a degree sequences can be performed in polynomial time has acquired more and more relevance, in order to spot the hard core of the generic reconstruction problem.

Recently, but still before the result in [6], some necessary conditions had been given for a sequence to be the degree sequence of a k-uniform hypergraph. Such a sequence is called an k-*sequence*. Most of these generalized the Erdös and Gallai theorem, or were based on two well known theorems by Havel [11] and Hakimi [10]. On the other hand, few necessary conditions were known. Among them, one of prominent interest is provided in [1], which uses Dewdney's theorem in [5], and sets a lower bound on the length of a sequence related to an k-uniform hypergraph. This result has been algorithmically rephrased in [3,9], by using techniques borrowed from Discrete Tomography. This research area mainly focuses on the retrieval of geometrical information and the full reconstruction of an unknown object modelled by a binary or integer matrix by means of a set of projections along prescribed directions (usually the horizontal and the vertical ones). The possibility of representing a k-hypergraph by its incidence (binary) matrix whose horizontal and vertical projections turn out to be the constant vector of entries k and its degree sequence, respectively, provides evidence of the possibility of a fruitful use of tomographic tools in the detection and reconstruction of the hypergraphs related to a degree sequence.

In particular, the present research focuses on investigating the properties of the 3-sequences that originate from a generalization of the gadget used by Deza et al. in [6] for their NP-completeness proof (see [8] for a preliminary study). These are denoted by $\mathcal{D}_n$, according to their length n. The relevance of those sequences is mainly due to their uniqueness property, i.e., there exists a unique 3-hypergraph compatible with them, up to isomorphism. It is known that there exists an operator called a *trade* that allows one to travel amongst all hypergraphs having the same degree sequence [12]; here we prove that the 3-hypergraphs related to the elements of each $\mathcal{D}_n$ act as a sort of fixed point for this operator.

Furthermore, since each element of $\mathcal{D}_n$ can be related to an infinite number of integer sequences, we group them into equivalence classes and we choose a representative for each class. We compute a lower bound to the asymptotic growth of the representative's elements gaining information about the cardinality of $\mathcal{D}_n$ and obtaining clues for a strategy for reconstruction.

From this preliminary study, a series of open problems results, with the long term aim of characterizing and of reconstructing wider classes of unique degree sequences. An example is given in Example 2.

So, in the next section, we will provide definitions and the results useful for our study. Section 3 will be devoted to presenting the most relevant properties of the degree sequences in $\mathcal{D}_n$ and the asymptotic growth of the elements of their representatives. In Sect. 4, we consider the reconstruction problem of a simple subclass of $\mathcal{D}_n$ and the isomorphism problem on unique sequences. We conclude our work by pointing out some open questions in Sect. 5.

2 Basic Notions and Definitions

We recall the basic definition of hypergraphs and fix the notation used in the rest of the paper. A hypergraph H is defined as a pair $H = (V, E)$ such that V is the set of vertices and $E \subset \mathcal{P}(V) \setminus \{\emptyset\}$ is the set of hyperedges, or briefly edges when no ambiguities occur, where $\mathcal{P}(V)$ is the power set of V.

A hypergraph is *simple* if it is loopless, i.e., it does not allow singleton edges. The *degree* of a vertex $v \in V$ is the number of edges containing v. The degree sequence $d = (d_1, d_2, \ldots, d_n)$ of a hypergraph H is the list of its vertex degrees usually arranged in non-increasing order. Let us denote by $\sigma(d)$ the sum of the elements of d. When $\mathcal{H}$ is k-uniform, the sequence d is called k-graphic. Notice that the case $k = 2$ corresponds to graphs, and a 2-graphic sequence is simply called graphic.

The seminal book by Berge [2] contains some essential results about hypergraphs and also information about their applications.

The problem of characterizing the graphic sequences of simple graphs was solved by Erdös and Gallai [7], in 1960, and algorithmically by Havel [11] and Hakimi [10], while only recently Deza et al. in [6] have shown the NP-completeness of the characterization of k-graphic sequences, with $k \geq 3$. The *3-Hypergraph Degree Sequence Problem* is the question of determining whether a given sequence d is the degree sequence of a 3-hypergraph.

In their proof, the authors mapped the instances of the known NP-complete problem *3-Partition* into instances of the 3-Hypergraph Degree Sequence problem. The mapping has the property that there is a 1-to-1 correspondence between the solutions of the instances I of *3-partition* and the 3-hypergraphs having the prescribed degree sequence d, computed from I.

To compute d, Deza et al. used an intermediate step in which they constructed a "gadget", and computed its degree sequence d' from I. This sequence turns out to have very interesting properties, as shown in the next sections, and it constitutes the focus of our research.

Hereafter, we define the procedure *Gen-pi(s)* that generalizes the gadget computation presented in [6], regardless the specific characteristics of the length and element sum of the instance of *3-Partition*. The input of this computation is an integer sequence $s = (s_1, \ldots, s_n)$, and it returns a degree sequence, denoted d_s, to emphasize its dependence on s, of a 3-hypergraph. Following the notation in [6], let $\{0, 1\}_3^n$ be the set of all the binary sequences of length n having exactly three elements equal to 1 and let s^T be the transpose of the vector s.

Algorithm 1. *Gen-pi(s)*

---

set $E \leftarrow \emptyset$
for all $e \in \{0, 1\}_3^n$ **do**
    if $e \cdot s^T > 0$ **then**
        $E \leftarrow E \cup \{e\}$
    end if
end for
return $d_s = \Sigma E$

The class of degree sequences of length n generated by the algorithm *Gen-pi*, from an input sequence s is indicated as $\mathcal{D}_n$.

The action of *Gen-pi* on s will be clearer after introducing the notion of incidence matrix of a hypergraph. Given a hypergraph $H = (V, E)$ such that $|V_{\mathcal{H}}| = n$ and $|E_{\mathcal{H}}| = m$, its *incidence matrix* is a $m \times n$ binary matrix where $a_{i,j} = 1$ if and only if the edge e_i contains the vertex v_j, otherwise $a_{i,j} = 0$. So $\Sigma_{i=1}^m a_{i,j} = d_j$ is the degree of the vertex v_j, and when $\mathcal{H}$ is k−uniform we have $\Sigma_{j=1}^n a_{i,j} = k$ for each edge e_i.

We observe that, for k-hypergraphs, the property of being simple means that all the rows of the incidence matrix are different.

Example 1. *Let us consider the integer sequence $s = (3, 2, 0, -1, -2)$. The output of Gen-pi(s) is the degree sequence $d_s = (5, 4, 4, 3, 2)$. Figure 1 shows a 3-hypergraph $H(d_s)$ having d_s as degree sequence, together with its related incidence matrix $M(d_s)$.*

The matrix representation of the hypergraph $H(d_s)$ in Fig. 2 provides an immediate idea of the action of *Gen-pi* on s. In the sequel, we will consider only degree sequences obtained from sequences s such that $s_1 + s_2 > s_n$, i.e., without null columns or, equivalently, hypergraphs without isolated vertices.

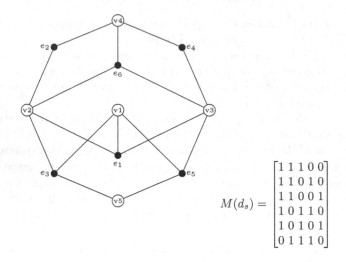

$$M(d_s) = \begin{bmatrix} 1 & 1 & 1 & 0 & 0 \\ 1 & 1 & 0 & 1 & 0 \\ 1 & 1 & 0 & 0 & 1 \\ 1 & 0 & 1 & 1 & 0 \\ 1 & 0 & 1 & 0 & 1 \\ 0 & 1 & 1 & 1 & 0 \end{bmatrix}$$

Fig. 1. The 3-hypergraph $H(d_s)$ and its related incidence matrix $M(d_s)$. The black nodes represent the edges of the 3-hypergraph and the white circles represent its vertices. The edge e_i refers to the i-th row of $M(d_s)$.

3   Properties of the Elements of $\mathcal{D}_n$

A relevant property of a degree sequence d_s directly follows from the action of *Gen-pi* on a generic integer sequence s:

Theorem 1. *There exists one only 3-hypergraph (up to isomorphism) having degree sequence* d_s.

Proof. The result can be obtained by contradiction. Let $d_s = (d_1, \ldots, d_n)$ and $s = (s_1, \ldots, s_n)$ be as in the algorithm *Gen-pi*. We observe that the number Max$= \Sigma_{i=1}^n s_i d_i$ is the maximum that can be realized by a sequence of $m = \Sigma_{i=1}^n d_s/3$ different triplets $(s_{i_1}, s_{j_1}, s_{k_1}), \ldots, (s_{i_m}, s_{j_m}, s_{k_m})$ of elements of s, containing all the edges $(i_1, j_1, k_1), \ldots, (i_m, j_m, k_m)$ of the 3-hypergraph $H(d_s)$. Any other 3-hypergraph $H(d_s)'$ having the same degree sequence would involve at least an edge $e = (i, j, k)$ not in $H(d_s)$, so that $s_i + s_j + s_k = t \leq 0$. As a consequence, the remaining $m - 1$ different edges $(i_1', j_1', k_1'), \ldots, (i_{m-1}', j_{m-1}', k_{m-1}')$ of $H(d_s)'$ must satisfy $\Sigma_{z=1}^{m-1} s_{i_z'} + s_{j_z'} + s_{k_z'} = \text{Max} + t$, and we reach a contradiction. $\qquad\square$

We say that the sequence d_s is *unique*, as well as the corresponding 3-hypergraph.

Note that the number of elements of $\mathcal{D}_n$ is finite, since each d_s has $\binom{n}{3}\frac{3}{n}$ as a maximum entry. So, by cardinality reasons, there exist an infinite number of non-increasing integer sequences s that generate at least one degree sequence d_s. An easy check reveals that for each degree sequence d_s there exists an infinite number of generating integer sequences.

As an example, all the length n sequences of positive integers generate the same constant degree sequence $d_s \in \mathcal{D}_n$ whose elements are $\binom{n}{3}\frac{3}{n} = \frac{(n-1)(n-2)}{2}$. Obviously, this sequence is maximal in $\mathcal{D}_n$ w.r.t the lexicographical order.

An easy check reveals that if a sequence s has two equal elements s_i and s_{i+1}, then also the elements d_i and d_{i+1} of the related d_s degree sequence are equal. The reverse is also true.

Property 1. *If there exists an index $i < n$ of d_s such that $d_i = d_{i+1}$, then there exists a sequence s' such that $s'_i = s'_{i+1}$ and $d_s = d_{s'}$.*

Proof. Let us construct s' from s: if $s_i = s_{i+1}$, then $s' = s$. So, let us consider the case $s_i \neq s_{i+1}$. Since $s_i > s_{i+1}$, then each triplet $s_{i+1} + s_j + s_k > 0$ implies the triplet $s_i + s_j + s_k > 0$, and each triplet $s_i + s_{j'} + s_{k'} \leq 0$ implies the triplet $s_{i+1} + s_{j'} + s_{k'} \leq 0$, with $j, j', k, k' \leq n$. Since $d_i = d_{i+1}$, then the reverse of both implications also holds, and furthermore, these inequalities also hold if s_i is replaced by an integer $s_{i+1} \leq \bar{s} \leq s_i$. For the inequalities that involve both s_i and s_{i+1}, i.e., those of the forms $s_i + s_{i+1} + s_k > 0$ or $s_i + s_{i+1} + s_{k'} \leq 0$, they are preserved when s_i and s_{i+1} are both replaced by the value $\bar{s} = \frac{s_i + s_{i+1}}{2}$. Unfortunately, $\bar{s}$ is not always an integer, and a final observation is required: the inequalities used in the *Gen-pi* procedure are preserved when all the elements of s are doubled, i.e., by abuse of notation, it holds $d_s = d_{2*s}$.

Putting things together, from s we can define the sequence s' as follows: first we initialize $s' = 2s$, then we set $s'_i = s'_{i+1} = \frac{s'_i + s'_{i+1}}{2} = \bar{s}'$. Doubling the elements of s, we find that $\bar{s}'$ is integer, and since $s'_i \leq \bar{s}' \leq s'_{i+1}$, all the inequalities that define d_s are preserved, as required. □

Property 2. *Let $s_i = s_{i+1}$ be two elements of an integer sequence s. The ith column and the $(i+1)^{st}$ column of the incidence matrix M_s of the hypergraph H_s generated by procedure Gen-pi(s) are equal.*

The proof directly follows from the fact that the elements d_i and d_{i+1} satisfy the same inequalities, and so they are present in the edges involving the same vertices.

Returning to the equivalence classes of the elements in $\mathcal{D}_n$, we propose as representative of the (non-void) class $[d]$, the sequence s_d that is minimal in lexicographic order. We stress the fact that s_d has the minimal first element among all the elements of $[d]$.

Unfortunately, equal elements of d do not always correspond to equal elements in the representative of s_d. As an example, exhaustive computation reveals that the representative of $d = (6, 5, 5, 4, 4)$ is $s_d = (2, 1, 1, 0, -1)$.

In order to compute and characterize the representative s_d of each equivalence class $[d]$, it is useful to understand the growth rate of the elements inside s_d according to length, with special attention to its first (and maximal) one.

So, for each length n and each $d \in \mathcal{D}_n$, we denote

$$M_n = \max_{s_d}\{s_1 : s_d = (s_1, s_2, \ldots, s_n)\}.$$

We call the sequences s_d where such a maximal first element is present *maximal* sequences. The first elements of the sequence $\{M_n\}_{n>2}$ up to $n = 8$, obtained by exhaustive computation are $1, 1, 2, 4, 6, 10$. For each n, there are several degree sequences whose representatives have the same maximal first element. The following table shows some of them according to the length parameter n:

Table 1. Table of the representatives s of the equivalence classes $[d_s]$ having maximal first element for the sequence lengths $n = 5, 6, 7$ and 8.

$n = 5$		$n = 7$	
s	d_s	s	d_s
$[2, 1, 1, 0, -1]$	$[6, 5, 5, 4, 4]$	$[6, 3, 2, 0, -1, -2, -3]$	$[15, 11, 10, 9, 8, 7, 6]$
$[2, 1, 1, -1, -2]$	$[5, 4, 4, 3, 2]$	$[6, 3, 2, 0, -2, -3, -4]$	$[13, 10, 9, 7, 6, 5, 4]$
$[2, 1, 0, 0, -1]$	$[6, 4, 4, 4, 3]$	$[6, 3, 1, 1, 0, -2, -3]$	$[15, 13, 10, 10, 9, 8, 7]$
$[2, 1, 0, -1, -2]$	$[4, 3, 2, 2, 1]$	$[6, 2, 1, 1, -1, -2, -3]$	$[15, 10, 9, 9, 8, 7, 5]$
$[2, 0, 0, -1, -1]$	$[5, 3, 3, 2, 2]$	$[6, 2, 1, 0, -1, -2, -3]$	$[15, 9, 8, 7, 7, 6, 5]$
$n = 6$		$n = 8$	
s	d_s	s	d_s
$[4, 2, 1, 0, -1, -2]$	$[10, 8, 7, 6, 6, 5]$	$[10, 4, 3, 2, 0, -1, -3, -6]$	$[21, 16, 15, 14, 13, 12, 10, 7]$
$[4, 1, 1, -1, -1, -2]$	$[10, 6, 6, 5, 5, 4]$	$[10, 4, 2, 2, 1, -2, -4, -5]$	$[21, 17, 14, 14, 13, 12, 9, 8]$

Let us investigate the asymptotic growth of the sequence $\{M_n\}_{n>3}$ by constructing a class of degree sequences $\mathcal{C}$ that provides a lower bound to it.

First we define the operator *Extend* that allow us to compute from a sequence $s = (s_1, \ldots, s_n)$ a sequence $s' = (s'_1, \ldots, s'_{n+2}) = Extend(s)$ as follows:

$$s'_i = \begin{cases} s_{i-1}, & \text{for } 2 \leq i \leq n+1 \\ s_1 + s_2 - s_n + 1, & \text{for } i = 1 \\ -(s_1 + s_2), & \text{for } i = n+2 \end{cases}$$

The following property holds

Property 3. *If s is the representative of the class $[d] \in D_n$, then $s' = Extend(s)$ is the representative of the respective class in D_{n+2}.*

Proof. Let $d_{s'}$ be the degree sequence generated by s'. We proceed by contradiction assuming there exists a representative $t = (t_1, \ldots, t_{n+2})$ of $[d_{s'}]$ different from s' and having $t_1 < s'_1$. We notice that the sequence $\tilde{t} = (t_2, \ldots, t_{n+1})$ generates d, since the elements $s'_2 + s'_3 \leq s'_{n+2}$, and such inequality must hold also in t, so also the elements $t_2, \ldots, t_{n+1}$ are not involved in any edge including t_{n+2}. On the other hand, by construction, $s'_2 + s'_3 = -s'_{n+2}$, so it holds $t_{n+2} \leq s'_{n+2}$.

Finally, we notice that the inequality $s'_1 + s'_{n+1} + s'_{n+2} > 0$ also holds since $s'_1 + s'_{n+1} + s'_{n+2} = 1$. In order to preserve the same inequality in t, i.e., $t_1 + t_{n+1} + t_{n+2} > 0$, having $t_{n+2} \leq s'_{n+2}$, we need $t_1 \geq s'_1$, and we reach a contradiction. $\square$

We underline that by iterating the application of the procedure *Extend* to a sequence s of length n, it produces longer sequences having the same length parity as s.

To illustrate the action of *Extend*, we depict its behaviour on the incidence matrices H_s and $H_{s'}$, with $s' = Extend(s)$. The basic idea is to extend H_s by adding an initial and a final column, as well as a set of starting rows, maximizing their number while leaving those in H_s unchanged:

$$
H_{s'} = \begin{pmatrix}
1 & 1 & 1 & 0 & \cdots\cdots & 0 \\
1 & 1 & 0 & 1 & \cdots\cdots & 0 \\
& \vdots & & & & \\
1 & \cdots\cdots\cdots\cdots & 1 & 1 \\
0 & & & & & 0 \\
\vdots & & \begin{bmatrix} & H_s & \end{bmatrix} & & & \vdots \\
0 & & & & & 0
\end{pmatrix}
$$

Let $s_0 = (s_{0,1}, \ldots, s_{0,n})$ be an integer sequence, and $s_k = (s_{k,1}, \ldots, s_{k,n+2k}) = Extend^k(s_0)$ be the sequence obtained by recursively applying the procedure *Extend* k-times to s_0. The following result holds

Theorem 2. *The integer sequence $\{s_{k,1}\}_k$ satisfies the recurrence relation:*

$$s_{k,1} = s_{k-1,1} + 2s_{k-2,1} + s_{k-3,1} + 1 \qquad k \geq 3 \tag{1}$$

with $s_{1,1}$ and $s_{2,1}$ being the first elements of $s_1 = Extend(s_0)$, and $s_2 = Extend^2(s_0)$, respectively.

Proof. The result immediately follows from the definition of the *Extend* operator:

$$
\begin{aligned}
s_{k,1} &= s_{k-1,1} + s_{k-1,2} - s_{k-1,n+2(k-1)} + 1 \\
s_{k-1,2} &= s_{k-2,1} \\
s_{k-1,n+2(k-1)} &= -s_{k-2,1} - s_{k-2,2}
\end{aligned}
$$

Replacing $s_{k-1,2}$ and $s_{k-1,n+2(k-1)}$ in the first equation we obtain the recurrence relation. □

We observe that, starting from two representative sequences $s_e = (1, 1, -1, -1)$ and $s_o = (2, 1, 0, -1, -2)$, having even and odd length, respectively, that are maximal w.r.t. the first element, the procedure *Extend* produces the two sequences $Extend(s_e) = (4, 1, 1, -1, -1, -2)$ and $Extend(s_o) = (6, 2, 1, 0, -1, -2, -3)$ that turn out to be two representatives with maximal first element of length 6 and 7, respectively. This property is not maintained when considering $Extend^2(s_e) = (8, 4, 1, 1, -1, -1, -2, -5)$, since $M_8 = 10$.

From this simple observation, we realize that the action of *Extend* on representatives with maximal first (and second) element deserves a deeper investigation in order to find an operator that allows us to pass from maximal representatives to maximal representatives. However, the *Extend* operator provides a lower bound to the growth rate of the M_n sequence.

Theorem 3. *The growth constant* λ *of the sequence* $\{s_{k,1}\}_k$ *is* $2.147 < \lambda < 2.148$.

Proof. The result follows from constructing the generating function for the recurrence relation (1) whose denominator is $q(x) = (1 - x - 2x^2 - x^3)(1 - x)$. The denominator has a unique minimal real root which can be computed numerically, $\rho = 0.466$. Therefore the asymptotic behaviour is controlled by λ that is the inverse of ρ, namely $f_{3,k} \sim (\lambda)^k$ $\qquad\qquad\square$

Corollary 1. *The growth constant* λ' *of* $\{M_n\}_{n>3}$ *is* $\lambda' \geq \lambda$.

4   Further Results on Unique Sequences

The property that each element d of $\mathcal{D}_n$ is unique, has a great relevance in the reconstruction of the related 3-hypergraph H_d. In particular, it is remarkable that, once one detects some edges of H_d, then those edges will not be modified till the end of the reconstruction or, equivalently, they will not be involved in possible backtracking steps, so limiting the complexity of the process.

As an example, this is the case of one or more elements of $d = (d_1, \ldots, d_n)$ that are maximal or minimal. Focusing on the first element d_1, if its value is $\binom{n-1}{2}$ this means that all edges involving the vertex v_1 are present, so they can be added and the process can proceed recursively on the next elements of d. Equivalently if $d_1 = n - 2$, then, by the definition of d, the edges involving v_1 are $(v_1, v_2, v_3), (v_1, v_2, v_4), \ldots, (v_1, v_2, v_n)$, that can be added to H_d.

Apart from those simple cases, some more are possible that may involve all the elements of d in a complex pattern of entanglements. In the intent of discovering a polynomial time way of managing these fixed patterns, a first and natural reconstruction algorithm is provided to deal with those simple cases that we address as *maximal instances*. This name is due to the fact that the sequence is maximal w.r.t. the lexicographical order among those having the same sum of elements.

The algorithm *Rec-max*, here not fully detailed for brevity sake, accepts a degree sequence d as input and produces the incidence matrix of a 3-hypergraph compatible with d, if d is a maximal sequence, otherwise it fails. We use the notation $M \oplus h$ to append the row h to the matrix M and $\deg(M)$ to denote the vector of column sums of the matrix M, i.e., the degree sequence of the related hypergraph.

The proof of the correctness of the algorithm is straightforward.

4.1   On the Isomorphism Properties of $\mathcal{D}_n$

The remarkable properties of the elements of $\mathcal{D}_n$ have an interesting consequence on the isomorphism problem restricted to the class. We recall that two (hyper)graphs H_1 and H_2 are isomorphic if and only if one can pass from the incidence matrix of H_1 to that of H_2 by a first round of column shifts, that corresponds to a mapping φ of the vertices of H_1 into those of H_2, followed by a

Algorithm 2. Rec-max(d)

set $d' = d$, $M_d = \emptyset$ and $i = 1$;
Step 1: create the maximal d'_i (w.r.t. the lexicographical order) length n binary sequences $h_1, \ldots, h_{d'_i}$ having exactly three elements 1, and such that the first of them lies in position i;
Step 2: set $M_d = M_d \oplus h_1 \oplus \cdots \oplus h_{d'_i}$;
Step 3: set $d' = d - \deg(M_d)$;
Step 4: **if** d' is the null vector **then**
$\qquad\qquad$ RETURN M_d;
$\qquad\quad$ **else if** d' has an element less than 0 or $i = n - 2$ **then**
$\qquad\qquad$ RETURN failure
$\qquad\quad$ **else** GoTo Step 1 updating $i = i + 1$.

second round of row shifts that allows to check the exact correspondence of the (hyper)edges.

Obviously, to check the isomorphism, φ has to preserve the vertices' degrees in H_1 and H_2, so in case of degree sequences with no equal elements φ equals Id, the identity mapping.

Actually, in case of equal elements in the degree sequence of H_1 and H_2, φ has, in general, to inspect the isomorphism for each possible permutations of the related columns (at present, no better strategy is available).

On the other hand, Property 2 states that, by construction, in each element $H \in \mathcal{D}_n$, the columns related to vertices having the same degree are equal, so again $\varphi = Id$ is a suitable mapping. This equality does not hold, in general, for each unique degree sequence (e.g., the degree sequence $d = (1,1,1,1,1,1)$ is unique w.r.t. 3-hypergraphs, but the check of the isomorphism of two related 3-hypergraphs H_1 and H_2 needs, in general, φ to be different from Id).

The following example reveals a new potential research line concerning the characterization and reconstruction of unique sequences that include, but are not restricted to, those generated by Gen-pi:

Example 2. Consider the degree sequence $\hat{d} = (25, 19, 19, 16, 16, 12, 10, 10, 5)$. Its uniqueness is witnessed by the related matrix $M_{\hat{d}}$ in Fig. 2, and whose construction seems to be not so far from that performed by Gen-pi. It remains an open problem to find a suitable meaning to the words "not so far" that could lead to a generalization of Gen-pi.

The following computations show that there does not exist an integer sequence s such that $\hat{d} = \hat{d}_s$. Recall that, by Property 1, among all the sequences in $[\hat{d}]$, if any, there exists one s such that $s_2 = s_3$, $s_4 = s_5$ and $s_7 = s_8$:
By the inequalities of Table 2 we obtain the result, in particular: by (1) and (3) we obtain $-2s_8 + s_6 + s_9 > 0$, by this last and (2) we obtain $2s_2 + 2s_5 + s_6 + s_9 > 0$ and finally using (5) we reach $2s_2 + s_9 > 0$, against (4).

v1	v2	v3	v4	v5	v6	v7	v8	v9	v1	v2	v3	v4	v5	v6	v7	v8	v9
1	1	1	0	0	0	0	0	0	0	1	1	1	0	0	0	0	0
				...									...				
1	1	0	0	0	0	0	0	1	0	1	1	0	0	0	0	1	0
1	0	1	1	0	0	0	0	0	0	1	0	1	1	0	0	0	0
				...									...				
1	0	1	0	0	0	0	0	1	0	1	0	1	0	0	0	1	0
1	0	0	1	1	0	0	0	0	0	1	0	0	1	1	0	0	0
				...									...				
1	0	0	1	0	0	0	0	1	0	1	0	0	1	0	0	1	0
1	0	0	0	1	1	0	0	0	0	0	1	1	1	0	0	0	0
				...									...				
1	0	0	0	1	0	0	0	1	0	0	1	1	0	0	0	1	0
1	0	0	0	0	1	1	0	0	0	0	1	0	1	1	0	0	0
				...									...				
1	0	0	0	0	0	1	0	1	0	0	1	0	1	0	0	1	0

Fig. 2. The incidence matrix $M(\hat{d})$ of a 3-hypergraph having degree sequence $\hat{d}$. On the left the part of the matrix related to the edges involving v_1, while on the right the remaining ones. The 3-hypergraph is unique as can be seen by the construction of the matrix.

Table 2. Inequalities related to some rows of the matrix $H_{\hat{d}}$. Edges are represented by triplets of vertex indices.

$$(1,6,9) \in H_{\hat{d}} \rightarrow s_1 + s_6 + s_9 > 0 \quad (1)$$
$$(2,5,8) \in H_{\hat{d}} \rightarrow s_2 + s_5 + s_8 > 0 \quad (2)$$
$$(1,7,8) \notin H_{\hat{d}} \rightarrow s_1 + 2s_8 \leq 0 \quad (3)$$
$$(2,3,9) \notin H_{\hat{d}} \rightarrow 2s_2 + s_9 \leq 0 \quad (4)$$
$$(4,5,6) \notin H_{\hat{d}} \rightarrow 2s_5 + s_6 \leq 0 \quad (5)$$

5   Conclusions and Open Problems

In this article, we consider the class $\mathcal{D}_n$ of degree sequences of 3-hypergraphs on n vertices that extend those defined in [6] and that are computed starting from a given integer sequence. First, we prove that each degree sequence $d \in \mathcal{D}_n$ is unique, i.e., the related 3-hypergraph H_d is unique up to isomorphism, then we define the representative integer sequence s_d which leads to the reconstruction of H_d. Some properties of s_d are shown, in particular we determine a lower bound to the growth rate of their maximal elements according to the length n, related to the number of edges of the 3-hypergraph H_d. This result is useful to generate and enumerate the elements of $\mathcal{D}_n$, establishing the size of the class. In this context, we point out two open problems:

i) define a variant of the *Extend* operator that allows to maintain the maximality property of the representatives;

ii) find the growth rate of the sequence $\{M_n\}_{n>2}$ and characterize the maximal representatives of each $\mathcal{D}_n$.

Furthermore, a simple algorithm is defined to reconstruct the maximal (in lexicographic order) sequences of $\mathcal{D}_n$ having prescribed sum of elements. From its definition, we realize that a generalization to include all the elements of $\mathcal{D}_n$ would involve some backtracking when elements less than zero appear in d'. We propose the following research line:

iii) find some properties related to the computation of d_s from s that prevent or, at least, restrict the backtracking in *Rec-max*. This will lead to the solution of the reconstruction problem related to $\mathcal{D}_n$. As an alternative, prove that it cannot be done in polynomial time.

Finally, the degree sequence $\hat{d}$ in Example 2 deserves attention: it admits a unique 3-hypergraph whose *structure* is close to that of the hypergraphs related to the elements of $\mathcal{D}_n$, but without being generated by an integer sequence. A final open problem is proposed:

iv) define a notion of structure of the elements of $\mathcal{D}_n$ and expand the class including unique degree sequences with similar structure. Investigate the properties of the new class.

References

1. Behrens, S., et al.: New results on degree sequences of uniform hypergraphs. Electron. J. Comb. **20**(4), P14 (2013)
2. Berge, C.: Hypergraphs. North-Holland, Amsterdam (1989)
3. Brlek, S., Frosini, A.: A tomographical interpretation of a sufficient condition on *h*-graphical sequences. In: Normand, N., Guédon, J., Autrusseau, F. (eds.) DGCI 2016. LNCS, vol. 9647, pp. 95–104. Springer, Cham (2016). https://doi.org/10.1007/978-3-319-32360-2_7
4. Colbourne, C.J., Kocay, W.L., Stinson, D.R.: Some NP-complete problems for hypergraph degree sequences. Discrete Appl. Math. **14**, 239–254 (1986)
5. Dewdney, A.K.: Degree sequences in complexes and hypergraphs. Proc. Amer. Math. Soc. **53**(2), 535–540 (1975)
6. Deza, A., Levin, A., Meesum, S.M., Onn, S.: Optimization over degree sequences. SIAM J. Disc. Math. **32**(3), 2067–2079 (2018)
7. Erdös, P., Gallai, T.: Graphs with prescribed degrees of vertices (in Hungarian). Math. Lapok. **11**, 264–274 (1960)
8. Frosini, A., Palma, G., Rinaldi, S.: Combinatorial properties of degree sequences of 3-uniform hypergraphs arising from saind arrays. In: Anselmo, M., Della Vedova, G., Manea, F., Pauly, A. (eds.) CiE 2020. LNCS, vol. 12098, pp. 228–238. Springer, Cham (2020). https://doi.org/10.1007/978-3-030-51466-2_20
9. Frosini, A., Picouleau, C., Rinaldi, S.: On the degree sequences of uniform hypergraphs. In: Gonzalez-Diaz, R., Jimenez, M.-J., Medrano, B. (eds.) DGCI 2013. LNCS, vol. 7749, pp. 300–310. Springer, Heidelberg (2013). https://doi.org/10.1007/978-3-642-37067-0_26

10. Hakimi, S.: On realizability of a set of integers as degrees of the vertices of a lineargraph. I. J. Soc. Indust. Appl. Math. **10**, 496–506 (1962)
11. Havel, V.: A remark on the existence of finite graphs (in Czech). Časopis. Pěst. Mat. **80**, 477–480 (1955)
12. Kocay, W., Li, P.C.: On 3-hypergraphs with equal degree sequences. Ars Combinatoria **82**, 145–157 (2006)

Power Sum Polynomials in a Discrete Tomography Perspective

Silvia M. C. Pagani[(⊠)] and Silvia Pianta

Dipartimento di Matematica e Fisica "N. Tartaglia", Università Cattolica del Sacro Cuore via Musei 41, 25121 Brescia, Italy
{silvia.pagani,silvia.pianta}@unicatt.it

Abstract. For a point of the projective space $PG(n, q)$, its Rédei factor is the linear polynomial in $n + 1$ variables, whose coefficients are the point coordinates. The power sum polynomial of a subset S of $PG(n, q)$ is the sum of the $(q-1)$-th powers of the Rédei factors of the points of S. The fact that many subsets may share the same power sum polynomial offers a natural connection to discrete tomography. In this paper we deal with the two-dimensional case and show that the notion of ghost, whose employment enables to find all solutions of the tomographic problem, can be rephrased in the finite geometry context, where subsets with null power sum polynomial are called ghosts as well. In the latter case, one can add ghosts still preserving the power sum polynomial by means of the multiset sum (modulo the field characteristic). We prove some general results on ghosts in $PG(2, q)$ and compute their number in case q is a prime.

Keywords: Discrete tomography · Ghost · Multiset sum · Power sum polynomial · Projective plane

1 Introduction

The aim of the present paper is to investigate the structure and the size of those point sets of $PG(2, q)$ associated to a given homogeneous polynomial of degree $q-1$, introduced by P. Sziklai in [8] and named *power sum polynomial*. It is a hard task in general to determine all sets sharing the same power sum polynomial. The ill-posedness of the aforementioned problem enables to construct a link to discrete tomography. There, linear algebra offers a straightforward way of dealing with all images agreeing with the same set of projections (see [5]). The key role is played by *ghosts*, which constitute the kernel of the linear system in which the problem may be translated.

Motivated by the description of the set of solutions of a tomographic problem, we investigate the recovery of the subsets of $PG(2, q)$ related to the same power sum polynomial in the following way. We look for an operation on the subsets of $PG(2, q)$ such that those with associated null polynomial, which we call ghosts as well, constitute the kernel of a suitable function mapping a subset to the

© Springer Nature Switzerland AG 2021
J. Lindblad et al. (Eds.): DGMM 2021, LNCS 12708, pp. 325–337, 2021.
https://doi.org/10.1007/978-3-030-76657-3_23

corresponding power sum polynomial. A good choice for the operation turns out to be the multiset sum modulo p (the field characteristic). Then, we investigate the algebraic and geometric properties of the set of ghosts and find out the size of the subset of ghosts when $q = p$, namely, for prime fields.

We remark that the "polynomial technique" for subsets of a projective space is a well investigated research field in the framework of finite geometries. It consists in the study of the interplay between subsets of a projective space and polynomials over finite fields (see for instance [1,2]). It is essentially listed in three steps: rephrasing the theorem to be proven into a relationship on points, reformulating the (new) theorem in terms of polynomials over finite fields, and performing the calculations. This technique was introduced in the 70's by L. Rédei [7] and inspired a series of results on blocking sets, directions and codes from the 90's onward, such as [3,9,10]. The advantage of choosing the power sum polynomial, w.r.t. the more studied Rédei polynomial, is the fact that the first one has a lower degree when the subset has more than $q - 1$ points.

The paper is organized as follows. Section 2 establishes the framework and gives the main definitions. In Sect. 3 the tomographic problem is recalled, together with a focus on the similarities with the treated problem. Section 4 gives some general results for ghosts in the projective plane, while in Sect. 5 we set $q = p$ and show the size of the set of ghosts in that case. Section 6 provides possible further work and concludes the paper.

2   The Power Sum Polynomial

Let $\mathbb{F}_q$ denote the Galois field of order $q = p^h$, p prime, $h \geq 1$, whilst $PG(n, q)$ will refer to the projective space of dimension n over $\mathbb{F}_q$. Let $\mathbf{X} = (X_1, \ldots, X_{n+1})$ be the vector of the variables.

Definition 1. *Let $P = (p_1, \ldots, p_{n+1})$ be a point of $PG(n, q)$. The* Rédei factor *corresponding to P is the linear polynomial $P \cdot \mathbf{X} = p_1 X_1 + \ldots + p_{n+1} X_{n+1}$.*

The zeros of a Rédei factor $P \cdot \mathbf{X}$ are the Plücker coordinates of the hyperplanes through P.

The well-known *Rédei polynomial* of a point set S is defined as the product of the Rédei factors corresponding to the points of S. We are interested in a different polynomial.

Definition 2. *Let $S = \{P_i : i = 1, \ldots, |S|\} \subseteq PG(n, q)$ be a point set. The* power sum polynomial *of S is defined as*

$$G^S(\mathbf{X}) := \sum_{i=1}^{|S|} (P_i \cdot \mathbf{X})^{q-1}.$$

The power sum polynomial is therefore a homogeneous polynomial of degree $q - 1$. Denote by $\mathbb{F}_q^{q-1}[\mathbf{X}]$ the set of homogeneous polynomials of degree $q - 1$ in the variables $\mathbf{X}$ with coefficients in $\mathbb{F}_q$. It is a vector space over $\mathbb{F}_q$ with dimension $\binom{n+q-1}{n}$. For $n = 2$, we will write

$$G^S(X, Y, Z) = \sum_{k=1}^{|S|} (a_k X + b_k Y + c_k Z)^{q-1}$$

$$= \sum_{k=1}^{|S|} \sum_{i=0}^{q-1} \sum_{j=0}^{q-1-i} \binom{q-1}{i,j} (a_k X)^{q-1-i-j} (b_k Y)^j (c_k Z)^i ,$$

where $\binom{q-1}{i,j} = \dfrac{(q-1)!}{i!j!(q-1-i-j)!}$ for $i + j \leq q - 1$.

The power sum polynomial of the union of disjoint subsets of $\mathrm{PG}(n, q)$ is clearly the sum of the corresponding power sum polynomials. If a hyperplane $(x_1, \ldots, x_{n+1})$ intersects S in m points, then, from the well-known fact that

$$\forall \alpha \in \mathbb{F}_q : \alpha^{q-1} = \begin{cases} 0 & \text{if } \alpha = 0, \\ 1 & \text{otherwise,} \end{cases}$$

it results that $G^S(x_1, \ldots, x_{n+1}) = |S| - m$.

As noted in [8], while there is an one-to-one correspondence between Rédei polynomials and point sets, the same power sum polynomial may refer to different point sets.

Example 1. Consider the Fano plane $\mathrm{PG}(2, 2)$ and the subset $S_1 = \{(0, 0, 1)\}$. The corresponding power sum polynomial is $G^{S_1}(X, Y, Z) = Z$, which is shared with other subsets of $\mathrm{PG}(2, 2)$, such as $S_2 = \{(1, 0, 1), (1, 0, 0)\}$ and $S_3 = \{(1, 0, 0), (0, 1, 0), (1, 1, 1)\}$.   □

We address the following problem: Given a homogeneous polynomial G of degree $q - 1$, find all sets $S \subseteq \mathrm{PG}(2, q)$ such that $G^S = G$. The choice of considering the two-dimensional case is justified in the next section.

3 Discrete Tomography

The problem stated at the end of the previous section belongs to the large class of the inverse problems, where an object has to be retrieved from the measurements of its features, broadly intended. In the same class one can find the *tomographic problem*, whose measurements are the projections of the unknown object along a set of given directions. Tomography originates from the work of J. Radon [6], who proved that, under some regularity hypotheses, a density function defined over $\mathbb{R}^2$ can be reconstructed from its projections taken along angles in the whole range $[0, \pi)$. From [6] on, the tomographic problem has mainly addressed the two-dimensional case (in diagnostics imaging, for instance, the 3D scan of the body is obtained by stacking 2D slices), as we will do in the present paper. The function to be recovered is also called an *image*.

When a finite set of directions is employed and the image is assumed to be defined on a finite subset of the lattice $\mathbb{Z}^2$, one deals with *discrete tomography*. Directions are lattice directions, i.e., with rational slopes, and projections are obtained by summing up the values of the points which are intersections between lines with given direction and the domain (see Fig. 1). Usually, the codomain of the image is (a subset of) $\mathbb{Z}$.

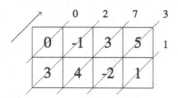

Fig. 1. A four by two integer-valued image and the projections along a direction.

In general, neither existence nor uniqueness of a solution applies to a discrete tomographic problem, which is ill-posed as most of the inverse problems. In fact, on the one side the available data may be inconsistent due to noise, so that no image exists with prescribed projections. On the other side, when dealing with consistent measurement (i.e., at least one solution exists) many images which agree with the projections could be allowed (see for instance [4]). The lack of uniqueness is caused by the presence of images whose projections along the set of considered directions are zero, and therefore are, in a sense, invisible to measurements. Such images are known as *ghosts* and constitute the kernel of a suitable linear system, obtained by ordering the projections in some way.

Linear algebra and [5] show how to treat ghosts in order to move among the solutions of a tomographic problem. In fact, every solution of a same problem may be obtained as the sum between a peculiar solution and a suitable ghost (see Fig. 2 for an example). We investigate their counterpart in the power sum polynomial problem, which will be called *ghosts* as well, in particular in the two-dimensional case.

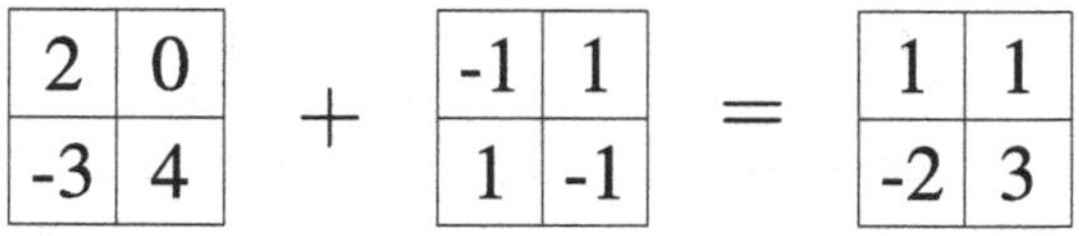

$$
\begin{array}{|c|c|}
\hline
2 & 0 \\
\hline
-3 & 4 \\
\hline
\end{array}
\;+\;
\begin{array}{|c|c|}
\hline
-1 & 1 \\
\hline
1 & -1 \\
\hline
\end{array}
\;=\;
\begin{array}{|c|c|}
\hline
1 & 1 \\
\hline
-2 & 3 \\
\hline
\end{array}
$$

Fig. 2. Adding a ghost (w.r.t. the coordinate directions in this case) to an image produces a new image, whose horizontal and vertical projections equal those of the starting image.

3.1   Connections and Differences

Our goal is to determine all point sets of $PG(2,q)$ sharing the same power sum polynomial. Arguing in analogy with the tomographic framework, the role of the grid (i.e., the domain) and that of an image are played by $PG(2,q)$ and a point set S, respectively. The counterpart of tomographic ghosts is defined as follows.

Definition 3. *Let $S \subseteq PG(2,q)$. We say that S is a ghost if $G^S(X, Y, Z) \equiv \mathbf{0}$.*

Note that, in the tomographic case, the set of directions has to be specified, while the definition of the power sum polynomial does not depend on other features. We study the structure of ghosts and look for an operation which maps a pair, consisting of a generic point set and a ghost, to another point set with the same power sum polynomial as the previous generic one. The investigation of ghosts therefore completes the search of subsets S such that $G^S = G$ for a given polynomial G.

4   Ghosts and Multiset Sum in PG(2, q)

As stated in the previous section, we have to study the structure of ghosts. We first prove some general results about classes of ghosts.

In order to prove that a point set $S \subseteq PG(2,q)$ has the zero polynomial as the corresponding one, we have to find when G^S vanishes identically. We look at the intersections of S with lines, which are the hyperplanes in the plane. In fact, if $|S \cap \ell| = m$ for a line ℓ, then $G^S(\ell) = |S| - m$, where, by an abuse of notation, we write that the polynomial is computed in ℓ instead of the Plücker coordinates of ℓ.

Therefore, as $\deg G^S < q$, we seek those point sets whose intersections with all lines of $PG(2,q)$ have size congruent to that of S modulo the field characteristic p. According to [8, Theorem 13.5], such point sets S are exactly those intersecting each line in a fixed (mod p) number of points. So, our ghosts are the *generalized Vandermonde sets* defined in [8].

Theorem 1. *Let $S \subseteq PG(2,q)$ be a ghost. Then $PG(2,q) \backslash S$ is a ghost.*

Proof. Since S has constant intersection size with lines, the same holds true for its complement $PG(2,q) \backslash S$.   □

As a consequence, both the empty set and the whole projective plane are ghosts.

Theorem 2. *A partial pencil $\mathcal{P}$ of $\lambda p + 1$ lines, $\lambda = 0, \ldots, p^{h-1}$, is a ghost. Consequently, a set of $q - \lambda p$ lines through a point P minus $\mathcal{P}$ is a ghost.*

Proof. Every line ℓ meets $\mathcal{P}$ in either one, $\lambda p + 1$ or $q + 1$ points. In all cases $m = 1 \mod p$ and

$$G^{\mathcal{P}}(\ell) = (\lambda p + 1)q + 1 - m = 0 \mod p.$$

□

Therefore, every line is a ghost, as well as any affine plane contained in $PG(2, q)$.

It would be desirable to define a binary operation in order to endow the subsets of $PG(2, q)$ with a group structure, so that the set of ghosts is stable under such an operation. Unfortunately, the usual set-theoretical union is not a good choice. Consider for instance the two lines ℓ_1, ℓ_2 of $PG(2, 2)$, whose points have coordinates satisfying $X = 0$ and $Y = 0$, respectively. Lines are ghosts of $PG(2, 2)$; their union consists of the five points $(0, 1, 0)$, $(0, 0, 1)$, $(0, 1, 1)$, $(1, 0, 0)$, $(1, 1, 0)$ and the corresponding power sum polynomial is

$$G^{\ell_1 \cup \ell_2} = Y + Z + (Y + Z) + X + (X + Y) = Y,$$

so the union of two ghosts is not a ghost in general. To deal with a deeper algebraic structure, we need to consider a different kind of operation.

4.1   Multiset Sum

The structure of all solutions of a same tomographic problem can be described by linear algebra as the sum between a peculiar solution and the vector subspace of ghosts (see [5]). Concerning power sum polynomials, the usual set-theoretical union is not enough and has to be replaced with the multiset sum.

Let A be a multiset, namely, a set whose elements can appear with multiplicity greater than one. Denote by $m_A(x)$ the multiplicity of the element x in A, which is zero if $x \notin A$. The *multiset sum* of two multisets A and B is a multiset C such that, for any element x,

$$m_C(x) = m_A(x) + m_B(x).$$

Denote the set of subsets S of $PG(2, q)$, where each point P is counted $m_S(P)$ mod p times, by $p^{PG(2,q)}$. Of course $p^{PG(2,q)}$ has p^{q^2+q+1} elements. Moreover, consider the binary operation $\uplus_p$, which is the multiset sum modulo p.

Lemma 1. $\left(p^{PG(2,q)}, \uplus_p\right)$ *is an abelian p-group.*

Proof. The operation $\uplus_p$ is clearly internal to $p^{PG(2,q)}$, associative and commutative. The identity element is the empty set and the inverse of a subset S, whose element P_i is counted λ_i times $(i = 1, \ldots, |S|)$, is the subset where P_i is counted $p - \lambda_i$ times.

Also, $p^{PG(2,q)}$ is a p-group since every non-identity element has period p. $\square$

It follows that the map

$$\varphi : \begin{cases} p^{PG(2,q)} \longrightarrow \mathbb{F}_q^{q-1}[X, Y, Z], \\ S \longmapsto G^S \end{cases}$$

preserves the group operation, i.e.,

$$\varphi\left(S_1 \uplus_p S_2\right) = G^{S_1 \uplus_p S_2} = G^{S_1} + G^{S_2} = \varphi(S_1) + \varphi(S_2).$$

Therefore φ is a homomorphism between abelian groups. In particular, if $A \subseteq B$, then $G^{B \backslash A} = G^B - G^A$.

The kernel of φ is the set of ghosts $\mathcal{G}$, which is indeed a subgroup of $p^{\mathrm{PG}(2,q)}$. Moreover

$$p^{\mathrm{PG}(2,q)} / \ker \varphi \cong \mathrm{im}\varphi.$$

We now extend the results of Theorems 1 and 2 to multisets. For a subset A of a multiset B, we define the complement of A in B as the multiset where each element $b \in B$ is counted $m_B(b) - m_A(b)$ times.

Theorem 3. *Let S be a ghost. Then the complement of S is a ghost. Moreover, lines and (multiset) sums of lines are ghosts of $PG(2,q)$.*

Proof. It results $G^{\mathrm{PG}(2,q) \backslash S} = G^{\mathrm{PG}(2,q)} - G^S = 0 - 0 = 0$. The statement about lines follows from Theorem 2 and the properties of $\uplus_p$. □

5   The Case $q = p$

From now on we focus on the case $q = p^h = p$ (i.e., $h = 1$), namely, we refer to prime fields. In order to get information about the size of the subgroup of ghosts, we argue on the size of the image of φ.

5.1   Surjectivity of φ

We now prove that, for $q = p$ prime, the dimension of $\mathrm{im}\varphi$ is $\binom{p+1}{2}$, namely, the function φ is onto.

If $p = 2$, it is trivial to prove that the images of points $(1,0,0)$, $(0,1,0)$, $(0,0,1)$ constitute a basis of $\mathbb{F}_2^1[X,Y,Z]$. For p odd, we will prove that the set consisting of the images of the $\binom{p+1}{2}$ points

$$\begin{aligned}
&(1,0,0), (1,0,1), (1,0,2), \ldots, (1,0,p-1), \\
&(1,1,0), (1,1,1), \ldots, (1,1,p-2), \\
&\ldots, \\
&(1,p-2,0), (1,p-2,1), \\
&(1,p-1,0)
\end{aligned} \tag{1}$$

is linearly independent. This can be done by showing that the matrix having as rows the components (w.r.t. the canonical basis of $\mathbb{F}_p^{p-1}[X,Y,Z]$) of the images of the above points is non-singular. Each entry of the matrix is the coefficient of a certain monomial in the development of $(X + bY + cZ)^{p-1}$, where $(1,b,c)$ is one of the above points. Since from the column corresponding to the monomial $X^{p-1-i-j}Y^jZ^i$ we can extract the coefficient $\binom{p-1}{i,j}$ and this does not interfere with the fact that the determinant is either zero or non-zero, we will omit the coefficients and simply write the value of b^jc^i. By reordering rows and columns in a suitable way we get the matrix in Table 1, which is a block lower triangular matrix. Its determinant is given by the product of the determinants of its blocks. The four blocks are: the one by one upper leftmost block (1), two equal blocks

$$\begin{pmatrix} 1 & 1 & 1 & \dots 1 \\ 2 & 2^2 & 2^3 & \dots 1 \\ \vdots & & & \\ p-1 & (p-1)^2 & (p-1)^3 & \dots 1 \end{pmatrix}$$

and the remaining block of order $\frac{(p-2)(p-1)}{2}$.

Table 1. The initial matrix.

	1	b	b^2	$\dots$ b^{p-1}	c	c^2	$\dots$ c^{p-1}	bc	b^2c	$\dots$ bc^{p-2}
$(1,0,0)$	1	0	$\dots$							
$(1,1,0)$	1	1	1	$\dots$ 1	0		$\dots$			
$(1,2,0)$	1	2	2^2	$\dots$ 1	0		$\dots$			
$\vdots$	$\vdots$									
$(1,p-1,0)$	1	$p-1$	$(p-1)^2$	$\dots$ 1	0		$\dots$			
$(1,0,1)$	1	0	0	$\dots$ 0	1	1	$\dots$ 1	0	$\dots$	
$(1,0,2)$	1	0	0	$\dots$ 0	2	2^2	$\dots$ 1	0	$\dots$	
$\vdots$	$\vdots$									
$(1,0,p-1)$	1	0	0	$\dots$ 0	$p-1$	$(p-1)^2$	$\dots$ 1	0	$\dots$	
$(1,1,1)$	1	1	1	$\dots$ 1	1	1	$\dots$ 1	1	1	$\dots$ 1
$\vdots$	$\vdots$									
$(1,p-2,1)$	1	$p-2$	$(p-2)^2$	$\dots$ 1	1	1	$\dots$ 1	$p-2$	$(p-2)^2$	$\dots$ $p-2$

The first three blocks are non-singular (in particular, the second and the third blocks are Vandermonde matrices), so the matrix has non-zero determinant if and only if the fourth block has.

The columns of the fourth block are indexed as $bc, b^2c, bc^2, \dots, bc^{p-2}$, so we can extract a factor bc from every row to get, after reordering rows and columns suitably, the matrix in Table 2.

In order to prove that the determinant of the matrix in Table 2 is non-zero, we apply some basic operations to its rows in order to obtain a block upper triangular matrix, whose blocks are Vandermonde matrices and which is singular if and only if the starting matrix is. The procedure consists of $p-2$ steps; in each of them we consider some rows as *pivotal* and substitute to a non-pivotal row a linear combination of it and the corresponding pivotal row, in order to have zeroes as entries in the columns corresponding to a certain power of c.

Table 2. The $\frac{(p-2)(p-1)}{2}$ lower leftmost block.

	1	b	b^2	$\cdots$	b^{p-3}	c	bc	$\cdots$	$b^{p-4}c$	c^2	$\cdots$	c^{p-3}
$(1,1,1)$	1	1	1	$\cdots$								
$(1,2,1)$	1	2	2^2	$\cdots$	2^{p-3}	1	2	$\cdots$	2^{p-4}	1	$\cdots$	1
$(1,3,1)$	1	3	3^2	$\cdots$	3^{p-3}	1	3	$\cdots$	3^{p-4}	1	$\cdots$	1
$\vdots$	$\vdots$											
$(1,p-2,1)$	1	$p-2$	$(p-2)^2$	$\cdots$	$(p-2)^{p-3}$	1	$p-2$	$\cdots$	$(p-2)^{p-4}$	1	$\cdots$	1
$(1,1,2)$	1	1	1	$\cdots$	1	2	2	$\cdots$	2	2^2	$\cdots$	2^{p-3}
$(1,2,2)$	1	2	2^2	$\cdots$	2^{p-3}	2	$2\cdot2$	$\cdots$	$2^{p-4}\cdot2$	2^2	$\cdots$	2^{p-3}
$\vdots$	$\vdots$											
$(1,p-3,2)$	1	$p-3$	$(p-3)^2$	$\cdots$	$(p-3)^{p-3}$	2	$(p-3)\cdot2$	$\cdots$	$(p-3)^{p-4}\cdot2$	2^2	$\cdots$	2^{p-3}
$(1,1,3)$	1	1	1	$\cdots$	1	3	3	$\cdots$	3	3^2	$\cdots$	3^{p-3}
$\vdots$	$\vdots$											
$(1,1,p-2)$	1	1	1	$\cdots$	1	$p-2$	$p-2$	$p-2$	$\cdots$	$(p-2)^2$	$\cdots$	$(p-2)^{p-3}$

Denote by $(1,b,c)^{(n)}$ the row corresponding to the image of the point $(1,b,c)$ at the n-th step of the procedure. It is defined recursively as follows:

$$(1,b,c)^{(0)} \quad : \quad \text{the row corresponding to } \varphi(1,b,c),$$
$$(1,b,c)^{(1)} := (1,b,c)^{(0)} - (1,b,1)^{(0)}, \quad c \geq 2,$$
$$(1,b,c)^{(n)} := \frac{(1,b,c)^{(n-1)}}{c-(n-1)} - (1,b,n)^{(n-1)}, \quad n \geq 2, c \geq n+1,$$

where the subtraction is intended to be element-wise. The third entry is at least $n+1$ since at Step n the pivotal rows are $(1,b,n)^{(n-1)}$, $b=1,\ldots,p-1-n$, so only rows such that $c \geq n+1$ will be modified.

After each step, the matrix becomes block upper triangular consisting of two blocks: the upper leftmost block is a Vandermonde matrix and the other block has to be treated at the next step. Rows in the matrix in Table 2 are ordered such that the first $p-2$ rows $((1,b,1), b=1,\ldots,p-2)$ are the pivotal ones in the first step, the following $p-3$ ones $((1,b,2), b=1,\ldots,p-3)$ will be the pivotal rows in the second step after the operations, and so on.

At Step n, the row $(1,b,c)^{(n)}$ has the entry corresponding to the column $b^\lambda c^\mu$ equal to:

- for $n=1$: $b^\lambda\left(c^\mu - 1\right)$;
- for $n=2n'$, $n' \geq 1$:

$$b^\lambda \sum_{i_1=0}^{\mu-2}\left(2^{\mu-1-i_1}-1\right)\sum_{i_2=0}^{i_1-2}\left(4^{i_1-1-i_2}-3^{i_1-1-i_2}\right)\sum_{i_3=0}^{i_2-2}\cdots$$
$$\cdot\sum_{i_{n'}=0}^{i_{n'-1}-2}(c-n)\left(n^{i_{n'-1}-1-i_{n'}}-(n-1)^{i_{n'-1}-1-i_{n'}}\right)c^{i_{n'}};$$

– for $n = 2n' + 1$, $n' \geq 1$:

$$b^\lambda \sum_{i_1=0}^{\mu-2} \left(2^{\mu-1-i_1} - 1\right) \sum_{i_2=0}^{i_1-2} \left(4^{i_1-1-i_2} - 3^{i_1-1-i_2}\right) \sum_{i_3=0}^{i_2-2} \cdots$$

$$\cdot \sum_{i_{n'}=1}^{i_{n'-1}-2} \left((2n')^{i_{n'-1}-1-i_{n'}} - (2n'-1)^{i_{n'-1}-1-i_{n'}}\right) \left(c^{i_{n'}} - n^{i_{n'}}\right).$$

The above formulas may be proven by induction on n. Note that, for $\mu < n$, there are empty summations and therefore the entry is zero. Moreover, all elements of row $(1, b, c)^{(n)}$ are divisible by $c - n$ even over the integers.

In particular, the pivotal rows at the next step, say $n+1$, have a Vandermonde submatrix in the columns corresponding to the power $b^\lambda c^n$, $\lambda = 0, \ldots, p - 2 - n$ (it is obtained by substituting $\mu = n$ and $c = n + 1$ in the above formulas; note that, for $\mu = n$, all summations reduce to a single term).

At the final step $p - 2$, the resulting matrix is block triangular and each block is a Vandermonde matrix with no repeated rows, so non-singular. We can conclude that the images of points in (1) are linearly independent and then $\mathrm{im}\varphi$ has dimension $\binom{p+1}{2}$.

In the next example we show how the procedure works.

Example 2. Set $p = 7$ and consider the $\frac{5 \cdot 6}{2} = 15$ points

$$(1,1,1), (1,1,2), (1,1,3), (1,1,4), (1,1,5), (1,2,1), (1,2,2), (1,2,3), (1,2,4),$$
$$(1,3,1), (1,3,2), (1,3,3), (1,4,1), (1,4,2), (1,5,1),$$

which are those, among the $\binom{7+1}{2} = 28$ points whose images constitute a basis for $\mathrm{im}\varphi$, having no zero coordinate. This means that the above points are those whose images constitute the block on which the procedure is applied.

The initial matrix is

	1	b	b^2	b^3	b^4	c	bc	b^2c	b^3c	c^2	bc^2	b^2c^2	c^3	bc^3	c^4
$(1,1,1)$	1	1	1	1	1	1	1	1	1	1	1	1	1	1	1
$(1,2,1)$	1	2	2^2	2^3	2^4	1	2	2^2	2^3	1	2	2^2	1	2	1
$(1,3,1)$	1	3	3^2	3^3	3^4	1	3	3^2	3^3	1	3	3^2	1	3	1
$(1,4,1)$	1	4	4^2	4^3	4^4	1	4	4^2	4^3	1	4	4^2	1	4	1
$(1,5,1)$	1	5	5^2	5^3	5^4	1	5	5^2	5^3	1	5	5^2	1	5	1
$(1,1,2)$	1	1	1	1	1	2	2	2	2	2^2	2^2	2^2	2^3	2^3	2^4
$(1,2,2)$	1	2	2^2	2^3	2^4	2	$2 \cdot 2$	$2^2 \cdot 2$	$2^3 \cdot 2$	2^2	$2 \cdot 2^2$	$2^2 \cdot 2^2$	2^3	$2 \cdot 2^3$	2^4
$(1,3,2)$	1	3	3^2	3^3	3^4	2	$3 \cdot 2$	$3^2 \cdot 2$	$3^3 \cdot 2$	2^2	$3 \cdot 2^2$	$3^2 \cdot 2^2$	2^3	$3 \cdot 2^3$	2^4
$(1,4,2)$	1	4	4^2	4^3	4^4	2	$4 \cdot 2$	$4^2 \cdot 2$	$4^3 \cdot 2$	2^2	$4 \cdot 2^2$	$4^2 \cdot 2^2$	2^3	$4 \cdot 2^3$	2^4
$(1,1,3)$	1	1	1	1	1	3	3	3	3	3^2	3^2	3^2	3^3	3^3	3^4
$(1,2,3)$	1	2	2^2	2^3	2^4	3	$2 \cdot 3$	$2^2 \cdot 3$	$2^3 \cdot 3$	3^2	$2 \cdot 3^2$	$2^2 \cdot 3^2$	3^3	$2 \cdot 3^3$	3^4
$(1,3,3)$	1	3	3^2	3^3	3^4	3	$3 \cdot 3$	$3^2 \cdot 3$	$3^3 \cdot 3$	3^2	$3 \cdot 3^2$	$3^2 \cdot 3^2$	3^3	$3 \cdot 3^3$	3^4
$(1,1,4)$	1	1	1	1	1	4	4	4	4	4^2	4^2	4^2	4^3	4^3	4^4
$(1,2,4)$	1	2	2^2	2^3	2^4	4	$2 \cdot 4$	$2^2 \cdot 4$	$2^3 \cdot 4$	4^2	$2 \cdot 4^2$	$2^2 \cdot 4^2$	4^3	$2 \cdot 4^3$	4^4
$(1,1,5)$	1	1	1	1	1	5	5	5	5	5^2	5^2	5^2	5^3	5^3	5^4

The upper leftmost 5 by 5 block is a Vandermonde matrix and the corresponding rows will be the pivotal ones at Step 1. The other rows will be manipulated so that the first five entries are zero. Therefore, matrix at Step 1 is

	1	b	b^2	b^3	b^4	c	bc	b^2c	b^3c	c^2	bc^2	b^2c^2	c^3	bc^3	c^4
$(1,1,1)$	1	1	1	1	1	*	...								
$(1,2,1)$	1	2	2^2	2^3	2^4	*	...								
$(1,3,1)$	1	3	3^2	3^3	3^4	*	...								
$(1,4,1)$	1	4	4^2	4^3	4^4	*	...								
$(1,5,1)$	1	5	5^2	5^3	5^4	*	...								
$(1,1,2)^{(1)}$	0	0	0	0	0	1	1	1	1	3	3	3	7	7	15
$(1,2,2)^{(1)}$	0	0	0	0	0	1	2	2^2	2^3	3	$2\cdot3$	$2^2\cdot3$	7	$2\cdot7$	15
$(1,3,2)^{(1)}$	0	0	0	0	0	1	3	3^2	3^3	3	$3\cdot3$	$3^2\cdot3$	7	$3\cdot7$	15
$(1,4,2)^{(1)}$	0	0	0	0	0	1	4	4^2	4^3	3	$4\cdot3$	$4^2\cdot3$	7	$4\cdot7$	15
$(1,1,3)^{(1)}$	0	0	0	0	0	2	2	2	2	8	8	8	26	26	80
$(1,2,3)^{(1)}$	0	0	0	0	0	2	$2\cdot2$	$2^2\cdot2$	$2^3\cdot2$	8	$2\cdot8$	$2^2\cdot8$	26	$2\cdot26$	80
$(1,3,3)^{(1)}$	0	0	0	0	0	2	$3\cdot2$	$3^2\cdot2$	$3^3\cdot2$	8	$3\cdot8$	$3^2\cdot8$	26	$3\cdot26$	80
$(1,1,4)^{(1)}$	0	0	0	0	0	3	3	3	3	15	15	15	63	63	255
$(1,2,4)^{(1)}$	0	0	0	0	0	3	$2\cdot3$	$2^2\cdot3$	$2^3\cdot3$	15	$2\cdot15$	$2^2\cdot15$	63	$2\cdot63$	255
$(1,1,5)^{(1)}$	0	0	0	0	0	4	4	4	4	24	24	24	124	124	624

where the asterisk means any value. In Step 2 we focus on the 10 by 10 lower rightmost block. Rows corresponding to $(1,b,2)^{(1)}$, $b=1,2,3,4$, form a Vandermonde matrix in the first four columns and they will be the pivotal ones. The (sub-)matrix at Step 2 is

	c	bc	b^2c	b^3c	c^2	bc^2	b^2c^2	c^3	bc^3	c^4
$(1,1,2)^{(1)}$	1	1	1	1	*	...				
$(1,2,2)^{(1)}$	1	2	2^2	2^3	*	...				
$(1,3,2)^{(1)}$	1	3	3^2	3^3	*	...				
$(1,4,2)^{(1)}$	1	4	4^2	4^3	*	...				
$(1,1,3)^{(2)}$	0	0	0	0	1	1	1	6	6	25
$(1,2,3)^{(2)}$	0	0	0	0	1	2	2^2	6	$2\cdot6$	25
$(1,3,3)^{(2)}$	0	0	0	0	1	3	3^2	6	$3\cdot6$	25
$(1,1,4)^{(2)}$	0	0	0	0	2	2	2	14	14	70
$(1,2,4)^{(2)}$	0	0	0	0	2	$2\cdot2$	$2^2\cdot2$	14	$2\cdot14$	70
$(1,1,5)^{(2)}$	0	0	0	0	3	3	3	24	24	141

At Step 3 we consider the 6 by 6 lower rightmost submatrix, whose pivotal rows will be $(1, b, 3)^{(2)}$, $b = 1, 2, 3$, and the remaining rows are modified as follows:

	c^2	bc^2	b^2c^2	c^3	bc^3	c^4
$(1,1,3)^{(2)}$	1	1	1	$*$	$\dots$	
$(1,2,3)^{(2)}$	1	2	2^2	$*$	$\dots$	
$(1,3,3)^{(2)}$	1	3	3^2	$*$	$\dots$	
$(1,1,4)^{(3)}$	0	0	0	1	1	10
$(1,2,4)^{(3)}$	0	0	0	1	2	10
$(1,1,5)^{(3)}$	0	0	0	2	2	22

The matrix at Step 4 is the following one:

	c^3	bc^3	c^4
$(1,1,4)^{(3)}$	1	1	$*$
$(1,2,4)^{(3)}$	1	2	$*$
$(1,1,5)^{(4)}$	0	0	1

At Step $5 = 7 - 2$ the one by one matrix Equation (1) is left. Therefore, the initial matrix has been transformed into a block triangular matrix, whose blocks are Vandermonde matrices with no rows repeated, and so non-singular. This means that also the initial matrix is non-singular. Combined with the fact that even the other three blocks of the matrix, whose rows are the images of the 28 points as in (1), are non-singular, it results that $\mathrm{im}\varphi$ has dimension 28. $\qquad\square$

The knowledge of dim $\mathrm{im}\varphi$ enables us to compute the size of the ghost subgroup $\mathcal{G}$.

Theorem 4. $|\mathcal{G}| = p^{\binom{p+1}{2}+1}$.

Proof. $\mathrm{im}\varphi$ has $p^{\binom{p+1}{2}}$ elements. So, there are

$$\frac{p^{p^2+p+1}}{p^{\binom{p+1}{2}}} = p^{\binom{p+1}{2}+1}$$

ghosts in $\mathrm{PG}(2, p)$. $\qquad\square$

6 Conclusions

In this paper we have addressed the problem of classifying the subsets of $\mathrm{PG}(2, q)$ associated to a same power sum polynomial. By exploiting some similarities with the classical tomographic problem, we have shown that two sets with a same power sum polynomial differ, in the multiset sum sense, by a set whose power sum polynomial is the null one. In analogy to tomography, such a set has been called ghost. We have then counted the number of ghosts of $\mathrm{PG}(2, p)$ by arguing on the image of the function φ, which maps a (multi)set of $\mathrm{PG}(2, p)$ to the corresponding power sum polynomial.

The present paper may be considered as a starting point in the study of the connection between (multi)sets of points in $PG(2, q)$ and their corresponding power sum polynomials within the framework of the polynomial technique for finite geometries. Therefore several future developments may be taken into account.

An immediate related research could be the study of the behavior of the power sum polynomials under the intersection of two sets.

Secondly, we aim to study the size of $im\varphi$ even when h is greater than one. This would give the exact number of ghosts in every projective plane.

Moreover, the subgroup of ghosts deserves a deeper investigation. In particular, we would like to find, if possible, a set of generators for the ghosts.

Further, our investigation has dealt with the planar case, due to its connections with the tomographic problem. Another development is the extension to higher dimensions, to treat the problem in full generality.

References

1. Blokhuis, A.: Polynomials in Finite Geometries and Combinatorics, pp. 35–52. London Mathematical Society Lecture Note Series, Cambridge University Press (1993). https://doi.org/10.1017/CBO9780511662089.003
2. Buekenhout, F. (ed.): Handbook of Incidence Geometry, North Holland (1995). https://doi.org/10.1016/B978-0-444-88355-1.X5000-2
3. De Beule, J.: Direction problems in affine spaces. Galois Geometries and Applications, pp. 79–94 (2014)
4. Fishburn, P., Lagarias, J., Reeds, J., Shepp, L.: Sets uniquely determined by projections on axes II. Discrete case. Discrete Math. **91**(2), 149–159 (1991). https://doi.org/10.1016/0012-365X(91)90106-C
5. Hajdu, L., Tijdeman, R.: Algebraic aspects of discrete tomography. J. Reine Angew. Math. **534**, 119–128 (2001). https://doi.org/10.1515/crll.2001.037
6. Radon, J.: Über die Bestimmung von Funktionen durch ihre Integralwerte längs gewisser Mannigfaltigkeiten. Ber. Verh. Sächs. Akad. Wiss. Leipzig Math.-Phys. Kl. (69), 262–277 (1917)
7. Rédei, L.: Lacunary Polynomials over Finite Fields. North Holland, Amsterdam (1973)
8. Sziklai, P.: Polynomials in finite geometry, manuscript. http://web.cs.elte.hu/sziklai/polynom/poly08feb.pdf
9. Szőnyi, T., Gács, A., Weiner, Z.: On the spectrum of minimal blocking sets in $PG(2, q)$. J. Geom. **76**(1–2), 256–281 (2003). https://doi.org/10.1007/s00022-003-1702-2
10. Szőnyi, T., Weiner, Z.: Stability of k mod p multisets and small weight codewords of the code generated by the lines of PG(2,q). J. Combin. Theory Ser. A **157**, 321–333 (2018). https://doi.org/10.1016/j.jcta.2018.02.005

On the Reconstruction of 3-Uniform Hypergraphs from Step-Two Degree Sequences

Andrea Frosini[1], Giulia Palma[2(✉)], and Simone Rinaldi[2]

[1] Dipartimento di Matematica e Informatica,
Università di Firenze, Firenze, Italy
andrea.frosini@unifi.it
[2] Dipartimento di Ingegneria dell'Informazione e Scienze Matematiche,
Università di Siena, Siena, Italy
{giulia.palma2,rinaldi}@unisi.it

Abstract. A nonnegative integer sequence is k-graphic if it is the degree sequence of a k-uniform simple hypergraph. The problem of deciding whether a given sequence π admits a 3-uniform simple hypergraph has recently been proved to be NP-complete, after long years of research. Thus, it is helpful to find which classes of instances are polynomially solvable in order to restrict the NP-hard core of the problem and design algorithms for real-life applications. Several necessary and few sufficient conditions for π to be k-graphic, with $k \geq 3$, appear in the literature. Frosini et al. defined a polynomial time algorithm to reconstruct k-uniform hypergraphs having regular or almost regular degree sequences. Our study fits in this research line defining some conditions and a polynomial time algorithm to reconstruct 3-uniform hypergraphs having step-two degree sequences, i.e., $\pi = (d, \ldots, d, d-2, \ldots, d-2)$. Our results are likely to be easily generalized to $k \geq 4$ and to other families of similar degree sequences.

Keywords: 3-Uniform hypergraphs · Degree sequences · Reconstruction problem

1 Introduction

The *degree sequence* of a simple hypergraph is the list of its vertex degrees, usually arranged in decreasing order. Given a nonnegative integer sequence π, the possibility of an efficient test for the existence of a simple hypergraph having degree sequence π remained unsolved for many years (see [1,3]). In 2018, Deza et al. in [5] proved that, assuming that $P \neq NP$, such an efficient characterization does not exist even for the simplest case of 3-uniform simple hypergraphs, briefly 3-hypergraphs. Furthermore, computing the number of different hypergraphs sharing the same degree sequence, usually addressed as Uniqueness Problem,

© Springer Nature Switzerland AG 2021
J. Lindblad et al. (Eds.): DGMM 2021, LNCS 12708, pp. 338–347, 2021.
https://doi.org/10.1007/978-3-030-76657-3_24

constitutes a challenging and still unsolved problem also in the simplest cases of homogeneous or almost homogeneous ones.

On the other hand, the degree sequences of simple graphs has been efficiently characterized by Erdős-Gallai in [6]. Relying on this result, several polynomial time algorithms have been defined to reconstruct the incidence matrix of a graph from a given degree sequence (if such a graph exists).

Riding this wave, necessary and sufficient conditions for a nonnegative integer sequence π to be a k-sequence, i.e., to be the degree sequence of a k-hypergraph, with $k \geq 3$, can be found in the successive literature, and they mainly rely on a result by Dewdney [4], based on a recursive decomposition of π. This characterization does not yield to an efficient algorithm and the question to determine a more practical characterization remains open. Brlek and Frosini in [2] defined a P-time reconstruction algorithm for the case of homogenous k-sequences. Later, this result was extended to step-one k-sequences (see [7]). A remarkable fact is that in both cases, all the k-sequences satisfy a simple necessary and sufficient condition. Here, we push further the investigation by focusing on the class of step-two 3-sequences, and we prove that such a condition is not sufficient any more. We also stress that changing from step-one to step-two 3-sequences is critical for the reconstruction of the related 3-hypergraph.

In particular, we first define a series of step-two 3-sequences that we indicate as basic, and that are used as starting point in the final reconstruction algorithm. At one time, we point out the few step-two sequences that are not 3-sequences. Finally, we show how to identify a 3-hypergraph solution related to the given 3-sequence by using as edges the cyclic shifts of 3-dense Lyndon words.

So, in the next section, we give the definitions useful for our study, we point out some relevant previous results and we introduce the main problem. Section 3 is devoted to the consistency problem concerning 3-hypergraphs having step-two degree sequences. Then, in Sect. 4, the reconstruction problem is solved, and the related polynomial time algorithm is provided. We conclude the article pointing out some open problems concerning the realizability of the step-two degree sequences, in Sect. 5.

2   Definitions, Previous Results and Introduction of the Problem

A *hypergraph* H is defined as a couple (V, E), where $V = (v_1, \ldots, v_n)$ is a finite set of n vertices, and E is the set of hyperedges, i.e., a collection of subsets of V, $\{e_1, e_2, \ldots, e_m\}$ where each e_i is non-empty (see [1]). A hypergraph is *simple* if it is loopless, i.e., there is no singleton among the hyperedges, and without parallel hyperedges, i.e., $e \neq e'$ for any pair $e, e' \in E$. Moreover, a hypergraph is said to be k-*uniform*, briefly k-hypergraph, if each hyperedge has cardinality k. The *degree* of a vertex $v \in V$ is the number of hyperedges $e \in E$ such that $v \in e$. The *degree sequence* of H is the list of its vertex degrees, usually written in non-increasing order. In this context, we consider the problem of characterizing the degree sequences $\pi = (d_1, d_2, \ldots, d_n)$ of a 3-hypergraph H, briefly 3-sequences, and reconstruct one of the related 3-hypergraph. We recall that each hypergraph can be represented through its incidence matrix, i.e., the binary matrix $A = (a_{i,j})$, with $i = 1 \ldots |E|$, $j = 1 \ldots |V|$, such that $a_{i,j} = 1$ if and only if the vertex v_j belongs to the hyperedge e_i. So, our study focuses on the characterization of a subset of 3-sequences and the reconstruction of the incidence matrix of their related 3-hypergraphs.

We underline that if a matrix A is the incidence matrix of a hypergraph H, then its rows' and columns' sums correspond to the sequence of the cardinalities of the elements of E and to the degree sequence π, respectively.

Frosini et al. in [7] defined a P-time algorithm to reconstruct k-hypergraphs having d-regular or almost d-regular degree sequences, i.e., the cases where π equals $(d, \ldots, d)$ or $(d, \ldots, d, d-1, \ldots, d-1)$, respectively. The algorithm relies on the notion of Lyndon words, that we recall hereafter. Concerning the related decision problem, they provided a condition that characterizes the d-regular and almost d-regular degree sequences in terms of the inequality

$$d \leq \frac{k}{n} \cdot \binom{n}{k}. \tag{1}$$

In the following, relying on these results, we investigate step-two 3-sequences, i.e., the sequences of the form $\pi = (d, \ldots, d, d-2, \ldots, d-2)$, and we show that their characterization and the reconstruction of (one of) the related 3-hypergraphs cannot be obtained as a simple generalization of the results in [7].

Finally, we recall a couple of standard notions of combinatorics of words that will be useful in the sequel. Following the notation in [8], the *necklace* of a binary word w is the equivalence class of all its cyclic shifts. If w is an aperiodic word, i.e., it can not be expressed as power of one if its proper sub-words, then it is called a *Lyndon word*. It holds the property that a word w of length n is a Lyndon word if and only if its necklace has n different elements. We are interested in fixed-density words (resp. Lyndon words), in which an additional parameter h, called *density*, that represents the number of 1s in the words, is added. The set of necklaces (resp. Lyndon words) with density h is represented by $N(n, h)$ (resp. $L(n, h)$).

For sake of brevity, we denote by $\mathcal{P}$, the set of all step-two 3-sequences whose maximum value satisfies Equation (1), and we use the exponential notation (x^i, y^j) to indicate binary words starting with i times the digit x, followed by j times the digit y.

3   The Characterization of Step-Two 3-Sequences

We face the consistency problem on the instance $\pi = (d, \ldots, d, d-2, \ldots, d-2)$. Given a step-two degree sequence $\pi = (d^g, (d-2)^{n-g})$, we define the *complement* of π, denoted by $\overline{\pi}$, the vector $\overline{\pi} = ((d_{max} - d + 2)^{n-g}, (d_{max} - d)^g)$, where $d_{max} = \frac{k}{n} \cdot \binom{n}{k}$.

A direct consequence of the definition of complement of a degree sequence is the following proposition:

Proposition 1. *A degree sequence π is k-graphic if and only if its complement is k-graphic.*

Proof. Since π is k-graphic, then there exists a $m \times n$ incidence matrix $M(\pi)$ associated with π. Given the incidence matrix M_n associated with the homogeneous regular vector $(d, d, \ldots, d)$ of length n (here d is d_{max}, the reconstructed hypergraph is the regular one and it is trivially unique), we remove all the rows of the submatrix $M(\pi)$, and we obtain the incidence matrix of a k-hypergraph associated with the mirror image of the vector $\overline{\pi}$ (i.e., the vector $\overline{\pi}$ read from right to left). By flipping horizontally the columns of the obtained matrix, we find the incidence matrix of a k-hypergraph having $\overline{\pi}$ as degree sequence. Hence, $\overline{\pi}$ is k-graphic.

Now, we consider the reconstruction problem for step-two degree sequences of $\mathcal{P}$ of length n, starting with the smallest value of n. There are no 3-hypergraphs having step-two degree sequences of length $n \leq 4$, indeed an easy exhaustive check reveals it. Concerning step-two sequences in $\mathcal{P}$, whose length is $n \geq 5$, we prove the following result:

Theorem 1. *Every step-two sequence $\pi = (2^3, 0^p)$, with $p \geq 2$, is not a 3-graphic sequence.*

Proof. As a 3-sequence, $\pi = (2^3, 0^p)$ admits a unique $2 \times (p+3)$ incidence matrix M of a 3-hypergraph. Since the two rows of M have to be equal, then the related 3-hypergraph on $p + 3$ vertices and two hyperedges is not simple, against the definition of the 3-graphicality of a sequence.

Let us call *basic sequences* the step-two sequences $(d^g, (d-2)^{n-g}) \in \mathcal{P}$, with $2 \leq d \leq 4$, different from those considered in Theorem 1 and the sequences $(5^3, 3^p)$, with $p \geq 3$. We choose to indicate the first class of sequences as *basic* since they are the smallest step-two 3-sequences in terms of their first element d. Furthermore, the reconstruction algorithm for step-two 3-sequences that we will present in the sequel uses a decomposition strategy that reaches, as a final step, one of the elements in the class.

We include as basic sequences also the step-two sequences $(5^3, 3^p)$, with $p \geq 3$, since their reconstruction constitutes the only exception in our strategy, and they need to be treated separately in the following proposition.

Proposition 2. *Every basic sequence is a 3-graphic sequence.*

Proof. We obtain the result by defining, for each type of basic sequences, a class of 3-hypergraphs satisfying it. We consider separately the following cases of the basic sequences. Note that they are the only admissible cases to have as elements' sum a multiple of three:

1. $(2^{3g}, 0^p)$, with $g > 1$, $p \geq 1$ and such that $3g + p > 4$;
2. $(3^g, 1^{3p})$, with $g \geq 1$, $p \geq 1$ and such that $g + 3p > 4$,
3. $(4^{3g}, 2^{3p})$, with $g \geq 1$, $p \geq 1$ and such that $3g + 3p > 4$;

4. $(5^3, 3^p)$, with $p \geq 3$.

Case 1 can be reconstructed just using the cyclic shifts of the Lyndon word $(111)(0)^{n-3}$ where the first 1-element is in position i, with $1 \leq i < 3g - 1$ and i is not a multiple of 3, and the Lyndon word $10^{3(g-1)}110^{n-3g}$.

$$\begin{bmatrix} 2\,2\,2\,2\,2\,2\,0\,0\,0 \\ \hline 1\,1\,1\,0\,0\,0\,0\,0\,0 \\ 0\,1\,1\,1\,0\,0\,0\,0\,0 \\ 0\,0\,0\,1\,1\,1\,0\,0\,0 \\ 1\,0\,0\,0\,1\,1\,0\,0\,0 \end{bmatrix}$$

Case $d = 2 : (2^6, 0^3)$.

We specify that this is only one of the possible ways of reconstructing this class of sequences and that it is generally valid as g and p vary.

Case 2 is divided into 3 subcases:

2.1. If $g = 1$, then $p \geq 2$ (otherwise $n = 4$). We use the Lyndon word $1^3 0^{n-3}$ and its cyclic shift $10^{n-3}11$. We use the cyclic shifts $0^5 1110^{n-8}$, ..., $0^{n-5} 11100$, if necessary. Then we use the Lyndon word 100110^{n-5}.

2.2. If $g = 2$, we use the Lyndon word $1^3 0^{n-3}$ and its cyclic shift $110^{n-3}1$. We use the cyclic shifts $0^4 1110^{n-7}$, ..., $0^{n-4} 1110$, if necessary. Then we use the Lyndon word 11010^{n-4}.

2.3. If $g \geq 3$ we use the Lyndon word $1^3 0^{n-3}$ and its cyclic shifts, where the first 1-element is in position i, with $1 \leq i \leq g - 1$; thus, its cyclic shift $110^{n-3}1$. We use the cyclic shifts $0^{g+2} 1110^{n-g-5}$, ..., $0^{n-4} 1110$, if necessary. Then we use the Lyndon word $10^{g-2} 1010^{n-g-2}$.

$$\begin{bmatrix} 3\,1\,1\,1\,1\,1\,1 \\ \hline 1\,1\,1\,0\,0\,0\,0 \\ 1\,0\,0\,0\,0\,1\,1 \\ 1\,0\,0\,1\,1\,0\,0 \end{bmatrix} \qquad \begin{bmatrix} 3\,3\,1\,1\,1\,1\,1\,1 \\ \hline 1\,1\,1\,0\,0\,0\,0\,0 \\ 1\,1\,0\,0\,0\,0\,0\,1 \\ 0\,0\,0\,0\,1\,1\,1\,0 \\ 1\,1\,0\,1\,0\,0\,0\,0 \end{bmatrix} \qquad \begin{bmatrix} 3\,3\,3\,1\,1\,1\,1\,1\,1 \\ \hline 1\,1\,1\,0\,0\,0\,0\,0\,0 \\ 0\,1\,1\,1\,0\,0\,0\,0\,0 \\ 1\,1\,0\,0\,0\,0\,0\,0\,1 \\ 0\,0\,0\,0\,0\,1\,1\,1\,0 \\ 1\,0\,1\,0\,1\,0\,0\,0\,0 \end{bmatrix}$$

Case $d = 3$, $g = 1$: $(3, 1^6)$ Case $d = 3$, $g = 2$: $(3^2, 1^6)$ Case $d = 3$, $g = 3$: $(3^3, 1^6)$

Case 3 is divided into 3 subcases:

3.1. $(4^{3g'}, 2^{3p'})$, with $g', p' > 0$, $n > 6$, since the length of the sequence is a multiple of 3, we can use the cyclic shifts of the periodic word $(10^{n/3-1})^3$, and the matrix associated with the sequence $(3^{3g'}, 1^{3p'})$ that we have already reconstructed in Case 2.

3.2. $(4^{3g'+1}, 2^{3p'+1})$, with $g', p' \geq 0$, we use the cyclic shifts of the Lyndon word $(111)0^{n-3}$, with the first 1-element in position i, with $1 \leq i \leq 3g'$, then the cyclic shifts of the same Lyndon word with the first 1-element in position $3g' + 3, \ldots, n - 2$. Now, we have to reconstruct the matrix associated with the sequence $S = (2^2, 1^{3g'-1}, 0^{3p'+1})$. An easy check reveals that this can be done arranging the cyclic shifts of the Lyndon word $10101(0)^{n-5}$. If the length of the sequence S is even, this is sufficient. Instead, we have also to use the Lyndon word $110^{n-3}1$.

3.3. $(4^{3g'+2}, 2^{3p'+2})$, with $g', p' \geq 0$, we use the cyclic shifts of the Lyndon word $(111)0^{n-3}$, with the first 1-element in position i, with $1 \leq i \leq 3g' + 2$; the cyclic shift of the same Lyndon word with the first 1-element in position $3g' + 4, \ldots, n-1$. Now, we have to reconstruct the matrix associated with the sequence $S = (2, 1^{3g'+1}, 0^{3p'+2})$.

Concerning case 4, the hypergraph can be reconstructed using the matrix obtained from the matrix of all cyclic shifts of the word $1^3 0^{n-3}$, except the word $0^{n-3}1^3$. Now, we have to reconstruct the matrix associated with the sequence $G = (2^3, 0^{n-6}, 1^3)$. This can be done using the Lyndon words $110^{n-5}10^2$, $0110^{n-5}10$, and $1010^{n-4}1$.

$$
\begin{bmatrix}
4\,4\,4\,4\,4\,4\,2\,2\,2\,2\,2\,2 \\
\hline
3\,3\,3\,3\,3\,3\,1\,1\,1\,1\,1\,1 \\
1\,0\,0\,0\,1\,0\,0\,0\,1\,0\,0\,0 \\
0\,1\,0\,0\,0\,1\,0\,0\,0\,1\,0\,0 \\
0\,0\,1\,0\,0\,0\,1\,0\,0\,0\,1\,0 \\
0\,0\,0\,1\,0\,0\,0\,1\,0\,0\,0\,1
\end{bmatrix}
\qquad
\begin{bmatrix}
5\,5\,5\,3\,3\,3 \\
\hline
1\,1\,1\,0\,0\,0 \\
0\,1\,1\,1\,0\,0 \\
0\,0\,1\,1\,1\,0 \\
1\,0\,0\,0\,1\,1 \\
1\,1\,0\,0\,0\,1 \\
1\,1\,0\,1\,0\,0 \\
1\,0\,1\,0\,0\,1
\end{bmatrix}
$$

Case $d = 4$, (3.1), $g' = p' = 2$: $(4^6, 2^6)$. Case $d = 5$: $(5^3, 3^3)$.

□

We underline that the 3-hypergraphs compatible with the basic sequences defined in the above proposition, are not unique.

Remark 1. The basic sequences can be reconstructed using three different necklaces at most.

The procedure that reconstructs the incidence matrix associated with a basic element π is denoted by **RecBasic**(π).

4   Reconstructing a 3-Hypergraph from an Element of $\mathcal{P}$

We recall that Sawada [9] presented a constant amortized time (CAT) algorithm **FastFixedContent** for the exhaustive generation of necklaces $N(n, h)$ of fixed length and density, and a slight modification of it, here denoted **GenLyndon**(n, h), for the CAT generation of the Lyndon words $L(n, h)$. This latter constructs a generating tree of the words, and since the tree has height h, the computational cost of generating k words of $L(n, h)$ is $O(k \cdot h \cdot n)$.

Our reconstruction algorithm is denoted **RecP**(π), and described in Algorithm 1.

Proposition 3. *Let π be a sequence in $\mathcal{P}$ of length $n > 4$. There is a sufficient number of Lyndon words required in the run of* **RecP**(π).

Algorithm 1 RecP(π)

Input: The sequence $\pi = (d, \ldots, d, d - 2, \ldots, d - 2) \in \mathcal{P}$ of length $n \geq 5$.
Output: An incidence matrix M of the 3–hypergraph whose degree sequence is π.
We solve **RecP(π_i)**, where π_i is the minimum between π and $\overline{\pi}$ in the lexicographical order.
Step 1: We initialize $g = 0$, and the vector $D = \pi_i$.
Step 2: While D is not a basic sequence, do $D = D - (3^n)$ and $g = g + 1$.
Step 3: $B = $ **RecBasic(D)**, using the Lyndon words $\ell_1, \ldots, \ell_r$, where $r = 2$ or 3 depending on D.
If $g + r > L(n, 3)$, then give **Failure**.
Step 4: By applying **GenLyndon(n, 3)**, generate the sequence of Lyndon words, and let $u_1, \ldots, u_q$, be the set of these Lyndon words after removing $\ell_1, \ldots, \ell_r$.
Step 5: We take $u_1, \ldots u_g$, and we generate the matrix of the necklaces $M(u_1), \ldots, M(u_g)$. Juxtapose them to B.

Proof. We need to consider the cases $n = 5$ and $n = 6$ separately.

For $n = 5$ there are only four sequences in $\mathcal{P}$: $\pi_1 = (3^2, 1^3)$ and its complement $\overline{\pi_1} = (5^3, 3^2)$, and $\pi_2 = (4, 2^4)$ and its complement $\overline{\pi_2} = (4^4, 2)$. π_1 and π_2 are basic sequences. Since $L(5, 3) = 2$ and the procedure **RecBasic(π_1)** requires just two Lyndon words, then these words are sufficient to reconstruct the matrices associated with π_1. Concerning π_2, the procedure **RecBasic(π_2)** would need three Lyndon words, however, in this special situation, the two Lyndon words are sufficient:

$$\begin{bmatrix} 1\,1\,1\,0\,0 \\ 1\,0\,0\,1\,1 \\ \hline 1\,1\,0\,1\,0 \\ 1\,0\,1\,1\,0 \end{bmatrix}$$

We observe that the sequence $\overline{\pi_1}$ is not a basic sequence and does not fall within the sequences considered in Theorem 1. However, it can be reconstructed since it is the complement of a 3-graphic sequence.

For $n = 6$ there are only seven sequences in $\mathcal{P}$: $\pi_1 = (3^3, 1^3)$ and its complement $\overline{\pi_1} = (9^3, 7^3)$, $\pi_2 = (4^3, 2^3)$ and its complement $\overline{\pi_2} = (8^3, 6^3)$, $\pi_3 = (5^3, 3^3)$ and its complement $\overline{\pi_3} = (7^3, 5^3)$, and $\pi_4 = \overline{\pi_4} = (6^3, 4^3)$. Sequences π_1, π_2, π_3 are basic sequences and can be reconstructed with a maximum of three Lyndon words. Since $L(6, 3) = 4$, the words are sufficient. Concerning π_4 four words are sufficient too, indeed it takes two words to apply the procedure **RecBasic(π_1)**, and then another word to obtain a matrix having vertical projections $(3, 3, 3, 3, 3, 3)$.

Now, let $\pi \in \mathcal{P}$ be a generic sequence of length $n > 6$, such that π is less than $\overline{\pi}$ in the lexicographic order, and denote by $l(B)$ the number of words needed to reconstruct the basic sequence $B = B(\pi)$ associated with π, i.e., the basic sequence obtained in last reconstruction step of **RecBasic(π)**. The number of

Lyndon words of length n and density 3 is

$$L(n,3) = \begin{cases} \frac{1}{n}\binom{n}{3} & \text{if } n \text{ is not a multiple of 3,} \\ \frac{1}{n}[\binom{n}{3} - \frac{n}{3}] \text{ otherwise.} \end{cases}$$

We show that there is a sufficient number of Lyndon words. A simple computation leads to the following inequality:

$$\frac{\lceil \frac{d_{max}}{2} \cdot \frac{3}{n}\binom{n}{3}\rceil - d_B}{3} + l(B) \le L(n) \tag{2}$$

where:

- $\lceil \frac{d_{max}}{2} \cdot \frac{3}{n}\binom{n}{3}\rceil$ is the maximal degree of a step-two degree sequence of length n in $\mathcal{P}$ up to complement;
- d_B is the greatest degree of a basic sequence (hence either $d_B = 2$, or $d_B = 3$, or $d_B = 4$, or $d_B = 5$).

From the inequality (2) we get:

- if n is a multiple of 3:

$$6l(B) - 2d_B \le \frac{(n-1)(n-2)}{2}.$$

  In the worst case (i.e., $d_B = 4$ and $n = 7$), we get: $10 \le 15$.
- Otherwise, if n is not a multiple of 3, we get:

$$6l(B) - 2d_B + 2 \le \frac{(n-1)(n-2)}{2}.$$

  In the worst case (i.e., $d_B = 4$ and $n = 9$), we get: $12 \le 28$.

□

Remark 2. Observe that **Failure** occurs if the number of Lyndon words of $L(n,3)$ is not sufficient to reconstruct the matrix. However, due to the statement above, **RecP** cannot give **Failure**.

Remark 3. The reason why $\pi = (5^3, 3^p)$, with $p \ge 1$, is a basic sequence, is that if we apply **RecP** to π, we would obtain the sequence $(2,2,2,0^p)$, which is not reconstructable (see Theorem 1).

The next theorem is an immediate consequence of Proposition 3 and Remark 3.

Theorem 2. *Given a sequence $\pi \in \mathcal{P}$, π is 3-graphical if and only if π can be reconstructed with the **RecP**.*

Example 1. Let us apply our algorithm to reconstruct the step-two sequence $\pi = (7,7,7,7,7,5,5)$ of length $n = 7$.

- In Step 1 we initialize $g = 0$, and $D = (7,7,7,7,7,5,5)$.
- From Step 2, we get the basic sequence $D = (4,4,4,4,4,2,2)$, and $g = 1$.
- In Step 3, we apply the procedure $B = $**RecBasic**$(D)$, using the Lyndon words $\ell_1 = 0000111$, $\ell_2 = 0001101$, $\ell_3 = 0010101$. Then $r = 3$. Since $g + r = 4 < L(7,3) = 5$, **RecP** does not give **Failure**.
- From Step 4, by applying **GenLyndon**(7,3) and removing ℓ_1, ℓ_2, ℓ_3, we get $u_1 = 0001011$ and $u_2 = 0010011$.
- In Step 5, we get the matrix depicted below.

$$\begin{bmatrix} 0 & 0 & 0 & 1 & 0 & 1 & 1 \\ 1 & 0 & 0 & 0 & 1 & 0 & 1 \\ 1 & 1 & 0 & 0 & 0 & 1 & 0 \\ 0 & 1 & 1 & 0 & 0 & 0 & 1 \\ 1 & 0 & 1 & 1 & 0 & 0 & 0 \\ 0 & 1 & 0 & 1 & 1 & 0 & 0 \\ 0 & 0 & 1 & 0 & 1 & 1 & 0 \\ \hline 1 & 1 & 1 & 0 & 0 & 0 & 0 \\ 0 & 1 & 1 & 1 & 0 & 0 & 0 \\ 0 & 0 & 1 & 1 & 1 & 0 & 0 \\ 0 & 0 & 0 & 1 & 1 & 1 & 0 \\ 0 & 0 & 0 & 0 & 1 & 1 & 1 \\ 1 & 1 & 0 & 0 & 0 & 0 & 1 \\ \hline 1 & 0 & 1 & 0 & 1 & 0 & 0 \\ \hline 1 & 1 & 0 & 1 & 0 & 0 & 0 \end{bmatrix}$$

The validity of **RecP**(π) is a simple consequence of Theorem 2. Clearly, the obtained matrix is the incidence matrix of a 3-hypergraph having degree sequence π, indeed by construction, all the rows are distinct. Moreover, the algorithm always terminates since, at each iteration, we add as many rows as possible to the final solution. Concerning the complexity analysis, we need to generate $O(m)$ different Lyndon words and shift each of them $O(n)$ times. Thus, since the algorithm **GenLyndon**$(n,3)$ requires $O(f \cdot h \cdot n)$, that is $O(3 \cdot f \cdot n)$ steps to generate f words of $L(n,3)$, the whole process takes polynomial time.

As a consequence of Theorem 2 we have a simple characterization of the step-two sequences which are 3-graphic.

Corollary 1. *A sequence $\pi \in \mathcal{P}$ is 3-graphic if and only if nor π neither $\bar{\pi}$ are of the form $(2^3, 0^p)$, $p \geq 1$.*

This implies that for any $n \geq 5$ there are two sequences of $\mathcal{P}$ which are not 3-graphic. As an example, with $n = 7$ the two sequences are $(2^3, 0^4)$ and $(15^4, 13^3)$.

5   Conclusions and Future Developments

We defined a polynomial algorithm that reconstructs the incidence matrix of a 3-uniform hypergraph realizing a step-two sequence of integers, when such a hypergraph exists.

The algorithm we furnished is tuned for 3-uniform hypergraphs, but we believe that it admits a generalization to k-uniform ones. Thus, a possible direction for further research is to generalize our algorithm to obtain a polynomial algorithm for the reconstruction of k-uniform hyeprgraphs having a step-two degree sequence. The characterization of the degree sequences of k-uniform hyperpgraphs, $k \geq 3$, is an NP-hard problem. Therefore, under the assumption that $P \neq NP$ there is no hope to find a good characterization; but to find a compact nice looking characterization should be of great interest in order to design algorithms for real-life applications.

The problem addressed in our article appears particularly interesting to us since, for almost $d-$regular sequences, all the sequences that satisfy Eq. 1 are 3-graphic, while this is not true anymore for step-two sequences. Therefore, it is significant to understand what happens as the step increases and to evaluate how many and which sequences satisfy Eq. 1, but are not 3-graphic.

Another line of research concerns the characterization and the study of integer sequences having span two, i.e., $\pi = (d, \ldots, d, d-1, \ldots, d-1, d-2, \ldots, d-2)$.

References

1. Berge, C.: Hyperpgraphs. North-Holland, Amsterdam (1989)
2. Brlek, S., Frosini, A.: A tomographical interpretation of a sufficient condition on h-graphical sequences. In: Normand, N., Guédon, J., Autrusseau, F. (eds.) DGCI 2016. LNCS, vol. 9647, pp. 95–104. Springer, Cham (2016). https://doi.org/10.1007/978-3-319-32360-2_7
3. Colbourn, C.J., Kocay, W.L., Stinson, D.R.: Some NP-complete problems for hypergraph degree sequences. Discrete Appl. Math. **14**, 239–254 (1986)
4. Dewdney, A.K.: Degree sequences in complexes and hypergraphs. Proc. Amer. Math. Soc. **53**(2), 535–540 (1975)
5. Deza, A., Levin, A., Meesum, S.M., Onn, S.: Optimization over degree sequences. SIAM J. Fisc. Math. **32**(3), 2067–2079 (2018)
6. Erdös, P., Gallai, T.: Graphs with prescribed degrees of vertices (Hungarian). Mat. Lapok **11**, 264–274 (1960)
7. Frosini, A., Picouleau, C., Rinaldi, S.: On the degree sequences of uniform hypergraphs. In: Gonzalez-Diaz, R., Jimenez, M.-J., Medrano, B. (eds.) DGCI 2013. LNCS, vol. 7749, pp. 300–310. Springer, Heidelberg (2013). https://doi.org/10.1007/978-3-642-37067-0_26
8. Ruskey, F., Sawada, J.: An efficient algorithm for generating necklaces with fixed density. Siam J. Comput. **29**, 671–684 (1999)
9. Sawada, J.: A fast algorithm to generate necklaces with fixed content. Theor. Comput. Sci. **301**, 477–489 (2003)

Hierarchical and Graph-Based Models, Analysis and Segmentation

Towards Interactive Image Segmentation by Dynamic and Iterative Spanning Forest

Isabela Borlido Barcelos[1], Felipe Belém[2], Paulo Miranda[3],
Alexandre Xavier Falcão[2], Zenilton K. G. do Patrocínio Jr.[1],
and Silvio Jamil F. Guimarães[1(✉)]

[1] Laboratory of Image and Multimedia Data Science (ImScience), Pontifical Catholic University of Minas Gerais, Belo Horizonte 31980–110, Brazil
{sjamil,zenilton}@pucminas.br
[2] Laboratory of Image Data Science (LIDS), University of Campinas,
São Paulo 13083–852, Brazil
{afalcao,felipe.belem}@ic.unicamp.br
[3] Institute of Mathematics and Statistics, University of Campinas,
São Paulo 05508–090, Brazil
pmiranda@vision.ime.usp.br

Abstract. Interactive image segmentation aims to partition the image into background and foreground objects by taking into account seeds inserted by users. Nowadays, many methods are capable of generating segmentations with few user interactions, especially region-based techniques. However, such methods are highly sensitive to seed displacement and quantity, and delineation errors are often propagated to the final segmentation result. Recently, a novel superpixel segmentation framework, named Dynamic and Iterative Spanning Forest (DISF), was proposed, which achieved top delineation performance while assessing many seed-based state-of-the-art methods' drawbacks. In this work, we propose interactive DISF (iDISF), an interactive segmentation framework, by modifying each step of DISF to consider user-validated information. DISF uses the Image Foresting Transform (IFT) framework for computing an optimum-path forest rooted in a seed set in the delineation step. To consider path and image gradient variation, we propose three new connectivity functions for the IFT. Finally, we also propose two new seed removal strategies for detecting relevant seeds for subsequent iterations. Results show segmentation improvements for minimal user effort—*i.e.*, a single click—and show theoretical advances that may benefit recent optimum-path-based interactive methods from scribbles.

Keywords: Graph-based image segmentation · Dynamic and iterative spanning forest · Interactive segmentation

The authors thank Conselho Nacional de Desenvolvimento Científico e Tecnológico – CNPq – (PQ 310075/2019-0), Coordenação de Aperfeiçoamento de Pessoal de Nível Superior – CAPES – (Grant COFECUB 88887.191730/2018-00) and Fundação de Amparo à Pesquisa do Estado de Minas Gerais – FAPEMIG – (Grants PPM-00006-18).

J. Lindblad et al. (Eds.): DGMM 2021, LNCS 12708, pp. 351–364, 2021.
https://doi.org/10.1007/978-3-030-76657-3_25

1   Introduction

Image segmentation consists of partitioning an image into significant regions, and its applications covers high- and low-level tasks, such as: medical image analysis [12,21]; iris detection [23]; and face recognition [5]. Since the accurate delineation is of utmost importance, the presence of noise and smooth edges imposes a major difficulty for automatic segmentation processes, making interactive approaches necessary [13,14]. In interactive image segmentation, the method receives information from the user—as a region of interest around the object [16] or internal and external line markers (*i.e.*, scribbles) [18,19]—to better delineate the desired object [17]. For interactive segmentation, some approaches offer an effective object delineation. In contour-based methods [7], the process of manually setting the object borders may achieve effective delineation, at the expense of being an exhausting activity for medium to large image sets. Whereas for strategies based on Graph cut [20], although popular, their performance degrades when the region of interest does not delimit the object tightly. The most popular approaches, using deep-learning techniques, have the ability of generalizing and locating the object with minimum user effort, but may not maintain their performance for unseen objects and requires a large amount of annotated images for parameter optimization [17].

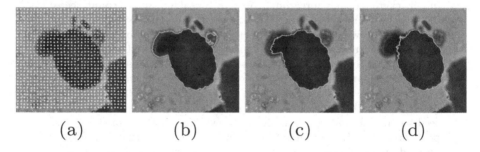

    (a)   (b)   (c)   (d)

Fig. 1. (a) Initial seed oversampling with $N_0 = 1000$ (yellow) and a user-defined seed (cyan). (b) One can run DISF and try to select the superpixel that best represents the object, but this does not solve the problem. In iDISF, even with a single object seed (a), the result can improve (c). However, with the proposed functions for dynamic arc-weight estimation, it's possible to achieve even better results (d). (Color figure online)

Among the state-of-the-art methods for interactive segmentation, there are the ones based on the Image Foresting Transform (IFT) [9]. The IFT is an elegant framework for developing image operators based on connectivity and solve the segmentation problem by computing an *optimum-path forest* from the image graph. By assigning seed labels, each object (and background) is formally defined by the trees with the same label [9]. In interactive segmentation, *Dynamic Trees* [3] surpassed state-of-the-art methods by computing dynamically the arc weights based on mid-level image properties of the growing trees. In superpixel segmentation [2,4,10,22], the *Dynamic and Iterative Spanning Forest* (DISF) [2] is a novel three-staged seed-based superpixel algorithm that exploits Dynamic Trees for assessing major drawbacks in seed-based state-of-the-art methods in terms of speed, segmentation consistency, and object delineation.

Since DISF was not conceived for interactive segmentation purposes, in this work we explore its decoupled structure to propose an interactive segmentation approach, named *interactive* DISF (iDISF). From scribbles inside and outside the desired object, the new method performs an oversampling of background seeds and calculates an optimum-path forest rooted at the seeds, removing the less relevant ones at each iteration. For seed removal, we propose two strategies, in which the first considers the distance from superpixels to the scribbles, and the second takes into account the superpixel position on the object's border. Alongside this, we propose three connectivity functions with a dynamic trees approach considering the local gradient and the contrast of the image.

Figure 1 illustrates the ineffectiveness of the simplest solution and how iDISF works to circumvent the problem. By starting from a seed oversampling (yellow seed in Fig. 1a) and a selected pixel inside the object (cyan seed in Fig. 1a), *a posteriori* superpixel selection in DISF reduce user effort to the choice of a single superpixel, but it does not solve the problem. In iDISF, even with a single user-selected seed the segmentation can be improved (Fig. 1c), achieving even better results with a proposed connectivity function (Fig. 1d).

The concepts to understand those IFT-based methods are described in Sect. 2 and iDISF is presented in Sect. 3. In Sect. 4, we demonstrate that iDISF can considerably increase segmentation accuracy with minimum user effort and that our novel connectivity functions can improve IFT-based methods. Finally, Sect. 5 provides a conclusion and future work.

2   Theoretical Background

2.1   Graph Notions

An image **I** is a pair $(\mathcal{I}, I)$ where $\mathcal{I}$ is a set of pixels, and I is a mapping that assigns to each pixel $p \in \mathcal{I}$ a value $I(p)$ in some arbitrary feature space (*e.g.*, the CIELAB color space). An image may be interpreted as a graph $G = (\mathcal{I}, \mathcal{A}, I)$, where the set $\mathcal{A} \subset \mathcal{I} \times \mathcal{I}$ of arcs derives from a given *adjacency relation* (*e.g.*, 4-neighborhood) and the nodes in $\mathcal{I}$ are weighted by function I. The methods exploit optimum connectivity between a seed set $\mathcal{S} \subset \mathcal{I}$ and the remaining pixels through paths in G, but irrespective to any specific seed. Let Π_G be the set of all possible paths in G. A *path* $\pi_p \in \Pi_G$ with terminus p is then a sequence $\pi_p = \langle p_1, p_2, .., p_k = p \rangle$ of adjacent nodes, being *trivial* when $\pi_p = \langle p \rangle$. A path $\pi_q = \pi_p \cdot \langle p, q \rangle$ is the extension of a path π_p with an arc $(p, q) \in \mathcal{A}$.

2.2   Image Foresting Transform and Dynamic Trees

Given a path-cost function f_* (we will use $f_* \in \{f_1, f_2, f_3, f_4\}$ along the text) and an image graph G, the IFT algorithm [9] generates an *optimum-path forest* $\mathcal{P}$, by assigning one *optimum-path* π_q^* to each node $q \in \mathcal{I}$—*i.e.*, $\mathcal{P}$ is an acyclic map that assigns to every node q its predecessor $\mathcal{P}(q) = p$ in the optimum path $\pi_q^* = \text{argmin}_{\pi_q \in \Pi_G}\{f_*(\pi_q)\}$, or a marker *nil* $\notin \mathcal{I}$ when $\langle q \rangle = \text{argmin}_{\pi_q \in \Pi_G}\{f_*(\pi_q)\}$ indicates that q is a root of the map. When f_* does not satisfy the conditions in [6], the predecessor map $\mathcal{P}$ is still a spanning forest with properties suitable for image segmentation [22]. In this work, the methods constrain the optimum paths to start from a seed set $\mathcal{S} \subset \mathcal{I}$ by setting

$$f_*(\langle q \rangle) = \begin{cases} 0, & \text{if } q \in \mathcal{S} \\ +\infty & \text{otherwise.} \end{cases} \tag{1}$$

For a given seed set $\mathcal{S}$, the IFT starts from all nodes being trivial paths with costs defined by Eq. 1 and grows optimum-path trees rooted at $\mathcal{S}$, such that each seed $s \in \mathcal{S}$ conquers its most strongly connected nodes $p \in \mathcal{I}$. During this process, let $\mathcal{T}_s$ be a growing optimum-path tree rooted at s. For every node $p \in \mathcal{T}_s$, we can store its root node $R(p) = s$. When a seed s tries to conquer a node q, adjacent to p, through the optimum path π_p, the algorithm evaluates if $f_*(\pi_p \cdot \langle p, q \rangle) < f_*(\pi_q)$ and then substitutes π_q by $\pi_p \cdot \langle p, q \rangle$ when it is true.

For interactive image segmentation based on Dynamic Trees [3], the user selects seeds inside and outside the objects and $f_*(\pi_p \cdot \langle p, q \rangle)$ may be defined as

$$f_1(\pi_p \cdot \langle p, q \rangle) = \max\{f_1(\pi_p), \|\mu_{\mathcal{T}_{R(p)}} - I(q)\|_2\}, \tag{2}$$

where $\mu_{\mathcal{T}_{R(p)}} = \frac{1}{|\mathcal{T}_{R(p)}|}\sum_{\forall t \in \mathcal{T}_{R(p)}} I(t)$ its the mean color vector of a tree $\mathcal{T}_{R(p)}$, rooted at $R(p) \in \mathcal{S}$. The *Dynamic Tress* approach has been evaluated with various dynamic arc-weight functions, being more effective than methods based on watershed transform, graph cut, and deep extreme cut [8].

2.3   Dynamic and Iterative Spanning Forest (DISF)

The DISF framework consists of three steps [2]: (a) seed oversampling, increasing the chances to select relevant seeds for an initial seed set $\mathcal{S}$; (b) superpixel delineation by the IFT algorithm for the path-cost function f_1 (Eq. 2); and (c) seed removal from $\mathcal{S}$ based on a superpixel-based seed relevance criterion, increasing the chances of retaining relevant seeds in $\mathcal{S}$. Steps (b) and (c) are repeated until the desired number of superpixels is achieved. For step (a), DISF adopts grid sampling [1] with a considerably higher number N_0 of samples (seeds) than the desired number $N_f \ll N_0$ of superpixels. In grid sampling, seeds are separated by the same distance d (Eq. 3) along with the main directions of the image domain.

$$d = \sqrt{|\mathcal{I}|/N_0} \tag{3}$$

To avoid seeds in border pixels, they are also moved to a position with the lowest gradient value among the 8 neighbors. One can see that, as N_0 increases, the space between seeds and the probability of missing a critical region are significantly reduced.

To achieve N_f superpixels, DISF retains $N_{i+1} = \max\{N_0 \exp^{-i}, N_f\}$ relevant seeds in $\mathcal{S}$, after each iteration $i \in \mathbb{N}^*$ to be used for the next iteration $i+1$, being the remaining ones discarded. The relevance of a seed depends on properties of its superpixel (size and color contrast). Since each superpixel is a tree, a *tree-adjacency relation* can be formally defined as: $\mathcal{B}_1 = \{(\mathcal{T}_s, \mathcal{T}_t) \mid \exists(x,y) \in \mathcal{A}, x \in \mathcal{T}_s, y \in \mathcal{T}_t$ and $s \neq t\}$. Then, the *relevance value* of a seed $s \in \mathcal{S}$ is defined by

$$V(s) = \frac{|\mathcal{T}_s|}{|\mathcal{I}|} \min_{\forall (\mathcal{T}_s, \mathcal{T}_t) \in \mathcal{B}_1} \{\|\mu_{\mathcal{T}_s} - \mu_{\mathcal{T}_t}\|_2\}. \tag{4}$$

3   Interactive DISF

DISF can provide accurate superpixel delineation [2], but no information assigns superpixels to objects of interest. For interactive segmentation, we modified DISF to consider scribbles drawn by the user and, from an oversampling of background seeds, remove the less relevant ones considering their position concerning those of the scribbles. We named this variant the *interactive* DISF (iDISF).

The three steps of DISF are slightly modified in iDISF. In step (a), the user selects seeds inside and outside the desired object and this slightly affects seed oversampling. For step (b), we introduce three novel connectivity functions that add gradient information to Eq. 2. In step (c), the seeds sampled by user scribbles truly indicate that their superpixel is part of the object or background (thus, relevant and must not be removed). We also propose two approaches for seed removal, in which one considers the distance to the object seeds, and the other the superpixel position in objects' border. The next sections provide the details of those steps (a)–(c) in iDISF.

3.1   Seed Oversampling

In iDISF, the user draws scribbles inside and outside the object of interest. From these pixels, $\mathbf{S_o}$ is the set of seeds labeled as object and $\mathbf{S_b}$ as background, where $\mathbf{S_o} \cap \mathbf{S_b} = \{\varnothing\}$, for $|\mathbf{S_o}| > 0$, and $|\mathbf{S_b}| \geq 0$. Therefore, it is not mandatory for the user to label background seeds. Let $\mathcal{S}$ denote the seed set in which, initially, is set to $\mathcal{S} \leftarrow \mathbf{S_o} \cup \mathbf{S_b}$. Moreover, consider L as a label map in which assigns a label to every node $p \in \mathcal{I}$, such that $L(p) = 1$ indicates an object seed, and $L(p) = 0$, a background one. Then, given a seed $s \in \mathbf{S_o}$, $L(s) = 1$, and it is analogous for $\mathbf{S_b}$.

Afterward, a seed oversampling is performed by an initial number of N_0 seeds, given by the user, using grid sampling [1]. The motivation is to represent the background regions whose pixels are not included in $\mathbf{S_b}$ and have distinct characteristics (*e.g.*, color or texture) from any pixel in $\mathbf{S_b}$. In iDISF, to maintain a minimum distance between the seeds, and to avoid overlap between seeds, each seed p, sampled by grid strategy, is included in $\mathcal{S}$ only if $\|p - q\|_2 > \frac{d}{2}$ for all $q \in \{\mathbf{S_o} \cup \mathbf{S_b}\}$. Thus, N_G seeds sampled by grid strategy are included in $\mathcal{S}$.

3.2   Superpixel Delineation

Superpixel delineation can be based on the IFT algorithm using the connectivity function f_1 (Eq. 2), as introduced for Dynamic Trees [3]. However, the arc weight $w(p,q) = \|\mu_{T_{R(p)}} - I(q)\|_2$ does not account for local color contrast around q and mean color contrast inside the tree $T_{R(p)}$. For that, we propose in Eqs. 5–7 three novel functions to evaluate non-trivial paths (Eq. 1 is still valid for trivial paths). They provide different ways to estimate arc weights dynamically.

$$f_2(\pi_p \cdot \langle p,q \rangle) = \max\{f_2(\pi_p), w(p,q) \times g_1(p,q)\}, \tag{5}$$

$$f_3(\pi_p \cdot \langle p,q \rangle) = \max\{f_3(\pi_p), w(p,q) + g_1(p,q)\}, \tag{6}$$

$$f_4(\pi_p \cdot \langle p,q \rangle) = \max\{f_4(\pi_p), w(p,q) \times g_2(p,q)\}. \tag{7}$$

The main idea of the proposed functions is to control the growth of each superpixel $T_{R(p)}$ in regions that might have higher/lower color contrast (*i.e.*, texture). Let $\|\nabla I(q)\|_2$ be a normalized image-gradient magnitude at a pixel q, and consider $C_\mathcal{I}$ and $C_{T_{R(p)}}$ the standard deviation divided by the mean value of $\|\nabla I(q)\|_2$ for all node q in $\mathcal{I}$ and in $T_{R(p)}$, respectively. Since $C_\mathcal{I}$ and $C_{T_{R(p)}}$ provide the relative dispersion of the gradient, they are used to measure the texture in $\mathcal{I}$ and $T_{R(p)}$, respectively. Therefore, more homogeneous regions have smaller values. Functions $\beta_* \in \{\beta_1, \beta_2\}$ are created to favor superpixel growth in regions of higher contrast, and α to controls the increasing/decreasing weight according to the contrast on the image, $C_\mathcal{I}$. The α in β_* favor superpixel growth in regions of lower contrast but whose images have high contrast.

$$\beta_1(p) = \max\{1, c_2\alpha, C_{T_{R(p)}}\}, \tag{8}$$

$$\beta_2(p) = \max\left\{1, c_2\alpha, \frac{C_{T_{R(p)}}}{|T_{R(p)}|}\right\}, \tag{9}$$

$$\alpha = \max\{c_1, C_\mathcal{I}\}, \tag{10}$$

where $0 < c_1 \leq 1$ and $0 < c_2 \leq 1$ are constants that control the minimum values of α and β_*, respectively. Finally, functions $g_*(p,q)$ are defined as

$$g_*(p,q) = \frac{\|\nabla I(q)\|_2^{\frac{1}{\alpha}}}{\beta_*(p)}, \tag{11}$$

Figure 2 illustrates examples of segmentation with different values of c_1 and c_2 when f_2 is the connectivity function. The main idea in c_1 is to provide a minimum $C_\mathcal{I}$ to prevent low gradient images from having a high increase in $\|\nabla I(q)\|_2$. Moreover, c_2 prevents regions with low contrast from receiving a high decrease in weight due to a high α.

3.3   Seed Removal

In this work, we evaluated two approaches for seed removal. The first one, named *seed removal by relevance*, consider the minimum Euclidean distance from a tree's pixel coordinates to any seed in $\mathbf{S}_o$. The second approach, named *seed removal by class*, distinguish the trees into two classes and calculates different relevance and removal conditions according to the class of the seed. In both approaches, we consider that the user draws the object scribbles inside the desired object, and the background scribbles outside the object. Therefore, the seeds in $\mathbf{S}_o$ and $\mathbf{S}_b$ are never removed.

Seed Removal by Relevance. Given an adjacency relation $\mathcal{B}_1$, and a function $D(\mathcal{T}_s, \mathbf{S}_o) = \min_{\forall \{p,q | p \in \mathcal{T}_s, q \in \mathbf{S}_o\}} \{\|p - q\|_2\}$ that returns the minimum Euclidean distance between the coordinates of the nodes in $\mathcal{T}_s$ and the seeds in $\mathbf{S}_o$. The relevance in DISF is modified (Eq. 12) to assign the maximum relevance for any s in $\{\mathbf{S}_o \cup \mathbf{S}_b\}$, and to include the distance $D(\mathcal{T}_s, \mathbf{S}_o)$ in the relevance criterion for seeds sampled by grid strategy. The motivation is to remove the seeds whose superpixel is closer to the object's scribbles and that may have some node inside the object.

$$
V_1(s) = \begin{cases} +\infty, & \text{if } s \in \{\mathbf{S}_o \cup \mathbf{S}_b\} \\ \frac{|\mathcal{T}_s| D(\mathcal{T}_s, \mathbf{S}_o)}{|\mathcal{I}|} \min_{\forall (\mathcal{T}_s, \mathcal{T}_t) \in \mathcal{B}_1} \{\|\mu_{\mathcal{T}_s} - \mu_{\mathcal{T}_t}\|_2\}, & \text{otherwise} \end{cases} \tag{12}
$$

Afterward, it includes each seed s in $\mathcal{S}$ in a queue Q with priority $V_1(s)$ and selects $N_{i+1} = \max\{(|\mathbf{S}_o| + |\mathbf{S}_b| + N_G) \exp^{-i}, |\mathbf{S}_o| + |\mathbf{S}_b| + N_f\}$ seeds with higher priority for iteration $i+1$. In which N_G is the number of seeds sampled (grid) in the first iteration, and N_f is the number of desired superpixels whose root was sampled by grid strategy. When N_{i+1} is equal to $|\mathcal{S}|$, the algorithm stops.

Seed Removal by Class. Similarly to $\mathcal{B}_1$, let $\mathcal{B}_2 = \{(\mathcal{T}_s, \mathcal{T}_t) \mid \exists (x, y) \in \mathcal{A}, x \in \mathcal{T}_s, y \in \mathcal{T}_t$ and $L(s) \neq L(t)\}$ be an tree-adjacency relation between object and background superpixels (*i.e.*, the object border). Then, according to L, it is possible to classify the superpixels in two categories: (1) the ones adjacent to superpixels with same label; and (2) those adjacents to at least one superpixel with different label. The classifying function M maps each seed s into one of the latter categories (*i.e.*, $M(s) \in \{1, 2\}$).

The *relevance by class* (Eq. 13) criterion assigns different relevance values for each seed $s \in \mathcal{S}$ with respect to M, such that, if $M(s) = 1$, the DISF's original strategy is applied; otherwise, we evaluate the color dissimilarity between the superpixels in the object border, only.

$$
V_2(s) = \begin{cases} \frac{|\mathcal{T}_s|}{|\mathcal{I}|} \min_{\forall (\mathcal{T}_s, \mathcal{T}_t) \in \mathcal{B}_1} \{\|\mu_{\mathcal{T}_s} - \mu_{\mathcal{T}_t}\|_2\}, & \text{if } M(s) = 1 \\ \min_{\forall (\mathcal{T}_s, \mathcal{T}_t) \in \mathcal{B}_2} \{\|\mu_{\mathcal{T}_s} - \mu_{\mathcal{T}_t}\|_2\}, & \text{otherwise} \end{cases} \tag{13}
$$

For instance, if a background superpixel is at the border, its relevance depends on how similar it is to its object neighbors (*e.g.*, less relevant if it is more similar). Note that any $s \in \{\mathbf{S}_o \cup \mathbf{S}_b\}$ is not applicable for being removed.

First, let $\mathbb{S}_x \subset \mathcal{S}$ be the set of seeds s whose $M(s) = x$, and let $\mu(\mathbb{S}_x)$ and $\sigma(\mathbb{S}_x)$ be its relevance mean and its relevance standard deviation values, respectively. Then, for non-user seeds (*i.e.*, sampling by grid strategy), two priority

Table 1. Comparison among DISF and iDISF with functions f_1–f_4 on the Parasites dataset

Method	N_0	FB	FOP	FR	SC	SSC	BGM	BCE	Mean N_f/iter
DISF f_1	200	0.64402	0.82283	0.96328	0.94813	0.94692	0.96964	0.92954	4.66234
iDISF *superpixels* f_1	200	0.68015	0.84884	0.96654	0.95137	0.95109	0.97134	0.93501	8.92207
iDISF *superpixels* f_2	200	0.72323	0.87560	0.96787	0.95469	0.95368	0.97338	0.93870	9.49350
iDISF *superpixels* f_3	200	0.69894	0.85186	**0.97017**	0.95605	0.95544	0.97407	0.94133	10.12987
iDISF *superpixels* f_4	200	**0.72912**	**0.88432**	0.96994	**0.95746**	**0.95653**	**0.97496**	**0.94259**	8.96103
iDISF *iterations* f_1	200	0.78286	0.90849	0.97764	0.96594	0.96599	0.97954	0.95622	5.57143
iDISF *iterations* f_2	500	**0.78545**	**0.90035**	**0.97794**	**0.96681**	**0.96654**	**0.98014**	**0.95699**	7.11688
iDISF *iterations* f_3	200	0.77231	0.89956	0.97559	0.96324	0.96361	0.97819	0.95261	5.12987
iDISF *iterations* f_4	500	0.77236	0.89263	0.97712	0.96504	0.96515	0.97910	0.95448	7.14286

queues—Q_1 and Q_2—are created in which the seed s is included in $Q_{M(s)}$ with priority $V_{M(s)}(s)$.

For each seed $s \in Q_x$, with $M(s) = x$, in non-decreasing order, s is maintained in $\mathcal{S}$ if $V_x(s) \geq \mu(\mathbb{S}_x) + \sigma(\mathbb{S}_x)$. Additionally, for each $s \in Q_1$, s is maintained if any of its neighboring seeds has been discarded, irrespective of $V(s)$. This latter condition delays the growth of unrepresentative background superpixels. Any s that satisfies none of the conditions imposed its discarded. Differently from DISF and iDISF with removal by relevance strategy, the stopping criteria for iDISF with removal by class are designed taking into account the desired number of iterations or $Q_1 = Q_2 = \{\varnothing\}$.

4   Experimental Results

In this section, we describe the experimental setup for iDISF evaluation using the f_1–f_4 connectivity functions, being f_2–f_4 proposed in this work. To distinguish the proposed seed removal approaches, we named them by iDISF *superpixels* and iDISF *iterations*. The iDISF *superpixels* perform *seed removal by relevance* with a controllable number of superpixels. On the other hand, the iDISF *iterations* accomplish *seed removal by class* with a controllable number of iterations.

4.1   Experimental Setup

To evaluate our seed removal proposals and connectivity functions, we performed two experiments. For the former, we have selected a private dataset, called Parasites, with 77 optical microscopy images of helminth eggs (Fig. 1 and second row of Fig. 2). The eggs might be connected to similar parts of the background. In the second one, we compare iDISF with a Dynamic Trees approach in an IFT framework in a public image dataset, GeoStar [11], which comes with 151 natural images (second row of Fig. 2) and user-drawn scribbles.

For the experiments, DISF and iDISF are executed with $N_0 \in \{50, 100, 200, 500, 1000\}$, $N_f \in \{2, 5, 10, 25, 50, 100, 200, 250, 500, 750, 1000\}$ (for DISF and iDISF *superpixels*), and a range from 1 to 10 iterations (for iDISF *iterations*) for each connectivity function, f_1–f_4. To evaluate our proposals with minimum user effort, we added a one-pixel scribble inside the object to each image in the Parasites dataset. For DISF, we perform *a posteriori* object selection given the one-pixel scribble of each image. Finally, we empirically set $c_1 = 0.7$ and $c_2 = 1$.

For quantitative evaluation, we considered classic segmentation metrics [15]: (i) *segmentation covering* (SC); (ii) *swapped segmentation covering* (SSC); (iii) *bipartite graph matching* (BGM); (iv) *bidirectional consistency error* (BCE); (v) *f-measure for regions* (FR); (vi) *f-measure for boundaries* (FB); and (vii) *f-measure for object and parts* (FOP).

4.2   Evaluation with Minimal User Effort

Table 1 shows the best DISF and iDISF variants' performances and their respective parameters.One can see that both iDISF can be significantly more effective than DISF with *a posteriori* object selection by the user irrespective the initial number of grid samples. Moreover, iDISF with IFT gradient-aware connectivity functions obtained the best results, indicating that f_1 might not capture finer texture details. Figure 3 illustrates the qualitative results of DISF and iDISF *superpixels* with all connectivity functions. iDISF is capable of effectively separate the egg from its attached impurity, differently from DISF with *a posteriori* superpixel selection. Moreover, one can see that methods considering f_1 function tends to generate leakings, while f_2–f_4, by considering the gradient variation, reduces such error significantly.

Table 2. Comparison among an IFT and iDISF with functions f_1–f_4 on the GeoStar dataset.

Method	N_0	FB	FOP	FR	SC	SSC	BGM	BCE	Mean N_f/iter
IFT f_1	0	0.41209	0.50484	0.89773	0.86954	0.85711	0.91464	0.83485	1
IFT f_2	0	0.39377	0.49406	0.89615	0.86789	0.85213	0.91038	0.83030	1
IFT f_3	0	0.39152	0.49988	0.89579	0.86723	0.85167	0.90995	0.82977	1
IFT f_4	0	0.38934	0.48727	0.89316	0.86340	0.84985	0.90872	0.82808	1
iDISF *superpixels* f_1	200	**0.43147**	**0.55688**	**0.92063**	**0.90115**	**0.88234**	**0.93477**	**0.86514**	18.27333
iDISF *superpixels* f_2	200	0.41226	0.53870	0.91785	0.89842	0.87744	0.93195	0.86015	18.64667
iDISF *superpixels* f_3	200	0.41444	0.54113	0.91792	0.89801	0.87774	0.93196	0.86090	18.04667
iDISF *superpixels* f_4	200	0.41276	0.54266	0.91741	0.89779	0.87701	0.93163	0.85908	15.84667
iDISF *iterations* f_1	200	**0.43435**	0.54570	0.92030	0.90112	0.88224	0.93468	0.86613	5.34000
	500	0.43316	**0.56060**	0.92186	0.90340	**0.88386**	0.93585	0.86724	6.22667
iDISF *iterations* f_2	500	0.42163	0.55928	0.92299	**0.90511**	0.88341	**0.93587**	**0.86747**	6.32666
iDISF *iterations* f_3	500	0.41930	0.55234	0.92163	0.90330	0.88240	0.93516	0.86649	6.15333
iDISF *iterations* f_4	1000	0.42131	0.56012	**0.92301**	0.90500	0.88320	0.93577	0.86724	7.34667

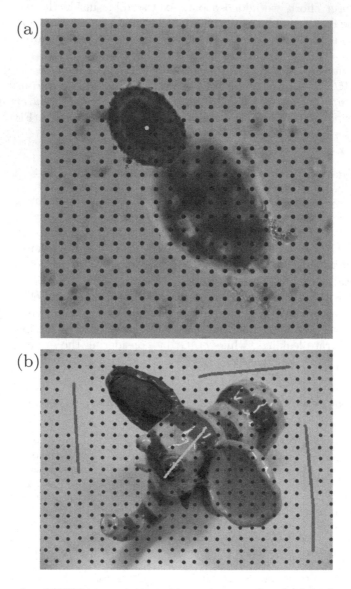

Fig. 2. Examples of iDISF segmentation with varying c_1 and c_2. (a) Initial oversampling with grid samples (blue) and object (cyan) and background (red) scribbles. (b) On the first row, none background scribbles was labeled, $c_2 = 1$ and c_1 varies as 0.1 (left), 0.5 (middle) and 1 (right). On the second row, $c_1 = 1$ and c_2 varies as 0.1 (left), 0.5 (middle) and 1 (right). All results are computed with connectivity function f_2, $N_G = 500$, and *seed removal by class* with 7 iterations. (Color figure online)

4.3   Evaluation with Scribbles

For the GeoStar dataset, with internal and external scribbles, we compare iDISF with functions f_1–f_4 to an IFT with a Dynamic Trees approach, which showed to be more effective than methods based on watershed transformation, graph cut, and deep extreme cut [3]. Therefore, the baseline here is the IFT with function f_1. Table 2 illustrates the quantitative results for the best IFT and iDISF variants' performances and their respective parameters. In this case, both iDISF

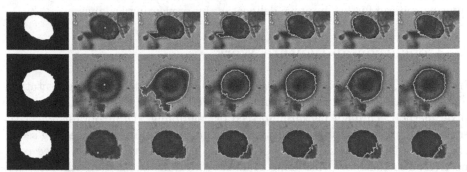

(a) Ground-truth and seed (b) DISF f_1 (c) iDISF f_1 (d) iDISF f_2 (e) iDISF f_3 (f) iDISF f_4

Fig. 3. Examples of object delineation on the Parasites dataset. The first column (a) contains the ground-truth image on the left and the seed labeled as object on the right (cyan). (b) shows DISF result with function f_1, and (c)–(f) iDISF results for f_1–f_4, respectively. (Color figure online)

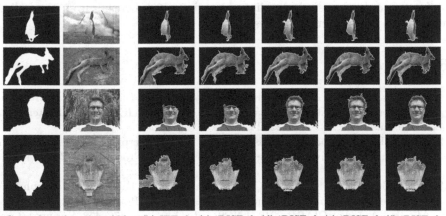

(a) Ground-truth and scribbles (b) IFT f_1 (c) iDISF f_1 (d) iDISF f_2 (e) iDISF f_3 (f) iDISF f_4

Fig. 4. Examples of object delineation on the GeoStar dataset. The first column (a) contains the ground-truth image on the left and the dataset scribbles on the right, in which the scribbles in cyan mark the objects and the scribbles in red mark the background. (b) shows resuls in a IFT framework with function f_1, and (c)–(f) contains results of iDISF with functions f_1–f_4, respectively. (Color figure online) (*iterations* and *superpixels*) obtain better results than IFT, which shows that a strategy of iterative segmentation can improve the final segmentation significantly. Moreover, like in Parasites dataset, the inclusion of gradient variation in the connectivity function has shown to be important for delineation.

Figure 4 illustrates the qualitative results of IFT and iDISF *iterations* with the connectivity functions f_1–f_4. Due to the background seed oversampling, iDISF increases the seed competition and, therefore, is capable of preventing major object and background leaks, and, therefore, achieving better delineation than the baseline. As stated before, the images show that inclusion of gradient information can further improve delineation by preventing leakings.

5   Conclusion and Future Work

In this work, we propose *interactive DISF* (iDISF), an interactive segmentation approach based on the *Dynamic and Iterative Spanning Forest* (DISF) superpixel method. Through background seed oversampling, iDISF obtains effective segmentation by iteratively computing optimum-path forests and selecting seeds that favors improving subsequent iterations. Moreover, we propose three connectivity functions for considering gradient variations along the path, and two seed removal strategies for a proper selection of relevant seeds (given an user-validated information). Results show that iDISF is capable of effectively segment the object, possibly requiring minimal user effort (*i.e.*, single click). In a real medical application, iDISF achieves effective object segmentation by considering the gradient variation during its computation. While in a natural image dataset, iDISF surpasses the baselines by a significant margin. For future work, we intend to develop seed sampling strategies which benefits from the indirect information of the user's scribbles.

References

1. Achanta, R., Shaji, A., Smith, K., Lucchi, A., Fua, P., Süsstrunk, S.: SLIC superpixels compared to state-of-the-art superpixel methods. IEEE Trans. Pattern Anal. Mach. Intell. **34**(11), 2274–2282 (2012)
2. Belém, F.C., Guimarães, S.J.F., Falcão, A.X.: Superpixel segmentation using dynamic and iterative spanning forest. IEEE Signal Proces. Lett. **27**, 1440–1444 (2020)
3. Bragantini, J., Martins, S.B., Castelo-Fernandez, C., Falcão, A.X.: Graph-based image segmentation using dynamic trees. In: Vera-Rodriguez, R., Fierrez, J., Morales, A. (eds.) CIARP 2018. LNCS, vol. 11401, pp. 470–478. Springer, Cham (2019). https://doi.org/10.1007/978-3-030-13469-3_55
4. de Castro Belém, F., Guimarães, S.J.F., Falcão, A.X.: Superpixel segmentation by object-based iterative spanning forest. In: Iberoamerican Congress on Pattern Recognition, pp. 334–341. Springer (2018)
5. Chen, J., Bai, X.: Thermal face segmentation based on circular shortest path. Infrared Phys. Technol. **97**, 391–400 (2019)

6. Ciesielski, K.C., Falcão, A.X., Miranda, P.A.V.: Path-value functions for which Dijkstra's algorithm returns optimal mapping. J. Math. Imaging Vis. (2018). https://doi.org/10.1007/s10851-018-0793-1

7. Condori, M.A.T., Mansilla, L.A.C., Miranda, P.A.V.: Bandeirantes: a graph-based approach for curve tracing and boundary tracking. In: Angulo, J., Velasco-Forero, S., Meyer, F. (eds.) ISMM 2017. LNCS, vol. 10225, pp. 95–106. Springer, Cham (2017). https://doi.org/10.1007/978-3-319-57240-6_8

8. Falcão, A., Bragantini, J.: The role of optimum connectivity in image segmentation: can the algorithm learn object information during the process? In: Couprie, M., Cousty, J., Kenmochi, Y., Mustafa, N. (eds.) DGCI 2019. LNCS, vol. 11414, pp. 180–194. Springer, Cham (2019). https://doi.org/10.1007/978-3-030-14085-4_15

9. Falcão, A.X., Stolfi, J., de Alencar Lotufo, R.: The image foresting transform: theory, algorithms, and applications. IEEE Trans. Pattern Anal. Mach. Intell. **26**(1), 19–29 (2004)

10. Galvão, F.L., Guimarães, S.J.F., Falcão, A.X.: Image segmentation using dense and sparse hierarchies of superpixels. Pattern Recogn. **108**, 107532 (2020). https://doi.org/10.1016/j.patcog.2020.107532. http://www.sciencedirect.com/science/article/pii/S0031320320303356

11. Gulshan, V., Rother, C., Criminisi, A., Blake, A., Zisserman, A.: Geodesic star convexity for interactive image segmentation. In: IEEE Computer Society Conference on Computer Vision and Pattern Recognition, pp. 3129–3136. IEEE (2010)

12. Huang, Q., Huang, Y., Luo, Y., Yuan, F., Li, X.: Segmentation of breast ultrasound image with semantic classification of superpixels. Med. Image Anal. **61**, 101657 (2020)

13. Miranda, P.A.V., Mansilla, L.A.C.: Oriented image foresting transform segmentation by seed competition. IEEE Trans. Image Process. **23**(1), 389–398 (2013). https://doi.org/10.1109/TIP.2013.2288867

14. Peng, B., Zhang, L., Zhang, D.: A survey of graph theoretical approaches to image segmentation. Pattern Recogn. **46**(3), 1020–1038 (2013)

15. Pont-Tuset, J., Marques, F.: Measures and meta-measures for the supervised evaluation of image segmentation. In: Proceedings of the IEEE Conference on Computer Vision and Pattern Recognition, pp. 2131–2138 (2013)

16. Rajchl, M., Lee, M.C., Oktay, O., Kamnitsas, K., Passerat-Palmbach, J., Bai, W., Damodaram, M., Rutherford, M.A., Hajnal, J.V., Kainz, B., et al.: Deepcut: object segmentation from bounding box annotations using convolutional neural networks. IEEE Trans. Med. Imaging **36**(2), 674–683 (2016). https://doi.org/10.1109/TMI.2016.2621185

17. Ramadan, H., Lachqar, C., Tairi, H.: A survey of recent interactive image segmentation methods. Comput. Vis. Media **6**(4), 355–384 (2020). https://doi.org/10.1007/s41095-020-0177-5

18. Rauber, P.E., Falcão, A.X., Spina, T.V., de Rezende, P.J.: Interactive segmentation by image foresting transform on superpixel graphs. In: 2013 XXVI Conference on Graphics, Patterns and Images, pp. 131–138. IEEE (2013)

19. Sung, M.C., Chang, L.W.: Using multi-layer random walker for image segmentation. In: International Workshop on Advanced Image Technology (IWAIT), pp. 1–4. IEEE (2018)

20. Tang, M., Gorelick, L., Veksler, O., Boykov, Y.: Grabcut in one cut. In: Proceedings of the IEEE International Conference on Computer Vision (ICCV), pp. 1769–1776. IEEE (2013)

21. Tiwari, A., Srivastava, S., Pant, M.: Brain tumor segmentation and classification from magnetic resonance images: Review of selected methods from 2014 to 2019. Pattern Recogn. Lett. **131**, 244–260 (2020)
22. Vargas-Muñoz, J.E., Chowdhury, A.S., Alexandre, E.B., Galvão, F.L., Miranda, P.A.V., Falcão, A.X.: An iterative spanning forest framework for superpixel segmentation. IEEE Trans. Image Process. **28**(7), 3477–3489 (2019)
23. Vyas, R., Kanumuri, T., Sheoran, G.: Non-parametric iris localization using pupil's uniform intensities and adaptive masking. In: 2017 14th IEEE India Council International Conference (INDICON), pp. 1–4. IEEE (2017)

Stability of the Tree of Shapes to Additive Noise

Nicolas Boutry[✉] and Guillaume Tochon

EPITA Research and Development Laboratory (LRDE), Le Kremlin-Bicêtre, France
{nicolas.boutry,guillaume.tochon}@lrde.epita.fr

Abstract. The tree of shapes (ToS) is a famous self-dual hierarchical structure in mathematical morphology, which represents the inclusion relationship of the shapes (*i.e.* the interior of the level lines with holes filled) in a grayscale image. The ToS has already found numerous applications in image processing tasks, such as grain filtering, contour extraction, image simplification, and so on. Its structure consistency is bound to the cleanliness of the level lines, which are themselves deeply affected by the presence of noise within the image. However, according to our knowledge, no one has measured before how resistant to (additive) noise this hierarchical structure is. In this paper, we propose and compare several measures to evaluate the stability of the ToS structure to noise.

Keywords: Tree of shapes · Additive noise · Stability · Tree distance

1 Introduction

The tree of shapes (ToS) is a morphological hierarchical representation which encodes the inclusion relationship between the shapes (*i.e.*, the interior of the level lines with holes filled) in a given image. Initially proposed by [20], it has then been extensively studied, both for its theoretical [1,10,13] and computational [8, 14] properties, as well as its wide range of applications in image processing (such as image segmentation [7] and simplification [26], object detection and extraction [19], morphological filtering [24], shape-space representations [18,25]), morphological attribute profiles computation [16], feature extraction for image retrieval [4], and curve matching [22]).

It is usually considered as the fusion between the max-tree [23] and its dual, the min-tree. Being a morphological representation, the ToS structure relies on the existence of a total ordering between the pixel values (to order by inclusion the level lines of the image), hence its existence for grayscale images (although it has been successfully adapted to multichannel images [9]). Therefore, the potential presence of noise within the image appears as a critical question regarding the cleanliness of the level lines, and thus the stability and the significance of the ToS representation with respect to the image it is built upon. More specifically, the fact that level lines would tend to degrade as the signal-to-noise ratio in the image decreases comes with the intuition that the resulting ToS also loses

ⓒ Springer Nature Switzerland AG 2021
J. Lindblad et al. (Eds.): DGMM 2021, LNCS 12708, pp. 365–377, 2021.
https://doi.org/10.1007/978-3-030-76657-3_26

relevance (at least for most of the shapes it contains). Measuring 1) how far from the "clean" ToS lies the "noisy" ToS (*i.e.*, the ToS built on the image corrupted by noise) and 2) how much the shapes of the "noisy" ToS have been degraded by the presence of noise could give some useful insights on the image content as well as the credibility that should be given to further processing applied on the ToS (such as image segmentation or object recognition resulting from it).

Unfortunately, the ToS stability has not been really explored yet in the mathematical morphology community. A study of the robustness of common hierarchical structures in terms of pixel classification performances of some morphological attribute profiles has been carried out in [16], leading to the conclusion that the ToS, the min/max trees and the ω-tree show superior performance compared to the α-tree depending of the choice of some threshold. In [4], it is shown that the MSER feature extractor can be significantly improved using the tree of shapes for some image retrieval tasks, due to an addition estimated to between 20% and 40% of features. This tree-based measure is robust to noise since it is derived from the MSER methodology, but no ToS robustness measure is given. In [22], an approach based on curve matching is proposed which is rotation-invariant and seems to be robust to noise (although this property is not emphasized in the cited work). Clearly, these three papers are not motivated by the idea of giving a complementary toolbox of robustness/stability measures as deeply as we do in the present paper.

When it comes to measuring distances between tree graphs, possible solutions encompasses tree-edit distances [3], graph distances [6], treelets based on graph kernels [12], Reeb graphs distances [2], or interleaving distances between merge trees [21]. However, the ToS structure is richer than a simple graph since all nodes also bear an image-related meaning that should also be taken into account to evaluate the similarity between such structures (for instance, two ToS might contain very different shapes even though their graph structure is the same). In this paper, the aim is to proceed to a very exploratory research, which concerns the definition of mathematical tools able to measure *how much a ToS is perturbed* when we add noise in the image it comes from (see Fig. 1). The plan is the following: Sect. 2 recalls the mathematical background necessary to understand the paper, Sects. 3, 4, 5 and 6 present spectral, topological, and geometrical measures of the stability of a ToS to added noise, and Sect. 7 concludes the paper with a summary of the different measures presented in the paper and their properties (variance, type of convergence, monotonicity, slope around 0, and so on). In the appendix (Sect. A), we add material about preliminary results on natural images, we show elementary measures computed on synthetic/natural images, and we show the evolution of a tree of shape on natural/synthetic images when noise amplitude increases.

2   Mathematical Background About the Tree of Shapes

We begin by briefly recalling the way the ToS is computed (see Fig. 2). Starting from a given image $I : \Omega \to \mathbb{R}$, we can compute for each possible level

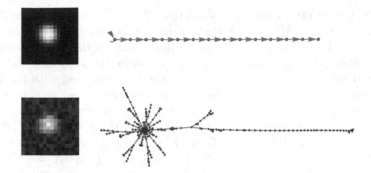

Fig. 1. In the raster scan order, a synthetic image without noise, its tree of shapes, the noisy version of this image, and its tree of shapes (we used an additive noise). The aim of this paper is to provide measures of these perturbations at the hierarchical level, taking account (or not) of the shapes of the respective trees.

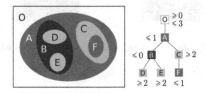

Fig. 2. An example of tree of shapes computation from [14]

$\ell \in \mathbb{R}$ the upper threshold sets: $[I \geq \ell] = \{x \in \Omega \; ; \; I(x) \geq \ell\}$, and the lower threshold sets: $[I < \ell] = \{x \in \Omega \; ; \; I(x) < \ell\}$. Using these sets, we compute their connected components in Ω and we saturate them (using the cavity fill-in operator), to obtain the upper shapes: $\mathcal{S}_\geq = \{Sat(\Gamma) \; ; \; \Gamma \in \mathcal{CC}([I \geq \ell])\}$, and the lower shapes: $\mathcal{S}_< = \{Sat(\Gamma) \; ; \; \Gamma \in \mathcal{CC}([I < \ell])\}$. By merging the sets $\mathcal{S}_\geq$ and $\mathcal{S}_<$, and assuming that the image is well-composed [5], we obtain the so-called *tree of shapes* $\mathcal{T}$ of I. It is known to be self-dual and contrast-invariant, and this way it represents the inclusion relationship between the shapes in the image.

3   Proposed Spectral Measures

After a brief recall in matter of Hausdorff and spectral distances, we propose four candidate spectral distances to measure stability of the ToS to additive noise.

3.1   Mathematical Preliminaries

The Hausdorff Distance. Let N be some positive integer. Let h be a mapping from $\mathbb{R} \times \mathbb{R}^N$ to $\mathbb{R}$ such as for any $v \in \mathbb{R}$ and $E \in \mathbb{R}^N$: $h(v, E) = \min_{e \in E} |v - e|$. We can define the Hausdorff distance [15] on two sets E_1, E_2 of N scalars by:

$$\mathcal{HA}(E_1, E_2) = \max\left(\max_{e_1 \in E_1} h(e_1, E_2), \max_{e_2 \in E_2} h(e_2, E_1)\right).$$

The Co-Spectral Distance. Given two trees $\mathcal{T}$ and $\mathcal{T}'$, let us assume that we want to measure their difference based on their spectrum, more exactly on the spectrum of the Laplacian [11] of the adjacency matrix of these trees considered as graphs. Starting from a tree $\mathcal{T}$, we can compute its adjacency matrix $\mathcal{A}$, which is defined as $\mathcal{A}_{i,j} = \mathcal{A}_{j,i} = 1$ if the node j is connected with the node i in $\mathcal{T}$, otherwise $\mathcal{A}_{i,j} = \mathcal{A}_{j,i} = 0$. Using this adjacency matrix, we can compute the Laplacian of $\mathcal{A}$, denoted $\mathcal{L}$, which is defined by the formula: $\mathcal{L} = \mathcal{D} - \mathcal{A}$, where $\mathcal{D}$ is the degree matrix [11] of $\mathcal{A}$. Based on these definitions[1], the co-spectral distance between two trees $\mathcal{T}$ and $\mathcal{T}'$ with the same number of nodes N is defined as $\sum_{i \in [1,N]} (\lambda_i - \lambda'_i)^2$, where $\Lambda = \{\lambda_i\}_{i \in [1,N]}$ (resp. $\Lambda' = \{\lambda'_i\}_{i \in [1,N]}$) is the spectrum of the Laplacian matrix $\mathcal{L}$ (resp. $\mathcal{L}'$) of $\mathcal{T}$ (resp. $\mathcal{T}'$).

Fig. 3. The original image (maximum pixel intensity $= 100$) and noisy versions ($\zeta = 20$, $\zeta = 50$ and $\zeta = 100$).

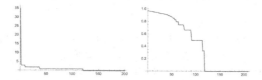

Fig. 4. Spectra of a ToS when the adjacency is weighted by zeros/ones (left) and by the IoU (right). We can see that the spectrum on the right side has more relief thanks to the IoU.

3.2 Proposed Spectral-Based Distances

We start from the ToS $\mathcal{T}$ corresponding to a given image $I : \Omega \to \mathbb{N}$ (see Fig. 3). Given some $\zeta \in \mathbb{N}$, we add an independent noise n which follows a uniform discrete law on $[\![0, \zeta]\!]$ so that $I' = n + I$, and we compute its ToS $\mathcal{T}'$. ζ is said to be the noise amplitude.

In this section, we use the following methodology:

$$I \to \mathcal{T} \to \mathcal{A} \to \mathcal{L} \to \Lambda .$$

---

[1] We recall that the spectrum of the Laplacian of a graph does not depend on the enumeration of its nodes, which explains why we can establish measures on the spectra of two graphs to compute a "distance" between them.

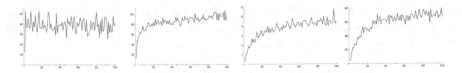

Fig. 5. From left to right, computation of μ_1, μ_2, μ_3 and μ_4 as a function of the noise amplitude ranging from 0 to 100 (the amplitude of the initial image).

First Approach. Let $\mathcal{T}$ and $\mathcal{T}'$ be two ToS computed on images I and I', respectively. We consider the adjacency matrices deduced from these ToS by setting $\mathcal{A}_{i,j}$ to 1 when $\mathcal{S}_j$ is parent of $\mathcal{S}_i$ (otherwise we write 0), and we compute the spectra of their respective Laplacians Λ and Λ' (see Fig. 4 (left)). Then, we compute the Hausdorff distance applied to Λ and Λ' to define our first measure:

$$\mu_1(\Lambda, \Lambda') = \mathcal{HA}(\Lambda, \Lambda').$$

Observation: We can observe in Fig. 5 (left) that this distance does not increase with the amplitude of the additive noise, leading to the conclusion that this measure does not seem well-suited for stability estimation. Indeed, we expect the function to be increasing with the noise amplitude to represent that the higher the noise, the more perturbed (thus the farther away) the "noisy" ToS.

Second Approach. Due to the failure of μ_1, we reconsider the Hausdorff formula for stability estimation and investigate the following measure:

$$\mu_2(\Lambda, \Lambda') = \mathcal{HA}_{mod}(\Lambda, \Lambda') = \max\left(\sum_{\lambda \in \Lambda} h(\lambda, \Lambda'), \sum_{\lambda' \in \Lambda'} h(\lambda', \Lambda)\right).$$

The aim of μ_2 is to consider all the eigenvalues that have been inserted in the new spectrum: the more they differ from the initial ones (or the more numerous they are), the higher their impact. At the same time, we want to ensure that if no new eigenvalue is introduced, then the stability measure is zero.

Observation: The first thing we can remark (see Fig. 5 left middle) is that it is monotonic (in mean). A second remark is that the slope of μ_2 seems to be high for low noise amplitudes. This is normal since the structure of the ToS is modified (many new branches are inserted and initial branches are elongated) even for very small values of ζ, in particular for synthetic images like the one under study.

Third Approach. We did not consider in the previous approaches the shapes of the two trees $\mathcal{T}$ and $\mathcal{T}'$. We propose then to add this information in the adjacency matrix by setting $\mathcal{A}_{i,j}$ as the intersection over union (IoU) value of $\mathcal{S}_i$

and $\mathcal{S}_j$ when $\mathcal{S}_j$ is parent of $\mathcal{S}_i$, otherwise we set it to zero. We then obtain the spectra depicted in Fig. 4 (right). Following, we compute between the two new spectra Λ_{IoU} and Λ'_{IoU} the formula:

$$\mu_3(\mathcal{T}, \mathcal{T}') = \mathcal{HA}_{mod}(\Lambda_{IoU}, \Lambda'_{IoU}).$$

Observation: We observe in Fig. 5 (right) that μ_3 increases slower than μ_2 and then is less sensible to the high perturbation that the ToS undergoes. Furthermore, the monotonic behavior of μ_2 is preserved.

Fourth Approach. Being in the case where we can have spectra of different sizes, we propose now to extend the usual definition of co-spectral distance in the following manner. We start from the discrete spectra Λ_{IoU} and Λ'_{IoU} and then interpolate them in a linear way. We obtain $\mathscr{I}_{IoU}$ and $\mathscr{I}'_{IoU}$, respectively. Since they are continuous, we can easily resize them to obtain two signals with a same support $[1, \max(N, N')]$ where N is the size of Λ and N' is the size of Λ'. This leads to our new measure:

$$\mu_4(\mathcal{T}, \mathcal{T}') = \int_1^{\max(N,N')} \left(\mathscr{I}_{IoU}(t) - \mathscr{I}'_{IoU}(t)\right)^2 dt.$$

Observation: The results of this measure can be observed in the rightmost panel of Fig. 5, which shows a much satisfying measure in the sense that it keeps increasing as long as $\mathcal{T}$ is perturbed.

Conclusion About the Spectral Approaches: Taking into account the IoU in the coefficients of the adjacency matrix allows us to obtain a more stable Laplacian spectrum but needs more computations. We recommend then μ_2 for fast computations, and μ_3 or μ_4 for more stable stability evaluation.

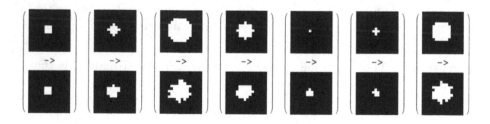

Fig. 6. Example of matching obtained using the HMA: the (initial) shapes extracted from $\mathcal{T}$ depicted in the first row are associated by the HMA to the (noisy) shapes extracted from $\mathcal{T}'$ depicted in the second row.

4   Proposed Elongation Measure

Let us assume that we have as usual our two trees T and T' computed from two images $I : \Omega \to \mathbb{R}$ and $I' : \Omega \to \mathbb{R}$, respectively. We now apply the Hungarian Matching Algorithm [17] (HMA) on these two tree of shapes based on the IoU measure computed on their shapes (we do not consider the structure of the trees but only their shapes since the tree structure can be induced by the set of shapes of each image). In this way, the pairing cost between two shapes $S \in T$ and $S' \in T'$ is $\mathrm{IoU}(S, S')$, and the final optimal matching $\mathcal{H}$ is defined as the injective function f from T to T' which maximizes the total cost:

$$\mathcal{H} = \arg\min_{f \text{injective}} \sum_{S \in T} (1 - \mathrm{IoU}(S, f(S))).$$

Note that usually, we use the HMA on sets of same cardinality, or on squared matrices. In our case, we use a bipartite graph $\mathcal{A}$ with the rows corresponding to the initial tree T, the columns corresponding to the final (noisy) tree T', and each $\mathcal{A}_{i,j}$ represents the (position-sensitive) pairing cost between the ith shape of T and the jth shape of T' (see Fig. 6). Now that we have our matching between the shapes of T and a subset $\mathcal{H}(T)$ of T', we can compute the elongation measure ℓ of T' relatively to T for any shape in T' in the following manner. When S' equals Ω, we define $\ell(S') = 0$ (since this shape never moves in the ToS). Then, for any shape $S' \in \mathcal{H}(T)\backslash\{\Omega\}$, we compute its inverse image $\mathcal{H}^{-1}(S')$ by the HMA. Now, we compute its parent $\mathrm{Par}_T(\mathcal{H}^{-1}(S'))$. We are ensured that this parent exists since in the tree of shapes, any shape has a parent (Ω is its own parent). Then, we compute its image $\mathfrak{P}(S') := \mathcal{H}(\mathrm{Par}_T(\mathcal{H}^{-1}(S')))$ in T'. After having defined:

$$\ell_0(S') = \mathrm{depth}_{T'}(S') - \mathrm{depth}_{T'}(\mathfrak{P}(S')) - 1,$$

the value:

$$\ell(S') = \max(0, \ell_0(S'))$$

is then a measure of how much the tree has been elongated from a local point of view for a given shape of $\mathcal{H}(T)$.

Let us remark that it can happen that $\ell_0(S')$ is negative. Indeed, the HMA ensures optimality in matters of costs, but does not guarantee that $\mathfrak{P}(S')$ is a parent of S' by following the procedure described above (even if we observed experimentally in simple cases that it is almost always the case). Let us recall however that, even if the HMA is optimal, the given pairing solution is not always unique, so this measure depends on the matching result. In the case where we want to obtain a scalar measure of the elongation, we proceed the following way. First we estimate the number of shapes of $\mathcal{H}(T)$ whose elongation is positive: $N' = \mathrm{Card}(\{S' \in \mathcal{H}(T) ; \ell_0(S') \geq 0\})$. Then, if N' is equal to zero, we consider that the total elongation is zero. Otherwise, it is equal to:

$$\ell_{tot}(T, T') = \frac{1}{N'} \sum_{S' \in \mathcal{H}(T) \text{ s.t. } \ell_0(S') \geq 0} \ell_0(S').$$

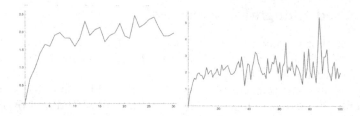

Fig. 7. Total elongation ℓ_{tot} as a function of the noise amplitude. On the left side, we show the elongation for noises whose amplitude goes from 0 to 30, and on the right side, from 0 to 100. We can remark that when the noise amplitude reaches about 30% of the signal amplitude, the measure converges in mean but its variance increases dramatically.

Observation: Despite its high variance as soon as we reach values next to $\zeta = 30$ (see Fig. 7), and then its non effectiveness as measure of the stability of the tree of shapes for high noise amplitudes, this measure remains the only proposed topological measure and is thus worthy of existence.

5   Proposed Deformation Measure $\mathcal{M}$

Let us assume that we have the same hypotheses as usual on the trees $\mathcal{T}$ and $\mathcal{T}'$ and that we apply the HMA on their sets of shapes. We have a matching between $\mathcal{T}$ and $\mathcal{H}(\mathcal{T})$ from which we can compute the first part of our measure:

$$\mathcal{M}_1(\mathcal{T}, \mathcal{T}') := \sum_{\mathcal{S} \in \mathcal{T}} d_{\mathrm{IoU}}(\mathcal{S}, \mathcal{H}(\mathcal{S})),$$

with $d_{\mathrm{IoU}}(\mathcal{S}, \mathcal{S}') = 1 - \mathrm{IoU}(\mathcal{S}, \mathcal{S}')$ their Jaccard index. The term $\mathcal{M}_1(\mathcal{T}, \mathcal{T}')$ measures how much the shapes of $\mathcal{T}$ have been deformed due to the added noise. Now, we have also to consider the intermediary shapes which have been added in-between the initial shapes. For this aim, we define for $A, B \in \mathcal{T}'$ with $A \supset B$:

$$\mathrm{Interm}_{\mathcal{T}'}(A, B) = \{\mathcal{S} \in \mathcal{T}' \; ; \; A \supset \mathcal{S} \supset B\}$$

Using this last notation, we can define the set of intermediary shapes in $\mathcal{T}'$:

$$\mathcal{I} = \{\mathcal{S}' \in \mathcal{T}' \; ; \; \exists A', B' \in \mathcal{H}(\mathcal{T}), A' \supset \mathcal{S}' \supset B'\} = \bigcup_{A, B \in \mathcal{H}(\mathcal{T})} \mathrm{Interm}_{\mathcal{T}'}(A, B).$$

Since the common domain Ω to I and I' (parent of every shape $\mathcal{S}$ in $\mathcal{T}$ and every shape $\mathcal{S}'$ in $\mathcal{T}'$) always belong to $\mathcal{H}(\mathcal{T})$, we can simplify $\mathcal{I}$:

$$\mathcal{I} = \bigcup_{\mathcal{S}' \in \mathcal{H}(\mathcal{T})} \mathrm{Parents}_{\mathcal{T}'}(\mathcal{S}') \setminus \mathcal{H}(\mathcal{T}),$$

where $\mathrm{Parents}_{\mathcal{T}'}(\mathcal{S}')$ is the set of strict parents of $\mathcal{S}'$ in $\mathcal{T}'$.

Once we have processed the matching shapes in $\mathcal{T}'$, some shapes remain (the ones which do not match at all with the initial tree). For them, we consider the measure:

$$\mathcal{M}_2(\mathcal{T},\mathcal{T}') = \sum_{\mathcal{S}'\in\mathcal{T}'\setminus\mathcal{H}(\mathcal{T})\setminus\mathcal{I}} \min_{\mathcal{S}\in\mathcal{T}} d_{\mathrm{IoU}}(\mathcal{S}',\mathcal{S}).$$

This measure shows how much the little shapes perturb the global structure of the tree. We finally conclude with the total measure of the deformation between $\mathcal{T}$ and $\mathcal{T}'$:

$$\mathcal{M}_{\mathrm{tot}}(\mathcal{T},\mathcal{T}') := \alpha_1\mathcal{M}_1(\mathcal{T},\mathcal{T}') + \alpha_2\mathcal{M}_2(\mathcal{T},\mathcal{T}'),$$

with $\{\alpha_i\}_{i\in[1,2]}$ a set of non-zero positive parameters. We can remark that when $\mathcal{T} = \mathcal{T}'$, we obtain $\mathcal{M}_{\mathrm{tot}}(\mathcal{T},\mathcal{T}') = 0$.

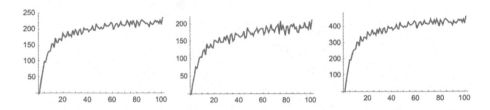

Fig. 8. From left to right, $\mathcal{M}_1$, $\mathcal{M}_2$, and $\mathcal{M}_{\mathrm{tot}}$ with α_1 and α_2 are arbitrarily chosen equal to one.

Observation: In Fig. 8, we can remark that $\mathcal{M}_1$ and $\mathcal{M}_2$ seem to have the same behavior when ζ increases. The main remark on the total measure is that it does not suffer from the limitation of the elongation measure presented in the previous section, and then seems to us to be a good candidate to quantify noise in a hierarchical structure like the ToS.

6   Proposed Measure β Based on d_{IoU}-Matching

Assuming we have the same notations as usual, we propose now to compute measures based on d_{IoU} (and not anymore using the HMA). In other words, we do what we call d_{IoU}-*matching*. Let us first define the following function $\xi : \mathcal{T} \to \mathcal{T}'$ as:

$$\xi(\mathcal{S}) := \arg\min_{\mathcal{S}'\in\mathcal{T}'} d_{\mathrm{IoU}}(\mathcal{S},\mathcal{S}')$$

This mapping represents the closest shape (in the d_{IoU} sense) of $\mathcal{S}$ in $\mathcal{T}'$. For the sake of simplicity, let us define for any tree $\mathcal{T}_0$:

$$\mathcal{P}_{\mathcal{T}_0}(k) := \left\{\mathcal{S}\in\mathcal{T}_0 \text{ s.t. } \mathrm{depth}_{\mathcal{T}_0}(\mathcal{S}) = k\right\}$$

Once we have this mapping and this notation, we can define a measure based on the intersection over union using the same mapping ξ:

$$\beta(k) := \frac{1}{\operatorname{Card}(\mathcal{P}_{\mathcal{T}}(k))} \sum_{\mathcal{S} \in \mathcal{P}_{\mathcal{T}}(k)} d_{\mathrm{IoU}}(\mathcal{S}, \xi(\mathcal{S})).$$

Note that we tested also its dual version by switching $\mathcal{T}$ and $\mathcal{T}'$ in this formula but the results were not relevant as a stability estimator.

β represents how much the shapes at depth k in $\mathcal{T}$ are perturbed, and we effectively observed that the higher the noise amplitude, the more d_{IoU} tends to one, since the IoU surely tends to zero.

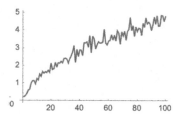

Fig. 9. The measure $\beta^{\vee}$ as a function of the noise amplitude ζ for $\zeta \in [1, 100]$.

In order to obtain (like in the previous sections) a quantification of the perturbation of the tree of shapes relatively to the noise amplitude, we propose then to compute the sum of the terms β over the possible depths of the components. We obtain then:

$$\beta^{\vee} = \sum_{k \in [0, \mathrm{depth}(\mathcal{T})]} \beta(k)$$

which is depicted in Fig. 9.

	Type	Asymp. conv.	Monotonic	$\mathfrak{s}$	HMA-based	d_{IoU}-based	Variance
μ_1	Spectral	Yes	No	1	No	No	High
μ_2	Spectral	Yes	Yes	3	No	No	Small
μ_3	Spectral	Yes	Yes	13	No	No	Small
μ_4	Spectral	Yes	Yes	20	No	No	Small
ℓ_{tot}	Topological	No	No	46	Yes	No	High
$\mathcal{M}_1$	Geometrical	Yes	Yes	10	Yes	No	Small
$\mathcal{M}_2$	Geometrical	Yes	Yes	10	Yes	No	Small
$\beta^{\vee}$	Geometrical	No	Yes	26	No	Yes	Small

Fig. 10. Summary of the different properties of each approach on the toy-image. Note that the representative slowness $\mathfrak{s}$ in the interval of low noise amplitudes is measured by the value of ζ at which the curve reaches the 50% of its maximum on the interval $[1, 100]$ for the first time.

Observation: In fact, this is the first measure we found among all our experiments which increases very slowly at the beginning, and furthermore which increases as long as ζ does so. Nevertheless, at excessively high noise amplitudes ($\zeta \geq 180$), the variance of β becomes too high and is not representative anymore.

7   Conclusion

Thanks to our exploratory research, we have been able to propose many measures of how much a tree of shapes is robust to noise. These measures can be geometrical (based on the shapes and the deformation of their contours), topological (based on the depth of the tree of shapes of an image), or spectral (based on the eigenvalues of the Laplacian of the tree of shapes). Furthermore, theses experiments are symptomatic of the difficulty to efficiently measure a phenomenon which can seem intuitively simple. Indeed, considering the behavior on the interval $[1, 100]$ where 100 is the amplitude of the signal, the only measure which has a low variance, is monotonic, and which does not converges asymptotically (three quality criteria according to us), is the last proposed measure β, that we estimate being the default measure any user should choose in a general context. Figure 10 summarizes the main properties of each of our formulas. As future work, we will investigate our measures on different types of hierarchical representations and noise models, on a larger benchmark of natural images to strengthen the provability of our observations.

A   Appendix

A.1   Preservation of the Behavior of our Measures on Natural Images

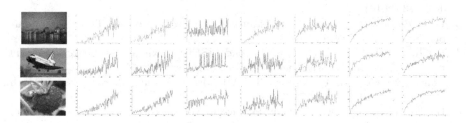

Fig. 11. From left to right, the studied image and the computations of μ_1, μ_2 , μ_3, μ_4, ℓ, $\mathcal{M}$ and β on three natural images.

The main difference with synthetic images is that natural images show a stronger variance (see Fig. 11). Conversely, the behavior of our measures are preserved except for μ_1 which becomes relevant on natural images.

A.2   Preservation of the Behavior of our Measures on Natural Images

Fig. 12. Images and their depth, number of nodes, and maximal degrees as a function of the noise amplitude.

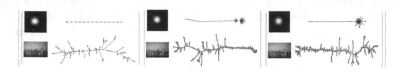

Fig. 13. Ramifications appear in the tree of shapes as long as we add noise to the represented image.

As we can observe in Figs. 12 and 13, elementary measures such as depth, numbers of nodes, and maximal degrees are not sufficient to measure the robustness of the ToS structure to noise.

References

1. Ballester, C., Caselles, V., Monasse, P.: The tree of shapes of an image. ESAIM: Control, Optim. Calc. Var. **9**, 1–18 (2003)
2. Bauer, U., Ge, X., Wang, Y.: Measuring distance between Reeb graphs. In: Proceedings of the Thirtieth Annual Symposium on Computational Geometry, pp. 464–473 (2014)
3. Bille, P.: A survey on tree edit distance and related problems. Theoret. Comput. Sci. **337**(1–3), 217–239 (2005)
4. Bosilj, P., Kijak, E., Lefèvre, S.: Beyond MSER: maximally stable regions using tree of shapes. In: British Machine Vision Conference (2015)
5. Boutry, N., Géraud, T., Najman, L.: A tutorial on well-composedness. J. Math. Imaging Vis. **60**(3), 443–478 (2018)
6. Bunke, H., Shearer, K.: A graph distance metric based on the maximal common subgraph. Pattern Recogn. Lett. **19**(3–4), 255–259 (1998)
7. Cardelino, J., Randall, G., Bertalmio, M., Caselles, V.: Region based segmentation using the tree of shapes. In: 2006 International Conference on Image Processing, pp. 2421–2424. IEEE (2006)
8. Carlinet, E., Crozet, S., Géraud, T.: The tree of shapes turned into a max-tree: a simple and efficient linear algorithm. In: 2018 25th IEEE International Conference on Image Processing (ICIP), pp. 1488–1492. IEEE (2018)
9. Carlinet, E., Géraud, T.: MToS: a tree of shapes for multivariate images. IEEE Trans. Image Process. **24**(12), 5330–5342 (2015)

10. Caselles, V., Monasse, P.: Geometric Description Of Images as Topographic Maps. Springer, New York (2009)
11. Chung, F.R., Graham, F.C.: Spectral Graph Theory. No. 92, American Mathematical Soc. (1997)
12. Gaüzere, B., Brun, L., Villemin, D.: Two new graphs kernels in chemoinformatics. Pattern Recogn. Lett. **33**(15), 2038–2047 (2012)
13. Géraud, T., Carlinet, E., Crozet, S.: Self-duality and digital topology: links between the morphological tree of shapes and well-composed gray-level images. In: Proceedings of the International Symposium on Mathematical Morphology, pp. 573–584. Springer (2015)
14. Géraud, T., Carlinet, E., Crozet, S., Najman, L.: A quasi-linear algorithm to compute the tree of shapes of n-D images. In: Proceedings of the International Symposium on Mathematical Morphology, pp. 98–110. Springer, Berlin (2013)
15. Huttenlocher, D.P., Klanderman, G.A., Rucklidge, W.J.: Comparing images using the Hausdorff distance. IEEE Trans. Pattern Anal. Mach. Intell. **15**(9), 850–863 (1993)
16. Koç, S.G., Aptoula, E., Bosilj, P., Damodaran, B.B., Dalla Mura, M., Lefevre, S.: A comparative noise robustness study of tree representations for attribute profile construction. In: 2017 25th Signal Processing and Communications Applications Conference (SIU), pp. 1–4. IEEE (2017)
17. Kuhn, H.W.: The hungarian method for the assignment problem. Naval Res. logistics Q. **2**(1–2), 83–97 (1955)
18. Lê Duy Huynh, N.B., Géraud, T.: Connected filters on generalized shape-spaces. Pattern Recogn. Lett. **128**, 348–354 (2019)
19. Lê Duy Huynh, Xu.Y., Géraud, T.: Morphology-based hierarchical representation with application to text segmentation in natural images. In: Proceedings of the International Conference on Pattern Recognition, pp. 4029–4034
20. Monasse, P., Guichard, F.: Fast computation of a contrast-invariant image representation. IEEE Trans. Image Process. **9**(5), 860–872 (2000)
21. Morozov, D., Beketayev, K., Weber, G.: Interleaving distance between merge trees. Discret. Comput. Geom. **49**(22–45), 52 (2013)
22. Pan, Y., Birdwell, J.D., Djouadi, S.M.: Preferential image segmentation using trees of shapes. IEEE Trans. Image Process. **18**(4), 854–866 (2009)
23. Salembier, P., Oliveras, A., Garrido, L.: Antiextensive connected operators for image and sequence processing. IEEE Trans. Image Process. **7**(4), 555–570 (1998)
24. Xu, Y., Géraud, T., Najman, L.: Morphological filtering in shape spaces: Applications using tree-based image representations. In: Proceedings of the International Conference on Pattern Recognition, pp. 485–488. IEEE (2012)
25. Xu, Y., Géraud, T., Najman, L.: Connected filtering on tree-based shape-spaces. IEEE Trans. Pattern Anal. Mach. Intell. **38**(6), 1126–1140 (2015)
26. Xu, Y., Géraud, T., Najman, L.: Hierarchical image simplification and segmentation based on Mumford-Shah-salient level line selection. Pattern Recogn. Lett. **83**, 278–286 (2016)

An Algebraic Framework for Out-of-Core Hierarchical Segmentation Algorithms

Jean Cousty[1]($\boxtimes$), Benjamin Perret[1], Harold Phelippeau[2], Stela Carneiro[1], Pierre Kamlay[2], and Lilian Buzer[1]

[1] LIGM, Univ Gustave Eiffel, CNRS, ESIEE Paris, 77454 Marne-la-Vallée, France
[2] Thermo Fisher Scientific, Bordeaux, France

Abstract. Binary partition hierarchies and minimum spanning trees are key structures for numerous hierarchical analysis methods, as those involved in computer vision and mathematical morphology. In this article, we consider the problem of their computation in an out-of-core manner, *i.e.*, by minimizing the size of the data structures that are simultaneously needed at the different computation steps. Out-of-core algorithms are necessary when the data are too large to fit entirely in the main memory of the computer, which can be the case with very large images in 2-, 3-, or higher dimension space. We propose a new algebraic framework composed of four main operations on hierarchies: edge-addition, select, insert, and join. Based on this framework, we propose and establish the correctness of an out-of-core calculus for binary partition hierarchies and for minimum spanning trees. First applications to image processing suggest the practical efficiency of this calculus.

1 Introduction

Image segmentation, one of the oldest problems in computer vision, consists in partitioning an image into meaningful regions that can be used to perform higher-order tasks. However, the objects of interest do not all appear at the same scale and can be nested within each other making segmentation a ill-posed problem. In this article, we are interested in the more general problem of hierarchical (or multi-scale) segmentation: find a series of partitions ordered from fine to coarse describing how the finer details are grouped to form higher level regions. This problem generalizes the one of hierarchical clustering, studied in classification, with the additional benefit of considering class-connectivity in a graph framework [2]. Nowadays, state of the art image analysis procedures involving segmentation [4,11] include two steps performed independently: i/learning contour cues and regional attributes with machine or deep learning strategies; and ii/computing and processing hierarchical segmentation trees. One of the trends in the computer vision community is the integration of these two tasks with for instance the introduction of hierarchy processing layers [1] into deep learning architectures for use in an end-to-end fashion. Binary partition hierarchies [17] built from an altitude ordering (a total order on the pairs of adjacent pixels

© Springer Nature Switzerland AG 2021
J. Lindblad et al. (Eds.): DGMM 2021, LNCS 12708, pp. 378–390, 2021.
https://doi.org/10.1007/978-3-030-76657-3_27

or vertices) and minimum spanning trees [3,12] constitute key structures for numerous hierarchical analysis methods.

While the algorithms for building and processing hierarchical segmentations are well established in the regular case of a single image of a standard size, there is a lack of efficient scalable algorithms. Scalability is needed to analyze large images at native resolution or large datasets. Indeed, giga- or tera-bytes images, which become common with the improvements in sensor resolution, cannot fit within the main memory of a computer and recent (deep) learning methods require browsing several times datasets of millions of images. The classical sequential hierarchy algorithms are not adapted to these situations and therefore need to be completely redesigned [5–7,10] to cope with these situations.

In [5,7,10], the authors investigate parallel algorithms to compute the min and max trees – hierarchical structures used for various applications, such as attribute filtering and segmentation – of terabytes images. In [6], the computation of minimum spanning trees of streaming images is considered. In [9], the authors investigate the parallel computation of quasi-flat zones hierarchies. From a high-level point of view, these approaches work independently on small pieces of the space, "join" the information found on adjacent pieces, and "propagate" (or "insert") this joint information into other pieces.

In this article, we envision the problem of computing binary partition hierarchies (and associated minimum spanning trees) under the out-of-core constraint, that is minimizing the size of the data structures loaded simultaneously into the principal memory of the computer. In other words, out-of-core algorithms target the processing of data or images that are too large to fit into the main memory at once without necessarily using any parallelization. With respect to these objectives, our main contributions in this article are the following:

- the formalization of a representation of a hierarchy of partitions, called a distribution, that is suited to out-of-core hierarchical analysis (Sect. 2);
- the introduction and the algebraic study of a fundamental external binary operation on hierarchies called edge-addition. This operation allows us to provide new characterizations of binary partition hierarchies and to introduce the notion of a partial binary partition hierarchy that is useful in the context of distributions (Sect. 3);
- three operations on hierarchies called select, join, and insert (Sects. 2 and 4) to handle distributions;
- an out-of-core calculus for (distributions of) binary partition hierarchies and minimum spanning trees (over a causal partition) of the space (Sect. 5); this calculus only involves the three introduced operations applied to partial hierarchies and it requires $O(K)$ calls to these operations, where K is the number of parts onto which the data structures are distributed;
- a discussion and a proof of concept of the efficiency of this calculus to design out-of-core algorithms for binary partition hierarchies and minimum spanning trees (Sect. 6).

Due to the space limitation in this article, formal proofs of the properties are omitted and will be given in a forthcoming extended version. However, a preliminary version of these proofs is available on https://perso.esiee.fr/~coustyj/hoocc/.

2   Distributed Hierarchies of Partitions: Select Operation

In this section, we first remind the definitions of a partial partition and of a hierarchy of partial partitions [15,16]. Then, we introduce the select operation. Intuitively, this operation consists in "selecting" the part of a given hierarchy made of the regions which hit a given set. Finally, we define the notion of a distribution of a hierarchy over a partition of the space. Such distribution is a set of hierarchies where each element is the result of "selecting" the part of the given hierarchy associated with one of the class of the given partition. These notions allow one to define a component forest and a border tree as presented in [5,10] and to set up the formal problem that we tackle in this article.

Let V be a set. A *(partial) partition of V* is a set of pairwise disjoint subsets of V. Any element of a partition is called a *region* of this partition. The *support (or ground)* of a partition $\mathbf{P}$, denoted by $gr(\mathbf{P})$, is the union of the regions of $\mathbf{P}$. A partition whose support is V is called a *complete partition of V*. Let $\mathbf{P}$ and $\mathbf{Q}$ be two partitions of V. We say that $\mathbf{Q}$ is a *refinement of $\mathbf{P}$* if any region of $\mathbf{Q}$ is included in a region of $\mathbf{P}$. A *hierarchy on V* is a series $(\mathbf{P}_0, \ldots, \mathbf{P}_\ell)$ of partitions of V such that, for any λ in $\{0, \ldots, \ell - 1\}$, the partition $\mathbf{P}_\lambda$ is a refinement of $\mathbf{P}_{\lambda+1}$. Let $\mathcal{H} = (\mathbf{P}_0, \ldots, \mathbf{P}_\ell)$ be a hierarchy. The integer ℓ is called the *depth of $\mathcal{H}$* and, for any λ in $\{0, \ldots, \ell\}$, the partition $\mathbf{P}_\lambda$ is called the *λ-scale of $\mathcal{H}$*. In the following, if λ is an integer in $\{0, \ldots, \ell\}$, we denote by $\mathcal{H}[\lambda]$ the λ-scale of $\mathcal{H}$. For any λ in $\{1, \ldots, \ell\}$, any region of the λ-scale of $\mathcal{H}$ is also called a *region of $\mathcal{H}$*. The hierarchy $\mathcal{H}$ is *complete* if $\mathcal{H}[0] = \{\{x\} \mid x \in V\}$ and if $\mathcal{H}[\ell] = \{V\}$. The hierarchy $\mathcal{H}$ is *binary* if, for any $\lambda \in \{1, \ldots, \ell\}$, we have $|\mathbf{P}_\lambda| \in \{|\mathbf{P}_{\lambda-1}| - 1, |\mathbf{P}_{\lambda-1}|\}$. We denote by $\mathcal{H}_\ell(V)$ the set of all hierarchies on V of depth ℓ, by $\mathbf{P}(V)$ the set of all partitions on V, and by $2^{|V|}$ the set of all subsets of V.

In the following, the symbol ℓ stands for any positive integer.

Let V be a set. The operation *sel* is the map from $2^{|V|} \times \mathbf{P}(V)$ to $\mathbf{P}(V)$ which associates to any subset X of V and to any partition $\mathbf{P}$ of V the subset $\mathrm{sel}(X, \mathbf{P})$ of $\mathbf{P}$ which contains every region of $\mathbf{P}$ that contains an element of X. The operation *select* is the map from $2^{|V|} \times \mathcal{H}_\ell(V)$ in $\mathcal{H}_\ell(V)$ which associates to any subset X of V and to any hierarchy $\mathcal{H}$ on V the hierarchy $\mathrm{select}(X, \mathcal{H}) = (\mathrm{sel}(X, \mathcal{H}[0]), \ldots, \mathrm{sel}(X, \mathcal{H}[\ell]))$.

Definition 1 (Distributed hierarchy). *Let V be a set, let $\mathbf{P}$ be a complete partition on V and let $\mathcal{H}$ be a hierarchy on V. The* distribution *of $\mathcal{H}$ over $\mathbf{P}$ is the set $\{\mathrm{select}(R, \mathcal{H}) \mid R \in \mathbf{P}\}$.*

The operation *select* and the notion of a distribution of a hierarchy are illustrated in Fig. 1. The following property asserts that the distribution of any hierarchy $\mathcal{H}$ over a partition of the space is indeed a representation of this hierarchy, which means that one can retrieve $\mathcal{H}$ from its distribution.

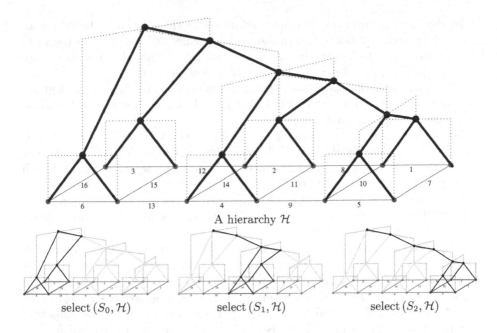

Fig. 1. Illustration of a distribution of a hierarchy. The hierarchies are represented in black-bold in the form of trees. First row: the binary partition hierarchy of the weighted graph depicted in red; and second row: its distribution over the partition $\{S_0, S_1, S_2\}$, where S_0 (resp. S_1, S_2) contains the four leftmost (resp. centermost, rightmost) vertices of the graph.

Property 2 (reconstruction). *Let V be a set, let $\mathbf{P}$ be a complete partition on V, let $\mathcal{H}$ be a hierarchy on V of depth ℓ, and let δ be the distribution of $\mathcal{H}$ over $\mathbf{P}$. Then, for any λ in $\{0, \ldots, \ell\}$, we have $\mathcal{H}[\lambda] = \cup\{\mathcal{D}[\lambda] \mid \mathcal{D} \in \delta\}$.*

3 Binary Partition Hierarchies: Edge-Addition Operations

The binary partition hierarchy plays a central role when one has to deal with hierarchies of partitions, notably in the framework of mathematical morphology [3,12]. It allows one to obtain the set representation of a hierarchy from its saliency map or from its ultrametric distance representations [3,13]. It can also be efficiently post-processed to obtain other hierarchies useful in image analysis applications like the quasi-flat zones and the constrained connectivity hierarchies [18], the hierarchical watersheds [13], the various HGB hierarchies [8], and to simplify hierarchies by removing non-significant regions [14] or to obtain marker-based segmentations of an image [17]. When the binary partition hierarchy is built from the edge-weights of a graph, its regions are in bijection with the vertices and edges of a minimum spanning tree of this graph [3,12] and thus embeds this minimum spanning tree.

In this section, we provide a new definition of this hierarchy based on an (elementary) external binary operation on hierarchies. We investigate the algebraic properties of this operation. As shown in Sect. 5, this allows us to design calculus to obtain distributions of binary partition hierarchies.

The binary partition hierarchy depends on an ordering of the edges of a graph that structures the set on which the hierarchy is built. Let us start this section with basic notions on graphs.

We define a graph as a pair $X = (V, E)$ where V is a finite set and E is composed of unordered pairs of distinct elements in V, i.e., E is a subset of $\{\{x, y\} \subseteq V \mid x \neq y\}$. Each element of V is called a *vertex* of X, and each element of E is called an edge of X.

In the remaining part of this article, we assume that $G = (V, E)$ is a graph and that $\ell = |E|$. Let us now give the definitions of some basic operators on graphs that are used in the sequel.

We denote by $\epsilon^{\times}$ the operator that maps to any subset X of V the subset of E made of the edges of G composed of two vertices in X, i.e. $\epsilon^{\times}(X) = \{\{x, y\} \in E \mid x \in X, y \in X\}$. We denote by $\delta^{\bullet}$ the operator that maps any subset F of E to the subset of V made of every vertex of V that belongs to an edge in F, i.e., $\delta^{\bullet}(F) = \cup F$. We denote by $\delta^{\times}$ the operator that maps to any subset X of V the subset of E made of every edge of G that contains a vertex of X, i.e., $\delta^{\times}(X) = \{\{x, y\} \in E \mid \{x, y\} \cap X \neq \emptyset\}$. Finally, we denote by γ the operator that maps to any two subsets X and Y of V the subset of E made of every edge of E that contains exactly one vertex of X and exactly one vertex of Y, i.e., $\gamma(X, Y) = \epsilon^{\times}(X \cup Y) \setminus \epsilon^{\times}(X) \setminus \epsilon^{\times}(Y)$. The set $\gamma(X, Y)$ is called the *common neighborhood of X and Y*

A *total order on E* is a sequence $(u_1, \ldots, u_\ell)$ such that $\{u_1, \ldots, u_\ell\} = E$. Let $\prec$ be a total order on E. Let k in $\{1, \ldots, \ell\}$, we denote by $u_k^{\prec}$ the k-th element of E for the order $\prec$. Let u be an edge in E. The *rank of u for $\prec$*, denoted by $r^{\prec}(u)$, is the unique integer k such that $u = u_k^{\prec}$. Let v be an edge of G. We write $u \prec v$ if the rank of u is less than the one of v ,i.e., $r^{\prec}(u) < r^{\prec}(v)$.

In the remaining part of this article, the symbol $\prec$ denotes a total order on E.

Let $\mathbf{P}$ be a partition of V. We consider the *class map associated with $\mathbf{P}$* (see [15]), denoted by $\mathrm{Cl}_{\mathbf{P}}$, as the map from V to the set of all subsets of V such that $\mathrm{Cl}_{\mathbf{P}}(x) = \emptyset$ if x does not belong to the support of $\mathbf{P}$ and $\mathrm{Cl}_{\mathbf{P}}(x) = R$ where R is the unique region of $\mathbf{P}$ which contains x, otherwise.

Let $\mathcal{X}$ be a hierarchy on V. Let $\{x, y\}$ be an edge in E and let k be the rank of $\{x, y\}$ for $\prec$. The *update of $\mathcal{X}$ with respect to $\{x, y\}$*, denoted by $\mathcal{X} \oplus \{x, y\}$, is the hierarchy defined by:

$$\mathcal{X} \oplus \{x, y\}[\lambda] = \mathcal{X}[\lambda], \text{ for any } \lambda \text{ in } \{0, \ldots, k-1\}; \text{ and}$$
$$\mathcal{X} \oplus \{x, y\}[\lambda] = \mathcal{X}[\lambda] \setminus \{R_x, R_y\} \cup \{R_x \cup R_y\}$$

where $R_x = \mathrm{Cl}_{\mathcal{X}[\lambda]}(x)$ and $R_y = \mathrm{Cl}_{\mathcal{X}[\lambda]}(y)$, for any λ in $\{k, \ldots, \ell\}$.

Property 3 *Let u and v be two edges in E and let $\mathcal{H}$ be a hierarchy. The following statements hold true:*

$$\mathcal{H} \oplus u = \mathcal{H} \oplus u \oplus u$$

$$\mathcal{H} \oplus u \oplus v = \mathcal{H} \oplus v \oplus u$$

Definition 4 (edge-addition) *Let $E' \subseteq E$ and let $\mathcal{H}$ be a hierarchy. We set $\mathcal{H} \boxplus E' = \mathcal{H} \oplus u_1 \oplus \ldots \oplus u_{|E'|}$ where $E' = \{u_1, \ldots, u_{|E'|}\}$. The binary operation $\boxplus$ is called the* edge-addition.

Let X be a set, we denote by $\perp_X$ the hierarchy defined by $\perp_X [\lambda] = \{\{x\} \mid x \in X\}$, for any λ in $\{0, \ldots \ell\}$.

Definition 5 (Binary partition hierarchy) *The* binary partition hierarchy *(for $\prec$), denoted by $\mathcal{B}^\prec$, is the hierarchy $\perp_V \boxplus E$. Let A be a subset of V, we define the (partial)* binary partition hierarchy *(for $\prec$) on A, denoted by $\mathcal{B}_A^\prec$, as the hierarchy $\perp_A \boxplus \epsilon^\times(A)$.*

Property 6 *The binary partition hierarchy for $\prec$ is complete and binary.*

The notion of a binary partition hierarchy is illustrated in Figs. 1 and 2. In Fig. 1, the hierarchy $\mathcal{H}$ (in bold in the first row) is the binary partition hierarchy of the graph shown in red for the order $\prec$ associated with the ranks given below the edges. In Fig. 2, the two hierarchies $\mathcal{X}$ and $\mathcal{Y}$ represented in the first row in blue and in green are the partial binary partition hierarchies associated to the 8 leftmost (resp. to the 4 rightmost) vertices of the graph.

The definition of a binary partition hierarchy given in Definition 5 differs from the one given in [3]. In [3], the edges of E are considered in increasing order of rank, whereas in Definition 5 the edges are added in any order since, from Property 3, the chosen order does not change the resulting hierarchy. From the particular case where the edges are picked in increasing order in Definition 5, it can be observed that the two definitions are equivalent.

In this article, we consider the problem of computing the distribution of the binary partition hierarchy over a partition of the space under the out-of-core constraint. To this end, we finish this section by analyzing some algebraic properties of the external operation $\boxplus$.

Let $\mathcal{X}$ and $\mathcal{Y}$ be two hierarchies in $\mathcal{H}_\ell(V)$. We say that $\mathcal{Y}$ is *smaller than* $\mathcal{X}$, and we write $\mathcal{Y} \sqsubseteq \mathcal{X}$, if, for any λ in $\{0, \ldots, \ell\}$, the partition $\mathcal{Y}[\lambda]$ is a refinement of $\mathcal{X}[\lambda]$. The set $\mathcal{H}_\ell(V)$ equipped with the relation $\sqsubseteq$ is a lattice whose supremum and infimum are denoted by $\sqcup$ and $\sqcap$ respectively. We refer to [16] for a general study of the lattices of (partial) partitions and of hierarchies. Let $u = \{x, y\}$ be an edge in E, we denote by $\mathcal{H}_u$ the hierarchy $\perp_u \oplus u$. We have:

for any λ in $\{0, \ldots, r^\prec(u) - 1\}, \mathcal{H}_u[\lambda] = \{\{x\}, \{y\}\}$; and
for any λ in $\{r^\prec(u), \ldots, \ell\}, \mathcal{H}_u[\lambda] = \{u\}$.

Property 7 *Let $\mathcal{X}$ and $\mathcal{Y}$ be two hierarchies in $\mathcal{H}_\ell(V)$, let u be an edge in E, and let F and T be two subsets of E. Then, the following equalities hold true:*

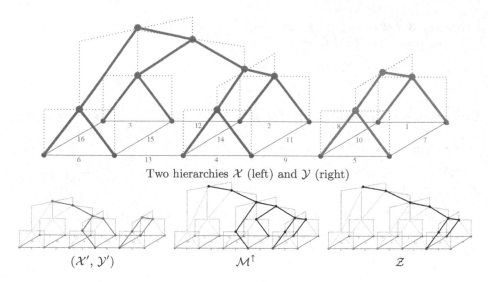

Two hierarchies $\mathcal{X}$ (left) and $\mathcal{Y}$ (right)

$(\mathcal{X}', \mathcal{Y}')$ $\mathcal{M}^{\uparrow}$ $\mathcal{Z}$

Fig. 2. Illustration of partial binary partition hierarchies and of the operations select and join. First row: the two partial binary partition hierarchies $\mathcal{X}$ and $\mathcal{Y}$ (represented in blue and in green, respectively) on the sets A and B of the 8 leftmost vertices and of the 4 rightmost vertices, respectively. Second row: (left) two hierarchies $\mathcal{X}'$ and $\mathcal{Y}'$ which are equal to select $(\gamma_B^{\bullet}(A), \mathcal{X})$ and to select $(\gamma_A^{\bullet}(B), \mathcal{Y})$ respectively; (center) the hierarchy $\mathcal{M}^{\uparrow} = \mathrm{join}\,(\mathcal{X}', \mathcal{Y}')$; and (right) the hierarchy $\mathcal{Z} = \mathrm{select}\,(\gamma_A^{\bullet}(B), \mathcal{M}^{\uparrow})$.

1. $\mathcal{H} \oplus u = \mathcal{H} \sqcup \mathcal{H}_u$;
2. $\mathcal{H} \boxplus F = \mathcal{H} \sqcup (\sqcup\{\mathcal{H}_u \mid u \in F\})$;
3. $(\mathcal{X} \sqcup \mathcal{Y}) \boxplus F = \mathcal{X} \sqcup (\mathcal{Y} \boxplus F) = (\mathcal{X} \boxplus F) \sqcup \mathcal{Y}$;
4. $\mathcal{X} \boxplus (F \cup I) = (\mathcal{X} \boxplus F) \boxplus I = (\mathcal{X} \boxplus I) \boxplus F$; and
5. $(\mathcal{X} \sqcap \mathcal{Y}) \boxplus F = (\mathcal{X} \boxplus F) \sqcap (\mathcal{Y} \boxplus F)$.

4   Join and Insert Operations

In this section, we introduce the two main binary operations, called join and insert, that allow us to compute (the distribution of) binary partition hierarchies in an out-of-core manner. Following the high-level schemes presented in [5,7,9, 10], the basic idea is to work independently on small pieces of the space, "join" the information found on adjacent pieces and "propagate" (or "insert") this information into other pieces. After presenting the definition of these operations, we investigate some basic properties which are then used in the following section to establish an out-of-core calculus for distributed binary partition hierarchies.

The *extent* of a hierarchy is the union of the set of its regions and its *ground* is the support of its 0-scale. The ground and the extent of a hierarchy $\mathcal{H}$ are denoted by $gr(\mathcal{H})$ and $ex(\mathcal{H})$, respectively.

Definition 8 (join) *Let $\mathcal{X}$ and $\mathcal{Y}$ be two hierarchies in $\mathcal{H}_\ell(V)$. The join of $\mathcal{X}$ and $\mathcal{Y}$, denoted by $join(\mathcal{X}, \mathcal{Y})$, is the hierarchy defined by $join(\mathcal{X}, \mathcal{Y}) = (\mathcal{X} \sqcup \mathcal{Y}) \boxplus F$, where F is the common neighborhood of the grounds of $\mathcal{X}$ and of $\mathcal{Y}$, i.e., $F = \gamma(gr(\mathcal{X}), gr(\mathcal{Y}))$.*

The operation join is illustrated in Fig. 2. The next property indicates that the binary partition hierarchy for $\prec$ on a set $A \cup B$ can be obtained by first considering the binary partition hierarchies on A and on B and then considering the join of these two hierarchies.

Property 9 *Let A and B be two subsets of V. Then, we have $join(\mathcal{B}_A^\prec, \mathcal{B}_B^\prec) = \mathcal{B}_{A \cup B}^\prec$.*

Definition 10 (Insert) *Let $\mathcal{X}$ and $\mathcal{Y}$ be two hierarchies.*
We say that $\mathcal{X}$ is insertable in $\mathcal{Y}$ if, for any λ in $\{0, \ldots, \ell\}$, for any region Y of $\mathcal{Y}[\lambda]$, Y is either included in a region of $\mathcal{X}[\lambda]$ or is included in $V \setminus gr(\mathcal{X}[\lambda])$.
Let $\mathcal{X}$ be insertable in $\mathcal{Y}$. The insertion of $\mathcal{X}$ into $\mathcal{Y}$ is the hierarchy $\mathcal{Z}$, such that, for any λ in $\{0, \ldots, \ell\}$, $\mathcal{Z}[\lambda] = \mathcal{X}[\lambda] \cup \{R \in \mathcal{Y}[\lambda] \mid R \cap gr(\mathcal{X}[\lambda]) = \emptyset\}$. The insertion of $\mathcal{X}$ into $\mathcal{Y}$ is denoted by $insert(\mathcal{X}, \mathcal{Y})$.

For instance, in Fig. 2, the hierarchy $\mathcal{Z}$ is insertable in the hierarchy $\mathcal{Y}$. The insertion of $\mathcal{Z}$ into $\mathcal{Y}$ is the hierarchy depicted in Fig. 1(right-bottom).

Property 11 *Let $\mathcal{X}$ and $\mathcal{Y}$ be two hierarchies. The following statements hold true:*

1. *the hierarchy $\mathcal{X}$ is insertable in $\mathcal{Y}$ if and only if $select(gr(\mathcal{X}), \mathcal{Y}) \sqsubseteq \mathcal{X}$.*
2. *if $\mathcal{X}$ is insertable in $\mathcal{Y}$, then we have $insert(\mathcal{X}, \mathcal{Y}) = \mathcal{X} \sqcup \mathcal{Y}$.*

The next lemma states that the selection is distributive over the insertion. Thus, when one needs to calculate a selection of the insertion of a hierarchy into another one, then it is enough to insert a selection of the first hierarchy in the selection of the second. In other words, the insertion can be performed in subparts of the hierarchy instead of on the whole hierarchy, a desirable property in the context of out-of-core computation.

Lemma 12 *Let A be a subset of V and let $\mathcal{X}$ and $\mathcal{Y}$ be two hierarchies such that $\mathcal{X}$ is insertable in $\mathcal{Y}$. Then, we have:*

$$select(A, insert(\mathcal{X}, \mathcal{Y})) = insert(select(A, \mathcal{X}), select(A, \mathcal{Y}))$$

The following lemma shows that when one needs to add a certain set of edges to a hierarchy, the addition can be performed "locally" on a subpart of the hierarchy (of the regions containing an element of the edges to be added) before being inserted in the initial hierarchy.

Lemma 13 *Let $\mathcal{X}$ be a hierarchy in $\mathcal{H}_\ell(V)$ and let F be a subset of E. Then, we have $\mathcal{X} \boxplus F = insert(select(\delta^\bullet(F), \mathcal{X}) \boxplus F, \mathcal{X})$.*

Due to Lemmas 12 and 13, we deduce the following property. It provides us with an out-of-core calculus to obtain the distribution of a binary partition hierarchy over a bi-partition $\{A, B\}$. This calculus is out-of-core in the sense that it does not consider the binary partition hierarchy on the whole set $A \cup B$.

Let A and B be any two subsets of V. We set $\gamma_B^\bullet(A) = \delta^\bullet(\gamma(A, B)) \cap A$. In other words, $\gamma_B^\bullet(A)$ contains every vertex in A which adjacent to a vertex in B.

Property 14 *Let A and B be two subsets of V.*
Let $\mathcal{M}^\uparrow = join\left(select\left(\gamma_B^\bullet(A), \mathcal{B}_A^\prec\right), \; select\left(\gamma_A^\bullet(B), \mathcal{B}_B^\prec\right)\right)$. Then, we have:

$$select\left(B, \mathcal{B}_{A \cup B}^\prec\right) = insert\left(select\left(\gamma_A^\bullet(B), \mathcal{M}^\uparrow\right), \mathcal{B}_B^\prec\right).$$

Property 14 is illustrated in Figs. 1 and 2. The first row of Fig. 2 represents two hierarchies $\mathcal{X} = \mathcal{B}_A^\prec$ and $\mathcal{Y} = \mathcal{B}_B^\prec$, where A (resp B) is the set of the 8 leftmost (resp. 4 rightmost) vertices of the depicted graph. The hierarchies $\mathcal{X}' = select\left(\gamma_B^\bullet(A), \mathcal{B}_A^\prec\right)$, $\mathcal{Y}' = select\left(\gamma_A^\bullet(B), \mathcal{B}_B^\prec\right)$, $\mathcal{M}^\uparrow = join\left(select\left(\gamma_B^\bullet(A), \mathcal{B}_A^\prec\right), \; select\left(\gamma_A^\bullet(B), \mathcal{B}_B^\prec\right)\right)$, and $\mathcal{Z} = select\left(\gamma_A^\bullet(B), \mathcal{M}^\uparrow\right)$ are depicted from left to right in the second row of Fig. 2. It can be observed that, as established by Property 14, the insertion of $\mathcal{Z}$ into $\mathcal{X} = \mathcal{B}_A^\prec$ is indeed equal to $select\left(B, \mathcal{B}_{A \cup B}^\prec\right)$ (shown in Fig. 1(bottom-right)). The next lemma is the key ingredient to extend the above calculus to partitions with more than two sets.

Lemma 15 *Let A, B and C be three pairwise disjoint subsets of V such that $\gamma(A, C) = \emptyset$. Let $\mathcal{M}^\uparrow = join\left(select\left(\gamma_B^\bullet(A), \mathcal{B}_A^\prec\right), \; select\left(\gamma_A^\bullet(B), \mathcal{B}_B^\prec\right)\right)$ and let $\mathcal{M}^\downarrow = join\left(select\left(\gamma_B^\bullet(A), \mathcal{B}_A^\prec\right), \; select\left(\gamma_A^\bullet(B), \mathcal{B}_{B \cup C}^\prec\right)\right)$. Then, we have :*

1. $select\left(A, \mathcal{B}_{A \cup B \cup C}^\prec\right) = insert\left(select\left(\gamma_B^\bullet(A), \mathcal{M}^\downarrow\right), \mathcal{B}_A^\prec\right)$; and
2. $\mathcal{M}^\downarrow = insert\left(select\left(\gamma_A^\bullet(B), \mathcal{B}_{A \cup B \cup C}^\prec\right), \mathcal{M}^\uparrow\right)$.

5   Out-of-Core Calculus

In this section, after providing the definition of a causal partition, we introduce an out-of-core calculus (with an associated algorithm) to efficiently obtain the distribution of a binary partition hierarchy on a causal partition. The main result of this section (Theorem 17) states the correctness of this calculus.

A causal partition of V is a series $(S_0, \ldots, S_k)$ of subsets of V such that: (i) $\{S_0, \ldots, S_k\}$ is a partition of V; (ii) $\delta^\times(S_0) = \epsilon^\times(S_0) \cup \gamma(S_0, S_1)$; (iii) $\delta^\times(S_i) = \epsilon^\times(S_i) \cup \gamma(S_i, S_{i-1}) \cup \gamma(S_i, S_{i+1})$, for any i in $\{1, \ldots, k-1\}$; and (iv) $\delta^\times(S_k) = \epsilon^\times(S_k) \cup \gamma(S_k, S_{k-1})$. Any element of a causal partition δ is called a *slice of δ*.

Let us now present induction formulae to compute the distribution of a binary partition hierarchy over a causal partition of the space.

Definition 16 *Let $(S_0, \ldots, S_k)$ be a causal partition of V. We set*

1. $\mathcal{B}_0^\uparrow = \mathcal{B}_{S_0}^\prec$;
2. $\forall i \in \{1, \ldots, k\}, \mathcal{M}_i^\uparrow = join\left(select\left(\gamma_{S_i}^\bullet(S_{i-1}), \mathcal{B}_{i-1}^\uparrow\right), \; select\left(\gamma_{S_{i-1}}^\bullet(S_i), \mathcal{B}_{S_i}^\prec\right)\right);$

3. $\forall i \in \{1, \dots, k\}$, $\mathcal{B}_i^\uparrow = insert(select\left(\gamma_{S_{i-1}}^\bullet(S_i), \mathcal{M}_i^\uparrow\right), \mathcal{B}_{S_i}^\prec)$;

4. $\mathcal{B}_k^\downarrow = \mathcal{B}_k^\uparrow$ and $\mathcal{M}_k^\downarrow = \mathcal{M}_k^\uparrow$;

5. $\forall i \in \{0, \dots, k-1\}$, $\mathcal{B}_i^\downarrow = insert(select\left(\gamma_{S_{i-1}}^\bullet(S_i), \mathcal{M}_{i+1}^\downarrow\right), \mathcal{B}_i^\uparrow)$; and

6. $\forall i \in \{1, \dots, k-1\}$, $\mathcal{M}_i^\downarrow = insert(select\left(\gamma_{S_{i-1}}^\bullet(S_i), \mathcal{B}_i^\downarrow\right), \mathcal{M}_i^\uparrow)$.

On the one hand, the three first formulae of Definition 16 provide us with a causal (inductive) calculus: i-th term of the series relies on the $(i-1)$-th term of the series. In particular, $\mathcal{M}_i^\uparrow$ depends on $\mathcal{B}_{i-1}^\uparrow$ and that $\mathcal{B}_i^\uparrow$ depends on $\mathcal{M}_i^\uparrow$, for any i in $\{1, \dots, k\}$. On the other hand, the three last formulae of Definition 16 provide us with an anti-causal calculus since the terms of index i rely on the ones of index $(i+1)$: $\mathcal{B}_i^\downarrow$ depends on $\mathcal{M}_{i+1}^\downarrow$ and $\mathcal{M}_i^\downarrow$ depends on $\mathcal{B}_i^\downarrow$. Note also that $\mathcal{M}_i^\downarrow$ depends on $\mathcal{M}_i^\uparrow$ and that $\mathcal{B}_i^\downarrow$ depends on $\mathcal{B}_i^\uparrow$. Hence, the anti-causal calculus must be performed after the causal ones. One can also observe that the computation of each term requires only the knowledge of hierarchies whose grounds are included in two consecutive slices of the given causal partition. In this sense, we can say that the above formulae form an out-of-core calculus. Algorithm 1, presented hereafter, is a sequential implementation of the above induction formulae. Given a partition of V into $k+1$ slices, it can be observed that Algorithm 1 performs $O(k)$ calls to select, insert, join, and to the algorithm PlayingWithKruskal [12] applied to a single slice of the given partition.

Algorithm 1: Out-of-core binary partition hierarchy.

Data: A graph (V, E), a total order $\prec$ on E, and a causal partition $(S_0, \dots, S_k)$ of V

Result: $\{\mathcal{B}_0^\downarrow, \dots, \mathcal{B}_k^\downarrow\}$: the distribution of the binary partition hierarchy $\mathcal{B}_V^\prec$ over $\{S_0, \dots, S_k\}$.

1 $\mathcal{B}_0^\uparrow := \mathcal{B}_{S_0}^\prec$ // Def. 16.1, call PlayingWithKruskal algorithm

2 **foreach** i *from 1 to k* **do**   // Causal traversal of the slices

3 | Call PlayingWithKruskal algorithm to compute $\mathcal{B}_{S_i}^\prec$

4 | $\mathcal{M}_i^\uparrow := join\left(select\left(\gamma_{S_i}^\bullet(S_{i-1}), \mathcal{B}_{i-1}^\uparrow\right), select\left(\gamma_{S_{i-1}}^\bullet(S_i), \mathcal{B}_{S_i}^\prec\right)\right)$ // Def. 16.2

5 | $\mathcal{B}_i^\uparrow := insert(select\left(\gamma_{S_{i-1}}^\bullet(S_i), \mathcal{M}_i^\uparrow\right), \mathcal{B}_{S_i}^\prec)$ // Def. 16.3

6 $\mathcal{B}_k^\downarrow := \mathcal{B}_k^\uparrow$; $\mathcal{M}_k^\downarrow := \mathcal{M}_k^\uparrow$ // Def. 16.4

7 **foreach** i *from $k-1$ to 0* **do**   // Anticausal traversal of the slices

8 | $\mathcal{B}_i^\downarrow := insert(select\left(\gamma_{S_{i+1}}^\bullet(S_i), \mathcal{M}_{i+1}^\downarrow\right), \mathcal{B}_i^\uparrow)$ // Def. 16.5

9 | **if** $i > 0$ **then** $\mathcal{M}_i^\downarrow := insert(select\left(\gamma_{S_{i-1}}^\bullet(S_i), \mathcal{B}_i^\downarrow\right), \mathcal{M}_i^\uparrow)$;   // Def. 16.6

Let us finish this section by Theorem 17 that establishes that the formulae of Definition 16 indeed allows one to compute a distributed binary partition hierarchy, hence, establishing the correctness of Algorithm 1.

Theorem 17 *Let $(S_0, \ldots, S_k)$ be a causal partition of V. Let i be any element in $\{0, \ldots, k\}$. The following statements hold true:*

1. $\mathcal{B}_i^{\uparrow} = select\left(S_i, \mathcal{B}_{\cup\{S_j \mid j \in \{0,\ldots i\}\}}^{\prec}\right);$

2. $\mathcal{B}_i^{\downarrow} = select(S_i, \mathcal{B}_V^{\prec});$ and

3. *the set $\left\{\mathcal{B}_i^{\downarrow} \mid i \in \{0, \ldots, k\}\right\}$ is the distribution of $\mathcal{B}_V^{\prec}$ over $\{S_0, \ldots, S_k\}$.*

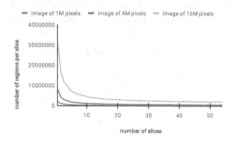 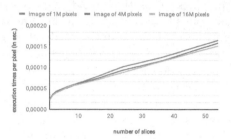

Fig. 3. Left: average size of the hierarchies in distributions of binary partition hierarchies over causal partitions; and right: execution times of the proposed calculus from three images of 1000×1000, 2000×2000, and of 4000×4000 pixels.

6   Discussion and Conclusion

In this article, an algebraic framework to study binary partition hierarchies is presented. Based on this framework, a new calculus allowing us to obtain the binary partition hierarchy associated to a given total order of the edges of a graph is introduced: the result of this calculus is a distributed representation of the binary partition hierarchy where the hierarchy is "split into subparts", each of them containing the hierarchical information associated to one predefined subset, called a slice, of the space. The proposed calculus only comprises binary operations that depend on two partial hierarchies. Thus, if the size of the partial hierarchies is limited compared to the size of the full hierarchy, computing the result of each operation only requires loading data structures of limited size into memory. In this sense, the proposed calculus is out-of-core. In order to test the practical efficiency of this calculus, an implementation in Python is designed and made available online at https://perso.esiee.fr/~coustyj/hoocc/. Figure 3 shows the first results obtained for images of 1K × 1K, 2K × 2K, and of 4K × 4K pixels: on the left, the average sizes of the partial hierarchies are displayed as functions of the number of slices into which the image domain is partitioned and, on the right, the execution times per pixel are displayed as functions of the number of slices. It can be seen that the sizes of the partial hierarchies decrease as an inverse function of the number of slices whereas the execution times increase linearly with respect to the number of slices as expected from the theoretical analysis. For instance, compared to the standard case where the image is considered as a single slice, for an image of 16M pixels, the average size of the partial structures is divided by approximately 10, when the space is partitioned into 16 slices while the execution time is multiplied by approximately 3. A detailed description of the algorithms for computing the result of the operations select, join, and insert will be provided in forthcoming articles. Directions for future work also include extensions of the calculus to partitions of the space which are not necessarily causal (*e.g.* grid-like partitions), out-of-core algorithms for image processing based on binary partition hierarchies like (hierarchical) watersheds or marker-based segmentation, and the adaptation of the proposed algorithm to parallel computations.

Acknowledgements. This work was partly supported by the ANR under grant ANR-20-CE23-0019.

References

1. Chierchia, G., Perret, B.: Ultrametric fitting by gradient descent. In: NeurIPS, pp. 3181–3192 (2019)
2. Cousty, J., Najman, L., Kenmochi, Y., Guimarães, S.: Hierarchical segmentations with graphs: quasi-flat zones, minimum spanning trees, and saliency maps. JMIV **60**(4), 479–502 (2018)
3. Cousty, J., Najman, L., Perret, B.: Constructive links between some morphological hierarchies on edge-weighted graphs. In: ISMM, pp. 86–97 (2013)
4. Farabet, C., Couprie, C., Najman, L., LeCun, Y.: Learning hierarchical features for scene labeling. TPAMI **35**(8), 1915–1929 (2012)
5. Gazagnes, S., Wilkinson, M.H.: Distributed component forests in 2-D: hierarchical image representations suitable for tera-scale images. IJPRAI **33**(11), 1940012 (2019)
6. Gigli, L., Velasco-Forero, S., Marcotegui, B.: On minimum spanning tree streaming for hierarchical segmentation. PRL **138**, 155–162 (2020)
7. Götz, M., Cavallaro, G., Geraud, T., Book, M., Riedel, M.: Parallel computation of component trees on distributed memory machines. TPDS **29**(11), 2582–2598 (2018)
8. Guimarães, S., Kenmochi, Y., Cousty, J., Patrocinio, Z., Najman, L.: Hierarchizing graph-based image segmentation algorithms relying on region dissimilarity: the case of the Felzenszwalb-Huttenlocher method. MMTA **2**(1), 55–75 (2017)
9. Havel, J., Merciol, F., Lefèvre, S.: Efficient tree construction for multiscale image representation and processing. J. Real-Time Image Proc. **16**(4), 1129–1146 (2016). https://doi.org/10.1007/s11554-016-0604-0
10. Kazemier, J.J., Ouzounis, G.K., Wilkinson, M.H.: Connected morphological attribute filters on distributed memory parallel machines. In: ISMM, pp. 357–368 (2017)
11. Maninis, K.K., Pont-Tuset, J., Arbeláez, P., Van Gool, L.: Convolutional oriented boundaries: from image segmentation to high-level tasks. TPAMI **40**(4), 819–833 (2017)
12. Najman, L., Cousty, J., Perret, B.: Playing with kruskal: algorithms for morphological trees in edge-weighted graphs. In: ISMM, pp. 135–146 (2013)
13. Najman, L., Schmitt, M.: Geodesic saliency of watershed contours and hierarchical segmentation. TPAMI **18**(12), 1163–1173 (1996)

14. Perret, B., Cousty, J., Guimarães, S.J.F., Kenmochi, Y., Najman, L.: Removing non-significant regions in hierarchical clustering and segmentation. PRL **128**, 433–439 (2019)
15. Ronse, C.: Partial partitions, partial connections and connective segmentation. JMIV **32**(2), 97–125 (2008)
16. Ronse, C.: Ordering partial partitions for image segmentation and filtering: Merging, creating and inflating blocks. JMIV **49**(1), 202–233 (2014)
17. Salembier, P., Garrido, L.: Binary partition tree as an efficient representation for image processing, segmentation, and information retrieval. TIP **9**(4), 561–576 (2000)
18. Soille, P.: Constrained connectivity for hierarchical image partitioning and simplification. TPAMI **30**(7), 1132–1145 (2008)

Fuzzy-Marker-Based Segmentation Using Hierarchies

Gabriel Barbosa da Fonseca[1,2(✉)], Benjamin Perret[2], Romain Negrel[3], Jean Cousty[2], and Silvio Jamil F. Guimarães[1]

[1] Laboratory of Image and Multimedia Data Science (ImScience), Pontifical Catholic University of Minas Gerais, Belo Horizonte 31980-110, Brazil
`sjamil@pucminas.br`
[2] LIGM, Univ Gustave Eiffel, CNRS, F-77454 Marne-la-Vallée, France
`{gabriel.fonseca,benjamin.perret,jean.cousty}@esiee.fr`
[3] ESIEE Paris, Université Gustave Eiffel, Marne-la-Vallée, France
`romain.negrel@esiee.fr`

Abstract. This article extends a classical marker-based image segmentation method proposed by Salembier and Garrido in 2000. In the original approach, the segmentation relies on two sets of pixels which play the role of object and background markers. In the proposed extension, the markers are not represented by crisp sets, but by fuzzy ones, *i.e.*, functions of the image domain into the real interval $[0, 1]$ indicating the degree of membership of each pixel to the markers. We show that when the fuzzy markers are indicator functions of crisp sets, the proposed method produces the same result as the original one. We present a linear-time algorithm for computing the result of the proposed method given two fuzzy markers and we establish the correctness of this algorithm. Additionally, we discuss possible applications of the proposed approach, such as adjusting marker strength in interactive image segmentation procedures and optimizing marker locations with gradient descent methods.

1 Introduction

Image segmentation is one of the fundamental tasks in image processing, and can be described as partitioning an image into distinct regions that comprise similar objects. Many argue that image segmentation is in fact a multi-scale problem, where regions at a coarser detail level are formed by a merging of regions at a finer detail level [1,11,25]. Following this idea, authors often utilize hierarchies constructed from a given image to produce a segmentation [1,4,20,23], and efficient algorithms for constructing such hierarchies have been proposed [16] (Fig. 1).

The authors thank Conselho Nacional de Desenvolvimento Científico e Tecnológico – CNPq – (PQ 310075/2019-0), Coordenação de Aperfeiçoamento de Pessoal de Nível Superior – CAPES – (Grant COFECUB 88887.191730/2018-00) and Fundação de Amparo à Pesquisa do Estado de Minas Gerais – FAPEMIG – (Grants PPM-00006-18).

J. Lindblad et al. (Eds.): DGMM 2021, LNCS 12708, pp. 391–403, 2021.
https://doi.org/10.1007/978-3-030-76657-3_28

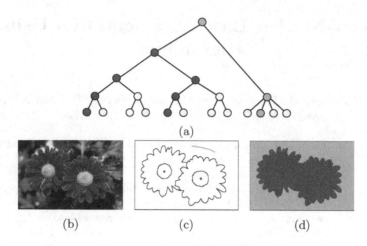

(a)

(b)   (c)   (d)

Fig. 1. A tree representation for a marker-based image segmentation with respect to a hierarchy on (a). On (b), the original image, on (c) a representation of the partitions with markers, and a segmentation on (d).

Even with the recent advances in the area of image processing, automatic segmentation can still be a very challenging task. The introduction of prior knowledge in the form of markers (often called seeds) that carry information about the location of objects of interest can drastically improve segmentation results. Many works study different approaches for producing marker-based segmentations, ranging from the classic graph-based works relying on watersheds [7,23], graph cuts [5,19], random walks [12], geodesics [2] and shortest paths [10], to the more recent works based on convolutional neural networks [13,14,24]. Also, when markers are provided by users over an interaction system, they can take different shapes. These shapes include scribbles [2,7,12], bounding boxes [19], points [13], and image regions [15,20]. In classical graph-based methods, markers usually have a binary nature, *i.e.*, a region of the image is either marked and represents an object with total certainty, or is unmarked. With that, distinct markers in the same image guide the segmentation with equal power.

Different approaches were proposed for controlling the way markers propagate when guiding a segmentation. In [23], the shape of the surface where markers propagate is modified to simulate a flooding by a viscous fluid, but different markers still guide the segmentation with equal power. In [6], the costs for propagating a marker to a region are dynamically estimated during propagation time. In this work, we are interested in the scenario where markers have a fuzzy nature instead of a binary one, and where different markers propagate and guide a segmentation with different power.

For that, we propose an extension of a classical method introduced in [20], which uses two sets of pixels as object and background markers and propagate them over a hierarchy of partitions to produce a two-class segmentation, *i.e.*, a partition of the image into a region classified as object and a region classified as background. In this extension, markers are not binary, but represented by positive real functions, referred to as fuzzy markers. The value of each element in a fuzzy marker represents its membership degree to the marker and dictates the influence of the marker at its location.

To present our extension, we rely on a characterization of [20] based on connection values (also called "fuzzy connectedness" [22] or "degree of connectivity" [18]). In our extension, we produce a segmentation based on fuzzy connection values to the object and background markers, similarly to the method proposed in [21]. However, contrary to [21], the fuzzy connection values proposed in this work also take the membership degree of elements to fuzzy markers into consideration.

The main contributions of this article are fourfold. First, we propose an extension of a classical segmentation method, where markers are fuzzy instead of being crisp. Second, we show that this method is indeed equivalent to the original one when markers are crisp (with Property 5). Third, we provide an efficient algorithm for computing in linear-time the proposed segmentation. Finally, we establish the correctness of the proposed algorithm (with Corollary 9). Due to space limitation, proofs of the presented properties will be included in an forthcoming extended version of this article. Furthermore, we discuss possible applications of the proposed method, such as adjusting the strength of markers when performing interactive segmentation and optimizing seed locations using gradient descent methods.

This article is organized as follows. In Sect. 2, we present the marker-based segmentation proposed in [20]. In Sect. 3, we define an extension of the marker-based segmentation, where markers are represented by fuzzy sets. We introduce an efficient algorithm for computing the fuzzy-marker-based segmentation in Sect. 4. Finally, in Sect. 5 we present the conclusion, possible applications of the proposed method, and directions for future works.

2   Marker-Based Segmentation Using Hierarchies

In this section, we present a classical approach for performing a marker-based image segmentation proposed by Salembier and Garrido in [20]. This method relies on extracting relevant regions of a hierarchy with respect to provided markers. First, we provide fundamental notions to define indexed hierarchies. Then, we present the notion of a marker-based segmentation from a hierarchy as introduced in [20]. Finally, we present a characterization of this segmentation based on ultrametric distances associated to given hierarchies. This characterization allows us to extend in Sect. 3 the method proposed in [20] to the case where the markers are fuzzy instead of being crisp.

In this article, the symbol V denotes a finite nonempty set. A *hierarchy* $\mathcal{H}$ (on V) is a set of subsets of V such that:

1. V is an element of $\mathcal{H}$;
2. for every element x of V, the singleton $\{x\}$ belongs to $\mathcal{H}$; and

3. for any two elements X and Y of $\mathcal{H}$ if the intersection of X and Y is nonempty, then X either includes Y or is included in Y.

Let $\mathcal{H}$ be any hierarchy, any element of $\mathcal{H}$ is called a *region of $\mathcal{H}$*. For any region X of $\mathcal{H}$, we define the *index of X* as a positive real value denoted by $\omega(X)$. An *indexed hierarchy (on V)* is a pair $(\mathcal{H}, \omega)$, where $\mathcal{H}$ is a hierarchy and where ω is a function from $\mathcal{H}$ to $\mathbb{R}^+$ such that:

1. $\omega(X) = 0$ if and only if X is a singleton; and
2. for any two regions X and Y of $\mathcal{H}$, if X is included in Y, then we have $\omega(X) < \omega(Y)$.

Let $\mathcal{H}$ be a hierarchy and let X and Y be two regions of $\mathcal{H}$. The region X is a *parent of Y* and Y is a *child of X* if Y is included in X and if any region of $\mathcal{H}$ which is proper superset of Y is also a superset of X. A region R of $\mathcal{H}$ is called a *leaf (resp. root) of $\mathcal{H}$* if it is not the parent (resp. child) of any region of $\mathcal{H}$. It can be observed that V is the only root of $\mathcal{H}$ and that the set of leaves of $\mathcal{H}$ is precisely the set of all singletons on V. It can also be noticed that any non-root region X of $\mathcal{H}$ has a unique parent, which is denoted by $par(X)$ in the following. A hierarchy $\mathcal{H}$ is considered a *binary hierarchy* if any non-leaf region has exactly two distinct children.

In the remaining part of this article, the pair $(\mathcal{H}, \omega)$ denotes an indexed binary hierarchy, that is an indexed hierarchy such that $\mathcal{H}$ is a binary hierarchy.

Let us now provide the definition of the method introduced in [20]. The main idea is to produce a partition by classifying as object any region of the hierarchy that intersects a given object marker and that does not intersect the background marker.

Let O and B be two subsets of V. The *marker-based segmentation (of V) for (O, B) (with respect to $(\mathcal{H}, \omega)$)*, denoted by $S_{\mathcal{H}}(O, B)$ is the union of the regions of $\mathcal{H}$ which contain an element of O but no element of B:

$$S_{\mathcal{H}}(O, B) = \cup\{X \in \mathcal{H} \mid X \cap O \neq \emptyset, X \cap B = \emptyset\}. \tag{1}$$

Our main interest in this article is to extend the notion of a segmentation for a pair of subsets of the space called markers to the case where the markers are fuzzy. In order to present this extension, we will rely on an alternative characterization of the marker-based segmentation based on a (ultrametric) distance induced by the hierarchy.

We define $d_{\mathcal{H}}$ as the function from $V \times V$ to $\mathbb{R}^+$ such that, for any two elements x and y of V, the value $d_{\mathcal{H}}(x, y)$ is the index of the smallest region of $\mathcal{H}$ which contains both x and y:

$$d_{\mathcal{H}}(x, y) = \min\{\omega(X) \mid X \in \mathcal{H}, x \in X, y \in X\}. \tag{2}$$

It is well known that the function $d_{\mathcal{H}}$ is a distance which is furthermore ultrametric [3]. Let x be any element of V and let Y be any subset of V. We define the *connection value of x to Y* (for the hierarchy $\mathcal{H}$), denoted by $\mathcal{C}_{\mathcal{H}}(x, Y)$, as the shortest distance between x and an element of Y:

$$\mathcal{C}_{\mathcal{H}}(x, Y) = \min\{d_{\mathcal{H}}(x, y) \mid y \in Y\}. \tag{3}$$

Property 1. *Let O and B be two subsets of V. The following statement holds true:*

$$S_{\mathcal{H}}(O, B) = \{x \in V \mid C_{\mathcal{H}}(x, O) < C_{\mathcal{H}}(x, B)\}.$$

In other words, Property 1 states that a point x belongs to the marker-based segmentation $S_{\mathcal{H}}(O, B)$ if the shortest distance between x and an element of O is smaller than the shortest distance between x and an element of B.

3   Fuzzy-Marker-Based Segmentation

The marker-based segmentation defined in [20] and presented in Sect. 2, relies on two subsets which play the role of markers. In this section we define an extension in which markers are not crisp sets anymore but fuzzy sets.

Fuzzy sets can be seen as membership functions defined on a set, in which each element has a membership degree. On a crisp set, the relationship between an element and a set is binary, *i.e.*, either the element belongs to the set or not. With a fuzzy set, elements can be considered partially included in a set, with higher membership degrees meaning a "stronger" inclusion in the set.

A *fuzzy set* (on V) is defined as a function from V to the real interval $[0, 1]$. Let μ be a fuzzy set. The function μ is also referred to as a membership function of V, where for any element x of V, the value $\mu(x)$ represents the membership degree of x in μ.

In the remaining of this article, fuzzy markers will be represented by fuzzy sets. We now extend the notion of connection value to fuzzy sets.

Definition 2 (Fuzzy Connection Value). *Let x be an element of V, and let μ be a fuzzy set. The fuzzy connection value of x to μ (for $(\mathcal{H}, \omega)$), denoted by $C_{\mathcal{H}}^{f}(x, \mu)$, is defined by:*

$$C_{\mathcal{H}}^{f}(x, \mu) = \min\{\alpha(\mu(y))(1 - \mu(y) + d_{\mathcal{H}}(x, y)) \mid y \in V\},$$

where α is a decreasing function such that $\alpha(1) = 1$ and $\alpha(0)$ is strictly greater than the maximal value of $d_{\mathcal{H}}$, i.e., $\alpha(0) > \max\{d_{\mathcal{H}}(x, y) \mid x \in V, y \in V\}$.

The notion of a fuzzy connection value is an extension of the one of connection value. The information given by the membership degree of an element to a fuzzy set is incorporated to the fuzzy connection value by the term $\alpha(\mu(y))$. Observe that for two elements x and y in V, such that $\mu(y)$ is equal to 1, we have that $\alpha(\mu(y))(1 - \mu(y) + d_{\mathcal{H}}(x, y))$ is equal to $d_{\mathcal{H}}(x, y)$. Since α is a decreasing function, as the value $\mu(y)$ decreases, the overall fuzzy connection value increases. In other words, we want that elements with low membership degrees lead to greater fuzzy connection values than elements with high membership degrees. The term $1 - \mu(y)$ and the constraint on $\alpha(0)$ allow us to show the link between connection values and fuzzy connection values, which is formally introduced in Property 3.

The following property makes the link between connection values and fuzzy connection values explicit by means of indicator functions.

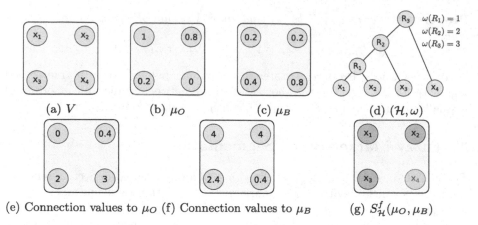

(e) Connection values to μ_O (f) Connection values to μ_B (g) $S_{\mathcal{H}}^f(\mu_O, \mu_B)$

Fig. 2. Example of a fuzzy-marker-based segmentation. Graphical representations of: (a) a set V; (b,c) fuzzy sets μ_O and μ_B; (d) an indexed hierarchy $(\mathcal{H}, \omega)$ on V; (e,f) fuzzy connection values to μ_O and μ_B, respectively; (g) the segmentation $S_{\mathcal{H}}^f(\mu_O, \mu_B)$ highlighted in blue. The α function is given by $\alpha(x) = (6 - 5x)$.

Let A be a subset of V. The *indicator function of A* is the fuzzy set $\mathbb{1}_A$, such that for any x in V, the value $\mathbb{1}_A$ is equal to 1 if x belongs to A and is equal to 0 otherwise.

Property 3. *Let A be a non-empty subset of V, and $\mathbb{1}_A$ be the indicator function of A. Then, the following statement holds true:*

$$\forall x \in V, \quad \mathcal{C}_{\mathcal{H}}(x, A) = \mathcal{C}_{\mathcal{H}}^f(x, \mathbb{1}_A).$$

We are now ready to extend the notion of a segmentation of a set V for a pair of subsets of V to a segmentation of V for a pair of fuzzy sets of V.

Definition 4 (Fuzzy-Marker-Based Segmentation). *Let μ_O and μ_B be two fuzzy sets. The* fuzzy-marker-based segmentation *(of V) for (μ_O, μ_B) (with respect to $(\mathcal{H}, \omega)$), denoted by $S_{\mathcal{H}}^f(\mu_O, \mu_B)$ is defined by*

$$S_{\mathcal{H}}^f(\mu_O, \mu_B) = \left\{ x \in V \mid \mathcal{C}_{\mathcal{H}}^f(x, \mu_O) < \mathcal{C}_{\mathcal{H}}^f(x, \mu_B) \right\},$$

In other words, the fuzzy-marker-based segmentation of V for (μ_O, μ_B) is the set that contains every element of V with a fuzzy connection value to μ_O smaller than to μ_B.

An example of a fuzzy-marker-based segmentation for a pair of fuzzy sets is illustrated in Fig. 2. A set V is illustrated in Fig. 2a, followed by the fuzzy markers μ_O and μ_B in Figs. 2b and 2c, respectively. We can observe in Fig. 2d a tree representation of a given indexed hierarchy. The fuzzy connection values of the elements of V to the fuzzy sets μ_O and μ_B can be observed in Figs. 2e and 2f, respectively. For the elements x_1, x_2, x_3, the fuzzy connection value to the fuzzy set μ_O is smaller than to the fuzzy set μ_B. Consequently, $\{x_1, x_2, x_3\}$ is the segmentation $S_{\mathcal{H}}^f(\mu_O, \mu_B)$, as shown in Fig. 2g.

The following property shows the relationship between the fuzzy-marker-based segmentation proposed in this article and the marker-based segmentation proposed in [20].

Property 5. *Let O and B be two subsets of V, and $\mathbb{1}_O$ and $\mathbb{1}_B$ be the indicator functions of O and B, respectively. Then, the following statement holds true:*

$$S_{\mathcal{H}}^f(\mathbb{1}_O, \mathbb{1}_B) = S_{\mathcal{H}}(O, B).$$

In other words, Property 5 states that when the fuzzy markers indeed represent crisp sets, the marker-based segmentation is the same as the fuzzy-marker-based segmentation.

4 Efficient Computation of Fuzzy-Marker-Based Segmentation

In this section, we present an efficient algorithm for computing the fuzzy connection values for every element of a set to a given fuzzy set on this space. Before presenting the algorithm, we give some definitions and properties that allow us to introduce the algorithm and prove its correctness. Finally, we give a brief discussion about the complexity of the proposed algorithm.

Let X and Y be two distinct regions of $\mathcal{H}$. The region X *is a sibling of* Y and X *and* Y *are siblings* if the parent of X and the parent of Y are the same. Since $\mathcal{H}$ is a binary hierarchy, every region on $\mathcal{H}$ that has a parent has exactly one sibling. If X is a non-root region of $\mathcal{H}$, we denote by $sib(X)$ *the unique sibling of* X.

We define the *restriction of a fuzzy set μ to a subset X of V*, denoted by $\mu{\downarrow}_X$, as the fuzzy set such that $\mu{\downarrow}_X (x) = \mu(x)$ for any x in X, and $\mu{\downarrow}_X (x) = 0$ for any x in $V \setminus X$, i.e., $\mu{\downarrow}_X = \inf(\mu, \mathbb{1}_X)$.

Property 6. *Let μ be a fuzzy set and let x be an element in V. Then, the fuzzy connection value from x to μ is given by:*

$$\mathcal{C}_{\mathcal{H}}^f(x, \mu) = \min\left\{\mathcal{C}_{\mathcal{H}}^f(x, \mu{\downarrow}_{V\setminus\{x\}}), \alpha(\mu(x))(1 - \mu(x))\right\}.$$

From Property 6, we see that the connection value of x to μ can be obtained from the connection value of x to the restriction of μ to $V\setminus\{x\}$ and from $\alpha(\mu(x))(1 - \mu(x))$. The computation of the later term can be done in constant time. We will now present an efficient way to compute the former one.

Let μ be a fuzzy set. We denote by $max(\mu)$ the maximum of $\mu(x)$ for all x in V.

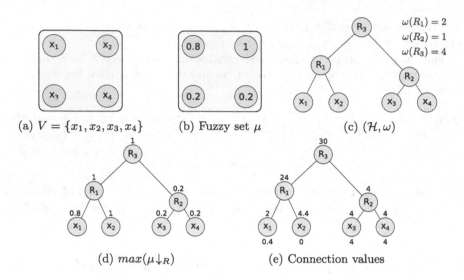

Fig. 3. A set V on (a). A fuzzy set μ on (b). A representation of an indexed hierarchy $(\mathcal{H}, \omega)$ on (c). On (d), the value $max(\mu \downarrow_R)$ for each region of $\mathcal{H}$ in blue. On (e), the regional connection value of each region of $\mathcal{H}$ in blue and the fuzzy connection value of each element of V (the leaves) in red. For this example, we set $\alpha(x) = 6 - 5x$.

Definition 7 (Regional Connection Value). *Let μ be a fuzzy set and let R be a region of $\mathcal{H}$. The regional connection value of R, denoted by $T_{\mathcal{H}}^{\mu}(R)$, is defined by:*

$$T_{\mathcal{H}}^{\mu}(R) = \alpha(0)(1 + \omega(R)), \text{ if } R = V \text{ (i.e., } R \text{ is the root of } \mathcal{H}\text{); and}$$
$$T_{\mathcal{H}}^{\mu}(R) = \min\left(T_{\mathcal{H}}^{\mu}(par(R)), \alpha(max(\mu \downarrow_{sib(R)}))(1 - max(\mu \downarrow_{sib(R)}) + \omega(par(R)))\right),$$
$$\text{if } R \neq V.$$

The regional connection value $T_{\mathcal{H}}^{\mu}(R)$ of each region can be obtained by browsing the hierarchy $\mathcal{H}$ in root-to-leaves order, *i.e.*, an order in which a parent region is processed before its children. Algorithm 1 presents the different steps to compute these values in linear time with respect to the number of regions in $\mathcal{H}$.

In Fig. 3 we have illustrations of a set V in Fig. 3a, a fuzzy set μ in Fig. 3b, and an indexed hierarchy $(\mathcal{H}, \omega)$ in Fig. 3c. In Fig. 3e, we have the values of the regional connection values of every region of the hierarchy in blue. For example, by using the equations from Definition 7, the regional connection value of the root of $\mathcal{H}$ (region R_3) is given by $\alpha(0)(1 + \omega(R_3))$. By setting $\alpha = 6 - 5x$, as in the example illustrated in Fig. 3, we have that $T_{\mathcal{H}}^{\mu}(R_3) = (6)(1 + 4) = 30$. For the left child of the root (region R_1), its regional connection value is given by the minimum between the values $T_{\mathcal{H}}^{\mu}(R_3)$ and $\alpha(max(\mu \downarrow_{R_2}))(1 - max(\mu \downarrow_{R_2}) + \omega(R_3)) = (6 - 5 * 0.2)(1 - 0.2 + 4) = 24$. As the value $T_{\mathcal{H}}^{\mu}(R_3) = 30$, we have $T_{\mathcal{H}}^{\mu}(R_1) = min(30, 24) = 24$.

As we will see with Property 8 and Corollary 9, the regional connection values allows us to compute the fuzzy connection values of every element x in V. With the proposed algorithm, we can efficiently compute the fuzzy connection values from all elements of a set to two distinct fuzzy sets. Having the fuzzy connection values to a pair of fuzzy sets, one can compute the fuzzy-marker-based segmentation of a given set with Definition 4.

The following property links the regional connection value to fuzzy connection values. This property allows us to show the correctness of the proposed algorithm for computing fuzzy connection values.

Property 8. *Let μ be a fuzzy set and let x be an element of V. Then, the fuzzy connection value of x to the restriction of μ to $V \setminus \{x\}$ is equal to $T_{\mathcal{H}}^{\mu}(\{x\})$:*

$$\mathcal{C}_{\mathcal{H}}^{f}(x, \mu \downarrow_{V \setminus \{x\}}) = T_{\mathcal{H}}^{\mu}(\{x\}).$$

Then, from Properties 6 and 8 we can derive Corollary 9, which presents how to obtain the fuzzy connection values for all x in V to μ by using the regional connection values.

Corollary 9. *Let x be an element of V, and μ be a fuzzy set. For any x in V, the fuzzy connection value $\mathcal{C}_{\mathcal{H}}^{f}(x, \mu)$ is given by:*

$$\mathcal{C}_{\mathcal{H}}^{f}(x, \mu) = \min \left\{ T_{\mathcal{H}}^{\mu}(\{x\}), \alpha(\mu(x))(1 - \mu(x)) \right\}.$$

A set V is illustrated in Fig. 3a, followed by a fuzzy set μ illustrated in Fig. 3b and a hierarchy $(\mathcal{H}, \omega)$ illustrated in Fig. 3c. In Fig. 3e, we can observe the fuzzy connection values (in red) of the elements of the set V to μ with respect to $(\mathcal{H}, \omega)$. By observing Fig. 3e and using the equation from Corollary 9, it can be seen that the fuzzy connection value of the leaf x_1 is equal to the minimum between $T_{\mathcal{H}}^{\mu}(\{x_1\})$ and $\alpha(\mu(x_1))(1 - \mu(x_1))$. By setting $\alpha(x) = 6 - 5x$, we have that $T_{\mathcal{H}}^{\mu}(\{x_1\}) = 2$ and $\alpha(0.8)(1 - 0.8) = 0.4$, thus the fuzzy connection value of x_1 to μ is equal to 0.4.

The computation of the value $\max(\mu \downarrow_R)$ for a function μ and every regions R in $\mathcal{H}$ can be done with a single pass on the hierarchy in leaves-to-root order, *i.e.*, in which a region is processed after its children. The computation of $T_{\mathcal{H}}^{\mu}(\{x\})$ for every x in V is done with a single pass on the hierarchy, in linear time with respect to the number of regions of $\mathcal{H}$. Then, after computing $T_{\mathcal{H}}^{\mu}(\{x\})$ for every x in V, computing the fuzzy connection values $\mathcal{C}_{\mathcal{H}}^{f}(x, \mu)$ for all x in V is done in linear time with respect to the number of elements in V, as it can be observed in Algorithm 2. As the number of regions in $\mathcal{H}$ is at most $2|V| - 1$, this leads to an overall complexity of $O(|V|)$.

Algorithm 1: Regional connection values

    Input   : A hierarchy $(\mathcal{H}, \omega)$, a fuzzy set μ, and the values $\max(\mu\downarrow_R)$ for every region R of $\mathcal{H}$

    Output: The array T containing the values $T_{\mathcal{H}}^{\mu}(\{x\})$ for all x in V

1 $T[V] := \alpha(0)(1 + \omega(V))$;

2 **foreach** *non-root region R in $\mathcal{H}$ in root-to-leaves order* **do**

3      $c_{sib} := \alpha(max(\mu\downarrow_{sib(R)}))(1 - max(\mu\downarrow_{sib(R)}) + \omega(par(R)))$;

4      $T[R] := \min\left(T[par(R)], c_{sib}\right)$;

5 **end**

Algorithm 2: Fuzzy connection values

    Input   : A set V, a fuzzy set μ on V

    Output: The array Out containing the values $\mathcal{C}_{\mathcal{H}}^{f}(x, \mu)$ for all x in V

1 **Compute** regional connection values with **Algorithm 1**;

2 **foreach** *element x in V* **do**

3      $Out[x] := \min\left(T[x], \alpha(\mu(x))(1 - \mu(x))\right)$;

4 **end**

---

5   Conclusion and Discussions

In this work, we propose an extension of a classical method of marker-based image segmentation to the case where markers are represented by fuzzy sets. We also propose an algorithm for computing the fuzzy-marker-based segmentation in linear time, and prove its correctness. With the proposed extension, a new extent of applications becomes possible, such as:

Interactive Segmentation with Adjustable Marker Strength: In an interactive segmentation framework, markers usually represent the location of objects of interest, and distinct markers guide the segmentation with equal power. By using the method proposed in this article, it is possible to add extra information to the markers, in the form of marker strength. With a fuzzy marker, we can adjust its membership degree from 0 to 1, where 1 denotes the maximum strength, and 0 the minimum.

An example of such application is illustrated in Fig. 4. The segmentations produced in the illustration are computed with a watershed by area hierarchy [8] created over a 4-adjacency graph weighted by a SED gradient [9]. We can see on the second row of Fig. 4 that when the markers are crisp (with strength of 1 on all marker locations) there is a leakage of the object segmentation to the background. We show that we can decrease this leakage by adjusting the strength of the object markers, without the need of changing the locations of any markers. The code for performing a fuzzy-marker-based interactive segmentation with adjustable marker strength developed using the HIGRA library [17] is available at https://higra.readthedocs.io/en/latest/notebooks.html.

Marker Optimization: Given a ground-truth segmentation, we can also ask ourselves what are the optimal markers to obtain this segmentation. It can be

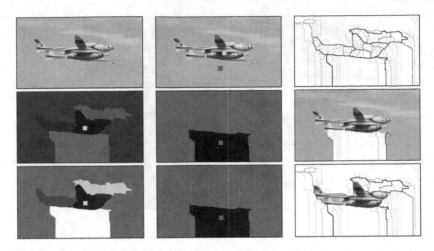

Fig. 4. Example of fuzzy-marker-based interactive segmentation. On the top row, from left to right: an image, object and background makers (in blue and red, respectively) superimposed over the image, and a saliency map representation of a hierarchy on the image. On the second and third row we can observe the map of fuzzy connection values from the regions of the image to the object marker, followed by the map of fuzzy connection values to the background marker, and finally the produced segmentation. On the maps of connection values, darker colors represent smaller values. On the second row, the markers are crisp, and on the third row, the value of the object markers is set to 0.5.

seen that the function $\mathcal{C}_{\mathcal{H}}^{f}$ given in Definition 2 is sub-differentiable with respect to μ, this allows us to use a classical gradient descent algorithm to optimize markers for a given ground-truth.

Given an image and a pair of fuzzy markers, we can produce a segmentation for the given markers. With the produced segmentation, we can compute a segmentation loss with respect to the ground-truth, using for instance a pixel-wise cross-entropy loss. We search for a pair of markers that minimize the loss function, *i.e.*, that produce a segmentation which is closer to the ground-truth. We can find a solution to this minimization problem with gradient descent, thus optimizing markers for a given ground-truth segmentation.

An example of a marker optimization is shown in Fig. 5. In the example, we use a pair of object and background markers, such that the background marker intersects the object of interest. Due to part of the background marker being located over the object, the produced segmentation has regions of the object wrongly classified as background. After the optimization process, we can observe that the produced markers lead to a better object segmentation, *i.e.*, a segmentation which is more similar to the considered ground-truth. For the example in Fig. 5, we use a watershed by area hierarchy created over a 4-adjacency graph weighted by a SED gradient. During the optimization process we also introduce a set of regularization functions to reduce the size and increase the smoothness

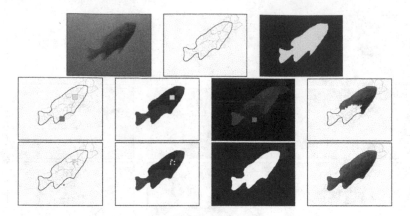

Fig. 5. Illustration of the marker optimization. On the top row, from left to right: an image; a representation of a saliency map of a hierarchy for the image; and the ground-truth of the object segmentation. On the second and third row: the markers (object in blue, background in red) over the saliency maps, followed by the map of fuzzy connection values to the object marker, the map of fuzzy connection values to the background marker, and the fuzzy-marker-based segmentation. On the second row we have the initial crisp markers, and on the bottom row we have the optimized markers. On the maps of fuzzy connection values, darker color represents lower values.

of the markers. Finally, we apply a grayscale dilation on the markers after the optimization for better visualization.

In future works, we will extend the studies on learning marker locations, using the proposed optimization framework to train a marker proposal convolutional neural network. Ideally, the markers proposed by the network would be able to recover the segmentation ground-truth, and follow shape constraints that characterize easily editable markers.

References

1. Arbeláez, P., Maire, M., Fowlkes, C., Malik, J.: Contour detection and hierarchical image segmentation. PAMI **33**(5), 898–916 (2011)
2. Bai, X., Sapiro, G.: Geodesic matting: a framework for fast interactive image and video segmentation and matting. IJCV **82**(2), 113–132 (2009)
3. Benzecri, J.P., et al.: L'analyse des données. 1. la taxinomie, pp. 195, 196. Dunod, Paris (1973)
4. Beucher, S.: Watershed, hierarchical segmentation and waterfall algorithm. In: ISMM, pp. 69–76. Springer (1994)
5. Boykov, Y.Y., Jolly, M.P.: Interactive graph cuts for optimal boundary & region segmentation of objects in N-D images. In: ICCV, vol. 1, pp. 105–112. IEEE (2001)
6. Bragantini, J., Martins, S.B., Castelo-Fernandez, C., Falcão, A.X.: Graph-based image segmentation using dynamic trees. In: Vera-Rodriguez, R., Fierrez, J., Morales, A. (eds.) CIARP 2018. LNCS, vol. 11401, pp. 470–478. Springer, Cham (2019). https://doi.org/10.1007/978-3-030-13469-3_55

7. Cousty, J., Bertrand, G., Najman, L., Couprie, M.: Watershed cuts: minimum spanning forests and the drop of water principle. PAMI **31**(8), 1362–1374 (2008)
8. Cousty, J., Najman, L.: Incremental algorithm for hierarchical minimum spanning forests and saliency of watershed cuts. In: Soille, P., Pesaresi, M., Ouzounis, G.K. (eds.) ISMM 2011. LNCS, vol. 6671, pp. 272–283. Springer, Heidelberg (2011). https://doi.org/10.1007/978-3-642-21569-8_24
9. Dollár, P., Zitnick, C.L.: Fast edge detection using structured forests. PAMI **37**(8), 1558–1570 (2014)
10. Falcão, A.X., Stolfi, J., de Alencar Lotufo, R.: The image foresting transform: theory, algorithms, and applications. PAMI **26**(1), 19–29 (2004)
11. Gómez, D., Yanez, J., Guada, C., Rodríguez, J.T., Montero, J., Zarrazola, E.: Fuzzy image segmentation based upon hierarchical clustering. Knowl.-Based Syst. **87**, 26–37 (2015)
12. Grady, L.: Random walks for image segmentation. PAMI **28**(11), 1768–1783 (2006)
13. Jang, W.D., Kim, C.S.: Interactive image segmentation via backpropagating refinement scheme. In: CVPR, pp. 5297–5306. IEEE (2019)
14. Li, Z., Chen, Q., Koltun, V.: Interactive image segmentation with latent diversity. In: CVPR, pp. 577–585. IEEE (2018)
15. Malmberg, F., Nordenskjöld, R., Strand, R., Kullberg, J.: Smartpaint: a tool for interactive segmentation of medical volume images. CMBBE **5**(1), 36–44 (2017)
16. Najman, L., Cousty, J., Perret, B.: Playing with Kruskal: algorithms for morphological trees in edge-weighted graphs. In: Hendriks, C.L.L., Borgefors, G., Strand, R. (eds.) ISMM 2013. LNCS, vol. 7883, pp. 135–146. Springer, Heidelberg (2013). https://doi.org/10.1007/978-3-642-38294-9_12
17. Perret, B., Chierchia, G., Cousty, J., Guimarães, S.J.F., Kenmochi, Y., Najman, L.: Higra: Hierarchical graph analysis. SoftwareX **10**, 1–6 (2019)
18. Rosenfeld, A.: Fuzzy digital topology. Inf. Control **40**(1), 76–87 (1979)
19. Rother, C., Kolmogorov, V., Blake, A.: "Grabcut": interactive foreground extraction using iterated graph cuts. TOG **23**(3), 309–314 (2004)
20. Salembier, P., Garrido, L.: Binary partition tree as an efficient representation for image processing, segmentation, and information retrieval. TIP **9**(4), 561–576 (2000)
21. Udupa, J.K., Saha, P.K., Lotufo, R.A.: Fuzzy connected object definition in images with respect to co-objects. In: Medical Imaging 1999: Image Processing, vol. 3661, pp. 236–245. International Society for Optics and Photonics (1999)
22. Udupa, J.K., Samarasekera, S.: Fuzzy connectedness and object definition: theory, algorithms, and applications in image segmentation. Graph. Models Image Process. **58**(3), 246–261 (1996)
23. Vachier, C., Meyer, F.: The viscous watershed transform. JMIV **22**(2–3), 251–267 (2005)
24. Xu, N., Price, B., Cohen, S., Yang, J., Huang, T.S.: Deep interactive object selection. In: CVPR, pp. 373–381. IEEE (2016)
25. Yin, S., Qian, Y., Gong, M.: Unsupervised hierarchical image segmentation through fuzzy entropy maximization. Pattern Recogn. **68**, 245–259 (2017)

Graph-Based Supervoxel Computation from Iterative Spanning Forest

Carolina Jerônimo[1], Felipe Belém[3], Sarah A. Carneiro[2],
Zenilton K. G. Patrocínio Jr.[1], Laurent Najman[2], Alexandre Falcão[3],
and Silvio Jamil F. Guimarães[1,2(✉)]

[1] Laboratory of Image and Multimedia Data Science (ImScience),
Pontifical Catholic University of Minas Gerais, Belo Horizonte 31980-110, Brazil
{zenilton,sjamil}@pucminas.br
[2] LIGM, CNRS, ESIEE Paris, Université Gustave Eiffel,
77454 Marne-la-Vallée, France
{sarah.alcar,laurent.najman}@esiee.fr
[3] Laboratory of Image Data Science (LIDS), University of Campinas,
Campinas 13083-852, Brazil
{felipe.belem,afalcao}@ic.unicamp.br

Abstract. Supervoxel segmentation leads to major improvements in video analysis since it generates simpler but meaningful primitives (*i.e.*, supervoxels). Thanks to the flexibility of the Iterative Spanning Forest (ISF) framework and recent strategies introduced by the Dynamic Iterative Spanning Forest (DISF) for superpixel computation, we propose a new graph-based method for supervoxel generation by using iterative spanning forest framework, so-called ISF2SVX, based on a pipeline composed by four stages: (a) graph creation; (b) seed oversampling; (c) IFT-based superpixel delineation; and (d) seed set reduction. Moreover, experimental results show that ISF2SVX is capable of effectively describing the video's color variation through its supervoxels, while being competitive for the remaining metrics considered.

Keywords: Graph-based method · Supervoxel computation · Iterative spanning forest

1 Introduction

In image and video applications, it is often necessary to separate the objects from its background for subsequent analysis. One approach generates groups of connected elements (*i.e.*, superpixels or supervoxels) which shares a common property (*e.g.*, color and texture). By generating numerous groups, the object

The authors thank Conselho Nacional de Desenvolvimento Científico e Tecnológico – CNPq – (PQ 310075/2019-0), Coordenação de Aperfeiçoamento de Pessoal de Nível Superior – CAPES – (Grant COFECUB 88887.191730/2018-00) and Fundação de Amparo à Pesquisa do Estado de Minas Gerais – FAPEMIG – (Grants PPM-00006-18).

© Springer Nature Switzerland AG 2021
J. Lindblad et al. (Eds.): DGMM 2021, LNCS 12708, pp. 404–415, 2021.
https://doi.org/10.1007/978-3-030-76657-3_29

can be effectively defined by its comprising regions, being the major premise of superpixel and supervoxel segmentation algorithms. Such methods are applied in many contexts such as: (i) object detection [11,13]; (ii) cloud connectivity [14,20,22]; and (iii) long-range tracking [16].

For early video processing, one can interpret it as a three-dimensional spatiotemporal volume and segment its objects. The *graph-based supervoxel* (GB) [9] is a image segmentation method based on graphs, presenting good boundary adherence but it is so computationally expensive. The *hierarchical GB* (GBH) [12] considers the GB strategy for computing a hierarchical iterative method; while the *stream GBH* (sGBH) [24] extends the latter for online video segmentation. The authors in [17] proposed the method *cp-HOScale* that improves GBH by computing the whole hierarchy without increasing the computational cost. Although GBH overcomes GB speed performance drawback, it does not guarantee the generation of the desired number of supervoxels. Analogous for GB and GBH, MeanShift [15] and *Segmentation by Weighted Aggregation* (SWA) [7] optimize the normalized cuts criterion, in which SWA performs it hierarchically. However, while MeanShift presents a fair delineation performance, SWA does not guarantee to produce the exact number of supervoxels. Three properties are desirable in video supervoxel segmentation: (i) spatiotemporal boundary adherence; (ii) computational efficiency; and (iii) ability to control the number of supervoxels generated. However, no supervoxel segmentation algorithm has all these characteristics [21].

Recent advances in superpixel segmentation (*e.g.*, deep learning strategies) are strongly related to the image dimensionality and, thus, an extension for video might not guarantee the same performance as the one reported. Thus, the improvements in both categories are often self-contained, significantly limiting their possible improvements. As an example, while GB [9] and GBH [12] equivalents are considered state-of-the-art methods in video segmentation, in superpixel segmentation, they were surpassed by a large set of newer and more effective approaches [2,19]. Finally, although the authors in [2] discuss how hierarchical superpixel methods might propagate errors to coarser levels, one can see that such a statement holds for hierarchical supervoxel segmentation, which is often considered to be a desirable property [21].

Inspired by the *Iterative Spanning Forest* (ISF) [19], a recent superpixel segmentation framework, in this work, we propose a supervoxel segmentation framework for video segmentation, named *ISF for Supervoxels* (ISF2SVX). Similar to ISF, our approach is composed of independent steps: (i) graph construction; (ii) seed sampling; (iii) supervoxel generation; and (iv) seed recomputation. In step (i), the video volume is converted to a directed graph representation which will be used as input for determining the seeds in step (ii). Then, for several iterations, ISF2SVX generates supervoxels through the *Image Foresting Transform* (IFT) [8] using improved seed sets—in steps (iii) and (iv), respectively. Figure 1 illustrates examples of results obtained by ISF2SVX by changing the strategy for seed sampling (grid and random) for 10 and 500 supervoxels.

(a) 10 supervoxels

(b) 500 supervoxels

Fig. 1. Examples of video segmentations for a video extracted of the GATech. The original frames are illustrated in the first row. We illustrate examples of the proposed method changing the seed sampling. We illustrate results for (a) 10 and (b) 500 supervoxels, considering grid and random seed sampling (second and third rows). Each resulting region is colored by its mean color.(Color figure online)

This paper is organized as follows. In Sect. 2, important concepts used in this work such as graphs and IFT are clarified. In Sect. 3, the methodology for the proposed segmentation approach is explained. In Sect. 4, we describe the experiments performed and compare the achieved results to other methods. Finally, some concluding notes and suggestions for future work are presented in Sect. 5.

2   Theoretical Background

In this section, we explain the necessary concepts and techniques related to our proposal. We first introduce some graph notions to present the core delineation algorithm of our proposal: *Image Foresting Transform* (IFT) [8] framework.

2.1   Graph

A *video* V can be represented as a pair $\mathsf{V} = (\mathcal{V}, \mathbf{I})$ in which $\mathcal{V} \subseteq \mathbb{N}^3$ denotes the set of *volume elements* (*i.e.*, voxels), and $\mathbf{I}$ maps every $v \in \mathcal{V}$ to a feature vector $\mathbf{I}(v) \in \mathbb{R}^m$. One can see that, for $m = 3$, V is a colored video (*e.g.*, RGB or CIELAB colorspaces). It is possible to create a *simple graph* (*i.e.*, no loops and no parallel edges) $\mathsf{G} = (\mathcal{N}, \mathcal{E}, \mathbf{I})$, derived from V, in which $\mathcal{N} \subseteq \mathcal{V}$ denotes the *vertex* set and $\mathcal{E} \subset \mathcal{N}^2$, the *edge* set. Two nodes $v_i, v_j \in \mathcal{N}$ are said to be *adjacent* if $(v_i, v_j) \in \mathcal{E}$. In this work, the elements in $\mathcal{E}$ are *arcs* (*i.e.*, G is a *digraph*). Consider $\pi_{s \rightsquigarrow t} = \langle s = v_1, v_2, \ldots, v_n = t \rangle$ to be a finite sequence of adjacent nodes (*i.e.*, a *path*) in which $(v_i, v_{i+1}) \in \mathcal{E}$ for $1 \leq i < n$. For simplicity, we may omit the path *origin* voxel by writing π_t. For $n = 1$, $\pi_t = \langle t \rangle$ is said to be *trivial*. We denote the *extension* of a path π_s by an arc $(s, t) \in \mathcal{E}$ as $\pi_s \cdot \langle s, t \rangle$ with the two instances of s being merged into one.

2.2   IFT

The *Image Foresting Transform* (IFT) [8] is a framework for the development of image processing operators based on connectivity and has been used to reduce image processing tasks as optimum-path forest computations over the image graph. As indicated by the authors [8], the IFT is independent of the input's dimensions and, therefore, the relation between pixels (or voxels) in such dimensionality can effectively be represented by an *adjacency relation* between them. In this work, we consider the IFT version restricted to a *seed set* $\mathcal{S} \subset \mathcal{N}$.

For a given arc $(s, t) \in \mathcal{E}$, it is possible to assign a non-negative *arc-cost* value $\mathbf{w}_*(s, t) \in \mathbb{R}^+$ through an *arc-cost function* $\mathbf{w}_*$. A common approach is to compute the ℓ_2-norm between the nodes' features—*i.e.*, $\|\mathbf{I}(s) - \mathbf{I}(t)\|_2$ for $s, t \in \mathcal{N}$. Consider Π_G the set of all possible paths in G. Then, a *connectivity function* $\mathbf{f}_*$ maps every path in Π_G to a *path-cost value* $\mathbf{f}_*(\pi_t) \in \mathbb{R}^+$. One of the most effective connectivity functions for object delineation is the $\mathbf{f}_{\max}$ function:

$$\mathbf{f}_{\max}(\langle t \rangle) = \begin{cases} 0 & \text{if } t \in \mathcal{S}, \\ +\infty & \text{otherwise} \end{cases} \tag{1}$$

$$\mathbf{f}_{\max}(\pi_s \cdot \langle s, t \rangle) = \max\{\mathbf{f}_{\max}(\pi_s), \mathbf{w}_*(s, t)\}$$

A path π_t^* is said to be *optimum* if, for any other path $\tau_t \in \Pi_\mathsf{G}$, $\mathbf{f}_*(\pi_t^*) \leq \mathbf{f}_*(\tau_t)$.

Let $\mathbf{C}$ be a *cost map* in which assigns, to every path $\pi_t \in \Pi_\mathsf{G}$, its respective path-cost value $\mathbf{f}_*(\pi_t)$. The IFT algorithm minimizes $\mathbf{C}(t) = \min_{\forall \pi_t \in \Pi_\mathsf{G}} \{\mathbf{f}_*(\pi_t)\}$ whenever $\mathbf{f}_*$ satisfies certain conditions [5]. First, the IFT assigns path-costs to all trivial paths accordingly and, then, it computes optimum paths in a non-decreasing order, from the seeds to the remaining nodes in the graph. Therefore, independently if $\mathbf{f}_*$ suffices the desired properties in [5], the IFT always generates a spanning forest and, consequently, each supervoxel is an unique tree. During the segmentation process, a *predecessor map* $\mathbf{P}$ is generated and defined. Such map assigns any node $t \in \mathcal{N}$ to its *predecessor* s in the optimum path $\pi_s^* \cdot \langle s, t \rangle$, or to a distinctive marker $nil \notin \mathcal{N}$—in such case, t is said to be a *root* of $\mathbf{P}$. In this work, every seed is a root of $\mathbf{P}$. One may see that $\mathbf{P}$ is a representation of an *optimum-path forest*, and it allows to recursively obtain the optimum-path root $\mathbf{R}(t)$ of t and its root's label $\mathbf{L}(\mathbf{R}(t))$.

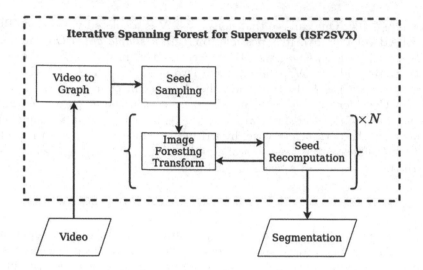

Fig. 2. Diagram of the proposed methodology for video supervoxel segmentation.

3   A Strategy for Supervoxel Computation Based on Iterative Spanning Forest

In this section, we present our approach for supervoxel computation based on *Iterative Spanning Forest* (ISF) [19] superpixel framework. Our proposal, so-called ISF2SVX, adopts a four step methodology: (a) graph creation; (b) seed sampling; (c) IFT-based supervoxel delineation; and (d) seed set recomputation. This pipeline is illustrated in Fig. 2. Although our framework permits conceiving uncountable distinct variants, in this work, we assess some of the latest findings proposed by the ISF-based *Dynamic and Iterative Spanning Forest* (DISF) [2] method, which has proven to be more effective than state-of-the-art superpixel segmentation methods.

3.1   Video to Graph

Differently from 2D and 3D images, the presence of the same object in between frames imposes a major challenge for generating temporally coherent supervoxels. Therefore, it is recommended that the arcs and their respective arc-costs should reflect such condition. In this work, we operate on a video $V = (\mathcal{V}, \mathbf{I})$ whose graph $G = (\mathcal{N}, \mathcal{E}, \mathbf{I})$ is modeled as a single volume of nodes, and the outgoing arcs $(s, t) \in \mathcal{E}$ of a node s, for any $t \in \mathcal{N}$ which $s \neq t$, are defined, for instance, by an *adjacency relation* (*e.g.*, 26-adjacency). Another possibility to transform the video into a graph may consider motion information, like optical flow, in order to guide the edge creation, however this strategy is out-of-the-scope of this work since our main aim here is to study the behaviour of a simple adjacency relation.

3.2   Seed Sampling

For a given graph G, the second step generates the seed set $\mathcal{S} \subset \mathcal{N}$ for the first iteration of the IFT algorithm. In [2], the authors pointed out the drawbacks of initially sampling a number $N_0 \in \mathbb{N}$ of seeds approximate to the desired number $N_f \in \mathbb{N}$ of superpixels (or supervoxels). The relevance of a seed—which promotes effective delineation—is related to its location in a graph and, therefore, a strategy of oversampling may overcome the latter by increasing the probability of such seed being inserted in $\mathcal{S}$.

Most methods adopt a grid sampling scheme [1] (hereinafter named GRID) by selecting equally distanced seeds within the graph. Given a desired number $N_0 \in \mathbb{N}$ of seeds, and by computing an approximate supervoxel size $s = \frac{N}{N_0}$, one can determine the stride d between seeds as $\sqrt[n]{s}$, where n denotes the data dimensionality—i.e., $n = 2$ and $n = 3$, for 2D images, and for 3D images and videos, respectively. Finally, for avoiding seeds in high contrast regions (i.e., probable object boundaries), the seeds are shifted to the lowest gradient position defined in an 8- or 26- neighborhood, for $n = 2$ or $n = 3$, respectively.

Considering an oversampling GRID strategy, d decreases sharply and, therefore, the proximity between seeds favors extreme competition, which often leads to better object delineation [2]. However, due to the excessive number of seeds, one can presume that a random selection of N_0 initial seeds can result in an even distribution in the graph, without compromising the selection of relevant ones. In this work, we propose such random oversampling strategy, named RND.

3.3   Supervoxel Generation

Once seeds are sampled, the supervoxels are generated using the IFT algorithm considering a connectivity function $\mathbf{f}_*$ and an arc-cost function $\mathbf{w}_*$. In this work, we consider the $\mathbf{f}_{\max}$ connectivity function for computing the path-costs.

In [19], the authors recall an arc-cost function $\mathbf{w}_1(p, q) = (\alpha \| \mathbf{I}(\mathbf{R}(p)) - \mathbf{I}(q) \|_2)^\beta + \| p - q \|_2$ in which $\alpha \in \mathbb{R}_*^+$ permits the user to control the regularity of the superpixels, and to control their adherence to boundaries through a factor $\beta \in \mathbb{R}_*^+$. However, superpixel and supervoxel regularity tends to prejudice the object delineation performance [2].

In DISF, the arc-costs are computed dynamically considering mid-level superpixel features, using a function first proposed in [3]. Let $\mathcal{T}_x \subset \mathcal{N}$ be an optimum-path growing tree rooted in a node $x \in \mathcal{N}$, and let $\mu(\mathcal{T}_x)$ be its mean feature vector. Then, the arc-cost function $\mathbf{w}_2$ can be formally defined as $\mathbf{w}_2(p, q) = \| \mu(\mathcal{T}_{\mathbf{R}(p)}) - \mathbf{I}(q) \|_2$. The function $\mathbf{w}_2$ has proven to be more effective than classic arc-cost functions for both superpixel segmentation [2] and for interactive image segmentation [3].

However, since the arc-costs are computed dynamically, $\mathbf{w}_2$ may generate discrepant segmentations, especially in regions with distinct colors, but equal gradient variation. Moreover, the order of arc evaluation during the IFT may also affect the aforementioned results. Therefore, in this work, we address such instability by computing a root-based arc-cost function $\mathbf{w}_3 = \| \mathbf{I}(\mathbf{R}(p)) - \mathbf{I}(q) \|_2$.

Although it is not a dynamic estimation, root-based functions often present top delineation performance [19].

3.4   Seed Recomputation

In ISF2SVX, the fourth step aims to update the seed set $\mathcal{S}$ in order to improve the supervoxel delineation for subsequent iterations. Such update can be performed by including, shifting or removing the seeds in $\mathcal{S}$, but respecting as much as possible the desired final number of supervoxels $N_f \in \mathbb{N}$ in the last iteration. Since, in this work, an oversampling strategy is presented, it is important to note that $N_0 \gg N_f$.

Since the presence of relevant seeds is expected—due to oversampling —, in [2], the authors proposed a new methodology for seed recomputation: removing irrelevant seeds based on a certain criterion. The motivation for that consists in promoting the growth of relevant superpixels (or supervoxels), by removing the irrelevant ones and maintaining the competition among the primers. At each iteration $i \in \mathbb{N}$, $\mathbf{M}(i) = \max\{N_0 \exp^{-i}, N_f\}$ relevant seeds are maintained for the subsequent iteration $i + 1$, while the remaining ones are discarded. In DISF, the stopping criterion is reaching the desired number of superpixels, which is often less than 10—a common value for many iterative methods.

The $\mathbf{M}(i)$ relevant seeds may be selected by a combination of their sizes and contrast [2] in which the former indicates the supervoxel's growth ability, and the latter, whether the supervoxel is located in a homogeneous region (thus, probably irrelevant). Let $\mathcal{B}$ be a *tree adjacency relation*, which defines the immediate neighbors of any supervoxel. Then, with the use of a priority queue, a relevance of a seed s can be measured by a function $\mathbf{V}(s) = \frac{|\mathcal{T}_s|}{|\mathcal{N}|} \min_{\forall(\mathcal{T}_s,\mathcal{T}_r)\in\mathcal{B}}\{\|\mu(\mathcal{T}_s) - \mu(\mathcal{T}_r)\|_2\}$, which $\mathcal{T}_r$ is an *adjacent supervoxel* of $\mathcal{T}_s$.

4   Experimental Analysis

We evaluated all methods considering the Chen [4] and Segtrack [18] datasets, both containing groundtruth annotations. Using the LIBSVX [23] library, we selected five classic evaluation metrics: (a) 3D boundary recall (BR); (b) 3D segmentation accuracy (SA); (c) 3D undersegmentation error (UE); (d) Explained variation (EV); and (e) Mean duration. BR measures the quality of the spatiotemporal boundary delineation, while SA measures the fraction of groundtruth segments which are correctly classified (*i.e.*, higher is better for both). UE calculates the fraction of object supervoxels overlapping background voxels—and vice-versa—(*i.e.*, lower is better). EV measures the method's ability to describe the video's color variations through its supervoxels (*i.e.*, higher is better). Finally, the mean supervoxel duration measures if a supervoxel perpetuates throughout the frames, indicating a temporal coherence to the object which it compounds (*i.e.*, higher is better).

In this work, we propose two ISF2SVX variants. One, named ISF2SVX-GRID-DYN, oversamples using GRID and computes supervoxels considering the arc-cost function w_2. The other, ISF2SVX-RND-ROOT, oversamples using RND, and considers the w_3 function. We compared our approaches with different state-of-the-art methods: (i) GB [9]; (ii) GBH [12]; (iii) SWA [7]; (iv) Mean-Shift [15]; and (v) cp-HOScale [17]. The number of supervoxels varied from 200 to 900 and, for the baselines, the recommended parameter settings were used.

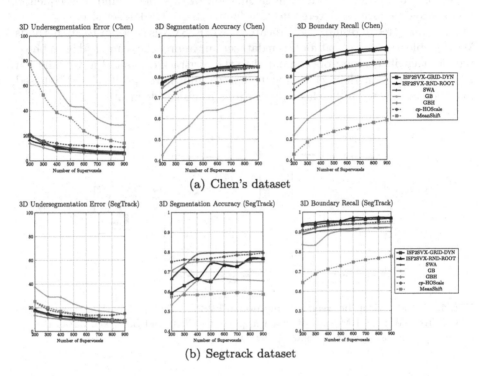

(a) Chen's dataset

(b) Segtrack dataset

Fig. 3. A comparison between our method ISF2SVX, and the methods cp-HOScale GB, GBH, SWA, and MeanShift when applied to Chen and SegTrack datasets. The comparison is based on the following metrics: (i) 3D undersegmentation error; (ii) 3D segmentation accuracy; (iii) 3D boundary recall.

4.1 Quantitative Analysis

Considering the undersegmentation error, it is possible to observe in the plots in Fig. 3 that both of the ISF2SVX variations managed to be compatible or even better than the compared works. Although for a smaller number of supervoxels, the GBH method performs slightly better in both Chen and Segtrack dataset, one can notice that after 700 supervoxels our methods are consistent with GBH. This may indicate that the construction of the hierarchy is not so beneficial to prevent supervoxel leaks, since region merging errors can occur and make the leak even more evident.

When taking into account the segmentation accuracy (Fig. 3), we notice that ISF2SVX has produced well-defined segmentations as well as some of the compared baseline methods. Although it is possible to detect an instability related to the segmentation accuracy in the Segtrack dataset, this instability is given by a single video whose object of interest is significantly small and more than one supervoxel composes it. Thus, the calculation of this metric and, consequently, the average performance were affected due to the small dataset size.

Boundary recall results also indicate that ISF2SVX is superior to all methods. We are able to observe that for Chen dataset our approaches can yield even better metrics that its competitors compared to the Segtrack dataset (Fig. 3). This can be associated with the fact that path-based, and more specifically IFT-based, methods are known to be effective solutions in object delineation [2,19].

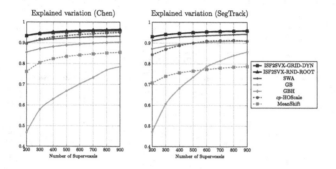

Fig. 4. A comparison between our method ISF2SVX, and the methods cp-HOScale GB, GBH, SWA, and MeanShift when applied to Chen and SegTrack datasets. The comparison is based on the explained variation.

In Fig. 4 is possible to observe that, since IFT minimizes the accumulated cost of the path, the internal variation of supervoxels tends to be greatly minimized. Thus, we can see better explained variation metric results for our variations when compared to previous studies. In addition, as there are no regularity constraints, the competition between seeds becomes more intense and, therefore, leads to a lower probability of incorporating dissimilar voxels. Furthermore, as one can see in Fig. 5, ISF2SVX, in both variants, outperforms the other methods in terms of mean duration. It is worth to mention that this metric tries to capture the temporal coherence of the supervoxels.

4.2   Qualitative Analysis

In Fig. 6, we compare the variant ISF2SVX-GRID-DYN with the baselines in a single video. As one may see, our approach manages to generate large supervoxels in non-significant regions (*e.g.*, the grass), while effectively delineates even small important regions (*e.g.*, the player's head). In contrast, for 100 supervoxels, all baselines generates too many small and irrelevant supervoxels. For 20

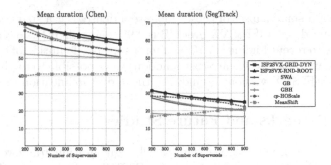

Fig. 5. A comparison between our method ISF2SVX, and the methods *cp*-HOScale GB, GBH, SWA, and MeanShift when applied to Chen and SegTrack datasets. The comparison is based on the mean duration.

supervoxels, such quantity is severely reduced at the expense of degrading the object delineation performance.

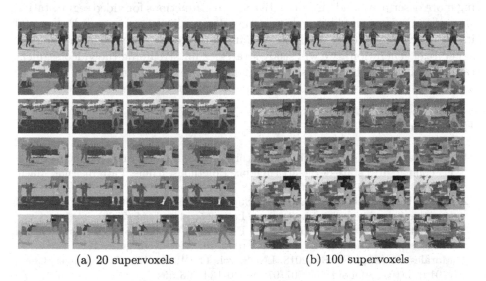

(a) 20 supervoxels   (b) 100 supervoxels

Fig. 6. Example extracted from Chen dataset. The first row are the original frames, the following rows, from top to bottom are results with 20 and 100 supervoxels obtained from GB, GBH, SWA, *cp*-HOScale, and ISF2SVX-GRID-DYN.

When comparing the mean execution time in such video—given an interval of [200, 900] supervoxels —, ISF2SVX obtains a speed-up of 5.9 and 7.1 against the second and third fastest baselines (*i.e.*, GBH and *cp*-HOScale, respectively). Although *cp*-HOScale manages to compute the whole hierarchy—thus is capable of obtaining many segmentations without requiring any recomputation —, it is unlikely that, in an application, the user would need to manipulate all the levels, and not a small subset of those (*i.e.*, use a dense over a sparse hierarchy). Finally, the speed-up of ISF2SVX over GB is 0.95, being slightly slower than GB. However, due to recent findings [6,10] and for a suitable definition of components (*e.g.*, GRID sampling and w_3 arc-cost function), it is possible to further improve the speed of ISF2SVX without prejudicing the object delineation performance.

5 Final Remarks and Future Studies

In this paper, we propose a new supervoxel segmentation framework, named *Iterative Spanning Forest for Supervoxels* (ISF2SVX), which was inspired by the *Iterative Spanning Forest* (ISF) superpixel segmentation framework. Our approach not only benefits from recent improvements in superpixel segmentation, but also permits the development of effective video segmentation algorithms through the definition of its components. Results show that ISF2SVX variants outperforms state-of-the-art methods with a great margin in two datasets, especially in terms of delineation and color description (by its supervoxels).

For further works, we would like to study the behaviour of ISF2SVX considering more descriptive and discriminative arc-cost functions for video segmentation since the ones presented here relies on the color gradient between an element and its conquering tree (or root). Moreover, instead of early video segmentation (or supervoxel generation), we will study strategies for streaming the video segmentation method. Furthermore, strategies for seed oversampling location will be an interesting direction since we may learn good positions to the set of seeds.

References

1. Achanta, R., Shaji, A., Smith, K., Lucchi, A., Fua, P., Süsstrunk, S.: Slic superpixels compared to state-of-the-art superpixel methods. Trans. Pattern Anal. Mach. Intell. **34**(11), 2274–2282 (2012)
2. Belém, F., Guimarães, S., Falcão, A.: Superpixel segmentation using dynamic and iterative spanning forest. Sig. Process. Lett. **27**, 1440–1444 (2020)
3. Bragantini, J., Martins, S.B., Castelo-Fernandez, C., Falcão, A.X.: Graph-based image segmentation using dynamic trees. In: Vera-Rodriguez, R., Fierrez, J., Morales, A. (eds.) CIARP 2018. LNCS, vol. 11401, pp. 470–478. Springer, Cham (2019). https://doi.org/10.1007/978-3-030-13469-3_55
4. Chen, A., Corso, J.: Propagating multi-class pixel labels throughout video frames. In: Western New York Image Processing Workshop, pp. 14–17 (2010)
5. Ciesielski, C., Falcão, A., Miranda, P.: Path-value functions for which Dijkstra's algorithm returns optimal mapping. J. Math. Imag. Vis. **60**(7), 1025–1036 (2018)
6. Condori, M.A., Cappabianco, F.A., Falcão, A.X., Miranda, P.A.: An extension of the differential image foresting transform and its application to superpixel generation. J. Vis. Commun. Image Represent. **71**, 102748 (2020)
7. Corso, J., Sharon, E., Dube, S., El-Saden, S., Sinha, U., Yuille, A.: Efficient multilevel brain tumor segmentation with integrated Bayesian model classification. Trans. Med. Imag. **27**(5), 629–640 (2008)
8. Falcão, A., Stolfi, J., Lotufo, R.: The image foresting transform: theory, algorithms, and applications. Trans. Pattern Anal. Mach. Intell. **26**(1), 19–29 (2004)

9. Felzenszwalb, P., Huttenlocher, D.: Efficient graph-based image segmentation. Int. J. Comput. Vis. **59**(2), 167–181 (2004)
10. Gonçalves, H.M., de Vasconcelos, G.J., Rangel, P.R., Carvalho, M., Archilha, N.L., Spina, T.V.: cudaIFT: 180x faster image foresting transform for waterpixel estimation using CUDA. In: VISIGRAPP (4: VISAPP). pp. 395–404 (2019)
11. Griffin, B.A., Corso, J.J.: Video object segmentation using supervoxel-based gerrymandering (2017). arXiv:1704.05165
12. Grundmann, M., Kwatra, V., Han, M., Essa, I.: Efficient hierarchical graph-based video segmentation. In: Computer Vision and Pattern Recognition (CVPR), pp. 2141–2148. IEEE (2010)
13. Oneata, D., Revaud, J., Verbeek, J., Schmid, C.: Spatio-temporal object detection proposals. In: Fleet, D., Pajdla, T., Schiele, B., Tuytelaars, T. (eds.) ECCV 2014. LNCS, vol. 8691, pp. 737–752. Springer, Cham (2014). https://doi.org/10.1007/978-3-319-10578-9_48
14. Papon, J., Abramov, A., Schoeler, M., Worgotter, F.: Voxel cloud connectivity segmentation-supervoxels for point clouds. In: Proceedings of the IEEE Conference on Computer Vision and Pattern Recognition, pp. 2027–2034 (2013)
15. Paris, S., Durand, F.: A topological approach to hierarchical segmentation using mean shift. In: 2007 IEEE Conference on Computer Vision and Pattern Recognition, pp. 1–8. IEEE (2007)
16. Sheng, H., Zhang, X., Zhang, Y., Wu, Y., Chen, J., Xiong, Z.: Enhanced association with supervoxels in multiple hypothesis tracking. IEEE Access **7**, 2107–2117 (2018)
17. Souza, K., Araújo, A., Patrocínio, Z., Jr., Guimarães, S.: Graph-based hierarchical video segmentation based on a simple dissimilarity measure. Pattern Recogn. Lett. **47**, 85–92 (2014)
18. Tsai, D., Flagg, M., Nakazawa, A., Rehg, J.: Motion coherent tracking using multi-label MRF optimization. Int. J. Comput. Vis. **100**(2), 190–202 (2012)
19. Vargas-Muñoz, J., Chowdhury, A., Alexandre, E., Galvão, F., Miranda, P., Falcão, A.: An iterative spanning forest framework for superpixel segmentation. Trans. Image Process. **28**(7), 3477–3489 (2019)
20. Verdoja, F., Thomas, D., Sugimoto, A.: Fast 3d point cloud segmentation using supervoxels with geometry and color for 3d scene understanding. In: 2017 IEEE International Conference on Multimedia and Expo (ICME), pp. 1285–1290. IEEE (2017)
21. Wang, B., Chen, Y., Liu, W., Qin, J., Du, Y., Han, G., He, S.: Real-time hierarchical supervoxel segmentation via a minimum spanning tree. Trans. Image Process. **29**, 9665–9677 (2020)
22. Wu, F., Wen, C., Guo, Y., Wang, J., Yu, Y., Wang, C., Li, J.: Rapid localization and extraction of street light poles in mobile LiDAR point clouds: a supervoxel-based approach. IEEE Trans. Intell. Trans. Syst. **18**(2), 292–305 (2016)
23. Xu, C., Corso, J.: LibSVX: a supervoxel library and benchmark for early video processing. Int. J. Comput. Vis. **119**(3), 272–290 (2016)
24. Xu, C., Xiong, C., Corso, J.J.: Streaming hierarchical video segmentation. In: Fitzgibbon, A., Lazebnik, S., Perona, P., Sato, Y., Schmid, C. (eds.) ECCV 2012. LNCS, vol. 7577, pp. 626–639. Springer, Heidelberg (2012). https://doi.org/10.1007/978-3-642-33783-3_45

Graph-Based M-tortuosity Estimation

Adam Hammoumi[1]([✉]), Maxime Moreaud[1,2], Elsa Jolimaitre[1],
Thibaud Chevalier[3], Alexey Novikov[4], and Michaela Klotz[4]

[1] IFP energies nouvelles, Rond-point de l'échangeur de Solaize BP3,
69360 Solaize, France
adam.hammoumi@ifpen.fr
[2] MINES ParisTech, PSL-Research University, CMM, 77305 Fontainebleau, France
[3] IFP energies nouvelles, 1 et 4 avenue de Bois-Préeau,
92852 Rueil-Malmaison, France
[4] LSFC Laboratoire de synthèse et fonctionnalisation des céramiques UMR 3080
CNRS/SAINT-GOBAIN CREE, Saint-Gobain Research Provence,
550 avenue Alphonse Jauffret, Cavaillon, France

Abstract. The sinuosity of a porous microstructure may be quantified
by geometric tortuosity characterization, namely the ratio of geodesic
and euclidean distances. The assessment of geometric tortuosity, among
other descriptors, is of importance for rigorous characterization of complex materials. This paper proposes a new way of calculation, based on a
graph structure, of the topological descriptor *M-tortuosity* introduced in
[3]. The original *M-tortuosity* descriptor is based on a geodesic distance
computation algorithm. A pore network partition [7] method is used to
extract pores and construct a graph from the void of a porous microstructure. Through this scheme, pores are the nodes, distances between pores
are the arcs between nodes and the goal boils down to the determination
of the shortest paths between nodes. Solving this on a graph requires a
tree search formulation of the problem. Our results have shown a dras-
tic time complexity decrease while preserving good agreement with the
original results. The added value of our method consists in its simplicity
of implementation and its reduced execution time.

Keywords: Geometric tortuosity · Pore network partition · Shortest
path search

1 Introduction

Let us consider a particle travelling through a medium where there are different phases, void and matter, for instance. Its path is tortuous and ambiguous, by virtue of its environment topology and morphology. A multi-scale context of chaotic assembly of aggregates and grains may take the phenomenon
to another level of complexity, forcing the particle to cross areas of high and
low density. This particle may be a part of the diffusion flow phenomenon in a
porous media, or the electrical transport of ionic charges by conduction through
a solid phase. Under the urge of understanding these transport phenomena, the

© Springer Nature Switzerland AG 2021
J. Lindblad et al. (Eds.): DGMM 2021, LNCS 12708, pp. 416–428, 2021.
https://doi.org/10.1007/978-3-030-76657-3_30

characterization and the analysis of complex microstructures is a fundamental step. *Porosity ϵ, specific surface area s_p* and *tortuosity τ* are among the most common descriptors of a porous medium. They are respectively providing information about volume fraction of void, accessible surface to the fluid and yielding estimation of meandering paths throughout the microstructure. In the history of tortuosity estimation, the focus has frequently been the characterization of microstructures to estimate their hydraulic and electrical transport properties [6]. Considering the diversity of the tortuosity concept, ad hoc models are created to address this issue. Typically, by including the transport process being studied, different conceptions and treatments of the question of tortuosity arise. Until now, unifying these definitions is still a puzzling task [12]. This work is based on the framework of geometric tortuosity, where only the morphological aspects of the microstructure are considered. The *M-tortuosity* descriptor proposed in [3] allows a global characterization of the microstructure by providing topological and morphological information, all contained in one scalar $\hat{\tau}$. The latter has proven to be an effective asset to discriminate microstructures by their tortuosities. The original algorithm of the *M-tortuosity* lies on multiple computation of geodesic distance maps, which can lead to important time complexity. An estimation of the aforementioned descriptor, allowing faster computation time is proposed. We call it *GM-tortuosity*. Our procedure can be divided in two steps. First, the pore network structure is extracted by means of a pore network partition *PNP* method. It involves a distance map computation in order to extract maxima points, a filtering process to drop irrelevant points and a geodesic distance transform with source propagation applied to the filtered pore network to generate partitions. In summary, the *PNP* method splits the pore space into multiple pores of different sizes. Second, the generated pore network is transformed to a graph where the pores represent nodes and connections between pores are weighted by corresponding euclidean distances. To find distance separating two distant pores, the shortest pathway computation algorithm Djikstra's [4] is used. Finally, we follow the guidelines of the original *M-tortuosity* algorithm with the only difference that the newly calculated distances replace the geodesic distances.

2   Related Work

Chaniot *et al.* [3] have proposed a novel descriptor for geometric tortuosity. The following ideas recall the most important features and methods of this work. The geometric tortuosity between two points can be defined as the ratio of tortuous and straight distance separating them. The formalism of the geodesic distance transform *GDT* allows a precise computation of geodesic distances [5]. *GDT* is based on the propagation of distances from source points. That is, for a given point, it would be possible to have the distance that separates it from each point of the network allowing the generation of a distance map. Hence, the tortuosity between two points can be correctly estimated via *GDT*. By considering all the points of the network, a deterministic *M-tortuosity* methodology was proposed: first, on a binary function $I \colon \mathbb{R}^3 \to \{0,1\}$, a bounded set of features points, pore network elements, is defined as $X = \{x \in \mathbb{R}^3/I(x) = 1\}$. For each $(x, y) \in X^2$ where $x \neq y$ and $x \neq c$, such that c is the center of inertia of the network, the geometric tortuosity is defined as $\tau_{x,y} = \frac{D_G(y,x;X)}{D(y,x)}$ where D_G is the geodesic distance between points x and y, calculated using [5] and D is the associated euclidean distance. Given the former x, it is possible to define an arithmetic mean of the geometric tortuosities C_x, where the respective weights are the geodesic distances $D_G(y, x; X)$.

$$C_x = \frac{\int_{X\setminus\{x\}} D_G(y, x; X)\tau_{x,y} \, dy}{\int_{X\setminus\{x\}} D_G(y, x; X) \, dy} \tag{1}$$

Subsequently, a scalar τ_M can be defined as the arithmetic mean of $\{C_x\}_{x \in X\setminus\{c\}}$, where the respective weights are the euclidean distances from c, $D(x, c)$.

$$\tau_M = \frac{\int_{X\setminus\{x\}} D(x, c)C_x \, dx}{\int_{X\setminus\{x\}} D(x, c) \, dx} \tag{2}$$

The evaluation of τ_M provides information about the total geometric tortuosity in X. Besides the definition of the geometric tortuosity, and the use of the geodesic distance transform, Eqs. (1 and 2) offer an overview of the proposed methodology. Since, it is not practical to consider all the points of the network, given the high cost of the geodesic distance map computation time. It is convenient to privilege an adequate sampling method. However, by considering a single source point, all the estimated distances will only refer to that point, and the distance between the other sampled points will not be considered, causing a poor representation of the network. Yet, repeating this process for the case where each sampled point is considered as a source point will allow the exploration of new paths and effectively estimate the tortuosity of the network. This work has proposed a deterministic and an estimated *M-tortuosity* descriptor. Our method presents faster computation methodology of the *M-tortuosity* estimator.

Recent work [8] has addressed the question of geometrical tortuosity and constrictivity by means of image analysis and morphological operations. The developed descriptors were performed on membranes of electrolysis cells *X-ray* tomography images. Experimental and numerical procedures were conducted in order to understand how these parameters are related to the measured transport properties, which allowed the identification of discriminating parameters. It has been reported that the constrictivity is strongly correlated with the transport properties of the studied membrane materials. The relevant part to us in this work is related to the used methodology to characterize the pore structure by means of image analysis. This method was initially proposed in [9] and goes as follows: first, regions of interest are extracted from tomography images. They undergo filtering and segmentation operation which enables the computation of the porosity of the material. Second, a skeletonization is performed on the resulting image to probe the porous network. Now that the pores were identified by their inertia centers, the porous network is simplified to a string of connected pores, where each pore is carrying information about its distance from nearest pore. Finally, the porous network forms a graph and the distances between pores are estimated by the Djikstra's shortest path search algorithm, yielding tortuous distances between two distant pores. The ratio of the latter and the euclidean distance between the two points defines a local tortuosity estimation. Repeating this process for a large number of data permits the computation of an overall tortuosity estimation calculated by averaging over all local tortuosity values. This method and ours share common operations, namely the graph-based organisation of pores and distances. Since this method may seem similar considering the sequence of steps on overall, our method is based on the framework of the *M-tortuosity*, making use of the *PNP* approach to generate the pore network and yielding an approximation of the total geometric tortuosity.

3   Graph-Based M-tortuosity Method

Inside complex 3D microstructures, it is difficult to visualize geodesic distances between points. Hence the interest of using toy cases, which will allow us to illustrate in a concrete way the obtained results. A simple two-phase scheme, Fig. 1a, is used to demonstrate the concepts behind the used techniques. In this section, we recall the important steps of the pore network partition method. The latter will allow the extraction of the pore network in form of uniquely labeled pores. Then, we explain the mapping of pores and their adjacency properties into a graph representation. At the end, a pseudo algorithm is given in order to summarize all the steps performed.

3.1   Pore Network Generation

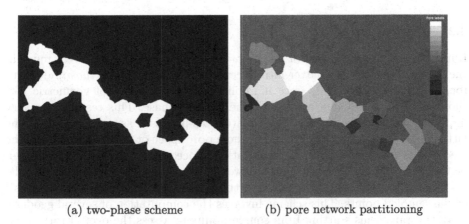

(a) two-phase scheme   (b) pore network partitioning

Fig. 1. The pore network partition method applied to a simple two-phase scheme. Pores in (b) are uniquely labeled and identified by different colors, corresponding to their size. (Color figure online)

Pore network models are one of the basis of physical and chemical simulation processes in porous materials. The quality of the predictions is sensitively linked to the representation of the pore network [15]. For real microstructures, the shape of pores is irregular and their morphology is complex and stochastic. Morphological models [13] allow a realistic representation by providing complex pore networks. For this modelling process, the *PNP* method introduced in [7] is an effective way of partitioning the pore network by providing coherent pore size distributions. The simplistic example shown in Fig. 1a is represented as a binary image, containing solid and void phase. The randomly drawn object in white corresponds to the pore network. By applying the *PNP* method to the porous phase surface, the latter gets divided into distinct pores, Fig. 1b. The resulting pores are labeled on the basis of their proximity to the solid phase i.e. their size. That is, the ability of discerning pores will allow an accurate representation of a wide range of pores: narrow, wide and in-between. This result is achieved throughout the following steps: first, a distance map, replacing the value of each foreground pixel by its value to the nearest background element is generated by applying a distance transform operation [2] on the microstructure. Then, a local maxima extraction operation is performed on the distance map to store equidistant points from two opposite walls of the solid phase. After the withdrawal of the rest of the points, disks of $r_i = label_{maxima_i}$, or spheres in the case of 3D modelling, are drawn around points $maxima_i$. Included elements are labeled by the same maxima label. If the center of a disk overlap with another disk of different label, the lowest value disk is removed along with its maxima point. The process of the creation of disks is repeated here again on the filtered maxima points. At last, a geodesic distance transform with source propagation is applied to generate pores. This procedure can be generalized to complex microstructures. Related results are discussed in Sect. 4.2.

3.2   GM-tortuosity

Building upon the formalism of the *M-tortuosity*, our steps proceed very much in the same way as [3]. The latter introduces a deterministic *M-tortuosity* descriptor, which is not practical since it implies the computation of geometric tortuosities of all the points composing the studied phase. This results in a very important computation time. An estimator was proposed as an alternative. The latter starts by either considering the whole pore space or only the points on the medial axis obtained from a skeletonization procedure (the used algorithm can be found in [10]). This operation is followed by a sampling strategy on the resulting pore space/skeleton. Geometric tortuosities can be computed on the basis of the sampled points. This scheme involves the computation of several geodesic distance transforms starting from source points to assess the overall tortuosity. In our method, we revised the way geodesic distances are computed by transposing this computation process to a shortest path algorithm on a graph. Let $I \colon \mathbb{Z}^3 \to \{0,1\}$ be a binary volume on a digitized $3D$ euclidean space. $I = S \cup P$ can be seen as the union of the solid S and the pore phase P. After the application of the *PNP* method, the modified pore phase $\hat{P}$ can be written as a set of distinct pores $\hat{P} = \{p_i\}_{i \in [0, N-1]}$, where N is the total number of pores.

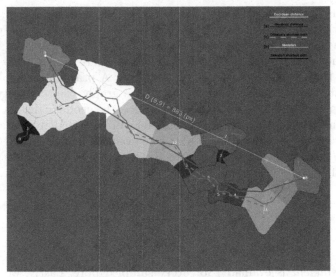

(a) Euclidean and geodesic paths between pores

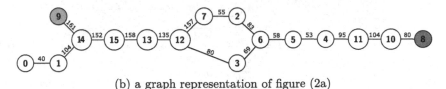

(b) a graph representation of figure (2a)

Fig. 2. Graph representation of pores and their adjacency of the two-phase scheme Fig. 1a.

Following the guidelines of the original *M-tortuosity*, a uniform sampling strategy is applied to the $\hat{P}$ set. The sampled points are extracted by a uniform distribution $U([0, N-1])$ allowing random selection of pores. Considering p_i and p_j as two pores localized by their center of inertia $c_{p_i} = \{x_{p_i}, y_{p_i}, z_{p_i}\}$ and $c_{p_j} = \{x_{p_j}, y_{p_j}, z_{p_j}\}$, respectively. The geometric tortuosity between the aforementioned pores can be defined as:

$$\tau_{i,j} = \frac{D_G(c_{p_i}, c_{p_j})}{D_E(c_{p_i}, c_{p_j})} \tag{3}$$

where $D_G(c_{p_i}, c_{p_j})$ and $D_E(c_{p_i}, c_{p_j})$ denote geodesic and euclidean distance between p_i and p_j, respectively. The computation of the euclidean distance is straightforward:

$$D_E(c_{p_i}, c_{p_j}) = \sqrt{(x_{p_i} - x_{p_j})^2 + (y_{p_i} - y_{p_j})^2 + (z_{p_i} - z_{p_j})^2} \tag{4}$$

The geodesic distance is computed through the Djikstra's algorithm. For that, a graph of the size $|\hat{P}|$ is constructed. The nodes of the graph are the pores. To establish the adjacency matrix, the neighbors of each pore are identified, and their euclidean distances from the source pore are stored. Two pores are neighbors if they have a common contact surface. A practical example is given in Fig. 2. After transforming the porosity into a graph, the tortuosity between two distant points can be estimated. For instance, the geodesic distance between p_8 and p_9 in the graph is defined as the sum of intermediate distances. i.e., $D_G(p_8, p_9) = 1145$. Therefore $\tau_{8,9} = 1.29$. The evaluation of the global tortuosity requires the computation of the *GM-coefficient*. Given N sampled pores, for $n \in [0, N - 1]$:

$$\hat{C}_n^{-1} = \frac{\sum\limits_{m=0, m \neq n}^{N-1} \frac{1}{D_G(p_m, p_n)}}{\sum\limits_{m=0, m \neq n}^{N-1} \frac{1}{D_G(p_m, p_n) \cdot \tau_{m,n}}} \tag{5}$$

The *GM-coefficient* is evaluated for every sampled pore. Lastly, we recall the definition of the *GM-tortuosity*:

$$GM = \frac{\sum\limits_{n=0}^{N-1} \frac{1}{D_E(p_n, c)}}{\sum\limits_{n=0}^{N-1} \frac{1}{D_E(p_n, c) \cdot \hat{C}_n^{-1}}} \tag{6}$$

where c is the center of inertia of the porous phase. The computation of $\hat{\tau}_M$ is the final step of the *GM-tortuosity* procedure. The definitions given above are all based on the original work [3]. The following pseudo code summarizes the aforementioned steps (cf. Algorithm 1).

GM-tortuosity Properties. Consider two points of the porosity phase $(x, y) \in P^2$. We can express the distance between the two points by

$$d(x, y) = \inf_{\Gamma \in P_{x,y}} \int_0^{\ell(\Gamma)} ds \tag{7}$$

where Γ is a path parameterized by its arc length $s \in [0, \ell(\Gamma)]$ and $P_{x,y}$ is the set of all differentiable paths. Assuming that P is a convex space, $\forall (x, y) \in P^2$ $\exists! \Gamma$ such that $d(x, y) = d_E(x, y)$. However, this property is not valid for concave spaces, and $d(x, y)$ may be expressed as

$$d_M(x, y) = \inf\left(\sum_{i=0}^{M} d_E(x_i, y_i)\right) \quad \Gamma_{x_i, y_i} \subset P \tag{8}$$

where $x_0 = x$ and $y_M = B$. In this case, $d_M(x, y)$ is the geodesic distance between points x and y, which is defined as the minimum distance formed by the sum over intermediate paths Γ_{x_i, y_i}. We denote the total geodesic path Γ_M.

Algorithm 1. *GM-tortuosity* estimation

$I \leftarrow$ Binary volume
Computation of the *PNP* function:
$\hat{I} \leftarrow (I \setminus P) \cup \hat{P}$
Construct graph $g(|\hat{P}|)$
Construct adjacency list:
for $i \in [0, N-1]$ **do**
    for $j \in [0, N-1]$ **do**
        if $p_i \cap p_j \neq 0$ **then**
            store $(p_i, p_j, D_E(c_{p_i}, c_{p_j}))$
        end if
    end for
end for
Computation of sampled pores $\hat{P}_s = \{p_k\}_{k \in [0, N-1]}$
for $p_n \in \hat{P}_s$ **do**
    Computation of shortest paths between sampled pores and source pore p_n:
    Geodesic distances $D_G(p_n, \cdot)$
    for $p_m \in \hat{P}_s$ with $p_m \neq p_n$ **do**
        Computation of geometric tortuosity   ▷ Eq. (3)
    end for
    Computation of *GM-Coefficient*   ▷ Eq. (5)
end for
Computation of *GM-tortuosity*   ▷ Eq. (6)

Considering the way the total geodesic path Γ_{GM} is calculated using the *GM-tortuosity*, it can be seen as a graph path. i.e., as a sequence $[x, x_1, x_2, x_3, ..., y]$ such that (x, x_1), (x_1, x_2),..., (x_{M-1}, y) are the edges, or intermediate paths, of the graph and the x_i are distinct pore inertia centers. Notice that, Γ_{GM}, is one of the possible paths given in Eq. (8), whereas Γ_M is by definition the shortest possible path. Therefore, we have the property:

$$d_M(x, y) \leq d_{GM}(x, y) \quad \forall (x, y) \in P^2 \tag{9}$$

such that $d_{GM}(x, y)$ is the geodesic distance formed by the edges of the graph.

Equation (6) provides a direct relationship between the geodesic distance and the *M/GM-tortuosity* descriptors, therefore their values can be predicted from d_M and d_{GM}. From Eq. (9), the following property can be derived:

$$\forall P \subset I \quad GM(P) \geq M(P) \tag{10}$$

4   Experiments and Results

4.1   Morphological Analysis of *GM-tortuosity*

Some of the implications of the morphological aspects of *GM-tortuosity* can be understood from Fig. 2. Provided that the difference between the two methods lies in the way the geodesic distance is calculated, Fig. 2 distinguishes between three different approaches in the assessment of the path between point 9 and point 8. First, the geodesic distance transform propagation between the two endpoints, in the domain constrained by the porosity phase, will produce the shortest geodesic path between the two points (path a). Second, the propagation of the same geodesic distance in a much more restricted domain, which is that of the morphological skeleton of the pore phase of width = 1 pixel, will produce path b. In this case, the propagation is constrained by the morphology of the skeleton, which alters the shortest path by forcing the algorithm to pass through unnecessary skeleton points. The latter corresponds to the longest geodesic distance between the two points. These two approaches may be combined with the *M-tortuosity* computation constrained by the domain boundaries (Porosity phase or skeleton). On the other hand, the procedure proposed in this work suggests constructing a path by linking only the centers of inertia of the intermediate partitions between the two endpoints (path c). To justify this choice, we refer to the concept of the morphological skeleton [14]. The latter is formed along the centers of balls of maximum radius that can be contained between two points of the same domain. For instance, as explained above concerning the *PNP* method, the maxima filtered points correspond to the maximum values of the distance map after a filtering operation. These maxima points, which will subsequently be the centers of inertia of the partitions, can be approached by the centers of balls, having the maximum radius and contained in the same phase. Therefore, they can be considered as sampled points of the skeleton. According to this reasoning, *GM-tortuosity* produces intermediate distances in terms of values between the calculation of geodesic distance in the whole porosity space and the one based in the skeleton. To validate this proposal, the *M-tortuosity* was calculated by considering the whole porosity and skeleton domain in Fig. 2. These values are then compared with the result found by our approach. The following results are obtained: *M-tortuosity* = 1,041, *M-tortuosity* in the skeleton = 1,21 and *GM-tortuosity* = 1.049. Our method overestimates the values of one approach and underestimates the ones of the other.

4.2   Stochastic Models of Complex Microstructures

We evaluate our method on multi-scales Cox Boolean models [11,13]. This modeling process is based on the intrinsic structure of the microstructures of porous materials allowing to model grains by elements, such as spheres or platelets. The model is characterized by the volume fraction of grains inside the matrix for 1-scale models, and by the volume fraction of grains inside and outside the aggregates, modeled by spheres, for 2-scales models. These different levels of size hierarchy define the scales of porosity of the material. For this study, we are interested in two cases. The first one, defining a one scale model, requires the definition of the size of particles as well as their volume fraction within the matrix. For the two scale model, the parameters of the aggregates must be defined in addition to those of the particles. In particular, we distinguish between R the radius of spheres, and l, h, t the length, height and thickness of the platelets. V_v the volume fraction of particles inside the matrix in the case of a 1-scale model.

V_{inc} denotes the volume fraction of the aggregates of size R_{inc}, $V_{v_{in}}$ and $V_{v_{out}}$ are the volume fractions of the grains inside and outside the aggregates, respectively. For comparisons with the *M-tortuosity* original results, we generate four microstructures, 1-scale and 2-scales models of spheres and platelets. We fix the dimensions of the spheres and the platelets, $R = 10$, $l = 10$, $h = 5$, $t = 5$ for all microstructures. $V_v = 0.4$ and $V_v = 0.3$ are fixed for the 1-scale model of spheres and platelets, respectively. $R_{inc} = 30$, $V_{v_{inc}} = 0.5$, $V_{v_{in}} = 0.3$ and $V_{v_{out}} = 0.5$ are the parameters for the 2-scales model of spheres. The 2-scales model of platelets is characterized by: $R_{inc} = 30$, $V_{v_{inc}} = 0.4$, $V_{v_{in}} = 0.3$ and $V_{v_{out}} = 0.7$. Microstructures are simulated within a box of 100^3 voxels. All microstructures volumes were generated using "plug im!" (2020) [1]. The calculation of *GM-tortuosity* is made available in the plug im! platform.

Fig. 3. Stochastic boolean models representing mono/multi scales microstructures. (a) and (b) are 1-scale microstructures of spheres and platelets, respectively. (c) and (d) are multi-scale microstructures of spheres and platelets, respectively. The parameters of these microstructures are given in Subsect. 4.2.

4.3   Results

Time Complexity. First, the time complexities of both methods are examined. Second, the mean computation time of the two methods on the examples from Fig. 3 is given. Lastly, additional example of computation time on a bigger microstructure of the size 300^3 voxels is provided.

As specified above, the generation of the pore network by means of the PNP method requires the use of two distance transforms. For both operations, a raster scanning algorithm was used. The latter only requires two image/volume scans that are forward and backward raster scans. Assuming that the total number of pixels/voxels is n, the PNP method requires $2(n) + 2(n)$ operations, which leads us to a linear time complexity $O(n)$. On the other hand, the implementation of the Djikstra's shortest path is based on a priority queue algorithm. Assuming that there are v vertices, or partitions, present in the graph constructed from the PNP output, the queue contains $O(v)$ vertices and each pop operation takes a time $O(\log v)$. Therefore, the time complexity needed to execute the main loop is linearithmic $O(v \log(v))$. In practice, n denotes millions of pixels/voxels whether v is of the order of thousands of elements. The M-tortuosity implementation used for comparison with our results is based on a raster-scanning algorithm as well [3]. Provided that the algorithm performs a sampling strategy over the porosity space followed by a geodesic distance transform operation, the time needed for the computation is $N \times O(n)$, where N is the number of the sampled points. The algorithm requires a linear time complexity $O(n)$. We conclude that the time complexity in the case of the original M-tortuosity algorithm exceeds significantly the time complexity of the GM-tortuosity algorithm.

The *M-tortuosity* and *GM-tortuosity* simulations were conducted on a (CPU: i7 2.6 GHz, RAM: 16 GB) personal computer. In the provided benchmark, N is taken equal to 64. Given the experiments in Fig. 3, the mean *M-tortuosity* computation time for a given microstructure is about 58 s, whereas for the *GM-tortuosity*, the mean computation time marks a very important decrease to reach 5 s. Another example is a rendered 1-scale boolean model of spheres of the size 300^3 voxels, which is illustrated in Fig. 4 as well as its pore network, generated by the *PNP* method. The computation time for the original and our method are 1832 s and 107 s, respectively. It is specified that the recorded 107 s includes the computation time of the *PNP* method as well. Concerning *GM-tortuosity*, computation time is highly governed by *PNP* construction. For M-tortuosity, the number of points has a strong impact on performance, while it is negligible for *GM-tortuosity*.

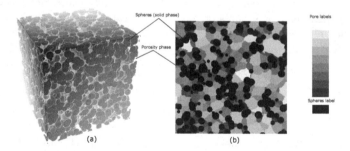

Fig. 4. (a) Rendered 1-scale boolean model of spheres of the size 300^3 voxels. (b) its corresponding pore network for one plane of the microstructure.

GM-tortuosity. Based on the generated microstructures illustrated in Subsect. 4.2, a comparison is conducted between the results of the *M-tortuosity* original estimator, available in [1], and the results of our *GM-tortuosity* estimator. The main advantage of the *M-tortuosity* method lies in its discriminating ability between microstructures. *GM-tortuosity* should produce higher tortuosity result as the graph construction strategy forces the passage through each center of inertia of the pores, and consequently increases the estimated geodesic distance. We need to check the consistency between the results of the two methods in the sense that the values of our method must follow the same behavior as those of the original method. Thus, the maximum and minimum values must match and the rest of the results should follow the same logic. The validation of this proposition would imply that the *GM-tortuosity* is a reliable descriptor in distinguishing between microstructures based on their geometric tortuosities. Considering 64 uniformly sampled points, or pores in our method, Table 1 depicts the mean geometric tortuosities related to *M-tortuosity* and *GM-tortuosity* for four simulations of four types of microstructures (Fig. 3: (a), (b), (c), (d)), evaluated in the pore and solid phase. The 2-scales model of platelets has the highest coefficient for both methods. Similarly, the 1-scale model of platelets presents the lowest coefficient for both methods. In addition, the results related to the evaluation of the *M-tortuosity* on the skeleton, given in Table 1, proves the validity of our proposal in Sect. 4.1. Indeed, the *GM-tortuosity* underestimates/overestimates the results of *M-tortuosity* evaluated in the skeleton/microstructure, respectively.

Table 1. *M-tortuosity* and *GM-tortuosity* mean values of four models of every microstructure shown in Fig. 3.

Model	1-S spheres (a)	2-S spheres (c)	1-S platelets (b)	2-S platelets (d)
Tortuosity in pore phase				
M-tortuosity	1.022	1.021	1.015	1.087
M-tortuosity (Skeleton)	1.202	1.215	1.175	1.211
GM-tortuosity	1.195	1.180	1.127	1.204
Tortuosity in solid phase				
M-tortuosity	1.33	1.32	1.26	1.05
M-tortuosity (Skeleton)	1.53	1.52	1.71	1.22
GM-tortuosity	1.45	1.41	1.39	1.17

5 Conclusion

Besides having a fast computation time, it was shown that the *GM-tortuosity* estimator provides reliable results for the given examples, allowing to preserve the discriminating aspect of the original descriptor. It also has the advantage of being able to use a very large number of source points with no major impact on calculation times. Further studies will have to be carried out concerning the comparison of the two descriptors, in particular as a function of pore morphology, affecting more or less strongly the estimation of geodesic distance on the graph.

References

1. "plug im!" an open access and customizable software for signal and image processing (2020). https://www.plugim.fr
2. Borgefors, G.: Distance transformations in digital images. Comput. Vis. Graph. Image Process. **34**(3), 344–371 (1986). https://doi.org/10.1016/S0734-189X(86)80047-0. http://www.sciencedirect.com/science/article/pii/S0734189X8 6800470

3. Chaniot, J., Moreaud, M., Sorbier, L., Fournel, T., Becker, J.M.: Tortuosimetric operator for complex porous media characterization. Image Anal. Stereol. **38**(1), 25–41 (2019). https://doi.org/10.5566/ias.2039. https://www.ias-iss.org/ojs/IAS/article/view/2039

4. Cormen, T.H., Leiserson, C.E., Rivest, R.L., Stein, C.: Introduction to Algorithms, 3rd edn. The MIT Press, Cambridge (2009)

5. Criminisi, A., Sharp, T., Rother, C., P'erez, P.: Geodesic image and video editing. ACM Trans. Graph. **29**(5) (2010). https://doi.org/10.1145/1857907.1857910

6. Ghanbarian, B., Hunt, A.G., Ewing, R.P., Sahimi, M.: Tortuosity in porous media: a critical review. Soil Sci. Soc. Am. J. **77**(5), 1461–1477 (2013). https://doi.org/10.2136/sssaj2012.0435. https://acsess.onlinelibrary.wiley.com/doi/abs/10.2136/sssaj2012.0435

7. Hammoumi, A., Moreaud, M., Jolimaitre, E., Chevalier, T., Novikov, A., Klotz, M.: Efficient pore network extraction method based on the distance transform. In: Masrour, T., El Hassani, I., Cherrafi, A. (eds.) A2IA 2020. LNNS, vol. 144, pp. 1–13. Springer, Cham (2021). https://doi.org/10.1007/978-3-030-53970-2_1

8. Holzer, L., et al.: The influence of constrictivity on the effective transport properties of porous layers in electrolysis and fuel cells. J. Mater. Sci. **48**(7), 2934–2952 (2013). https://doi.org/10.1007/s10853-012-6968-z

9. Keller, L.M., Holzer, L., Wepf, R., Gasser, P.: 3D geometry and topology of pore pathways in Opalinus clay: implications for mass transport. Appl. Clay Sci. **52**(1), 85–95 (2011). https://doi.org/10.1016/j.clay.2011.02.003. http://www.sciencedirect.com/science/article/pii/S0169131711000573

10. Lee, T., Kashyap, R., Chu, C.: Building skeleton models via 3-D medial surface axis thinning algorithms. CVGIP: Graph. Models Image Process. **56**(6), 462–478 (1994). https://doi.org/10.1006/cgip.1994.1042. http://www.sciencedirect.com/science/article/pii/S104996528471042X

11. Matheron, G.: Random Sets and Integral Geometry. Wiley, New York (1974)

12. Moldrup, P., Olesen, T., Komatsu, T., Schjønning, P., Rolston, D.: Tortuosity, diffusivity, and permeability in the soil liquid and gaseous phases. Soil Sci. Soc. Am. J. **65**(3), 613–623 (2001). https://doi.org/10.2136/sssaj2001.653613x. https://acsess.onlinelibrary.wiley.com/doi/abs/10.2136/sssaj2001.653613x

13. Moreaud, M., Chaniot, J., Fournel, T., Becker, J.M., Sorbier, L.: Multi-scale stochastic morphological models for 3D complex microstructures. In: 2018 17th Workshop on Information Optics (WIO), pp. 1–3 (2018). https://doi.org/10.1109/WIO.2018.8643455

14. Serra, J.: Image Analysis and Mathematical Morphology. Academic Press, Inc., Cambridge (1983)

15. Xiong, Q., Baychev, T.G., Jivkov, A.P.: Review of pore network modelling of porous media: experimental characterisations, network constructions and applications to reactive transport. J. Contam. Hydrol. **192**, 101–117 (2016). https://doi.org/10.1016/j.jconhyd.2016.07.002. http://www.sciencedirect.com/science/article/pii/S016977221630122X

A Maximum-Flow Model for Digital Elastica Shape Optimization

Daniel Martins Antunes[1]([✉]) [ID], Jacques-Olivier Lachaud[1] [ID],
and Hugues Talbot[2] [ID]

[1] Université Savoie Mont-Blanc, Le Bourget-du-lac, France
{daniel.martins-antunes,jacques-olivier.lachaud}@univ-smb.fr
[2] CentraleSupelec, Université Paris-Saclay, Inria, Gif-sur-Yvette, France
hugues.talbot@centralesupelec.fr

Abstract. The Elastica is a curve regularization model that integrates
the squared curvature in addition to the curve length. It has been shown
to be useful for contour smoothing and interpolation, for example in the
presence of thin elements.

In this article, we propose a graph-cut based model for optimizing the
discrete Elastica energy using a fast and efficient graph-cut model. Even
though the Elastica energy is neither convex nor sub-modular, we show
that the final shape we achieve is often very close to the globally optimal
one.

Our model easily adapts to image segmentation tasks. We show that
compared to previous works and state-of-the-art algorithms, our proposal
is simpler to implement, faster, and yields comparable or better results.

Keywords: Multigrid convergence · Digital estimators · Curvature ·
Discrete optimization

1 Introduction

The Elastica model and associated energy is a curve regularisation problem with
a rich history [12,15], which involves the squared curvature. It was introduced in
computer vision in [16] as a means to regularize edges or segmentation boundaries with an explicit curvature component. Being associated with second-order
derivatives, the notion of curvature is difficult to compute and optimize in a
regularisation framework.

Multiple attempts have been made by prominent researchers to introduce
an explicit curvature term in curve regularizers using various methods and
approaches. Even early active contour models [10] had an elasticity component
equivalent to a notion of curvature. Similarly, with level-set methods, local curvature can readily be estimated as in [13]. However these approaches typically
optimise a non-convex local energy by gradient descent, and most were non-
geometric, i.e. discretization-dependent. In contrast [5] proposed a geometric
level-set segmentation method, but using only the perimeter as a regularizer, and

© Springer Nature Switzerland AG 2021
J. Lindblad et al. (Eds.): DGMM 2021, LNCS 12708, pp. 429–440, 2021.
https://doi.org/10.1007/978-3-030-76657-3_31

was still non-convex. All the numerous approaches based on total-variation mini-mization for image restoration or segmentation [19] and even the Mumford-Shah functional [17] use this quite elementary regularizer, probably because optimiz-ing it in an exact manner was already a challenge for a long time [2,4,6], whether in the discrete or continuous cases.

In more recent works, optimizing the Elastica, which is a geometric regular-izer, has seen renewed interest. In [14], authors successfully use a computational geometry approach to perform image restoration with the Elastica as regular-izer. In [7], authors compute an approximate discrete version of the Elastica and optimize it with discrete calculus. In [18], an efficient, discrete approxi-mation of curvature is optimized with a specific solver using local submodular approximations [8]. However, in these works, the poor quality of the curvature approximation may limit accuracy and hence the quality of the result.

In a previous work [1] we proposed to formulate a digital flow that approxi-mates an Elastica-related flow using a multigrid-convergent curvature estimator, within a discrete variational framework. We also presented an application of this model as a post-processing step to a segmentation framework.

In this work, we propose a novel approach that still uses a multigrid-convergent estimation of curvature, but we optimize using a maximal flow algo-rithm.

1.1 Multigrid Convergent Estimators

Geometric measurements in digital objects can be tricky. Intuitively, a good estimator should converge to its continuous counterpart value as the grid res-olution is refined. The criterion that formalizes this intuition is the multigrid convergence property.

Definition 1 *(Multigrid convergence). Let $\mathcal{F}$ a family of shapes in the plane and Q a global measurement (e.g., perimeter, area) on members of $\mathcal{F}$. Additionally, denote $D_h(S)$ a digitization of shape S in a digital grid of resolution h. The estimator $\hat{Q}$ of Q is* multigrid convergent *for the family $\mathcal{F}$ if and only if for every shape $S \in \mathcal{F}$, there exists $h_S > 0$ such that*

$$\forall h \leq h_S, \quad |\hat{Q}(D_h(S), h) - Q(S)| \leq \tau_S(h),$$

where $\tau_S : \mathbb{R}_+ \setminus \{0\} \to \mathbb{R}_+$ tends to zero as h tends to zero and is the speed of convergence *of $\hat{Q}$ towards Q for S.*

Tangent and curvature are examples of local properties computed along the boundary of some shape S in the plane. We need a slightly different definition of multigrid convergence in order to map points of the Euclidean boundary to those in the digital contour.

Definition 2 *(Multigrid convergence for local geometric quantities). Let $\mathcal{F}$ a family of shapes in the plane and Q a local measurement along the boundary ∂S of $S \in \mathcal{F}$. Additionally, denote $D_h(S)$ a digitization of S in a digital grid* *of resolution h and $\partial_h S$ its digital contour. The estimator $\hat{Q}$ of Q is* multigrid convergent *for the family $\mathcal{F}$ if and only if for every shape $S \in \mathcal{F}$, there exists $h_S > 0$ such that the estimate $\hat{Q}(D_h(S), p, h)$ is defined for all $p \in \partial_h S$ with $0 < h < h_S$, and for any $x \in \partial S$,*

$$\forall p \in \partial_h S \text{ with } \|p - x\|_\infty \leq h, \quad |\hat{Q}(D_h(S), p, h) - Q(S, x)| \leq \tau_S(h),$$

where $\tau_S : \mathbb{R}_+ \setminus \{0\} \to \mathbb{R}_+$ has null limit at 0. This function defines the speed of convergence *of $\hat{Q}$ towards Q for S.*

We now recall the notion of Elastica, as well as its digital counterpart.

1.2 Digital Elastica

Letting κ be the curvature function along some Euclidean shape $S \subset \mathbb{R}^2$. The elastica energy of S with parameters $\boldsymbol{\theta} = (\alpha \geq 0, \beta \geq 0)$ is defined as

$$E_\theta(S) = \int_{\partial S} \alpha + \beta \kappa(s)^2 ds.$$

Similarly, the digital elastica $\hat{E}_\theta$ of some digitization $D_h(S)$ of S is defined as

$$\hat{E}_\theta(D_h(S)) = \sum_{\dot{e} \in \partial_h S} \hat{s}(\dot{e}) \left(\alpha + \beta \hat{\kappa}^2(D_h(S), \dot{e}, h) \right), \tag{1}$$

where $\dot{e}$ denotes the center of the linel e and the estimators of length $\hat{s}$ and curvature $\hat{\kappa}$ are multigrid convergent.

1.3 A Multigrid-Convergent Estimation of Curvature

Let S be an arbitrary shape. The following definition yields a curvature estimation at every point of its boundary.

Definition 3 *(Integral Invariant Curvature Estimator). Let $D_h(S)$ a digitization of $S \subset \mathbb{R}^2$ and $B_r(p)$ the Euclidean disk of radius r centered at p. The integral invariant curvature estimator is defined for every point $p \in \partial_h S$ as*

$$\hat{\kappa}_r(D_h(S), p, h) := \frac{3}{r^3} \left(\frac{\pi r^2}{2} - \widehat{Area}\left(D_h(B_r(p)) \cap D_h(S), h\right) \right). \tag{2}$$

where $\widehat{Area}(D, h)$ estimates the area of D by counting its grid points and then scaling them by h^2. This estimator is multigrid convergent for the family of compact shapes in the plane with 3-smooth boundary. It converges with speed $O(h^{\frac{1}{3}})$ for radii chosen as $r = \Theta(h^{\frac{1}{3}})$ [11].

In the expression above, we will substitute an arbitrary subset D of $\mathbb{Z}^2$ to $D_h(S)$ since the continuous shape S is unknown. In the following we omit the grid step h to simplify expressions (or, putting it differently, we assume that the shape of interest is rescaled by $1/h$ and we set $h = 1$).

2   Digital Elastica Minimization via Graph Cuts

In this section we present a graph cut model that converges to the optimum digital shape under digital elastica regularization. Moreover, the model is easily adapted to image segmentation tasks.

2.1   Balance Coefficient

In the core of the model is the notion of balance coefficient. We are going to extend Eq. 2 to the whole digital domain. In fact, since we are more interested in the balance of intersected and non-intersected points, we slightly change Eq. 2 and give it another name. We define the *balance coefficient* as

$$u_r(D, p) = \left(\frac{\pi r^2}{2} - \widehat{\mathrm{Area}}(B_r(p) \cap D) \right)^2 .$$

The balance coefficient at p gives us as an *approximation* of the new squared curvature value when the shape is perturbed a little around p. Therefore, it is reasonable to evolve the shape towards the zero level set of the balance coefficient function (see Fig. 1).

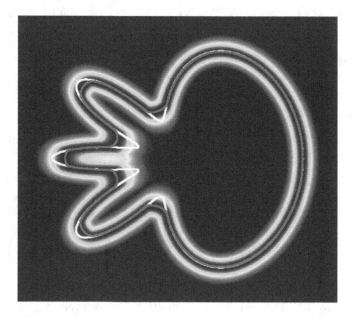

Fig. 1. Balance coefficient zero level set. We notice that evolving the initial contour (colored in white) to the zero level set of the balance coefficient (colored in magenta) regularizes the shape with respect to the squared curvature. (Color figure online)

2.2   Model Overview

Let $D^{(0)}$ be the initial digital set. The GraphFlow model produces a sequence of digital shapes $D^{(k)}$ by executing two main steps:

Candidate selection: We associate to $D^{(k)}$ a set of neighbor shapes $\mathcal{P}(D^{(k)})$. For each $D' \in \mathcal{P}(D^{(k)})$ we construct the capacitated graph $\mathcal{G}_{D'}$ such that its minimum cut Q minimizes the candidate function.

$$cand(D) = data(D) + \sum_{p \in D'} \sum_{q \in \mathcal{N}_4(p)} u_r(D, p) + u_r(D, q). \tag{3}$$

The proposition $mincut(Q, G)$ indicates that Q is a minimum cut of G.
Validation: Each minimum cut Q computed in the previous step induces a solution candidate D_Q. We group the solution candidates in the set $sol(D^{(k)})$ and we choose the one that minimizes the validation energy

$$val(D) = data(D) + \hat{E}_\theta(D). \tag{4}$$

The candidate function is computed on a band around the contour of $D^{(k)}$. Let $d_D : \Omega \to \mathbb{R}$ be the signed Euclidean distance transformation with respect to shape D. The value $d_D(p)$ gives the Euclidean distance between $p \notin D$ and the closest point in D. For points $p \in D$, $d_D(p)$ gives the negative of the distance between p and the closest point not in D.

Definition 4 *(Optimization band). Let $D \subset \Omega \subset \mathbb{Z}^2$ a digital set and natural number $n > 0$. The optimization band $O_n(D)$ is defined as*

$$O_n(D) := \{p \in \Omega \mid -n \leq d_D(p) \leq n\}.$$

We use as the neighborhood of shapes the a-probe set

Definition 5 *(a-probe set). Let $D \subset \Omega \subset \mathbb{Z}^2$ a digital set and a some natural number. The a-probe set of D is defined as*

$$\mathcal{P}_a(D) = D \cup \bigcup_{a' < a} D^{+a'} \cup D^{-a'},$$

where D^{+a} (D^{-a}) denotes a dilation(erosion) by a disk of radius a.

2.3   Shape Optimization

We are going to evolve the initial contour ∂D of some digital shape D to the zero level set of its balance coefficient by computing the minimum cut of the candidate graphs. In this experiment, the data term in both candidate and validation functions equal to zero.

Definition 6 *Candidate graph. Let $D \subset \Omega \subset \mathbb{Z}^2$ a digital set and natural number $n > 0$. We define $\mathcal{G}_D(\mathcal{V}, \mathcal{E}, c)$ as the candidate graph of D with optimization band n such that*

$$\mathcal{V} = \{\, v_p \mid p \in O_n(D) \,\} \cup \{s, t\}$$
$$\mathcal{E} = \{\, \{v_p, v_q\} \mid p \in O_n(D) \text{ and } q \in \mathcal{N}_4(p) \,\} \cup \mathcal{E}_{st}$$
$$\mathcal{E}_{st} = \{\, \{s, v_p\} \mid d_D(p) = -n \,\} \cup \{\, \{v_p, t\} \mid d_D(p) = n \,\}.$$

The vertices s, t are virtual vertices representing the source and target vertices as it is usual in a minimum cut framework. In particular, after the minimum cut is computed, vertices connected to the source will define the new digital shape. The innermost (resp. outermost) pixels of the optimization band are connected to the source (resp. target), and we identify such vertices as

$$\mathcal{V}_s := \{v_p \in \Omega \mid d_D(p) = -n\}$$
$$\mathcal{V}_t := \{v_p \in \Omega \mid d_D(p) = n\}.$$

The set $\mathcal{E}_{st}$ comprises all the edges having the source as their starting point or the target as their endpoint. Next, we describe how to set the edges' capacities.

Edge e	$c(e)$	For
$\{v_p, v_q\}$	$u_r(D, p) + u_r(D, q)$	$\{v_p, v_q\} \in \mathcal{E}_u$
$\{v_p, s\}$	M	$v_p \in \mathcal{V}_s$
$\{v_p, t\}$	M	$v_p \in \mathcal{V}_t$

Here M is twice the highest value of the balance coefficient plus one, i.e.

$$M = 1 + \max_{p \in O_n(D)} 2 * u_r(D, p).$$

2.4   Image Segmentation

The GraphFlow model is suitable for image processing tasks. We present our experiments using the data term employed by Boykov-Jolly (BJ) graph cut model described in [3]. Let $x \in [0, 1]^{|D|}$ represent the label of each pixel (0 for background and 1 for foreground). Then, we define the data term as

$$data(D) = \gamma_r \sum_{p \in D} \psi_p(x_p) + \gamma_b \sum_{p \in D'} \sum_{q \in \mathcal{N}_4(p)} \psi_{p,q}(x_p, x_q),$$

where $\gamma_r \geq 0$ and $\gamma_b \geq 0$ are parameters controlling the influence of the data and space coherence terms, respectively. Given the image $I : \Omega \to [0, 1]^3$, the unary and pairwise terms are defined as

$$\psi_p(x_p) = \begin{cases} -\ln H_{bg}\big(I(p)\big), & \text{if } x_p = 0 \\ -\ln H_{fg}\big(I(p)\big), & \text{if } x_p = 1, \end{cases}$$

$$\psi_{p,q}(x_p, x_q) = \begin{cases} \exp\left(-\dfrac{1}{|(p,q)|}\dfrac{(I(p)-I(q))^2}{2\sigma^2}\right), & q \in \mathcal{N}_4(p) \\ 0, & \text{otherwise.} \end{cases}$$

The terms H_{bg} and H_{fg} are mixed Gaussian distribution constructed from foreground and background seeds given by the user.

2.5 Elastica GraphFlow Algorithm

The GraphFlow algorithm implements a local-search strategy to minimize (4) with a search space given by the solution of the candidates set defined in the previous section. We choose to let the model flow even in the case where the next shape $D^{(k+1)}$ has a higher energy than the previous shape $D^{(k)}$. It is a simple strategy to escape local minima. In the implementation presented here, the only stopping condition is the number of iterations. This strategy could be tailored to a specific application.

---

input : An image I or a digital set D; the optimization band n; the probe set
parameter a; parameter vector $\boldsymbol{\theta} = (\alpha, \beta)$; parameter vector
$\boldsymbol{\gamma} = (\gamma_r, \gamma_b)$; the maximum number of iterations maxIt;

if *Image I is given* **then**
| $D^{(0)} \longleftarrow graphcut_B J(I)$;
end
else
| $D^{(0)} \longleftarrow D$;
| $(\gamma_r, \gamma_b) \longleftarrow (0,0)$;
end

$k \longleftarrow 0$;
while $k <$ maxIt **do**
| //Candidate selection
| $sol(D^{(k)}) \longleftarrow \bigcup_{D' \in \mathcal{P}_a(D^{(k)})} \big\{ (Q, D_Q) \mid mincut(Q, \mathcal{G}_{D'}) \big\}$;
| //Candidate validation
| $D^{(k+1)} \longleftarrow \underset{(Q,S)\in sol(D^{(k)})}{\arg\min}\ data(S) + \hat{E}_{\boldsymbol{\theta}}(S)$;
| $k \longleftarrow k+1$;
end

Algorithm 1: GraphFlow algorithm.

The Elastica GraphFlow algorithm has two fundamental steps. In the candidate selection, we build the solution of the candidates set from the minimum cuts of the candidate graphs. Next, in the validation step, we choose the digital set with minimum value for (4). If we interpret the balance coefficient minimization as the best move one can make towards digital elastica minimization, the solution of the candidates set can be seen as the neighboring shapes with highest potential to minimize the elastica energy for the given a-probe set.

Some results of Algorithm 1 are shown in the next section.

3   Results and Discussion

We first present some of our own results, then some comparison with our previous work and with the reference implementation of [20].

3.1   Results

The GraphFlow algorithm produces a flow that is in accordance with expectations for a flow guided by the elastica energy. In particular, it grows and shrinks in accordance with the α coefficient in the digital elastica (see Fig. 2) for a-probe sets such that $a > 0$. If we use a 0-probe set, we recover a flow similar to the curve-shortening flow [9].

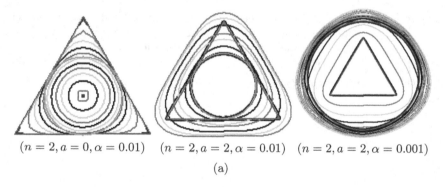

$(n = 2, a = 0, \alpha = 0.01)$ $(n = 2, a = 2, \alpha = 0.01)$ $(n = 2, a = 2, \alpha = 0.001)$

(a)

Fig. 2. GraphFlow results. The GraphFlow algorithm can shrink and grow in accordance with length penalization and it converges to a shape close to the theoretical global optimum (green curve) in the free elastica problem. We are using $n = 2, a = 2$ and shapes are displayed every 10 iterations. (Color figure online)

The solutions are very similar to those achieved in [1], but with the advantage of producing smoother flows and up to 100× faster. In Fig. 3, we show that our algorithm can easily be used for segmentation, and that it produces results that are parametrically smoother than those of the BJ model.

In Fig. 4, we can observe that the Elastica GraphFlow presents the completion property, i.e., it tends to return a segmentation with fewer disconnected components.

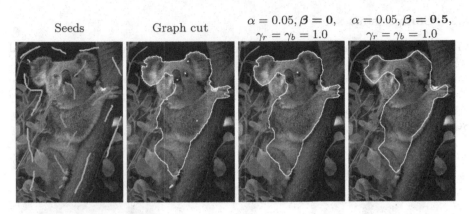

Fig. 3. GraphFlow segmentation. Given foreground (green) and background (gray) seeds in picture (a); Graph cut produces picture (b) which is used as input of the Graph-Flow algorithm; in pictures (c) and (d) we display the output of our Elastica GraphFlow algorithm with and without the squared curvature term in the regularization. (Color figure online)

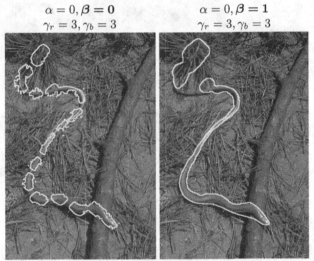

Fig. 4. GraphFlow and completion property. The oversegmented picture on the left was obtained with no squared curvature regularization, while the picture on the right was obtained by setting $\beta = 1.0$. (Color figure online)

Table 1. Running time for the image segmentation problem.

Exp-comparison running time			
Model	Minimum	Maximum	Average
SLCR	2.87 min	52.24 min	18.4 min
Previous work	60 s	297 s	156 s
This work	11 s	150 s	75 s

3.2   Comparisons

In this section we compare our proposed results for segmentation with several state-of-the-art results in Fig. 5: the reference BJ segmentation method; the linear relaxation methods by Schoeneman *et al.*; and our previous results. We observe that our results are smoother than BJ, as in the previous figure. They are very similar to our previous results, and much better than the reference SCLR method from literature.

Fig. 5. Segmentation results comparison. Top row, Graph-Cut BJ results; second row: references SCLR results; third row: previous results from [1]; last row: proposed method.

All the experiments were executed on a 32-core 2.4 GHz CPU. According to Table 1, our proposed method is significantly faster than our previous work, and much faster than the reference SLCR method.

4   Conclusion

In this article, we described a graph cut model for optimizing the Elastica energy that is suitable for both discrete curve evolution and image segmentation. The evolution produced by the Elastica GraphFlow responds to the length penalization term α, i.e., the shape tends to grow (shrink) for lower (higher) values of α and to the curvature penalization β. Furthermore we observe a convergence to a shape that appears close to the expected, theoretical global optimum of the Elastica model in the cases where this globally optimal curve can be computed. Our Elastica GraphFlow algorithm is significantly faster and simpler to implement than the previous models presented in the literature.

In our method, we need to use a family of shapes (a-probes). In future work, we plan to use a dynamic family with the help a parameter-free estimator. In the same vein, a multi-resolution approach would be very useful, in particular for dealing with very thin portions of the image. We also note that a globally optimal solution with multigrid convergent estimators is yet to be proposed.

References

1. Antunes, D., Lachaud, J.O., Talbot, H.: An elastica-driven digital curve evolution model for image segmentation. J. Math. Imaging Vis. **63**, 1–17 (2021)
2. Appleton, B., Talbot, H.: Globally minimal surfaces by continuous maximal flows. IEEE Trans. Pattern Anal. Mach. Intell. **28**(1), 106–118 (2005)
3. Boykov, Y.Y., Jolly, M.P.: Interactive graph cuts for optimal boundary & region segmentation of objects in N-D images. In: Proceedings Eighth IEEE International Conference on Computer Vision, ICCV 2001, vol. 1, pp. 105–112 (2001)
4. Boykov, Y., Veksler, O., Zabih, R.: Fast approximate energy minimization via graph cuts. IEEE Trans. Pattern Anal. Mach. Intell. **23**(11), 1222–1239 (2001)
5. Caselles, V., Kimmel, R., Sapiro, G.: Geodesic active contours. Int. J. Comput. Vision **22**(1), 61–79 (1997)
6. Chambolle, A.: An algorithm for total variation minimization and applications. J. Math. Imaging Vis. **20**(1), 89–97 (2004)
7. El-Zehiry, N.Y., Grady, L.: Fast global optimization of curvature. In: 2010 IEEE Computer Society Conference on Computer Vision and Pattern Recognition, pp. 3257–3264 (2010)
8. Gorelick, L., Veksler, O., Boykov, Y., Ben Ayed, I., Delong, A.: Local submodular approximations for binary pairwise energies. In: Computer Vision and Pattern Recognition, vol. 1, p. 4 (2014)
9. Huisken, G., et al.: Flow by mean curvature of convex surfaces into spheres. J. Differ. Geom. **20**(1), 237–266 (1984)
10. Kass, M., Witkin, A., Terzopoulos, D.: Snakes: active contour models. Int. J. Comput. Vision **1**, 321–331 (1988)

11. Lachaud, J.-O., Coeurjolly, D., Levallois, J.: Robust and convergent curvature and normal estimators with digital integral invariants. In: Najman, L., Romon, P. (eds.) Modern Approaches to Discrete Curvature. LNM, vol. 2184, pp. 293–348. Springer, Cham (2017). https://doi.org/10.1007/978-3-319-58002-9_9
12. Levien, R.: The elastica: a mathematical history. Electrical Engineering and Computer Sciences University of California at Berkeley (2008)
13. Malladi, R., Sethian, J.A., Vemuri, B.C.: Shape modeling with front propagation: a level set approach. IEEE Trans. Pattern Anal. Mach. Intell. **17**(2), 158–175 (1995)
14. Masnou, S., Morel, J.M.: Level lines based disocclusion. In: Proceedings 1998 International Conference on Image Processing, ICIP 1998 (Cat. No.98CB36269), vol. 3, pp. 259–263 (1998)
15. Matsutani, S., et al.: Euler's elastica and beyond. J. Geom. Symmetry Phys. **17**, 45–86 (2010)
16. Mumford, D.: Elastica and computer vision. In: Bajaj, C.L. (ed.) Algebraic Geometry and Its Applications, pp. 491–506. Springer, New York (1994). https://doi.org/10.1007/978-1-4612-2628-4_31
17. Mumford, D., Shah, J.: Optimal approximation by piecewise smooth functions and associated variational problems. Commun. Pure Appl. Math. **42**(5), 577–685 (1989)
18. Nieuwenhuis, C., Toeppe, E., Gorelick, L., Veksler, O., Boykov, Y.: Efficient squared curvature. In: 2014 IEEE Conference on Computer Vision and Pattern Recognition, pp. 4098–4105 (2014)
19. Rudin, L.I., Osher, S., Fatemi, E.: Nonlinear total variation based noise removal algorithms. Physica D **60**(1–4), 259–268 (1992)
20. Schoenemann, T., Kahl, F., Cremers, D.: Curvature regularity for region-based image segmentation and inpainting: a linear programming relaxation. In: 2009 IEEE 12th International Conference on Computer Vision, pp. 17–23 (2009)

Image Segmentation by Relaxed Deep Extreme Cut with Connected Extreme Points

Débora E. C. Oliveira, Caio L. Demario, and Paulo A.V. Miranda(✉)(iD)

Institute of Mathematics and Statistics, University of São Paulo,
CEP 05508–090, São Paulo, SP, Brazil
{emilli.costa,caiolopes,pmiranda}@ime.usp.br

Abstract. In this work, we propose a hybrid method for image segmentation based on the selection of four extreme points (leftmost, rightmost, top and bottom pixels at the object boundary), combining Deep Extreme Cut, a connectivity constraint for the extreme points, a marker-based color classifier from automatically estimated markers and a final relaxation procedure with the boundary polarity constraint, which is related to the extension of Random Walks to directed graphs as proposed by Singaraju et al. Its second constituent element presents theoretical contributions on how to optimally convert the 4 point boundary-based selection into connected region-based markers for image segmentation. The proposed method is able to correct imperfections from Deep Extreme Cut, leading to considerably improved results, in public datasets of natural images, with minimal user intervention (only four mouse clicks).

Keywords: Deep Extreme Cut · Connected extreme points · Iterative relaxation

1 Introduction

Image segmentation is one of the most fundamental and challenging problems in image processing and computer vision. It is the task of partitioning an input image into objects of interest by assigning distinct labels to their composing pixels [2]. Hence, image segmentation can be interpreted as a classification problem at the pixel level. Not surprisingly, machine learning-based methods are among the most prominent solutions to the segmentation problem, especially after the advent of deep learning techniques [13]. In this context, a state-of-the-art method in grabcut-style [25] for interactive segmentation with minimal user involvement is Deep Extreme Cut [16]. It considers extreme points (leftmost, rightmost, top and bottom pixels at the object boundary) as 4 clicks for each object (Fig. 1a), the CNN produces a probability map for the object (Fig. 1b), and the final segmented mask is obtained by thresholding (Figs. 1c and d). The CNN was trained

Thanks to CNPq (308985/2015-0, 313554/2018-8, 465446/2014-0), CAPES (88887.136422/2017-00) and FAPESP (2014/12236-1, 2014/50937-1, 2016/21591-5).

© Springer Nature Switzerland AG 2021
J. Lindblad et al. (Eds.): DGMM 2021, LNCS 12708, pp. 441–453, 2021.
https://doi.org/10.1007/978-3-030-76657-3_32

to minimize the standard cross entropy loss, using the ResNet-101 [13] as the backbone of its architecture, removing the fully connected layers as well as the max pooling layers in the last two stages to preserve acceptable output resolution for dense prediction [16].

The user input based on extreme points is usually a good choice for single-shot algorithms, since from the 4 indicated points we can infer object's bounding box, and consequently, all points outside the box are known to belong to the background. However, considerable residual errors all around the boundary are quite common in Deep Extreme Cut (Figs. 1e and f). The interactive correction of all these residual errors spread along the entire boundary in order to generate a truly correct segmentation result would require a very large interaction effort on the part of the user, reducing or even completely annulling the method's gains. In order to improve the results of the method in comparison benchmarks, normally a high threshold is considered to eliminate false positives, but this can result in the loss of important object's parts for some images (e.g., the left wing in Figs. 1c and d). Note that there is no single threshold that works correctly for all parts of the object.

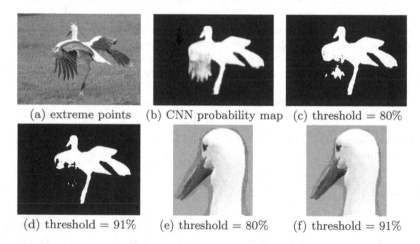

(a) extreme points   (b) CNN probability map   (c) threshold = 80%

(d) threshold = 91%   (e) threshold = 80%   (f) threshold = 91%

Fig. 1. Example of segmentation of a natural image by Deep Extreme Cut. (a) The selection of extreme points with 4 clicks. (b) The probability map for the object produced by the CNN with weights fine-tuned on SBD + PASCAL train [8,12]. (c) The result by Deep Extreme Cut with its default threshold value of 80%. (d) The result by Deep Extreme Cut with threshold value of 91% as required for optimal performance on the GrabCut dataset. (e-f) Notice, that Deep Extreme Cut has poor boundary adherence.

In this work, we propose a hybrid method for image segmentation, based on the selection of four extreme points, by a combination of the following parts:

Deep Extreme Cut [16] (Fig. 1), a connectivity constraint for the extreme points motivated by the good results presented in [17,18] (Fig. 2a), a marker-based color classifier [21] from automatically estimated markers (Fig. 2b), and a final relaxation procedure with the boundary polarity constraint [6], which is related to the extension of Random Walks to directed graphs as proposed by Singaraju et al. [26] (Figs. 2c and d).

After computing Deep Extreme Cut for the selected extreme points, we optimally convert the 4 point boundary-based selection into connected region-based markers for image segmentation with a strong theoretical basis (Figs. 2a and b). This allows us to use the arc-weight estimation based on markers from [21] to enhance the object's boundary. We then combine the object membership map from [21] with the Deep Extreme Cut result to use as a starting point for the final refinement by the relaxation procedure from [6]. The proposed method is able to correct imperfections from Deep Extreme Cut, leading to considerably improved results (Figs. 2c and d), in public datasets of natural images, with minimal user intervention (only four mouse clicks).

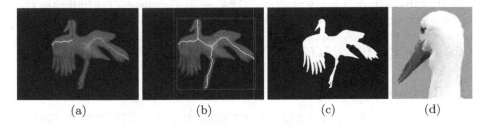

     (a)   (b)   (c)   (d)

Fig. 2. The result of segmentation of a natural image by the proposed hybrid method. (a) The optimal conversion of the 4 point boundary-based selection into a connected region-based marker for the object. (b) The final internal (yellow) and external (blue) region-based markers. (c-d) The segmentation result by the proposed method with 4 clicks. Note the improvements in the boundary adherence compared to Fig. 1 (Colour figure online).

The next section presents the required notations and the relaxation procedure from [6]. The proposed method is then presented in Sect. 3. The experiments and conclusions are shown in Sects. 4 and 5, respectively.

2 Background

A 2D image can be interpreted as a weighted digraph $G = \langle \mathcal{V}, \mathcal{A}, \omega \rangle$, whose vertices $\mathcal{V} \subset \mathbb{Z}^2$ are the image pixels in its image domain and whose arcs are the ordered pairs $\langle s, t \rangle \in \mathcal{A}$ of neighboring pixel, according to the 8-neighborhood. The set of adjacent nodes to s is denoted by $\mathcal{N}(s) = \{t \in \mathcal{V} : \langle s, t \rangle \in \mathcal{A}\}$. We use $\omega_{st} \geq 0$ as a short for $\omega(\langle s, t \rangle)$, where $\omega : \mathcal{A} \to \mathbb{R}$ is the arc-weight function. The digraph G is symmetric if for any of its arcs $\langle s, t \rangle \in \mathcal{A}$, the pair $\langle t, s \rangle$ is also an arc of G, but we can have $\omega_{st} \neq \omega_{ts}$.

A *path (in G)* of length $\ell \geq 0$ is a sequence $\pi = \langle v_0, \ldots, v_\ell \rangle$ of adjacent vertices (*i.e.*, $\langle v_j, v_{j+1} \rangle \in \mathcal{A}$ for any $j < \ell$) with no repeated vertices ($v_i \neq v_j$ for $i \neq j$). The set of vertices along a path π can be indicated as $\mathcal{V}(\pi) = \{v_0, v_1, \ldots, v_\ell\}$. For a given path $\pi = \langle v_0, \ldots, v_\ell \rangle$, we use the notation $(\pi)^{-1} = \langle v_\ell, \ldots, v_0 \rangle$ to indicate the reverse path. A path $\pi_t = \langle v_0, \ldots, v_\ell \rangle$ is a path with *terminus* at a node $t = v_\ell$. When we want to explicitly indicate the origin of the path, the notation $\pi_{s \rightsquigarrow t} = \langle v_0, \ldots, v_\ell \rangle$ may also be used, where $v_0 = s$ stands for the origin and $v_\ell = t$ for the destination node. A path is *trivial* when $\pi_t = \langle t \rangle$. A path $\pi_t = \pi_s \cdot \langle s, t \rangle$ indicates the extension of a path π_s by an arc $\langle s, t \rangle$ and, more generally, $\pi_{r \rightsquigarrow s} \cdot \pi_{s \rightsquigarrow t} = \langle r, \ldots, s, \ldots, t \rangle$ indicates the concatenation of two paths. $\Pi(G)$ indicates the set of all possible paths in a graph G. A *connectivity function* $f : \Pi(G) \to \mathbb{R}$ computes a cost value $f(\pi_t)$ for any path π_t, usually based on arc weights. A path π_t is *optimum* if $f(\pi_t) \leq f(\tau_t)$ for any other path $\tau_t \in \Pi(G)$. We use $\mathring{\pi}_t$ to indicate an optimum path.

For the sake of simplicity, we will focus on object/background segmentation, as indicated by two seed sets $\mathcal{S}_o \subset \mathcal{V}$ and $\mathcal{S}_b \subset \mathcal{V}$, respectively. It can be a crisp mapping $\mathcal{L} : \mathcal{V} \to \{0, 1\}$, that assigns to each vertex a label (1 for object and 0 for background), or a fuzzy mapping $\mathcal{L} : \mathcal{V} \to [0, 1]$, where $\mathcal{L}(s)$ indicates the object membership degree of s. Both cases are subject to $\mathcal{L}(s) = 1$ for $s \in \mathcal{S}_o$ and $\mathcal{L}(s) = 0$ for $s \in \mathcal{S}_b$. Notice that, a crisp $\mathcal{L}$ is the *characteristic function* χ_P of the subset $P = \{s \in \mathcal{V} : \mathcal{L}(s) = 1\}$ of $\mathcal{V}$.

2.1   Relaxation Procedure for Directed Graphs

As proposed by Demario and Miranda [6], extending the work of Malmberg et al. [15], for an initial computed segmentation $\mathcal{L}^0$, a sequence of fuzzy segmentations $\mathcal{L}^1, \mathcal{L}^2, \ldots, \mathcal{L}^N$ by iterative relaxation is obtained as follows:

$$\mathcal{L}^{i+1}(s) = \begin{cases} \frac{\sum_{t \in \mathcal{N}(s)} W^i(s,t) \cdot \mathcal{L}^i(t)}{\sum_{t \in \mathcal{N}(s)} W^i(s,t)} & \text{if } s \notin \mathcal{S}_o \cup \mathcal{S}_b \\ \mathcal{L}^i(s) & \text{otherwise} \end{cases} \tag{1}$$

where the values of $W^i(s, t)$ reflect the behavior of diodes:

$$W^i(s, t) = \begin{cases} \omega_{st} & \text{if } \mathcal{L}^i(s) \geq \mathcal{L}^i(t) \\ \omega_{ts} & \text{if } \mathcal{L}^i(s) < \mathcal{L}^i(t) \end{cases} \tag{2}$$

The final crisp $\mathcal{L}$ is then obtained, by assigning $\mathcal{L}(s) = 1$ to all $s \in \mathcal{V}$ with $\mathcal{L}^N(s) \geq 0.5$ and $\mathcal{L}(s) = 0$ otherwise. In [15], $\mathcal{L}^0$ is taken as IFT [9] with the additive path cost [5], while Oriented Image Foresting Transform (OIFT) [22] is used as $\mathcal{L}^0$ for [6].

This solution $\mathcal{L}^N$ converges to an extension of Random Walks (RW) [10] to directed graphs, as proposed by Singaraju et al. [26], for sufficiently high values of N, which minimizes the following energy function:

$$\mathcal{F}(\mathcal{L}) = \sum_{\langle s,t \rangle \in \mathcal{A}} [\omega_{st} \mathcal{I}(\mathcal{L}_s \geq \mathcal{L}_t) + \omega_{ts} \mathcal{I}(\mathcal{L}_s < \mathcal{L}_t)] \cdot (\mathcal{L}_s - \mathcal{L}_t)^2 \tag{3}$$

where $\mathcal{L}$ is a fuzzy segmentation satisfying the seeds, $\mathcal{L}_s$ is a short for $\mathcal{L}(s)$, and $\mathcal{I}(A)$ is an indicator function of the event A, such that $\mathcal{I}(A) = 1$, if A is true and $\mathcal{I}(A) = 0$ otherwise.

For limited values of N, we get a lower amount of relaxation leading to a hybrid result. In [6,15], the hybrid approach is used to fix contour irregularities from the original methods, improving the perceived quality of the segmentation results, by increasing the circularity[1] and accuracy in relation to OIFT and IFT.

In this work, we consider the second version presented in [6], which is inspired by the cost propagation of OIFT, where a background node t (i.e., $\mathcal{L}^i(t) < 0.5$) processes ω_{st} of anti-parallel (reverse) arcs to its neighbor nodes s, by setting:

$$W^i(s,t) = \begin{cases} \omega_{st} & \text{if } \mathcal{L}^i(t) < 0.5 \\ \omega_{ts} & \text{if } \mathcal{L}^i(t) \geq 0.5 \end{cases} \tag{4}$$

3   Relaxed Deep Extreme Cut with Connected Points

Now, let's address the problem of how to connect the four extreme points. Since extreme points are selected over the boundary of the desired object, they correspond to a particular kind of anchor point, as used by boundary tracking approaches [20]. We will denote these anchor points by the set $A = \{a_1, a_2, \ldots, a_k\}$. In this work, we are interested in the case where $k = 4$.

Consider an energy map $E : \mathcal{V} \to \mathbb{R}$, such that high values of $E(t)$ indicate pixels most likely to belong to the object of interest. Thus, a good connection between the anchor points should pass through the regions with the highest values on the map E and should be composed of as few as possible pixels, in order to minimize the risk of marking as object's seed some background pixels.

For $X \subseteq \mathcal{V}$, $0 \leq E(t) < E_{max}$ and $\bar{E}(t) = E_{max} - E(t)$, let $\varepsilon(X) = \sum_{t \in X} \bar{E}(t)$. We want to find a set $T \subseteq \mathcal{V}$, subject to the constraints that $\{a_1, a_2, \ldots, a_k\} \subseteq T$ and T is a 8-connected component, such that $\varepsilon(T)$ is minimum. In other words, let $R = \{X \subseteq \mathcal{V} : A \subseteq X \wedge X \text{ is 8-connected component}\}$ and $\varepsilon_{min} = \min_{X \in R} \varepsilon(X)$. We want to find a set $T \in R$, such that $\varepsilon(T) = \varepsilon_{min}$.

For a set $A = \{a_1\}$ with a single anchor point ($k = 1$), we have the trivial solution $T = \{a_1\}$. For the case $A = \{a_1, a_2\}$ with $k = 2$, the solution can be obtained by computing an optimum path $\mathring{\pi}_{a_1 \leadsto a_2}$ in G with 8-neighborhood with the following connectivity function f_i with $i = 1$:

$$f_i(\langle t \rangle) = \begin{cases} \bar{E}(t) & \text{if } t = a_i \\ +\infty & \text{otherwise} \end{cases}$$

$$f_i(\pi_s \cdot \langle s, t \rangle) = f_i(\pi_s) + \bar{E}(t) \tag{5}$$

The solution is then obtained as $T = \mathcal{V}(\mathring{\pi}_{a_1 \leadsto a_2})$. Note that $f_1(\mathring{\pi}_{a_1 \leadsto a_2}) = \varepsilon(T)$. The optimum path $\mathring{\pi}_{a_1 \leadsto a_2}$ can be efficiently computed by using the Image Foresting Transform [5].

---

[1] Circularity was measured by the isoperimetric quotient. That is, the ratio of the object area to the area of a circle with the same perimeter.

The case $A = \{a_1, a_2, a_3\}$ with $k = 3$ is fairly more complex. We can conclude that T, in this case, can be described as the union of three paths $\pi_{a_1 \rightsquigarrow p}$, $\pi_{a_2 \rightsquigarrow p}$ and $\pi_{a_3 \rightsquigarrow p}$ with a junction point at a vertex p, as shown in Fig. 3a (that is, $T = \bigcup_{i=1}^3 \mathcal{V}(\pi_{a_i \rightsquigarrow p})$ and $\mathcal{V}(\pi_{a_i \rightsquigarrow p}) \cap \mathcal{V}(\pi_{a_j \rightsquigarrow p}) = \{p\}$ for $i \neq j$). The problem is that p and the paths are unknown. Note that we have $\varepsilon_{min} = \left[\sum_{i=1}^3 f_i(\pi_{a_i \rightsquigarrow p})\right] - 2 \cdot \bar{E}(p)$.

Proposition 1. *In the following, let $\mathring{\pi}_{a_i \rightsquigarrow x}$ be an optimum path by the cost function f_i from a_i to any pixel x. Let $\hat{\varepsilon}(p') = \left[\sum_{i=1}^3 f_i(\mathring{\pi}_{a_i \rightsquigarrow p'})\right] - 2 \cdot \bar{E}(p')$ for $p' \in V$. The junction point for $k = 3$ can be computed as $p^* = \arg\min_{p' \in V} \hat{\varepsilon}(p')$ and $T^* = \bigcup_{i=1}^3 \mathcal{V}(\mathring{\pi}_{a_i \rightsquigarrow p^*})$ is optimal. That is, $\varepsilon(T^*) = \varepsilon_{min}$.*

Proof. Let $\tilde{R} = \{X \in R : X \text{ is minimal with property } X \in R\}$ and $T(p') \in \tilde{R}$ be the best tree with junction point at p' (i.e., $\varepsilon(T(p'))$ is lower than or equal to $\varepsilon(X)$ for any other tree $X \in \tilde{R}$ with junction point at p') and let $T(a_i, p')$ be the path interconnecting a_i and p' in the tree $T(p')$.

By the definition of optimum paths, we have that $f_i(T(a_i, p')) \geq f_i(\mathring{\pi}_{a_i \rightsquigarrow p'})$ for $i = 1, \ldots, 3$. Consequently, we have that:

$$\varepsilon(T(p')) = \left[\sum_{i=1}^3 f_i(T(a_i, p'))\right] - 2 \cdot \bar{E}(p') \geq \left[\sum_{i=1}^3 f_i(\mathring{\pi}_{a_i \rightsquigarrow p'})\right] - 2 \cdot \bar{E}(p') = \hat{\varepsilon}(p') \tag{6}$$

Therefore:

$$\varepsilon(T(p')) \geq \hat{\varepsilon}(p') \tag{7}$$

We have two cases:

Case (1): If $\mathcal{V}(\mathring{\pi}_{a_i \rightsquigarrow p'}) \cap \mathcal{V}(\mathring{\pi}_{a_j \rightsquigarrow p'}) = \{p'\}$ for $i \neq j$, then $\bigcup_{i=1}^3 \mathcal{V}(\mathring{\pi}_{a_i \rightsquigarrow p'})$ forms a tree in $\tilde{R}$ with junction point at p'. But since $T(p')$ is the best tree by definition, we have $\varepsilon(T(p')) \leq \varepsilon(\bigcup_{i=1}^3 \mathcal{V}(\mathring{\pi}_{a_i \rightsquigarrow p'})) = \hat{\varepsilon}(p')$. Finally, combining with Eq. 7, we obtain $\varepsilon(T(p')) = \hat{\varepsilon}(p')$.

Case (2): If $\mathcal{V}(\mathring{\pi}_{a_i \rightsquigarrow p'}) \cap \mathcal{V}(\mathring{\pi}_{a_j \rightsquigarrow p'}) \neq \{p'\}$ for $i \neq j$, then the paths intersect at more than one point. Let $q' \neq p'$ be the first crossing point between $\mathring{\pi}_{a_i \rightsquigarrow p'}$ and $\mathring{\pi}_{a_j \rightsquigarrow p'}$ with $i \neq j$. Therefore, we have:

$$\mathring{\pi}_{a_i \rightsquigarrow p'} = \langle a_i, \ldots, q' \ldots, p' \rangle = \mathring{\pi}_{a_i \rightsquigarrow q'} \cdot \tau_{q' \rightsquigarrow p'} \tag{8}$$

$$\mathring{\pi}_{a_j \rightsquigarrow p'} = \langle a_j, \ldots, q' \ldots, p' \rangle = \mathring{\pi}_{a_j \rightsquigarrow q'} \cdot \rho_{q' \rightsquigarrow p'} \tag{9}$$

From the optimality of the paths and the additive nature of the cost function f_i, we have that the costs added by the suffixes $\tau_{q' \rightsquigarrow p'}$ and $\rho_{q' \rightsquigarrow p'}$ are the same, that is:

$$\Delta = \varepsilon(\mathcal{V}(\tau_{q' \rightsquigarrow p'})) = \varepsilon(\mathcal{V}(\rho_{q' \rightsquigarrow p'})) \tag{10}$$

Therefore, in the computation of $\hat{\varepsilon}(p')$, we have that the value $\Delta - \bar{E}(p')$ is counted twice. So in that case we may conclude that: $\varepsilon_{min} \leq \varepsilon(T(q')) \leq f_i(\mathring{\pi}_{a_i \rightsquigarrow q'}) + f_j(\mathring{\pi}_{a_j \rightsquigarrow q'}) + f_k(\mathring{\pi}_{a_k \rightsquigarrow p'} \cdot (\tau_{q' \rightsquigarrow p'})^{-1}) - 2 \cdot \bar{E}(q') < \hat{\varepsilon}(p')$. Thus, we have $\varepsilon_{min} < \hat{\varepsilon}(p')$. Combining with Eq. 7 we finally have: $\varepsilon_{min} < \hat{\varepsilon}(p') \leq \varepsilon(T(p'))$.

Since for an optimal solution $T(p)$ at pixel p, we have $\varepsilon_{min} = \varepsilon(T(p))$, therefore we can conclude that **Case (2)** can only occur for $p' \neq p$.

Therefore, for $p' = p$, we have the first case and consequently we have $\varepsilon_{min} = \varepsilon(T(p)) = \hat{\varepsilon}(p)$, which corresponds to an optimal solution.

Given that in Proposition 1 the best solution (with minimum $\hat{\varepsilon}(p')$) is selected among all possible values of p' and given that for $p' = p$ we have an optimal solution, so we have that $\varepsilon(T^*) = \varepsilon_{min}$, as we wanted to prove.

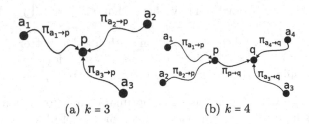

(a) $k = 3$ (b) $k = 4$

Fig. 3. Illustration of the expected connection behavior of the anchor points for an optimal solution. (a) The case for $k = 3$. Note that this definition already includes the particular cases where the junction point p coincides with one of the anchor points. (b) The case for $k = 4$.

Now let's handle the case where $A = \{a_1, a_2, a_3, a_4\}$ with $k = 4$. For $k = 4$, we can always re-enumerate the anchor points so that the optimal solution takes the form of the scheme shown in Fig. 3b. We look for the junction points p and q such that $T = \left[\bigcup_{i=1}^{2} \mathcal{V}(\pi_{a_i \rightsquigarrow p}) \right] \cup \left[\bigcup_{i=3}^{4} \mathcal{V}(\pi_{a_i \rightsquigarrow q}) \right] \cup \mathcal{V}(\pi_{p \rightsquigarrow q})$. So we have:

$$\varepsilon(T) = \varepsilon_{min} = \sum_{i=1}^{2} \varepsilon(\mathcal{V}(\pi_{a_i \rightsquigarrow p})) + \sum_{i=3}^{4} \varepsilon(\mathcal{V}(\pi_{a_i \rightsquigarrow q})) + \varepsilon(\mathcal{V}(\pi_{p \rightsquigarrow q})) - 2 \cdot \bar{E}(p) - 2 \cdot \bar{E}(q) \tag{11}$$

Consider the following algorithm:

Algorithm 1. CONNECTION OF ANCHOR POINTS FOR $k = 4$

INPUT: Image graph $G = \langle \mathcal{V}, \mathcal{A}, \omega \rangle$, energy map $\bar{E}$ and set of anchor points $A = \{a_1, a_2, a_3, a_4\}$.
OUTPUT: Connected set of object seeds T_c.

1. *Set $e^{\bullet} \leftarrow \infty$.*
2. **For each $p' \in \mathcal{V}$, do**
3. | *Compute $T' = T^*$ as stated in Proposition 1, for $k = 3$ taking as anchor points $\{p', a_3, a_4\}$.*
4. | $e' = \varepsilon(T') + \left[\sum_{i=1}^{2} f_i(\hat{\pi}_{a_i \rightsquigarrow p'}) \right] - 2 \cdot \bar{E}(p')$.
5. | **If $e' < e^{\bullet}$ then**
6. | | *Set $e^{\bullet} \leftarrow e'$.*
7. | | *Set $p^{\bullet} \leftarrow p'$.*
8. | | *Set $T^{\bullet} \leftarrow T'$.*
9. *Set $T_c \leftarrow T^{\bullet} \cup \mathcal{V}(\hat{\pi}_{a_1 \rightsquigarrow p^{\bullet}}) \cup \mathcal{V}(\hat{\pi}_{a_2 \rightsquigarrow p^{\bullet}})$.*
10. *Return T_c.*

Proposition 2. *The set $\mathcal{T}_c$ computed by Alg. 1 is optimal. That is, $\varepsilon(\mathcal{T}_c) = \varepsilon_{min}$.*

Proof. For $p' = p$, we have that $C' = \bigcup_{i=1}^{2} \mathcal{V}(\mathring{\pi}_{a_i \rightsquigarrow p})$ is an optimal solution to the connection problem of the points $\{a_1, a_2, p\}$ by arguments similar to those presented for the proof of Proposition 1 for $k = 3$. Therefore, $\varepsilon(C') \leq \varepsilon(\bigcup_{i=1}^{2} \mathcal{V}(\pi_{a_i \rightsquigarrow p}))$.

$\mathcal{T}'$ in Line 3 of Algorithm 1 is an optimal solution to the connection problem of the points $\{p, a_3, a_4\}$ by Proposition 1. Hence, $\varepsilon(\mathcal{T}') \leq \varepsilon(\mathcal{V}(\pi_{p \rightsquigarrow q}) \cup \mathcal{V}(\pi_{a_3 \rightsquigarrow q}) \cup \mathcal{V}(\pi_{a_4 \rightsquigarrow q}))$. Therefore, from Eq. 11 we have that $\varepsilon_{min} = \varepsilon(\bigcup_{i=1}^{2} \mathcal{V}(\pi_{a_i \rightsquigarrow p})) + \varepsilon(\mathcal{V}(\pi_{p \rightsquigarrow q}) \cup \mathcal{V}(\pi_{a_3 \rightsquigarrow q}) \cup \mathcal{V}(\pi_{a_4 \rightsquigarrow q})) - \bar{E}(p) \geq \varepsilon(C') + \varepsilon(\mathcal{T}') - \bar{E}(p)$. Due to the impossibility of the existence of a solution with a value better than ε_{min}, we have by exclusion that $e' = \varepsilon(C') + \varepsilon(\mathcal{T}') - \bar{E}(p) = \varepsilon_{min}$ for $p' = p$. To conclude the proof it is sufficient to note that $e'(p') \geq \varepsilon_{min}$ for $p' \neq p$.

Although Algorithm 1 guarantees an optimal connection result, its computational cost is high with a computational complexity of $O(n^2 \cdot \log n)$, where n is the number of pixels. In order to have a more efficient algorithm for practical use, we adopt a strategy similar to that used by the α-expansion algorithm for the approximate energy minimization with Graph Cuts [1]. We propose an iterative algorithm by fixing at each iteration one of the junction points (p or q) and calculating the other via a minimization by Proposition 1 for the case with $k = 3$. This process is alternately repeated until convergence or until it reaches a maximum iteration limit. In practice, we noticed that this algorithm generated optimal results in all tested cases, being considerably more efficient, with complexity $O(n \cdot \log n)$. This solution is adopted in the experiments.

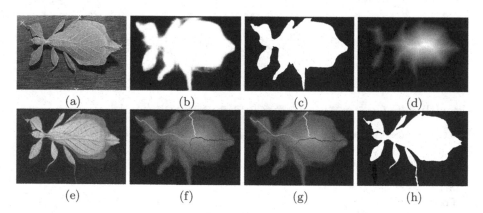

Fig. 4. The segmentation result of a natural image by the proposed method. (a) The extreme points with 4 clicks. (b) The probability map for the object produced by the CNN. (c) The result by Deep Extreme Cut. (d) Euclidean distance transform (EDT) of the segmentation by Deep Extreme Cut from (c). (e) The mean color distance of each pixel in Lab space in relation to the five nearest pixels in $\mathcal{S}_b$. (f) The optimal connection for $\beta = 3.5$ over the map $E(t)$. Note that the red path passes incorrectly through the background. (g) A better optimal connection is obtained for $\beta = 7.0$ over the map $E(t)$. (h) The final segmentation using the markers from (g). (Colour figure online)

The energy map is computed as $E(t) = W(t)^\beta$, where $W(t)$ is the weighted mean of the probability map for the object produced by the CNN (Figs. 4a-b), the square value of the Euclidean distance transform (EDT)^2 of the segmentation by Deep Extreme Cut (Figs. 4c-d), and the mean color distance in Lab space of t compared to the five nearest pixels in S_b (Fig. 4e), using the following weights 20%, 30% and 50%, respectively. The power β favors longer paths (Figs. 4f-h). When β tends to infinity, we have that the connection paths between the extreme points converge to the connection by Riverbed [20] as proposed in [17,18]. The seeds in S_b are taken as pixels in an external frame of the bounding box (see Fig. 2b) and S_o is computed as T_c.

The object membership map $\hat{M}(t)$ from [21], using the markers in S_o and S_b, is then computed and its is combined with the segmentation by Deep Extreme Cut $\mathcal{L}_{DEC}(t)$ by a weighted mean, in order to generate a starting point $\mathcal{L}^0$ for the final refinement by the relaxation procedure. $\mathcal{L}^0$ is computed as follows:

$$\mathcal{L}^0(t) = (1 - q_1)q_2 \times \mathcal{L}_{DEC}(t) + (1 - (1 - q_1)q_2) \times \hat{M}(t) \qquad (12)$$

where q_1 and q_2 are measures of the quality of the maps $\hat{M}(t)$ and $\mathcal{L}_{DEC}(t)$, respectively. The measure q_1 is estimated as the square root of the sharpness measure^3 of the membership map $\hat{M}(t)$ and we use $q_2 = \left(\frac{\sum_{s \in S_o} \mathcal{L}_{DEC}(s)}{|S_o|} \right)^3$, which penalizes the result of $\mathcal{L}_{DEC}(t)$ when object seeds pass through its exterior.

4   Experimental Results

In the experiments, the total number of iterations of the relaxation procedure was computed as $N = diag \times 0.04 \times [1 + q_1]$, where $diag$ is the bounding box diagonal size in pixels. The arc-weights were computed similarly to what was proposed in [21] with some adjustments from [6], with the polarity as described in [19] for colored images, basically using $\mathcal{L}^0$ as a reference map to define the orientations. Other issues and minor details can be consulted directly in the source code of the method which is publicly available.^{4}

Regarding the error rate measure^5, Deep Extreme Cut [16] has already been shown to surpass several methods from the literature, including Deconvolution [23], Deep GrabCut [30], MILCut [29], Box Prior [14], Extreme Clicking [24], One Cut [28] and Kernel cut [27]. Our hybrid method has even better results, increasing the gains in relation to all these approaches.

We used four datasets of natural images: The GrabCut dataset of 50 images [25]; the new NatImg21 dataset of 64 natural images; the new BirdsInsects, possessing objects with thin and elongated parts, which is composed of

---

^{2} The squared EDT values are quantized into 256 levels (8 bits) prior to the weighted mean computation.

^{3} The sharpness measure is the complement of fuzziness as defined in [15].

^{4} The source code is available on the website: http://www.vision.ime.usp.br/~pmir anda/downloads.html.

^{5} The error rate is the percentage of misclassified pixels within the bounding boxes.

Table 1. Experimental results.

	Dice coeff.	Jaccard index	Gen. Jac. (GB $\delta = 1.0$)	Error Rate	Dice coeff.	Jaccard index	Gen. Jac. (GB $\delta = 1.0$)	Error Rate
Grabcut(5)	91.01%	85.39%	79.34%	8.09%	86.48%	78.58%	69.40%	11.82%
Grabcut(10)	91.60%	86.29%	80.97%	7.40%	86.68%	78.89%	69.91%	11.59%
Grabcut(20)	91.61%	86.32%	80.98%	7.39%	87.05%	79.37%	70.74%	11.12%
Grabcut(40)	91.65%	86.38%	81.10%	7.34%	87.30%	79.68%	71.24%	10.78%
DEXTR	96.66%	94.36%	94.05%	2.30%	93.38%	88.54%	86.58%	5.16%
Proposed($\beta = 3.5$)	**98.32%**	**96.79%**	**96.54%**	**1.31%**	95.77%	92.34%	90.75%	3.30%
Proposed($\beta = 7.0$)	98.29%	96.74%	96.49%	1.34%	**95.89%**	**92.55%**	**91.01%**	**3.21%**

(a) Grabcut dataset | | | | (b) NatImg21 dataset

	Dice coeff.	Jaccard index	Gen. Jac. (GB $\delta = 1.0$)	Error Rate	Dice coeff.	Jaccard index	Gen. Jac. (GB $\delta = 1.0$)	Error Rate
Grabcut(5)	82.65%	73.69%	63.13%	11.27%	82.36%	73.54%	68.02%	13.38%
Grabcut(10)	83.22%	74.55%	64.99%	10.89%	82.57%	73.88%	68.75%	13.06%
Grabcut(20)	83.58%	75.08%	65.95%	10.64%	82.64%	74.00%	68.92%	12.99%
Grabcut(40)	83.82%	75.41%	66.35%	10.42%	82.67%	74.05%	69.00%	12.95%
DEXTR	92.31%	86.01%	83.77%	3.73%	96.53%	93.73%	93.47%	2.56%
Proposed($\beta = 3.5$)	95.73%	92.00%	91.28%	2.01%	97.29%	94.98%	94.77%	2.04%
Proposed($\beta = 7.0$)	**95.89%**	**92.29%**	**91.61%**	**1.94%**	**97.34%**	**95.03%**	**94.82%**	**2.02%**

(c) BirdsInsects dataset | | | | (d) Geostar dataset

120 images; and the GeoStar dataset from [11], which is composed of 151 images. NatImg21 and BirdsInsects are publicly available to the community.[6]

Besides computing the error rate, we also computed the similarity by Dice Coefficient [7] and the Jaccard metric (ratio of Intersection over Union) between the ground truth and the segmented results for different methods. For the Grabcut method, we used its implementation available in the OpenCV in Python with different numbers of iterations (5,10,20 and 40). For Deep Extreme Cut (DEXTR), we used their pre-computed results, as available in their web page for the GrabCut dataset[7] and, for the other datasets, we used their code in Python as available in the same web page and computed the extreme points automatically from the ground truth images.

We also computed the average accuracy utilizing the General Balanced (GB) metric [4] in Eq. 13, with $\delta = 1.0$, where TP, FP and FN stand for the true positive, false positive, and false negative sets of pixels, respectively and $\mathcal{O}$ is the set of pixels representing the object of interest in the ground-truth image. GB with $\delta = 1.0$ is a balanced and general version of the Jaccard metric [4]. Note that the differences between the proposed method and Deep Extreme Cut were emphasized and became more pronounced in the new measure (Table 1). The gains were less pronounced in the last dataset, as it has a great tolerance for errors around the boundary in its ground truths, hiding the segmentation

---

[6] The datasets are available on the website: http://www.vision.ime.usp.br/~pmiranda/downloads.html.

[7] https://cvlsegmentation.github.io/dextr/.

flaws of the methods. Nevertheless, the proposed method performed better in all tested datasets.

$$\text{GB} = \frac{\delta|\text{TP}|}{\delta|\text{TP}| + |\text{FN}| + \text{FP}_\text{S}}, \qquad \text{FP}_\text{S} = \sum_{i=0}^{+\infty} \frac{|\text{FP}|^{i+1}}{|\mathcal{O}|^i} \qquad (13)$$

5 Conclusion

We have successfully improved Deep Extreme Cut [16], correcting imperfections in its segmented results. This demonstrates the importance of graph optimization methods as a natural complement to machine learning techniques.

Our procedure to connect the extreme points can also be used as a general strategy to convert region-based methods to the grabcut-style. For instance, some preliminary results, not shown in this paper, indicate that Dynamic Trees [3] in grabcut-style can lead to far better results compared to the Grabcut method. Therefore, as future work, we intend to convert region-based methods to the grabcut-style and to adapt our hybrid strategy in other domains, including 3D medical applications.

References

1. Boykov, Y., Veksler, O., Zabih, R.: Fast approximate energy minimization via graph cuts. IEEE Trans. Pattern Anal. Mach. Intell. **23**(11), 1222–1239 (2001)
2. Boykov, Y., Jolly, M.P.: Interactive graph cuts for optimal boundary & region segmentation of objects in N-D images. Int. Conf. Comput. Vision (ICCV) **1**, 105–112 (2001)
3. Bragantini, J., Martins, S.B., Castelo-Fernandez, C., Falcão, A.X.: Graph-based image segmentation using dynamic trees. In: Vera-Rodriguez, R., Fierrez, J., Morales, A. (eds.) CIARP 2018. LNCS, vol. 11401, pp. 470–478. Springer, Cham (2019). https://doi.org/10.1007/978-3-030-13469-3_55
4. Cappabianco, F.A.M., Ribeiro, P.F.O., de Miranda, P.A.V., Udupa, J.K.: A general and balanced region-based metric for evaluating medical image segmentation algorithms. In: 2019 IEEE International Conference on Image Processing (ICIP), pp. 1525–1529, September 2019. https://doi.org/10.1109/ICIP.2019.8803083
5. Ciesielski, K., Falcão, A., Miranda, P.: Path-value functions for which Dijkstra's algorithm returns optimal mapping. J. Math. Imag. Vision **60**(7), 1025–1036 (2018)
6. Demario, C.L., Miranda, P.A.V.: Relaxed oriented image foresting transform for seeded image segmentation. In: 2019 IEEE International Conference on Image Processing (ICIP), pp. 1520–1524 (2019)
7. Dice, L.: Measures of the amount of ecologic association between species. Ecology **26**, 297–302 (1945)
8. Everingham, M., Van Gool, L., Williams, C.K.I., Winn, J., Zisserman, A.: The PASCAL Visual Object Classes Challenge 2012 (VOC2012) Results. http://www.pascal-network.org/challenges/VOC/voc2012/workshop/index.html
9. Falcão, A., Stolfi, J., Lotufo, R.: The image foresting transform: theory, algorithms, and applications. Trans. PAMI **26**(1), 19–29 (2004)

10. Grady, L.: Random walks for image segmentation. Trans. PAMI **28**(11), 1768–1783 (2006)
11. Gulshan, V., Rother, C., Criminisi, A., Blake, A., Zisserman, A.: Geodesic star convexity for interactive image segmentation. In: Proceedings of Computer Vision and Pattern Recognition, pp. 3129–3136 (2010)
12. Hariharan, B., Arbelaez, P., Bourdev, L., Maji, S., Malik, J.: Semantic contours from inverse detectors. In: International Conference on Computer Vision (ICCV) (2011)
13. He, K., Zhang, X., Ren, S., Sun, J.: Deep residual learning for image recognition. In: IEEE Conference on Computer Vision and Pattern Recognition (CVPR), pp. 770–778, June 2016. https://doi.org/10.1109/CVPR.2016.90
14. Lempitsky, V., Kohli, P., Rother, C., Sharp, T.: Image segmentation with a bounding box prior. In: 2009 IEEE 12th International Conference on Computer Vision, pp. 277–284 (2009)
15. Malmberg, F., Nyström, I., Mehnert, A., Engstrom, C., Bengtsson, E.: Relaxed image foresting transforms for interactive volume image segmentation. In: Proceedings of SPIE, vol. 7623, pp. 7623–7623-11 (2010)
16. Maninis, K.K., Caelles, S., Pont-Tuset, J., Gool, L.V.: Deep extreme cut: from extreme points to object segmentation. In: IEEE Conference on Computer Vision and Pattern Recognition, pp. 616–625 (2018)
17. Mansilla, L.A.C., Miranda, P.A.V.: Oriented image foresting transform segmentation: Connectivity constraints with adjustable width. In: 2016 29th SIBGRAPI Conference on Graphics, Patterns and Images (SIBGRAPI), pp. 289–296, October 2016. https://doi.org/10.1109/SIBGRAPI.2016.047
18. Mansilla, L.A.C., Miranda, P.A.V., Cappabianco, F.A.M.: Oriented image foresting transform segmentation with connectivity constraints. In: 2016 IEEE International Conference on Image Processing (ICIP), pp. 2554–2558, September 2016
19. Mansilla, L., Miranda, P.: Image segmentation by oriented image foresting transform: handling ties and colored images. In: 18th International Conference on Digital Signal Processing, pp. 1–6. Santorini, GR, July 2013
20. Miranda, P., Falcão, A., Spina, T.: Riverbed: a novel user-steered image segmentation method based on optimum boundary tracking. IEEE Trans. Image Process. **21**(6), 3042–3052 (2012)
21. Miranda, P., Falcão, A., Udupa, J.: Synergistic arc-weight estimation for interactive image segmentation using graphs. Comput. Vision Image Underst. **114**(1), 85–99 (2010)
22. Miranda, P., Mansilla, L.: Oriented image foresting transform segmentation by seed competition. IEEE Trans. Image Process. **23**(1), 389–398 (2014)
23. Noh, H., Hong, S., Han, B.: Learning deconvolution network for semantic segmentation. In: 2015 IEEE International Conference on Computer Vision (ICCV), pp. 1520–1528 (2015)
24. Papadopoulos, D.P., Uijlings, J.R.R., Keller, F., Ferrari, V.: Extreme clicking for efficient object annotation. In: 2017 IEEE International Conference on Computer Vision (ICCV), pp. 4940–4949 (2017)
25. Rother, C., Kolmogorov, V., Blake, A.: "grabcut": interactive foreground extraction using iterated graph cuts. ACM Trans. Gr. **23**(3), 309–314 (2004)
26. Singaraju, D., Grady, L., Vidal, R.: Interactive image segmentation via minimization of quadratic energies on directed graphs. In: International Conference on Computer Vision and Pattern Recognition, pp. 1–8, June 2008

27. Tang, M., Ayed, I.B., Marin, D., Boykov, Y.: Secrets of grabcut and kernel k-means. In: 2015 IEEE International Conference on Computer Vision (ICCV), pp. 1555–1563 (2015)
28. Tang, M., Gorelick, L., Veksler, O., Boykov, Y.: Grabcut in one cut. In: 2013 IEEE International Conference on Computer Vision, pp. 1769–1776 (2013)
29. Wu, J., Zhao, Y., Zhu, J., Luo, S., Tu, Z.: Milcut: a sweeping line multiple instance learning paradigm for interactive image segmentation. In: 2014 IEEE Conference on Computer Vision and Pattern Recognition, pp. 256–263 (2014)
30. Xu, N., Price, B., Cohen, S., Yang, J., Huang, T.: Deep grabcut for object selection. In: Kim, T.-K., Stefanos Zafeiriou, G.B., Mikolajczyk, K. (eds.) Proceedings of the British Machine Vision Conference (BMVC), pp. 182.1-182.12. BMVA Press, September 2017

Learning-Based Approaches
to Mathematical Morphology

On Some Associations Between Mathematical Morphology and Artificial Intelligence

Isabelle Bloch[1]([⊠])[iD], Samy Blusseau[2][iD], Ramón Pino Pérez[3][iD],
Élodie Puybareau[4][iD], and Guillaume Tochon[4][iD]

[1] Sorbonne Université, CNRS, LIP6, 75005 Paris, France
isabelle.bloch@sorbonne-universite.fr
[2] Centre for Mathematical Morphology, Mines ParisTech, PSL Research University,
Fontainebleau, France
samy.blusseau@mines-paristech.fr
[3] CRIL-CNRS, Université d'Artois, Lens, France
ramon.pinoperez@univ-artois.fr
[4] EPITA Research and Development Laboratory (LRDE),
Le Kremlin-Bicêtre, France
{elodie.puybareau,guillaume.tochon}@lrde.epita.fr

Abstract. This paper aims at providing an overview of the use of mathematical morphology, in its algebraic setting, in several fields of artificial intelligence (AI). Three domains of AI will be covered. In the first domain, mathematical morphology operators will be expressed in some logics (propositional, modal, description logics) to answer typical questions in knowledge representation and reasoning, such as revision, fusion, explanatory relations, satisfying usual postulates. In the second domain, spatial reasoning will benefit from spatial relations modeled using fuzzy sets and morphological operators, with applications in model-based image understanding. In the third domain, interactions between mathematical morphology and deep learning will be detailed. Morphological neural networks were introduced as an alternative to classical architectures, yielding a new geometry in decision surfaces. Deep networks were also trained to learn morphological operators and pipelines, and morphological algorithms were used as companion tools to machine learning, for pre/post processing or even regularization purposes. These ideas have known a large resurgence in the last few years and new ones are emerging.

Keywords: Mathematical morphology · Artificial intelligence ·
Lattice · Logics · Spatial reasoning · Fuzzy sets · Neural networks ·
Deep learning

I. Bloch—This work was partly done while I. Bloch was with LTCI, Télécom Paris, Institut Polytechnique de Paris, France.

J. Lindblad et al. (Eds.): DGMM 2021, LNCS 12708, pp. 457–469, 2021.
https://doi.org/10.1007/978-3-030-76657-3_33

1   Introduction

This paper aims at highlighting the usefulness of mathematical morphology in artificial intelligence (AI). To this end, we restrict ourselves to the deterministic setting and to increasing operators. The underlying structure of complete lattices allows applying basic operators to various settings, and this is the bridge we establish between the so far rather disconnected domains of mathematical morphology and AI formalisms. Concrete operators depending on structuring elements will provide simple and intuitive examples. On the AI side, we address several of its components, from purely symbolic approaches to machine learning.

In Sect. 2, we illustrate links between mathematical morphology and logics. In propositional logics, considering the lattice of formulas, morphological operators will act on formulas (and on their models). Such examples will be described in several logics. We will then show how they can be used in typical reasoning problems in AI, to define revision operators, merging operators, or explanatory relations. In Sect. 3, mathematical morphology will be shown to be helpful for spatial reasoning. Spatial reasoning aims at modeling spatial entities and spatial relations to reason about them. A typical example is the problem of model-based image understanding. Models of a scene usually involve spatial relations to provide information on the structure of the scene and on the spatial arrangement of the objects it contains. Moreover, such relations allow disambiguating objects with similar shapes and appearances, and are more robust to deformations or pathological cases. Mathematical morphology is then useful, combined with fuzzy sets, to model such spatial relations, taking into account their intrinsic vagueness (e.g. left to, close to), and to compute them efficiently. These models can then be used in spatial reasoning processes. Finally in Sect. 4, we will move to machine learning methods in AI, and combine mathematical morphology and deep learning. Interactions between mathematical morphology and deep learning have been investigated since the 1980s across several aspects. Besides, they have known a large resurgence in the last few years, and new ones are emerging. We shall give an overview of this trend.

This paper is based on a tutorial taught by some of the authors at ECAI 2020. It is an overview, relying on the existing literature and previous works by the authors, and paving the way for future research at the cross-road of mathematical morphology and artificial intelligence.

2   Mathematical Morphology and Logics

In this section, we first show how basic morphological operators can be applied to logical formulas, and then use them to address typical reasoning problems in artificial intelligence.

Propositional Logic. Let us start with propositional logic, as originally proposed in [13]. We assume that the language is defined by a set V of variables (here assumed to be finite), denoted by $a, b...$, and the standard connectives

$\land$ (conjunction), $\lor$ (disjunction), $\neg$ (negation), $\rightarrow$ (implication). Formulas are built on this language. The consequence relation is denoted by $\vdash$. Tautology and antilogy are denoted by $\top$ and $\bot$, respectively. A world or interpretation is an assignment of truth values to all variables. Ω denotes the set of all worlds. The set of models of a formula φ is denoted by $[\![\varphi]\!]$, and is a subset of Ω, defined as the set of worlds in which the formula is true. Since a formula is semantically characterized by its set of models, it is equivalent to consider the lattice of the set of formulas (up to syntactic equivalence) and the lattice $(\mathcal{P}(\Omega), \subseteq)$ where $\mathcal{P}(\Omega)$ is the set of subsets of Ω. Algebraic morphological operators are then defined as in any lattice. Now, the notion of structuring element can be formalized in this setting as a binary relation between worlds, and we will note for a relation B: $\omega' \in B_\omega$ iff (ω, ω') belongs to the relation. The inverse relation is denoted by $\check{B}$ $(\omega \in \check{B}_{\omega'}$ iff $\omega' \in B_\omega)$. Then, the morphological dilation of a formula φ with a structuring element B is simply defined via the semantic as:

$$[\![\delta_B(\varphi)]\!] = \delta_B([\![\varphi]\!]) = \{\omega \in \Omega \mid \check{B}_\omega \cap [\![\varphi]\!] \neq \emptyset\},$$

In a similar way, the morphological erosion is defined as:

$$[\![\varepsilon_B(\varphi)]\!] = \varepsilon_B([\![\varphi]\!]) = \{\omega \in \Omega \mid B_\omega \subseteq [\![\varphi]\!]\}.$$

Let us illustrate these ideas on a simple example, where the relation B between worlds is defined as a neighborhood relation, *e.g.* based on a threshold on a Hamming distance d_H between worlds. Let $\varphi = (a \land b \land c) \lor (\neg a \land \neg b \land c)$, and B be the ball of radius 1 of the Hamming distance: $B_\omega = \{\omega' \in \Omega \mid d_H(\omega, \omega') \leq 1\}$, *i.e.* B_ω comprises all worlds where at most one variable is instantiated differently from ω. The dilation of φ by B is then $\delta(\varphi) = (\neg a \lor b \lor c) \land (a \lor \neg b \lor c)$.

Many other operators from mathematical morphology can be defined in a similar way, and can be used to filter sets of models, making the operations described next more robust to noise or outliers for instance, to segment the main parts of a formula, etc.

Reasoning in AI Using Morphological Operators. With the help of these operators of dilation and erosion, we can successfully tackle some important reasoning aspects in AI, such as revision, merging (or fusion), abduction (explanatory reasoning), mediation..., which can find very concrete solutions in this morphological framework [14,15,17,18]. Another interesting feature is that these solutions, while simple and tractable, satisfy the properties usually required in such reasoning problems. Let us mention three examples.

Belief Revision. Let us assume that we have a set of preferences or beliefs, represented by one formula φ, as in [32]. If a new information is available, modeled as a formula ψ, the initial set should be revised to account for this new information [4]. This revision operation, denoted by $\varphi \circ \psi$ is usually assumed to induce a minimal change on the initial set of beliefs. A very simple concrete form for $\circ$ is to dilate the initial preferences or beliefs, until they become consistent with

ψ, i.e. $\varphi \circ \psi = \delta^n(\varphi) \wedge \psi$, with $n = \min\{k \in \mathbb{N} \mid \delta^k(\varphi) \wedge \psi$ is consistent$\}$. Taking the minimum size of dilation achieving consistency directly models the idea of minimal change. Interestingly enough, it can be proved that this particular revision operator satisfies the AGM postulates [4] in Katsuno and Mendelzon's model [32], which are widely considered as the postulates every revision operator should satisfy. Hence, the dilation based approach allows for a concrete computation of the revision, satisfying both the AGM postulates and the minimal change constraint.

Belief Merging. Let us now consider that m sets of beliefs or preferences are represented by logical formulas $\{\varphi_1...\varphi_m\}$ (which is a multi-set). Again, dilation can be the basis of the definition of several operators for logical fusion of these belief sets, under integrity constraints encoded as a propositional formula μ. Typically, each set of beliefs can be dilated until their conjunction with the constraint is consistent, e.g. $\Delta_\mu^{\max}(\varphi_1, ..., \varphi_m) = \delta^n(\varphi_1) \wedge \delta^n(\varphi_2) \wedge ... \wedge \delta^n(\varphi_m) \wedge \mu$, where $n = \min\{k \in \mathbb{N} \mid \delta^k(\varphi_1) \wedge ... \wedge \delta^k(\varphi_m) \wedge \mu$ is consistent$\}$, or $\Delta_\mu^\Sigma(\varphi_1, ..., \varphi_m) = \bigvee_{(n_1,...,n_m)} \delta^{n_1}(\varphi_1) \wedge \delta^{n_2}(\varphi_2) \wedge ... \wedge \delta^{n_m}(\varphi_m) \wedge \mu$ where $\sum_{i=1}^m n_i$ is minimal with $\delta^{n_1}(\varphi_1) \wedge \delta^{n_2}(\varphi_2) \wedge ... \wedge \delta^{n_m}(\varphi_m) \wedge \mu$ consistent. These two definitions correspond to merging operators introduced in [34], and satisfy the rationality postulates such operators should satisfy. An example is illustrated, by representing models as sets, in Fig. 1 (left).

Abduction. Abductive reasoning belongs to the now popular domain of explainable AI, and aims at finding the "best"explanation γ to an observation α, according to a knowledge base Σ. In a logical setting, Σ is a set of formulas, and γ and α are formulas. The problem is then expressed as: $\Sigma \cup \{\gamma\} \vdash \alpha$. Again, we expect γ to satisfy a number or properties, expressed as rationality postulates in [42,43]. The general idea of using mathematical morphology in this context is to find the most central models in Σ or in $\Sigma \wedge \alpha$, satisfying α. Note that reducing the set of models amounts to adding formulas to Σ. This idea can be implemented in a simple way by using successive erosions, *e.g.* as:

$$\alpha \vartriangleright^\ell \gamma \overset{def}{\Leftrightarrow} \gamma \vdash_\Sigma \varepsilon_\ell(\Sigma \wedge \alpha) \; ; \quad \text{or} \quad \alpha \vartriangleright^{\ell c} \gamma \overset{def}{\Leftrightarrow} \gamma \vdash_\Sigma \varepsilon_{\ell c}(\Sigma, \alpha) \wedge \alpha$$

where ε_ℓ denotes the last erosion before obtaining an empty set, and $\varepsilon_{\ell c}$ denotes the last erosion which is still consistent with α. An example is illustrated in Fig. 1 (right). A useful feature of this approach is that different types of explanations can be obtained by appropriate choices of the structuring element used in the erosions. For instance if $\Sigma = \{a \rightarrow c, b \rightarrow c\}$ and $\alpha = c$, then depending on the meaning of a, b, c, we can seek a disjunctive explanation of c $(a \vee b)$, or a conjunctive one $(a \wedge b)$, or an exclusive disjunction $((a \wedge \neg b) \vee (\neg a \wedge b))$ [14].

Other Logics. The ideas described above have been extended to several other logics, more expressive than propositional logic. We just mention some of these extensions here:

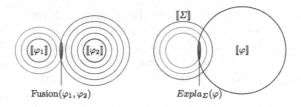

Fig. 1. Fusion based on dilations (left) and explanation based on erosions (right).

- Dilation of formulas expressed in first order logic has been used in [28,29] for application to merging, along with an efficient implementation using decision diagrams.
- In modal logic, the two modalities modeling possibility ◇ and necessity □ can be expressed as dilation and erosion, respectively [16]. From a semantic point of view, the accessibility relation, expressing a binary relation between worlds, can be considered as a structuring element. It is then natural to consider that □φ is satisfied in a world ω if φ is satisfied in all worlds accessible from ω, corresponding to the erosion by this structuring element, while ◇φ is satisfied in ω if φ is satisfied in at least one world accessible from ω, hence corresponding to the dilation.
- In description logic δ and ε can be naturally included in the logic as binary predicates [6,31], and are thus involved in ontological reasoning.
- A generalization of all the above was developed in the abstract framework of satisfaction systems and institutions, which encompasses many logics [1–3]. This allows extending the revision operator described above in propositional logic to a revision operator acting in any logic, based on a notion of relaxation (slightly different from a dilation). Similarly, abductive reasoning can be performed based on notions of cutting and retraction, similar to erosions.

3   Mathematical Morphology for Spatial Reasoning

In this section, we illustrate how mathematical morphology can be used for spatial reasoning in various settings. Spatial reasoning is defined as the domain of knowledge representation on spatial entities and spatial relationships, and reasoning on them. Spatial entities can be represented as abstract formulas in a logical (symbolic, qualitative) setting, as regions or keypoints in a quantitative setting, or as fuzzy regions in the semi-qualitative setting of fuzzy sets (*i.e.* still a deterministic setting). On the symbolic side, spatial relations can be represented as formulas, connectives, modalities or predicates in a logical setting. On the numerical side, they are best represented using fuzzy models, in order to account for their intrinsic imprecision (*e.g.* "to the right of", "close to").

Spatial Reasoning in a Qualitative Setting Using Morpho-Logic. Let us first consider the qualitative setting, using various logics. As a first example, let us consider abductive reasoning, as introduced in Sect. 2. Image understanding can be expressed as an explanatory process, which aims at providing the "best" explanation γ to the observations $\mathcal{O}$ according to a knowledge base $\mathcal{K}$ [6], *i.e.* such that $\mathcal{K} \cup \{\gamma\} \vdash \mathcal{O}$. Observations can be images, or results of some image analysis process (e.g. segmentation of some structures in the images). The knowledge base $\mathcal{K}$ models expert knowledge on the domain, on the structures present in the scene and on their relations (contrast, spatial relations...). It can be expressed in description logic for instance. A solution to the abduction problem consists in translating knowledge and observation in a lattice of concepts, and by applying erosions in this lattice to find the best explanation according to a minimality criterion [6]. Another algorithmic solution relies on tableau methods [54], an algorithmic way to satisfiability problems, where formulas are developed in different branches until inconsistencies are found. In the example in Fig. 2 (left), a MRI brain image with a tumor can be interpreted, at a higher level, using an anatomical knowledge base, as "Peripheral Small Deforming Tumoral Brain".

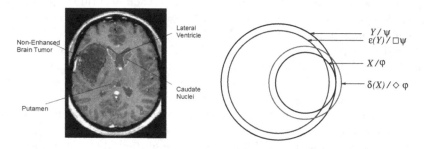

Fig. 2. Left: pathological brain with a tumor. Finding a high level interpretation of the image can be formalized as an abduction problem, where the knowledge base contains expert knowledge and the observation is the image and segmentation results. Right: Tangential part from morphological operators (X and Y are models of formulas φ and ψ, respectively).

Let us now consider modal morpho-logic, where the two modalities are defined as erosion and dilation (*i.e.* $\Box \equiv \varepsilon$ and $\Diamond \equiv \delta$), and consider the domain of mereotopology, specifically the Region Connection Calculus (RCC) formalism [45]. In this theory, several topological relations are defined from a connection predicate, in first order logic. Modal morpho-logic leads to simpler and decidable expressions of some of these relations. Let us provide a few examples, where φ and ψ are formulas representing abstract spatial entities:

- φ is a tangential part of ψ iff $\varphi \to \psi$ and $\Diamond \varphi \wedge \neg \psi \not\to \bot$ (or $\varphi \to \psi$ and $\varphi \wedge \neg \Box \psi \not\to \bot$). A simple model in the 2D space of such a relation is illustrated in Fig. 2 (right).
- φ is a non tangential part of ψ iff $\Diamond \varphi \to \psi$ (or $\varphi \to \Box \psi$).
- φ and ψ are externally connected (adjacent) iff $\varphi \wedge \psi \to \bot$ and $\Diamond \varphi \wedge \psi \not\to \bot$ (or $\varphi \wedge \Diamond \psi \not\to \bot$).

Further links between mathematical morphology and RCC can be found in [10, 12, 35].

Semi-qualitative Framework Using Fuzzy Modeling of Spatial Relations. When imprecision on knowledge and on data has to be taken into account, a semi-qualitative framework is best appropriate, such as fuzzy sets theory. In this theory, every piece of information becomes a matter of degree, and the membership of an element to a set, the degree to which some elements satisfy a relation, the truth value of a logical formula, etc. are values in $[0, 1]$. Note that what is most important is the ranking between different values, rather than their absolute value. This theory offers many tools for information representation and processing [26]. One of the problems to be addressed when reasoning on both qualitative or symbolic knowledge and on numerical data is the so-called semantic gap, between abstract concepts and concrete information extracted from data (e.g. images). The notion of linguistic variable [55] is then useful to establish links between a concept and its representation in a specific concrete domain. Here again, mathematical morphology, extended to handle fuzzy sets [8], can play an important role that we illustrate here on the modeling of spatial relations [7].

The main idea is to model the semantics of a spatial relation R expressed in a linguistic way as a fuzzy structuring element ν_R (i.e. a function from the spatial domain into $[0, 1]$). Then, dilating a reference object, possibly fuzzy, defined by its membership function μ, provides a fuzzy region of space where the value $\delta_{\nu_R}(\mu)(x)$ at each point x represents the degree to which relation R to μ is satisfied at this point. The degree to which another object satisfies relation R to μ can then be computed from the aggregation, using some fuzzy fusion operator, of the values $\delta_{\nu_R}(\mu)(y)$ for all points y of this second object. Several relations can be modeled according to this principle, such as topological relations, directions, distances, as well as more complex relations (between, along, parallel, aligned...). A useful feature of fuzzy representations is that a relation and its degree of satisfaction can be represented as a number, a fuzzy number, an interval, a distribution, a spatial fuzzy set, etc., in a same unifying framework.

This framework can be the basis of spatial reasoning, for example for structural model based image understanding. Indeed, modeling explicitly the structural knowledge we may have on a scene helps recognizing individual structures (disambiguating them in case of similar shape and appearance), as well as their global organization. This can be achieved using various methods, each having two main components: knowledge representation and reasoning. Fuzzy models of objects and relations can enhance qualitative representations (logical formulas, knowledge bases, ontologies), and then be used in logical reasoning, including morpho-logic (for instance, the set of models of a formula becomes a fuzzy set). They can serve as attributes in structural representations such as graphs, hypergraphs, conceptual graphs. Reasoning then relies on matching, sequential graph traversal to guide the exploration of an image, constraint satisfaction problems, etc. (see *e.g.* [9, 11] and the references therein). From the interpretation results, it is then possible to go back to the initial language of the domain to provide linguistic descriptions of the image content, as in the previous example on brain image interpretation.

4   Mathematical Morphology and Neural Networks

We now move to a different branch of AI, namely machine learning based on neural networks, which has tremendously grown in the past years. As a matter of fact, there has been an increasing effort devoted to the combination of mathematical morphology and neural network frameworks. More specifically, morphological operations can be integrated as efficient pre- or post-processing tools within machine/deep learning processing frameworks. Further, the structural similarity between neuron operations (weighted linear combination of input values, potentially mapped by non-linear activation functions) and elementary morphological operations such as erosion and dilation makes it tempting to substitute the former by the latter, resulting in morphological perceptrons and morphological layers. This section reviews these major lines of research.

Mathematical Morphology as a Pre/Post-Processing Step. A well-known drawback of deep convolutional neural networks (CNNs) is their poor ability to segment very thin structures, and their sensibility to noise and contrast [20,25]. Integrating mathematical morphology as pre-processing has improved the results of several classical CNN architectures on such problems. For example, the use of the morphological top-hat helps to enhance the very small structures of medical images. In [53], the top-hat is used in one of the three channels of a VGG input to segment small white matter hyperintensities, while in [22], the top-hat results are directly fed to the networks and proved their usefulness by guiding the networks (here ConvNet and Mask R-CNN) to focus on chosen parts of the image (here knee meniscus tear). In [24], a geodesic reconstruction is performed to inject topological (global) information into Whole Slide Imaging images (that are huge and heavy) before doing patches. This step improved the skin segmentation results of U-Net.

Mathematical morphology can also be helpful in post-processing steps. For example, to avoid heavy computation time and memory use, some 3D segmentation tasks can be treated as successive 2D segmentations. Without any spatial context, successive segmented slices can be disconnected or show aberrations. Mathematical morphology can hence be used as a regularization step to remove these abnormalities [44].

Morphological Neural Networks for Images. Morphological neural networks were introduced in the late 1980s with a definition of neurons as weighted rank filters [52] or, in a less general form, as performing dilations or erosions [23]. Replacing the linear perceptron's dot product by the non-linear max-plus and min-plus operators has induced a new geometry of decision surfaces, which we may refer to as *bounding box* geometry [46,49,50,56], and alternative (or complementary) strategies to gradient descent in networks training. Hybrid approaches mixing linear and morphological layers have also been developed for an even richer geometry [21,30,41,51,57], and dilation layers showed interesting pruning properties when located after linear layers [21,57]. The latter studies, however, only consider dense layers and are therefore little suited to image analysis, in contrast to convolutional networks which can handle large images.

Yet, translation invariant morphological architectures are a crucial issue. Indeed, the success of a morphological framework involving elementary operations (dilation and erosion) and their combinations (closing, opening, top-hat, etc.) often comes with a tedious trial-and-error setting to derive the optimal sequence of operations and their respective structuring elements. Deep CNNs are a potential solution to automatically learn this optimal sequence and the structuring elements. As a matter of fact, the weights of each layer filter could be interpreted as (non necessary flat) structuring elements, provided that the conventional convolution operation has been replaced by erosion or dilation, and that the layer has a way to learn which operation to use. Strictly speaking, one should then talk in that case about morphological (and no longer convolutional) neural network architectures.

In the early 2010s, when deep learning optimization tools were not as well diffused as today, attempts were made to overcome the non-differentiability of the min and max operations of erosions and dilations in convolutional-like approaches relying on stochastic gradient descent [37]. Based on the Counter-Harmonic Mean (CHM) [5,19], the so-called *PConv* layer is not only smooth, but it can approximate non-flat dilations, erosions and classical convolutions, depending on their parameter p, which can be trained along with the kernel parameters. Recently, this idea was successfully applied to digit recognition tasks [38]. Even lately, another smooth approximation of min and max, *i.e.* the so-called LogSumExp function (also known as multivariate softplus), has been investigated in [48] to learn binary structuring elements, and extended to grayscale structuring elements in [47]. Finally, a last smooth version of min and max operations based on the α-softmax [36] function has been proposed and shown to outperform the classical *PConv* layer in learning non flat structuring elements [33].

On the other hand, deep neural networks including non-smooth operators (the ReLU activation function, max and min pooling layers, to name a few) have been efficiently trained with stochastic gradient descent for years. Indeed, these operators are actually differentiable almost everywhere, and a descent direction can be defined even in their zero-measure non-smooth regions. Therefore, it is natural that translation-invariant morphological layers were recently optimized just as usual convolutional ones [27,39,40], that is, with stochastic gradient descent and back-propagation. In the latter studies, deep architectures including morphological layers were applied to classification, image denoising and restoration as well as edge detection. Two remarkable results are that the morphological trainable max-pooling improves significantly the classical max-pooling, and that independent morphological layers converge without constraint towards an adjunction.

From this brief review, some challenges clearly appear. First, the morphological networks we just mentioned are all way more shallow than the most popular classical architectures, and none of them compete with state-of-the-art CNNs on tasks like segmentation or classification on large-scale image datasets. Besides, successful architectures including morphological layers almost always contain classical convolutional layers. These two observations tend to indicate that insights are still needed regarding the representation power of purely morphological networks, as well as their optimization when many layers are stacked.

5   Conclusion

In this paper, we highlighted some links between mathematical morphology and different components of artificial intelligence, including symbolic AI for knowledge representation and reasoning in various logics, fuzzy sets for reasoning under uncertainty, and machine learning based on neural networks, which are all very active topics in AI. Future work on mathematical morphology and symbolic AI is planned in two ways: extend the "toolbox" of morpho-logic with other morphological operations to enrich both knowledge representation and reasoning, and enhance mathematical morphology with the inference power of logics. Both directions can be endowed with an uncertainty modeling layer, based on fuzzy sets theory. Similarly, a current trend in deep learning is to introduce knowledge in neural networks, which could be, in the future, modeled as morpho-logic or fuzzy spatial relations.

References

1. Aiguier, M., Atif, J., Bloch, I., Hudelot, C.: Belief revision, minimal change and relaxation: a general framework based on satisfaction systems, and applications to description logics. Artif. Intell. **256**, 160–180 (2018)
2. Aiguier, M., Atif, J., Bloch, I., Pino Pérez, R.: Explanatory relations in arbitrary logics based on satisfaction systems, cutting and retraction. Int. J. Approx. Reason. **102**, 1–20 (2018)
3. Aiguier, M., Bloch, I.: Logical dual concepts based on mathematical morphology in stratified institutions: applications to spatial reasoning. J. Appl. Non-Classical Logics **29**(4), 392–429 (2019)
4. Alchourron, C., Gardenfors, P., Makinson, D.: On the logic of theory change. J. Symb. Log. **50**(2), 510–530 (1985)
5. Angulo, J.: Pseudo-morphological image diffusion using the counter-harmonic paradigm. In: Blanc-Talon, J., Bone, D., Philips, W., Popescu, D., Scheunders, P. (eds.) ACIVS 2010. LNCS, vol. 6474, pp. 426–437. Springer, Heidelberg (2010). https://doi.org/10.1007/978-3-642-17688-3_40
6. Atif, J., Hudelot, C., Bloch, I.: Explanatory reasoning for image understanding using formal concept analysis and description logics. IEEE Trans. Syst. Man. Cybern. Syst. **44**(5), 552–570 (2014)
7. Bloch, I.: Fuzzy spatial relationships for image processing and interpretation: a review. Image Vis. Comput. **23**(2), 89–110 (2005)

8. Bloch, I.: Duality vs. adjunction for fuzzy mathematical morphology and general form of fuzzy erosions and dilations. Fuzzy Sets Syst. **160**, 1858–1867 (2009)

9. Bloch, I.: Fuzzy sets for image processing and understanding. Fuzzy Sets Syst. **281**, 280–291 (2015)

10. Bloch, I.: Topological relations between bipolar fuzzy sets based on mathematical morphology. In: Angulo, J., Velasco-Forero, S., Meyer, F. (eds.) ISMM 2017. LNCS, vol. 10225, pp. 40–51. Springer, Cham (2017). https://doi.org/10.1007/978-3-319-57240-6_4

11. Bloch, I.: On linguistic descriptions of image content. In: Rencontres Francophones sur la Logique Floue et ses Applications (LFA), pp. 57–64. Sète, France (2020)

12. Bloch, I., Heijmans, H., Ronse, C.: Mathematical morphology. In: Aiello, M., Pratt-Hartman, I., van Benthem, J. (eds.) Handbook of Spatial Logics, pp. 857–947. Springer, Cham (2007). Chapter 13

13. Bloch, I., Lang, J.: Towards mathematical morpho-logics. In: 8th International Conference on Information Processing and Management of Uncertainty in Knowledge based Systems IPMU 2000, vol. III, pp. 1405–1412. Madrid, Spain (2000)

14. Bloch, I., Lang, J., Pino Pérez, R., Uzcátegui, C.: Morphologic for knowledge dynamics: revision, fusion, abduction. Technical report (2018). arXiv:1802.05142

15. Bloch, I., Pino Pérez, R., Uzcategui, C.: Mediation in the framework of morpho-logic. In: European Conference on Artificial Intelligence, pp. 190–194 (2006)

16. Bloch, I.: Modal logics based on mathematical morphology for qualitative spatial reasoning. J. Appl. Non-Classical Log. **12**(3–4), 399–423 (2002)

17. Bloch, I., Pino Pérez, R., Uzcátegui, C.: Explanatory relations based on mathematical morphology. In: Benferhat, S., Besnard, P. (eds.) ECSQARU 2001. LNCS (LNAI), vol. 2143, pp. 736–747. Springer, Heidelberg (2001). https://doi.org/10.1007/3-540-44652-4_65

18. Bloch, I., Pino Pérez, R., Uzcátegui, C.: A unified treatment for knowledge dynamics. In: KR, pp. 329–337 (2004)

19. Bullen, P.S.: Handbook of Means and Their Inequalities. Springer Science & Business Media, Berlin (2013)

20. Bunne, C., Rahmann, L., Wolf, T.: Studying invariances of trained convolutional neural networks (2018). arXiv preprint arXiv:1803.05963

21. Charisopoulos, V., Maragos, P.: Morphological perceptrons: geometry and training algorithms. In: Angulo, J., Velasco-Forero, S., Meyer, F. (eds.) ISMM 2017. LNCS, vol. 10225, pp. 3–15. Springer, Cham (2017). https://doi.org/10.1007/978-3-319-57240-6_1

22. Couteaux, V., et al.: Automatic knee meniscus tear detection and orientation classification with mask-RCNN. Diagn. Interv. Imaging **100**(4), 235–242 (2019)

23. Davidson, J.L., Ritter, G.X.: Theory of morphological neural networks. In: Digital Optical Computing II, vol. 1215, pp. 378–388. ISOP (1990)

24. Decencière, E., et al.: Dealing with topological information within a fully convolutional neural network. In: Blanc-Talon, J., Helbert, D., Philips, W., Popescu, D., Scheunders, P. (eds.) ACIVS 2018. LNCS, vol. 11182, pp. 462–471. Springer, Cham (2018). https://doi.org/10.1007/978-3-030-01449-0_39

25. Dodge, S., Karam, L.: Understanding how image quality affects deep neural networks. In: 2016 Eighth International Conference on Quality of Multimedia Experience (QoMEX), pp. 1–6. IEEE (2016)

26. Dubois, D., Prade, H.: Fuzzy Sets and Systems: Theory and Applications. Academic Press, New-York (1980)

27. Franchi, G., Fehri, A., Yao, A.: Deep morphological networks. Pattern Recogn. **102**, 107246 (2020)

28. Gorogiannis, N., Hunter, A.: Implementing semantic merging operators using binary decision diagrams. Int. J. Approx. Reason. **49**(1), 234–251 (2008)
29. Gorogiannis, N., Hunter, A.: Merging first-order knowledge using dilation operators. In: Hartmann, S., Kern-Isberner, G. (eds.) FoIKS 2008. LNCS, vol. 4932, pp. 132–150. Springer, Heidelberg (2008). https://doi.org/10.1007/978-3-540-77684-0_11
30. Hernández, G., Zamora, E., Sossa, H., Téllez, G., Furlán, F.: Hybrid neural networks for big data classification. Neurocomputing **390**, 327–340 (2020)
31. Hudelot, C., Atif, J., Bloch, I.: Fuzzy spatial relation ontology for image interpretation. Fuzzy Sets Syst. **159**(15), 1929–1951 (2008)
32. Katsuno, H., Mendelzon, A.O.: Propositional knowledge base revision and minimal change. Artif. Intell. **52**, 263–294 (1991)
33. Kirszenberg, A., Tochon, G., Puybareau, É., Angulo, J.: Going beyond p-convolutions to learn grayscale morphological operators. In: IAPR International Conference on Discrete Geometry and Mathematical Morphology (DGMM). Springer, Uppsala, Sweden (2021)
34. Konieczny, S., Pino Pérez, R.: Logic based merging. J. Philos. Log. **40**(2), 239–270 (2011)
35. Landini, G., Galton, A., Randell, D., Fouad, S.: Novel applications of discrete mereotopology to mathematical morphology. Signal Process. Image Commun. **76**, 109–117 (2019)
36. Lange, M., Zühlke, D., Holz, O., Villmann, T., Mittweida, S.G.: Applications of Lp-norms and their smooth approximations for gradient based learning vector quantization. In: European Symposium on Artificial Neural Networks, Computational Intelligence and Machine Learning, pp. 271–276 (2014)
37. Masci, J., Angulo, J., Schmidhuber, J.: A learning framework for morphological operators using counter–harmonic mean. In: Hendriks, C.L.L., Borgefors, G., Strand, R. (eds.) ISMM 2013. LNCS, vol. 7883, pp. 329–340. Springer, Heidelberg (2013). https://doi.org/10.1007/978-3-642-38294-9_28
38. Mellouli, D., Hamdani, T.M., Sanchez-Medina, J.J., Ayed, M.B., Alimi, A.M.: Morphological convolutional neural network architecture for digit recognition. IEEE Trans. Neural Netw. Learn. Syst. **30**(9), 2876–2885 (2019)
39. Mondal, R., Dey, M.S., Chanda, B.: Image restoration by learning morphological opening-closing network. Math. Morphol. Theory Appl. **4**(1), 87–107 (2020)
40. Nogueira, K., Chanussot, J., Dalla Mura, M., Schwartz, W.R., Dos Santos, J.A.: An introduction to deep morphological networks (2019). arXiv:1906.01751
41. Pessoa, L.F., Maragos, P.: Neural networks with hybrid morphological/rank/linear nodes: a unifying framework with applications to handwritten character recognition. Pattern Recogn. **33**(6), 945–960 (2000)
42. Pino Pérez, R., Uzcátegui, C.: Jumping to explanations versus jumping to conclusions. Artif. Intell. **111**(1–2), 131–169 (1999)
43. Pino Pérez, R., Uzcátegui, C.: Preferences and explanations. Artif. Intell. **149**(1), 1–30 (2003)
44. Puybareau, É., Tochon, G., Chazalon, J., Fabrizio, J.: Segmentation of gliomas and prediction of patient overall survival: a simple and fast procedure. In: Crimi, A., et al. (eds.) BrainLes 2018. LNCS, vol. 11384, pp. 199–209. Springer, Cham (2019). https://doi.org/10.1007/978-3-030-11726-9_18
45. Randell, D., Cui, Z., Cohn, A.: A spatial logic based on regions and connection. In: 3rd International Conference on Knowledge Representation and Reasoning (KR), pp. 165–176 (1992)

46. Ritter, G.X., Sussner, P.: An introduction to morphological neural networks. In: 13th IEEE International Conference on Pattern Recognition, vol. 4, pp. 709–717 (1996)
47. Shen, Y., Zhong, X., Shih, F.Y.: Deep morphological neural networks (2019). arXiv preprint arXiv:1909.01532
48. Shih, F.Y., Shen, Y., Zhong, X.: Development of deep learning framework for mathematical morphology. Int. J. Pattern Recogn. Artif. Intell. **33**(06), 1954024 (2019)
49. Sussner, P., Esmi, E.L.: Morphological perceptrons with competitive learning: Lattice-theoretical framework and constructive learning algorithm. Inf. Sci. **181**(10), 1929–1950 (2011)
50. Sussner, P.: Morphological perceptron learning. In: IEEE International Symposium on Intelligent Control (ISIC), pp. 477–482 (1998)
51. Sussner, P., Campiotti, I.: Extreme learning machine for a new hybrid morphological/linear perceptron. Neural Netw. **123**, 288–298 (2020)
52. Wilson, S.S.: Morphological networks. In: Visual Communications and Image Processing IV, vol. 1199, pp. 483–495. ISOP (1989)
53. Xu, Y., Géraud, T., Puybareau, É., Bloch, I., Chazalon, J.: White matter hyperintensities segmentation in a few seconds using fully convolutional network and transfer learning. In: Crimi, A., Bakas, S., Kuijf, H., Menze, B., Reyes, M. (eds.) Brainlesion: Glioma, Multiple Sclerosis, Stroke and Traumatic Brain Injuries, pp. 501–514. Springer, Cham (2018)
54. Yang, Y., Atif, J., Bloch, I.: Abductive reasoning using tableau methods for highlevel image interpretation. In: Hölldobler, S., Krötzsch, M., Peñaloza, R., Rudolph, S. (eds.) KI 2015. LNCS (LNAI), vol. 9324, pp. 356–365. Springer, Cham (2015). https://doi.org/10.1007/978-3-319-24489-1_34
55. Zadeh, L.A.: The concept of a linguistic variable and its application to approximate reasoning. Inf. Sci. **8**, 199–249 (1975)
56. Zamora, E., Sossa, H.: Dendrite morphological neurons trained by stochastic gradient descent. Neurocomputing **260**, 420–431 (2017)
57. Zhang, Y., Blusseau, S., Velasco-Forero, S., Bloch, I., Angulo, J.: Max-plus operators applied to filter selection and model pruning in neural networks. In: Burgeth, B., Kleefeld, A., Naegel, B., Passat, N., Perret, B. (eds.) ISMM 2019. LNCS, vol. 11564, pp. 310–322. Springer, Cham (2019). https://doi.org/10.1007/978-3-030-20867-7_24

Going Beyond p-convolutions to Learn Grayscale Morphological Operators

Alexandre Kirszenberg[1], Guillaume Tochon[1(✉)], Élodie Puybareau[1], and Jesus Angulo[2]

[1] EPITA Research and Development Laboratory (LRDE),
Le Kremlin-Bicêtre, France
{alexandre.kirszenberg,guillaume.tochon,elodie.puybareau}@lrde.epita.fr
[2] Centre for Mathematical Morphology, Mines ParisTech,
PSL Research University, Paris, France
jesus.angulo@mines-paristech.fr

Abstract. Integrating mathematical morphology operations within deep neural networks has been subject to increasing attention lately. However, replacing standard convolution layers with erosions or dilations is particularly challenging because the min and max operations are not differentiable. Relying on the asymptotic behavior of the counter-harmonic mean, p-convolutional layers were proposed as a possible workaround to this issue since they can perform pseudo-dilation or pseudo-erosion operations (depending on the value of their inner parameter p), and very promising results were reported. In this work, we present two new morphological layers based on the same principle as the p-convolutional layer while circumventing its principal drawbacks, and demonstrate their potential interest in further implementations within deep convolutional neural network architectures.

Keywords: Morphological layer · p-convolution · Counter-harmonic mean · Grayscale mathematical morphology

1 Introduction

Mathematical morphology deals with the non-linear filtering of images [15]. The elementary operations of mathematical morphology amount to computing the minimum (for the erosion) or maximum (for the dilation) of all pixel values within a neighborhood of some given shape and size (the structuring element) of the pixel under study. Combining those elementary operations, one can define more advanced (but still non-linear) filters, such as openings and closings, which have many times proven to be successful at various image processing tasks such as filtering, segmentation or edge detection [18]. However, deriving the optimal combination of operations and the design (shape and size) of their respective structuring element is generally done in a tedious and time-consuming trial-and-error fashion. Thus, delegating the automatic identification of the right sequence

© Springer Nature Switzerland AG 2021
J. Lindblad et al. (Eds.): DGMM 2021, LNCS 12708, pp. 470–482, 2021.
https://doi.org/10.1007/978-3-030-76657-3_34

of operations to use and their structuring element to some machine learning technique is an appealing strategy.

On the other hand, artificial neural networks are composed of units (or neurons) connected to each other and organized in layers. The output of each neuron is expressed as the linear combination of its inputs weighed by trainable weights, potentially mapped by a non-linear activation function [6]. Convolutional neural networks (CNNs) work in a similar fashion, replacing neurons with convolutional filters [8].

Because of the similarity between their respective operations, there has been an increasing interest in past years to integrate morphological operations within the framework of neural networks, and two major lines of research have emerged. The first one, tracing back to the end of the 80s, replaces the multiplication and addition of linear perceptron units with addition and maximum [4,14,20], resulting in so-called non-linear morphological perceptrons [19] (see [3,21] for recent works in this domain). The second line, mainly motivated by the rise of deep CNNs, explores the integration of elementary morphological operations in such networks to automatically learn their optimal shape and weights, the major issue being that the min and max operations are not differentiable. A first workaround is to replace them by smooth differentiable approximations, making them suited to the conventional gradient descent learning approach via back-propagation [8]. In their seminal work, Masci *et al.* [10] relied on some properties of the counter-harmonic mean [2] (CHM) to provide p-convolutional (*PConv*) layers, the value of the trainable parameter p dictating which of the elementary morphological operation the layer ultimately mimicks. The CHM was also used as an alternative to the standard max-pooling layer in classical CNN architectures [11]. LogSumExp functions (also known as multivariate softplus) were proposed as replacements of min and max operations to learn binary [17] and grayscale [16] structuring elements. An alternative approach was followed in [5,12]: the non-linear morphological operations remained unchanged, and the backpropagation step was instead adapted to handle them in the same way the classical max-pooling layer is handled in standard CNNs. Finally, morphological operations were recreated and optimized as combinations of depthwise and pointwise convolution with depthwise pooling [13].

Looking at all recently proposed approaches (apart from [10], all other aforementionned works date back to no later than 2017) and the diversity of their evaluation (image classification on MNIST database [9], image denoising and restoration, edge detection and so on), it seems to us that the magical formula for integrating morphological operations within CNNs has yet to be derived. For this reason, we would like to draw the attention in this work back to the *PConv* layer proposed in [10]. As a matter of fact, very promising results were reported but never investigated further. Relying on the CHM framework, we propose two possible extensions to the *PConv* layer, for which we demonstrate potential interest in further implementations within deep neural network architectures.

In Sect. 2, we review the work on p-convolutions [10], presenting their main advantages, properties and limitations. In Sect. 3, we propose two new morphological layers, namely the $\mathcal{L}$Morph layer (also based on the CHM) and the $\mathcal{S}$Morph layer (based on the regularized softmax). Both proposed layers are compatible with grayscale mathematical morphology and nonflat structuring elements. In Sect. 4, we showcase some results from our implementations, and proceed to compare these results to those of the p-convolution layer. Finally, Sect. 5 provides a conclusion and some perspectives from our contributions.

2   p-convolutions: Definitions, Properties and Pitfalls

In this section, we detail the notion of p-convolution as presented in [10].

2.1   Grayscale Mathematical Morphology

In mathematical morphology, an image is classically represented as a 2D function $f : E \rightarrow \mathbb{R}$ with $x \in E$ being the pixel coordinates in the 2D grid $E \subseteq \mathbb{Z}^2$ and $f(x) \in \mathbb{R}$ being the pixel value. In grayscale mathematical morphology, $i.e.$ when both the image f and the structuring element b are real-valued (and not binary), erosion $f \ominus b$ and dilation $f \oplus b$ operations can be written as:

$$(f \ominus b)(x) = \inf_{y \in E} \{f(y) - b(x - y)\} \tag{1}$$

$$(f \oplus b)(x) = \sup_{y \in E} \{f(y) + b(x - y)\} \tag{2}$$

This formalism also encompasses the use of flat (binary) structuring elements, which are then written as

$$b(x) = \begin{cases} 0 & \text{if } x \in B \\ -\infty & \text{otherwise} \end{cases}, \tag{3}$$

where $B \subseteq E$ is the support of the structuring function b.

2.2   The Counter-Harmonic Mean and the p-convolution

Let $p \in \mathbb{R}$. The counter-harmonic mean (CHM) of order p of a given non negative vector $\mathbf{x} = (x_1, \ldots, x_n) \in (\mathbb{R}^+)^n$ with non negative weights $\mathbf{w} = (w_1, \ldots, w_n) \in (\mathbb{R}^+)^n$ is defined as

$$CHM(\mathbf{x}, \mathbf{w}, p) = \frac{\sum_{i=1}^{n} w_i x_i^p}{\sum_{i=1}^{n} w_i x_i^{p-1}} . \tag{4}$$

The CHM is also known as the Lehmer mean [2]. Asymptotically, one has $\lim_{p \rightarrow +\infty} CHM(\mathbf{x}, \mathbf{w}, p) = \sup_i x_i$ and $\lim_{p \rightarrow -\infty} CHM(\mathbf{x}, \mathbf{w}, p) = \inf_i x_i$.
The p-convolution of an image f at pixel x for a given (positive) convolution kernel $w : W \subseteq E \rightarrow \mathbb{R}^+$ is defined as:

$$PConv(f, w, p)(x) = (f *_p w)(x) = \frac{(f^{p+1} * w)(x)}{(f^p * w)(x)} = \frac{\sum_{y \in W(x)} f^{p+1}(y) w(x - y)}{\sum_{y \in W(x)} f^p(y) w(x - y)} \tag{5}$$

where $f^p(x)$ denotes the pixel value $f(x)$ raised at the power of p, $W(x)$ is the spatial support of kernel w centered at x, and the scalar p controls the type of operation to perform.

Based on the asymptotic properties of the CHM, the morphological behavior of the *PConv* operation with respect to p has notably been studied in [1]. More precisely, when $p > 0$ (resp. $p < 0$), the operation is a pseudo-dilation (resp. pseudo-erosion), and when $p \to \infty$ (resp. $-\infty$), the largest (resp. smallest) pixel value in the local neighborhood $W(x)$ of pixel x dominates the weighted sum (5) and the $PConv(f, w, p)(x)$ acts as a non-flat grayscale dilation (resp. a non-flat grayscale erosion) with the structuring function $b(x) = \frac{1}{p} log(w(x))$:

$$\lim_{p \to +\infty} (f *_p w)(x) = \sup_{y \in W(x)} \left\{ f(y) + \frac{1}{p} \log (w(x - y)) \right\} \tag{6}$$

$$\lim_{p \to -\infty} (f *_p w)(x) = \inf_{y \in W(x)} \left\{ f(y) - \frac{1}{p} \log (w(x - y)) \right\} \tag{7}$$

In practice, Eqs. (6) and (7) hold true for $|p| > 10$. The flat structuring function (3) can be recovered by using constant weight kernels, *i.e.*, $w(x) = 1$ if $x \in W$ and $w(x) = 0$ if $x \notin W$ and $|p| \gg 0$. As stated in [10], the *PConv* operation is differentiable, thus compatible with gradient descent learning approaches via back-propagation.

2.3   Limits of the p-convolution Layer

In order for the *PConv* layer to be defined on all its possible input parameters, w and f must be strictly positive. Otherwise, the following issues can occur:

- If $f(x)$ contains null values and p is negative, $f^p(x)$ is not defined;
- If $f(x)$ contains negative values and p is a non-null, non-integer real number, $f^p(x)$ can contain complex numbers;
- If $w(x)$ or $f^p(x)$ contain null values, $\frac{1}{(f^p * w)(x)}$ is not defined.

As such, before feeding an image to the p-convolution operation, it first must be rescaled between $[1, 2]$:

$$f_r(x) = 1.0 + \frac{f(x) - \min_{x \in E} f(x)}{\max_{x \in E} f(x) - \min_{x \in E} f(x)} \tag{8}$$

Moreover, if several *PConv* layers are concatenated one behind the other (to achieve (pseudo-) opening and closing operations for instance), a rescaling must be performed before each layer. Particular care must also be taken with the output of the last *PConv* layer, since it must also be rescaled to ensure that the range of the output matches that of the target. This is done by adding a trainable scale/bias $1 \times 1 \times 1$ convolution layer at the end of the network.

Last but not least, a notable drawback of the *PConv* layer when it comes to learning a specific (binary or non-flat) structuring element is that it tends to be hollow and flattened out in the center (see further presented results in Sect. 4).

3   Proposed $\mathcal{L}$Morph and $\mathcal{S}$Morph Layers

As exposed in the previous Sect. 2.3, the *PConv* layer has a few edge cases and drawbacks. For this reason, we now propose two new morphological layers, based upon the same fundamental principle as the *PConv* operation, with the intent of making them compatible with general grayscale mathematical morphology.

3.1   Introducing the $\mathcal{L}$Morph Operation

Our main objective is still to circumvent the non-differentiability of min and max functions by replacing them with smooth and differentiable approximations. In the previous Sect. 2, we presented the CHM, whose asymptotic behavior is exploited by the *PConv* layer [10]. Relying on this behavior once more, we now propose to define the following $\mathcal{L}$Morph (for $\mathcal{L}$ehmer-mean based Morphological) operation:

$$\mathcal{L}\mathrm{Morph}(f, w, p)(x) = \frac{\sum_{y \in W(x)} (f(y) + w(x - y))^{p+1}}{\sum_{y \in W(x)} (f(y) + w(x - y))^{p}} \tag{9}$$

where $w : W \to \mathbb{R}^+$ is the structuring function and $p \in \mathbb{R}$. Defined as such, we can identify $\mathcal{L}\mathrm{Morph}(f, w, p)$ with the CHM defined by the Eq. (4): all weights w_i (resp. entries x_i) of Eq. (4) correspond to 1 (resp. $f(y) + w(x - y)$) in the Eq. (9), from which we can deduce the following asymptotic behavior:

$$\lim_{p \to +\infty} \mathcal{L}\mathrm{Morph}(f, w, p)(x) = \sup_{y \in W(x)} \{f(y) + w(x - y)\} = (f \oplus w)(x) \tag{10}$$

$$\lim_{p \to -\infty} \mathcal{L}\mathrm{Morph}(f, w, p)(x) = \inf_{y \in W(x)} \{f(y) + w(x - y)\} = (f \ominus -w)(x) \tag{11}$$

By changing the sign of p, one can achieve either pseudo-dilation (if $p > 0$) or pseudo-erosion (if $p < 0$). Figure 1 displays examples of applying the $\mathcal{L}$Morph function with a given non-flat structuring element for different values of p. In practice $|p| > 20$ is sufficient to reproduce non-flat grayscale dilation or non-flat grayscale erosion. Note however that the applied structuring function is $-w$ in the case of an erosion.

Relying on the CHM like the *PConv* layer brings over some shared limitations: the input image f must be positive and rescaled following Eq. (8), and the structuring function w must be positive or null.

3.2   Introducing the $\mathcal{S}$Morph Operation

Deriving a morphological layer based on the asymptotic behavior of the CHM has a major drawback in that the input must be rescaled within the range $[1, 2]$. In order to circumvent this issue, we now propose to leverage the α-softmax function [7], which is defined as:

$$\mathcal{S}_\alpha(\mathbf{x}) = \frac{\sum_{i=1}^{n} x_i e^{\alpha x_i}}{\sum_{i=1}^{n} e^{\alpha x_i}} \tag{12}$$

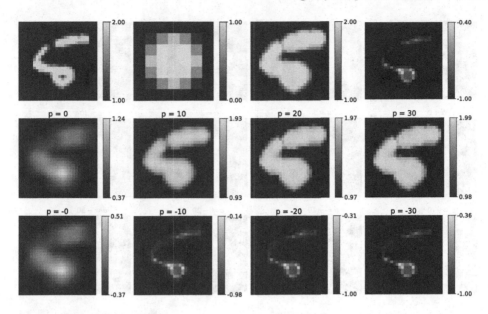

Fig. 1. Top row: input image, non-flat structuring element, target dilation, target erosion. Middle row: $\mathcal{L}$Morph pseudo-dilation for increasing value of p. Bottom row: $\mathcal{L}$Morph pseudo-erosion for increasing value of $|p|$.

for some $\mathbf{x} = (x_1, \ldots, x_n) \in \mathbb{R}^n$ and $\alpha \in \mathbb{R}$. In fact, $\mathcal{S}_\alpha$ has the desired properties that $\lim_{\alpha \to +\infty} \mathcal{S}_\alpha(\mathbf{x}) = \max_i x_i$ and $\lim_{\alpha \to -\infty} \mathcal{S}_\alpha(\mathbf{x}) = \min_i x_i$. This function is less restrictive than the CHM since it does not require the elements of $\mathbf{x}$ to be strictly positive. A major benefit is that it is no longer necessary to rescale its input.

Exploiting this property, we define in the following the $\mathcal{S}$Morph (standing for Smooth Morphological) operation:

$$\mathcal{S}\mathrm{Morph}(f, w, \alpha)(x) = \frac{\sum_{y \in W(x)} (f(y) + w(x - y)) e^{\alpha(f(y) + w(x-y))}}{\sum_{y \in W(x)} e^{\alpha(f(y) + w(x-y))}}, \tag{13}$$

where $w : W \to \mathbb{R}$ plays the role of the structuring function. We can see from the properties of $\mathcal{S}_\alpha$ that the following holds true:

$$\lim_{\alpha \to +\infty} \mathcal{S}\mathrm{Morph}(f, w, \alpha)(x) = (f \oplus w)(x) \tag{14}$$

$$\lim_{\alpha \to -\infty} \mathcal{S}\mathrm{Morph}(f, w, \alpha)(x) = (f \ominus -w)(x) \tag{15}$$

As such, just like the *PConv* and $\mathcal{L}$Morph layers, the proposed $\mathcal{S}$Morph operation can alternate between a pseudo-dilation ($\alpha > 0$) and a pseudo-erosion ($\alpha < 0$). Furthermore, when $\alpha \gg 0$ (resp. $\alpha \ll 0$), this function approximates the grayscale dilation (resp. the grayscale erosion).

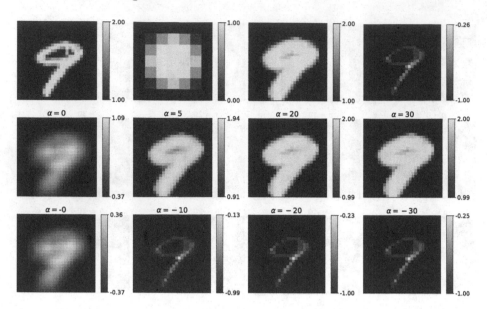

Fig. 2. Top row: input image, non-flat structuring element, target dilation, target erosion. Middle row: $\mathcal{S}$Morph pseudo-dilation for increasing value of α. Bottom row: $\mathcal{S}$Morph pseudo-erosion for increasing value of $|\alpha|$.

Figure 2 showcases examples of applying the $\mathcal{S}$Morph function with a given non-flat structuring element for different values of α. We can see that, as $|\alpha|$ increases, the operation better and better approximates the target operation.

4   Conducted Experiments

4.1   Experimental Protocol

In the following, we evaluate the ability of the proposed $\mathcal{L}$Morph and $\mathcal{S}$Morph layers to properly learn a target structuring element and compare with results obtained by the $PConv$ layer. To do so, we apply in turn dilation $\oplus$, erosion $\ominus$, closing $\bullet$ and opening $\circ$ to all 60000 digits of the MNIST dataset [9], with the target structuring elements displayed by Fig. 3. For dilation and erosion (resp. closing and opening), each network is composed of a single (resp. two) morphological layer(s) followed by a scale/bias $Conv$ $1 \times 1 \times 1$ to rescale the output into the range of the target images. Note that for both $PConv$ and $\mathcal{L}$Morph networks, the image also has to be rescaled in the range $[1, 2]$ before passing through the morphological layer. We train all networks with a batch size of 32, optimizing for the mean squared error (MSE) loss with the Adam optimizer (with starting learning rate $\eta = 0.01$). The learning rate of the optimizer is scheduled to decrease by a factor of 10 when the loss plateaus for 5 consecutive epochs. Convergence is reached when the loss plateaus for 10 consecutive epochs. For the $PConv$ layer, the filter is initialized with 1s and $p = 0$. For $\mathcal{L}$Morph, the filter is initialized with a folded normal distribution with standard deviation $\sigma = 0.01$, and $p = 0$. For the $\mathcal{S}$Morph layer, the filter is initialized with a centered normal distribution with standard deviation $\sigma = 0.01$ and $\alpha = 0$. In all instances, the training is done simultaneously on the weights and the parameter p or α.

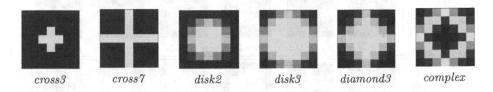

cross3   cross7   disk2   disk3   diamond3   complex

Fig. 3. 7×7 target grayscale structuring elements.

In order to assess the performance of the morphological networks for all scenarios (one scenario being one morphological operation $\oplus$, $\ominus$, $\bullet$ and $\circ$ and one target structuring element among those presented by Fig. 3), we computed the root mean square error (RMSE) between the filter learned at convergence and the target filter. The loss at convergence as well as the value of the parameter p or α also serve as quantitative criteria.

4.2   Obtained Results

Figure 4 gathers the structuring elements learned by the $PConv$, $\mathcal{L}$Morph and $\mathcal{S}$Morph layers for dilation and erosion, and the value of their respective parameter. Looking at the sign of the parameter, all three networks succeed at finding the correct morphological operation. The magnitude of the parameter at convergence also confirms that the operation applied by all networks can be considered as dilation or erosion (and not simply pseudo-dilation or pseudo-erosion). However, looking at the shape of the learned structuring element, it is clear that the $PConv$ layer suffers from the hollow effect mentioned in Sect. 2.3, while both $\mathcal{L}$Morph and $\mathcal{S}$Morph layers accurately retrieve the target structuring element. This is confirmed by the RMSE values between those structuring elements and their respective targets, as presented by Table 1. More particularly, $\mathcal{L}$Morph always achieves the lowest RMSE value for dilation tasks, while $\mathcal{S}$Morph succeeds better on erosion. In any case, the loss at convergence of the $\mathcal{S}$Morph network is almost consistently lower than that of the $\mathcal{L}$Morph network by one or two orders of magnitude, and by two to three with respect to the $PConv$ network.

Figure 5 displays the structuring elements learned by the $PConv$, $\mathcal{L}$Morph and $\mathcal{S}$Morph layers for all 6 target structuring elements for closing and opening operations. This time, since each network comprises of two morphological layers (and a scale/bias $Conv$ $1 \times 1 \times 1$ layer), it is worth mentioning that the two filters evolve independently from each other. Nevertheless, the two morphological layers are expected to learn filters having exactly the same shape, with parameter p or α of opposite signs, once training converges.

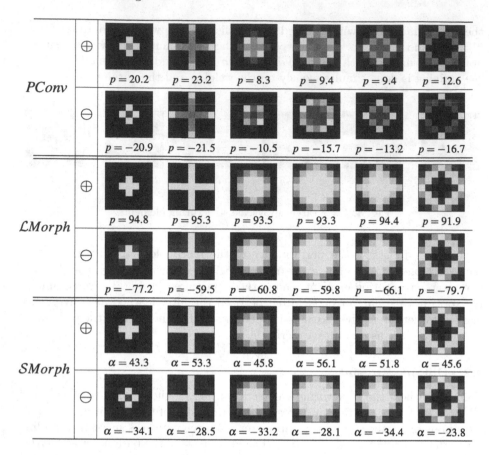

Fig. 4. Learned structuring element (with corresponding p or α at convergence) for $PConv$, $\mathcal{L}$Morph and $\mathcal{S}$Morph layers on dilation $\oplus$ and erosion $\ominus$ tasks.

As can be seen on Fig. 5, the $PConv$ network always succeeds at learning the right morphological operation: the first (resp. second) layer converges to $p > 0$ (resp. $p < 0$) for the closing, and the opposite behavior for the opening. However, quite often $|p| < 10$, indicating that the layer is applying pseudo-dilation or pseudo-erosion only. In addition, the learned structuring element suffers again from the hollow effect for the opening, and does not find the correct shape for the closing. The $\mathcal{L}$Morph network succeeds in learning the correct operation and shape for the closing operation with large target structuring elements (all but *cross3* and *disk2*). For the opening operation however, it consistently fails at retrieving the shape of the target structuring element. This counter-performance is up to now unexplained. The $\mathcal{S}$Morph network also struggles with small structuring elements for both the opening and closing, but perfectly recovers large ones. The edge case of small target structuring elements could come from the scale/bias $Conv$ $1 \times 1 \times 1$, which over-compensates for the gain or loss of aver-

Table 1. MSE loss at convergence and RMSE between the learned structuring element displayed by Fig. 4 and the target for $PConv$, $\mathcal{L}$Morph and $\mathcal{S}$Morph layers on dilation $\oplus$ and erosion $\ominus$ tasks. Best (lowest) results are in bold.

			cross3	cross7	disk2	disk3	diamond3	complex	
$PConv$	$\oplus$	LOSS	0.5	0.8	5.1	6.4	6.8	3.3	$\times 10^{-4}$
		RMSE	0.41	1.42	2.22	3.09	2.80	2.38	
	$\ominus$	LOSS	2.4	0.62	13	2.6	5.2	1.2	$\times 10^{-5}$
		RMSE	0.82	1.55	2.82	3.77	3.63	2.76	
$\mathcal{L}Morph$	$\oplus$	LOSS	0.84	1.2	0.78	1.2	1.2	2.1	$\times 10^{-5}$
		RMSE	**0.02**	**0.00**	**0.02**	**0.05**	**0.01**	**0.48**	
	$\ominus$	LOSS	1.1	**0.37**	1.3	**0.37**	0.58	**1.1**	$\times 10^{-6}$
		RMSE	**0.44**	0.63	0.37	0.29	0.38	0.48	
$\mathcal{S}Morph$	$\oplus$	LOSS	**1.8**	**4.0**	**2.5**	**4.9**	**3.9**	**1.7**	$\times 10^{-7}$
		RMSE	0.10	0.13	0.08	0.06	0.09	0.51	
	$\ominus$	LOSS	210	4.1	**0.8**	4.1	**4.3**	9.5	$\times 10^{-7}$
		RMSE	0.78	**0.15**	**0.09**	**0.05**	**0.09**	0.51	

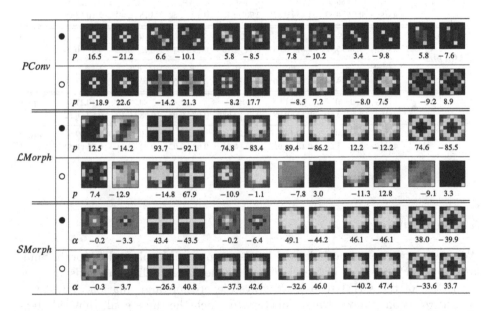

Fig. 5. Learned structuring elements (with corresponding p or α value for each layer) for $PConv$, $\mathcal{L}$Morph and $\mathcal{S}$Morph layers on closing ● and opening ○ tasks.

age pixel intensities. Thus, when back-propagating the error during the learning phase, all filters behind the scale/bias $Conv$ $1 \times 1 \times 1$ layer might not learn the right operation with a parameter p or α not converging toward the correct sign domain.

Table 2 presents the MSE loss at convergence and RMSE value between the learned filters and the target structuring elements for closing and opening scenarios. Except for the aforementioned edge case, the SMorph layer consistently achieves the lowest loss value and RMSE for opening, while the best results for closing are obtained either by LMorph or SMorph layers. Overall, apart from small structuring elements for closing or opening operations, the proposed SMorph layer outperforms its $PConv$ and LMorph counterparts. Last but not least, it should also be noted that the SMorph layer is also numerically more stable. As a matter of fact, raising to the power of p in the $PConv$ and LMorph layers faster induces floating point accuracy issues.

Table 2. MSE loss at convergence and RMSE between the learned structuring elements displayed by Fig. 5 and the target for $PConv$, LMorph and SMorph layers on closing • and opening ○ tasks. Best (lowest) results are in bold.

			cross3	cross7	disk2	disk3	diamond3	complex	
$PConv$	•	LOSS	0.09	5.1	1.7	1.0	5.4	4.9	$\times 10^{-3}$
		RMSE	**0.83 | 0.81**	3.94 | 3.97	2.82 | 3.07	4.44 | 4.64	4.34 | 4.51	3.68 | 3.65	
	○	LOSS	**0.86**	0.76	4.8	3.9	4.4	2.3	$\times 10^{-4}$
		RMSE	**0.91 | 0.32**	1.45 | 0.87	2.70 | 2.49	3.10 | 2.52	3.15 | 2.43	2.10 | 2.08	
LMorph	•	LOSS	**0.5**	0.13	**2.0**	0.18	6.2	0.14	$\times 10^{-4}$
		RMSE	3.58 | 5.12	**0.01 | 0.63**	**0.14 | 1.16**	**0.07 | 0.40**	0.75 | 0.95	**0.47 | 0.99**	
	○	LOSS	8.7	36	2.0	2.7	4.6	2.5	$\times 10^{-3}$
		RMSE	3.17 | 4.74	2.90 | 1.20	1.99 | 1.61	2.81 | 5.56	1.66 | 3.54	2.72 | 3.98	
SMorph	•	LOSS	1700	**0.31**	4400	**0.42**	**0.37**	**0.17**	$\times 10^{-6}$
		RMSE	2.10 | 3.59	**0.14 | 0.08**	2.78 | 3.68	**0.08 | 0.01**	**0.08 | 0.10**	**0.52 | 0.52**	
	○	LOSS	4700	**0.14**	**0.22**	**0.25**	**0.24**	**0.09**	$\times 10^{-6}$
		RMSE	3.40 | 1.63	**0.18 | 0.02**	**0.05 | 0.01**	**0.02 | 0.02**	**0.02 | 0.01**	**0.51 | 0.52**	

5   Conclusion

We present two new morphological layers, namely LMorph and SMorph. Similarly to the $Pconv$ layer of Masci *et al.* [10], the former relies on the asymptotic properties of the CHM to achieve grayscale erosion and dilation. The latter instead relies on the α-softmax function to reach the same goal, thus sidestepping some of the limitations shared by the $PConv$ and LMorph layers (namely being restricted to strictly positive inputs, rescaled in the range $[1, 2]$, and positive structuring functions). In order to evaluate the performances of the proposed morphological layers, we applied in turn dilation, erosion, closing and opening on the whole MNIST dataset images, with 6 target structuring elements of various sizes and shapes, and the morphological layers were trained to retrieve those structuring elements. Qualitative and quantitative comparisons demonstrated that the SMorph layer overall outperforms both the $PConv$ and LMorph layers.

Future work includes investigating the edge cases uncovered for both proposed layers, as well as integrating them into more complex network architectures and evaluating them on concrete image processing applications.

References

1. Angulo, J.: Pseudo-morphological image diffusion using the counter-harmonic paradigm. In: Blanc-Talon, J., Bone, D., Philips, W., Popescu, D., Scheunders, P. (eds.) ACIVS 2010. LNCS, vol. 6474, pp. 426–437. Springer, Heidelberg (2010). https://doi.org/10.1007/978-3-642-17688-3_40
2. Bullen, P.S.: Handbook of Means and Their Inequalities, vol. 560. Springer, Heidelberg (2013). https://doi.org/10.1007/978-94-017-0399-4
3. Charisopoulos, V., Maragos, P.: Morphological perceptrons: geometry and training algorithms. In: Angulo, J., Velasco-Forero, S., Meyer, F. (eds.) ISMM 2017. LNCS, vol. 10225, pp. 3–15. Springer, Cham (2017). https://doi.org/10.1007/978-3-319-57240-6_1
4. Davidson, J.L., Ritter, G.X.: Theory of morphological neural networks. In: Digital Optical Computing II, vol. 1215, pp. 378–388. International Society for Optics and Photonics (1990)
5. Franchi, G., Fehri, A., Yao, A.: Deep morphological networks. Pattern Recogn. **102**, 107246 (2020)
6. Hassoun, M.H., et al.: Fundamentals of Artificial Neural Networks. MIT Press, Cambridge (1995)
7. Lange, M., Zühlke, D., Holz, O., Villmann, T., Mittweida, S.G.: Applications of lp-norms and their smooth approximations for gradient based learning vector quantization. In: ESANN, pp. 271–276 (2014)
8. LeCun, Y., Bengio, Y., Hinton, G.: Deep learning. Nature **521**(7553), 436–444 (2015)
9. LeCun, Y., Cortes, C., Burges, C.J.: The MNIST database of handwritten digits **10**(34), 14 (1998). http://yann.lecun.com/exdb/mnist/
10. Masci, J., Angulo, J., Schmidhuber, J.: A learning framework for morphological operators using counter–harmonic mean. In: Hendriks, C.L.L., Borgefors, G., Strand, R. (eds.) ISMM 2013. LNCS, vol. 7883, pp. 329–340. Springer, Heidelberg (2013). https://doi.org/10.1007/978-3-642-38294-9_28
11. Mellouli, D., Hamdani, T.M., Ayed, M.B., Alimi, A.M.: Morph-CNN: a morphological convolutional neural network for image classification. In: Liu, D., Xie, S., Li, Y., Zhao, D., El-Alfy, E.S. (eds.) International Conference on Neural Information Processing, vol. 10635, pp. 110–117. Springer, Heidelberg (2017). https://doi.org/10.1007/978-3-319-70096-0_12
12. Mondal, R., Dey, M.S., Chanda, B.: Image restoration by learning morphological opening-closing network. Math. Morphol. Theor. Appl. **4**(1), 87–107 (2020)
13. Nogueira, K., Chanussot, J., Dalla Mura, M., Schwartz, W.R., Santos, J.A.D.: An introduction to deep morphological networks. arXiv preprint arXiv:1906.01751 (2019)
14. Ritter, G.X., Sussner, P.: An introduction to morphological neural networks. In: Proceedings of 13th International Conference on Pattern Recognition, vol. 4, pp. 709–717. IEEE (1996)
15. Serra, J.: Image Analysis and Mathematical Morphology. Academic Press, Cambridge (1983)

16. Shen, Y., Zhong, X., Shih, F.Y.: Deep morphological neural networks. arXiv preprint arXiv:1909.01532 (2019)
17. Shih, F.Y., Shen, Y., Zhong, X.: Development of deep learning framework for mathematical morphology. Int. J. Pattern Recogn. Artif. Intell. **33**(06), 1954024 (2019)
18. Soille, P.: Morphological Image Analysis: Principles And Applications. Springer, Heidelberg (2013). https://doi.org/10.1007/978-3-662-05088-0
19. Sussner, P.: Morphological perceptron learning. In: Proceedings of the 1998 IEEE International Symposium on Intelligent Control (ISIC), pp. 477–482. IEEE (1998)
20. Wilson, S.S.: Morphological networks. In: Visual Communications and Image Processing IV, vol. 1199, pp. 483–495. International Society for Optics and Photonics (1989)
21. Zhang, Y., Blusseau, S., Velasco-Forero, S., Bloch, I., Angulo, J.: Max-plus operators applied to filter selection and model pruning in neural networks. In: Burgeth, B., Kleefeld, A., Naegel, B., Passat, N., Perret, B. (eds.) ISMM 2019. LNCS, vol. 11564, pp. 310–322. Springer, Cham (2019). https://doi.org/10.1007/978-3-030-20867-7_24

Scale Equivariant Neural Networks with Morphological Scale-Spaces

Mateus Sangalli$^{(\boxtimes)}$, Samy Blusseau$^{(\boxtimes)}$, Santiago Velasco-Forero, and Jesús Angulo

Centre for Mathematical Morphology, Mines Paristech, PSL Research University, Fontainebleau, France
{mateus.sangalli,samy.blusseau}@mines-paristech.fr

Abstract. The translation equivariance of convolutions can make convolutional neural networks translation equivariant or invariant. Equivariance to other transformations (e.g. rotations, affine transformations, scalings) may also be desirable as soon as we know *a priori* that transformed versions of the same objects appear in the data. The semigroup cross-correlation, which is a linear operator equivariant to semigroup actions, was recently proposed and applied in conjunction with the Gaussian scale-space to create architectures which are equivariant to discrete scalings. In this paper, a generalization using a broad class of liftings, including morphological scale-spaces, is proposed. The architectures obtained from different scale-spaces are tested and compared in supervised classification and semantic segmentation tasks where objects in test images appear at different scales compared to training images. In both classification and segmentation tasks, the scale-equivariant architectures improve dramatically the generalization to unseen scales compared to a convolutional baseline. Besides, in our experiments morphological scale-spaces outperformed the Gaussian scale-space in geometrical tasks.

Keywords: Morphological scale-space · Neural networks · Scale equivariance

1 Introduction

Convolutional Neural Network (CNN) models achieve state-of-the-art performance in many image analysis tasks. An important property of CNNs is that a translation applied to its inputs is equivalent to a translation applied to its features maps, as illustrated in Fig. 1. This property is a particular case of group equivariance [3]. An operator is equivariant with respect to a group if applying a group action in the input and then the operator, amounts to applying the operator to the original input and then an action of the same group to the outputs. In addition to translations, group actions can model many interesting classes of spatial transformations such as rotations, scalings, affine transformations, and so on.

© Springer Nature Switzerland AG 2021
J. Lindblad et al. (Eds.): DGMM 2021, LNCS 12708, pp. 483–495, 2021.
https://doi.org/10.1007/978-3-030-76657-3_35

Group equivariant CNNs [3] are a generalization of CNNs that, in addition to being equivariant to translations, are also equivariant to other groups of transformations. Many of these networks focus on equivariance to rotations, in different kinds of data [3,10,12,13]. A group equivariant neural network may also be used to obtain invariance, with reduction operations. An operator is invariant to some transformation if applying the operator to an input or to its transformed version produces the same output. Invariance is often crucial in image analysis tasks. For example, in a digit classification task a translation should not change the label of the digit, as illustrated by Fig. 1(a). The same holds for re-scaled versions of the same digit (Fig. 1(b)).

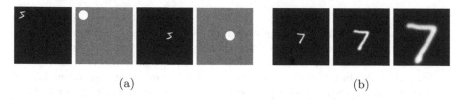

(a)   (b)

Fig. 1. Image from MNIST [7] and their translated versions with a illustration of their respective feature maps in a CNN (a), and images from MNIST Large Scale [6] at different scales (b).

Worrall and Welling [14] introduce neural networks equivariant to the action of semigroups, instead of groups. Semigroup actions can model non-invertible transformations and in [14] the authors focused on equivariance to downsampling in discrete domains. Focusing on downsampling is a way to address equivariance to scalings without creating spurious information through interpolation. In their architecture, the first layer is based on a Gaussian scale-space operator, and subsequent layers of this network are equivariant to the action of a semigroup of scalings and translations. Effectively, these operators are equivariant to rescaling of a discrete image. There are other scale-spaces with similar mathematical properties to the Gaussian scale-space, in particular the morphological scale-spaces.

In this paper we generalize the approach on scale-equivariant neural networks [14] by finding a sufficient condition in which scale-spaces, in the sense of [5], can be used as the first layer, or the so-called *lifting*, in an equivariant network, and investigate several architectures built on morphological scale-spaces. We observe that the morphological scale-spaces networks compare favorably to the Gaussian one in tasks of classification and segmentation of images at scales previously unseen by the network, in contrast to [14], in which the experiments test the overall performance of the network, but where the train and test sets objects follow the same scale distribution. The rest of the paper is organized as follows. In Sect. 2 we give a short review of the existing approaches related to equivariance in CNNs. Then the general mathematical framework we use is exposed in Sect. 3, before we focus specifically on the semigroup of scalings and translations in Sect. 4. In the latter, we also review the algebraic basis of scale-spaces [5] and apply this definition to generalize the scale-equivariant architecture of [14]. In Sect. 5 we test scale-equivariant architectures obtained from different scale-spaces. In particular we test the models in classification and segmentation of images where the objects in the test set appear at scales unseen in the training set. We opt to test the models in tasks where the scale of objects can be easily controlled and measured rather than tasks with real data, where even though objects may appear at different scales, it is difficult to explicitly differentiate between the training and test set scales'. We focus on simple experiments which depend on the shapes of the objects, rather than textures. In those experiments, the equivariant models improved the generalization accuracy dramatically over the convolutional models and the morphological scale-spaces' models performed well even in comparison to the Gaussian scale-spaces.

2   Related Work

In [15], scale equivariance is obtained by applying filters in the different scales using decomposed kernels to reduce the model complexity. The authors applied the equivariant model to multiscale classification and reconstruction of handwritten digit images and were able to surpass the regular CNN models in the generalization to unseen scales, even when using data augmentation. In [4] a locally scale invariant neural network architecture is defined. Filters are defined as linear combinations of a basis of steerable filters and max-pooling the result over different scales. The invariant network was successfully applied to the tasks of classification of re-scaled and distorted images of hand-written digits and clothing. Both of these approaches reduce computational cost and avoid creating spurious information through interpolation by using a decomposition into steerable filters, but applying these models to large scale variations can increase their cost significantly. In [6], a scale invariant architecture is proposed, in which input images are processed in different scales, with foveated images operators, i.e. images are processed with a higher resolution close to the center and a smaller resolution as the operators gets farther from the center. The MNIST Large Scale dataset, which is used later in Sect. 5 was introduced there, and the foveated networks achieved very high generalization performance in unseen scales when compared to regular CNNs. A disadvantage of this approach is that it assumes that the objects of interest are at the center of the image.

In [14], instead of treating scaling as an invertible operation, such as it would behave in a continuous domain, it is considered the action of downsampling the input image in a discrete domain. Because of that, the obtained operators are equivariant to a *semigroup*, and not a group. The semigroup action in the input consists of applying a scale-dependent Gaussian blur to the inputs and then downsampling, which is the way to re-scale discrete signals while avoiding aliasing. The convolutional filters are efficiently applied to feature maps defined on a semigroup formed by scalings and translations by means of dilated convolutions, without relying on interpolation. These operators are also easily scalable to large scale variations, since applying it at larger scales has the same computational cost. The semigroup equivariant models were applied to classification and semantic segmentation of datasets of large images, achieving results which are competitive with the literature. The Gaussian scale-space may not be appropriate when blurring affects the geometrical features that characterize the objects to analyse. With this in mind, the models in this paper are an extension of the ones in [14], but we generalize the approach to allow the usage of other scale-spaces [5], with a focus on morphological scale-spaces, as a step to the generalization of this model to more complicated data, as well as a way to shed some light into the workings of morphological scale-spaces in the context of these images.

3   General Setting

In the scope of image processing, the notion of equivariance of an operator means that a transformed version of an image should produce an "equivalently" transformed version of the original output by the operator, as illustrated in Fig. 1(a). We are specifically interested in *linear* operators, as they are the elementary operations in common neural networks.

3.1   Group Equivariance

Let $(G, .)$ be a discrete group and $\mathcal{F} = \mathbb{R}^G$ the set of functions mapping G to $\mathbb{R}$. Consider the family of operators $R_g, g \in G$ defined on $\mathcal{F}$ by

$$\forall g \in G, \forall f \in \mathcal{F}, \quad R_g(f) : u \in G \mapsto f(u.g^{-1}). \tag{1}$$

This family of operators is a right group action of G on $\mathcal{F}$ (as $R_{g_1} \circ R_{g_2} = R_{g_2 g_1}$). For illustration, the group G could be for example the group of translations of $\mathbb{Z}^2$, identified to $(\mathbb{Z}^2, +)$ itself, and in that case $\mathcal{F}$ could be seen as the set of "infinite" images (or classical images periodised over all $\mathbb{Z}^2$). In turn, the action R_g would be the translation of a function by a vector g.

Bearing in mind the final purpose of defining CNN layers, we focus on linear endomorphism of the vector space $\mathcal{F}$. Suppose such an operator H is equivariant with respect to R_g, that is,

$$\forall f \in \mathcal{F}, \quad H\left(R_g(f)\right) = R_g\left(H(f)\right). \tag{2}$$

Then, using linearity and the fact that the basis $(\mathbb{1}_{\{g\}})_{g \in G}$ spans $\mathcal{F}$, we get $H(f) = \sum_{g \in G} f(g) R_g(h)$ where $h = H(\mathbb{1}_{\{e_G\}})$ is the response of the filter H to the impulse on e_G, the neutral element of G. This writes in the more familiar form

$$\forall u \in G, \quad H(f)(u) = \sum_{g \in G} f(g) h(u.g^{-1}) = \sum_{g \in G} f(g^{-1}.u) h(g). \tag{3}$$

We end with a classical result, at the basis of linear filtering and convolutional neural networks: linearity and equivariance implies for an operator to be written as a convolution with a kernel h that represents it (and conversely). We shall note $H(f) := f \star_G h$ although this operation is commutative only if G is.

3.2 Semigroup Equivariance

Let's first stress the interest of extending the equivariance setting to semigroups. Recall that $(G, \cdot)$ is a semigroup if the law $\cdot$ is associative, but in general they do not have a neutral element or inverse elements. A semigroup induces a semigroup action $(\varphi_g)_{g \in G}$ on a set X, as soon as this family is homomorphic to the semigroup, that is, if either $\forall g, h \in G, \varphi_g \circ \varphi_h = \varphi_{g \cdot h}$. or $\forall g, h \in G, \varphi_g \circ \varphi_h = \varphi_{h \cdot g}$.

In image processing, important examples of semigroup actions are scale-spaces. As we will present in more details in Sect. 4.2, semigroup actions on images may be the convolution with a Gaussian kernel (Gaussian scale-space) or the application of a morphological operator such as erosion, dilation, opening or closing (morphological scale-spaces). Scale-spaces highlight the multi-scale nature of images and have shown great efficiency as image representations [8]. Besides, they are naturally complementary to the scaling operation. For example, the Gaussian blurring acts as a low-pass filter and allows the subsampling (or downscaling on a discrete domain) of an image to avoid aliasing artifacts.

Hence, equivariance of linear operators to semigroups seems highly desirable, as it is natural to expect that the same information at different scales produce the same responses up to some shift due to scale change. However, the derivation of Sect. 3.1 cannot be reproduced here since it includes group inversions, which precisely lack in semigroups. Still, we notice that Eq. 3 can also be written $H(f)(u) = \sum_{g \in G} f(u.g)h(g)$ if we change the function h for its conjugate, that is $h(u) = H(\mathbb{1}_{\{e_G\}})(u^{-1})$, and thanks to a change in variables. Now this expression can be applied to semigroups, considering the semigroup right action $R_u(f)(g) = f(u.g)$. We get that operators H defined by

$$\forall u \in G, \quad H(f)(u) = \sum_{g \in G} R_u(f)(g)h(g) \tag{4}$$

are indeed equivariant to the semigroup action $R_t, t \in G$, since

$$H(R_t(f))(u) = \sum_{g \in G} R_u(R_t(f))(g)h(g) = \sum_{g \in G} R_{tu}(f)(g)h(g) = R_t(H(f))(u).$$
$$\tag{5}$$

This class of semigroup equivariant operators is the semigroup cross-correlation proposed in [14]. We also write $f \star_G h$, remarking however that, contrary to the group case, this operation is not symmetrical in f and h even when the law $\cdot$ on G is commutative.

3.3 Lifting

So far we have considered functions defined on a general semi-group, but the input to CNNs are images defined on a discrete set X, which may be different from the semigroup we seek equivariance to. In particular this will be the case in Sect. 4, when we consider the semigroup product of translations and scalings. In theory the issue is easily overcome, as changing the range of the sum from

$g \in G$ to $x \in X$ in (4) does not change the equivariance property. In practice, that means defining a semigroup action $R_u(f)(x)$. We propose to split this task into two steps. First, we define the semigroup in which (4) holds as it is, like in Sect. 4. Second, we introduce a *lifting* operator Λ to map a function f defined on X into a function Λf defined on G, as it is done in Sect. 4.2. The operator H becomes then

$$\forall u \in G, \quad H(f)(u) = \sum_{g \in G} R_u(\Lambda f)(g)h(g).$$

Since f and $H(f)$ now lie in different spaces, a more general definition of equivariance is necessary: H is equivariant to G if there exists two actions R_u and R'_u such that $H(R'(f)) = R(H(f))$. Now we easily check that it is sufficient for the lifting operator to be equivariant to G, that is, $R_u \circ \Lambda = \Lambda \circ R'_u$, to have equivariance of H. Indeed, in that case by omitting parentheses for readability,

$$H(R'_t f)(u) = \sum_{g \in G} R_u \Lambda R'_t f(g)h(g) = \sum_{g \in G} R_u R_t \Lambda f(g)h(g) = R_t(Hf)(u). \quad (6)$$

The advantage of this two-step approach is the richness of operators induced by the variety of possible liftings, as exposed in Sect. 4.2.

4   Scale and Translations Semigroup

4.1   Scale Cross-Correlation

We focus on the semigroup product of scalings and translations $G = \mathcal{S} \times \mathbb{Z}^2 = (\mathcal{S} \times \mathbb{Z}^2, \cdot)$, with $\mathcal{S} = \{\gamma^{-i} | i \in \mathbb{N}\}$, and $\gamma > 1$ is an integer. The operation $\cdot$ is defined as $(s,x) \cdot (t,y) = (st, s^{-1}y + x)$. Assuming $\gamma = 2$, the operator (4) applied to a signal f at a point (s,x) becomes [14]

$$H(f)(2^{-k}, x) = (f \star_G h)(2^{-k}, x) = \sum_{l \geq 0} \sum_{y \in \mathbb{Z}^2} f(2^{-k-l}, 2^k y + x)h(2^l, y). \quad (7)$$

The operations here were defined for single channel images on G, but they can easily be applied to multichannel images. Let the input $f = (f_1, \ldots, f_n) \in (\mathbb{R}^n)^G$ be a signal with n channels. Assuming the output has m channels, the filter is of the form $h : G \to \mathbb{R}^{n \times m}$. In this case [14], we compute the operator H at channel $o \in \{1, \ldots, m\}$ as $(f \star_G h)_o(2^{-k}, x) := \sum_{c=1}^{n} (f_c \star_G h_{c,o})(2^{-k}, x)$.

Note that in (7), because of the multiplicative constant 2^k in the spatial component, the receptive field (i.e. the region that the network "sees" of the input image at any given output position) of networks consisting of scale semigroup correlations are large. Indeed, a network consisting of L scale cross-correlations with filters with dimensions $P \times K \times K$ (i.e. the support of the filter is a grid $\{(p, k_1, k_2) | p \in \{2^0, 2^1, \ldots, 2^{P-1}\}, 1 \leq k_1, k_2 \leq K\}$) has as a receptive field at each scale s a square of sides $K + 2^{s+P-1}(L-1)(K-1)$, which is large when compared to a convolutional network with L layers and $K \times K$ filters, which has a receptive field of size $L(K-1)+1$. In other words, an architecture built on scale cross-correlations attains the same receptive field as a deep CNN with much smaller depth and number of parameters.

As anticipated in Sect. 3.3, we now have equivariant operators on functions $f : G \to \mathbb{R}^n$ and we need to apply this to images supported by a grid. In the next sections we use the notion of scale-space to define lifting operators that map images to functions on the semigroup of scales and translations. With that we aim to define a neural network architecture that is equivariant with respect to re-scaling of 2D images.

4.2 Scale-Spaces as Lifting Operators

Following the definitions of [5], a family $\{S(t) : \mathcal{F}^{\mathbb{R}^2} \to \mathcal{F}^{\mathbb{R}^2} | t > 0\}$ of operators on images is a *scaling* if:

$$S(1) = \mathrm{id}, \qquad \forall t, s > 0 \ \ S(t)S(s) = S(ts) \tag{8}$$

where id is the identity transform. A scaling can be seen as an action of the group of continuous scalings, which is isomorphic to $(\mathbb{R}_*^+, \times)$. An example is the family $S^{p,q}$, $p, q \geq 0$, given by $(S^{p,q}(t)f)(\mathbf{x}) = t^q f\left(\frac{1}{t^p}\mathbf{x}\right)$, where p and q control rates of the spatial and contrast scaling, respectively.

Let S be a scaling and $\dotplus$ a commutative operation such that $(\mathbb{R}_*^+, \dotplus)$ is a semigroup. Then a $(S, \dotplus)$ *scale-space* is a family $\{T(t)|t > 0\}$ of operators such that, for all $t, s > 0$ [5]:

$$T(t)T(s) = T(t\dotplus s), \qquad T(t)S(t) = S(t)T(1). \tag{9}$$

The property $T(t)S(s) = S(s)T(t/s)$, for all $t, s > 0$, is a direct consequence of the second property [5]. Here, in addition to (9), we assume that the scale-space $T(t)$ is translation-equivariant for all $t > 0$ (i.e. $T(t)(L_z f) = L_z(T(t)f)$ where $L_z(f)(x) = f(x + z)$). Thanks to the second property in (9), a $(S^{p,0}, \dotplus)$ scale-space T defines an operator on images $f : \mathbb{R}^2 \to \mathbb{R}^C$

$$\forall (s, x) \in \mathcal{S} \times E \quad (\Lambda f)(s, x) = (T(s^{-\frac{1}{p}})f)(x) \tag{10}$$

such that, for all $(t, z) \in \mathcal{S} \times \mathbb{Z}^2$, $R_{(t,z)} \circ \Lambda = \Lambda \circ R'_{(t,z)}$, where $R'_{(s,x)}$ is applied to an image on a continuous domain, $f : \mathbb{R}^2 \to \mathbb{R}^C$, as $(R'_{(s,x)}f)(y) = f(s^{-1}y + x)$. So, in order for Λ to be our lifting operator we assume that the input f is a function on a continuous domain. In practice, we discretize Λ and the input images. With this, the morphological scale-spaces, as well as the Gaussian scale-space, being $(S^{\frac{1}{2},0}, \dotplus)$ scale-spaces, can be used as the lifting operators.

Gaussian Scale-Space Lifting: The Gaussian scale-space is a $(S^{\frac{1}{2}}, +)$ scale-space defined by the family $T_{\mathcal{G}}(t)$. For all images $f \in \mathbb{R}^{\mathbb{R}^2}$ and points $x \in \mathbb{R}^2$, $T_{\mathcal{G}}$ can be computed as the convolution $(T_{\mathcal{G}}(t)f)(x) = (f * \mathcal{G}_t)(x)$ where $\mathcal{G}_t(x) = (2\pi t)^{-1} \exp\left(-\frac{\|x\|^2}{2t}\right)$. This was the scale-space considered in [14]. There, it was assumed that image f has a maximum spatial frequency content. They model this by assuming that there exists a signal f_0 and a constant $s_0 > 0$ such that $f = (f_0 * \mathcal{G}_{s_0})$.

Quadratic Morphological Scale-Spaces: Morphological operators can form many different types of scale-spaces [5]. In this paper we consider quadratic morphological scale-spaces. The families of quadratic erosions and dilations by the structuring functions $q_t(x) = -\frac{\|x\|^2}{4ct}$ $t > 0$, given by

$$(T_{\varepsilon_q}(t)f)(x) = \varepsilon_t(f)(x) = \inf_{y \in \mathbb{R}^2} \left(f(x+y) - q_t(y) \right), \quad \text{and,} \tag{11}$$

$$(T_{\delta_q}(t)f)(x) = \delta_t(f)(x) = \sup_{y \in \mathbb{R}^2} \left(f(x-y) + q_t(y) \right), \tag{12}$$

form $(S^{\frac{1}{2},0}, +)$ scale-spaces that can be regarded as morphological counterparts to the Gaussian scale-space [11]. Here, to increase flexibility, we consider a parameter $c > 0$ learned by gradient descent with the rest of the parameters of the network. It is also shown in [5] that the openings $T_{\alpha_q}(t) = T_{\delta_q \circ \varepsilon_q} = \delta_t \circ \varepsilon_t$ and closings $T_{\beta_q}(t) = T_{\varepsilon_q \circ \delta_q} = \varepsilon_t \circ \delta_t$ by those structuring elements form $(S^{\frac{1}{2},0}, \vee)$ scale-spaces, where $\vee$ is the pairwise maximum, $a \vee b = \max\{a, b\}$.

5   Experiments

5.1   Image Classification

The MNIST Large Scale dataset is built upon the MNIST dataset [7] and was introduced to evaluate the ability of CNNs to generalize to scales not seen in the training set [6]. The dataset contains three training sets, tr1, tr2 and tr4, which consist of 50000 samples from the MNIST dataset upscaled by factors one, two and four respectively. The remaining 10000 samples from the original MNIST are used as validation sets, upscaled to match tr1, tr2 and tr4. In our experiments we use tr2 as the training set. The test set is re-scaled to the scales $2^{\frac{i}{4}}, i = -4, \ldots, 12$.

The scale-equivariant architecture used consists of the lifting layer, truncated at five scales, followed by $L = 5$ scale cross-correlations layers[1] and a global max-pooling, before a dense with a softmax activation. The filters have dimension $1 \times 3 \times 3$, and each layer has $16, 16, 32, 32, 64$ feature maps and $1, 2, 1, 2, 1$ strides, respectively, using Batch Normalization and ReLU activations. The architecture is similar to the one used in [6]. By taking a max-pooling over all scales and spatial positions we attempt to make the model invariant to the action of the semigroup G. This makes so that the output is at least as high as the output of the same model after the action of the G. So if at a certain scale the features are indicative of a certain class, the output should also be indicative of that class.

---

[1] In our experiments, we use the implementation of the cross-correlation layer from https://github.com/deworrall92/deep-scale-spaces.

We compare liftings that use scale-spaces: the Gaussian scale-space $T_{\mathcal{G}}$, the quadratic dilation and closing scale-spaces[2] T_{δ_q} and T_{β_q} and the scale-space $T_{\mathrm{id}}(t) = \mathrm{id}$ for all $t > 0$. We also compare with a CNN with size similar to the equivariant models'. For this proposal, one can measure the Euclidean distance between the features obtained from the same images at different scales to quantify the quality of invariance of the model. The distance is normalized by the norm of the inputs. For the purposes of this experiment, we consider the features of the whole dataset as a single vector.

Figure 2(a) shows the accuracies of the models when tested with different scales and in Fig. 2(b) the distances between the features obtained from the same input image at different scales. As expected the equivariant models outperform the CNN model in terms of generalization and even at the training scale. The difference between the peaks of the equivariant models and the CNN models may be related to the difference in the receptive field. In this experiment, the equivariant models performed similarly, with performance peaks at scales one, two and four which are one scaling upwards or downwards from one another. The distances in Fig. 2(b) are mostly consistent with the accuracies and smaller in the equivariant models.

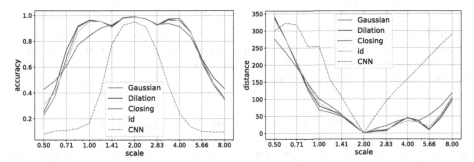

(a) Accuracies of the models at different scales.   (b) Distances between the features in the training scale and all the test scales.

Fig. 2. Results on MNIST Large Scale experiment averaged from five initializations of the models.

5.2 Image Segmentation

In this section we perform an experiment on image segmentation where the objects in the image are re-scaled independently of one another. Unlike in the classification problem, it would be difficult to obtain the images from the dataset by means of data augmentation. In this problem the network benefits from being

---

[2] We do not use erosion and opening scale-spaces because the bright objects in a dark background would be erased by the anti-extensive operators. For a dataset without a well-defined polarity, self-dual operators could be used instead.

locally invariant to re-scaling[3]. We apply a max-pooling over the scale dimension, i.e. the operator $M(f)(x) = \max\{f(s,x)|s \in \{2^0, 2^1, 2^2, \ldots, 2^N\}\}$. This means that the activations used by the softmax layer are the highest for each scale, and should in practice make a locally invariant-model.

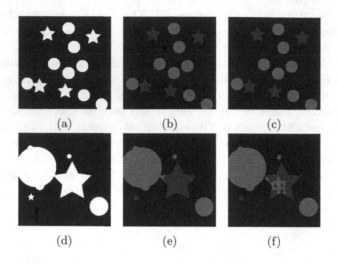

Fig. 3. Example images from the (a) training and (d) test sets of the segmentation experiments, and segmentation results using equivariant models with the (b)(e) proposed dilation and (c)(f) Gaussian scale-spaces. Pixels in red are classified as *disk*, those in blue as *star*. (Color figure online)

The dataset we use in this experiment consists of 224×224 synthetic binary images of shapes such as Fig. 3 which are divided in three classes: disks, stars and the background. In the training set, only one scale is present, like in Fig. 3(a). We construct test sets where each shape is re-scaled by a factor uniformly sampled from the interval $[2^{-i}, 2^i]$, in which we use $i = 1, 2$. Figure 3(b) shows an example of a test image. The train set contains 10000 images and the test sets contain 500 images each. The experiment is repeated ten times, each time generating a different training/test set pair.

The architecture chosen for the equivariant models consists simply of six layers of semigroup cross-correlations. Because the scale cross-correlation has a naturally large receptive field, subsampling is not necessary. The output of the network is a three-channel image with the scores for each class, and the class is chosen as the coordinate with the greatest score. To quantitatively evaluate the models, the Intersection over Union(IoU), or Jaccard index, between the ground truth image and the predictions is used. As baselines, we compare the models to a CNN with the same number of layers and a similar size and number of parameters, and also to a U-Net [9] architecture.

---

[3] We say that an operator ψ on images is locally invariant w.r.t. to $R'_{(s,x)}$ if $\forall (s,x) \in G$ $\psi R'_{(s,x)} f = R'_{(1, s^{-1}x)} \psi f$.

In Table 1 we compare the IoU obtained from different models. We see that CNN performs badly, compared to the equivariant models, even in the training set scale. This is partially attributed to the fact that the receptive field of the CNN is not as large, although having the same number of layers and a similar number of parameters. As expected, the equivariant models outperformed the CNN architectures and the U-Net architecture in the generalization to other scales.

Table 1. IoU between the ground truth images and the predictions obtained from equivariant models with different liftings, trained on images where objects only appear at scale one.

Scales	Gaussian Lifting	Dilation Lifting	Closing Lifting	Id Lifting	U-Net	CNN
1.	0.9929 ± 0.0006	0.9929 ± 0.0006	0.9929 ± 0.0005	0.9927 ± 0.0008	0.9927 ± 0.0006	0.9083 ± 0.0006
$[\frac{1}{2}, 2]$	0.92 ± 0.06	0.97 ± 0.01	0.89 ± 0.06	0.91 ± 0.03	0.86 ± 0.02	0.68 ± 0.01
$[\frac{1}{4}, 4]$	0.88 ± 0.07	0.93 ± 0.02	0.86 ± 0.05	0.88 ± 0.03	0.70 ± 0.04	0.627 ± 0.008

To analyse why the dilation is suited to this particular dataset, we can analyse the effect of applying a discrete re-scaling, i.e. a subsampling to the objects processed by the scale-spaces. In Fig. 4 we see the difference between a Gaussian and dilation lifting followed by a subsampling operator. Indeed, the persistence of concavities of the star shapes makes it easier to distinguish the objects in the last images.

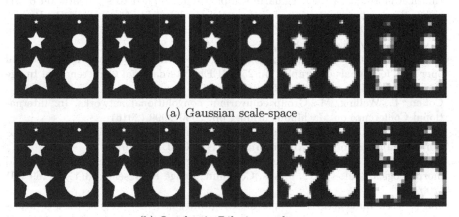

(a) Gaussian scale-space

(b) Quadratic Dilation scale-space

Fig. 4. The same image after being processed by the Gaussian (a) and quadratic dilations (b) scale-spaces and being subsampled by factors 2^i $i = 0, 1, 2, 3, 4$.

6 Conclusions and Future Work

In this paper we presented a generalization of the scale-equivariant models of Worrall and Welling [14] based on the general definition of scale-spaces given in [5]. The models obtained from this approach with different scale-spaces are evaluated in experiments designed to test invariance to change in scales. In our experiments, the generalization to new scales of the models based on scale-spaces surpassed the CNN baselines, including a U-Net model. We see that changing the type of scale-space in the architecture of [14] can induce a change in the performance of the model. In the datasets used, where the geometric information of the images is important, the dilation scale-space model compared very favorably to the Gaussian one. Regarding future works, we note that only the second property of the scale-space definition (9) is necessary in order for operators to be equivariant, which means that some operators that fulfill it but are not scale-spaces, such as the top-hat transform, can be used as lifting operators. It would also be interesting to compare these models with the invariance obtained from data augmentation and possibly combine the two approaches in a complementary way. Additionally, future works will explore the use of the proposed morphological scale-spaces on other types of data like 3D point clouds [1], graphs [2], and high dimensional images.

References

1. Asplund, T., Serna, A., Marcotegui, B., Strand, R., Luengo Hendriks, C.L.: Mathematical morphology on irregularly sampled data applied to segmentation of 3D point clouds of urban scenes. In: Burgeth, B., Kleefeld, A., Naegel, B., Passat, N., Perret, B. (eds.) ISMM 2019. LNCS, vol. 11564, pp. 375–387. Springer, Cham (2019). https://doi.org/10.1007/978-3-030-20867-7_29
2. Blusseau, S., Velasco-Forero, S., Angulo, J., Bloch, I.: Tropical and morphological operators for signals on graphs. In: 25th IEEE International Conference on Image Processing (ICIP), pp. 1198–1202 (2018)
3. Cohen, T., Welling, M.: Group equivariant convolutional networks. In: International Conference on Machine Learning, pp. 2990–2999 (2016)
4. Ghosh, R., Gupta, A.K.: Scale steerable filters for locally scale-invariant convolutional neural networks. arXiv preprint arXiv:1906.03861 (2019)
5. Heijmans, H.J., van den Boomgaard, R.: Algebraic framework for linear and morphological scale-spaces. J. Vis. Commun. Image Represent. $13(1-2)$, 269–301 (2002)
6. Jansson, Y., Lindeberg, T.: Exploring the ability of CNNs to generalise to previously unseen scales over wide scale ranges. arXiv preprint arXiv:2004.01536 (2020)
7. LeCun, Y., Cortes, C., Burges, C.: MNIST handwritten digit database (2010)
8. Lowe, D.G.: Object recognition from local scale-invariant features. In: Proceedings of the Seventh IEEE International Conference on Computer Vision, vol. 2, pp. 1150–1157 (1999)
9. Ronneberger, O., Fischer, P., Brox, T.: U-Net: convolutional networks for biomedical image segmentation. In: Navab, N., Hornegger, J., Wells, W.M., Frangi, A.F. (eds.) MICCAI 2015. LNCS, vol. 9351, pp. 234–241. Springer, Cham (2015). https://doi.org/10.1007/978-3-319-24574-4_28

10. Thomas, N., et al.: Tensor field networks: rotation-and translation-equivariant neural networks for 3D point clouds. arXiv preprint arXiv:1802.08219 (2018)

11. Van Den Boomgaard, R., Smeulders, A.: The morphological structure of images: the differential equations of morphological scale-space. IEEE Trans. Pattern Anal. Mach. Intell. **16**(11), 1101–1113 (1994)

12. Weiler, M., Geiger, M., Welling, M., Boomsma, W., Cohen, T.S.: 3D steerable CNNs: learning rotationally equivariant features in volumetric data. In: Advances in Neural Information Processing Systems, pp. 10381–10392 (2018)

13. Weiler, M., Hamprecht, F.A., Storath, M.: Learning steerable filters for rotation equivariant CNNs. In: Proceedings of the IEEE Conference on Computer Vision and Pattern Recognition, pp. 849–858 (2018)

14. Worrall, D., Welling, M.: Deep scale-spaces: equivariance over scale. In: Advances in Neural Information Processing Systems, pp. 7364–7376 (2019)

15. Zhu, W., Qiu, Q., Calderbank, R., Sapiro, G., Cheng, X.: Scale-equivariant neural networks with decomposed convolutional filters. arXiv preprint arXiv:1909.11193 (2019)

Multivariate and PDE-Based Mathematical Morphology, Morphological Filtering

Eigenfunctions of Ultrametric Morphological Openings and Closings

Jesús Angulo[✉]

MINES ParisTech, PSL-Research University,
CMM-Centre de Morphologie Mathématique, Fontainebleau, France
jesus.angulo@mines-paristech.fr

Abstract. This paper deals with the relationship between spectral analysis in min–max algebra and ultrametric morphological operators. Indeed, morphological semigroups in ultrametric spaces are essentially based on that algebra. Theory of eigenfunctionals in min–max analysis is revisited, including classical applications (preference analysis, percolation and hierarchical segmentation). Ultrametric distance is the fix point functional in min–max analysis and from this result, we prove that the ultrametric distance is the key ingredient to easily define the eigenfunctions of ultrametric morphological openings and closings.

Keywords: Ultrametric space · Ultrametric image processing · (min, max)-analysis · Morphological semigroups

1 Introduction

Given a square matrix A, one of the fundamental linear algebra problems, known as spectral analysis of A, is to find a number λ, called eigenvalue, and a vector v, called eigenvector, such that $Av = \lambda v$. This problem is ubiquitous in both mathematics and physics. In the infinity dimensional generalization, the problem is also relevant for linear operators. An example of its interest is the case of the spectral analysis of Laplace operator. The spectrum of the Laplace operator consists of all eigenvalues λ_i for which there is a corresponding eigenfunction ϕ_i with: $-\Delta\phi_i = \lambda_i\phi_i$. Then, for instance, given any initial heat distribution $f(x)$ in a bounded domain Ω, the solution of the heat equation at time t, $u(x,t) = (f * k_t)(x) = P_t f(x)$, can be written either by its convolution form or by its spectral expansion, i.e.,

$$P_t f(x) = \frac{1}{(4\pi t)^{d/2}} \int_\Omega f(y) e^{-|x-y|^2/4t} dy = \sum_{i=1}^{+\infty} e^{-t\lambda_i} \langle u_0, \phi_i \rangle \phi_i(x), \qquad (1)$$

with the heat kernel as $k_t(x,y) = \sum_{i=1}^{+\infty} e^{-t\lambda_i} \phi_i(x)\overline{\phi_i}(y)$.

From the 60s and 70s of last century, different applied mathematics areas have been studying a more general eigenproblem where the addition and multiplication in matrix and vectors operations are replaced by other pairs of operations.

J. Lindblad et al. (Eds.): DGMM 2021, LNCS 12708, pp. 499–511, 2021.
https://doi.org/10.1007/978-3-030-76657-3_36

If one replaces the addition and multiplication of vectors and matrices by operations of maximum and sum, the corresponding "linear algebra" is called max–plus algebra, which has been extensively studied, including the eigenproblem, see for instance the book [11] for exhaustive list of references. But that problem is out of the scope of this paper. Readers interested on max–plus matrix algebra and spectral analysis from the perspective of mathematical morphology are referred to the excellent survey by Maragos [13]. In the case where one replaces respectively by operations of maximum and minimum, we work on the so-called max–min algebra (also known as bottleneck algebra [5]). Spectral analysis in max–min algebra is also well studied from their first interpretation in the field of hierarchical clustering [8]. Eigenvectors of max-min matrices and their connection with paths in digraphs were widely investigated by Gondran and Minoux, see overview papers [9,12], by Cechlárova [5,6] and by Gavalec [7]. Spectral analysis in max–min algebras is also relatively classic in fuzzy reasoning [15]. This eigenproblem in distributive lattice was studied in [16]. Procedures and efficient algorithms to compute the maximal eigenvector of a given max-min matrix has been also considered [5]. Max-min algebra is also very relevant in several morphological frameworks, such as fuzzy logic, viscous morphology or geodesic reconstruction, see our overview in [1].

In this work, we are interested on relating the notion of spectral analysis in max–min algebra to ultrametric morphological operators [2]. Indeed, morphological semigroups in ultrametric spaces are essentially based on that algebra. The expansion provided by (1) for the diffusion process using the Laplacian eigenfunctions can be similarly formulated in ultrametric spaces [3]. The interpretation of ultrametric Laplace eigenfunctions depends obviously on the hierarchical organisation of ultrametric balls according to the ultrametric distance. We show here that the ultrametric distance is also the key ingredient to define the eigenfunctions of ultrametric morphological openings and closings. The theoretical results of this paper are mainly based on Gondran and Minoux theory, where the discrete case was considered in [9] and later, in [10] the continuous (and infinite dimensional) one. The later study also considered the preliminary interest in nonlinear physics such as percolation. This paper is another step forwards in our program of revisiting classical image/data processing on ultrametric representations. The rest of its contents is organised as follows. Section 2 provides a short reminder of the main definitions and properties of ultrametric morphological operators. Gondran and Minoux theory of min–max analysis of operators and matrices is reviewed is Sect. 3. Section 4 discusses our contribution to the study of the eigensystem of ultrametric morphological operators. Finally, Sect. 5 closes the paper with some conclusions and perspectives.

2   Reminder About Ultrametric Morphological Openings and Closings

Before going further, let us recall basic facts on ultrametric morphological openings and closings. One can refer to [2] for details.

An ultrametric space (X, d) is a metric space in which the triangle inequality is strengthened to $d(x, z) \leq \max\{d(x, y), d(y, z)\}$. Notably, that implies that the set of ball of radius r provides a partition of X into disjoint balls. Given a separable and complete ultrametric space (X, d), let us consider the family of non-negative bounded functions f on (X, d), $f : X \to [0, M]$. The complement (or negative) function of f, denoted f^c, is obtained by the involution $f^c(x) = M - f(x)$. The set of non-negative bounded functions on ultrametric space is a lattice with respect to the pointwise maximum $\vee$ and minimum $\wedge$.

Definition 1. *The canonical isotropic ultrametric structuring function is the parametric family $\{b_t\}_{t>0}$ of functions $b_t : X \times X \to (-\infty, M]$ given by*

$$b_t(x, y) = M - \frac{d(x, y)}{t}. \tag{2}$$

Definition 2. *Given an ultrametric structuring function $\{b_t\}_{t>0}$ in (X, d), for any non-negative bounded function f the ultrametric dilation $D_t f$ and the ultrametric erosion $E_t f$ of f on (X, d) according to b_t are defined as*

$$D_t f(x) = \sup_{y \in X} \{f(y) \wedge b_t(x, y)\}, \quad \forall x \in X, \tag{3}$$

$$E_t f(x) = \inf_{y \in X} \{f(y) \vee b_t^c(x, y)\}, \quad \forall x \in X. \tag{4}$$

We can easily identify that the ultrametric dilation is a kind of product in $(\max, \min)$-algebra of function f by b_t. Considering the classical algebraic definitions of morphological operators [17] for the case of ultrametric semigroups $\{D_t\}_{t\geq0}$, resp. $\{E_t\}_{t\geq0}$, they have the properties of increasingness and commutativity with supremum, resp. infimum, which involves that D_t is a dilation and E_t is an erosion. In addition, they are extensive, resp. anti-extensive, operators and, by the supremal semigroup property, both are idempotent operators, i.e., $D_t D_t = D_t$ and $E_t E_t = E_t$, which implies that D_t is a closing and E_t is an opening. Finally, these semigroups are just the so-called granulometric semigroup [17] and therefore $\{D_t\}_{t\geq0}$ is an anti-granulometry and $\{E_t\}_{t\geq0}$ is a granulometry, which involve interesting scale-space properties useful for filtering and decomposition.

Let (X, d) be a discrete ultrametric space. Choose a sequence $\{c_k\}_{k=0}^{\infty}$ of positive reals such that $c_0 = 0$ and $c_{k+1} > c_k \geq 0$, $k = 0, 1, \cdots$. Then, given $t > 0$, the following sequence $\{b_{k,t}\}_{k=0}^{\infty}$ is defined, such that

$$b_{k,t} = M - t^{-1}c_k. \tag{5}$$

Let us define $\forall k$, $\forall x \in X$, the ultrametric dilation and erosion of radius k on the associated partition as

$$Q_k^{\vee} f(x) = \sup_{y \in B_k(x)} f(y), \tag{6}$$

$$Q_k^{\wedge} f(x) = \inf_{y \in B_k(x)} f(y), \tag{7}$$

where $B_k(x)$ is the ultrametric ball of radius k and center x. Using now (6) and (7), it is straightforward to see that the ultrametric dilation and ultrametric erosion of f by $b_{k,t}$ can be written as

$$D_t f(x) = \sup_{0 \leq k \leq \infty} \left\{ Q_k^\vee f(x) \wedge b_{k,t} \right\}, \tag{8}$$

$$E_t f(x) = \inf_{0 \leq k \leq \infty} \left\{ Q_k^\wedge f(x) \vee (M - b_{k,t}) \right\}. \tag{9}$$

From this formulation, one does not need to compute explicitly the ultrametric distance between all-pairs of points x and y since $D_t f(x)$ and $E_t f(x)$ are obtained by working on the supremum and infimum mosaics $Q_k^\vee f(x)$ and $Q_k^\wedge f(x)$ from the set of partitions, which is usually finite, i.e., $k = 0, 1, \cdots, K$.

3   Eigen-Functionals in (min, max)-Analysis

In this section, the main elements of the Gondran and Minoux theory [9, 10] of eigenvalues and eigen-functionals of diagonally dominant endomorphophisms in min–max analysis is reviewed. The theory is the background to the specific problem of the study of eigenfunction of ultrametric morphological operators.

3.1   Inf-Diagonal Dominant Kernel (idd-kernel) in (min, max)-Algebra and Its Powers

Let us first introduce the axiomatic definition of an *inf-diagonal dominant kernel (idd-kernel)*.

Definition 3. *A proper lower semi-continuous (with closed and bounded lower-level sets) functional $\alpha : X \times X \to \overline{\mathbb{R}}$ is called an idd-kernel if the following two conditions are satisfied*

1. *Boundedness and diagonal uniformity: there exists a finite value 0_α such that*

$$\alpha(x, x) = 0_\alpha, \quad \forall x \in X;$$

2. *Inf-diagonal dominance, i.e.,*

$$\alpha(x, y) \geq 0_\alpha, \quad \forall x, y \in X;$$

which is equivalent to $\forall x \in X$:

$$\alpha(x, x) \leq \inf_{y \neq x} \left\{ \alpha(x, y) \right\}. \tag{10}$$

Let us denote by $\mathcal{A}$ the set of idd-kernels in X. Using the (min, max)-associativity property, the successive (min, max)-powers of an idd-kernel $\alpha \in \mathcal{A}$ in X may be defined recursively as:

$$\alpha^n(x, y) = \min_{z \in X} \left\{ \alpha^{n-1}(x, z) \vee \alpha(z, y) \right\}, \quad \forall n \in \mathbb{N}, n > 2. \tag{11}$$

Let $\alpha \in \mathcal{A}$ be a idd-kernel. Considering for instance $n = 2$, one has

$$\alpha^2(x, y) = \min_{z \in X} \{\alpha(x, z) \vee \alpha(z, y)\}.$$

In particular, taking $z = y$ above and using the inf-diagonal dominance, we have

$$\alpha^2(x, y) \leq \alpha(x, y) \vee \alpha(y, y) = \alpha(x, y), \quad x, y \in X,$$

which provides a non-increasing behaviour. Indeed, there exists a stronger convergence result to a fix-point which is easily seen for the fact that the sequence $\alpha(x, y), \alpha^2(x, y) \cdots \alpha^n(x, y)$ is non-increasing, together with the fact that it is bounded from below by $\alpha(x, x) = 0_\alpha$. More formally:

Proposition 1 ([10]). *The endomorphism α^* defined by limit*

$$\alpha^*(x, y) = \lim_{n \to \infty} \alpha^n(x, y) \tag{12}$$

always exists (i.e., limit is convergent) and satisfies the relationships

$$\alpha^* = (\alpha^*)^2 = \alpha \cdot_{\min,\max} \alpha^* = \alpha^* \cdot_{\min,\max} \alpha.$$

3.2 (min, max)-Eigenfunctions of α and α^*

Let us introduce now the (min, max)-product of a function $f \in \mathcal{F}(X, \overline{\mathbb{R}})$ and an idd-kernel $\alpha \in \mathcal{A}$ as follows

$$(f *_{\min,\max} \alpha)(x) = \inf_{y \in X} \{f(y) \vee \alpha(x, y)\}. \tag{13}$$

Definition 4. *Given an idd-kernel $\alpha \in \mathcal{A}$ and $\lambda \in \mathbb{R}$, a ψ is called a (min, max)-eigenfunction of α for the eigenvalue λ if and only if*

$$\psi *_{\min,\max} \alpha = \lambda \vee \psi.$$

The following two propositions provide on the one hand, the equivalence of the eigenfunctions of α and α^* and on the other hand an explicit way to compute the eigenfunctions from the "columns" of α^* [10]. We include the proof of the second proposition to justify the simplicity of the construction.

Proposition 2 ([10]). *Let $\lambda > 0_\alpha$. If ψ is a (min, max)-eigenfunction of α for the eigenvalue λ then $\psi(x) \geq \lambda, \forall x \in X$. In addition, one has*

- *$\lambda \vee \psi = 0_\alpha \vee \psi = \psi$;*
- *$\lambda \vee \psi = \psi *_{\min,\max} \alpha = \psi *_{\min,\max} \alpha^*$.*

Proposition 3 ([10]). *For $\alpha \in \mathcal{A}$ and $\lambda \in \mathbb{R}$ and for an arbitrary fixed $y \in X$, let ϕ_λ^y denote the functional in $f \in \mathcal{F}(X, \overline{\mathbb{R}})$ defined by*

$$\phi_\lambda^y(x) = \lambda \vee \alpha^*(x, y). \tag{14}$$

Then, ϕ_λ^y is a (min, max)-eigenfunction of α for the eigenvalue λ.

Proof. Since $\max\{\alpha(x,z); \phi_\lambda^y(z)\} = \max\{\alpha(x,z); \alpha(z,y); \lambda\}$, we obtain

$$(\phi_\lambda^y *_{\min,\max} \alpha)(x) = \inf_{z \in X} \{\alpha(x,z) \vee \phi_\lambda^y(z)\}$$

$$= \max\left\{\inf_{z \in X} \{\alpha(x,z) \vee \alpha^*(z,y)\}; \lambda\right\}$$

$$= \max\{\alpha \cdot_{\min,\max} \alpha^*(x,y); \lambda\}.$$

From properties of α^*, $\alpha \cdot_{\min,\max} \alpha^*(x,y) = \alpha^*(x,y)$, thus $\forall x$,

$$(\phi_\lambda^y *_{\min,\max} \alpha)(x) = \max\{\alpha^*(x,y); \lambda\} = \max\{\max(\alpha^*(x,y); \lambda); \lambda\}$$
$$= \max\{\phi_\lambda^y(x); \lambda\} = \phi_\lambda^y(x) \vee \lambda.$$

Finally, the following representation theorem provides us the interest of the theory.

Theorem 1 (Gondran and Minoux, 1998 [10]). *Let $\alpha \in \mathcal{A}$, $\lambda > 0_\alpha$, and for any $x, y \in X$, one computes $\phi_\lambda^y(x) = \lambda \vee \alpha^*(x,y)$. Then, the set*

$$G_\lambda = \{\phi_\lambda^y(x) : y \in X\},$$

is the unique minimal generator of the set of (min, max)*-eigenfunctions of λ. Consequently, for any* (min, max)*-eigenfunction ψ of α with eigenvalue λ, there exists a functional $h \in \mathcal{F}(X, \overline{\mathbb{R}})$ such that ψ can be expressed in terms of the $\phi_\lambda^y(x)$ as*

$$\psi(x) = \inf_{y \in X} \{h(y) \vee \phi_\lambda^y(x)\} = \langle h, \phi_\lambda^y\rangle_{\min,\max}. \tag{15}$$

3.3   Discrete Case of idd-kernels

Let X be a finite discrete space with $|X| = n$. The functional $\alpha(x_i, x_j)$ is represented by a matrix $A = (a_{ij}) \in M_n(\mathbb{R})$, i.e., $a_{ij} = \alpha(x_i, x_j)$. The corresponding eigenproblem is written as

$$A \cdot_{\min,\max} v = \lambda \vee v$$

where the matrix operations are given as follows. Given three matrices $A, B, C \in M_n(\mathbb{R})$, a scalar $\lambda \in \mathbb{R}$ and two vector $v, w \in \mathbb{R}^n$, one has matrix multiplication $A \cdot_{\min,\max} B = C \Leftrightarrow \bigwedge_{1 \leq k \leq n}(a_{ik} \vee b_{kj}) = c_{ij}$, multiplication of a matrix by a scalar $\lambda \vee A = B \Leftrightarrow \lambda \vee a_{ij} = b_{ij}$ and multiplication of a vector by a matrix $A \cdot_{\min,\max} v = w \Leftrightarrow \bigwedge_{1 \leq j \leq n}(a_{ij} \vee v_j) = w_i$. Thus

$$A^{(k)} = A \cdot_{\min,\max} A^{(k-1)},$$

is just the matrix product in the matrix algebra (min, max). The limit

$$A^* = \lim_{k \to \infty} A^{(k)} = \alpha^*(x_i, x_j)$$

is called the quasi-inverse of A in (min, max)-matrix algebra [9]. Obviously, (min, max)-eigenfunctions theory is valid for the discrete case.

3.4   Two Applications

We consider now two first applications of the (min, max)-spectral theory.

Preference Analysis in (max, min)-Algebra [12]. Given n objects, find a total ordering between them using the pairwise comparison preferences (or votes) given by K judges. The results of this kind of ranking can be represented by a preference $n \times n$-matrix $A = (a_{ij})$, where a_{ij} denotes the number of judges who prefer i to j. Note that by construction of the matrix A, $a_{ij} + a_{ji} = K$, $\forall i, j$, $i \neq j$. In the case of ties, it is assumed a $1/2$ contribution.

Starting from A, the method of partial orders is based on determining a hierarchy of preference relations on the objects with nested equivalence classes. More precisely, for any λ, the classes at level λ are defined as the strong connected components of the graph $G_\lambda = (X, E_\lambda)$, with node set X are the objects and the set of edges is $E_\lambda = \{(i, j) : a_{ij} \geq \lambda\}$.

Let us consider example with $n = 4$ and $K = 6$, given by the following matrix A and its quasi-inverse in the (max, min)-algebra A^*:

$$A = \begin{pmatrix} 0 & 3 & 4 & 3.5 \\ 3 & 0 & 4 & 1 \\ 2 & 2 & 0 & 5 \\ 2.5 & 5 & 1 & 0 \end{pmatrix}, \quad A^* = A^3 = \begin{pmatrix} 0 & 4 & 4 & 4 \\ 3 & 0 & 4 & 4 \\ 3 & 5 & 0 & 5 \\ 3 & 5 & 4 & 0 \end{pmatrix}.$$

There is thus three eigenvalues $\lambda_1 = 5$, $\lambda_2 = 4$ and $\lambda_3 = 3$. At level $\lambda_3 = 3$, G_3 is just a single connected component and the four objects are therefore not ordered. Level $\lambda_2 = 4$ leads to a quotient graph G_4, refined into two classes, where object 1 is preferred over the three others, but the objects 2, 3 and 4 are undistinguishable. The level $\lambda_1 = 5$ provides a differentiation order between them: 3 is preferred over 4, which, in turn, is preferred over 2. The (max, min)-approach can be compared with a $(+, \times)$-based spectral analysis of A, associated to the method proposed by Berge [4]. It consists in a "best mean ordering" of the objects according to the non-increasing values of the components of the real eigenvector v corresponding to the largest eigenvalue λ, $Av = \lambda v$. In the current example [12], the largest eigenvalue is $\lambda = 8.92$ and corresponding eigenvector is $v = (0.56\ 0.46\ 0.50\ 0.47)^T$. Thus A^* provides at the end the same order, but in addition various quotient graphs corresponding to the hierarchical partial orders.

Note that replacing objects and judges by drugs and effects on patients, the problem is relevant in medical analysis [15].

Percolation on Distribution of Particles in (max, min)-Algebra [10]. Let us consider a continuous (or discrete) distribution of particles in the space X and $\alpha(x, y)$ can be interpreted as a potential of interaction between particles located at x and y (instead of a "distance", it should be seen as an "affinity"); e.g., for a random function, we can for instance use the difference of intensities to define the affinity. The λ-percolation, or connectivity problem up to threshold λ, consists in finding for any pair of distinct points x and y in X a path with respect to the threshold λ.

Consider the dual (max, min)-eigensystem, i.e., sup-diagonal dominant kernel $\alpha(x, x) \geq \sup_{y \neq x}\{\alpha(x, y)\}$.

First, compute the limit $\alpha^*(x, y)$: as we show just below, that can obtained using the minimum spanning tree on the dual of the graph with the potential of interaction as affinity. The (max, min)-eigenfunctions for any eigenvalue λ are

$$\phi_\lambda^y(x) = \lambda \wedge \alpha^*(x, y).$$

Then, for any $y \in X$, there is a percolation path between x and y if $\phi_\lambda^y(x) = \lambda$.

4   Eigensystem on Ultrametric Morphological Operators

Eigenfunctional analysis in (min, max)-algebra is the natural framework in the case of ultrametric spaces.

4.1   (min, max)-Eigensystem in Ultrametric Space

Let (X, d) be a length space, i.e., a metric space in which the distance between two points $x, y \in X$ is given by the infimum of the lengths of paths which join them. It is easy to see that $d(x, y)$ is just an example of an idd-kernel. The corresponding limit of (min, max)-powers (11) can be written as [11]:

$$d^*(x, y) = \min_{\pi \in \text{path}(x, y)} \quad \max_{\pi \, : \, k=0, \cdots, p-1} d(z_k, z_{k+1}) \tag{16}$$

with $\pi = \{z_0 = x, z_1, \cdots, z_p = y\}$ is a path in X. We easily see that $d^*(x, y) \leq \max_z (d^*(x, z), d^*(z, y))$. Therefore $d^*(x, y)$ is the *sub-dominant ultrametric on* (X, d), defined as the largest ultrametric below the given dissimilarity $d(x, y)$.

Let us note $\Lambda = \{d^*(x, y) : x, y \in X, \ x \neq y\}$. Any $\lambda \in \Lambda$ is a (min, max)-eigenvalue of $d(x, y)$ and $d^*(x, y)$ and the corresponding *minimal generators of* (min, max)-*eigenfunctions* are given by

$$\phi_\lambda^y(x) = \lambda \vee d^*(x, y). \tag{17}$$

Application to Hierarchical Classification. Let us recall the pioneering result on the connection between hierarchical classification and spectral analysis in (min, max)-algebra.

Theorem 2 (Gondran, 1976 [8]). *At each level λ of a hierarchical classification (dendogram) associated to the sub-dominant ultrametric of a distance matrix $D_{i,j} = d(i, j)$, where $D_{i,:}$ denote the i-th column of D. Two objects (leafs of the tree) i and j belong to the same class at level λ (ultrametric ball of radius λ) if and only if the two (min, max)-eigenfunctions associated to their columns are equal, i.e.,*

$$\lambda \vee D_{i,:}^* = \lambda \vee D_{j,:}^*.$$

The set of distinct vectors of the form $\lambda \vee D_{i,:}^*$, forms the unique minimal generators of (min, max)-eigenvectors of D for eigenvalue λ. With $\lambda = 0$, the set of columns of D^* minimal generator of the eigenvectors associated to the neutral element of the "product": $0 \vee D_{i,:}^* = D_{i,:}^*$.

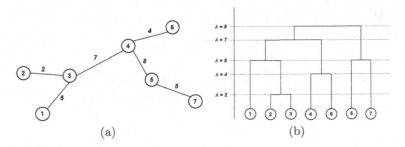

Fig. 1. (a) Example of minimum spanning tree and (b) associated dendrogram.

The following example, borrowed from [11] illustrates this theorem. Consider the distance matrix D and its pseudo-inverse D^*:

$$D = \begin{pmatrix} 0 & 7 & 5 & 8 & 10 & 8 & 10 \\ 7 & 0 & 2 & 10 & 9 & 9 & 10 \\ 5 & 2 & 0 & 7 & 11 & 10 & 9 \\ 8 & 10 & 7 & 0 & 8 & 4 & 11 \\ 10 & 9 & 11 & 8 & 0 & 9 & 5 \\ 8 & 9 & 10 & 4 & 9 & 0 & 10 \\ 10 & 10 & 9 & 11 & 5 & 10 & 0 \end{pmatrix}, \quad D^* = \begin{pmatrix} 0 & 5 & 5 & 7 & 8 & 7 & 8 \\ 5 & 0 & 2 & 7 & 8 & 7 & 8 \\ 5 & 2 & 0 & 7 & 8 & 7 & 8 \\ 7 & 7 & 7 & 0 & 8 & 4 & 8 \\ 8 & 8 & 8 & 8 & 0 & 8 & 5 \\ 7 & 7 & 7 & 4 & 8 & 0 & 8 \\ 8 & 8 & 8 & 8 & 5 & 8 & 0 \end{pmatrix}.$$

In this discrete setting, the matrix of ultrametric distances D^* can be computed efficiently using a minimum spanning tree (MST) algorithm. Figure 1(a) depicts the MST corresponding to the graph of D as adjacency matrix, matrix D^* is straightforward derived from it. For instance, $d*(2,7) = \max (d(2,3), d(3,4), d(4,5), d(5,8)) = 8$. The associated hierarchical classification represented by a dendrogram is given in Fig. 1(b). Taking for instance $\lambda = 5$, one has the three eigenvectors

$$v_1 = \lambda \vee D^*_{1,:} = \lambda \vee D^*_{2,:} = \lambda \vee D^*_{3,:} = (5\ 5\ 5\ 7\ 8\ 7\ 8)^T$$

$$v_2 = \lambda \vee D^*_{4,:} = \lambda \vee D^*_{6,:} = (7\ 7\ 7\ 5\ 8\ 5\ 8)^T$$

$$v_3 = \lambda \vee D^*_{5,:} = \lambda \vee D^*_{7,:} = (8\ 8\ 8\ 8\ 5\ 8\ 5)^T$$

which are the minimal generator $G_5 = \{v_1, v_2, v_3\}$.

Interpretation of (min, max)-Eigenfunction in an Ultrametric Space.
Using the fact that in an ultrametric space if $d^*(x,y) < r$ then $B_r(x) = B_r(y)$, one has that two points x and y that belonging to the ultrametric ball of radius λ have the same (min, max)-eigenfunction, i.e., if $y \in B_\lambda(x)$ (which implies that $x \in B_\lambda(y)$) then $\lambda \vee d^*(x, z) = \lambda \vee d^*(y, z)$, $z \in X \Leftrightarrow \phi^x_\lambda = \phi^y_\lambda$. In addition, we can easily see that

$$\phi^y_\lambda(x) = \lambda\, 1_{B_\lambda(y)}(x) + \underbrace{d^*(x,y)}_{\geq \lambda}\, 1_{X \setminus B_\lambda(y)}(x), \qquad (18)$$

with the following "normalization":

$$\inf_{y \in X} \phi_\lambda^y(x) = \lambda \mathbf{1}_X(x), \forall x \in X.$$

Figure 2 depicts two partitions of a discrete ultrametric space at levels λ and $\lambda + 1$ and the corresponding $(\min, \max)$-eigenfunction at point $y \in X$ and eigenvalue λ.

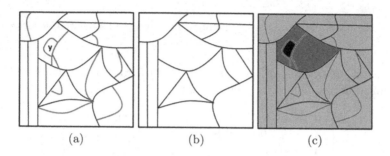

(a)   (b)   (c)

Fig. 2. $(\min, \max)$-eigenfunction at point $y \in X$ and eigenvalue λ: (a) and (b) depict two partitions of a discrete ultrametric space at levels λ and $\lambda + 1$; (c) eigenfunction $\phi_\lambda^y(x)$. Black corresponds to λ.

4.2 Eigenfunctions of Ultrametric Erosion-Opening and Dilation-Closing

The property $d^*(x,y) \cdot_{\min,\max} d^*(x,y) = d^*(x,y)$ is the basic ingredient in [2] to prove the existency of the supremal ultrametric morphological semigroups and the fact that ultrametric erosion and dilation are idempotent operators. Let us also notice that using expression (13), one has

$$E_t f(x) = \inf_{y \in X} \left\{ f(y) \vee t^{-1} d^*(x,y) \right\} = \left(f *_{\min,\max} t^{-1} d^* \right)(x)$$

$$= t^{-1} \left(t f *_{\min,\max} d^* \right)(x), \quad \forall x \in X, t > 0. \tag{19}$$

Proposition 4. *Given an ultrametric space (X, d^*), the corresponding $\phi_\lambda^y(x)$, $\forall y \in X$ and $\lambda \in \Lambda$, is an eigenfunction of the ultrametric erosion-opening with $t = 1$, i.e.,*

$$E_1 \phi_\lambda^y(x) = \lambda \vee \phi_\lambda^y(x). \tag{20}$$

For $t \neq 1$, one has the following scaling

$$E_t t^{-1} \phi_\lambda^y(x) = t^{-1} \left(\lambda \vee \phi_\lambda^y(x) \right).$$

Proof. We have $E_1\phi_\lambda^y(x) = (\phi_\lambda^y *_{\min,\max} d^*)(x)$, then

$$(\phi_\lambda^y *_{\min,\max} d^*)(x) = \inf_{z \in X} \{\phi_\lambda^y(z) \vee d^*(x,z)\} = \inf_{z \in X} \{\lambda \vee d^*(y,z) \vee d^*(z,x)\}$$
$$= \lambda \vee (d^*(x,y) *_{\min,\max} d^*(x,y)) = \lambda \vee d^*(x,y) = \lambda \vee \phi_\lambda^y(x).$$

When $t \neq 1$, from (19), we obtain

$$E_t t^{-1}\phi_\lambda^y(x) = t^{-1}(\phi_\lambda^y *_{\min,\max} d^*)(x) = t^{-1}E_1\phi_\lambda^y(x),$$

so finally, using (20), one has $E_t t^{-1}\phi_\lambda^y(x) = t^{-1}(\lambda \vee \phi_\lambda^y(x))$.

A similar result is obtained for the eigenfunctions of the ultrametric dilation-closing

$$D_1\bar{\phi}_\lambda^y(x) = \bar{\lambda} \wedge \bar{\phi}_\lambda^y(x). \tag{21}$$

where $\bar{\phi}_\lambda^y(x)$ are the corresponding (max, min)-eigenfunctions, obtained from the dual metric distance and the corresponding dual eigenvalues $\bar{\lambda}$.

Using the alternative representation of the discrete erosion (9) with $c_k = \lambda_k$ $E_t f(x) = \inf_{\lambda_k \in \Lambda}\{Q_{\lambda_k}^\wedge f(x) \vee \lambda_k\}$, we have the following result.

Proposition 5. *The ultrametric erosion-opening of a function f on (X, d^*) at $t = 1$ can be written as the expansion on the base of* (min, max)-*eigenfunctions* $\{\phi_{\lambda_k}^y\}_{\lambda_k \in \Lambda}$ *as follows:*

$$E_1 f(x) = \inf_{\lambda_k \in \Lambda}\{\langle f, \phi_{\lambda_k}^y\rangle_{\min,\max} \vee \lambda\}. \tag{22}$$

Scaling from (19) *provides the corresponding expansion for $E_t f(x)$.*

Proof. The (min, max)-scalar product of a function f on (X, d^*) and the (min, max)-eigenfunction ϕ_λ^y:

$$\langle f, \phi_\lambda^y\rangle_{\min,\max} = \inf_{y \in X}\{f(y) \vee \phi_\lambda^y(x)\}.$$

Now, using the expression (18), one has

$$\langle f, \phi_\lambda^y\rangle_{\min,\max} = \left[\inf_{x \in B_\lambda(y)}\{f(x)\} \vee \lambda\right] \wedge \inf_{x \notin B_\lambda(y)}\{f(x) \vee d^*(y,x)\}$$
$$= Q_\lambda^\wedge f(x) \vee \lambda.$$

Obviously, a similar expansion can be obtained from the dilation-closing using the (max, min)-eigenfunctions of ultrametric space X.

5   Conclusions and Perspectives

Morphological semigroups in ultrametric spaces can be seen as the min–max product (and its dual) of a function and a scaled version of the ultrametric distance. Ultrametric distance is a fixed point functional in min–max analysis and its eigenfunctions are defined in a direct way. From this viewpoint, min–max spectral analysis on ultrametric spaces describes the nested organization of ultrametric balls. Eigenfunctions of ultrametric distance are just the eigenfunctions of ultrametric erosion-opening and this spectral base provide an expansion of morphological operators.

This theory is related to Meyer's theory of watershed on node-or-edge-weighed graphs [14], since min–max algebra is also connected to analysis of paths on digraphs. A better understanding of the connection between the present spectral theory and Meyer's theory could help into the integrative use of ultrametric operators in segmentation and filtering.

References

1. Angulo, J.: Convolution in (max, min)-Algebra and Its Role in Mathematical Morphology. Adv. Imaging Electron Phys. **203**, 1–66 (2017)
2. Angulo, J., Velasco-Forero, S.: Morphological semigroups and scale-spaces on ultrametric spaces. In: Angulo, J., Velasco-Forero, S., Meyer, F. (eds.) ISMM 2017. LNCS, vol. 10225, pp. 28–39. Springer, Cham (2017). https://doi.org/10.1007/978-3-319-57240-6_3
3. Angulo, J., Velasco-Forero, S.: Hierarchical Laplacian and its spectrum in ultrametric image processing. In: Burgeth, B., Kleefeld, A., Naegel, B., Passat, N., Perret, B. (eds.) ISMM 2019. LNCS, vol. 11564, pp. 29–40. Springer, Cham (2019). https://doi.org/10.1007/978-3-030-20867-7_3
4. Berge, C.: La théorie des graphes et ses applications. Dunod (Libraire), Paris (1958)
5. Cechlárová, K.: Eigenvectors in bottleneck algebra. Linear Algebra Appl. **175**, 63–73 (1992)
6. Cechlárová, K.: Efficient Computation of the Greatest Eigenvector in Fuzzy Algebra. Tatra Mt. Math. Publ. **12**, 73–79 (1997)
7. Gavalec, M.: Monotone Eigenspace Structure in Max-Min Algebra. Linear Algebra Appl. **345**, 149–167 (2002)
8. Gondran, M.: Valeurs Propres et Vecteurs Propres en Classification Hiérarchique, R.A.I.R.O. Informatique Théorique **10**(3), 39–46 (1976)
9. Gondran, M., Minoux, M.: Linear Algebra in Dioïds. A Survey of Recent Results. Ann. Discrete Math. **19**, 147–164 (1984)
10. Gondran, M., Minoux, M.: Eigenvalues and Eigen-Functionals of Diagonally Dominant Endomorphophisms in Min-Max Analysis. Linear Algebra Appl. **282**, 47–61 (1998)
11. Gondran, M., Minoux, M.: Graphes, dioïdes et semi-anneaux. Editions TEC & DOC, Paris (2001)
12. Gondran, M., Minoux, M.: Dioïds and Semirings: Links to Fuzzy Sets and Other Applications. Fuzzy Sets Syst. **158**(12), 1273–1294 (2007)
13. Maragos, P.: Representations for Morphological Image Operators and Analogies with Linear Operators. Adv. Imaging Electron Phys. **177**, 45–187 (2013)

14. Meyer, F.: Topographical Tools for Filtering and Segmentation 1: Watersheds on Node- or Edge-weighted Graphs. Wiley, Hoboken (2019)
15. Sanchez, E.: Resolution of Eigen Fuzzy Sets Equations. Fuzzy Sets Syst. **1**, 69–74 (1978)
16. Tan, Y.J.: On the Eigenproblem of Matrices Over Distributive Lattices. Linear Algebra Appl. **374**, 87–106 (2003)
17. Serra, J.: Image Analysis and Mathematical Morphology. Vol II: Theoretical Advances. Academic Press, London (1988)

Measuring the Irregularity of Vector-Valued Morphological Operators Using Wasserstein Metric

Marcos Eduardo Valle[1]([✉]), Samuel Francisco[1,2], Marco Aurélio Granero[2], and Santiago Velasco-Forero[3]

[1] Universidade Estadual de Campinas, Campinas, SP, Brazil
valle@ime.unicamp.br
[2] Instituto Federal de Educação, Ciência e Tecnologia de São Paulo, São Paulo, SP, Brazil
[3] Center of Mathematical Morphology, Mines ParisTech, PSL Research University, Paris, France

Abstract. Mathematical morphology is a useful theory of nonlinear operators widely used for image processing and analysis. Despite the successful application of morphological operators for binary and gray-scale images, extending them to vector-valued images is not straightforward because there are no unambiguous orderings for vectors. Among the many approaches to multivalued mathematical morphology, those based on total orders are particularly promising. Morphological operators based on total orders do not produce the so-called false-colors. On the downside, they often introduce irregularities in the output image. Although the irregularity issue has a rigorous mathematical formulation, we are not aware of an efficient method to quantify it. In this paper, we propose to quantify the irregularity of a vector-valued morphological operator using the Wasserstein metric. The Wasserstein metric yields the minimal transport cost for transforming the input into the output image. We illustrate by examples how to quantify the irregularity of vector-valued morphological operators using the Wasserstein metric.

Keywords: Mathematical morphology · Vector-valued images · Total order · Irregularity issue · Optimal transportation

1 Introduction

Mathematical morphology (MM) is a nonlinear theory that uses geometric and topological concepts for image and signal processing. The theory of mathematical morphology is usually defined on an algebraic structure called complete lattices

This work was supported in part by the São Paulo Research Foundation (FAPESP) under grant no 2019/02278-2 and the National Council for Scientific and Technological Development (CNPq) under grant no 310118/2017-4.

which is satisfactory for binary and grayscale images [8,10]. In the case of vector-valued images, vector spaces endowed with a total order is one of the most comfortable frameworks for the extension of morphological processing [1,2,22]. Approaches that have been recently formulated using total orderings include [9,12,18–21]. Despite their successful applications for color and hyperspectral image processing, Chevallier and Angulo showed that the information contained in a total order is too weak to reproduce the natural topology of the value space [5]. As a consequence, morphological operators may introduce irregularities and aliasing on images.

Our motivation is to formulate quantitative measures to study the irregularity implied by a morphological operator on vector-valued images. We believe that this is the first work proposing a framework based on the Wasserstein metric to score this irregularity effect considering pairs of input/output images [11,23]. The paper is organized as follows: Sect. 2 introduced mathematical morphology concepts for vector-valued images and the difficulties produced by operators based on total orders. Section 3 presents our proposition of irregularity measure. Additionally, results on natural images show the goodness of proposed measures. The paper ends with Sect. 4 including conclusions and recommendations for future works.

2   Basic Concepts on Mathematical Morphology

Let us begin by presenting the basic concepts and the notations used in this paper. First, an image $\mathbf{I}$ corresponds to a mapping from a point set D to a value set $\mathbb{V}$, that is, $\mathbf{I} : D \to \mathbb{V}$. The set of all images from a domain D to $\mathbb{V}$ is denoted by $\mathcal{V} = \mathbb{V}^D$. Throughout the paper, we assume the point set D is finite and included in a space $\mathcal{E}$, where $(\mathcal{E}, +)$ is a group. Usually, we consider $\mathcal{E} = \mathbb{R}^2$ or $\mathcal{E} = \mathbb{Z}^2$ with the usual addition. Furthermore, we assume the value set $\mathbb{V}$ is a complete lattice equipped with a metric $d : \mathbb{V} \times \mathbb{V} \to [0, +\infty)$. Recall that a complete lattice $\mathbb{L}$ is a partially ordered set in which any subset $X \subset \mathbb{L}$ has both an infimum and a supremum [3]. The infimum and the supremum of X are denoted respectively by $\bigwedge X$ and $\bigvee X$.

2.1   Mathematical Morphology on Complete Lattices

Mathematical morphology (MM) is mainly concerned with image operators used to extract relevant geometric and topological information from an image [6,8,15]. The two elementary operators of MM are dilations and erosions. Many other operators, such as opening, closing, and the morphological gradient, are obtained by combining the elementary morphological operators.

Complete lattices provide an appropriate mathematical background for defining the elementary morphological operators [8,10]. Indeed, the elementary morphological operators are those that commute with the supremum and the infimum operations in a complete lattice. When the value set $\mathbb{V}$ is a complete lattice, the operators $\delta_S, \varepsilon_S : \mathcal{V} \to \mathcal{V}$ given by the following equations where $S \subseteq \mathcal{E}$ is finite are respectively a dilation and an erosion:

$$\delta_S(\mathbf{I})(p) = \bigvee_{\substack{s \in S \\ p-s \in D}} \mathbf{I}(p-s) \quad \text{and} \quad \varepsilon_S(\mathbf{I})(p) = \bigwedge_{\substack{s \in S \\ p+s \in D}} \mathbf{I}(p+s). \tag{1}$$

The set S is referred to as the structuring element (SE) [15]. The images $\delta_S(\mathbf{I})$ and $\varepsilon_S(\mathbf{I})$ are respectively the dilation and the erosion of $\mathbf{I}$ by the structuring element S.

Although there exist more general definitions, the elementary morphological operators given by (1) are widely used in practical situations. Combining dilations and erosions, we obtain many other morphological operators. In this paper, we focus on elementary operators defined by (1). We also consider openings γ_S and closings ϕ_S, which are obtained by the compositions $\gamma_S = \delta_S \circ \varepsilon_S$ and $\phi_S = \varepsilon_S \circ \delta_S$ [15].

2.2   Vector-Valued Mathematical Morphology

Let us now address morphological operators for vector-valued images. A vector-valued image is obtained by considering $\mathbb{V} \subseteq \bar{\mathbb{R}}^d$, where $\bar{\mathbb{R}} = \mathbb{R} \cup \{-\infty, +\infty\}$ and $d \geq 2$.

Vector-valued dilations and erosions can be defined using (1) whenever the vector-valued set $\mathbb{V}$ is a complete lattice. However, there are many different ordering schemes for vector-valued sets. Defining an appropriate ordering scheme is one of the main challenges of vector-valued MM. The following references provide a brief sense of interesting directions of research on vector-valued MM [1,2,4,5,7,9,18,19,21,22].

As examples of ordering schemes on vector-valued sets, let us recall the marginal and the lexicographical orderings [2]. The marginal ordering is defined as follows for all $\boldsymbol{u} = (u_1, \ldots, u_d) \in \mathbb{V}$ and $\boldsymbol{v} = (v_1, \ldots, v_d) \in \mathbb{V}$:

$$\boldsymbol{u} \leq_M \boldsymbol{v} \quad \Longleftrightarrow \quad u_i \leq v_i, \forall i = 1, \ldots, d, \tag{2}$$

where "$\leq$" denotes the usual ordering on $\mathbb{R}$. The marginal ordering is also called the component-wise ordering or the Cartesian product ordering. The lexicographical ordering is defined as follows:

$$\boldsymbol{u} \leq_L \boldsymbol{v} \quad \Longleftrightarrow \quad \exists i : u_i \leq v_i \text{ and } u_j = v_j, \forall j < i. \tag{3}$$

In contrast to the marginal ordering, the lexicographical ordering is a total ordering. Hence, one of the inequalities $\boldsymbol{u} \leq_L \boldsymbol{v}$ or $\boldsymbol{v} \leq_L \boldsymbol{u}$ holds for any $\boldsymbol{u}, \boldsymbol{v} \in \mathbb{V}$.

2.3   The False "Colors" Problem Versus the Irregularity Issue

One problem on vector-valued MM is the creation of "false colors" or, more generally, false values [14]. A morphological operator $\psi : \mathcal{V} \to \mathcal{V}$ introduces false values whenever there are values on $\psi(I)$ which do not belong to the original image $\mathbf{I}$. Formally, let $2^{\mathbb{V}}$ denote the power set of $\mathbb{V}$ and let $V : \mathcal{V} \to 2^{\mathbb{V}}$ be the mapping given by

$$V(\mathbf{I}) = \{\mathbf{I}(p) : p \in D\}, \quad \forall \mathbf{I} \in \mathcal{V}. \tag{4}$$

A morphological operator ψ introduces false colors if the set difference $V(\psi(\mathbf{I})) \setminus V(\mathbf{I})$ is not empty. The abnormal false values can be a problem in many applications such as when dealing with satellite data [14]. Using the marginal ordering, the dilation and the erosion given by (1) usually yield false colors.

A total ordering, such as the lexicographical ordering, circumvents the problem of the false values [14]. Using a total ordering, the supremum and the infimum of a finite set is an element of the set, i.e., they coincide with the maximum and minimum operations, respectively. As a consequence, if D is finite, the elementary morphological operators given by (1) only contain values of the input image $\mathbf{I}$. On the downside, a total ordering can be irregular in a metric space. According to Chevallier and Angulo, the irregularities follow because the topology induced by a total order may not reproduce the topology of a metric space [5]. Specifically, let the value set $\mathbb{V}$ be a totally ordered set as well as a metric space, with metric $d : \mathbb{V} \times \mathbb{V} \to [0, +\infty)$. Chevallier and Angulo showed that there exists $u, v, w \in \mathbb{V}$ such that $u \leq v \leq w$ but $d(u, w) < d(u, v)$ under mild conditions with respect to the connectivity of $\mathbb{V}$. In words, although w is closer to u than v, the inequalities $u \leq v \leq w$ suggest w is farther from u than v. Since the morphological operators are defined using the extrema operators, they do not take the metric of $\mathbb{V}$ into account.

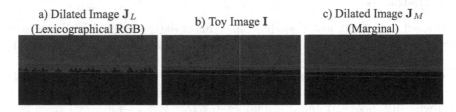

a) Dilated Image $\mathbf{J}_L$ (Lexicographical RGB)   b) Toy Image $\mathbf{I}$ c) Dilated Image $\mathbf{J}_M$ (Marginal)

Fig. 1. Illustrative example of the irregularity issue. Image with three colors and its corresponding dilation by a cross structuring element using the RGB lexicographial and marginal orderings. (Color figure online)

A visual interpretation of the irregularity issue is shown in Fig. 1, which is very similar to an example provided in [5]. Figure 1b) shows an image with three RGB colors, namely $u = (0, 0, 0)$, $v = (0, 0, 1)$, and $w = (0.005, 0, 0)$. The toy image $\mathbf{I}$ is obtained by replacing pure black values u by w with probability 0.3 from an image of size 32×64 with two stripes of colors blue and black. The dilations $\mathbf{J}_L$ and $\mathbf{J}_M$ by a cross as the structuring element using respectively the lexicographical RGB and the marginal ordering schemes are also depicted in Fig. 1. Visually, u and w are black colors while v is a pure blue. Using the Euclidean distance, we obtain $d(u, v) = 1$ and $d(u, w) = 0.005$. These distances agree with our color perception. However, using the lexicographical ordering, we obtain $u \leq_L v \leq_L w$. As a consequence, the following happens when we compute the dilation $\delta_S(\mathbf{I}) = \mathbf{J}_L$ using the lexicographical ordering: the blue pixel value v advances over the black u but it is overlaid by the black w, resulting in the irregularities shown in Fig. 1a). In contrast, the dilated image depicted in Fig. 1c) obtained using the marginal ordering does not present any visual irregularity.

Although we know the irregularity results from a divergence between the topologies induced by the metric and the total ordering, there is no consensual quantitative measure which agrees with our visual perception. A quantitative measure can help to choose an appropriate ordering scheme for vector-valued mathematical morphology. In the following section, we propose to measure the irregularity using the Wasserstein metric.

3   Measuring the Irregularity

In this section, we present a quantitative measure for the irregularity issue, called the *irregularity index*. We begin by presenting a global irregularity index. Then, we propose computing the irregularity index as an average of local irregularity indexes. Although we are interested in measuring the irregularity implied by a total ordering, we will not assume $\mathbb{V}$ is totally ordered. Indeed, the proposed irregularity measure is well defined whenever D is finite and $\mathbb{V}$ is a metric space.

3.1   The Wasserstein Metric and the Generalized Sum of Pixel-Wise Distances

The global irregularity index is defined in terms of the quotient of the Wasserstein metric and a generalized sum of pixel-wise distances. Let us begin by reviewing the generalized sum of pixel-wise distances.

Consider an image operator $\psi : \mathcal{V} \to \mathcal{V}$. Given an input image $\mathbf{I} \in \mathcal{V}$, let $\mathbf{J} = \psi(\mathbf{I})$ denote the output of the image operator. The generalized sum of pixel-wise distances of $\mathbf{I}$ and $\mathbf{J}$ is an operator $\mathcal{D}_p : \mathcal{V} \times \mathcal{V} \to [0, +\infty)$ given by

$$\mathcal{D}_p(\mathbf{I}, \mathbf{J}) = \left(\sum_{x \in D} d^p\big(\mathbf{I}(x), \mathbf{J}(x)\big) \right)^{\frac{1}{p}}, \quad p \geq 1. \tag{5}$$

The generalized sum of pixel-wise distances is one of the simplest measures that takes into account the metric d and the pixel locations. However, $\mathcal{D}_p$ is usually not properly scaled; possibly because its dimension is the same as the metric d. For example, using the Euclidean RGB distance and $p = 1$, the images shown in Fig. 1 yield the values $\mathcal{D}_1(\mathbf{I}, \mathbf{J}_L) = 34.12$ and $\mathcal{D}_1(\mathbf{I}, \mathbf{J}_M) = 66.05$. Note that $\mathcal{D}_1(\mathbf{I}, \mathbf{J}_L) \leq \mathcal{D}_1(\mathbf{I}, \mathbf{J}_M)$. Hence, the generalized sum of pixel-wise distances is not an appropriate quantitative measure for the irregularity issue.

Let us now review the Wasserstein metric, also known as the earth mover's distance or the Kantorovich-Rubinstein distance in certain contexts [11,23]. The Wasserstein metric, named after the Russian mathematician Leonid Vaseršteĭn, has been previously used by Rubner et al. for content-based image retrieval [11]. In the general case, the Wasserstein metric is used to compute distances between probability distributions. For discrete probabilities, however, it is formulated as a transportation problem.

The objective of a transportation problem is to minimize the cost to deliver items from n factories to m shops. In our context, the transportation problem minimizes the cost to transform the input image $\mathbf{I}$ into the output image $\mathbf{J}$. The cost is defined using the metric on the value set $\mathbb{V}$. Precisely, let $V(\mathbf{I}) = \{v_1, \ldots, v_n\}$ and $V(\mathbf{J}) = \{u_1, \ldots, u_m\}$ be the sets of color values of $\mathbf{I}$ and $\mathbf{J}$, respectively. Given $p \geq 1$, the cost to transform a value v_i of $\mathbf{I}$ into a value u_j of $\mathbf{J}$ is defined by

$$c_{ij} = d^p(v_i, u_j), \quad i = 1, \ldots, n, \ j = 1, \ldots, m. \tag{6}$$

The Wasserstein metric, denoted by $\mathcal{W}_p : \mathbb{V} \times \mathbb{V} \to [0, \infty)$ for $p \geq 1$, is given by

$$\mathcal{W}_p(\mathbf{I}, \mathbf{J}) = \left(\sum_{i=1}^{n} \sum_{j=1}^{m} c_{ij} x_{ij} \right)^{1/p}, \quad p \geq 1, \tag{7}$$

where x_{ij} solves the linear programming problem

$$\begin{cases} \text{minimize} & \displaystyle\sum_{i=1}^{n} \sum_{j=1}^{m} c_{ij} x_{ij} \\ \text{subject to} & \displaystyle\sum_{j=1}^{m} x_{ij} = \mathrm{Card}(\{x : \mathbf{I}(x) = v_i\}), \quad \forall i = 1, \ldots, n, \\ & \displaystyle\sum_{i=1}^{n} x_{ij} = \mathrm{Card}(\{x : \mathbf{J}(x) = u_j\}), \quad \forall j = 1, \ldots, m, \\ & x_{ij} \geq 0, \quad \forall i = 1, \ldots, n, \ \forall j = 1, \ldots, m. \end{cases} \tag{8}$$

In the transportation problem (8), the variable x_{ij} represents the amount of the pixel value v_i of $\mathbf{I}$ transformed to the pixel value u_j of $\mathbf{J}$. In some sense, the Wasserstein metric can be interpreted as the minimal cost to transform $\mathbf{I}$ into $\mathbf{J}$. Considering $p = 1$ and the Euclidean distance, we obtain $\mathcal{W}_1(\mathbf{I}, \mathbf{J}_L) = 6.18$ and $\mathcal{W}_1(\mathbf{I}, \mathbf{J}_M) = 65.94$ for the images shown in Fig. 1. Note that the inequality $\mathcal{W}_1(\mathbf{I}, \mathbf{J}_L) < \mathcal{W}_1(\mathbf{I}, \mathbf{J}_M)$ holds in this example. Like the generalized sum of pixel-wise distances, the Wasserstein metric is not an appropriate measure of the irregularity; possibly because it has the same dimension as the generalized sum of pixel-wise distances.

3.2   Global Irregularity Index

Although both generalized sum of pixel-wise distances and the Wasserstein metric are, *per se*, not appropriate to evaluate the irregularity issue, we advocate in this paper that their quotient yields a useful measure. First of all, note that the generalized sum of pixel-wise distances satisfies

$$D_p(\mathbf{I}, \mathbf{J}) = \left(\sum_{i=1}^{n} \sum_{j=1}^{m} c_{ij} y_{ij} \right)^{\frac{1}{p}}, \quad p \geq 1, \tag{9}$$

where

$$y_{ij} = \mathrm{Card}\left(\{x : \mathbf{I}(x) = v_i \text{ and } \mathbf{J}(x) = u_j, x \in D\}\right), \tag{10}$$

for all $i = 1, \ldots, n$ and $j = 1, \ldots, m$. Moreover, it is not hard to see that $y_{ij} \geq 0$,

$$\sum_{j=1}^{m} y_{ij} = \mathrm{Card}(\{x : \mathbf{I}(x) = v_i\}) \quad \text{and} \quad \sum_{i=1}^{n} y_{ij} = \mathrm{Card}(\{x : \mathbf{J}(x) = u_j\}) \tag{11}$$

for all $i = 1, \ldots, n$ and $j = 1, \ldots, m$. Therefore, the generalized sum of pixel-wise distances also measures the cost of transforming $\mathbf{I}$ into $\mathbf{J}$. Moreover, D_p and W_p have the same units and magnitudes. Because W_p is the minimal cost, the inequality $W_p(\mathbf{I}, \mathbf{J}) \leq D_p(\mathbf{I}, \mathbf{J})$ holds for any $\mathbf{I}$ and $\mathbf{J} = \psi(\mathbf{I})$. Using these remarks, we propose to measure the irregularity using the mapping $\Phi_p^g : V \times V \to [0, 1]$ given by the relative gap between D_p and W_p. Precisely, given images $\mathbf{I}, \mathbf{J} \in V$, we define the global irregularity index by means of the equation

$$\Phi_p^g(\mathbf{I}, \mathbf{J}) = \frac{D_p(\mathbf{I}, \mathbf{J}) - W_p(\mathbf{I}, \mathbf{J})}{D_p(\mathbf{I}, \mathbf{J})}, \quad \text{if} \quad D_p(\mathbf{I}, \mathbf{J}) \neq 0, \tag{12}$$

and $\Phi_p^g(\mathbf{I}, \mathbf{J}) = 0$ if $D_p(\mathbf{I}, \mathbf{J}) = 0$. Note that the larger the gap between $W_p(\mathbf{I}, \mathbf{J})$ and $D_p(\mathbf{I}, \mathbf{J})$, the larger the global irregularity index. Equivalently, we have

$$\Phi_p^g(\mathbf{I}, \mathbf{J}) = \begin{cases} 0, & \text{if } D_p(\mathbf{I}, \mathbf{J}) = 0, \\ 1 - \dfrac{W_p(\mathbf{I}, \mathbf{J})}{D_p(\mathbf{I}, \mathbf{J})}, & \text{otherwise.} \end{cases} \tag{13}$$

The irregularity index is symmetric and bounded, that is, $\Phi_p(\mathbf{I}, \mathbf{J}) = \Phi_p(\mathbf{J}, \mathbf{I})$ and $0 \leq \Phi_p(\mathbf{I}, \mathbf{J}) \leq 1$. Moreover, $\Phi_p^g(\mathbf{I}, \mathbf{J})$ is a dimensionless quantity. The more irregular is $\mathbf{J} = \psi(\mathbf{I})$, the larger the value of $\Phi_p^g(\mathbf{I}, \mathbf{J})$ is expect to be. For example, using $p = 1$ and the Euclidean distance, the irregularity index of the dilated images shown in Fig. 1a) and 1c) are respectively $\Phi_1^g(\mathbf{I}, \mathbf{J}_L) = 81.90\%$ and $\Phi_1^g(\mathbf{I}, \mathbf{J}_M) = 0.17\%$.

3.3   Average of Local Irregularity Indexes

Despite its mathematical formulation, computing the global irregularity index is not an easy task for natural images. Precisely, this irregularity index requires solving a linear programming problem with mn variables, where m and n are the number of distinct pixel values of the images $\mathbf{I}$ and $\mathbf{J}$, respectively. In practical situations, the dimension of the linear programming problem (8) is extremely large, making it impossible to be solved in real time. To circumvent this computational burden, we propose to compute the Wasserstein metric and the generalized sum of pixel-wise distances locally and aggregate the values into a single quantitative index.

Let $\{W_1, W_2, \ldots, W_k\}$, with $W_\ell \subseteq D$ for all $\ell = 1, \ldots, k$, be a family of possibly overlapping local windows such that $D \subseteq \cup_{i=1}^k W_i$. Also, let $\mathcal{D}_p(\mathbf{I}, \mathbf{J}|W_\ell)$ and $\mathcal{W}_p(\mathbf{I}, \mathbf{J}|W_\ell)$ denote the generalized sum of pixel-wise distances and the Wasserstein metric computed restricting the images $\mathbf{I}$ and $\mathbf{J}$ to the local window W_ℓ. The average of local irregularity indexes is defined by the following equation for all $\mathbf{I}, \mathbf{J} \in \mathcal{V}$:

$$\Phi_p^a(\mathbf{I}, \mathbf{J}) = 1 - \left(\prod_{\ell=1}^k \frac{\mathcal{W}_p(\mathbf{I}, \mathbf{J}|W_\ell)}{\mathcal{D}_p(\mathbf{I}, \mathbf{J}|W_\ell)} \right)^{\frac{1}{p}}. \tag{14}$$

We would like to emphasize that, because the irregularity index is given by a ratio, the geometric mean is used to aggregate the quotient of $\mathcal{W}_p(\mathbf{I}, \mathbf{J}|W_\ell)$ by $\mathcal{D}_p(\mathbf{I}, \mathbf{J}|W_\ell)$.

In our computational implementation, inspired by the structural similarity index (SSIM) [24], we used local square windows of size 8×8 with strides of 4 pixels. Using $p = 1$ and the Euclidean distance, the average of local irregularity indexes of the dilated images shown in Fig. 1a) and 1c) are respectively $\Phi_1^a(\mathbf{I}, \mathbf{J}_L) = 15.27\%$ and $\Phi_1^a(\mathbf{I}, \mathbf{J}_M) = 0.03\%$. Note that $\Phi_1^a(\mathbf{I}, \mathbf{J}_L)$ is significantly less than $\Phi_1^g(\mathbf{I}, \mathbf{J}_L) = 81.90\%$. However, the following experiment shows that the average of local irregularity indexes is highly correlated to the global irregularity index.

3.4   The Global Irregularity Index and the Average of Local Irregularity Indexes

Let us compare the global irregularity index and the average of local irregularity indexes using toy images similar to the one provided by Chevallier and Angulo [5] but with different probabilities. Precisely, we first construct an image of size 64×32 with two stripes of the same width but with the colors black $\boldsymbol{u} = (0, 0, 0)$ and blue $\boldsymbol{v} = (0, 0, 1)$ in the RGB color space. Then, a pure black pixel value is replaced by the black $\boldsymbol{w} = (0.005, 0, 0)$ with probability $\pi \in [0, 1]$. The resulting image $\mathbf{I}$ is dilated by a cross structuring element using both the lexicographical and the marginal ordering schemes. The simulation has been repeated 100 times for each probability $\pi \in [0, 1]$. Figure 2 shows the mean of both the global irregularity index and the average of local irregularity indexes by the probability π. Again, we used the Euclidean distance to compute the cost to transform pixel values. For a better interpretation of this graph, dilated images obtained using the lexicographical ordering from images generated with probabilities $\pi = 0.0, 0.25, 0.50, 0.75,$ and 1.0 are shown at the bottom of Fig. 2.

Note that both the global irregularity index and the average of the local indexes are very close to zero for the marginal ordering. In contrast, using the lexicographical RGB ordering, both irregularity indexes increase until close to $\pi = 0.3$ and then decrease. Furthermore, the irregularity indexes agree with the visual irregularity provided in the sample images at the bottom of Fig. 2. Finally, we would like to point out that the correlation between the global and average of local irregularity indexes is 98.77%. Therefore, although they have different scales, we believe both the global irregularity index and the average of the local index can be used as an effective measure for the irregularity issue.

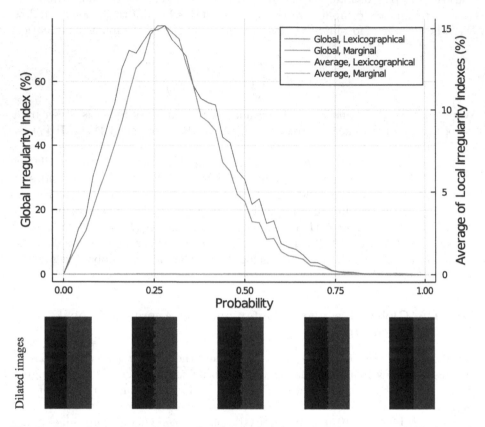

Fig. 2. Top: Irregularity indexes by the probability of replacing a pure black pixel value $u = (0,0,0)$ by the black $w = (0.005,0,0)$ in an image similar to Fig. 1b). **Bottom:** Lexicographical dilation of images obtained using $\pi = 0.0, 0.25, 0.5, 0.75$, and 1.0, respectively. (Color figure online)

3.5   Example with Natural Color Images

Let us now provide some examples with natural color images. Precisely, we compute the average of local irregularity indexes Φ_p^a, the generalized sum of pixel-wise distances $\mathcal{D}_p$, and the average of local Wasserstein metrics, denoted by $\mathcal{W}_p^a$, for several RGB images of size 256×256. Because of their computational burden, we refrained to compute the global irregularity index and the Wasserstein metric globally. In contrast to Φ_p^a, however, the local Wasserstein metrics are aggregated using the arithmetic mean. Moreover, for better scaling the $\mathcal{D}_p$ and $\mathcal{W}_p^a$, these quantities have been divided by the number of pixels of the processed image. Like in the previous examples, we used the Euclidean distance as the metric of the value set. The quantitative measures have been computed using the erosion, the dilation, the opening, and the closing by a 7×7 square structuring element. Two approaches based on the total ordering have been considered: One based on the RGB lexicographical order and the other based on a supervised reduced ordering [19]. In a supervised reduced ordering, the pixel values are ranked using a supervised machine learning technique trained on a set of background and foreground pixels. We also included the marginal approach for comparison purposes. Recall that the marginal approach is not based on a total order. Thus, it can circumvent the irregularity issue.

Figure 3 summarizes the outcome of this computational experiment. To facilitate the exposition, the average of the local irregularity indexes, the generalized sum of pixel-wise distances divided by the number of pixels of the image, and the average of local Wasserstein metric divided by the number of pixels of the local windows, are presented as a triple $(\Phi_p^a, \mathcal{D}_p, \mathcal{W}_p^a)$ below the output images in percentage.

The weakness of both the generalized sum of pixel-wise distances $\mathcal{D}_p$ and the Wasserstein metric $\mathcal{W}_p$ for measuring the irregularity is observed on the images shown in Fig. 3. As pointed out previously, $\mathcal{D}_p$ and $\mathcal{W}_p$ are not dimensionless measures; they are possibly measured using some photometric quantity such as lumen [17]. As a consequence, they may yield misleading values. For example, comparing the eroded boats and the dilated tulips depicted respectively in Fig. 3b) and 3f), one may conjecture that the larger $\mathcal{D}_p$ or $\mathcal{W}_p$, the more irregular is the image. However, opening the Lena image or closing the balloons images using the lexicographical ordering, yielded the irregular images shown in Fig. 3j) and 3n) whose quantities $\mathcal{D}_1$ and $\mathcal{W}_1^a$ are much smaller than the values obtained for the eroded boat images.

Let us now address the proposed irregularity index Φ_1^a, which is a dimensionless measure. Note that the three eroded versions of the boat image shown in Fig. 3b), 3c), and 3d) are quite similar. Accordingly, the corresponding irregularity indexes have similar magnitudes. In contrast, the dilated version of the tulips image, depicted in Fig. 3f), 3g), and 3h), are quite different. In particular, the irregularities of the red pixels in the flowers on Fig. 3f) obtained using the lexicographical ordering is noticeable. In agreement, the irregularity index Φ_1^a is much larger for the lexicographical ordering than for the marginal ordering. Similar remarks hold by comparing the irregularity indexes obtained for the images Fig. 3j), 3k), and 3l) or 3m), 3n), and 3o). In fact, except for the boat image, the irregularity index obtained for the lexicographical and supervised approaches are significantly larger than the irregularity index yielded by the marginal approach.

We invite the reader to carefully compare the images and the corresponding quantities shown in Fig. 3. We hope this example will help the reader to be convinced of the superiority of the proposed irregularity index, in particular, over both the generalized sum of pixel-wise distances and the Wasserstein metric.

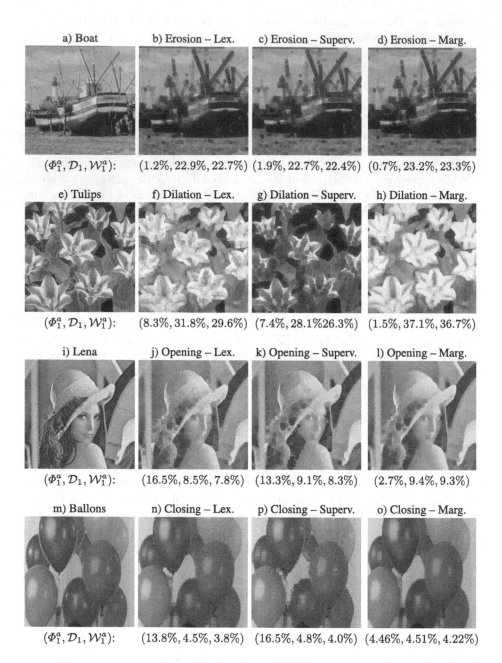

a) Boat   b) Erosion – Lex.   c) Erosion – Superv.   d) Erosion – Marg.

$(\Phi_1^a, \mathcal{D}_1, \mathcal{W}_1^a):$ $(1.2\%, 22.9\%, 22.7\%)$ $(1.9\%, 22.7\%, 22.4\%)$ $(0.7\%, 23.2\%, 23.3\%)$

e) Tulips   f) Dilation – Lex.   g) Dilation – Superv.   h) Dilation – Marg.

$(\Phi_1^a, \mathcal{D}_1, \mathcal{W}_1^a):$ $(8.3\%, 31.8\%, 29.6\%)$ $(7.4\%, 28.1\%26.3\%)$ $(1.5\%, 37.1\%, 36.7\%)$

i) Lena   j) Opening – Lex.   k) Opening – Superv.   l) Opening – Marg.

$(\Phi_1^a, \mathcal{D}_1, \mathcal{W}_1^a):$ $(16.5\%, 8.5\%, 7.8\%)$ $(13.3\%, 9.1\%, 8.3\%)$ $(2.7\%, 9.4\%, 9.3\%)$

m) Ballons   n) Closing – Lex.   p) Closing – Superv.   o) Closing – Marg.

$(\Phi_1^a, \mathcal{D}_1, \mathcal{W}_1^a):$ $(13.8\%, 4.5\%, 3.8\%)$ $(16.5\%, 4.8\%, 4.0\%)$ $(4.46\%, 4.51\%, 4.22\%)$

Fig. 3. Illustrative examples of the average of local irregularity indexes Φ_1^a computed for several color images using different morphological operators. For comparison purposes, we also included the sum of pixel-wise distances $\mathcal{D}_1$ and the average of local Wasserstein metric $\mathcal{W}_1^a$. (Color figure online)

4 Concluding Remarks

In this paper, we proposed two quantitative measures for the irregularity issue. Namely, the global irregularity index and the average of the local irregularity indexes, denoted respectively by Φ_p^g and Φ_p^a. Although Chevallier and Angulo provided a rigorous formulation of the irregularity issue [5], as far as we know, there is no effective quantitative measure for this problem. For example, the generalized sum of pixel-wise distance $\mathcal{D}_p$, which is closely related to the norm of (internal or external) gradient or top-hat operations, is not a dimensionless measure [15,16]. Similarly, the Wasserstein metric $\mathcal{W}_p$, which is eventually known as the earth mover's distance, is also not a dimensionless measure [11]. Hence, they may not be appropriately scaled. Accordingly, as can be observed in Fig. 3, both $\mathcal{D}_p$ and $\mathcal{W}_p$ are not appropriate for measuring the irregularity issue. In contrast, the dimensionless global irregularity index given by the relative gap between $\mathcal{D}_p$ and $\mathcal{W}_p$ yielded good quantitative values for the irregularity for images with few pixel values. In practical situations, however, the number of distinct pixel values makes it impossible to compute the global irregularity index. To circumvent this computational drawback, we proposed computing the geometric mean of the irregularity index on several small windows. Visual interpretations of the irregularity indexes are provided using both synthetic and natural images.

Finally, we would like to point out that the irregularity index can be used in the future to evaluate the performance of morphological operators. The irregularity index can also be used for the design of efficient morphological operators. For example, it can be used as the objective function for the design of vector-valued morphological operators based on uncertain reduced orderings [13].

References

1. Angulo, J.: Morphological Colour Operators in Totally Ordered Lattices Based on Distances: Application to Image Filtering, Enhancement and Analysis. Comput. Visi. Image Underst. **107**(1–2), 56–73 (2007)
2. Aptoula, E., Lefèvre, S.: A Comparative Study on Multivariate Mathematical Morphology. Pattern Recogn. **40**(11), 2914–2929 (2007)
3. Birkhoff, G.: Lattice Theory, 3rd edn. American Mathematical Society, Providence (1993)
4. Burgeth, B., Kleefeld, A.: An Approach to Color-Morphology Based on Einstein Addition and Loewner Order. Pattern Recogn. Lett. **47**, 29–39 (2014)
5. Chevallier, E., Angulo, J.: The Irregularity Issue of Total Orders on Metric Spaces and Its Consequences for Mathematical Morphology. J. Math. Imaging Vis. **54**(3), 344–357 (2015). https://doi.org/10.1007/s10851-015-0607-7
6. Dougherty, E.R., Lotufo, R.A.: Hands-on Morphological Image Processing. SPIE Press, Bellingham (2003)
7. van de Gronde, J., Roerdink, J.: Group-Invariant Colour Morphology Based on Frames. IEEE Trans. Image Process. **23**(3), 1276–1288 (2014)
8. Heijmans, H.J.A.M.: Mathematical Morphology: A Modern Approach in Image Processing Based on Algebra and Geometry. SIAM Rev. **37**(1), 1–36 (1995)

9. Lézoray, O.: Complete Lattice Learning for Multivariate Mathematical Morphology. J. Vis. Commun. Image Represent. **35**, 220–235 (2016). https://doi.org/10.1016/j.jvcir.2015.12.017

10. Ronse, C.: Why Mathematical Morphology Needs Complete Lattices. Sig. Proc. **21**(2), 129–154 (1990)

11. Rubner, Y., Tomasi, C., Guibas, L.J.: Earth Mover's Distance as a Metric for Image Retrieval. Int. J. Comput. Vis. **40**(2), 99–121 (11 2000). https://doi.org/10.1023/A:1026543900054. https://link.springer.com/article/10.1023/A:1026543900054

12. Sangalli, M., Valle, M.E.: Color mathematical morphology using a fuzzy color-based supervised ordering. In: Barreto, G.A., Coelho, R. (eds.) NAFIPS 2018. CCIS, vol. 831, pp. 278–289. Springer, Cham (2018). https://doi.org/10.1007/978-3-319-95312-0_24

13. Sangalli, M., Valle, M.E.: Approaches to multivalued mathematical morphology based on uncertain reduced orderings. In: Burgeth, B., Kleefeld, A., Naegel, B., Passat, N., Perret, B. (eds.) ISMM 2019. LNCS, vol. 11564, pp. 228–240. Springer, Cham (2019). https://doi.org/10.1007/978-3-030-20867-7_18

14. Serra, J.: The "false colour" problem. In: Wilkinson, M.H.F., Roerdink, J.B.T.M. (eds.) ISMM 2009. LNCS, vol. 5720, pp. 13–23. Springer, Heidelberg (2009). https://doi.org/10.1007/978-3-642-03613-2_2

15. Soille, P.: Morphological Image Analysis. Springer, Berlin (1999). https://doi.org/10.1007/978-3-662-03939-7

16. Soille, P., Vogt, J., Colombo, R.: Carving and Adaptive Drainage Enforcement of Grid Digital Elevation Models. Water Resour. Res. **39**(12), 1366 (2003)

17. Trussell, H.J., Vrhel, M.J.: Photometry and colorimetry. In: Fundamentals of Digital Imaging, pp. 191–244. Cambridge University Press, Cambridge (2008). https://doi.org/10.1017/CBO9780511754555.009

18. Valle, M.E., Valente, R.A.: Mathematical morphology on the spherical CIELab quantale with an application in color image boundary detection. J. Math. Imaging Vis. **57**(2), 183–201 (2017). https://doi.org/10.1007/s10851-016-0674-4. https://link.springer.com/article/10.1007/s10851-016-0674-4

19. Velasco-Forero, S., Angulo, J.: Supervised ordering in Rp: application to morphological processing of hyperspectral images. IEEE Trans. Image Process. **20**(11), 3301–3308 (2011). https://doi.org/10.1109/TIP.2011.2144611

20. Velasco-Forero, S., Angulo, J.: Random projection depth for multivariate mathematical morphology. IEEE J. Sel. Top. Sig. Process. **6**(7), 753–763 (2012). https://doi.org/10.1109/JSTSP.2012.2211336

21. Velasco-Forero, S., Angulo, J.: Mathematical morphology for vector images using statistical depth. In: Soille, P., Pesaresi, M., Ouzounis, G.K. (eds.) ISMM 2011. LNCS, vol. 6671, pp. 355–366. Springer, Heidelberg (2011). https://doi.org/10.1007/978-3-642-21569-8_31

22. Velasco-Forero, S., Angulo, J.: Vector ordering and multispectral morphological image processing. In: Celebi, M.E., Smolka, B. (eds.) Advances in Low-Level Color Image Processing. LNCVB, vol. 11, pp. 223–239. Springer, Dordrecht (2014). https://doi.org/10.1007/978-94-007-7584-8_7

23. Villani, C.: Topics in Optimal Transportation. American Mathematical Society, Providence (2003)

24. Wang, Z., Bovik, A.C., Sheikh, H.R., Simoncelli, E.P.: Image Quality Assessment: From Error Visibility to Structural Similarity. IEEE Trans. Image Process. **13**(4), 600–612 (2004)

An Equivalence Relation Between Morphological Dynamics and Persistent Homology in n-D

Nicolas Boutry[1(✉)], Thierry Géraud[1], and Laurent Najman[2]

[1] EPITA Research and Development Laboratory (LRDE),
Le Kremlin-Bicêtre, France
nicolas.boutry@lrde.epita.fr
[2] LIGM, Équipe A3SI, ESIEE, Université Gustave Eiffel, Champs-sur-Marne, France

Abstract. In Mathematical Morphology (MM), dynamics are used to compute markers to proceed for example to watershed-based image decomposition. At the same time, persistence is a concept coming from Persistent Homology (PH) and Morse Theory (MT) and represents the stability of the extrema of a Morse function. Since these concepts are similar on Morse functions, we studied their relationship and we found, and proved, that they are equal on 1D Morse functions. Here, we propose to extend this proof to n-D, $n \geq 2$, showing that this equality can be applied to n-D images and not only to 1D functions. This is a step further to show how much MM and MT are related.

Keywords: Mathematical Morphology · Morse Theory · Computational homology · Persistent homology · Dynamics · Persistence

1 Introduction

Fig. 1. Low sensibility of dynamics to noise (extracted from [15]).

In *Mathematical Morphology* [21–23], *dynamics* [14,15,24], defined in terms of continuous paths and optimization problems, represents a very powerful tool to measure the significance of extrema in a gray-level image (see Fig. 1). Thanks to dynamics, we can construct efficient markers of objects belonging to an image which do not depend on the size or on the shape of the object we want to segment (to compute watershed transforms [20,25] and proceed to image segmentation).

© Springer Nature Switzerland AG 2021
J. Lindblad et al. (Eds.): DGMM 2021, LNCS 12708, pp. 525–537, 2021.
https://doi.org/10.1007/978-3-030-76657-3_38

This contrasts with convolution filters very often used in digital signal processing or morphological filters [21–23] where geometrical properties do matter. Selecting components of high dynamics in an image is a way to filter objects depending on their contrast, whatever the scale of the objects.

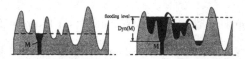

Fig. 2. The dynamics of a minimum of a given function can be computed thanks to a flooding algorithm (extracted from [15]).

Note that there exists an interesting relation between flooding algorithms and the computation of dynamics (see Fig. 2). Indeed, when we flood a local minimum in the topographical view of the $1D$ function, we are able to know the dynamics of this local minimum when water reaches some point of the function where water is lower than the height of the initial local minimum.

In *Persistent Homology* [6,10] well-known in *Computational Topology* [7], we can find the same paradigm: topological features whose *persistence* is high are "true" when the ones whose persistence is low are considered as sampling artifacts, whatever their scale. An example of application of persistence is the filtering of *Morse-Smale complexes* [8,9,16] used in *Morse Theory* [13,19] where pairs of extrema of low persistence are canceled for simplification purpose. This way, we obtain simplified topological representations of *Morse functions*. A discrete counterpart of Morse theory, known as *Discrete Morse Theory* can be found in [11–13,17].

As detailed in [5], pairing by persistence of critical values can be extended in a more general setting to pairing by *interval persistence* of critical points. The result is that they are able to do function matching based on their critical points and they are able to pair all the critical points of a given function (see Fig. 2 in [5]) where persistent homology does not succeed. However, due to the modification of the definition they introduce, this matching is not applicable when we consider usual threshold sets.

In this paper, we prove that the relation between Mathematical Morphology and Persistent Homology is strong in the sense that pairing by dynamics and pairing by persistence are equivalent (and then dynamics and persistence are equal) in n-D when we work with Morse functions. Note that this paper is the extension from 1D to n-D of [4].

The plan of the paper is the following: Sect. 2 recalls the mathematical background needed in this paper, Sect. 3 proves the equivalence between pairing by dynamics and pairing by persistence and Sect. 4 concludes the paper.

2   Mathematical Pre-requisites

We call *path* from $\mathbf{x}$ to $\mathbf{x}'$ both in $\mathbb{R}^n$ a continuous mapping from $[0, 1]$ to $\mathbb{R}^n$. Let Π_1, Π_2 be two paths satisfying $\Pi_1(1) = \Pi_2(0)$, then we denote by $\Pi_1 <> \Pi_2$ the *join* between these two paths. For any two points $\mathbf{x}^1, \mathbf{x}^2 \in \mathbb{R}^n$, we denote by $[\mathbf{x}^1, \mathbf{x}^2]$ the path:

$$\lambda \in [0, 1] \rightarrow (1 - \lambda).\mathbf{x}^1 + \lambda.\mathbf{x}^2.$$

Also, we work with $\mathbb{R}^n$ supplied with the Euclidean norm $\|.\|_2 : \mathbf{x} \rightarrow \|\mathbf{x}\|_2 = \sqrt{\sum_{i=1}^{n} \mathbf{x}_i^2}$.

We will use *lower threshold sets* coming from cross-section topology [2, 3, 18] of a function f defined for some real value $\lambda \in \mathbb{R}$ by:

$$[f < \lambda] = \left\{ x \in \mathbb{R}^n \mid f(x) < \lambda \right\},$$

and

$$[f \leq \lambda] = \left\{ x \in \mathbb{R}^n \mid f(x) \leq \lambda \right\}.$$

2.1   Morse Functions

We call *Morse functions* the real functions in $C^\infty(\mathbb{R}^n)$ whose Hessian is not degenerated at *critical values*, that is, where their gradient vanishes. A strong property of Morse functions is that their critical values are isolated.

Lemma 1 (Morse Lemma [1]). *Let $f : C^\infty(\mathbb{R}^n) \rightarrow \mathbb{R}$ be a Morse function. When $x^* \in \mathbb{R}^n$ is a critical point of f, then there exists some neighborhood V of x^* and some diffeomorphism $\varphi : V \rightarrow \mathbb{R}^n$ such that f is equal to a second order polynomial function of $\mathbf{x} = (x_1, \ldots, x_n)$ on V:*

$$\forall \mathbf{x} \in V, \ f \circ \varphi^{-1}(\mathbf{x}) = f(x^*) - x_1^2 - x_2^2 - \cdots - x_k^2 + x_{k+1}^2 + \cdots + x_n^2,$$

We call *k-saddle* of a Morse function a point $x \in \mathbb{R}^n$ such that the Hessian matrix has exactly k strictly negative eigenvalues (and then $(n - k)$ strictly positive eigenvalues); in this case, k is sometimes called the *index* of f at x. We say that a Morse function has *unique critical values* when for any two different critical values $x_1, x_2 \in \mathbb{R}^n$ of f, we have $f(x_1) \neq f(x_2)$.

2.2   Dynamics

From now on, $f : \mathbb{R}^n \rightarrow \mathbb{R}$ is a Morse function with unique critical values.

Let $\mathbf{x}^{\min}$ be a local minimum of f. Then we call *set of descending paths starting from* $\mathbf{x}^{\min}$ (shortly $(D_{\mathbf{x}^{\min}})$) the set of paths going from $\mathbf{x}^{\min}$ to some element $\mathbf{x}_< \in \mathbb{R}^n$ satisfying $f(\mathbf{x}_<) < f(\mathbf{x}^{\min})$.

The *effort* of a path $\Pi : [0, 1] \rightarrow \mathbb{R}^n$ (relatively to f) is equal to:

$$\max_{\ell \in [0,1], \ell' \in [0,1]} (f(\Pi(\ell)) - f(\Pi(\ell'))).$$

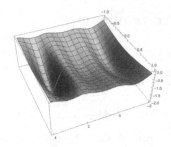

Fig. 3. Pairing by dynamics on a Morse function: the red and blue paths are both in $(D_{\mathbf{x}^{\min}})$ but only the blue one reaches a point $\mathbf{x}_<$ whose height is lower than $f(\mathbf{x}^{\min})$ with a minimal effort. (Color figure online)

A local minimum $\mathbf{x}^{\min}$ of f is said to be *matchable* if there exists some $\mathbf{x}_< \in \mathbb{R}^n$ such that $f(\mathbf{x}_<) < f(\mathbf{x}^{\min})$. We call *dynamics* of a matchable local minimum $\mathbf{x}^{\min}$ of f the value:

$$\mathrm{dyn}(\mathbf{x}^{\min}) = \min_{\Pi \in (D_{\mathbf{x}^{\min}})} \max_{\ell \in [0,1]} \left(f(\Pi(\ell)) - f(\mathbf{x}^{\min}) \right),$$

and we say that $\mathbf{x}^{\min}$ is *paired by dynamics* (see Fig. 3) with some 1-saddle $\mathbf{x}^{\mathrm{sad}} \in \mathbb{R}^n$ of f when:

$$\mathrm{dyn}(\mathbf{x}^{\min}) = f(\mathbf{x}^{\mathrm{sad}}) - f(\mathbf{x}^{\min}).$$

An *optimal* path Π^{opt} is an element of $(D_{\mathbf{x}^{\min}})$ whose effort is equal to $\min_{\Pi \in (D_{\mathbf{x}^{\min}})}(\mathrm{Effort}(\Pi))$. Note that for any local minimum $\mathbf{x}^{\min}$ of f, there always exists some optimal path Π^{opt} such that $\mathrm{Effort}(\Pi^{\mathrm{opt}}) = \mathrm{dyn}(\mathbf{x}^{\min})$.

Thanks to the uniqueness of critical values of f, there exists only one critical point of f which can be paired with $\mathbf{x}^{\min}$ by dynamics.

Dynamics are always positive, and the dynamics of an absolute minimum of f is set at $+\infty$ (by convention).

2.3 Topological Persistence

Let us denote by clo the closure operator, which adds to a subset of $\mathbb{R}^n$ all its accumulation points, and by $\mathcal{CC}(X)$ the connected components of a subset X of $\mathbb{R}^n$. We also define the *representative* of a subset X of $\mathbb{R}^n$ relatively to a Morse function f the point which minimizes f on X:

$$\mathrm{rep}(X) = \arg\min_{\mathbf{x} \in X} f(\mathbf{x}).$$

Definition 1. *Let f be some Morse function with unique critical values, and let $\mathbf{x}^{\mathrm{sad}}$ be the abscissa of some 1-saddle point of f. Now we define the following expressions. First,*

$$C^{\mathrm{sad}} = \mathcal{CC}([f \le f(\mathbf{x}^{\mathrm{sad}})], \mathbf{x}^{\mathrm{sad}})$$

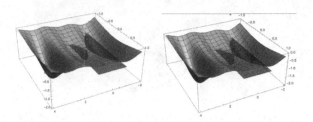

Fig. 4. Pairing by persistence on a Morse function: we compute the plane whose height is reaching $f(\mathbf{x}^{\mathrm{sad}})$ (see the left side), which allows us to compute C^{sad}, to deduce the components C_i^I whose closure contains $\mathbf{x}^{\mathrm{sad}}$, and to decide which representative is paired with $\mathbf{x}^{\mathrm{sad}}$ by persistence by choosing the one whose height is the greatest. We can also observe (see the right side) the *merge phase* where the two components merge and where the component whose representative is paired with $\mathbf{x}^{\mathrm{sad}}$ dies. (Color figure online)

denotes the component of the set $[f \leq f(\mathbf{x}^{\mathrm{sad}})]$ *which contains* $\mathbf{x}^{\mathrm{sad}}$. *Second, we denote by:*

$$\{C_i^I\}_{i \in I} = \mathcal{CC}([f < f(\mathbf{x}^{\mathrm{sad}})])$$

the connected components of the open set $[f < f(\mathbf{x}^{\mathrm{sad}})]$. *Third, we define*

$$\{C_i^{\mathrm{sad}}\}_{i \in I^{\mathrm{sad}}} = \{C_i^I \mid \mathbf{x}^{\mathrm{sad}} \in \mathrm{clo}(C_i^I)\}$$

the subset of components C_i^I *whose closure contains* $\mathbf{x}^{\mathrm{sad}}$. *Fourth, for each* $i \in I^{\mathrm{sad}}$, *we denote*

$$\mathrm{rep}_i = \arg\min_{x \in C_i^{\mathrm{sad}}} f(x)$$

the representative of C_i^{sad}. *Fifth, we define the abscissa*

$$\mathbf{x}^{\min} = \mathrm{rep}_{i_{\mathrm{paired}}}$$

with

$$i_{\mathrm{paired}} = \arg\max_{i \in I^{\mathrm{sad}}} f(\mathrm{rep}_i),$$

thus $\mathbf{x}^{\min}$ *is the representative of the component* C_i^{sad} *of minimal depth. In this context, we say that* $\mathbf{x}^{\mathrm{sad}}$ *is paired by persistence to* $\mathbf{x}^{\min}$ *(Fig. 4). Then, the persistence of* $\mathbf{x}^{\mathrm{sad}}$ *is equal to:*

$$\mathrm{Per}(\mathbf{x}^{\mathrm{sad}}) = f(\mathbf{x}^{\mathrm{sad}}) - f(\mathbf{x}^{\min}).$$

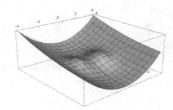

Fig. 5. Every optimal descending path goes through a 1-saddle. Observe the path in blue coming from the left side and decreasing when following the topographical view of the Morse function f. The effort of this path to reach the minimum of f is minimal thanks to the fact that it goes through the saddle point at the middle of the image. (Color figure online)

3   The n-D Equivalence

Let us make two important remarks that will help us in the sequel.

Lemma 2. *Let* $f : \mathbb{R}^n \to \mathbb{R}$ *be a Morse function and let* $\mathbf{x}^{\min}$ *be a local minimum of* f. *Then for any optimal path* Π^{opt} *in* $(D_{\mathbf{x}^{\min}})$, *there exists some* $\ell^* \in]0,1[$ *such that it is a maximum of* $f \circ \Pi^{\mathrm{opt}}$ *and at the same time* $\Pi^{\mathrm{opt}}(\ell^*)$ *is the abscissa of a 1-saddle point of* f.

Proof: This proof is depicted in Fig. 5. Let us proceed by counterposition, and let us prove that when a path Π in $(D_{\mathbf{x}^{\min}})$ does not go through a 1-saddle of f, it cannot be optimal.

Let Π be a path in $(D_{\mathbf{x}^{\min}})$. Let us define $\ell^* \in [0,1]$ as one of the positions where the mapping $f \circ \Pi$ is maximal:

$$\ell^* \in \arg\max_{\ell \in [0,1]} f(\Pi(\ell)),$$

and $x^* = \Pi(\ell^*)$. Let us prove that we can find another path Π' in $(D_{\mathbf{x}^{\min}})$ whose effort is lower than the one of Π.

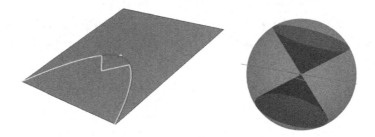

Fig. 6. How to compute descending paths of lower efforts. The initial path going through x^* (the little grey ball) is in red, the new path of lower effort is in green (the non-zero gradient case is on the left side, the zero-gradient case is on the right side). (Color figure online)

At x^*, f can satisfy three possibilities:

– When we have $\nabla f(x^*) \neq 0$ (see the left side of Fig. 6), then locally f is a plane of slope $\|\nabla f(x^*)\|$, and then we can easily find some path Π' in $(D_{\mathbf{x}^{min}})$ with a lower effort than Effort(Π). More precisely, let us fix some arbitrary small value $\varepsilon > 0$ and draw the closed topological ball $\bar{B}(x^*, \varepsilon)$, we can define three points:

$$\ell_{min} = \min\{\ell \mid \Pi(\ell) \in \bar{B}(x^*, \varepsilon)\},$$
$$\ell_{max} = \max\{\ell \mid \Pi(\ell) \in \bar{B}(x^*, \varepsilon)\},$$
$$x_B = x^* - \varepsilon.\frac{\nabla f(x^*)}{\|\nabla f(x^*)\|}.$$

Thanks to these points, we can define a new path Π':

$$\Pi|_{[0,\ell_{min}]} <> [\Pi(\ell_{min}), x_B] <> [x_B, \Pi(\ell_{max})] <> \Pi|_{[\ell_{max},1]}.$$

By doing this procedure at every point in $[0, 1]$ where $f \circ \Pi$ reaches its maximal value, we obtain a new path whose effort is lower than the one of Π.

– When we have $\nabla f(x^*) = 0$, then we are at a critical point of f. It cannot be a 0-saddle, that is, a local minimum, due to the existence of the descending path going through x^*. It cannot be a 1-saddle neither (by hypothesis). It is then a k-saddle point with $k \in [2, n]$ (see the right side of Fig. 6). Using Lemma 1, f is locally equal to a second order polynomial function (up to a change of coordinates φ s.t. $\varphi(x^*) = \mathbf{0}$):

$$f \circ \varphi^{-1}(\mathbf{x}) = f(x^*) - x_1^2 - x_2^2 - \cdots - x_k^2 + x_{k+1}^2 + \cdots + x_n^2.$$

Now, let us define for some arbitrary small value $\varepsilon > 0$:

$$\ell_{min} = \min\{\ell \mid \Pi(\ell) \in \bar{B}(\mathbf{0}, \varepsilon)\},$$
$$\ell_{max} = \max\{\ell \mid \Pi(\ell) \in \bar{B}(\mathbf{0}, \varepsilon)\},$$
$$\mathfrak{B} = \left\{\mathbf{x} \;\middle|\; \sum_{i \in [1,k]} x_i^2 \leq \varepsilon^2 \ \textbf{and} \ \forall j \in [k+1, n], x_j = 0\right\} \setminus \{\mathbf{0}\}.$$

This last set is connected since it is equal to a k-manifold (with $k \geq 2$) minus a point. Let us assume without constraints that $\Pi(\ell_{min})$ and $\Pi(\ell_{max})$ belong to $\mathfrak{B}$ (otherwise we can consider their orthogonal projections on the hyperplane of lower dimension containing $\mathfrak{B}$ but the reasoning is the same). Thus, there exists some path $\Pi_{\mathfrak{B}}$ joining $\Pi(\ell_{min})$ to $\Pi(\ell_{max})$ in $\mathfrak{B}$, from which we can deduce the path $\Pi' = \Pi|_{[0,\ell_{min}]} <> \Pi_{\mathfrak{B}} <> \Pi|_{[\ell_{max},1]}$ whose effort is lower than the one of Π since its image is inside $[f < f(x^*)]$.

Since we have seen that, in any possible case, Π is not optimal, it concludes the proof. □

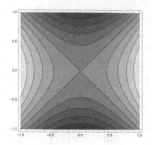

Fig. 7. A 1-saddle point leads to two open connected components. At a 1-saddle point whose abscissa is $\mathbf{x}^{\mathrm{sad}}$ (at the center of the image), the component $[f \leq f(\mathbf{x}^{\mathrm{sad}})]$ is locally the merge of the closure of two connected components (in orange) of $[f < f(\mathbf{x}^{\mathrm{sad}})]$ when f is a Morse function. (Color figure omline)

Proposition 1. *Let f be a Morse function from $\mathbb{R}^n$ to $\mathbb{R}$ with $n \geq 1$. When x^* is a critical point of index 1, then there exists $\varepsilon > 0$ such that:*

$$\mathrm{Card}\left(\mathcal{CC}(B(x^*,\varepsilon) \cap [f < f(x^*)])\right) = 2,$$

where Card *is the* cardinality operator.

Proof: The case $n = 1$ is obvious, let us then treat the case $n \geq 2$ (see Fig. 7). Thanks to Lemma 1 and thanks to the fact that $\mathbf{x}^{\mathrm{sad}}$ is the abscissa of a 1-saddle, we can say that (up to a change of coordinates and in a small neighborhood around $\mathbf{x}^{\mathrm{sad}}$) for any $\mathbf{x}$:

$$f(x) = f(\mathbf{x}^{\mathrm{sad}}) + \mathbf{x}^T \cdot \begin{bmatrix} -1 & \mathbf{0} \\ \mathbf{0} & \mathbb{I}_{n-1} \end{bmatrix} \cdot \mathbf{x},$$

where $\mathbb{I}_{n-1}$ is the identity matrix of dimension $(n-1) \times (n-1)$. In other words, around $\mathbf{x}^{\mathrm{sad}}$, we obtain that:

$$[f < f(\mathbf{x}^{\mathrm{sad}})] = \left\{ \mathbf{x} \ \middle| \ -x_1^2 + \sum_{i=2}^{n} x_i^2 < 0 \right\} = C_+ \cup C_-,$$

with:

$$C_+ = \left\{ \mathbf{x} \ \middle| \ x_1 > \sqrt{\sum_{i=2}^{n} x_i^2} \right\}, \quad C_- = \left\{ \mathbf{x} \ \middle| \ x_1 < -\sqrt{\sum_{i=2}^{n} x_i^2} \right\},$$

where C_+ and C_- are two open connected components of $\mathbb{R}^n$. Indeed, for any pair (M, M') of C_+, we have $x_1^M > \sqrt{\sum_{i=2}^{n}(x_i^M)^2}$ and $x_1^{M'} > \sqrt{\sum_{i=2}^{n}(x_i^{M'})^2}$, from which we define $N = (x_1^M, 0, \ldots, 0)^T \in C_+$ and $N' = (x_1^{M'}, 0, \ldots, 0)^T \in C_+$ from which we deduce the path $[M, N] <> [N, N'] <> [N', M']$ joining M to M' in C_+. The reasoning with C_- is the same. Since C_+ and C_- are two connected (separated) disjoint sets, the proof is done.   □

3.1   Pairing by Persistence Implies Pairing by Dynamics in n-D

Theorem 1. *Let f be a Morse function from $\mathbb{R}^n$ to $\mathbb{R}$. We assume that the 1-saddle point of f whose abscissa is $\mathbf{x}^{\text{sad}}$ is paired by persistence to a local minimum $\mathbf{x}^{\text{min}}$ of f. Then, $\mathbf{x}^{\text{min}}$ is paired by dynamics to $\mathbf{x}^{\text{sad}}$.*

Proof: Let us assume that $\mathbf{x}^{\text{sad}}$ is paired by persistence to $\mathbf{x}^{\text{min}}$, then we have the hypotheses described in Definition 1. Let us denote by C^{min} the connected component in $\{C_i\}_{i \in I^{\text{sad}}}$ satisfying that $\mathbf{x}^{\text{min}} = \text{rep}(C_{i_{\min}})$.

Since $\mathbf{x}^{\text{sad}}$ is the abscissa of a 1-saddle, by Proposition 1, we know that $\text{Card}(I^{\text{sad}}) = 2$, then there exists: $\mathbf{x}_< = \text{rep}(C^<)$ with $C^<$ the component C_i with $i \in I \setminus \{i_{\min}\}$, then $\mathbf{x}^{\text{min}}$ is matchable. Let us assume that the dynamics of $\mathbf{x}^{\text{min}}$ satisfies:

$$\text{dyn}(\mathbf{x}^{\text{min}}) < f(\mathbf{x}^{\text{sad}}) - f(\mathbf{x}^{\text{min}}). \quad \text{(HYP)}$$

This means that there exists a path $\Pi_<$ in $(D_{\mathbf{x}^{\text{min}}})$ such that:

$$\max_{\ell \in [0,1]} f(\Pi_<(\ell)) - f(\mathbf{x}^{\text{min}}) < f(\mathbf{x}^{\text{sad}}) - f(\mathbf{x}^{\text{min}}),$$

that is, for any $\ell \in [0,1]$, $f(\Pi_<(\ell)) < f(\mathbf{x}^{\text{sad}})$, and then by continuity in space of $\Pi_<$, the image of $[0,1]$ by $\Pi_<$ is in C^{min}. Because $\Pi_<$ belongs to $(D_{\mathbf{x}^{\text{min}}})$, there exists then some $\mathbf{x}_< \in C^{\text{min}}$ satisfying $f(\mathbf{x}_<) < f(\mathbf{x}^{\text{min}})$. We obtain a contradiction, (HYP) is then false. Then, we have $\text{dyn}(\mathbf{x}^{\text{min}}) \geq f(\mathbf{x}^{\text{sad}}) - f(\mathbf{x}^{\text{min}})$.

Because for any $i \in I^{\text{sad}}$, $\mathbf{x}^{\text{sad}}$ is an accumulation point of C_i in $\mathbb{R}^n$, there exist a path Π_m from $\mathbf{x}^{\text{min}}$ to $\mathbf{x}^{\text{sad}}$ such that:

$$\forall \ell \in [0,1], \Pi_m(\ell) \in C^{\text{sad}},$$

$$\forall \ell \in [0,1[, \Pi_m(\ell) \in C^{\text{min}}.$$

In the same way, there exists a path Π_M from $\mathbf{x}_<$ to $\mathbf{x}^{\text{sad}}$ such that:

$$\forall \ell \in [0,1], \Pi_M(\ell) \in C^{\text{sad}},$$

$$\forall \ell \in [0,1[, \Pi_M(\ell) \in C^<.$$

We can then build a path Π which is the concatenation of Π_m and $\ell \rightarrow \Pi_M(1 - \ell)$, which goes from $\mathbf{x}^{\text{min}}$ to $\mathbf{x}_<$ and goes through $\mathbf{x}^{\text{sad}}$. Since this path stays inside C^{sad}, we know that $\text{Effort}(\Pi) \leq f(\mathbf{x}^{\text{sad}}) - f(\mathbf{x}^{\text{min}})$, and then $\text{dyn}(\mathbf{x}^{\text{min}}) \leq f(\mathbf{x}^{\text{sad}}) - f(\mathbf{x}^{\text{min}})$.

By grouping the two inequalities, we obtain that $\text{dyn}(\mathbf{x}^{\text{min}}) = f(\mathbf{x}^{\text{sad}}) - f(\mathbf{x}^{\text{min}})$, and then by uniqueness of the critical values of f, $\mathbf{x}^{\text{min}}$ is then paired by dynamics to $\mathbf{x}^{\text{sad}}$. $\qquad \square$

3.2   Pairing by Dynamics Implies Pairing by Persistence in n-D

Theorem 2. *Let f be a Morse function from $\mathbb{R}^n$ to $\mathbb{R}$. We assume that the local minimum $\mathbf{x}^{\text{min}}$ of f is paired by dynamics to a 1-saddle of f of abscissa $\mathbf{x}^{\text{sad}}$. Then, $\mathbf{x}^{\text{sad}}$ is paired by persistence to $\mathbf{x}^{\text{min}}$.*

Proof: Let us assume that $\mathbf{x}^{\min}$ is paired to $\mathbf{x}^{\mathrm{sad}}$ by dynamics. Let us recall the usual framework relative to persistence:

$$C^{\mathrm{sad}} = \mathcal{CC}([f \leq f(\mathbf{x}^{\mathrm{sad}})], \mathbf{x}^{\mathrm{sad}}), \tag{1}$$

$$\{C_i^I\}_{i \in I} = \mathcal{CC}([f < f(\mathbf{x}^{\mathrm{sad}})]), \tag{2}$$

$$\{C_i^{\mathrm{sad}}\}_{i \in I^{\mathrm{sad}}} = \{C_i^I | \mathbf{x}^{\mathrm{sad}} \in \mathrm{clo}(C_i^I)\}, \tag{3}$$

$$\forall i \in I^{\mathrm{sad}}, \ \mathrm{rep}_i = \arg\min_{x \in C_i^{\mathrm{sad}}} f(x). \tag{4}$$

By Definition 1, $\mathbf{x}^{\mathrm{sad}}$ will be paired to the representative rep_i of C_i^{sad} which maximizes $f(\mathrm{rep}_i)$.

1. Let us show that there exists $i_{\min}$ such that $\mathbf{x}^{\min}$ is the representative of a component $C_{i_{\min}}^{\mathrm{sad}}$ of $\{C_i^{\mathrm{sad}}\}_{i \in I^{\mathrm{sad}}}$.

   (a) First, $\mathbf{x}^{\min}$ is paired by dynamics with $\mathbf{x}^{\mathrm{sad}}$ and $\mathrm{dyn}(\mathbf{x}^{\min})$ is greater than zero, then $f(\mathbf{x}^{\mathrm{sad}}) > f(\mathbf{x}^{\min})$, then $\mathbf{x}^{\min}$ belongs to $[f < f(\mathbf{x}^{\mathrm{sad}})]$, then there exists some $i_{\min} \in I$ such that $\mathbf{x}^{\min} \in C_{i_{\min}}$ (see Eq. (2) above).

   (b) Now, if we assume that $\mathbf{x}^{\min}$ is not the representative of $C_{i_{\min}}$, there exists then some $\mathbf{x}_<$ in $C_{i_{\min}}$ satisfying that $f(\mathbf{x}_<) < f(\mathbf{x}^{\min})$, and then there exists some Π in $(D_{\mathbf{x}^{\min}})$ whose image is contained in $C_{i_{\min}}$. In other words,

   $$\mathrm{dyn}(\mathbf{x}^{\min}) \leq \mathrm{Effort}(\Pi) < f(\mathbf{x}^{\mathrm{sad}}) - f(\mathbf{x}^{\min}),$$

   which contradicts the hypothesis that $\mathbf{x}^{\min}$ is paired with $\mathbf{x}^{\mathrm{sad}}$ by dynamics.

   (c) Let us show that $i_{\min}$ belongs to I^{sad}, that is, $\mathbf{x}^{\mathrm{sad}} \in \mathrm{clo}(C_{i_{\min}})$. Let us assume that:

   $$\mathbf{x}^{\mathrm{sad}} \notin \mathrm{clo}(C_{i_{\min}}). \quad \text{(HYP2)}$$

   Every path in $(D_{\mathbf{x}^{\min}})$ goes outside of $C_{i_{\min}}$ to reach some point whose image by f is lower than $f(\mathbf{x}^{\min})$ since $\mathbf{x}^{\min}$ has been proven to be the representative of $C_{i_{\min}}$. Then this path will intersect the boundary ∂ of $C_{i_{\min}}$. Since by (HYP2), $\mathbf{x}^{\mathrm{sad}}$ does not belong to the boundary ∂ of $C_{i_{\min}}$, any optimal path Π^* in $(D_{\mathbf{x}^{\min}})$ will go through one 1-saddle $\mathbf{x}^{\mathrm{sad}}_2 = \arg\max_{\ell \in [0,1]} f(\Pi^*(\ell))$ (by Lemma 2) different from $\mathbf{x}^{\mathrm{sad}}$ and verifying then $f(\mathbf{x}^{\mathrm{sad}}_2) > f(\mathbf{x}^{\mathrm{sad}})$. Thus, $\mathrm{dyn}(\mathbf{x}^{\min}) > f(\mathbf{x}^{\mathrm{sad}}) - f(\mathbf{x}^{\min})$, which contradicts the hypothesis that $\mathbf{x}^{\min}$ is paired with $\mathbf{x}^{\mathrm{sad}}$ by dynamics. Then, we have:

   $$\mathbf{x}^{\mathrm{sad}} \in \mathrm{clo}(C_{i_{\min}}).$$

2. Now let us show that $f(\mathbf{x}^{\min}) > f(\mathrm{rep}(C_i^{\mathrm{sad}}))$ for any $i \in I^{\mathrm{sad}} \setminus \{i_{\min}\}$. For this aim, we will prove that there exists some $i \in I^{\mathrm{sad}}$ such that $f(\mathrm{rep}(C_i^{\mathrm{sad}})) < f(\mathbf{x}^{\min})$ and we will conclude with Proposition 1. Let us assume that the representative r of each component C_i^{sad} except $C^{\min}$ satisfies $f(r) > f(\mathbf{x}^{\min})$, then any path Π of $(D_{\mathbf{x}^{\min}})$ will have to go outside C^{sad} to reach some point whose image by f is lower than $f(\mathbf{x}^{\min})$. We obtain the same situation as

before (see $(1.c)$), and then we obtain that the effort of Π will be greater than $f(\mathbf{x}^{\text{sad}}) - f(\mathbf{x}^{\text{min}})$, which leads to a contradiction with the hypothesis that $\mathbf{x}^{\text{min}}$ is paired with $\mathbf{x}^{\text{sad}}$ by dynamics. We have then that there exists $i \in I^{\text{sad}}$ such that $f(\text{rep}(C_i^{\text{sad}})) < f(\mathbf{x}^{\text{min}})$. Thanks to Proposition 1, we know then that $\mathbf{x}^{\text{min}}$ is the representative of the components of $[f < f(\mathbf{x}^{\text{sad}})]$ whose image by f is the greatest.
3. It follows that $\mathbf{x}^{\text{sad}}$ is paired with $\mathbf{x}^{\text{min}}$ by persistence.

4 Conclusion

We have proved that persistence and dynamics lead to same pairings in n-D, $n \geq 1$, which implies that they are equal whatever the dimension. Concerning the future works, we propose to investigate the relationship between persistence and dynamics in the discrete case [12] (that is, on complexes). We will also check under which conditions pairings by persistence and by dynamics are equivalent for functions that are not Morse. Furthermore, we will examine if the fast algorithms used in MM like watershed cuts, Betti numbers computations or attribute-based filtering are applicable to PH. Conversely, we will study if some PH concepts can be seen as the generalization of some MM concepts (for example, dynamics seems to be a particular case of persistence).

A   Ambiguities occurring when values are not unique

As depicted in Fig. 8, the abscissa of the blue point can be paired by persistence to the abscissas of the orange and/or the red points. The same thing appears when we want to pair the abscissa of the pink point to the abscissas of the green and/or blue points. This shows how much it is important to have unique critical values on Morse functions.

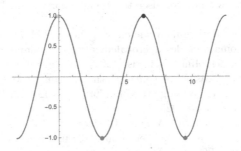

Fig. 8. Ambiguities can occur when critical values are not unique for pairing by dynamics and for pairing by persistence. (Color figure online)

References

1. Audin, M., Damian, M.: Morse Theory and Floer Homology. Universitext, 1st edn., pp. XIV, 96. Springer, London (2014)
2. Bertrand, G., Everat, J.-C., Couprie, M.: Topological approach to image segmentation. In: SPIE's 1996 International Symposium on Optical Science, Engineering, and Instrumentation, vol. 2826, pp. 65–76. International Society for Optics and Photonics (1996)
3. Beucher, S., Meyer, F.: The morphological approach to segmentation: the watershed transformation. In: Optical Engineering, New York, Marcel Dekker Incorporated, vol. 34, pp. 433–433 (1992)
4. Boutry, N., Géraud, T., Najman, L.: An equivalence relation between Morphological Dynamics and Persistent Homology in 1D. In: Burgeth, B., Kleefeld, A., Naegel, B., Passat, N., Perret, B. (eds.) Mathematical Morphology and Its Applications to Signal and Image Processing. Lecture Notes in Computer Science Series, vol. 11564, pp. 57–68. Springer, Cham (2019). https://doi.org/10.1007/978-3-030-20867-7_5
5. Dey, T.K., Wenger, R.: Stability of critical points with interval persistence. Discrete Comput. Geom. **38**(3), 479–512 (2007)
6. Edelsbrunner, H., Harer, J.: Persistent homology - a survey. Contemp. Math. **453**, 257–282 (2008)
7. Edelsbrunner, H., Harer, J.: Computational Topology: An Introduction. American Mathematical Society, Providence, USA (2010)
8. Edelsbrunner, H., Harer, J., Natarajan, V., Pascucci, V.: Morse-Smale complexes for piecewise linear 3-manifolds. In: Proceedings of the Nineteenth Annual Symposium on Computational Geometry, pp. 361–370 (2003)
9. Edelsbrunner, H., Harer, J., Zomorodian, A.: Hierarchical Morse-Smale complexes for piecewise linear 2-manifolds. Discrete Comput. Geom. **30**(1), 87–107 (2003)
10. Edelsbrunner, H., Letscher, D., Zomorodian, A.: Topological persistence and simplification. In: Foundations of Computer Science, pp. 454–463. IEEE (2000)
11. Forman, R.: A discrete Morse theory for cell complexes. In: Yau, S.-T. (ed.) Geometry. Topology for Raoul Bott. International Press, Somerville MA (1995)
12. Forman, R.: Morse Theory for Cell Complexes (1998)
13. Forman, R.: A user's guide to Discrete Morse theory. Sém. Lothar. Combin. **48**, 35 (2002)
14. Grimaud, M.: La géodésie numérique en Morphologie Mathématique. Application à la détection automatique des microcalcifications en mammographie numérique. Ph.D. thesis, École des Mines de Paris (1991)
15. Grimaud, M.: New measure of contrast: the dynamics. In: Image Algebra and Morphological Image Processing III, vol. 1769, pp. 292–306. International Society for Optics and Photonics (1992)
16. Günther, D., Reininghaus, J., Wagner, H., Hotz, I.: Efficient computation of 3D Morse-Smale complexes and persistent homology using discrete Morse theory. Vis. Comput. **28**(10), 959–969 (2012)
17. Jöllenbeck, M., Welker, V.: Minimal resolutions via Algebraic Discrete Morse Theory. American Mathematical Society (2009)
18. Meyer, F.: Skeletons and perceptual graphs. Signal Process. **16**(4), 335–363 (1989)
19. Willard Milnor, J., Spivak, M., Wells, R., Wells, R.: Morse Theory. Princeton University Press, Princeton, New Jersey (1963)

20. Najman, L., Schmitt, M.: Geodesic saliency of watershed contours and hierarchical segmentation. IEEE Trans. Pattern Anal. Mach. Intell. **18**(12), 1163–1173 (1996)
21. Najman, L., Talbot, H.: Mathematical Morphology: From Theory to Applications. Wiley (2013)
22. Serra, J.: Introduction to mathematical morphology. Comput. Vis. Graph. Image Process. **35**(3), 283–305 (1986)
23. Serra, J., Soille, P.: Mathematical Morphology and its Applications to Image Processing. Computational Imaging and Vision, vol. 2, p. 385. Springer, Dordrecht (2012)
24. Vachier, C.: Extraction de caractéristiques, segmentation d'image et Morphologie Mathématique. Ph.D. thesis, École Nationale Supérieure des Mines de Paris (1995)
25. Vincent, L., Soille, P.: Watersheds in digital spaces: an efficient algorithm based on immersion simulations. IEEE Trans. Pattern Anal. Mach. Intell. **13**(6), 583–598 (1991)

Sparse Approximate Solutions to Max-Plus Equations

Nikos Tsilivis[1(✉)], Anastasios Tsiamis[2], and Petros Maragos[1]

[1] School of ECE, National Technical University of Athens, Athens, Greece
maragos@cs.ntua.gr
[2] ESE Department, SEAS, University of Pennsylvania, Philadelphia, USA
atsiamis@seas.upenn.edu

Abstract. In this work, we study the problem of finding approximate, with minimum support set, solutions to matrix max-plus equations, which we call sparse approximate solutions. We show how one can obtain such solutions efficiently and in polynomial time for any ℓ_p approximation error. Subsequently, we propose a method for pruning morphological neural networks, based on the developed theory.

Keywords: Sparsity · Max-plus algebra · Submodular optimization

1 Introduction

In the last decades, the areas of signal and image processing have greatly benefited from the advancement of the theory of sparse representations [10]. Given a few linear measurements of an object of interest, sparse approximation theory provides efficient tools and algorithms for the acquisition of the sparsest (most zero elements) solution of the corresponding underdetermined linear system [10,20]. Based on the sparsity assumption of the initial signal, this allows perfect reconstruction from little data. Ideas stemming from this area have also given birth to *compressed sensing* techniques [5,9] that allow accurate reconstructions from limited random projections of the initial signal, with wide-ranging applications in photography, magnetic resonance imaging and others.

Yet, there is a variety of problems in areas such as scheduling and synchronization [2,7], morphological image and signal analysis [14,18,22] and optimization and optimal control [1,2,12] that do not admit linear representations. Instead, these problems share the ability to be described as a system of nonlinear equations, which involve maximum operations together with additions. The relevant theoretical framework has initially been developed in [2,4,7] and the appropriate algebra for this kind of problems is called *max-plus* algebra.

P. Maragos—The work of P. Maragos was co-financed by the European Regional Development Fund of the European Union and Greek national funds through the Operational Program Competitiveness, Entrepreneurship and Innovation, under the call RESEARCH–CREATE–INNOVATE (Project: "e-Prevention", code: T1EDK-02890).

© Springer Nature Switzerland AG 2021
J. Lindblad et al. (Eds.): DGMM 2021, LNCS 12708, pp. 538–550, 2021.
https://doi.org/10.1007/978-3-030-76657-3_39

Motivated by the sparsity in the linear setting, [23] introduced the notion of sparsity (signals with many $-\infty$ values, i.e. the identity element of this algebra) in max-plus algebra. Herein, we contribute to this theory, by studying the problem of sparse approximate solutions to matrix max-plus equations allowing the approximation error to be measured by any l_p norm. Indicatively, we also present a preliminary application of the theory to the pruning of morphological neural networks.

In particular, we make the following contributions: a) We pose a *generalized* problem of finding the sparsest approximate solution to matrix max-plus equations under a constraint which makes the problem more tractable, also known as the "lateness constraint". The approximation error is in terms of any ℓ_p norm, for $p < \infty$. b) We prove that for any ℓ_p norm, $p < \infty$, the problem has supermodular properties, which allows us to solve it approximately but efficiently via a greedy algorithm, with a derived approximation ratio. c) We investigate the ℓ_∞ case without the "lateness constraint", reveal its hardness and propose a heuristic method for solving it. d) We demonstrate how one may prune whole neurons from morphological neural networks using the developed theory.

2 Background Concepts

For max and min operations we use the well-established lattice-theoretic symbols of $\vee$ and $\wedge$, respectively. We use roman letters for functions, signals and their arguments and greek letters mainly for operators. Also, boldface roman letters for vectors (lowercase) and matrices (capital). If $\mathbf{M} = [m_{ij}]$ is a matrix, its (i,j) element is also denoted as m_{ij} or as $[\mathbf{M}]_{ij}$. Similarly, $\mathbf{x} = [x_i]$ denotes a column vector, whose i-th element is denoted as $[\mathbf{x}]_i$ or simply x_i.

2.1 Max-Plus Algebra

Max-plus arithmetic consists of the idempotent semiring $(\mathbb{R}_{\max}, \max, +)$, where $\mathbb{R}_{\max} = \mathbb{R} \cup \{-\infty\}$ is equipped with the standard maximum and sum operations, respectively. *Max-plus algebra* consists of vector operations that extend max-plus arithmetic to $\mathbb{R}_{\max}^n$. They include the pointwise operations of partial ordering $\mathbf{x} \leq \mathbf{y}$ and pointwise supremum $\mathbf{x} \vee \mathbf{y} = [x_i \vee y_i]$, together with a class of vector transformations defined below. Max-plus algebra is isomorphic to the *tropical algebra*, namely the min-plus semiring $(\mathbb{R}_{\min}, \min, +)$, $\mathbb{R}_{\min} = \mathbb{R} \cup \{\infty\}$ when extended to $\mathbb{R}_{\min}^n$ in a similar fashion. Vector transformations on $\mathbb{R}_{\max}^n$ (resp. $\mathbb{R}_{\min}^n$) that distribute over max-plus (resp. min-plus) vector superpositions can be represented as a max-plus $\boxplus$ (resp. min-plus $\boxplus'$) product of a matrix $\mathbf{A} \in \mathbb{R}_{\max}^{m \times n}(\mathbb{R}_{\min}^{m \times n})$ with an input vector $\mathbf{x} \in \mathbb{R}_{\max}^n(\mathbb{R}_{\min}^n)$:

$$[\mathbf{A} \boxplus \mathbf{x}]_i \triangleq \bigvee_{k=1}^n a_{ik} + x_k, \quad [\mathbf{A} \boxplus' \mathbf{x}]_i \triangleq \bigwedge_{k=1}^n a_{ik} + x_k \tag{1}$$

More details about general algebraic structures that obey those arithmetics can be found in [19]. In the case of a max-plus matrix equation $\mathbf{A} \boxplus \mathbf{x} = \mathbf{b}$, there is a solution if and only if the vector

$$\hat{\mathbf{x}} = (-\mathbf{A})^{\mathsf{T}} \boxplus' \mathbf{b} \tag{2}$$

satisfies it [4,7,19]. We call this vector the *principal solution* of the equation. It also satisfies the inequality $\mathbf{A} \boxplus \hat{\mathbf{x}} \leq \mathbf{b}$. The previous expressions are best understood through the notion of adjunctions [13,14]. A pair of lattice operators $\delta : \mathcal{L} \to \mathcal{M}, \varepsilon : \mathcal{M} \to \mathcal{L}$, where $\mathcal{L}, \mathcal{M}$ are complete lattices, is called an *adjunction* if it satisfies the relation $\delta(X) \leq Y \iff X \leq \varepsilon(Y)$ for all $X \in \mathcal{L}, Y \in \mathcal{M}$, and δ, ε are called the adjoint operator of each other. Loosely speaking, we can define an operator $\delta : \mathbb{R}_{\max}^n \to \mathbb{R}_{\max}^m$ for a matrix $\mathbf{A} \in \mathbb{R}_{\max}^{m \times n}$, such as $\delta(\mathbf{x}) = \mathbf{A} \boxplus \mathbf{x}$. It can be shown [19] that its adjoint operator ε is given by the formula $\varepsilon(\mathbf{y}) = (-\mathbf{A})^{\mathsf{T}} \boxplus' \mathbf{y}$, and the aforementioned expressions follow from this observation. Lastly, a vector $\mathbf{x} \in \mathbb{R}_{\max}^n$ is called *sparse* if it contains many $-\infty$ elements and we define its *support set*, supp$(\mathbf{x})$, to be the set of positions where vector $\mathbf{x}$ has finite values, that is supp$(\mathbf{x}) = \{i \mid x_i \neq -\infty\}$.

2.2   Submodularity

Let U be a universe of elements. A set function $f : 2^U \to \mathbb{R}$ is called *submodular* [17] if $\forall A \subseteq B \subseteq U$, $k \notin B$ holds:

$$f(A \cup \{k\}) - f(A) \geq f(B \cup \{k\}) - f(B). \tag{3}$$

A set function f is called *supermodular* if $-f$ is submodular. Submodular functions occur as models of many real world evaluations in a number of fields and allow many hard combinatorial problems to be solved fast and with strong approximation guarantees [3,16]. It has been suggested that their importance in discrete optimization is similar to convex functions' in continuous optimization [17].

The following definition captures the idea of how far a given function is from being submodular and generalizes the notion of submodularity.

Definition 1. *[8] Let U be a set and $f : 2^U \to \mathbb{R}^+$ be an increasing, non-negative, function. The submodularity ratio of f is*

$$\gamma_{U,k}(f) \triangleq \min_{L \subseteq U, S:|S| \leq k, S \cap L = \emptyset} \frac{\sum_{x \in S} f(L \cup \{x\}) - f(L)}{f(L \cup S) - f(L)} \tag{4}$$

Proposition 1. *[8] An increasing function $f : 2^U \to \mathbb{R}$ is submodular if and only if $\gamma_{U,k}(f) \geq 1$, $\forall U, k$.*

In [8], the authors used the submodularity ratio to analyze the properties of greedy algorithms in discrete optimization problems with functions that are only approximately submodular ($\gamma \in (0,1)$). They proved that the performance of the algorithms degrade gradually as a function of γ, thus allowing guarantees for a wider variety of objective functions.

3   Sparse Approximate Solutions to Max-Plus Equations

We consider the problem of finding the sparsest approximate solution to the max-plus matrix equation $\mathbf{A} \boxplus \mathbf{x} = \mathbf{b}, \mathbf{A} \in \mathbb{R}^{m \times n}, \mathbf{b} \in \mathbb{R}^m$. Such a solution should i) have minimum support set $\text{supp}(\mathbf{x})$, and ii) have small enough approximation error $\|\mathbf{b} - \mathbf{A} \boxplus \mathbf{x}\|_p^p$, for some $\ell_p, p < \infty$ norm. For this reason, given a prescribed constant ϵ, we formulate the following optimization problem:

$$\arg \min_{\mathbf{x} \in \mathbb{R}_{\max}^n} \; |\text{supp}(\mathbf{x})|, \text{ s.t. } \|\mathbf{b} - \mathbf{A} \boxplus \mathbf{x}\|_p^p \leq \epsilon, \; p < \infty$$

$$\mathbf{A} \boxplus \mathbf{x} \leq \mathbf{b}. \tag{5}$$

Note that we add an additional constraint $\mathbf{A} \boxplus \mathbf{x} \leq \mathbf{b}$, also known as the "lateness" constraint. This constraint makes problem (5) more tractable; it enables the reformulation of problem (5) as a set optimization problem in (13). In many applications this constraint is desirable–see [23]. However, in other situations, it might lead to less sparse solutions or higher residual error. A possible way to overcome this constraint is explored in Sect. 3.1.

Even with the additional lateness constraint, problem (5) is very hard to solve. For example, when $\epsilon = 0$, solving (5) is an $\mathcal{NP}$-hard problem [23]. Thus, we do not expect to find an efficient algorithm which solves (5) exactly. Instead, as we prove next, there is a polynomial time algorithm which finds an approximate solution, by leveraging its supermodular properties. First, let us show that the above problem can be formed as a discrete optimization problem over a set. We follow a similar procedure to [23], where the case $p = 1$ was examined. For the rest of this section, let $J = \{1, \dots, n\}$.

Lemma 1 *(Projection on the support set, ℓ_p case). Let $T \subseteq J$,*

$$X_T = \{\mathbf{x} \in \mathbb{R}_{max}^n : supp(\mathbf{x}) = T, \; \mathbf{A} \boxplus \mathbf{x} \leq \mathbf{b}\}. \tag{6}$$

and $\mathbf{x}|_T$ be defined as $\hat{\mathbf{x}}$ inside T and $-\infty$ otherwise, where $\hat{\mathbf{x}}$ is the principal solution defined in (2). Then, it holds:

- $\mathbf{x}|_T \in X_T$.
- $\|\mathbf{b} - \mathbf{A} \boxplus \mathbf{x}|_T\|_p^p \leq \|\mathbf{b} - \mathbf{A} \boxplus \mathbf{x}\|_p^p \; \forall \, \mathbf{x} \in X_T$.

Proof. – It suffices to show that $\mathbf{A} \boxplus \mathbf{x}|_T \leq \mathbf{b}$. For $j \in T$ it is $[\mathbf{x}|_T]_j = \hat{x}_j$ and for $j \in J \setminus T, [\mathbf{x}|_T]_j = -\infty \leq \hat{x}_j$. Thus,

$$\mathbf{x}|_T \leq \hat{\mathbf{x}} \iff \mathbf{A} \boxplus \mathbf{x}|_T \leq \mathbf{A} \boxplus \hat{\mathbf{x}} \implies \mathbf{A} \boxplus \mathbf{x}|_T \leq \mathbf{b}. \tag{7}$$

Hence, $\mathbf{x}|_T \in X_T$.
- Let $\mathbf{x} \in X_T$, then $\mathbf{A} \boxplus \mathbf{x} \leq \mathbf{b} \iff \mathbf{x} \leq \hat{\mathbf{x}}$, which implies (since both $\mathbf{x}, \mathbf{x}|_T$ have $-\infty$ values outside of T):

$$\mathbf{x} \leq \mathbf{x}|_T \iff \mathbf{b} - \mathbf{A} \boxplus \mathbf{x}|_T \leq \mathbf{b} - \mathbf{A} \boxplus \mathbf{x}. \tag{8}$$

Hence:

$$\|\mathbf{b} - \mathbf{A} \boxplus \mathbf{x}|_T\|_p^p = \sum_{j \in T} (\mathbf{b} - \mathbf{A} \boxplus \mathbf{x}|_T)_j^p \leq \sum_{j \in T} (\mathbf{b} - \mathbf{A} \boxplus \mathbf{x})_j^p = \|\mathbf{b} - \mathbf{A} \boxplus \mathbf{x}\|_p^p. \tag{9}$$

$\square$

The previous lemma informs us that we can fix the finite values of a solution of Problem (5) to be equal to those of the principal solution $\hat{\mathbf{x}}$. Indeed,

Proposition 2. *Let* $\mathbf{x}_{OPT}$ *be an optimal solution of (5), then we can construct a new one with values inside the support set equal to those of the principal solution* $\hat{\mathbf{x}}$.

Proof. Define

$$\mathbf{z} = \begin{cases} \hat{x}_j, & j \in \operatorname{supp}(\mathbf{x}_{OPT}) \\ -\infty, & \text{otherwise} \end{cases}, \tag{10}$$

then $\operatorname{supp}(\mathbf{x}_{OPT}) = \operatorname{supp}(\mathbf{z})$ and, from Lemma 1, $\|\mathbf{b} - \mathbf{A} \boxplus \mathbf{z}\|_p^p \leq \|\mathbf{b} - \mathbf{A} \boxplus \mathbf{x}_{OPT}\|_p^p$ and $\mathbf{A} \boxplus \mathbf{z} \leq \mathbf{b}$. Thus, $\mathbf{z}$ is also an optimal solution of (5). $\square$

Therefore, the only variable that matters in Problem (5) is the support set. To further clarify this, let us proceed with the following definitions:

Definition 2. *Let* $T \subseteq J$ *be a candidate support and let* $\mathbf{A}_j$ *denote the* j-*th column of* $\mathbf{A}$. *The error vector* $\mathbf{e} : 2^J \to \mathbb{R}^m$ *is defined as:*

$$\mathbf{e}(T) = \begin{cases} \mathbf{b} - \bigvee_{j \in T} (\mathbf{A}_j + \hat{x}_j), & T \neq \emptyset \\ \bigvee_{j \in J} \mathbf{e}(\{j\}), & T = \emptyset. \end{cases} \tag{11}$$

Observe that for any T, *it holds* $\bigvee_{j \in T} (\mathbf{A}_j + \hat{x}_j) \leq \bigvee_{j \in J} (\mathbf{A}_j + \hat{x}_j) \leq \mathbf{b}$, *which means that the above vector* $\mathbf{e}(T) = (e_1(T), e_2(T), \dots, e_m(T))^\intercal$ *is always non-negative. We also define the corresponding error function* $E_p : 2^J \to \mathbb{R}$ *as:*

$$E_p(T) = \|\mathbf{e}(T)\|_p^p = \sum_{i=1}^m (e_i(T))^p. \tag{12}$$

Problem (5) can now be written as:

$$\arg \min_{T \subseteq J} |T| \tag{13}$$
$$\text{s.t. } E_p(T) \leq \epsilon$$

The main results of this section are based on the following properties of E_p.

Theorem 1. *Error function* E_p *is decreasing and supermodular.*

Proof. Regarding the monotonicity, let $\emptyset \neq C \subseteq B \subset J$, then

$$\bigvee_{j \in C} (\mathbf{A}_j + \hat{x}_j) \leq \bigvee_{j \in B} (\mathbf{A}_j + \hat{x}_j) \iff \mathbf{e}(B) \leq \mathbf{e}(C), \tag{14}$$

thus raising the, non-negative, components of the two vectors to the p-th power and adding the inequalities together yields $E_p(B) \leq E_p(C)$. The case for $C = \emptyset$ easily follows from the definition of $\mathbf{e}$.

Let $S, L \subseteq U \subseteq J$, with $|S| \leq K$, $S \cap L = \emptyset$ and define $f(U) = -E_p(U)$, $\forall U$. Then:

$$\gamma_{U,K}(f) = \min_{L,S} \frac{\sum_{s_k \in S} f(L \cup \{s_k\}) - f(L)}{f(L \cup S) - f(L)}, \tag{15}$$

where $f(L) = \sum_{i=1}^{m} [b_i - \bigvee_{j \in L} (A_{ij} + \hat{x}_j)]^p$. Let now I_1 be the set:

$$I_1 = \{i \mid \bigvee_{j \in L \cup S} (A_{ij} + \hat{x}_j) = \bigvee_{j \in L} (A_{ij} + \hat{x}_j)\} \tag{16}$$

and for each $s_k \in S$, we define two sets of indices:

$$I_2(s_k) = \{i \mid \bigvee_{j \in L \cup \{s_k\}} (A_{ij} + \hat{x}_j) = \bigvee_{j \in L \cup S} (A_{ij} + \hat{x}_j) > \bigvee_{j \in L} (A_{ij} + \hat{x}_j)\} \tag{17}$$

and:

$$I_3(s_k) = \{i \mid \bigvee_{j \in L \cup S} (A_{ij} + \hat{x}_j) > \bigvee_{j \in L \cup \{s_k\}} (A_{ij} + \hat{x}_j) > \bigvee_{j \in L} (A_{ij} + \hat{x}_j)\}. \tag{18}$$

Then, if

$$\Sigma_1(L, S) = \sum_{s_k \in S} \sum_{i \in I_1, I_2(s_k)} \{-[b_i - \bigvee_{j \in L \cup \{s_k\}} (A_{ij} + \hat{x}_j)]^p + [b_i - \bigvee_{j \in L} (A_{ij} + \hat{x}_j)]^p\} \tag{19}$$

and

$$\Sigma_2(L, S) = \sum_{s_k \in S} \sum_{i \in I_3(s_k)} -[b_i - \bigvee_{j \in L \cup \{s_k\}} (A_{ij} + \hat{x}_j)]^p + [b_i - \bigvee_{j \in L} (A_{ij} + \hat{x}_j)]^p, \tag{20}$$

the ratio becomes:

$$\gamma_{U,K}(f) = \min_{L,S} \frac{\Sigma_1(L, S) + \Sigma_2(L, S)}{\Sigma_1(L, S)} \geq 1, \ \forall \, U, K \tag{21}$$

meaning (Proposition 1) that f is submodular or, equivalently, $E_p = -f$ is supermodular. $\square$

Setting $\tilde{E}_p(T) = \max(E_p(T), \epsilon)^1$ and leveraging the previous theorem, we are able to formulate problem (13), and thus the initial one (5), as a cardinality

---

[1] The new, truncated, error function remains supermodular; see [16].

Algorithm 1: Approximate solution of problem (5)

Input: A, b
Compute $\hat{\mathbf{x}} = (-\mathbf{A})^{\mathsf{T}} \boxplus' \mathbf{b}$
if $E_p(J) > \epsilon$ **then**
  | **return** Infeasible
Set $T_0 = \emptyset, k = 0$
while $E_p(T_k) > \epsilon$ **do**
  | $j = \arg\min_{s \in J \setminus T_k} E_p(T_k \cup \{s\})$
  | $T_{k+1} = T_k \cup \{j\}$
  | $k = k + 1$
end
$x_j = \hat{x}_j, j \in T_k$ and $x_j = -\infty$, otherwise
return $\mathbf{x}, T_k$

---

minimization problem subject to a supermodular equality constraint [24], which allows us to approximately solve it by the greedy Algorithm 1. The calculation of the principal solution requires $\mathcal{O}(nm)$ time and the greedy selection of the support set of the solution costs $\mathcal{O}(n^2)$ time. We call the solutions of problem (5) *Sparse Greatest Lower Estimates* of $\mathbf{b}$. Regarding the approximation ratio between the optimal solution and the output of Algorithm 1, the following proposition holds.

Proposition 3. *Let $\mathbf{x}$ be the output of Algorithm 1 after $k > 0$ iterations of the inner while loop and T_k the respective support set. Then, if T^* is the support set of the optimal solution of (5), the following inequality holds:*

$$\frac{|T_k|}{|T^*|} \leq 1 + \log\left(\frac{m\Delta^p - \epsilon}{E_p(T_{k-1}) - \epsilon}\right), \tag{22}$$

where $\Delta = \bigvee_{i,j}(b_i - A_{ij} - \hat{x}_j)$.

Proof. From [24], the following bound holds for the cardinality minimization problem subject to a supermodular and decreasing constraint, defined as function $f : 2^J \to \mathbb{R}$, by the greedy algorithm:

$$\frac{|T_k|}{|T^*|} \leq 1 + \log\left(\frac{f(\emptyset) - f(J)}{f(T_{k-1}) - f(J)}\right) \tag{23}$$

For our problem, it is $f = \tilde{E}_p$. Observe now that, since $k > 0$, $\tilde{E}_p(\emptyset) = E_p(\emptyset) \leq m\Delta^p$, $0 \leq \tilde{E}_p(J) = \epsilon$ and $\tilde{E}_p(T_{k-1}) > \epsilon$. Therefore, the result follows. $\qquad\square$

The ratio warns us to expect less optimal and, thus, less sparse vectors when increasing the norm p that we use to measure the approximation. It also hints towards an inapproximability result when $p \to \infty$, which is formalised next.

3.1 Sparse Vectors with Minimum ℓ_∞ Errors

Although in some settings the $\mathbf{A} \boxplus \mathbf{x} \leq \mathbf{b}$ constraint is needed [23], in other cases it could disqualify potentially sparsest vectors from consideration. Omitting the constraint, on the other hand, makes it unclear how to search for minimum error solutions for any ℓ_p $(p < \infty)$ norm. For instance, it has recently been reported that it is $\mathcal{NP}$-hard to determine if a given point is a local minimum for the ℓ_2 norm [15]. For that reason, we shift our attention to the case of $p = \infty$. It is well known [4,7] that problem $\min_{\mathbf{x} \in \mathbb{R}^n_{\max}} \|\mathbf{b} - \mathbf{A} \boxplus \mathbf{x}\|_\infty$ has a closed form solution; it can be calculated in $\mathcal{O}(nm)$ time by adding to the principal solution element-wise the half of its ℓ_∞ error. Note that this new vector does not necessarily satisfy $\mathbf{A} \boxplus \mathbf{x} \leq \mathbf{b}$, so it shows a way to overcome the aforementioned limitation.

First, let us demonstrate that problem (5), when considering the ℓ_∞ norm, becomes harder than before and non-approximable by the greedy Algorithm 1. Hence, consider now the following optimization problem:

$$\arg \min_{\mathbf{x} \in \mathbb{R}^n_{\max}} |\text{supp}(\mathbf{x})|$$
$$\text{s.t. } \|\mathbf{b} - \mathbf{A} \boxplus \mathbf{x}\|_\infty \leq \epsilon. \tag{24}$$

Thanks to a similar construction as in the previous section, this problem can be recast as a set-search problem.

Lemma 2 (Projection on the support set, ℓ_∞ case). *Let* $T \subseteq J$, $\mathbf{x}|_T$ *defined as* $\hat{\mathbf{x}}$ *inside* T *and* $-\infty$ *otherwise and* $\mathbf{x}^* = \mathbf{x}|_T + \frac{\|\mathbf{b} - \mathbf{A} \boxplus \mathbf{x}|_T\|_\infty}{2}$. *Then* $\forall \, \mathbf{z} \in \mathbb{R}^n_{max}$ *with* $\text{supp}(\mathbf{z}) = T$, *it holds:*

$$\|\mathbf{b} - \mathbf{A} \boxplus \mathbf{z}\|_\infty \geq \|\mathbf{b} - \mathbf{A} \boxplus \mathbf{x}^*\|_\infty = \frac{\|\mathbf{b} - \mathbf{A} \boxplus \mathbf{x}|_T\|_\infty}{2}. \tag{25}$$

Proof. (Sketch) By fixing the support set of the considered vectors equal to T, equivalently we omit the columns and indices of $\mathbf{A}$ and $\mathbf{x}$, respectively, that do not belong in T (since they will not be considered at the evaluation of the maximum). By doing so, we get a new equation with same vector $\mathbf{b}$ and restricted $\mathbf{A}, \mathbf{x}$. The vector $\mathbf{x}^*$ that minimizes the ℓ_∞ error of this equation is obtained from its principal solution plus the half of its ℓ_∞ error. But now observe that the new principal solution shares the same values with the original principal solution (follows from Lemma 1) inside T, which is exactly vector $\mathbf{x}|_T$. Extending $\mathbf{x}^*$ back to $\mathbb{R}^n_{\max}$ yields the result. □

So, a similar result to Proposition 2 holds.

Proposition 4. *Let* $\mathbf{x}_{OPT}$ *be an optimal solution of (24), then we can construct a new one with values inside the support set equal to those of the principal solution* $\hat{\mathbf{x}}$ *plus the half of its* ℓ_∞ *error.*

By defining $E_\infty(T) = \frac{\|\mathbf{b} - \mathbf{A} \boxplus \mathbf{x}|_T\|_\infty}{2}$, (24) becomes:

$$\arg \min_{T \subseteq J} |T|$$
$$\text{s.t. } E_\infty(T) \leq \epsilon \tag{26}$$

Unfortunately this problem does not admit an approximate solution by the greedy Algorithm 1 (to be precise, the modified version of Algorithm 1 when

E_p becomes E_∞), as its error function, although decreasing, is not supermodular. The following example also reveals that the submodularity ratio (4) of E_∞ is 0. Therefore, it is not even approximately supermodular and a solution by Algorithm 1 can be arbitrarily bad [8].

Example 1. Let $A = \begin{pmatrix} 0 & 5 & 2 \\ 4 & 1 & 0 \\ 0 & 1 & 0 \end{pmatrix}, \mathbf{b} = \begin{pmatrix} 3 \\ 1 \\ 0 \end{pmatrix}$, then principal solution $\hat{\mathbf{x}}$ is:

$$\hat{\mathbf{x}} = \begin{pmatrix} 0 & -4 & 0 \\ -5 & -1 & -1 \\ -2 & 0 & 0 \end{pmatrix} \boxplus' \begin{pmatrix} 3 \\ 1 \\ 0 \end{pmatrix} = \begin{pmatrix} -3 \\ -2 \\ 0 \end{pmatrix}.$$

We calculate now the error function on different sets:

- When $T = \{3\}$, then $\hat{\mathbf{x}}|_{\{3\}} = (-\infty, -\infty, 0)^\top$ and

$$E_\infty(\{3\}) = \tfrac{1}{2}\|\mathbf{b} - \bigvee_{j\in\{3\}}(\mathbf{A}_j + \hat{x}|_{\{3\},j})\|_\infty = \tfrac{1}{2}\| \begin{pmatrix} 3 \\ 1 \\ 0 \end{pmatrix} - \begin{pmatrix} 2 \\ 0 \\ 0 \end{pmatrix} \|_\infty = \tfrac{1}{2}.$$

- Likewise, when $T = \{1,3\}$, $E_\infty(\{1,3\}) = \tfrac{1}{2}\| \begin{pmatrix} 3 \\ 1 \\ 0 \end{pmatrix} - \begin{pmatrix} -3 \\ 1 \\ -3 \end{pmatrix} \vee \begin{pmatrix} 2 \\ 0 \\ 0 \end{pmatrix} \|_\infty = \tfrac{1}{2}.$

- $T = \{2,3\}$, $E_\infty(\{2,3\}) = \tfrac{1}{2}\| \begin{pmatrix} 3 \\ 1 \\ 0 \end{pmatrix} - \begin{pmatrix} 3 \\ -1 \\ -1 \end{pmatrix} \vee \begin{pmatrix} 2 \\ 0 \\ 0 \end{pmatrix} \|_\infty = \tfrac{1}{2}.$

- $T = \{1,2,3\}$, $E_\infty(\{1,2,3\}) = \tfrac{1}{2}\| \begin{pmatrix} 3 \\ 1 \\ 0 \end{pmatrix} - \begin{pmatrix} -3 \\ 1 \\ -3 \end{pmatrix} \vee \begin{pmatrix} 3 \\ -1 \\ -1 \end{pmatrix} \vee \begin{pmatrix} 2 \\ 0 \\ 0 \end{pmatrix} \|_\infty = 0.$

Let now $f = -E_\infty, L = \{3\}, S = \{1,2\}$, then, by (4), we have:

$$\frac{f(\{3\} \cup \{1\}) - f(\{3\}) + f(\{3\} \cup \{2\}) - f(\{3\})}{f(\{3\} \cup \{1,2\}) - f(\{3\})} = \frac{-1/2 + 1/2 - 1/2 + 1/2}{0 + 1/2} = 0,$$
$$(27)$$

meaning that f has submodularity ratio 0 or E_∞ is not even approximately supermodular.

Although the previous discussion denies from problem (24) a greedy solution with any guarantees, we propose next a practical alternative to get a sparse enough vector. We first obtain a sparse vector $\mathbf{x}_{p,\epsilon}$ by solving problem (5). Then, we add to this vector element-wise half of its ℓ_∞ error $\|\mathbf{b} - \mathbf{A} \boxplus \mathbf{x}_{p,\epsilon}\|_\infty/2$. Interestingly, this new solution minimizes the ℓ_∞ error among all vectors with the same support, as formalized in the following result.

Proposition 5. *Let* $\mathbf{x}_{SMMAE} \in \mathbb{R}^n_{max}$ *be defined as:*

$$\mathbf{x}_{SMMAE} = \mathbf{x}_{p,\epsilon} + \frac{\|\mathbf{b} - \mathbf{A} \boxplus \mathbf{x}_{p,\epsilon}\|_\infty}{2}, \qquad (28)$$

where $\mathbf{x}_{p,\epsilon}$ *is a solution of problem (5) with fixed* (p, ϵ). *Then* $\forall\ \mathbf{z} \in \mathbb{R}_{max}^n$ *with* $supp(\mathbf{z}) = supp(\mathbf{x}_{p,\epsilon})$, *it holds*

$$\|\mathbf{b} - \mathbf{A} \boxplus \mathbf{z}\|_\infty \geq \|\mathbf{b} - \mathbf{A} \boxplus \mathbf{x}_{SMMAE}\|_\infty = \frac{\|\mathbf{b} - \mathbf{A} \boxplus \mathbf{x}_{p,\epsilon}\|_\infty}{2} \quad (29)$$

and, also,

$$\|\mathbf{b} - \mathbf{A} \boxplus \mathbf{x}_{SMMAE}\|_\infty \leq \frac{\sqrt[p]{\epsilon}}{2}. \quad (30)$$

Proof. Observe that $\mathbf{x}_{p,\epsilon}$ is equal to the principal solution $\hat{\mathbf{x}}$ inside $supp(\mathbf{x}_{p,\epsilon})$. So the first inequality holds from Lemma 2. Regarding the second one, we have:

$$\|\mathbf{b} - \mathbf{A} \boxplus \mathbf{x}_{SMMAE}\|_\infty = \frac{\|\mathbf{b} - \mathbf{A} \boxplus \mathbf{x}_{p,\epsilon}\|_\infty}{2} = \frac{\bigvee_i (b_i - [\mathbf{A} \boxplus \mathbf{x}_{p,\epsilon}]_i)}{2}. \quad (31)$$

But, notice that:

$$(\bigvee_i b_i - [\mathbf{A} \boxplus \mathbf{x}_{p,\epsilon}]_i)^p = \bigvee_i (b_i - [\mathbf{A} \boxplus \mathbf{x}_{p,\epsilon}]_i)^p \leq \sum_i (b_i - [\mathbf{A} \boxplus \mathbf{x}_{p,\epsilon}]_i)^p \leq \epsilon, \quad (32)$$

so

$$\bigvee_i (b_i - [\mathbf{A} \boxplus \mathbf{x}_{p,\epsilon}]_i) \leq \sqrt[p]{\epsilon} \quad (33)$$

and the result follows from (31). Note that the bound tightens, as p increases. □

The above method provides sparse vectors that are approximate solutions of the equation with respect to the ℓ_∞ norm without the need of the lateness constraint. After computing $\mathbf{x}_{p,\epsilon}$, $\mathbf{x}_{SMMAE}$ requires $\mathcal{O}(m|supp(\mathbf{x}_{p,\epsilon})| + |supp(\mathbf{x}_{p,\epsilon})|) = \mathcal{O}((m + 1)|supp(\mathbf{x}_{p,\epsilon})|)$ time. We call $\mathbf{x}_{SMMAE}$ *Sparse Minimum Max Absolute Error (SMMAE)* estimate of $\mathbf{b}$.

4 Application in Neural Network Pruning

Recently, there has been a renewed interest in Morphological Neural Networks [6, 11, 21, 25] which consist of neural networks with layers performing morphological operations (dilations or erosions). While they are theoretically appealing because of the success that morphology operations had in traditional computer vision tasks and the universal approximation property that these networks possess, they have also shown an ability to be pruned and produce interpretable models. Herein, we propose a way to do this systematically, by formulating the pruning problem as a system of max-plus equations.

Let a morphological network be a multi-layered network that contains layers of linear transformations followed by max-plus operations. The authors of [25] call this sequence of layers as a *Max-plus block*. If $\mathbf{x} \in \mathbb{R}^d$ represents the input and k is the output's dimension, then a simple network of 1 Max-plus block (see Fig. 1) performs the following operations:

$$\begin{aligned} \mathbf{z} &= \mathbf{W}\mathbf{x}, \\ \mathbf{y} &= \mathbf{A} \boxplus \mathbf{z}, \end{aligned} \quad (34)$$

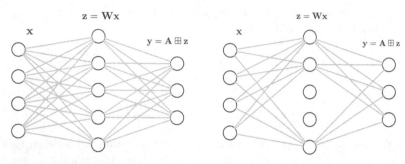

(a) A simple Max-plus block with $d = 4, n = 5, k = 3$.

(b) The same Max-plus block, after pruning two neurons from its first layer.

Fig. 1. Morphological neural networks.

where $\mathbf{W} \in \mathbb{R}^{n \times d}$ and $\mathbf{A} \in \mathbb{R}_{\max}^{k \times n}$. Suppose now that this network has been trained successfully, possibly with a redundant number n of neurons and we wish to maintain its accuracy while minimizing its size. For each training sample $(\mathbf{x}^{(i)}, \mathbf{y}^{(i)})$, it holds $\tilde{\mathbf{y}}^{(i)} = \mathbf{A} \boxplus \mathbf{z}^{(i)}$, where $\tilde{\mathbf{y}}^{(i)}$ is the network's prediction. We keep now fixed the prediction (that we wish to maintain) and the matrix $\mathbf{A}$ and we find a sparse approximate solution of this equation with respect to vector $\mathbf{z}^{(i)}$. Observe that if a value of $\mathbf{z}$ equals $-\infty$, then equivalently we can set the corresponding column of $\mathbf{A}$ to $-\infty$, thus pruning the whole unit. Of course, this naive technique would prune units that are important for other training samples. We propose overcoming this by finding sparse solutions for each sample, counting how many times each index $j \in \{1, \dots, n\}$ has been found inside the support set of a solution and then keeping only the k most frequent values.

The proposed method enables one to fully prune neurons from any layer that performs a max-plus operation, without harming its performance, and produce compact, interpretable networks. We support the above analysis by providing an experiment on MNIST and FashionMNIST datasets. Both datasets are balanced and contain 10 different classes.

Example 2. We train 2 networks for each dataset, containing 1 max-plus block with 64 and 128 neurons, respectively, inside the hidden layer, for 20 epochs with Stochastic Gradient Descent optimizing the Cross Entropy Loss. After the training, we pick at random 10000 samples from the training dataset (which account to 17% of the whole training data), we perform a forward pass over the network for each one of them to obtain predictions and then run Algorithm 1 with $p = 20$ and $\epsilon = 2^{20}$, so that we acquire sparse vectors $\mathbf{z}$ (and their support sets). Then, we simply find the 10 (same as the number of classes) most frequent indices inside the support sets of the solutions, keep the units that correspond to those indices and prune the rest of them. As can be seen in Table 1, all of the pruned networks record the same test accuracy as the full models, while having 54 and 118 *less* neurons, respectively. Note that trying to train from scratch networks with $n = 10$, under the same training setting, produces significantly worse results (around 60% for both datasets).

Table 1. Test set accuracy before and after pruning.

	MNIST		FashionMNIST	
	64	128	64	128
Full model	92.21	92.17	79.27	83.37
Pruned ($n = 10$)	92.21	92.17	79.27	83.37

5   Conclusions and Future Work

In this work, we developed the theory of sparsest approximate solutions to max-plus equations, tackled the hardness of finding one by exploiting problem's submodular structure and provided efficient algorithms for any ℓ_p approximation error. We briefly presented then a usage of the developed algorithms in a representative area of applications, the pruning of Morphological Neural Networks. It is a subject of future work to investigate the applications of sparsity in more areas of applications, perform further experiments on the proposed pruning technique in deeper and more general networks and develop a theory of sparsity in general nonlinear vector spaces called Complete Weighted Lattices [19].

References

1. Akian, M., Gaubert, S., Guterman, A.: Tropical polyhedra are equivalent to mean payoff games. Int. J. Algebra Comput. **22**(1), 1250001 (2012)
2. Baccelli, F., Cohen, G., Olsder, G.J., Quadrat, J.P.: Synchronization and Linearity: An Algebra for Discrete Event Systems. Wiley (1992)
3. Bach, F.: Learning with submodular functions: a convex optimization perspective (2013)
4. Butkovič, P.: Max-linear Systems: Theory and Algorithms. Monographs in Mathematics, 1st edn., p. 274. Springer, London (2010). https://doi.org/10.1007/978-1-84996-299-5
5. Candès, E., Romberg, J., Tao, T.: Stable signal recovery from incomplete and inaccurate measurements. Commun. Pure Appl. Math. **59**(8), 1207–1223 (2006)
6. Charisopoulos, V., Maragos, P.: Morphological perceptrons: geometry and training algorithms. In: Angulo, J., Velasco-Forero, S., Meyer, F. (eds.) ISMM 2017. LNCS, vol. 10225, pp. 3–15. Springer, Cham (2017). https://doi.org/10.1007/978-3-319-57240-6_1
7. Cuninghame-Green, R.: Minimax Algebra. Lecture Notes in Economics and Mathematical Systems, 1st edn., p. 258. Springer, Berlin, Heidelberg (1979). https://doi.org/10.1007/978-3-642-48708-8

8. Das, A., Kempe, D.: Approximate submodularity and its applications: Subset selection, sparse approximation and dictionary selection. J. Mach. Learn. Res. **19**(1), 74–107 (2018)
9. Donoho, D.: Compressed sensing. IEEE Trans. Inf. Theor. **52**, 1289–1306 (2006)
10. Elad, M.: Sparse and Redundant Representations: From Theory to Applications in Signal and Image Processing, 1st edn., p. 376. Springer, New York (2010)
11. Franchi, G., Fehri, A., Yao, A.: Deep morphological networks. Patter Recogn. **102**, 107246 (2020)
12. Gaubert, S., McEneaney, W., Qu, Z.: Curse of dimensionality reduction in max-plus based approximation methods: Theoretical estimates and improved pruning algorithms. In: Proceedings of IEEE Conference on Decision and Control and Europe Control Conference (2011)
13. Gierz, G., Hofmann, K.H., Keimel, K., Lawson, J.D., Mislove, M., Scott, D.S.: A Compendium of Continuous Lattices, 1st edn., p. 371. Springer, Berlin, Heidelberg (1980). https://doi.org/10.1007/978-3-642-67678-9
14. Heijmans, H.: Morphological Image Operators. Academic Press, Boston (1994)
15. Hook, J.: Max-plus linear inverse problems: 2-norm regression and system identification of max-plus linear dynamical systems with Gaussian noise (2019)
16. Krause, A., Golovin, D.: Submodular function maximization. In: Bordeaux, L., Hamadi, Y., Kohli, P. (eds.) Tractability, pp. 71–104. Cambridge University Press, Cambridge (2014). https://doi.org/10.1017/CBO9781139177801.004
17. Lovász, L.: Submodular Functions and Convexity. Mathematical Programming The State of the Art. Springer, Berlin, Heidelberg (1983). https://doi.org/10.1007/978-3-642-68874-4_10
18. Maragos, P.: Morphological filtering for image enhancement and feature detection. In: Image and Video Processing Handbook, 2nd edn. pp. 136–156 (2005)
19. Maragos, P.: Dynamical systems on weighted lattices: General theory. Math. Control Signals Syst. **29**, 21 (2017). https://doi.org/10.1007/s00498-017-0207-8
20. Natarajan, B.K.: Sparse approximate solutions to linear systems. SIAM J. Comput. **24**(2), 227–234 (1995)
21. Ritter, G.X., Urcid, G.: Lattice algebra approach to single-neuron computation. IEEE Trans. Neural Netw. **14**(2), 282–295 (2003)
22. Serra, J.: Image Analysis and Mathematical Morphology. Academic Press, Inc., Orlando, USA (1982)
23. Tsiamis, A., Maragos, P.: Sparsity in max-plus algebra and systems. Discrete Events Dyn. Syst. **29**, 163–189 (2019)
24. Wolsey, L.: An analysis of the greedy algorithm for the submodular set covering problem. Combinatorica **2**, 385–393 (1982)
25. Zhang, Y., Blusseau, S., Velasco-Forero, S., Bloch, I., Angulo, J.: Max-Plus Operators Applied to Filter Selection and Model Pruning in Neural Networks. In: Burgeth, B., Kleefeld, A., Naegel, B., Passat, N., Perret, B. (eds.) ISMM 2019. LNCS, vol. 11564, pp. 310–322. Springer, Cham (2019). https://doi.org/10.1007/978-3-030-20867-7_24

Author Index